Unity Cookbook

Fifth Edition

Over 160 recipes to craft your own masterpiece
in Unity 2023

Matt Smith

Shaun Ferns

Sinéad Murphy

<packt>

BIRMINGHAM—MUMBAI

Unity Cookbook
Fifth Edition

Copyright © 2023 Packt Publishing

Senior Publishing Product Manager: Larissa Pinto
Acquisition Editor – Peer Reviews: Gaurav Gavas and Jane D'Souza
Project Editor: Meenakshi Vijay
Content Development Editor: Davide Oliveri
Assitant Development Editor: Elliot Dallow
Copy Editor: Safis Editing
Technical Editor: Kushal Sharma
Proofreader: Safis Editing
Indexer: Subalakshmi Govindhan
Presentation Designer: Rajesh Shirsath
Developer Relations Marketing Executive: Sohini Ghosh

First published: June 2013
Second edition: October 2015
Third edition: August 2018
Fourth edition: September 2021
Fifth edition: November 2023

Production reference: 1281123

Published by Packt Publishing Ltd.
Grosvenor House
11 St Paul's Square
Birmingham
B3 1RB, UK.

ISBN 9781805123026

www.packt.com

We dedicate this book to Bobby O'Brien.

– Matt Smith & Sinéad Murphy

Dedicated to my amazing son, Sénan, a hero in the making.

– Shaun Ferns

Foreword

Not so long ago, developing professional quality games meant licensing an expensive game engine or writing your own from scratch. Then, you needed to hire a small army of developers to use it. Today, game engines like Unity have democratized game development to the point where you can simply download the tools and start making the game of your dreams right away.

Well... kinda. Having a powerful game creation tool is not the same thing as having the technical knowledge and skills to use it effectively.

I started coding games as a kid on my trusty ZX Spectrum, Commodore 64, and later the Amiga. I've been working as a professional game developer since 2003. When I first took the plunge into learning Unity development to create the Fungus storytelling tool, I found a huge amount of online documentation, tutorials, and forum answers available for Unity developers. This makes getting started with Unity development relatively easy, but the information can also be quite fragmented. Often, the last piece of the puzzle you need is buried 40 minutes into an hour-long tutorial video or on the 15th page of a forum thread. The hours you spend looking for these nuggets of wisdom is time that would be better spent working on your game.

The beauty of the *Unity Cookbook* is that Matt, Chico, Shaun, and Sinéad have distilled this knowledge into a neat collection of easy-to-follow recipes, and they have provided the scripts and complete working projects so that you can put it to use straight away.

In this latest edition, Matt, Shaun, and Sinéad have updated the recipes from the previous book and added new recipes to introduce many of the latest Unity features. There are some new chapters focusing on ProBuilder and mobile development; and other chapters include new topics such as non-horizontal AI NavMesh surfaces (think spiders crawling up walls and ceilings!), and advanced visual effects including vignettes and realistic animated water surfaces.

Getting started with Unity development is free and easy. When you're ready to take your skills to the next level, this book is an effective way to do just that. It covers a great deal in its hundreds of pages, and if you can master even half of what's here, you'll be well on the way to becoming a great Unity developer!

Chris Gregan

Chief Architect, Romero Games: www.romerogames.ie

Author of Fungus: fungusgames.com

Contributors

About the authors

Matt Smith is senior lecturer at **TU Dublin**, the Technological University of Dublin, Ireland, specialising in XR and interactive multimedia. He leads the university's DRIVE (Digital Realities, Interaction, and Virtual Environments) research group, and is currently supervising several PhD students in interaction design and XR technologies. In 1980, Matt started computer programming (on a ZX80). A few years later he submitted his first two games for the programming project component of his 'O'-level computing certificate (aged 16). In 1985, Matt wrote the lyrics, and was a member of the band that played (and sang, sorry about that by the way) the music on the B-side of the audio cassette carrying the computer game Confuzion (the game/song has a Wikipedia page...). In 2024, No Starch Press will publish his PHP Crash Course. Matt is still (pleasantly!) surprised at the popularity of his *Unity Cookbook* series – whose beginning was a book proposal sent to Packt Publishing over 10 years ago.

I'm grateful to my coauthors, Shaun and Sinéad, without whom this book wouldn't have been possible.

Many thanks to all my family. Thanks also to the editors, reviewers, and readers for their feedback. Thanks to my students, who continue to challenge and surprise me with their enthusiasm for multimedia and game development.

Shaun Ferns is an academic at **TU Dublin**, the Technological University of Dublin, Ireland, where he is a researcher in the DRIVE (Digital Realities, Interaction, and Virtual Environments) research group and an associate researcher at the Educational Informatics Lab (EILab) at OntarioTechU. Since 2016, he has been primarily researching and teaching multimedia development, and prior to that was involved in the delivery of several engineering programs. He is currently exploring the opportunities transmedia provides in improving user experience and engagement in cultural archive artifacts and serious games for the built environment. Shaun began to "play" with Unity when designing and building his house in 2010, developing an architectural walk-through to support the development of the design of the new home. Since then, he has been working on several Unity-based cultural projects and hopes to complete one soon! Shaun has taken up the challenge of playing the Irish tenor banjo and currently enjoys playing in Irish traditional music sessions with his friends.

When not practicing, he can be found wandering the cliffs and mountains around Donegal or swimming its Atlantic shores.

First and foremost, I am grateful to the students I have had the privilege of teaching and learning alongside over the past two decades. Your energy, excitement, and courage have challenged me to think more deeply and creatively. I am deeply inspired by your commitment to making the world a better place.

I am also indebted to the editors, reviewers, and readers who have provided feedback on this book. Your insights and suggestions have helped me to refine my arguments and improve the clarity and accessibility of my writing.

I am especially grateful to Matt, who has been a steadfast supporter, both professionally and personally. Your guidance, encouragement, and shared love of Game design and development through Unity have made this journey even more rewarding. Delighted to also welcome Sinéad to the authorship team. I hope this is the start of a long-lasting collaboration!

I am also grateful to my family and friends, who have provided me with unwavering support throughout this process. Finally, I would like to thank the many people who have made contributions to my education and professional development over the years. I am grateful for your mentorship and guidance.

Sinéad Murphy is currently Data Analytics Manager for the Irish NGO Trócaire. She has over 25 years of computing experience, including freelance IT training and database consulting, university lecturing in mathematics, IT skills, and programming at TU Dublin (Ireland) and Middlesex University (London). She is a published academic, with undergraduate and postgraduate degrees in mathematics, computing, and data science. She is passionate about the use of IT for understanding and visualising data, and using that understanding to make meaningful differences in the world. She is currently exploring the use of Python and Unity for data analytics and interactive visualisations.

Many thanks to my coauthors, Matt and Shaun, it's been a great experience working on my first book.

Thanks to my family for all their support. Thanks also to the editors and reviewers who provided feedback and suggestions.

About the reviewer

Jerry Medeiros is a seasoned professional with over a decade of hands-on experience in immersive technology, focusing on games and extended reality. With a robust background in artificial intelligence, Jerry brings a unique perspective to the intersection of technology and interactive experiences. Holding a degree in game development with a specialization in interaction design, as well as a Master of Computer Science with research expertise in artificial intelligence, Jerry is well versed in cutting-edge technologies and their applications. Additionally, an MBA in Innovation further underscores Jerry's commitment to driving creative and forward-thinking solutions in the tech industry.

Learn more on Discord

To join the Discord community for this book – where you can share feedback, ask questions to the author, and learn about new releases – follow the QR code below:

https://packt.link/unitydev

Table of Contents

Chapter 8: 2D Animation and Physics 245

Chapter 14: Shader Graphs and Video Players   501

Chapter 18: Virtual and Extended Reality (VR/XR)　　　　621

Preface

Game development is a broad and complex task. It is an interdisciplinary field, covering subjects as diverse as artificial intelligence, character animation, digital painting, and sound editing. All these areas of knowledge can materialize as the production of hundreds (or thousands!) of multimedia and data assets. A special software application—the game engine—is required to consolidate all these assets into a single product. Game engines are specialized pieces of software, which used to belong to an esoteric domain. They were expensive, inflexible, and extremely complicated to use. They were for big studios or hardcore programmers only. Then, along came Unity.

Unity represents the true democratization of game development. It is an engine and multimedia editing environment that is user-friendly and versatile. It has free and Pro versions; the latter includes even more features. Unity offers deployment to many platforms, including the following:

- **Mobile:** Android, iOS and Windows Phone
- **Web:** WebGL (and WebXR)
- **Desktop:** PC, Mac, and Linux platforms
- **Console:** Nintendo Switch, PS5/4/3, Xbox SeriesX/One/360, PlayStation Mobile, PlayStation Vita, and Wii U
- **Virtual Reality (VR)/Augmented Reality (AR):** Oculus Quest/2/3/Pro and Rift, Samsung Gear VR, HTC Vive Focus, Google Daydream, Microsoft Hololens, and the Apple Vision Pro

Today, Unity is used by a diverse community of developers all around the world. Some are students and hobbyists, but many are commercial organizations, ranging from garage developers to international studios, who use Unity to make a huge number of games—you might have already played some on one platform or another.

This book provides over 150 Unity game development recipes. Some recipes demonstrate Unity application techniques for multimedia features, including working with animations and using preinstalled package systems. Other recipes develop game components with C# scripts, ranging from working with data structures and data file manipulation to artificial intelligence algorithms for computer-controlled characters.

If you want to develop quality games in an organized and straightforward way, and you want to learn how to create useful game components and solve common problems, then both Unity and this book are for you.

Who this book is for

This book is for anyone who wants to explore a wide range of Unity scripting and multimedia features and find ready-to-use solutions for many game features. Programmers can explore multimedia features, and multimedia developers can try their hand at scripting. From intermediate to advanced users, from artists to coders, this book is for you, and everyone in your team! It is intended for everyone who has the basics of using Unity and a little programming knowledge in C#.

What this book covers

Chapter 1, Displaying Data with Core UI Elements, is filled with User Interface (UI) recipes to help you increase the entertainment and enjoyment value of your games through the quality of the visual elements displaying text and data. You'll learn a wide range of UI techniques for displaying text and images.

Chapter 2, Responding to User Events for Interactive UIs, teaches you about updating displays, and detecting and responding to user input actions, such as mouseovers. There are recipes for panels in visual layers, radio buttons and toggle groups, interactive text entry, directional radars, countdown timers, and custom mouse cursors.

Chapter 3, Inventory and Advanced UIs, relates to the many games that involve the player collecting items, such as keys to open doors and ammo for weapons, or choosing from a selection of items, such as from a collection of spells to cast. The recipes in this chapter offer a range of text and graphical solutions for displaying inventory status to the player, including whether they are carrying an item or not and the maximum number of items they are able to collect.

Chapter 4, Playing and Manipulating Sounds, suggests ways to use sound effects and soundtrack music to make your game more interesting. The chapter demonstrates how to manipulate sound at runtime through the use of scripts, Reverb Zones, and the Audio Mixer. It also includes recipes for real-time graphics visualizations of playing sounds, and a recipe to create a simple 140 bpm loop manager.

Chapter 5, Textures, Materials, and 3D Objects, contains recipes that will give you a better understanding of how to create, import, and modify 3D objects in Scenes. Recipes for this chapter include controlling how objects look by changing their textures and transparency, as well as creating GameObjects by creating and manipulating geometric primitives such as cubes and spheres.

Chapter 6, Creating 3D Environments with Terrains, contains recipes that will give you a better understanding of how to create and modify the large-scale geography of a Scene using the Unity terrain tools. You'll learn how to texture and height paint terrains, add holes, trees and vegetation, and also begin to explore the powerful, dynamic, realistic water features possible in HDRP (High Definition Render Pipeline) projects.

Chapter 7, Creating 3D Geometry with ProBuilder, contains recipes that will give you a better understanding of how to create and modify 3D objects within the Unity Editor using the powerful ProBuilder toolkit. As well as the basics of working with geometric meshes, you'll learn to extrude, texture, and vertex paint objects, gaining the skills to quickly prototype terrains and objects for complex game levels.

Chapter 8, 2D Animation and Physics, introduces some of Unity's powerful 2D animation and physics features. In this chapter, we present recipes to help you understand the relationships between the different animation elements in Unity, exploring the movement of different parts of the body and the use of sprite-sheet image files that contain sequences of sprite frame pictures. In this chapter, core Unity Animation concepts are presented, including Animation State Machines, Transitions, and Trigger events, as well as clipping via Sprite Masks. In addition, this chapter introduces the use of Tiles and Tilemaps for 2D games.

Chapter 9, Animated Character, focuses on character animation and demonstrates how to take advantage of Unity's Mecanim animation system. It covers a range of subjects, from basic character setup to controlling character animations with the old and new input systems.

Chapter 10, Saving and Loading Data, explores how games running on devices can benefit from persistent file-based data, and also communication with other networked applications. In this chapter, a range of recipes are presented that illustrate how to save and load data between Scenes, how to read data from text files, how to set up an online, database-driven leaderboard, and how to write Unity games that can communicate with such online systems.

Chapter 11, Controlling and Choosing Positions, presents a range of recipes for 2D and 3D user- and computer-controlled objects and characters, which can lead to games with a richer and more exciting user experience. Examples of these recipes include spawn-points, checkpoints, and physics-based approaches, such as applying forces when clicking on objects and firing projectiles into the Scene.

Chapter 12, Navigation Meshes and Agents, explores ways that Unity's NavMeshes and NavMesh Agents offer for the automation of object and character movement and pathfinding in your games. For example, recipes include ways to make objects follow predefined sequences of waypoints, or be controlled by mouse clicks for point-and-click control.

Chapter 13, Cameras, Lighting, and Visual Effects, presents recipes covering techniques for controlling and enhancing your game's cameras. It offers solutions to work with both single and multiple cameras, illustrates how to apply post-processing effects, such as vignettes and grainy grayscale videos. The chapter also introduces ways to work with Unity's powerful Cinemachine components. Other recipes in this chapter introduce visual effects including emissive materials and "cookie" textures, simulating objects casting shadows between the light source and the surfaces lights shine onto.

Chapter 14, Shader Graphs and Video Players, covers two powerful visual components in Unity: Shader Graphs and the Video Player. Both make it easy to add impressive visuals to your games with little or no programming. It includes recipes on how to simulate CCTV playback and download and play an online video, as well as an introduction to applying Shader Graphs in projects. Several recipes are presented for each of these features in this chapter.

Chapter 15, Particle Systems and Other Visual Effects, offers a hands-on approach to both using and re-purposing Unity's particle systems package, and also creating your own particle system from scratch.

Chapter 16, Mobile Games and Apps, provides an overview of and introduction to mobile projects in Unity. Since AR/VR/XR projects are mobile applications, this chapter acts as a foundation for those chapters too (*Chapters 17 and 18*).

Chapter 17, *Augmented Reality (AR)*, provides an overview of and introduction to AR projects in Unity. The recipes guide you through exploring the Unity AR examples, then creating and configuring your own AR projects.

Chapter 18, *Virtual and Extended Reality (VR/XR)*, provides an overview of and introduction to VR projects in Unity. Recipes include creating and configuring projects for VR, adding content, and building apps and deploying them onto devices or publishing them as WebXR via the web.

Chapter 19, *Advanced Topics: Gizmos, Automated Testing, and More*, explores a range of advanced topics, including creating your own gizmos to enhance design-time work in the Scene through visual grid guides with snapping. Automated code and runtime testing is also introduced, in addition to different approaches to saving and loading game data, and a final recipe introduces the new Python for Unity package, allowing scripting in the popular Python programming language.

Technical requirements to get the most out of this book

To complete the recipes in this book, there are some things that you will need.

For all chapters, you will need Unity 2023.1 or later, plus one of the following computer systems:

- Microsoft Windows 10 (64-bit)/GPU: DX10, DX11, and DX12-capable
- macOS GPU Metal-capable Intel or AMD

    - Mojave 10.14+ / Intel x64 with SSE2 instruction set support
    - Big Sur 11.0 / Apple Silicon M1 or later

- Linux Ubuntu 20.04 or Ubuntu 18.04 / Gnome desktop running on X11 / GPU: OpenGL 3.2+ or Vulkan-capable Nvidia or AMD

For each chapter, there is a folder in the book's GitHub repository that contains the asset files you will need; you can find these at `https://github.com/PacktPublishing/Unity-2023-Cookbook-Fifth-Edition`.

For recipes in some chapters, additional hardware/software will be helpful:

- *Chapter 4*, *Playing and Manipulating Sounds*

    To edit and create audio files yourself, you can download and install the free Audacity application for your computer system (Windows/Mac/Linux). You can find it at `https://www.audacityteam.org/download/`.

- *Chapter 5*, *Textures, Materials, and 3D Objects*

    To work with 3D objects in the Blender editor, you can download it for free at `www.blender.org`.

- *Chapter 10, Saving and Loading Data*

  - Since some of the recipes in this chapter make use of web servers and a database, for those recipes, you will require either the PHP 8 language (which comes with its own web server and SQLite database features) or an AMP package.

  - If you are installing the PHP language, refer to the installation guide and download links:

    - `https://www.php.net/manual/en/install.php`
    - `https://www.php.net/downloads`

  - If you do want to install a web server and database server application, a great choice is XAMPP. It is a free, cross-platform collection of everything you need to set up a database and web server on your local computer. The download page also contains FAQs and installation instructions for Windows, Mac, and Linux: `https://www.apachefriends.org/download.html`.

- *Chapter 15, Particle Systems and Other Visual Effects*

  If you wish to create your own image files, you will also need an image editor, such as Adobe Photoshop, which can be found at `www.adobe.com`, or GIMP, which is a free alternative and can be found at `www.gimp.org/`.

- *Chapter 16, Mobile Games and Apps*

  - If developing for Android, you'll need an Android mobile device.
  - If developing for Apple iOS, you'll need:

    - An Apple iOS mobile device.
    - A free Apple ID, which you can create on an Apple device or at `https://appleid.apple.com/`.
    - A Mac computer with the free Xcode program editor installed.
    - Note: If you don't have access to a Mac computer and Xcode, another way to develop for Apple iOS is to use Unity's Cloud Build services. Learn more about Cloud Build for iOS at `https://docs.unity3d.com/2020.1/Documentation/Manual/UnityCloudBuildiOS.html`.

- *Chapter 17, Augmented Reality (AR)*

  To get the most from this chapter's recipes, you will need an AR device. For this, you can use a dedicated device such as an AR headset, or you can use smartphone apps to begin experiencing AR.

- *Chapter 18, Virtual Reality (VR)*

  You will need a device to view VR apps. For this, you can use a dedicated device, such as a VR headset like the Meta Quest 1/2/3, Samsung Gear VR, or Apple Vision Pro. If you wish to use a smartphone for VR projects, there are many low-cost devices to choose from, such as Google Cardboard: https://developers.google.com/cardboard.

Download the example code files

You'll find the chapter figures, recipe assets, and completed Unity projects for each chapter at https://github.com/PacktPublishing/Unity-2023-Cookbook-Fifth-Edition.

You can either download these files as ZIP archives or use free Git software to download (clone) these files. These GitHub repositories will be updated with any improvements.

We also have other code bundles from our rich catalog of books and videos available at https://github.com/PacktPublishing/. Check them out!

Download the color images

We also provide a PDF file that has color images of the screenshots/diagrams used in this book. You can download it here: https://packt.link/gbp/9781805123026.

Conventions used

There are a number of text conventions used throughout this book.

CodeInText: Indicates code words in text, database table names, folder names, filenames, file extensions, pathnames, dummy URLs, user input, and Twitter handles. Here is an example: "The playerInventoryDisplay variable is a reference to an instance object of the PlayerInventoryDisplay class."

A block of code is set as follows:

```
public class PlayerInventoryDisplay : MonoBehaviour {
    public Text starText;
    public void OnChangeStarTotal(int numStars) {
        string starMessage = "total stars = " + numStars;
        starText.text = starMessage;
    }
}
```

When we wish to draw your attention to a particular part of a code block, the relevant lines or items are set in bold, like this:

```
public class PlayerInventoryDisplay : MonoBehaviour {
    public Text starText;
    public void OnChangeStarTotal(int numStars) {
        string starMessage = "total stars = " + numStars;
        starText.text = starMessage;
    }
}
```

Bold: Indicates a new term, an important word, or words that you see onscreen. For example, words in menus or dialog boxes appear in the text like this: "In the **Inspector** panel, set the font of Text-carrying-star to **Xolonium-Bold**, and set its color to yellow."

> Warnings or important notes appear like this.

> Tips and tricks appear like this.

Get in touch

Feedback from our readers is always welcome.

General feedback: If you have questions about any aspect of this book, mention the book title in the subject of your message and email us at customercare@packtpub.com.

Errata: Although we have taken every care to ensure the accuracy of our content, mistakes do happen. If you have found a mistake in this book, we would be grateful if you would report this to us. Please visit www.packt.com/submit-errata, selecting your book, clicking on the Errata Submission Form link, and entering the details.

Piracy: If you come across any illegal copies of our works in any form on the Internet, we would be grateful if you would provide us with the location address or website name. Please contact us at copyright@packt.com with a link to the material.

If you are interested in becoming an author: If there is a topic that you have expertise in and you are interested in either writing or contributing to a book, please visit authors.packtpub.com.

Share your thoughts

Once you've read *Unity Cookbook, Fifth Edition* we'd love to hear your thoughts! Scan the QR code below to go straight to the Amazon review page for this book and share your feedback.

https://packt.link/r/1805123025

Your review is important to us and the tech community and will help us make sure we're delivering excellent quality content.

Download a free PDF copy of this book

Thanks for purchasing this book!

Do you like to read on the go but are unable to carry your print books everywhere? Is your eBook purchase not compatible with the device of your choice?

Don't worry, now with every Packt book you get a DRM-free PDF version of that book at no cost.

Read anywhere, any place, on any device. Search, copy, and paste code from your favorite technical books directly into your application.

The perks don't stop there, you can get exclusive access to discounts, newsletters, and great free content in your inbox daily

Follow these simple steps to get the benefits:

1. Scan the QR code or visit the link below

https://packt.link/free-ebook/9781805123026

2. Submit your proof of purchase
3. That's it! We'll send your free PDF and other benefits to your email directly

1

Displaying Data with Core UI Elements

A key element that contributes to the entertainment and enjoyment of most games is the quality of the player's visual experience, and an important part of this is the **user interface (UI)**. UI elements involve ways for the user to interact with the game (such as buttons, cursors, and text boxes), as well as ways for the game to present up-to-date information to the user (such as the time remaining, current health, score, lives left, or location of enemies). This chapter is filled with UI recipes to give you a range of examples and ideas for creating game UIs.

> This chapter is all about the **Unity UI** system. This is based on GameObjects and their components, and the recommended system for runtime player visual UIs. There are other UI systems in Unity (UI Toolkit and IMGUI), but these are mostly used for Unity Editor design-time interfaces.

Every game and interactive multimedia application is different, and so this chapter attempts to fulfill two key roles:

- The first aim is to provide step-by-step instructions on how to create a range of Unity basic UI elements and, where appropriate, associate them with game variables in code.
- The second aim is to provide a rich illustration of how UI components can be used for a variety of purposes. This will help you get good ideas about how to make the Unity UI set of controls deliver the particular visual experience and interactions for the games that you are developing.

Basic UI components can provide static images and text to just make the screen look more interesting. By using scripts, we can change the content of these images and text objects so that the players' numeric scores can be updated, or we can show stickmen images to indicate how many lives the player has left. Other UI elements are interactive, allowing users to click on buttons, choose options, enter text, and so on.

More sophisticated kinds of UI can involve collecting and calculating data about the game (such as percentage time remaining or enemy hit damage; or the positions and types of key GameObjects in the scene and their relationship to the location and orientation of the player), and then displaying these values in a natural, graphical way (such as with progress bars or radar screens).

Core GameObjects, components, and concepts relating to Unity UI development include the following:

- **Canvas:** Every UI element is a child (or sub-child) of a **Canvas**. There can be multiple **Canvas** GameObjects in a single scene. If a **Canvas** is not already present, then one will automatically be created when a new UI GameObject is created, with that UI object as the child of the new **Canvas** GameObject.
- **EventSystem:** An **EventSystem** GameObject is required to manage the interaction events for UI controls. One will automatically be created with the first UI element. Unity only allows one **EventSystem** in any scene.
- **Visual UI controls:** The visible UI controls include **Button**, **Image**, **TextMeshPro**, and **Toggle**.
- **The Rect Transform component:** UI GameObjects are GameObjects that take up a rectangle on a 2D plane, and so have a **Rect Transform** component rather than a **Transform** component. The special **Rect Transform** component has some different properties from the scene's GameObject **Transform** component (with its straightforward *X/Y/Z* position, rotation, and scale properties). Associated with **Rect Transforms** are pivot points (reference points for scaling, resizing, and rotations) and anchor points.

The following diagram shows the four main categories of UI controls, each in a **Canvas** GameObject and interacting via an **EventSystem** GameObject. UI controls can have their own **Canvas**, or several UI controls can be in the same **Canvas**. The four categories are **display-only** and **interactive** UI controls, **non-visible** interactive components (such as ones to group a set of mutually exclusive radio buttons), and **C# script** classes to manage UI control behavior through logic written in the program code.

> Note that UI controls must be a child or descendant of a **Canvas**, otherwise, they will not work properly. Also, interactive UI controls will not work properly if the **EventSystem** GameObject is missing.

Both the **Canvas** and **EventSystem** GameObjects are automatically added to the **Hierarchy** panel as soon as the first UI GameObject is added to a scene:

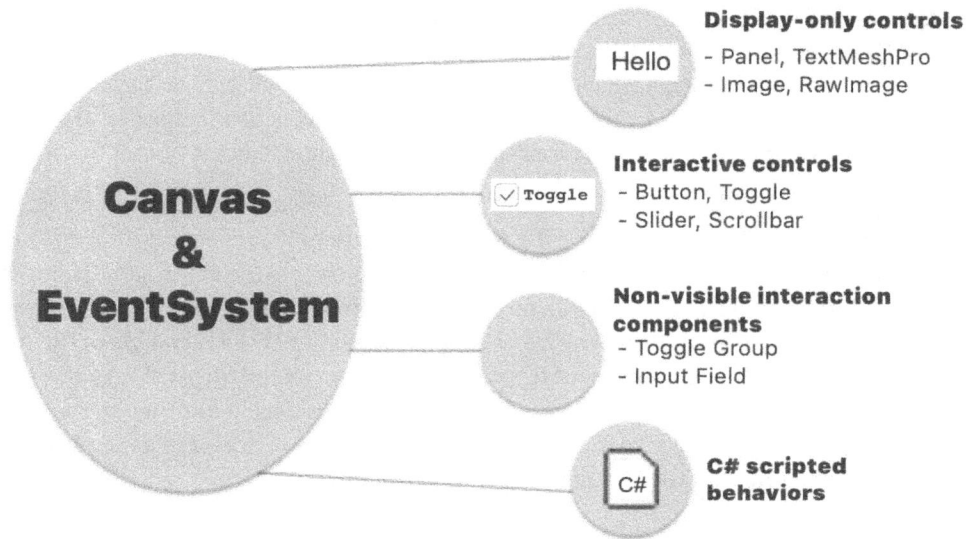

Figure 1.1: Canvas and EventSystem

Rect Transforms for UI GameObjects represent a rectangular area rather than a single point, which is the case for scene GameObject **transforms**. **Rect Transforms** describe how a UI element should be positioned, sized, scaled, and rotated. UI GameObjects have a **Pivot Point**, indicated in the **Scene** panel by a small blue circle. All transformations are made relative to this pivot point, for example, rotations of a GameObject are around this point. If a UI GameObject is moved, its pivot point is moved too.

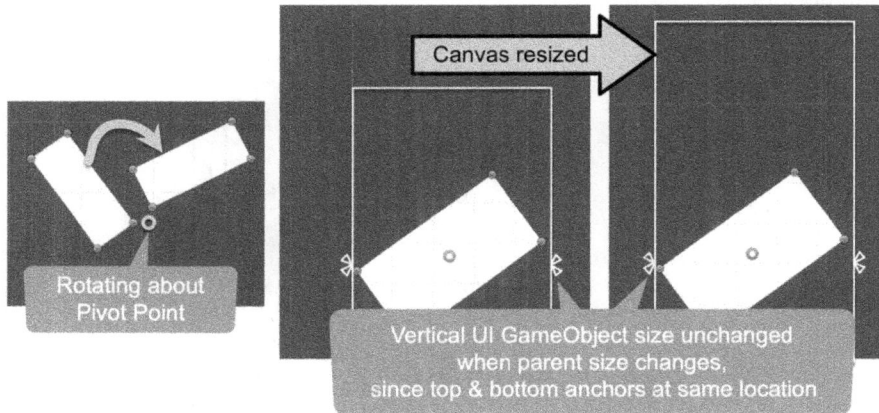

Figure 1.2: Transformations relative to pivot point and anchors

UI GameObjects also have **Anchors,** which determine the position and size of a UI GameObject relative to its parent. There are four anchors, indicating how a UI GameObject will be sized relative to its parent from the top, bottom, left, and right. Each anchor is indicated by a white triangle. UI GameObjects can be direct children of a canvas, or they can be child GameObjects of some other UI GameObject. The pivot point is relative to the immediate parent of each UI GameObject.

Rect Transforms have a width and height that can be changed without affecting the local **Scale** of the component. When the scale is changed for the **Rect Transform** of a UI element, this will also scale font sizes and borders on sliced images, and so on. If all four anchors are at the same point, resizing the **Canvas** will not stretch the **Rect Transform.** It will only affect its position. In this case, we'll see the **Pos X** and **Pos Y** properties and the **Width** and **Height** properties of the rectangle in the **Inspector** panel. However, if the anchors are not all at the same point, **Canvas** resizing will result in stretching the element's rectangle. So, instead of **Width,** we'll see the values for left and right – the position of the horizontal sides of the rectangle to the sides of the Canvas, where **Width** will depend on the actual Canvas width (and the same for top/bottom/height).

Unity provides a set of preset values for pivots and anchors, making the most common values very quick and easy to assign to an element's **Rect Transform.** The following screenshot shows the 3 x 3 grid of the **Anchor Presets** panel, which allows you to make quick choices about the left, right, top, bottom, middle, horizontal, and vertical values. Also, the extra column on the right offers horizontal stretch presets, while the extra row at the bottom offers vertical stretch presets. Pressing the *Shift + Alt* keys sets the pivot and anchors when a preset is clicked. Pressing just *Shift* sets the pivot but not the position, and pressing just *Alt* sets the position but not the pivot. *Figure 1.3* shows the **Anchor Presets** panel when both the *Shift + Alt* keys are pressed, whose icons illustrate the result of choosing vertical and horizontal positions for the anchors.

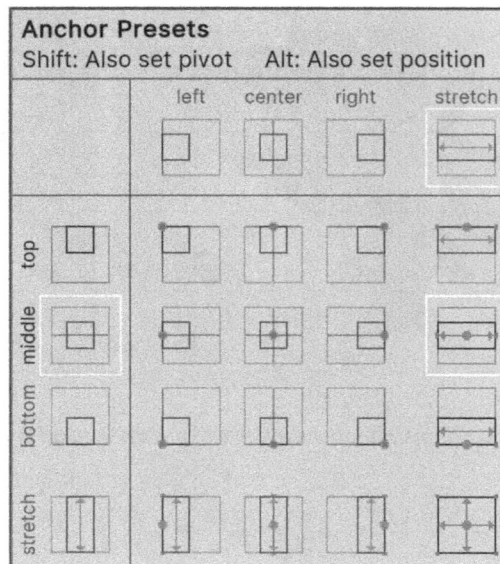

Figure 1.3: The Rect Transform Anchor Presets panel, when Shift and Alt pressed

There are three **Canvas** render modes:

- **Screen Space: Overlay:** In this mode, the UI elements are displayed without any reference to any camera (there is no need for any **Camera** in the scene). The UI elements are presented in front of (overlaying) any sort of camera display of the scene's contents.
- **Screen Space: Camera:** In this mode, the **Canvas** is treated as a flat plane in the frustum (viewing space) of a **Camera** scene – where this plane is always facing the camera. So, any scene objects in front of this plane will be rendered in front of the UI elements on the **Canvas**. The **Canvas** GameObject is automatically resized if the screen size, resolution, or camera settings are changed.
- **World Space:** In this mode, the **Canvas** acts as a flat plane in the frustum (viewing space) of a **Camera** scene – but the plane is not made to always face the **Camera**. The **Canvas** GameObject appears just as with any other objects in the scene, relative to where (if anywhere), in the camera's viewing frustum, the **Canvas** panel is located and oriented.

In this chapter, we are going to use the **Screen Space: Overlay** mode. However, all these recipes can be used with the other two modes as well.

Be creative! This chapter aims to act as a launching pad of ideas, techniques, and reusable C# scripts for your own projects. Get to know the range of Unity UI elements and try to work smart. Often, a UI component exists with most of the components that you may need for something in your game, but you may need to adapt it somehow. An example of this can be seen in the recipe that makes a UI Slider non-interactive, instead using it to display a red-green progress bar for the status of a countdown timer. We will take a detailed look at this in the *Displaying countdown times graphically with a UI Slider* section in *Chapter 2, Responding to User Events for Interactive UIs*.

Many of these recipes involve C# script classes that make use of the Unity scene-start event sequence of Awake() for all GameObjects, Start() for all GameObjects, and then Update() every frame to every GameObject. Therefore, you'll see many recipes in this chapter (and the whole book) where we cache references to GameObject components in the Awake() method, and then make use of these components in Start() and other methods once the scene is up and running.

> Since there is no 3D content in the recipes for this chapter, the projects created all use the 2D (Core) template. However, Unity UI works with all 2D and 3D projects.

In this chapter, we will cover the following recipes:

- Creating a Font Asset file for use with TextMeshPro
- Displaying a "Hello World" UI text message
- Displaying a digital clock
- Displaying a digital countdown timer
- Creating a message that fades away
- Displaying an image

Before you get started, in the Preface of the book you will find a list of the technical requirements needed to be able to complete all the recipes within this book. Overall requirements are listed first, then details of any additional hardware or software useful for individual chapters is also provided.

Creating a Font Asset file for use with TextMeshPro

The powerful TextMeshPro system is now core to Unity's UI system. While the TMP (TextMeshPro) Essentials comes with two fonts (Arial and Liberation Sans), and there are more available in the TMP examples, often you may have an existing font, or perhaps one from a client, that you will wish to use with your Unity project. In this recipe, we'll go through the steps of creating a Font Asset file, so you can then use whatever fonts you wish in your Unity projects.

Note: ensure you have a license appropriate for any fonts you use.

Getting ready

For this recipe, we have prepared the font that you need in a folder named Fonts in the 01_01 folder. Many thanks to Severin Meyer for making this font freely available at dafont.com.

How to do it...

To create a Font Asset file for use with TextMeshPro, follow these steps:

1. Create a new 2D (Core) project.

2. Import the provided Fonts folder, as described in the *Getting ready* section, so once you've copied these font files into your Unity project they'll be in the Assets folder in the **Project** panel.

3. Load the TextMeshPro essentials resources by choosing **Window | TextMeshPro | Import TMP Essential Resources**.

4. Now open the **Font Asset Creator,** by choosing **Window | TextMeshPro | Font Asset Creator**.

5. In the **Font Asset Creator** set the **Source Font File** to Xolonium-Bold file. Then click the **Generate Font Atlas** button.

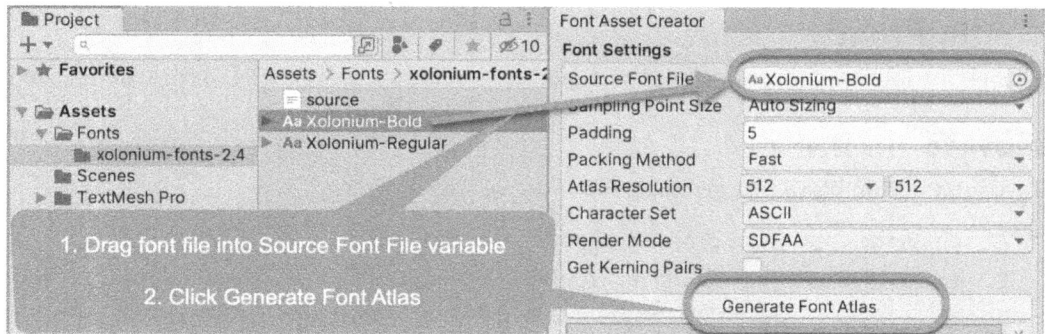

Figure 1.4: Using the Font Asset Creator tool

6. After a few seconds, the font atlas should have been created, and you can click the **Save** button. Select the Fonts folder as the destination for this new Xolonium-Bold SDF file.

7. When working with UI TextMeshPro text and button assets, you will now have the Xolonium-Bold font available in the list of fonts to choose from.

How it works...

The TestMeshPro feature requires Font Asset files in the SDF font atlas format. By following the steps in this recipe, you are able to create such files for any fonts you have on your computer.

> Signed Distance Field (SDF) fonts are a font encoding that makes text look crisp even after scaling and transformations.

Displaying a "Hello World" UI text message

The first traditional problem to be solved with new computing technology is to display the **Hello World** message, as shown in the following screenshot:

Figure 1.5: Displaying the "Hello World" message

In this recipe, you'll learn how to create a simple UI text object with this message, in large white text with a selected font, in the center of the screen.

Getting ready

This recipe follows on from the previous recipe, so make a copy of that and work on that copy.

How to do it...

To display a *Hello World* text message, follow these steps:

1. Open the copy of the project from the previous recipe.

2. In the **Hierarchy** panel, add a **TextMeshPro** GameObject to the scene by going to **GameObject | UI | Text - TextMeshPro**. Name this GameObject **Text-hello**. A **Canvas** and **EventSystem** GameObject will also be automatically added to the scene.

> Alternatively, you can use the **Create** menu immediately below the **Hierarchy** tab. To do so, go to **Create | UI | Text - TextMeshPro**.

3. Ensure that your new **Text-hello** GameObject is selected in the **Hierarchy** panel.

4. Now, in the **Inspector** panel, ensure the following properties are set (see screenshot):

    *   **Text Input** set to read **Hello World**
    *   **Font Asset** set to **Xolonium-Bold SDF**
    *   **Font Size** as per your requirements (large – this depends on your screen; try 50 or 100)
    *   **Vertex Color** set to white
    *   **Alignment** set to horizontal-center and vertical-center
    *   **Wrapping** set to **Enabled**

- **Overflow** set to **Overflow**

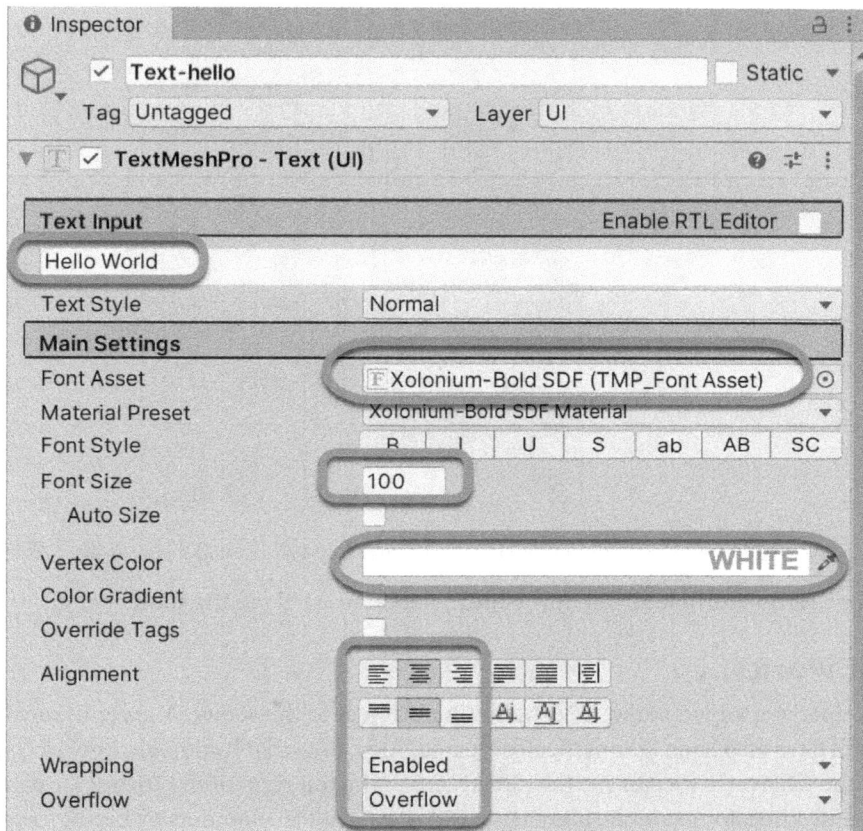

Figure 1.6: TMP text settings in the Inspector panel

5. In the **Inspector** panel, click **Rect Transform** to make a dropdown appear, and click on the **Anchor Presets** square icon (at the top left of the Rect Transform component in the **Inspector** panel). This should result in several rows and columns of preset position squares appearing.

Holding down *Shift + Alt*, click on the middle center one (**middle** row and **center column**).

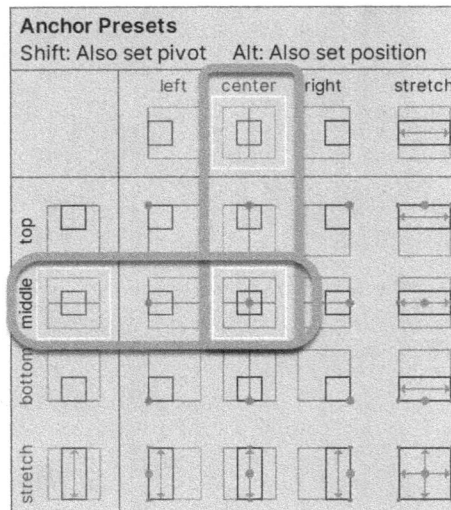

Figure 1.7: Selecting the horizontal center and vertical middle in the Rect Transform

6. Your **Hello World** text will now appear, centered nicely in the **Game** panel.

How it works...

In this recipe, you added a new **Text-hello** GameObject to the scene. A parent **Canvas** and UI **Event-System** will have also been automatically created. Also, note that by default, new UI GameObjects are added to the UI layer – we can see this illustrated at the top right of the **Inspector** panel in *Figure 1.6*. This is useful since, for example, it is easy to hide/reveal all UI elements by hiding/revealing this layer in the **Culling Mask** property of the **Camera** component of the **Main Camera** GameObject.

You set the text content and presentation properties and used the **Rect Transform** anchor presets to ensure that whatever way the screen is resized, the text will stay horizontally and vertically centered.

There's more...

Here are some more details you don't want to miss.

Styling substrings with rich text

Each separate **UI TextMeshPro** component can have its own color, size, boldness styling, and so on. However, if you wish to quickly add a highlighting style to the part of a string to be displayed to the user, you can apply HTML-style markups. The following are examples that are available without the need to create separate UI text objects:

*   Change the font to **Xolonium-Regular** (created by following the first recipe of this chapter)

- Embolden text with the b markup: I am **bold**
- Italicize text with the i markup: I am **<i>italic</i>**
- Set the text color with hex values or a color name: I am a **<color=green>**green text **</color>**, but I am <color=#FF0000>red</color>

> You can learn more by reading the Unity online manual's Rich Text page at https://docs.unity3d.com/Packages/com.unity.ugui@1.0/manual/StyledText.html

Exploring the TextMeshPro Examples and Extras

There are additional Text Mesh Pro resources and examples that are worth exploring to see a good range of the effects possible with this feature of Unity. You can load the TextMeshPro examples and extra resources by choosing **Window | TextMeshPro | Import TMP Examples and Extras**.

A new folder will be created in the **Project** panel: TextMeshpro | Examples & Extras. In this folder, you'll find scenes illustrating gradients, fonts, sprites, and more. There is also a Resources folder, offering additional fonts, gradient presets, and more.

Displaying a digital clock

Whether it is real-world time or an in-game countdown clock, many games are enhanced by some form of clock or timer display. The following screenshot shows the kind of clock we will be creating in this recipe:

Figure 1.8: Displaying a digital clock when the scene is run

The most straightforward type of clock to display is a string composed of integers for hours, minutes, and seconds, which is what we'll create in this recipe.

Getting ready

For this recipe, we have prepared the font that you will need in a folder named Fonts in the 01_01 folder. If you have not done so already, create a Xolonium-Bold SDF Font Asset file for the imported fonts, as described in the first recipe of this chapter.

How to do it...

To create a digital clock, follow these steps:

1. Create a new **Unity 2D project.**
2. Import the provided Fonts folder, containing your Xolonium-Bold SDF Font Asset file.

3. Add a **Text TMP** GameObject to the scene named **Text-clock** by choosing **GameObject | UI | Text - TextMeshPro**.

4. Ensure that the **Text-clock** GameObject is selected in the **Hierarchy** panel. Now, in the **Inspector** panel, ensure that the following properties are set:

   • **Font Type** set to **Xolonium Bold**
   • **Font Size** set to **20**
   • **Alignment** set to horizontal and vertical-center
   • **Overflow** settings set to **Overflow**
   • **Color** set to white

5. In **Rect Transform**, click on the **Anchor Presets** square icon, which will result in the appearance of several rows and columns of preset position squares. Holding down *Shift + Alt*, click on the **top** center item (top row, center column).

6. In the **Project** panel, create a folder named **_Scripts** and create a C# script class (menu: **Create | C# Script**) called ClockDigital in this new folder:

```csharp
using UnityEngine;
using TMPro;
using System;

public class ClockDigital : MonoBehaviour {
  private TextMeshProUGUI textClock;

  void Awake (){
    textClock = GetComponent<TextMeshProUGUI>();
  }

  void Update (){
    DateTime time = DateTime.Now;
    string hour = LeadingZero( time.Hour );
    string minute = LeadingZero( time.Minute );
    string second = LeadingZero( time.Second );

    textClock.text = hour + ":" + minute + ":" + second;
  }

  string LeadingZero (int n){
      return n.ToString().PadLeft(2, '0');
  }
}
```

> Note: In the **Project** panel, it can be useful to prefix important folders with an underscore character so that items appear first in a sequence.
>
> Since **scripts** and **scenes** are things that are most often accessed, prefixing their folder names with an underscore character, as in _Scenes and _Scripts, means they are always easy to find at the top in the **Project** panel.

7. Ensure the **Text-clock** GameObject is selected in the **Hierarchy** panel.

8. In the **Inspector** panel, add an instance of the ClockDigital script class as a component by clicking the **Add Component** button, selecting **Scripts,** and choosing the ClockDigital script class:

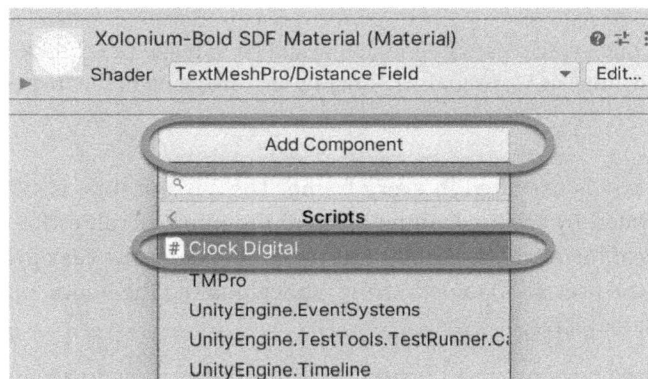

Figure 1.9: Adding scripted component Clock Digital in the Inspector

> Note: You can add script components through dragging and dropping.
>
> Script components can also be added to GameObjects via dragging and dropping. For example, with the **Text-clock** GameObject selected in the **Hierarchy** panel, drag your **ClockDigital** script onto it to add an instance of this script class as a component of the **Text-clock** GameObject.

9. When you run the scene, you will now see a digital clock that shows hours, minutes, and seconds in the top-center part of the screen.

How it works...

In this recipe, you added a **Text** GameObject to a scene. Then, you added an instance of the ClockDigital C# script class to that GameObject.

Notice that as well as the standard two C# packages (UnityEngine and System.Collections) that are written by default for every new script, you added the using statements to two more C# script packages, **TMPro** and System.

The **TMPro** package is needed since our code uses the TextMestPro text object, and the System package is needed since it contains the DateTime class that we need to access the clock on the computer where our game is running.

There is one variable, textClock, which will be a reference to the Text component, whose text content we wish to update in each frame with the current time in hours, minutes, and seconds.

The Awake() method (executed when the scene begins) sets the textClock variable to be a reference to the Text component in the GameObject, to which our scripted object has been added. Storing a reference to a component in this way is referred to as **caching** – this means that code that's executed later does not need to repeat the computationally expensive task of searching the GameObject hierarchy for a component of a particular type.

> Note that an alternative approach would be to make textClock a public variable. This would allow us to assign it via draging and dropping in the **Inspector** panel.

The Update() method is executed in every frame. The current time is stored in the time variable, and strings are created by adding leading zeros to the number values for the hours, minutes, and seconds properties of the variable. Finally, this method updates the text property (that is, the letters and numbers that the user sees) to be a string, concatenating the hours, minutes, and seconds with colon separator characters.

Although the code we have provided is useful for illustrating how to access the time component of a DateTime object individually, the Format(...) method of the String class can be used to format a DateTime object all in a single statement. For example, the preceding could be written more succinctly in a single statement; that is, **String.Format("HH:mm:ss", DateTime.Now)**. For more examples, see http://www.csharp-examples.net/string-format-datetime/.

The LeadingZero(...) method takes an integer as input and returns a string of this number with leading zeros added to the left if the value is less than **10**.

Displaying a digital countdown timer

As a game mechanic, countdown clocks are a popular feature in many games:

Figure 1.10: Countdown clock

This recipe, which will adapt the digital clock shown in the previous recipe, will show you how to display a digital countdown clock that will count down from a predetermined time to zero.

Getting ready

This recipe adapts to the previous one. So, make a copy of the project for the previous recipe and work on that copy.

For this recipe, we have prepared the `CountdownTimer` script that you need in a folder named `_Scripts` inside the `01_03` folder.

How to do it...

To create a digital countdown timer, follow these steps:

1. Import the provided `_Scripts` folder.

2. In the **Inspector** panel, remove the scripted component, **ClockDigital,** from the **Text-clock** GameObject. You can do this by choosing **Remove Component** from the 3-dot options menu icon for this component in the **Inspector** panel.

3. In the **Inspector** panel, add an instance of the `CountdownTimer` script class as a component by clicking the **Add Component** button, selecting **Scripts,** and choosing the `CountdownTimer` script class.

4. Create a `DigitalCountdown` C# script class that contains the following code, and add an instance as a scripted component to the **Text-clock** GameObject:

```
using UnityEngine;
using TMPro;

public class DigitalCountdown : MonoBehaviour {
    private TextMeshProUGUI textClock;
    private CountdownTimer countdownTimer;

    void Awake() {
        textClock = GetComponent<TextMeshProUGUI>();
        countdownTimer = GetComponent<CountdownTimer>();
    }
    void Start() {
        countdownTimer.ResetTimer( 30 );
    }

    void Update () {
        int timeRemaining = countdownTimer.GetSecondsRemaining();
        string message = TimerMessage(timeRemaining);
        textClock.text = message;
    }

    private string TimerMessage(int secondsLeft) {
```

```
            if (secondsLeft <= 0){
                return "countdown has finished";
            } else {
                return "Countdown seconds remaining = " + secondsLeft;
            }
        }
    }
```

5. When you run the scene, you will now see a digital clock counting down from 30. When the countdown reaches zero, a message stating **Countdown has finished** will be displayed.

How it works...

In this recipe, you added instances of the DigitalCountdown and CountdownTimer C# script classes to your scene's **UI TextMeshProUGUI** GameObject.

The Awake() method caches references to the TextMeshProUGUI and CountdownTimer components in the countdownTimer and textClock variables. The textClock variable will be a reference to the **UI** Text component, whose text content we wish to update in each frame with a *time-remaining* message (or a *timer-complete* message).

The Start() method calls the countdown timer object's CountdownTimerReset(...) method, passing an initial value of 30 seconds.

The Update() method is executed in every frame. This method retrieves the countdown timer's remaining seconds and stores this value as an integer (whole number) in the timeRemaining variable. This value is passed as a parameter to the TimerMessage() method, and the resulting message is stored in the string (text) variable message. Finally, this method updates the text property (that is, the letters and numbers that the user sees) of the textClock TextMeshProUGUI GameObject to be equal to the string message about the remaining seconds.

The TimerMessage() method takes an integer as input, and if the value is zero or less, a message stating the timer has finished is returned. Otherwise (if more than zero seconds remain), a message stating the number of remaining seconds is returned.

There's more...

Here are some more details you don't want to miss.

Automatically add components with [RequireComponent(...)]

The DigitalCountdown script class requires the same GameObject to also have an instance of the CountdownTimer script class. Rather than having to manually attach an instance of a required script, you can use the [RequireComponent(...)] C# attribute immediately before the class declaration statement. This will result in Unity automatically attaching an instance of the required script class.

If the script class or component cannot be found, an error will occur, and be reported in the **Console** panel.

For example, by writing the following code, Unity will add an instance of `CountdownTimer` as soon as an instance of the `DigitalCountdown` script class has been added as a component of a GameObject:

```
using UnityEngine;
using TMPro;


[RequireComponent (typeof (CountdownTimer))]
public class DigitalCountdown : MonoBehaviour {
```

You can learn more by reading the Unity documentation at `https://docs.unity3d.com/ScriptReference/RequireComponent.html`.

Creating a message that fades away

Sometimes, we want a message to only be displayed for a certain time, and then fade away and disappear. This recipe will describe the process for displaying a text message and then making it fade away completely after 5 seconds. It could be used for providing instructions or warnings to a player that disappears so as not to take up screen space.

Getting ready

This recipe adapts the previous one (*Displaying a digital countdown timer*). So, make a copy of the project for that recipe and work on that copy.

How to do it...

To display a text message that fades away, follow these steps:

1. In the **Inspector** panel, remove the scripted component, **DigitalCountdown**, from the **Text-clock** GameObject.
2. Select the **Text-clock** GameObject in the **Hierarchy** panel. Then, in the **Inspector** for the **Text-MeshPro – Text (UI)** component, set its **Text Input** default text to `hello world`.

3. Create a C# script class called `FadeAway` that contains the following code, and add an instance as a scripted component to the **Text-clock** GameObject:

```csharp
using UnityEngine;
using TMPro;

[RequireComponent(typeof(CountdownTimer))]
public class FadeAway : MonoBehaviour
{
    private CountdownTimer countdownTimer;
    private TextMeshProUGUI textUI;

    void Awake()
    {
        textUI = GetComponent<TextMeshProUGUI>();
        countdownTimer = GetComponent<CountdownTimer>();
    }

    void Start()
    {
        countdownTimer.ResetTimer(5);
    }

    void Update()
    {
        float alphaRemaining =
            countdownTimer.GetProportionTimeRemaining();
        print(alphaRemaining);
        Color c = textUI.color;
        c.a = alphaRemaining;
        textUI.color = c;
    }
}
```

4. When you run the scene, you will see that the `hello world` message on the screen slowly fades away, disappearing after 5 seconds.

How it works...

In this recipe, you added an instance of the `FadeAway` scripted class to the **Text-clock** GameObject. Due to the `RequireComponent(...)` attribute, an instance of the `CountdownTimer` script class was also **automatically** added.

The Awake() method caches references to the **TextMeshProUGUI** and **CountdownTimer** components in the countdownTimer and textUI variables.

The Start() method resets the countdown timer so that it starts counting down from 5 seconds.

The Update() method (executed every frame) retrieves the proportion of time remaining in our timer by calling the GetProportionTimeRemaining() method of the CountdownTimer script class. This method returns a value between **0.0** and **1.0**, which also happens to be the range of values for the alpha (transparency) property of the color property of a UI TextMeshPro GameObject.

> The flexible range of 0.0–1.0. It is often a good idea to represent proportions as values between 0.0 and 1.0. Either this will be just the value we want for something, or we can multiply the maximum value by our decimal proportion, and we get the appropriate value. For example, if we wanted the number of degrees of a circle for a given 0.0–0.1 proportion, we would just multiply by the maximum of 360, and so on.

The Update() method then retrieves the current color of the text being displayed (via textUI.color), updates its alpha property, and resets the text object to having this updated color value. The result is that each frame in the text object's transparency represents the current value of the proportion of the timer remaining until it fades to fully transparent when the timer gets to zero.

Displaying an image

There are many cases where we wish to display an image onscreen, including logos, maps, icons, and splash graphics. In this recipe, we will display an image centered at the top of the screen.

The following screenshot shows Unity displaying an image:

Figure 1.11: Displaying the Unity logo as an image

Getting ready

For this recipe, we have prepared the image that you need in a folder named **Images** in the 01_06 folder.

How to do it...

To display an image, follow these steps:

1. Create a new **Unity 2D** project.
2. Set the **Game** panel to **400 x 300**. Do this by displaying the **Game** panel, and then creating a new **Resolution** in the **Free Aspect** drop-down menu at the top of the panel.

3. Click the plus (+) symbol at the bottom of this menu, setting **Label** to **Core UI**, **Width** to 400, and **Height** to 300. Click **OK**; the **Game** panel should be set to this new resolution:

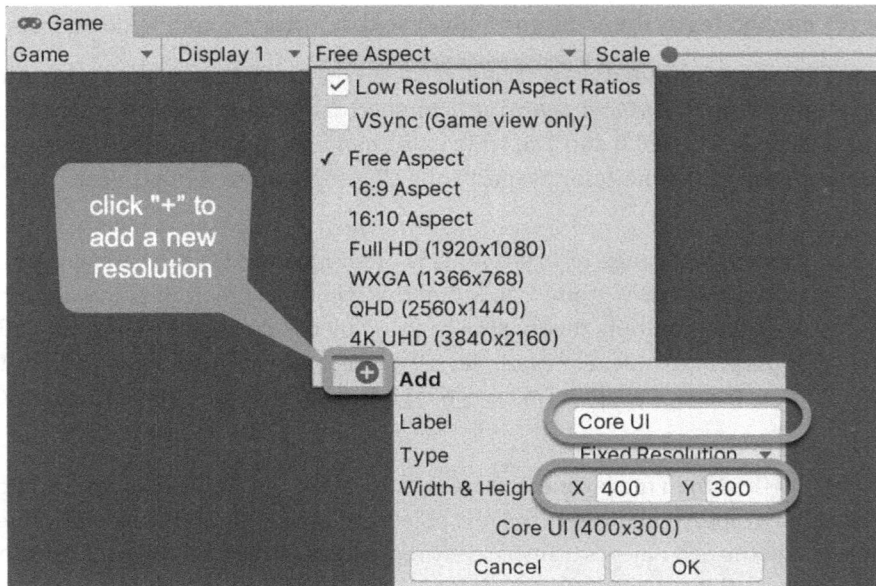

Figure 1.12: Adding a new screen resolution to the Game panel

> Alternatively, you can set the default **Game** panel's resolution by going to **Edit | Project Settings | Player** and then the width and height of **Resolution** and **Presentation** in the **Inspector** panel (having turned off the **Full-Screen** option).

4. Import the provided **Images** folder.

5. Select the unity_logo image asset file in the **Project** panel, and in the **Inspector** panel, ensure that its **Texture Type** is set to **Default**. If it has some other type, then choose **Default** from the drop-down list and click on the **Apply** button.

6. In the **Hierarchy** panel, add a **UI | RawImage** GameObject named **RawImage-logo** to the scene.

7. Ensure that the **RawImage-logo** GameObject is selected in the **Hierarchy** panel. In the **Inspector** panel for the **RawImage** component, click the file viewer circle icon at the right-hand side of the **Texture** property and select the unity_logo image:

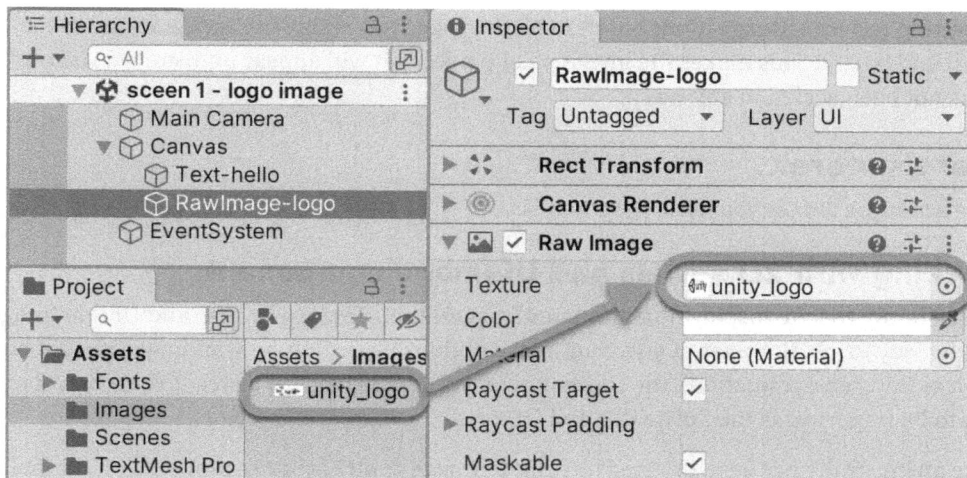

Figure 1.13: Setting a Texture for a Raw Image UI GameObject

> An alternative way of assigning this **Texture** is to drag the unity_logo image from your **Project** folder (**Images**) into the **Raw Image** public property **Texture**.

8. Click on the **Set Native Size** button to resize the image so that it is no longer stretched and distorted.

9. In **Rect Transform**, click on the **Anchor Presets** square icon, which will result in several rows and columns of preset position squares appearing. Holding down *Shift + Alt*, click on the top row and the center column.

10. The image will now be positioned neatly at the top of the **Game** panel and will be horizontally centered.

How it works...

In this recipe, you ensured that an image has its **Texture Type** set to **Default** – this is so that this image asset file type matches the type expected by the UI **RawImage** control. Default images are suitable for texturing 3D meshes, for example. For example, if its type was **Sprite** (2D and UI), then it could not be used with a UI **RawImage** control.

You also added a UI **RawImage** control to the scene. The **RawImage** control has been made to display the unity_logo image file. This image has been positioned at the top-center of the **Game** panel.

By setting the **Raw Image** component to **Native Size**, you resize the image in the scene to its pixel width and height. This makes the image pixel-perfect – it will appear undistorted since at native size it has not been scaled in any way.

There's more...

Here are some details you don't want to miss.

Working with 2D sprites and UI Image components

If you simply wish to display non-animated images, then **Texture** images and UI **RawImage** controls are the way to go. However, if you want more options regarding how an image should be displayed (such as tiling and animation), the **UI Image** control should be used instead. This control needs image files to be imported as the **Sprite (2D and UI)** type.

Once an image file has been dragged into the UI **Image** control's **Sprite** property, additional properties will be available, such as **Image Type**, and options to preserve the aspect ratio.

If you wish to prevent a UI **Sprite** GameObject from being distorted and stretched, go to the **Inspector** panel and check the **Preserve Aspect** option in its **Image** component.

See also

An example of tiling a sprite image can be found in the *Revealing icons for multiple object pickups by changing the size of a tiled image* recipe in *Chapter 3, Inventory UIs and Advanced UIs*.

Further reading

The following are some useful resources for learning more about working with core UI elements in Unity:

- The Unity manual provides a very good introduction to UI TextMesh Pro layouts and **Rect Transforms**:

    - https://docs.unity3d.com/Packages/com.unity.textmeshpro@4.0/manual/index.html

    - https://docs.unity3d.com/Packages/com.unity.ugui@1.0/manual/class-RectTransform.html

- Here is a comparison of the different Unity UI systems:

    - https://docs.unity3d.com/2023.2/Documentation/Manual/UI-system-compare.html

- In addition, Kodeco's (formerly Ray Wenderlich) *Unity* Introduction to TextMeshPro web tutorial also presents a helpful guide to TextMeshPro in Unity:

    - https://www.kodeco.com/22175776-introduction-to-textmesh-pro-in-unity

- To learn more about TextMeshPro, take a look at the following link:

  - `https://blogs.unity3d.com/2018/10/16/making-the-most-of-textmesh-pro-in-unity-2018/`

- Background on how TextMeshPro uses Signed Distance Functions:

  - `https://en.wikipedia.org/wiki/Signed_distance_function`

- Google offers a wide range of fonts you can use with your Unity projects:

  - `https://fonts.google.com/`

- Learn more about SDF fonts in the TextMeshPro documentation:

  - `https://docs.unity3d.com/Packages/com.unity.textmeshpro@4.0/manual/FontAssetsSDF.html`

Learn more on Discord

To join the Discord community for this book – where you can share feedback, ask questions to the author, and learn about new releases – follow the QR code below:

`https://packt.link/unitydev`

2

Responding to User Events for Interactive UIs

Almost all the recipes in this chapter involve different interactive UI controls. Although there are different kinds of interactive UI controls, the basic way to work with them, as well as to have scripted actions respond to user actions, is all based on the same idea: events triggering the execution of object method functions.

Then, for fun, and as an example of a very different kind of UI, the final recipe will demonstrate how to add sophisticated, real-time communication for the relative positions of objects in the scene to your game - in the form of a radar!

The UI can be used for three main purposes:

- To display **static (unchanging) values**, such as the name or logo image of the game, or word labels such as **Level** and **Score**, that tell us what the numbers next to them indicate (the recipes for these can be found in *Chapter 1, Displaying Data with Core UI Elements*).

- To display **values that change due to our scripts**, such as timers, scores, or the distance from our **Player** character to some other object (an example of this is the radar recipe at the end of this chapter, *Displaying a radar to indicate the relative locations of objects*).

- **Interactive** UI controls, whose purpose is to allow the player to communicate with the game scripts via their mouse or touchscreen. These are the ones we'll look at in detail in this chapter.

The core concept of working with Unity interactive UI controls is to *register an object's public method so that we're informed when a particular event occurs*. For example, we can add a UI dropdown to a scene named DropDown1, and then write a MyScript script class containing a NewValueAction() public method to perform an action. However, nothing will happen until we do two things:

- We need to add an *instance of the script class as a component* of a GameObject in the scene (which we'll name go1 for our example – although we can also add the script instance to the UI GameObject itself if we wish to).

- In the UI dropdown's properties, we need to *register the GameObject's public method* of its script component so that it responds to **On Value Changed** event messages.

The NewValueAction() public method of the MyScript script will typically retrieve the value that's been selected by the user in the dropdown and do something with it. For example, the NewValueAction() public method might confirm the value to the user, change the music volume, or change the game's difficulty. The NewValueAction() method will be executed each time GameObject go1 receives the NewValueAction() message. In the properties of DropDown1, we need to register go1's scripted component – that is, MyScript's NewValueAction() public method – as an event listener for **On Value Changed** events. We need to do all this at design time (that is, in the Unity Editor before running the scene):

Design Time

GameObject **DropDown1**

GameObject **go1**

Transform

MyScript
void Awake()
public NewValueAction()

component 2
component 3

NewValueAction()

UI Dropdown
Option A
Option B

Rect Transform

On Value Changed event listeners:
go1->MyScript.NewValueAction()

component 3

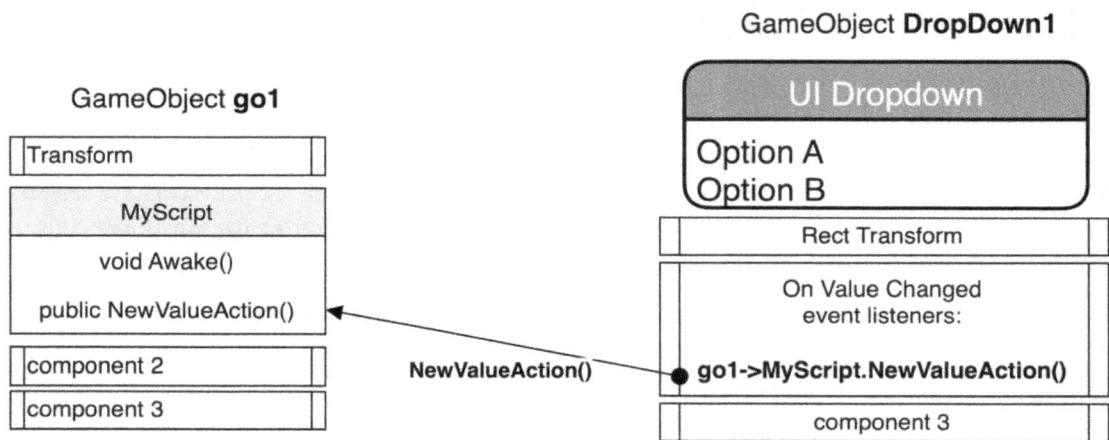

Figure 2.1: Graphical representation of the UI at design time

At **runtime** (when the scene in the application is running), the following will happen:

1. If the user changes the value in the drop-down menu of the DropDown1 GameObject (*step 1* in the following diagram), this will generate an **On Value Changed** event.

2. DropDown1 will update its display on the screen to show the user the newly selected value (*step 2a*). It will also send messages to all the GameObject components registered as listeners to **On Value Changed** events (*step 2b*).

3. In our example, this will lead to the `NewValueAction()` method in the go1 GameObject's scripted component being executed (*step 3*).

Run Time

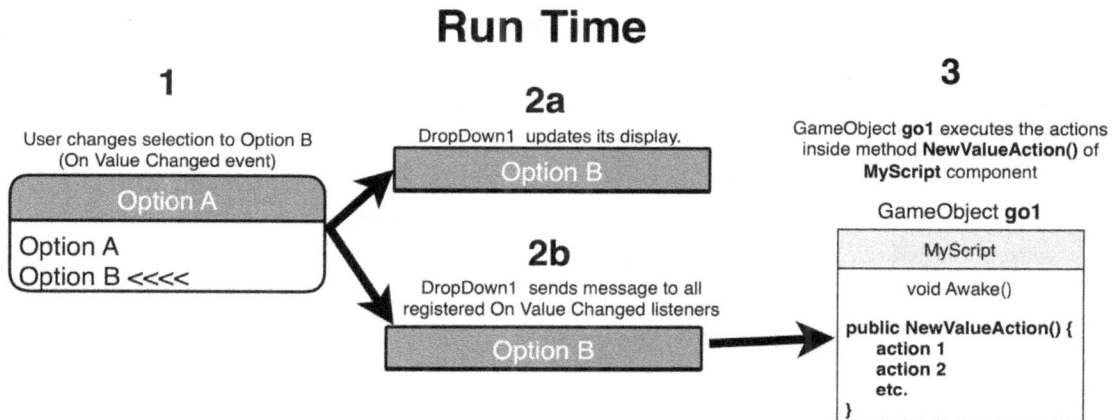

Figure 2.2: Graphical representation of the UI at runtime

Registering public object methods is a very common way to handle events such as user interaction or web communications, which may occur in different orders, may never occur, or may happen several times in a short period. Several software design patterns describe ways to work with these event setups, such as the *Observer* pattern and the *Publisher-Subscriber* design pattern (more details can be found at `https://unity.com/how-to/create-modular-and-maintainable-code-observer-pattern`).

Core GameObject components related to interactive Unity UI development include the following:

- **Visual UI controls:** The visible UI controls themselves include **Button, Image, Text,** and **Toggle.** These are the UI controls the user sees on the screen and uses their mouse/touchscreen to interact with. These are the GameObjects that maintain a list of object methods that have subscribed to user-interaction events.

- **Interaction UI controls:** These are non-visible components that are added to GameObjects; examples include **Input Field** and **Toggle Group.**

- **Panel:** UI objects can be grouped together (logically and physically) with UI **Panels. Panels** can play several roles, including providing a GameObject parent in the **Hierarchy** window for a related group of controls. They can provide a visual background image to graphically relate controls on the screen, and they can also have scripted resize and drag interactions added if desired.

In addition, the concept of sibling depth is important when multiple UI components are overlapping. The bottom-to-top display order (what appears on the top of what) for a UI element is determined initially by its place in the sequence in the **Hierarchy window**. At design time, this can be manually set by dragging GameObjects into the desired sequence in the **Hierarchy** window. At runtime, we can send messages to the **Rect Transforms** of GameObjects to dynamically change their **Hierarchy** position (and, therefore, the display order) as the game or user interaction demands. This is illustrated in the *Organizing images inside panels and changing panel depths via buttons* recipe in this chapter.

Often, a UI element exists with most of the components that you may need for something in your game, but you may need to adapt it somehow. An example of this can be seen in the *Displaying a countdown timer graphically with a UI Slider* recipe, which makes a UI Slider non-interactive so as to display a red-green progress bar for the status of a countdown timer.

In this chapter, we will cover the following recipes:

- Creating a UI Button to reveal an image
- Creating a UI Button to move between scenes
- Animating UI Button properties on mouseover
- Organizing image panels and changing panel depths via UI Buttons
- Displaying the value of an interactive UI Slider
- Displaying a countdown timer graphically with a UI Slider
- Setting custom mouse cursors for 2D and 3D GameObjects
- Setting custom mouse cursors for UI controls
- Interactive text entry with Input Field
- Detecting interactions with a single Toggle UI component
- Creating related radio buttons using UI Toggles
- Creating text UI Dropdown menus
- Creating image icon UI Dropdown menus
- Displaying a radar to indicate the relative locations of objects.

Creating a UI Button to reveal an image

In this recipe, we'll create a button that, when pressed, will make an image appear.

Getting ready

For this recipe, we have prepared the image that you need in a folder named Images in the 02_01 folder.

How to do it...

To create a UI Button to reveal an image, follow these steps:

1. Create a new **Unity 2D project**.
2. Import the **unity_logo** image into the **Project** folder.

3. Drag the **unity_logo** image into the scene and in the **Inspector** panel, set the **scale** property to **X 2 Y 2**, and uncheck the check box as per the screenshot to make this GameObject inactive. An inactive GameObject is not displayed, and nor does it have any active behavior (so no scripts run and it does not respond to any event messages).

4. Load the TextMeshPro Essential Resources, by choosing **Window | TextMeshPro | Import TMP Essential Resources**.

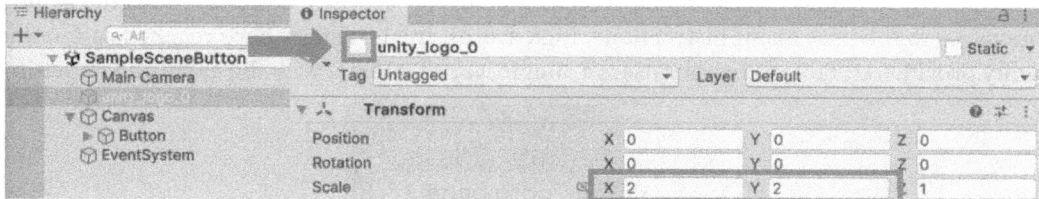

Figure 2.3: Inspector properties for an image in the scene

5. In the **Hierarchy** window, right-click and select **Create | UI | Button-TextMeshPro**.

6. Select **GameObject Button** in the **Hierarchy** window and click on the plus (+) button at the bottom of the **Inspector** window, to create a new **OnClick** event handler.

7. Drag the **unity_logo** image from the **Hierarchy** window over the **Object** slot immediately below the menu that says **Runtime Only**.

8. Select the **SetActive (bool)** method from the **GameObject** drop-down list (initially showing **No Function**) and click on the checkbox.

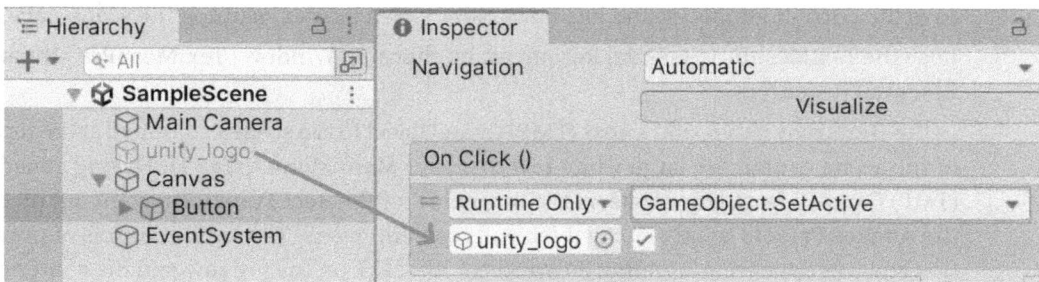

Figure 2.4: Settings for the OnClick event handler for the button

9. Save your changes and run the scene. When you click the button, the image will appear.

How it works...

In this recipe, you created a new scene, imported an image, and unchecked the **Visible in Runtime** checkbox for the image so that it would not be seen at runtime.

You added a UI Button and a new **OnClick** event action that executes the **GameObject.SetActive()** method of the **GameObject** drop-down list of the button, and you checked the box so that the image (unity_logo) appears when the button is clicked.

There's more...

As an alternative to having the image appear when clicking on a button, the same button could be used to make an image disappear by the checkbox on the image.

Creating a UI Button to move between scenes

The majority of games include menu screens that display settings, buttons to start the game playing, messages to the user about instructions, high scores, the level they have reached so far, and so on. Unity provides UI Buttons to offer users a simple way to interact with the game and its settings.

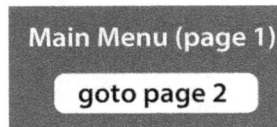

Figure 2.5: Example of a main menu UI Button

In this recipe, we'll create a very simple game consisting of two screens, each with a button to load the other one, as illustrated in the preceding screenshot.

How to do it...

To create a button-navigable multi-scene game, follow these steps:

1. Create a new **Unity 2D project**.
2. Save the current (empty) scene in a new folder called Scenes, naming the scene **page1**.
3. Load the TextMeshPro Essential Resources, by choosing **Window | TextMeshPro | Import TMP Essential Resources**.
4. In the **Hierarchy** panel, add a **Text** (TMP) GameObject to the scene positioned at the top center of the scene containing large white text that says **Main Menu (page 1)**. Having added a **Text** (TMP) GameObject to the scene, in the **Inspector** for the **Rect Transform** component, click on the **Anchor Presets** square icon, which will result in several rows and columns of preset position squares appearing. Holding down *Shift + Alt*, click on the top row and the center column.
5. Add a UI **Button-TextMeshPro** to the scene positioned in the middle-center of the screen.
6. In the **Hierarchy** window, click on the Tree View toggle triangle to display the **Text** child of this GameObject button. Select the Text GameObject and, in the **Inspector** window for the **Text Input** property, enter the text **goto page 2**:

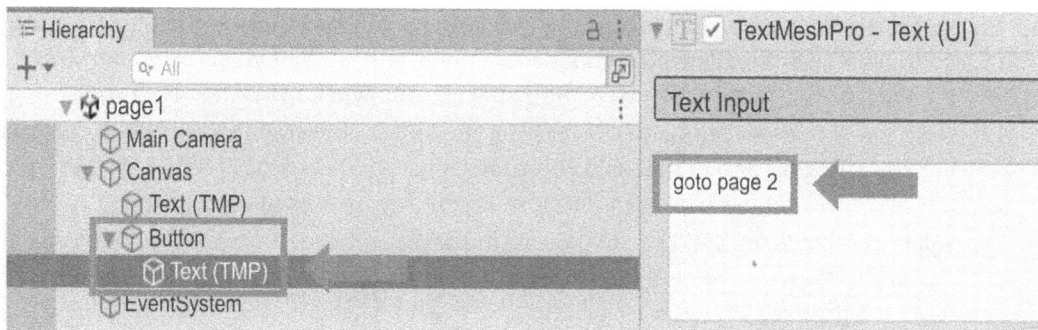

Figure 2.6: UI Button Text child

7. Create a second scene, named page2, with the UI text **Instructions (page 2)** and a **UI Button-TextMeshPro** with the text goto page 1. You can either repeat the preceding steps or duplicate the page1 scene file, name the duplicate **page2**, and then edit the UI TMP text and UI Button text appropriately.

8. Add both scenes to the build, which is the set of scenes that will end up in the actual application built by Unity. Open the **Build Settings** panel by choosing **File | Build Settings...**. Then drag the two scenes from the **Project** panel into the top section (**Scenes in Build**) of the **Build Settings** panel. Ensure the sequence goes page1 first, then page2 second – drag to rearrange them if necessary.

> We cannot tell Unity to load a scene that has not been added to the list of scenes in the build. This makes sense since when an application is built, we should never try to open a scene that isn't included as part of that application. The scene that appears first (index 0) in the **Build Settings** panel will be the first scene opened when the game is run.

9. Ensure you have the page1 scene open.

10. Create a new empty GameObject named SceneManager.

11. Create a C# script class called SceneLoader, in a new folder called _Scripts, that contains the following code. Then, add an instance of SceneLoader as a scripted component to the SceneManager GameObject:

```
using UnityEngine;
using UnityEngine.SceneManagement;

public class SceneLoader : MonoBehaviour {
    public void LoadOnClick(int sceneIndex) {
        SceneManager.LoadScene(sceneIndex);
    }
}
```

12. Select the **Button** GameObject in the **Hierarchy** panel and click on the plus (+) button at the bottom of the **Button (Script)** component in the **Inspector. This** will create a new **OnClick** event handler for this button (that is, a method to execute when the button is clicked).

13. Drag the SceneManager GameObject from the **Hierarchy** window over the **Object** slot immediately below the menu that says **Runtime Only**. This means that when the button receives an **OnClick** event, we can call a public method from a scripted object inside SceneManager.

14. Select the **LoadOnClick(int)** method from the **SceneLoader** drop-down list. Type 1 (the index of the scene we want to be loaded when this button is clicked) in the text box, below the method's drop-down menu.

    This integer, 1, will be passed to the method when the button receives an OnClick event message, as shown here:

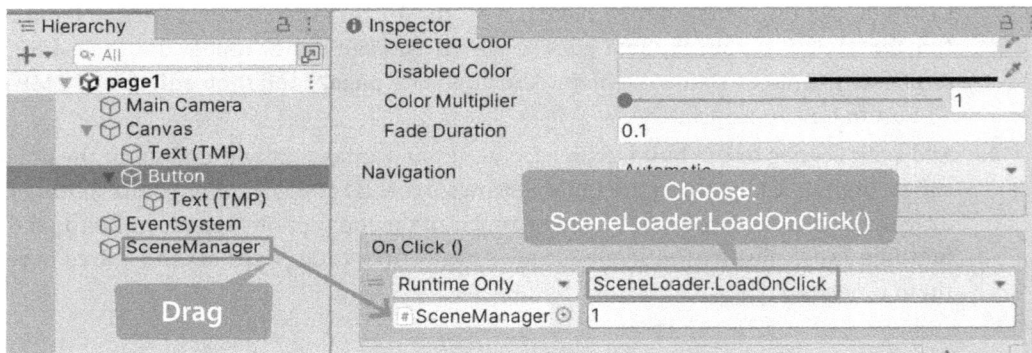

Figure 2.7: Button (Script) settings

Save the current scene (**page1**).

15. Open **page2** and follow the same steps to make the **page2** button load **page1**. That is, create a new empty SceneManager GameObject, add an instance of the SceneLoader script class to SceneManager, and then add an OnClick event action to the button that calls LoadOnClick and passes an integer of 0 so that **page1** is loaded.

16. Save **page2**.

17. When you run the **page1 scene,** you will be presented with your **Main Menu** text and a button that, when clicked, makes the game load the **page2 scene.** On **page2,** you'll have a button to take you back to **page1.**

How it works...

In this recipe, you created two scenes and added both of these scenes to the game's build. You added a UI Button and some UI Text to each scene.

When a UI Button is added to the **Hierarchy** window, a child UI Text object is also automatically created, and the content of the **Text Input** property of this UI Text child is the text that the user sees on the button.

Here, you created a script class and added an instance as a component to **GameObject SceneManager**. In fact, it didn't really matter where this script instance was added, so long as it was in one of the GameObjects of the scene. This is necessary since the OnClick event action of a button can only execute a method (function) of a component in a GameObject in the scene.

For the buttons for each scene, you added a new OnClick() event handler that invokes (executes) the LoadOnClick() method of the SceneLoader scripted component in **SceneManager**. This method inputs the integer index of the scene in the project's **Build Settings** so that the button on the **page1** scene gives integer 1 as the scene to be loaded and the button for **page2** gives integer 0.

There's more...

There are several ways in which we can visually inform the user that the button is interactive when they move their mouse over it. The simplest way is to add a **Color Tint** that will appear when the mouse is over the button – this is the default **Transition**. With **Button** selected in the **Hierarchy** window, choose a tint color (for example, red), for the **Highlighted Color** property of the **Button (Script)** component in the **Inspector** panel:

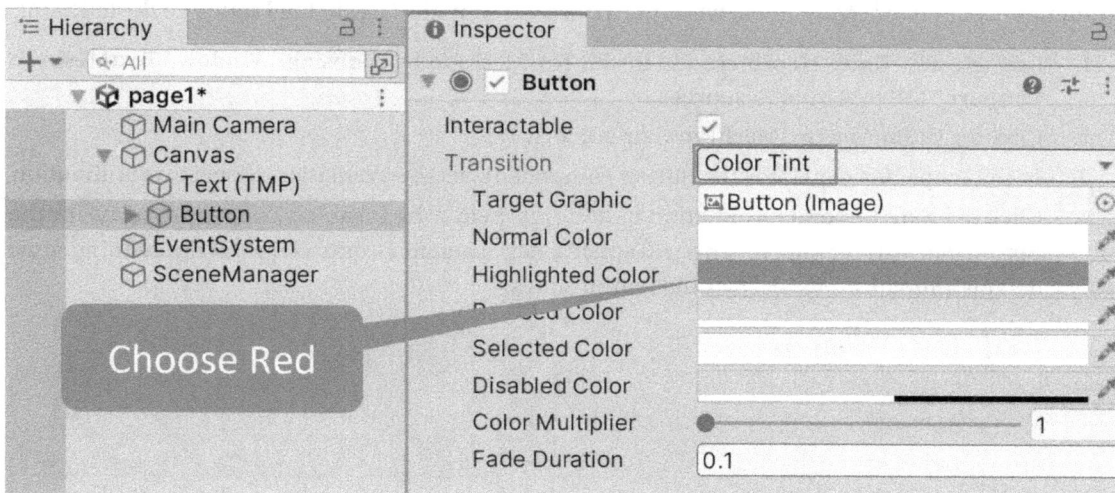

Figure 2.8: Adjusting the mouseover settings for buttons

Another form of visual **transition** to inform the user of an active button is **Sprite Swap**. In this case, the properties of different images for **Targeted/Highlighted/Pressed/Disabled** are available in the **Inspector** window. The default target graphic is the built-in Unity **Button (Image)** – this is the gray rounded rectangle default when GameObject buttons are created. Dragging in a very different-looking image for the highlighted sprite is an effective alternative to setting a **Color Tint**:

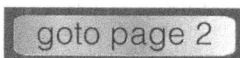

Figure 2.9: Example of an image as a button

We have provided a rainbow.png image in a folder named 02_02 that can be used for the **button** mouseover's **Highlighted sprite**. You will need to ensure this image asset has its **Texture Type** set to **Sprite (2D and UI)** in the **Inspector** window. The preceding screenshot shows the button with this rainbow background image.

Animating UI Button properties on mouseover

At the end of the previous recipe, we illustrated two ways to visually communicate buttons to users. The animation of button properties can be a highly effective and visually interesting way to reinforce to the user that the item their mouse is currently over is a clickable, active button. One common animation effect is for a button to become larger when the mouse is over it and then shrink back to its original size when the mouse is moved away. Animation effects are achieved by choosing the **Animation** option for the **Transition** property of a Button GameObject, and by creating an **Animation Controller** with triggers for the **Normal, Highlighted, Pressed,** and **Disabled** states.

How to do it...

To animate a button for enlargement when the mouse is over it (the **Highlighted** state), do the following:

1.   Create a new **Unity 2D project** and install TextMeshPro by choosing: **Window | TextMeshPro | Import TMP Essential Resources**.

2.   Create a **UI Button-TextMeshPro** GameObject.

3.   In the **Inspector** panel, for the **Button** component, set the **Transition** property to **Animation**.

4.   Click the **Auto Generate Animation** button (just below the **Disabled Trigger** property) for the **Button (Script)** component. This will create a new Animator Controller asset file defining some default animations for each of the button states.

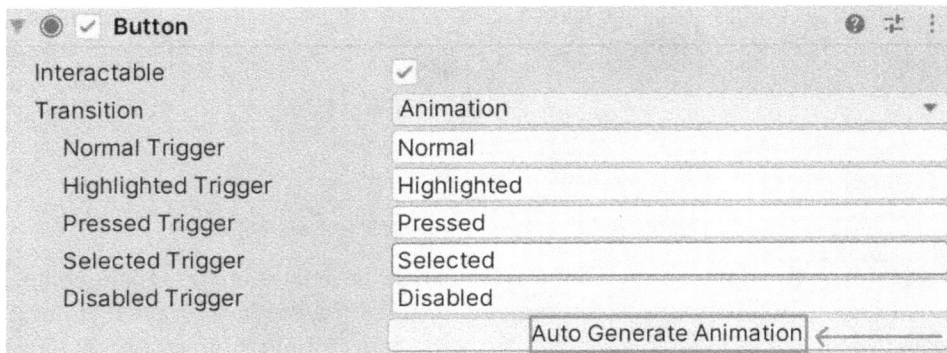

Figure 2.10: Auto Generate Animation

5.   Save the new controller (in a new folder called Animations), naming it button-animation-controller.

6.   Ensure that the Button GameObject is selected in the **Hierarchy** window. Open **Window | Animation | Animation**. In the **Animation** window, select the **Highlighted** clip from the drop-down menu:

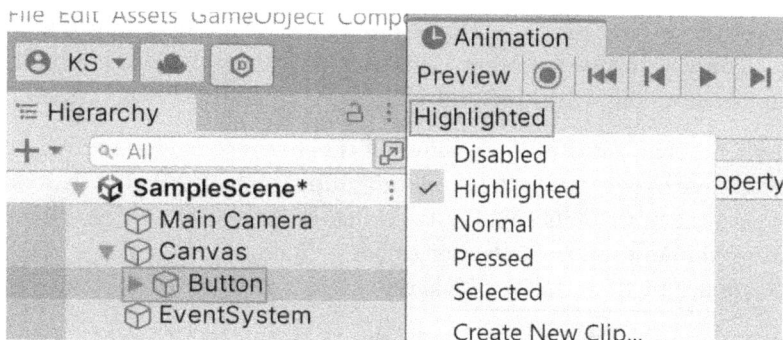

Figure 2.11: Selecting the Button GameObject in the Hierarchy window

7. In the **Animation** window, click on the red record circle button, and then click on the **Add Property** button, choosing to record changes to the **Rect Transform | Scale** property.

8. Two keyframes will have been created. Delete the second one at **1:00** (since we don't want a "bouncing" button):

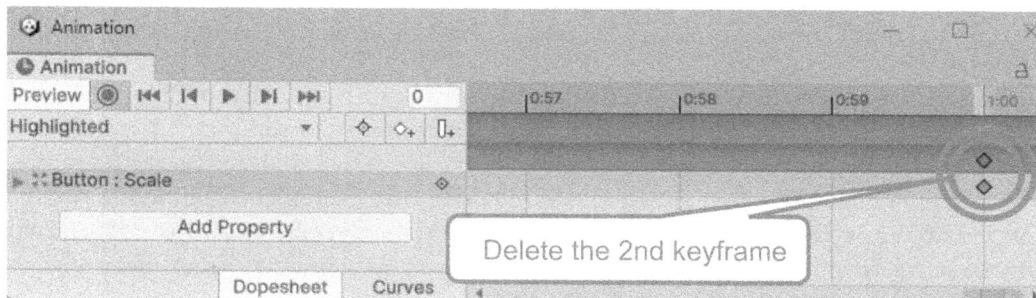

Figure 2.12: Deleting the keyframe

9. Select the frame at **1:00** by clicking one of the diamonds (both turn **blue** when selected), and then press the *Backspace/Delete* key.

10. Select the first keyframe at **0:00** (the only one now!). In the **Inspector** window, set the X and Y scale properties of the **Rect Transform** component to (1.2, 1.2).

11. Click on the red record circle button for the second time to stop recording the animation changes.

12. Save and run your scene. You will see that the button smoothly animates and becomes larger when the mouse is over it, and then smoothly returns to its original size when the mouse has moved away.

How it works...

In this recipe, you created a button and set its **Transition** mode to **Animation**. This makes Unity require an **Animation Controller** with four states: **Normal**, **Highlighted**, **Pressed**, and **Disabled**. You then made Unity automatically create an **Animation Controller** with these four states.

Then, you edited the animation for the **Highlighted** (mouseover) state, deleting the second keyframe, and making the only keyframe a version of the button that's larger so that its scale is 1.2. So, as is the case, if the GameObject has a **Scale** of 1 initially, when animating it will be scaled up to 1.2.

When the mouse is not hovering over the button, it's unchanged, and the **Normal** state settings are used. When the mouse moves over the button, the **Animation Controller** smoothly modifies the settings of the button to become those of its **Highlighted** state (that is, bigger). When the mouse is moved away from the button, the **Animation Controller** smoothly modifies the settings of the button to become those of its **Normal** state (that is, its original size).

The following web pages offer video and web-based tutorials on UI animations:

- The Unity documentation about UI button animations: `https://docs.unity3d.com/Packages/com.unity.ugui@1.0/manual/UIAnimationIntegration.html`
- Ray Wenderlich's great tutorial (part 2), including the available button animations, is available at `http://www.raywenderlich.com/79031/unity-new-gui-tutorial-part-2`

Organizing image panels and changing panel depths via buttons

UI Panels are provided by Unity to allow UI controls to be grouped and moved together, and also to visually group elements with an image background (if desired). The sibling depth is what determines which UI elements will appear above or below others. We can see the sibling depth explicitly in the **Hierarchy** window, since the top-to-bottom sequence of UI GameObjects in the **Hierarchy** window sets the sibling depth. So, the first item has a depth of 1, the second has a depth of 2, and so on. The UI GameObjects with larger sibling depths (further down the hierarchy, which means they're drawn later) will appear above the UI GameObjects with smaller sibling depths:

Figure 2.13: Example of organizing panels

In this recipe, we'll begin by creating two UI Panels, each showing a different playing card image, and we'll use one button to move between them. We'll then expand the recipe by creating one more UI Panel. We'll also add four triangle arrangement buttons to change the display order (move to bottom, move to top, move up one, and move down one).

Getting ready

For this recipe, we have prepared the images that you need in a folder named Images/ornamental_deck-png and Images /icons in the 02_04 folder.

How to do it...

To create the UI **Panels** whose layering can be changed by clicking buttons, follow these steps:

1. Create a new **Unity 2D project** and install TextMeshPro by choosing: **Window | TextMeshPro | Import TMP Essential Resources**.

2. Create a new **UI Panel** GameObject named Panel-jack-diamonds. Do the following to this panel:

    - For the **Image (Script)** component, drag the jack_of_diamonds playing card image asset file from the **Project** window into the **Source Image** property. Select the **Color** property and increase the **Alpha** value to 255 (so that this background image of the panel is no longer partly transparent).

    - For the **Rect Transform** property, position it in the middle-center part of the screen and set its **Width** to 200 and its **Height** to 300.

3. Create a **UI Button-TextMeshPro** GameObject named Button-move-to-front. In the **Hierarchy** window, make this button a child of **Panel-jack-diamonds**. Delete the **Text** child GameObject of this button (since we'll use an icon to indicate what this button does).

4. With the **Button-move-to-front** GameObject selected in the **Hierarchy** window, do the following in the **Inspector** window:

    - In **Rect Transform**, position the button at the top-center of the player card image so that it can be seen at the top of the playing card. Size the image to **Width** = 16 and **Height** = 16. Move the icon image down slightly, by setting **Pos Y** = -5 (to ensure we can see the horizontal bar above the triangle).

    - For the **Source Image** property of the **Image (Script)** component, select the arrangement triangle icon image; that is, **icon_move_to_front**.

    - Add an **OnClick** event handler by clicking on the plus (+) sign at the bottom of the **Button (Script)** component.

    - Drag Panel-jack-diamonds from the **Hierarchy** window over to the **Object** slot (immediately below the menu saying **Runtime Only**).

- Select the **RectTransform.SetAsLastSibling** method from the drop-down function list (initially showing **No Function**):

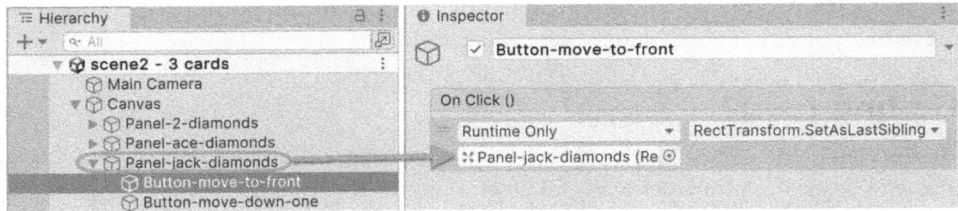

Figure 2.14: Addition of an OnClick event handler

5. Repeat *step 2* to create a second panel named `Panel-2-diamonds` with its own **move-to-front** button and a **Source Image** of `2_of_diamonds`. Move and position this new panel slightly to the right of `Panel-jack-diamonds`, allowing both **move-to-front** buttons to be seen.

6. Save your scene and run the game. You will be able to click the **move-to-front** button on either of the cards to move that card's panel to the front. If you run the game with the **Game** window not maximized, you'll actually see the panels changing the order in the list of the children of **Canvas** in the **Hierarchy** window.

How it works...

In this recipe, you created two UI Panels, each of which contains a background image of a playing card and a UI Button whose action will make its parent panel move to the front. You set the **Alpha** (transparency) setting of the background image's **Color** setting to **255** (no transparency).

You then added an **OnClick** event handler to the button of each **UI Panel**. This action sends a `SetAsLastSibling` message to the button's **panel** parent. When the **OnClick** message is received, the clicked panel is moved to the bottom (end) of the sequence of GameObjects in the **Canvas**, so this **panel** is drawn last from the **Canvas** objects. This means that it appears visually in front of all the other GameObjects.

The button's action illustrates how the **OnClick** function does not have to be calling a public method of a scripted component of an object, but it can be sending a message to one of the non-scripted components of the targeted GameObject. In this recipe, we send the **SetAsLastSibling** message to the **Rect Transform** component of the **panel** where the **button** is located.

There's more...

There are some details you don't want to miss.

Moving up or down by just one position, using scripted methods

While **Rect Transform** offers `SetAsLastSibling` (move to front) and `SetAsFirstSibling` (move to back), and even `SetSiblingIndex` (if we knew exactly what position in the sequence to type in), there isn't a built-in way to make an element move up or down just one position in the sequence of GameObjects in the **Hierarchy** window.

However, we can write two straightforward methods in C# to do this, and we can add buttons to call these methods, providing full control of the top-to-bottom arrangement of the UI controls on the screen. To implement four buttons (move-to-front/move-to-back/up one/down one), do the following:

1. Create a C# script class called `ArrangeActions` containing the following code and add an instance as a scripted component to each of your UI Panels:

```csharp
using UnityEngine;

public class ArrangeActions : MonoBehaviour {
    private RectTransform panelRectTransform;

    void Awake() {
        panelRectTransform = GetComponent<RectTransform>();
    }

    public void MoveDownOne() {
        print("(before change) " + gameObject.name + " sibling index
= " + panelRectTransform.GetSiblingIndex());

        int currentSiblingIndex = panelRectTransform.
GetSiblingIndex();
        if (currentSiblingIndex > 0) {}
            panelRectTransform.
SetSiblingIndex(currentSiblingIndex - 1);
        }

        print("(after change) " + gameObject.name + " sibling index
= " + panelRectTransform.GetSiblingIndex());
    }

    public void MoveUpOne() {
        print ("(before change) " + gameObject.name +  " sibling
index = " + panelRectTransform.GetSiblingIndex());

        int currentSiblingIndex = panelRectTransform.
GetSiblingIndex();
        int maxSiblingIndex = panelRectTransform.childCount - 1;
        if (currentSiblingIndex < maxSiblingIndex) {
            panelRectTransform.
SetSiblingIndex(currentSiblingIndex + 1);
        }
```

```
                    print ("(after change) " + gameObject.name +  " sibling
        index = " + panelRectTransform.GetSiblingIndex());
            }
        }
```

2. Add a second UI Button to each card panel, this time using the arrangement triangle icon image called **icon_move_to_back**, and set the **OnClick** event function for these buttons to **SetAsFirstSibling**.

3. Add two more **UI Buttons** to each card panel with the up and down triangle icon images; that is, **icon_up_one** and **icon_down_one**. Set the **OnClick** event handler function for the down-one buttons to call the MoveDownOne() method and set the function for the up-one buttons to call the MoveUpOne() method.

4. Copy one of the **UI Panels** to create a third card (this time showing the ace of diamonds). Arrange the three cards so that you can see all four buttons for at least two of the cards, even when those cards are at the bottom (see the screenshot at the beginning of this recipe).

5. Save the scene and run your game. You will now have full control over how to layer the three card **UI Panels**.

> Note that the MoveDownOne() and MoveUpOne() methods subtract and add 1 to the sibling depth of the panel the scripted object is a component of. The methods contain a test, so ensure we don't try to set a negative panel depth or a depth that is higher than the maximum index for the number of panels.

Displaying the value of an interactive UI Slider

A UI **Slider** is a graphical tool that allows a user to set the numerical value of an object.

Figure 2.15: Example of a UI Slider offering a range of 0 to 20

This recipe illustrates how to create an interactive **UI Slider** and execute a C# method each time the user changes the **UI Slider** value.

How to do it...

To create a **UI Slider** and display its value on the screen, follow these steps:

1. Create a new **Unity 2D project** and install TextMeshPro by choosing: **Window | TextMeshPro | Import TMP Essential Resources**.

2. Add a **UI Text-TextMeshPro** GameObject to the scene with a **Font** size of 30 and placeholder text, such as Slider value here (this text will be replaced with the slider value when the scene starts). Set **Overflow** to **Overflow**. Since we may change the font and message at a later date, it's useful to allow overflow to prevent some of the message being truncated.

3. In the **Hierarchy** window, add a UI Slider GameObject to the scene by going to **GameObject | UI | Slider**.

4. In the **Inspector** window, modify the settings for the position of the UI Slider GameObject's **Rect Transform** to the top-middle part of the screen.

5. In the **Inspector** window, modify the settings of **Position** for the UI Text's **Rect Transform** so that they're just below the slider (top, middle, then **Pos Y = -30**).

6. In the **Inspector** window, set the **UI Slider's Min Value** to 0 and **Max Value** to 20. Then, check the **Whole Numbers** checkbox:

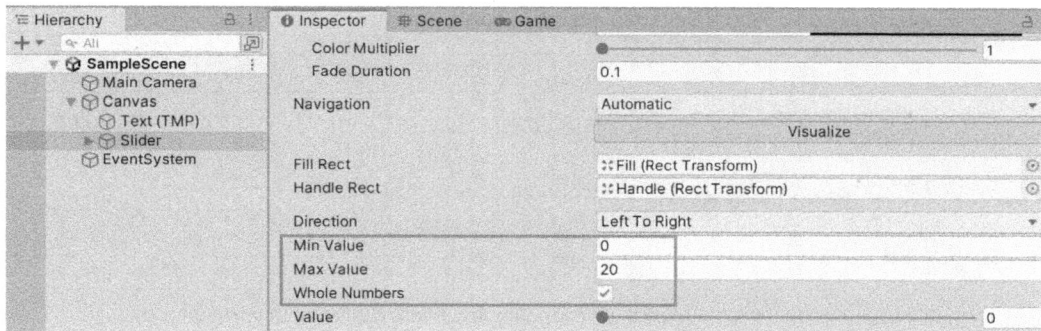

Figure 2.16: Setting the UI Slider's Min Value and Max Value

7. Create a C# script class called SliderValueToText containing the following code and add an instance as a scripted component to the Text(TMP)GameObject:

```csharp
using UnityEngine;
using UnityEngine.UI;
using TMPro;


public class SliderValueToText : MonoBehaviour {
    public Slider sliderUI;
    private TextMeshProUGUI textSliderValue;

    void Awake() {
        textSliderValue = GetComponent<TextMeshProUGUI>();
    }
```

```
void Start() {
    ShowSliderValue();
}

public void ShowSliderValue () {
    string sliderMessage = "Slider value = " + sliderUI.value;
    textSliderValue.text = sliderMessage;
}
}
```

8. Ensure that the Text(TMP) GameObject is selected in the **Hierarchy** window. Then, in the **Inspector** window, drag the **Slider** GameObject into the public **Slider UI** variable slot for the Slider Value To Text (Script) scripted component:

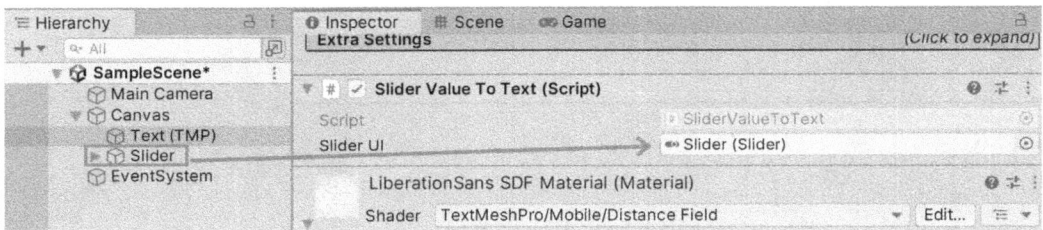

Figure 2.17: Dragging Slider into the Slider UI variable

9. Ensure that the Slider GameObject is selected in the **Hierarchy** window. Then, in the **Inspector** window, add an **OnValue Changed (Single) event** handler by clicking on the plus (+) sign at the bottom of the **Slider** component.

10. Drag the Text(TMP) GameObject from the **Hierarchy** window over to the **Object** slot (immediately below the menu that says **Runtime Only**), as shown in the following screenshot:

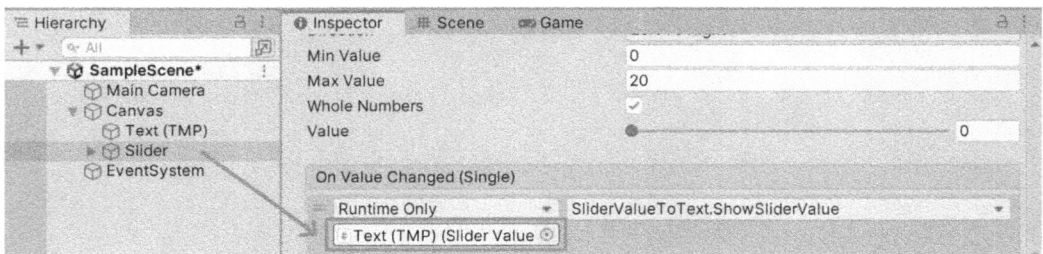

Figure 2.18: Dragging the Text GameObject into None (Object)

You have now told Unity which object a message should be sent to each time the slider is changed.

11. From the drop-down menu, select **SliderValueToText** and the **ShowSliderValue** method, as shown in the following screenshot. This means that each time the slider is updated, the ShowSliderValue() method, in the scripted object in the **Text(TMP)** GameObject, will be executed:

*Figure 2.19: Drop-down menu for **On Value Changed***

12. When you run the scene, you will see a **UI Slider**. Below it, you will see a text message in the form Slider value = <n>.

13. Each time the **UI Slider** is moved, the text value that's shown will be (almost) instantly updated. The values should range from 0 (the leftmost of the slider) to 20 (the rightmost of the slider).

How it works...

In this recipe, you created a **UI Slider** GameObject and set it to contain whole numbers in the range of 0 to 20.

You also added an instance of the SliderValueToText C# script class to the UI Text(TMP) GameObject.

The Awake() method caches references to the **Text** component in the textSliderValue variable.

The Start() method invokes the ShowSliderValue() method so that the display is correct when the scene begins (that is, the initial slider value is displayed).

The ShowSliderValue() method gets the value of the slider and then updates the text that's displayed to be a message in the form of Slider value = <n>.

Finally, you added the ShowSliderValue() method of the SliderValueToText scripted component to the **Slider** GameObject's list of **On Value Changed** event listeners. So, each time the slider value changes, it sends a message to call the ShowSliderValue() method so that the new value is updated on the screen.

Displaying a countdown timer graphically with a UI Slider

There are many cases where we wish to inform the player of how much time is left in a game or how much longer an element will take to download – for example, a loading progress bar, the time or health remaining compared to the starting maximum, or how much the player has filled up their water bottle from the fountain of youth.

In this recipe, we'll illustrate how to remove the interactive "handle" of a **UI Slider**, and then change the size and color of its components to provide us with an easy-to-use, general-purpose progress/proportion bar:

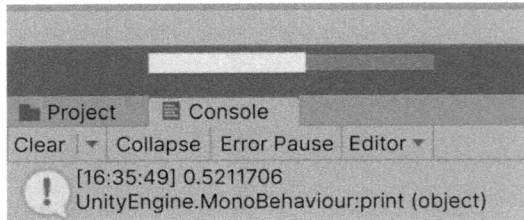

Figure 2.20: Example of a countdown timer with a **UI Slider**

In this recipe, we'll use our modified **UI Slider** to graphically present to the user how much time remains on a countdown timer.

Getting ready

For this recipe, we have prepared the script and images that you need in the 02_04 folder, respectively named _Scripts and Images.

How to do it...

To create a digital countdown timer with a graphical display, follow these steps:

1. Create a new **Unity 2D project** and install TextMeshPro by choosing: **Window | TextMeshPro | Import TMP Essential Resources**.

2. Import the CountdownTimer script and the red_square and green_square images into this project.

3. Add a UI Text(TMP) GameObject to the scene with a **Font** size of 30 and placeholder text such as a UI Slider value (this text will be replaced with the slider value when the scene starts). Check that **Overflow** is set to **Overflow**.

4. In the **Hierarchy** window, add a Slider GameObject to the scene by going to **GameObject | UI | Slider**.

5. In the **Inspector** window, modify the settings for the position of the Slider GameObject's **Rect Transform** to the top-center part of the screen.

6. Ensure that the Slider GameObject is selected in the **Hierarchy** window.

7. Deactivate the Handle Slide Area child GameObject (by unchecking it).

8. You'll see the "drag circle" disappear in the **Game** window (the user will not be dragging the slider since we want this slider to be display-only):

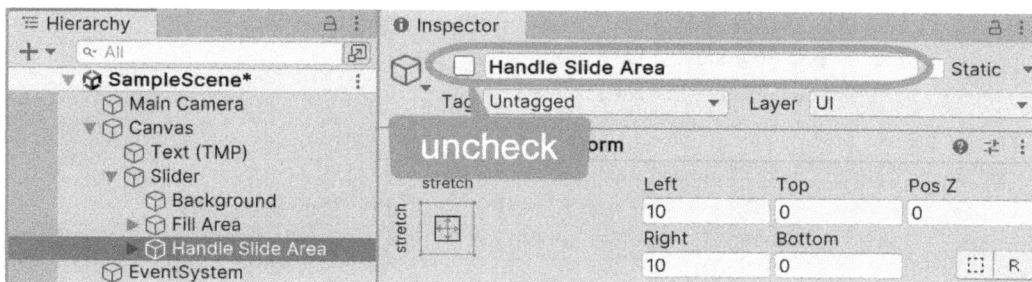

Figure 2.21: Ensuring Handle Slide Area is deactivated

9. Select the **Background** child and do the following:

    • Drag the red_square image into the **Source Image** property of the **Image** component in the **Inspector** window.

10. Select the Fill child of the Fill Area child and do the following:

    • Drag the green_square image into the **Source Image** property of the **Image** component in the **Inspector** window.

11. Select the Fill Area child and do the following:

    • In the **Rect Transform** component, use the **Anchors** preset position of left-middle.
    • Set **Width** to 155 and **Height** to 12:

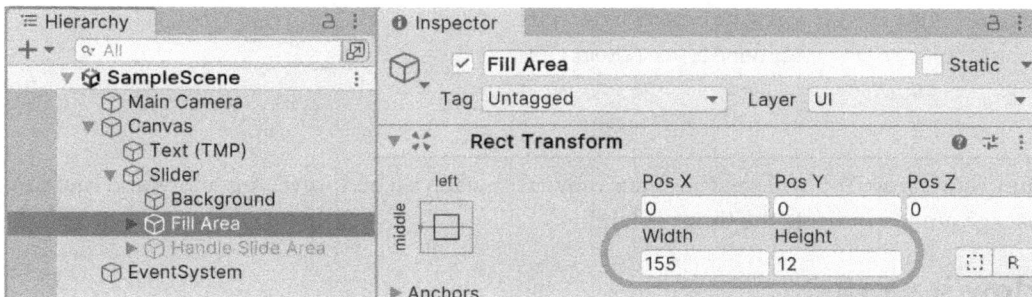

Figure 2.22: Selections in the Rect Transform component

12. Create a C# script class called SliderTimerDisplay that contains the following code and add an instance as a scripted component to the Slider GameObject:

```
using UnityEngine;
using UnityEngine.UI;
using TMPro;


[RequireComponent(typeof(CountdownTimer))]
```

```csharp
public class SliderTimerDisplay : MonoBehaviour {
    private CountdownTimer countdownTimer;
    private Slider sliderUI;

    void Awake() {
        countdownTimer = GetComponent<CountdownTimer>();
        sliderUI = GetComponent<Slider>();
    }

    void Start() {
        SetupSlider();
        countdownTimer.ResetTimer( 30 );
    }

    void Update () {
        sliderUI.value = countdownTimer.GetProportionTimeRemaining();
        print (countdownTimer.GetProportionTimeRemaining());
    }

    private void SetupSlider () {
        sliderUI.minValue = 0;
        sliderUI.maxValue = 1;
        sliderUI.wholeNumbers = false;
    }
}
```

Run your game. You will see the slider move with each second, revealing more and more of the red background to indicate the time remaining.

How it works...

In this recipe, you hid the **Handle Slide Area** child so that the **UI Slider** is for display only, which means it cannot be interacted with by the user. The **Background** color of the **UI Slider** was set to red so that, as the counter goes down, more and more red is revealed, warning the user that the time is running out.

The **Fill** property of the **UI Slider** was set to green so that the proportion remaining is displayed in green – the more green that's displayed, the greater the value of the slider/timer.

An instance of the provided CountdownTimer script class was automatically added as a component to the **UI Slider** via [RequireComponent(...)].

The Awake() method caches references to the **CountdownTimer** and **Slider** components in the countdownTimer and sliderUI variables.

The `Start()` method calls the `SetupSlider()` method and then resets the countdown timer so that it starts counting down from 30 seconds.

The `SetupSlider()` method sets up this slider for float (decimal) values between `0.0` and `1.0`.

In each frame, the `Update()` method sets the slider value to the float that's returned by calling the `GetProportionRemaining()` method from the running timer. At runtime, Unity adjusts the proportion of red/green that's displayed in the **UI Slider** so that it matches the slider's value.

Setting custom mouse cursors for 2D and 3D GameObjects

Cursor icons are often used to indicate the nature of the interactions that can be done with the mouse. Zooming, for instance, might be illustrated by a magnifying glass; shooting, on the other hand, is usually represented by a stylized target or reticle:

Figure 2.23: Mouse pointer represented as a stylized target

The preceding screenshot shows an example of the Unity logo with the cursor represented as a stylized target. In this recipe, we will learn how to implement custom mouse cursor icons to better illustrate your gameplay – or just to escape the Windows, macOS, and Linux default UI.

Getting ready

For this recipe, we have prepared the folders that you'll need in the `02_07` folder.

How to do it...

To make a custom cursor appear when the mouse is over a GameObject, follow these steps:

1. Create a new **Unity 2D project**.
2. Import the provided folder, called `Images`. Select the **unity_logo** image in the **Project** window. Then, in the **Inspector** window, change **Texture Type** to **Sprite (2D and UI)**. This is because we'll use this image for a 2D `Sprite` GameObject and it requires this **Texture Type** (it won't work with the **Default** type).
3. Go to **GameObject | 2D Object | Sprites | Square** to add the necessary GameObject to the scene. Name this `New Sprite`, if this wasn't the default name when it was created:

   - In the **Inspector** window, set the **Sprite** property of the **Sprite Renderer** component to the **unity_logo** image. In the GameObject's **Transform** component, set the scaling to (`3,3,3`) and, if necessary, reposition **Sprite** so that it's centered in the **Game** window when the scene runs.

- Go to **Component** | **Physics 2D** | **Box Collider 2D** to create a **Box Collider** and add it to the Sprite GameObject. This is needed for this GameObject to receive `OnMouseEnter` and `OnMouseExit` event messages.

4. Import the provided folder called `IconsCursors`. Select all three images in the **Project** window and, in the **Inspector** window, change **Texture Type** to **Cursor**. This will allow us to use these images as mouse cursors without any errors occurring.

5. Create a C# script class called `CustomCursorPointer` containing the following code and add an instance as a scripted component to the New Sprite GameObject:

```
using UnityEngine;

public class CustomCursorPointer : MonoBehaviour {
  public Texture2D cursorTexture2D;
  private CursorMode cursorMode = CursorMode.Auto;
  private Vector2 hotSpot = Vector2.zero;

  public void OnMouseEnter() {
    SetCustomCursor(cursorTexture2D);
  }

  public void OnMouseExit() {
    SetCustomCursor(null);
  }

  private void SetCustomCursor(Texture2D curText){
    Cursor.SetCursor(curText, hotSpot, cursorMode);
  }
}
```

The `OnMouseEnter()` and `OnMouseExit()` event methods have been deliberately declared as `public`. This will allow these methods to also be called from UI GameObjects when they receive the `OnPointerEnterExit` events.

6. With the **New Sprite** item selected in the **Hierarchy** window, drag the `CursorTarget` image into the public **Cursor Texture 2D** variable slot in the **Inspector** window for the **Custom Cursor Pointer (Script)** component:

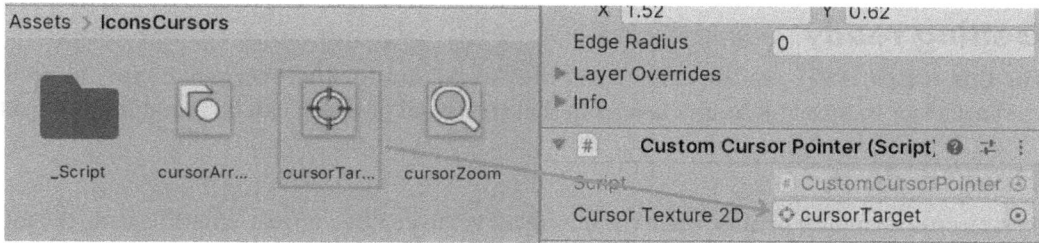

Figure 2.24: Cursor Texture 2D dragged to the variable slot

7. Save and run the current scene. When the mouse pointer moves over the Unity logo sprite, it will change to the custom **CursorTarget** image that you chose.

How it works...

In this recipe, you created a `Sprite` GameObject and assigned it the Unity logo image. You imported some cursor images and set their **Texture Type** to **Cursor** so that they can be used to change the image for the user's mouse pointer. You also added a **Box Collider** to the `Sprite` GameObject so that it would receive **OnMouseEnter** and **OnMouseExit** event messages.

Then, you created the `CustomCursorPointer` script class and added an instance object of this class to the `Sprite` GameObject. This script tells Unity to change the mouse pointer when an **OnMouseEnter** message is received – that is, when the user's mouse pointer moves over the part of the screen where the Unity logo's sprite image is being rendered. When an **OnMouseExit** event is received (the user's mouse pointer is no longer over the cube part of the screen), the system is told to go back to the operating system's default cursor. This event should be received within a few milliseconds of the user's mouse exiting from the collider.

Finally, you selected the `CursorTarget` image to be the custom mouse cursor image the user sees when the mouse is over the Unity logo image.

Setting custom mouse cursors for UI controls

The previous recipe demonstrated how to change the mouse pointer for 2D and 3D GameObjects receiving **OnMouseEnter** and **OnMouseExit** events. Unity UI controls do not receive **OnMouseEnter** and **OnMouseExit** events. Instead, UI controls can be made to respond to **PointerEnter** and **PointerExit** events if we add a special **Event Trigger** component to the UI GameObject:

Figure 2.25: Mouse pointer as a magnifying glass cursor

In this recipe, we'll change the mouse pointer to a custom magnifying glass cursor when it moves over a UI `Button` GameObject.

Getting ready

For this recipe, we'll use the same asset files as we did for the previous recipe, as well as the CustomCursorPointer C# script class from that recipe, all of which can be found in the 02_08 folder.

How to do it...

To set a custom mouse pointer when the mouse moves over a UI control GameObject, do the following:

1. Create a new **Unity 2D project** and install TextMeshPro by choosing: **Window | TextMeshPro | Import TMP Essential Resources**.

2. Import the provided IconsCursors folder. Select all three images in the **Project** window and, in the **Inspector** window, change **Texture Type** to **Cursor**. This will allow us to use these images as mouse cursors without any errors occurring.

3. Import the provided _Scripts folder containing the CustomCursorPointer C# script class.

4. Add a UI Button-TextMeshPro GameObject to the scene, leaving this named **Button.**

5. Add an instance of the CustomCursorPointer C# script class to the Button **GameObject.**

6. With the Button GameObject selected in the **Hierarchy** window, drag the CursorZoom image into the public **Cursor Texture 2D** variable slot in the **Inspector** window for the **Customer Cursor Pointer (Script)** component.

7. In the **Inspector** window, add an **Event Trigger** component to the Button GameObject by going to **Add Component | Event | Event Trigger**.

8. Add a **PointerEnter** event to your **Event Trigger** component, click on the plus (+) button to add an event handler slot, and drag the Button GameObject into the **Object** slot.

9. From the **Function** drop-down menu, choose **CustomCursorPointer** and then choose the **OnMouseEnter** method:

Figure 2.26: Event Trigger settings

10. Add a **Pointer Exit** event to your **Event Trigger** component, and make it call the OnMouseExit() method from **CustomCursorPointer** when this event is received.

11. Save and run the current scene. When the mouse pointer moves over our **UI Button,** it will change to the custom **CursorZoom** image that you chose.

How it works...

In this recipe, you imported some cursor images and set their **Texture Type** to **Cursor** so that they could be used to change the image for the user's mouse pointer. You also created a **UI Button** GameObject and added to it an **Event Trigger** component.

You then added an instance of the CustomCursorPointer C# script class to the Button GameObject and selected the magnifying-glass-style CursorZoom image.

After that, you created a **PointerEnter** event and linked it to invoke the **OnMouseEnter** method of the instance of the CustomCursorPointer script in the Button GameObject (which changes the mouse pointer image to the custom mouse cursor).

Finally, you created a PointerExit event and linked it to invoke the **OnMouseExit** method of the instance of the CustomCursorPointer C# script class to the Button GameObject (which resets the mouse cursor back to the system default).

Essentially, you have redirected PointerEnter/Exit events to invoke the OnMouseEnter/Exit methods of the CustomCursorPointer C# script class so that we can manage custom cursors for 2D, 3D, and UI GameObjects with the same scripting methods.

Interactive text entry with Input Field

While we often just wish to display non-interactive text messages to the user, there are times (such as name entry for high scores) where we want the user to be able to enter text or numbers into our game. Unity provides the UI Input Field component for this purpose. In this recipe, we'll create an input field that prompts the user to enter their name:

Name: [Matt] last entry = 'Matt'

Figure 2.27: Example of interactive text entry

Having interactive text on the screen isn't of much use unless we can *retrieve* the text that's entered to be used in our game logic, and we may need to know each time the user changes the text's content and act accordingly. In this recipe, we'll add an event handler C# script that detects each time the user finishes editing the text and updates an extra message onscreen, confirming the newly entered content.

Getting ready

For this recipe, we have prepared the required assets in a folder named IconsCursors in the 02_09 folder.

How to do it...

To create an interactive text input box for the user, follow these steps:

1. Create a new **Unity 2D project** and install TextMeshPro by choosing: **Window | TextMeshPro | Import TMP Essential Resources**.

2. Change the background of **Main Camera** to solid white. Do this in the **Inspector** for the Camera component by setting **Clear Flags** to **Solid Color**, and then choosing a white color for the **Background** property. The complete background of the **Game** panel should now be white.

3. Add a **UI Input Field - TextMeshPro** named `InputField` to the scene (importing TMP Essential Resources). Position this at the top center of the screen.

4. Add a **UI Text - TextMeshPro** GameObject to the scene, naming it `Text-prompt`. Position this to the left of **Input Field**. Change the **Text** property of this GameObject to `Name:`.

5. Create a new **UI Text - TextMeshPro** GameObject named **Text-display**. Position this to the right of the **Input Text** control, and make its text **red**.

6. Delete all of the content of the **Text** property of GameObject **Text-display** (so that, initially, the user won't see any text onscreen for this GameObject).

7. Create a new C# script class named `DisplayChangedTextContent` containing the following code:

```
using UnityEngine;
using TMPro;

public class DisplayChangedTextContent : MonoBehaviour
{
    public TMP_InputField inputField;
    private TextMeshProUGUI textDisplay;

    void Awake()
    {
        textDisplay = GetComponent<TextMeshProUGUI>();
    }

    public void DISPLAY_NEW_VALUE()
    {
        textDisplay.text = "last entry = '" + inputField.text + "'";
    }
}
```

8. Add an instance of the `DisplayChangedTextContent` C# script class to the `Text-display` GameObject.

9. With `Text-display` selected in the **Hierarchy** window, from the **Project** window, drag the **InputField** GameObject into the public **Input Field** variable of the **Display Changed Text Content (Script)** component:

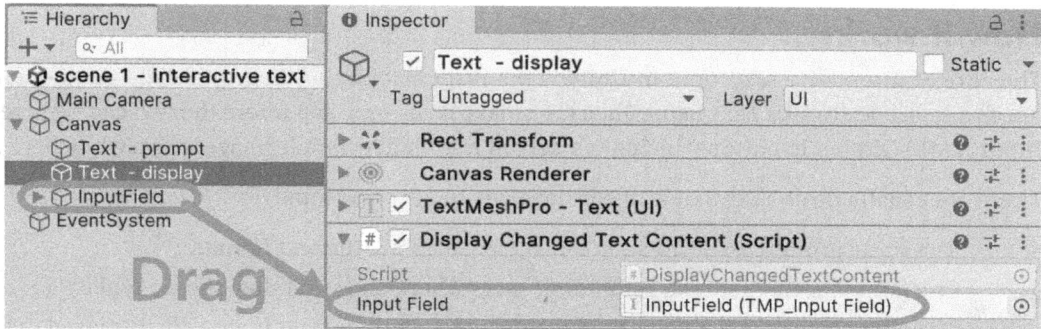

Figure 2.28: Setting the Input Field variable

10. Select the Input Field GameObject in the **Hierarchy**. Add an **On Value Changed (String)** event to the list of event handlers for the **Input Field (Script)** component. Click on the plus (+) button to add an event handler slot and drag the **Text-display** GameObject into the **Object** slot.

11. Then, from the **Function** drop-down menu, choose the **DisplayChangedTextContent** component and then choose the **DISPLAY_NEW_VALUE** method (we named this in capitals to make it easy to find in the menu!).

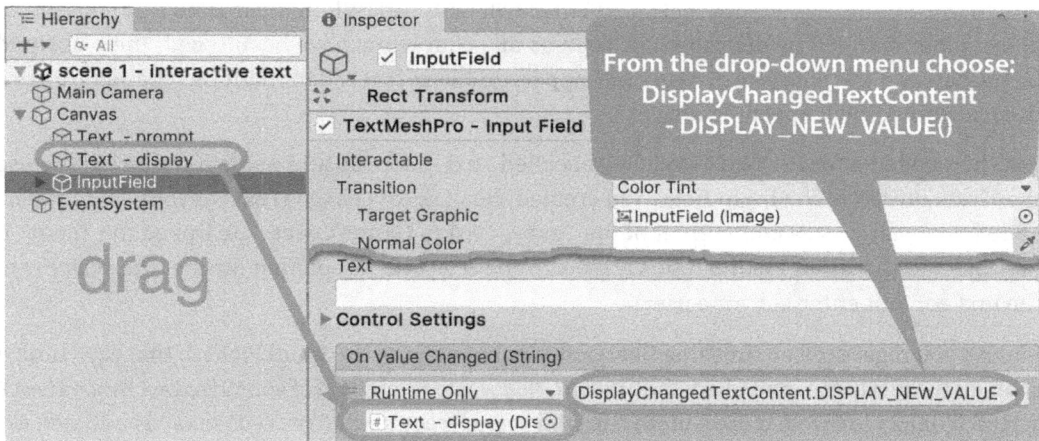

Figure 2.29: Making DISPLAY_NEW_VALUE the method to execute each time the Input Text changes

12. Save and run the scene. Each time the user changes the text in the **input field**, the **On Value Changed** event will fire, and you'll see a new content text message displayed in red on the screen.

How it works...

The core of interactive text input in Unity is the responsibility of the **Input Field** component. This needs a reference to a UI **Text** GameObject. To make it easier to see where the text can be typed, **Text Input (TMP)** (similar to buttons) includes a default rounded rectangle image with a white background.

There are usually three **Text** GameObjects involved in user text input:

- The static prompt text, which, in our recipe, displays the text **Name:**.
- The faint placeholder text, reminding users where and what they should type.
- The editable text object (with the font and color settings) that is actually displayed to the user, showing the characters as they type.

First, you created an `InputField` GameObject, which automatically provides two child `Text` GameObjects, named `Placeholder` and `Text`. These represent the faint placeholder text and the editable text, which you renamed `Text-input`. You then added a third `Text` GameObject, `Text-prompt`, containing **Name:**.

The built-in scripting that is part of **Input Field** components does lots of work for us. At runtime, a **Text-Input Input Caret** (text insertion cursor) GameObject is created, displaying the blinking vertical line to inform the user where their next letter will be typed. When there is no text content, the faint placeholder text will be displayed. As soon as any characters have been typed, the placeholder will be hidden and the characters typed will appear in black text. Then, if all the characters are deleted, the placeholder will appear again.

You then added a fourth **Text** GameObject called `Text-display` and made it red to tell the user what they last entered in the **input field**. You created the `DisplayChangedTextContent` C# script class and added an instance as a component of the `Text-display` **GameObject**. You linked the `InputField` GameObject to the **Input Field** public variable of the scripted component (so that the script can access the text content entered by the user).

Finally, you registered an **On Value Changed** event handler of the **Input Field** so that each time the user changes the text in the **input field**, the `DISPLAY_NEW_VALUE()` method of your `DisplayChangedTextContent` scripted object is invoked (executed), and the red text content of `Text-display` is updated to tell the user what the newly edited text consisted of.

> I suggest you use capital letters and underscores when naming methods you plan to register for event handlers, such as our input field **On Value Changed** event. This makes them much easier to find navigating component methods when setting up the event handler in the **Inspector**.

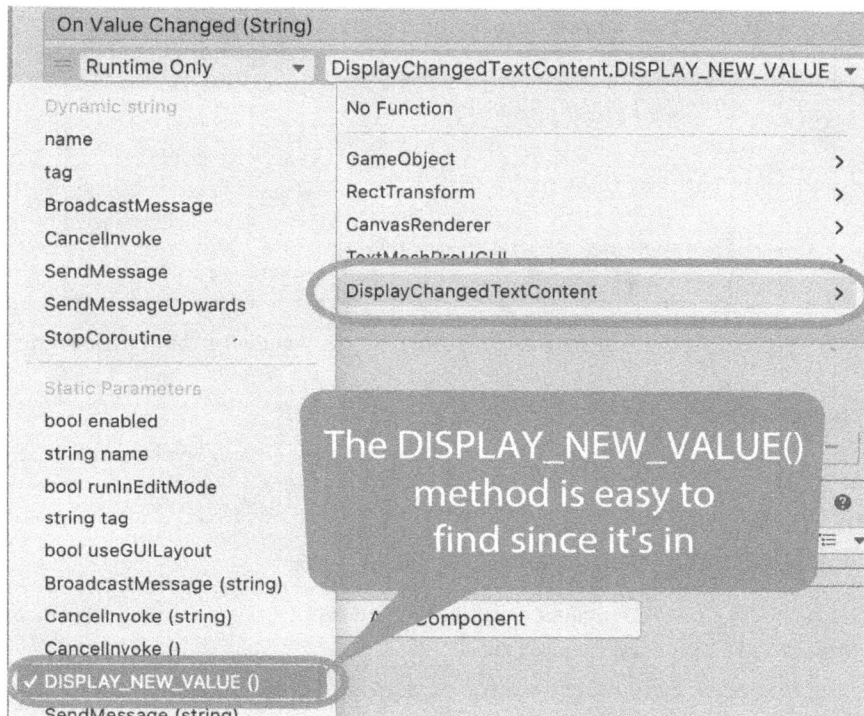

Figure 2.30: Making DISPLAY_NEW_VALUE the method to execute each time the Input Text changes

There's more...

Rather than the updated text being displayed with every key press, we can wait until *Tab* or *Enter* is pressed to submit a new text string by using an **On End Edit** event handler rather than **On Value Changed.**

Also, the content type of **Input Field (Script)** can be set (restricted) to several specific types of text input, including email addresses, integer or decimal numbers only, or password text (where an asterisk is displayed for each character that's entered). You can learn more about **input fields** by reading the Unity Manual page: https://docs.unity3d.com/Manual/script-InputField.html.

Detecting interactions with a single Toggle UI component

Users make choices and, often, these choices have *one of two* options (for example, sound on or off), or sometimes *one of several* possibilities (for example, difficulty level as easy/medium/hard). Unity **UI Toggles** allows users to turn options on and off; when combined with **toggle groups**, they restrict choices to one of the groups of items.

In this recipe, we'll explore the basic **Toggle** and a script to respond to a change in values.

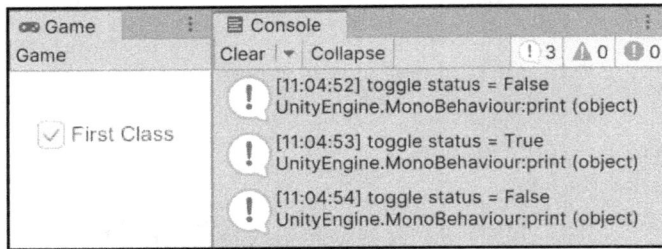

Figure 2.31: Example showing the button's status changing in the Console window

Getting ready

For this recipe, we have prepared the C# script `ToggleChangeManager` class in the `02_10` folder.

How to do it...

To display an on/off **UI Toggle** to the user, follow these steps:

1. Create a new **Unity 2D project** and install TextMeshPro by choosing: **Window | TextMeshPro | Import TMP Essential Resources**.

2. In the **Inspector** window, change the **Background** color of **Main Camera** to white.

3. Add a **UI Toggle** to the scene.

4. For the **Label** child of the `Toggle` GameObject, set the **Text** property to **First Class**.

5. Add an instance of the C# script class called `ToggleChangeManager` to the `Toggle` GameObject:

```
using UnityEngine;
using UnityEngine.UI;

public class ToggleChangeManager : MonoBehaviour {
    private Toggle toggle;

    void Awake () {
        toggle = GetComponent<Toggle>();
    }

    public void PrintNewToggleValue() {
        bool status = toggle.isOn;
        print ("toggle status = " + status);
    }
}
```

6. With the `Toggle` GameObject selected, add an **On Value Changed** event to the list of event handlers for the **Toggle (Script)** component, click on the plus (+) button to add an event handler slot, and drag `Toggle` into the **Object** slot.

Note. If this is the first time you have done this, it may seem strange to select a GameObject, and then drag this same GameObject into a property of one of its components. However, this is quite common, since the logic location for behavior like button actions and scripted actions is a component of the GameObject whose behavior is being set. So, it is correct to drag the Toggle GameObject into the Object slot for that toggle's On Value Changed event handler.

7. From the **Function** drop-down menu, choose **ToggleChangeManager** and then choose the **PrintNewToggleValue** method.

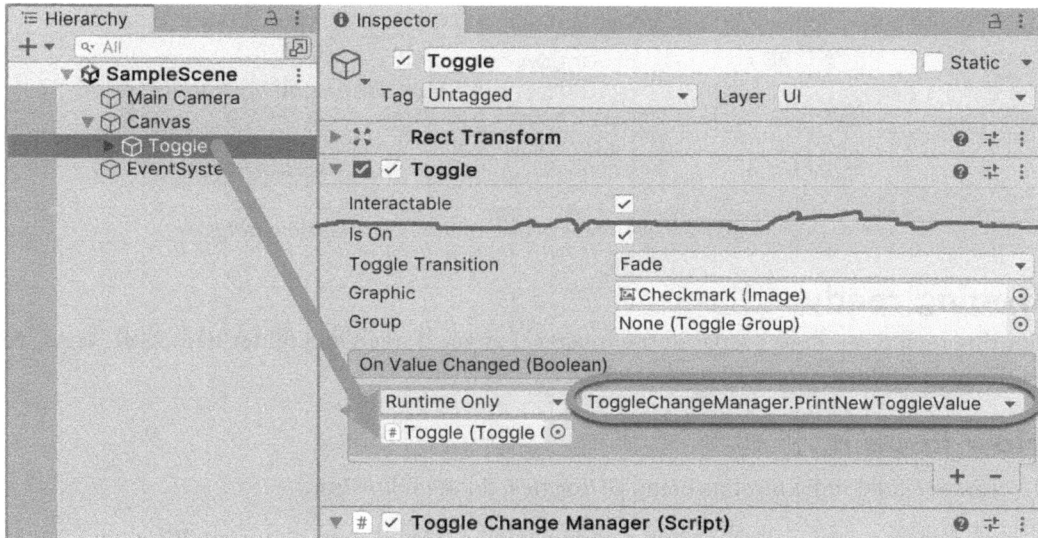

Figure 2.32: Setting the Toggle's On Value Changed event handler function

8. Save and run the scene. Each time you check or uncheck the Toggle GameObject, the **On Value Changed** event will fire, and you'll see a new text message printed into the **Console** window by our script, stating the new Boolean true/false value of Toggle.

How it works...

When you create a Unity **UI Toggle** GameObject, it comes with several child GameObjects automatically – Background, Checkmark, and the text's Label. Unless we need to style the look of a Toggle in a special way, all we must do is simply edit the text's Label so that the user knows what option or feature this Toggle is going to turn on/off.

The Awake() method of the ToggleChangeManager C# class caches a reference to the **Toggle** component in the GameObject where the script instance is located. When the game is running, each time the user clicks on the **Toggle** component to change its value, an **On Value Changed** event is fired. Then, we register the PrintNewToggleValue() method, which is to be executed when such an event occurs. This method retrieves, and then prints out to the **Console** window, the new Boolean true/false value of Toggle.

Creating related radio buttons using UI Toggles

Unity **UI Toggles** are also the base components if we wish to implement a group of mutually exclusive options in the style of *radio buttons*. We need to group related radio buttons together (**UI Toggles**) to ensure that when one radio button turns on (is selected), all the other radio buttons in the group turn off (are unselected).

We also need to change the visual look if we want to adhere to the usual style of radio buttons as circles, rather than the square **UI Toggle** default images:

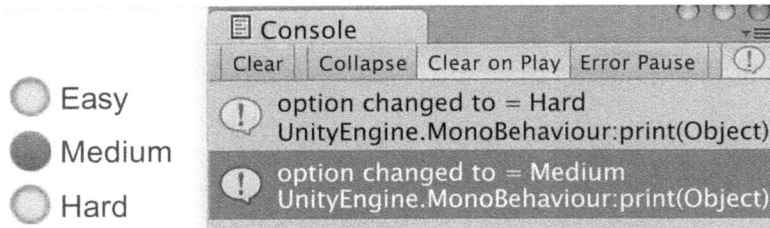

Figure 2.33: Example of three buttons with Console status

Getting ready

For this recipe, we have prepared the images that you'll need in a folder named UI Demo Textures in the 02_11 folder.

How to do it...

To create related radio buttons using UI Toggles, do the following:

1. Create a new Unity 2D project and install TextMeshPro by choosing: **Window | TextMeshPro | Import TMP Essential Resources**.
2. In the **Inspector** window, change the **Background** color of **Main Camera** to white.
3. Import the UI Demo Textures folder into the project.
4. Add a **UI Toggle** to the scene, naming this new GameObject Toggle-easy.
5. For the **Label** child of the Toggle-easy GameObject, set the **Text** property to **Easy**.
6. Select the Canvas GameObject and, in the **Inspector** window, add a **UI | Toggle Group** component.
7. With the Toggle-easy GameObject selected, in the **Inspector** window, drag the Canvas GameObject into the **Toggle Group** property of the **Toggle (Script)** component.

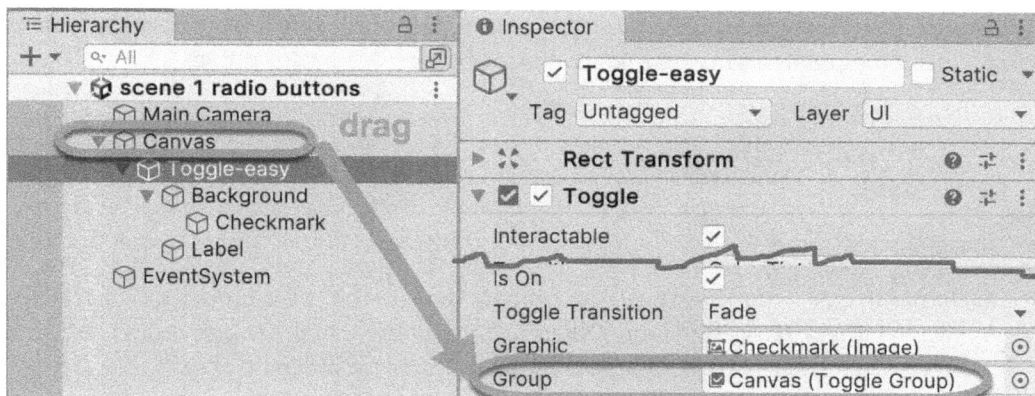

Figure 2.34: Assigning the Canvas Toggle Group to the Toggle-easy GameObject

8. Assign the `Toggle-easy` GameObject with a new Tag called **Easy**. Do this by selecting **Add Tag...** from the **Tag** drop-down menu in the **Inspector** window – this will open the **Tags & Layers** panel. Click the plus (+) button for a new Tag, and create a new Tag, `Easy`. Finally, select the `Toggle-easy` GameObject again in the **Hierarchy** window, and at the top of the **Inspector**, set its Tag to `Easy`.

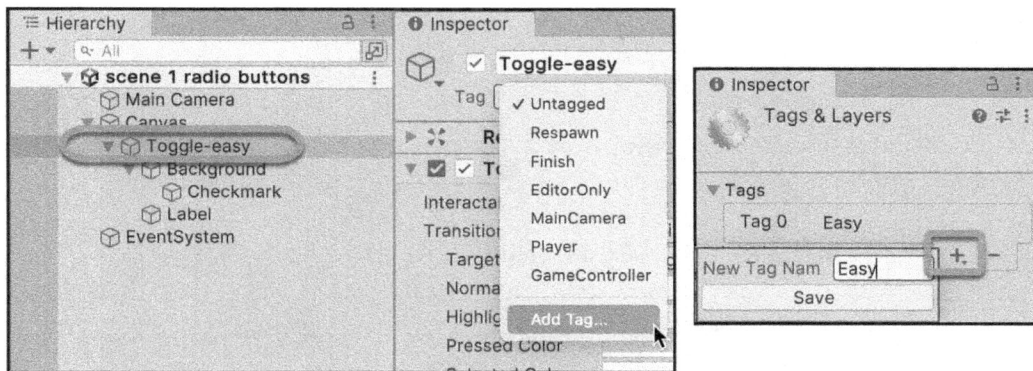

Figure 2.35: Creating new Tag called **Easy**

You can also create/edit **Tags & Layers** from the project **Settings** panel, accessed from the **Edit | Settings...** menu.

9. Select the **Background** child GameObject of Toggle-easy and, for its **Image** component, select the UIToggleBG image in the **Source Image** property (a circle outline).

> To make these toggles look more like radio buttons, the background of each is set to the circle outline image of UIToggleBG, and the checkmark (which displays the toggles that are on) is filled with the circle image called UIToggleButton.

10. In the **Inspector** for the **Toggle** component, ensure the **Is On** property is checked. Then select the **Checkmark** child GameObject of Toggle-easy. In the **Image** component, choose the UIToggleButton image for the **Source Image** property (a filled circle).

> Of the three choices (easy, medium, and hard) that we'll offer to the user, we'll set the easy option to be the one that is supposed to be initially selected. Therefore, we need its **Is On** property to be checked, which will lead to its checkmark image being displayed.

11. Add an instance of the RadioButtonManager C# script class to the Canvas GameObject:

```
using UnityEngine;
using System.Collections;
using UnityEngine.UI;

public class RadioButtonManager : MonoBehaviour {
  private string currentDifficulty = "Easy";

  public void PrintNewGroupValue(Toggle sender){
    // only take notice from Toggle just switched to On
    if(sender.isOn){
      currentDifficulty = sender.tag;
      print ("option changed to = " + currentDifficulty);
    }
  }
}
```

12. With the Toggle-easy GameObject selected, add an **On Value Changed** event to the list of event handlers for the **Toggle (Script)** component, click on the plus (+) button to add an event handler slot, and drag the Canvas GameObject into the **Object** slot.

13. Then, from the **Function** drop-down menu, choose **RadioButtonManager** and then choose the **PrintNewGroupValue** method.

14. Into the **Toggle** parameter slot, which is initially **None (Toggle)**, drag the `Toggle-easy` Game-Object. This means that the `Toggle-easy` GameObject calls the `PrintNewGroupValue(...)` method of a C# scripted component called `RadioButtonManager` in the `Canvas` GameObject, passing itself as a parameter.

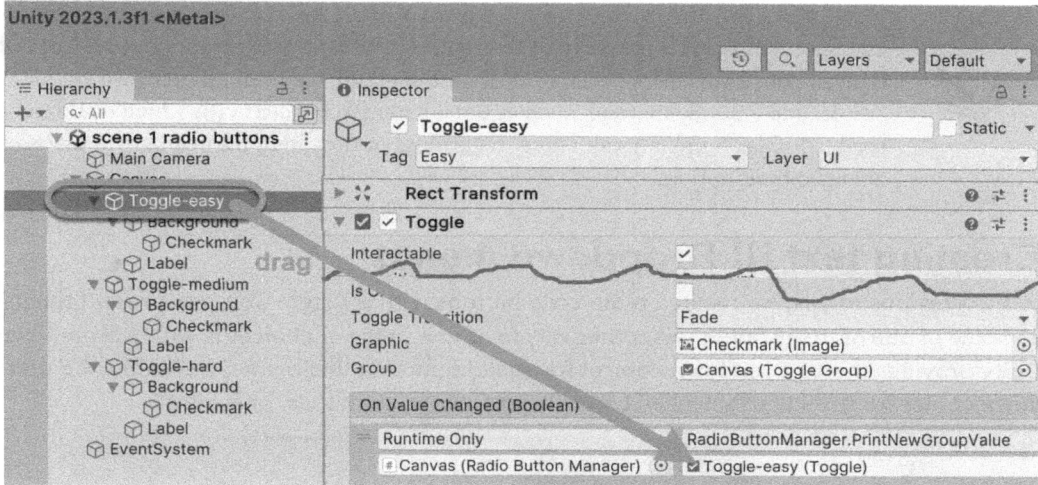

Figure 2.36: Dragging the Toggle-easy GameObject to the Toggle parameter slot

15. Duplicate the `Toggle-easy` GameObject, naming the copy `Toggle-medium`. Set its **Rect Transform** property's **Pos Y** to `-25` (so that this copy is positioned below the easy option) and uncheck the **Is On** property of the **Toggle** component. Create a new Tag, `Medium`, and assign the Tag to the new GameObject, `Toggle-medium`.

16. Duplicate the `Toggle-medium` GameObject, naming the copy `Toggle-hard`. Set its **Rect Transform** property's **Pos Y** to `-50` (so that this copy is positioned below the medium option). Create a new Tag, `Hard`, and assign the Tag to the new GameObject, `Toggle-hard`.

17. Save and run the scene. Each time you check one of the three radio buttons, the **On Value Changed** event will fire, and you'll see a new text message printed into the **Console** window by our script, stating the tag of whichever **Toggle** (radio button) was just set to true (**Is On**).

How it works...

By using the `UIToggleBG` and `UIToggleButton` images, we made the UI GameObject look like radio buttons – circles that when selected have a filled center. By adding a **Toggle Group** component to `Canvas`, and having each Toggle GameObject link to it, the three radio buttons can tell **Toggle Group** when they have been selected. Then, the other members of the group are deselected.

Each Toggle has an **On Value Changed** event handler that prints out the Tag for the GameObject. So, by creating the Tags `Easy`, `Medium`, and `Hard`, and assigning them to their corresponding GameObjects, we are able to print out a message corresponding to the radio button that has been clicked by the user.

If you had several groups of radio buttons in the same scene, one strategy is, for each group, to add the **Toggle Group** component to one of the Toggles and have all the others link to that one. So, all Toggles in each group link to the same **Toggle Group** component.

> **Note.** We store the current radio button value (the last one switched **On**) in the `currentDifficulty` property of the RadioButtonManager component of GameObject Canvas. Since variables declared outside a method are remembered, we could, for example, add a public method, such as `GetCurrentDifficulty()`, which could tell other scripted objects the current value, regardless of how long it's been since the user last changed their option.

Creating text UI Dropdown menus

In the previous recipe, we created radio-style buttons with a Toggle Group to present the user with a choice of one of many options. Another way to offer a range of choices is with a drop-down menu. Unity provides the **UI Dropdown** control for such menus. In this recipe, we'll offer the user a drop-down choice for the suit of a deck of cards (hearts, clubs, diamonds, or spades):

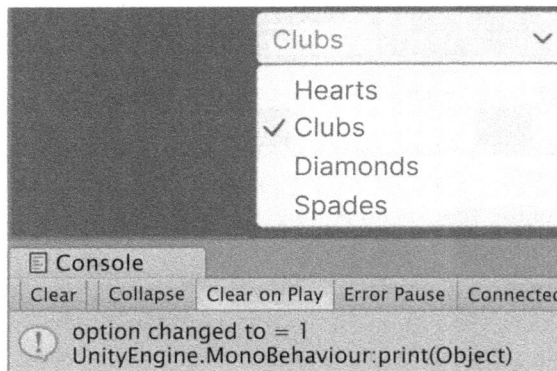

Figure 2.37: Checking the drop-down menu in the Console window

Note that the **UI Dropdown** that's created by default includes a scrollable area if there isn't space for all the options. We'll learn how to remove such GameObjects and components to reduce complexity when such a feature is not required.

How to do it...

To create a **UI Dropdown** control GameObject, follow these steps:

1. Create a new Unity 2D project and install TextMeshPro by choosing: **Window | TextMeshPro | Import TMP Essential Resources**.

2. Add a **UI Dropdown - TextMeshPro** to the scene.

3. In the **Inspector** window, for the **Dropdown (Script)** component, change the list of **Options** from **Option A**, **Option B**, and **Option C** to **Hearts**, **Clubs**, **Diamonds**, and **Spades**. You'll need to click the plus (+) button to add space for the fourth option, that is, **Spades**.

4. Add an instance of the C# script class called `DropdownManager` to the `Dropdown` GameObject:

```
using UnityEngine;
using TMPro;

public class DropdownManager : MonoBehaviour  {
    private TMP_Dropdown dropdown;

    private void Awake() {
        dropdown = GetComponent<TMP_Dropdown>();
    }

    public void PrintNewValue() {
        int currentValue = dropdown.value;
        print ("option changed to = " + currentValue);
    }
}
```

5. With the `Dropdown` GameObject selected, add an **On Value Changed** event to the list of event handlers for the **Dropdown (Script)** component, click on the plus (+) button to add an event handler slot, and drag **Dropdown** into the **Object** slot.

6. From the **Function** drop-down menu, choose **DropdownManager** and then choose the **Print-NewValue** method.

7. Save and run the scene. Each time you change **Dropdown**, the **On Value Changed** event will fire, and you'll see a new text message is printed to the **Console** window by our script, stating the **Integer** index of the chosen **Dropdown** value (0 for the first item, 1 for the second item, and so on).

8. Select the `Template` child GameObject of **Dropdown** in the **Project** window and, in its **Rect Transform panel**, reduce its height to 50. When you run the scene, you should see a scrollable area, since not all options fit within the template's height:

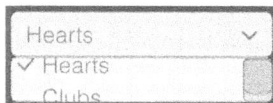

Figure 2.38: Example of a drop-down menu

9. Delete the `Scrollbar` child of the `Template` GameObject and remove the **Scroll Rect (Script)** component of it. When you run the scene now, you'll only see the first two options (**Hearts and Clubs**), with no way to access the other two options. When you are sure your template's height is sufficient for all its options, you can safely remove these scrollable options to simplify the GameObjects in your scene.

How it works...

When you create a Unity **UI DropDown-TextMeshPro** GameObject, it comes with several components and child GameObjects – `Label`, `Arrow`, and `Template` (as well as `ViewPort` and `Scrollbar`, and so on). Dropdowns work by duplicating the `Template` GameObject for each of the options listed in the **Dropdown (Script)** component. Both the **Text** and **Sprite** image values can be given for each option. The properties of the `Template` GameObject are used to control the visual style and behavior of the dropdown's thousands of possible settings.

First, you replaced the default options (**Option A, Option B**, and so on) in the **Dropdown (Script)** component. You then created a C# script class called `DropdownManager`, which, when attached to your **dropdown** and having its `PrintNewValue` method registered for **On Value Changed** events, means you can see the **Integer** index of the option each time the user changes their choice. Item index values start counting at zero (as is the case in many computing contexts), so 0 for the first item, 1 for the second item, and so on.

Since the default `Dropdown` GameObject that was created includes a **Scroll Rect (Script)** component and a `Scrollbar` child GameObject, when you reduced the height of `Template`, you could still scroll through the options. You then removed these items so that your dropdown didn't have a scrolling feature anymore.

Creating image icon UI Dropdown menus

In this recipe, you'll learn how to create **UI Dropdown** menus that show image icons next to the text for each menu item. We'll build on the previous recipe to offer the user a drop-down choice for the suit of a deck of cards (hearts, clubs, diamonds, or spades).

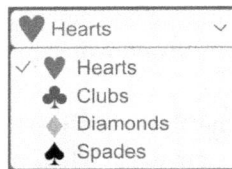

Figure 2.39: Example showing UI Dropdown menus with text and image

There are two pairs of items Unity uses to manage how text and images are displayed for a **UI Dropdown** control:

- The **Caption** `Text` and `Image` GameObjects (direct children of the Dropdown GameObject) are used to control how the currently selected item for the dropdown is displayed – this is the part of the dropdown we always see, regardless of whether it is being interacted with.

- The **Item** `Text` and `Image` GameObjects (children of the `Template` child of the `Dropdown` GameObject) define how each option is displayed as a row when the drop-down menu items are being displayed – the rows that are displayed when the user is actively working with the `Dropdown` GameObject.

So, we have to add an image in two places (for the **Caption Image** and the **Item Image** settings), in order to get a dropdown working fully with image icons for each option.

Getting ready

This recipe builds on the previous one. So, create a copy of that project and work on the copy.

For this recipe, we have prepared the image that you need in a folder named Images in the 02_13 folder.

How to do it...

To create image icon dropdown menus, follow these steps:

1. Open the copy you made of the previous recipe.
2. Import the provided Images folder.
3. In the **Inspector** window, for the **Dropdown TextMeshPro** component, for each item in the **Options** list – **Hearts, Clubs, Diamonds, and Spades** – drag the associated **Sprite** image from the card_suits folder into the **Project** window (hearts.png for **Hearts**, and so on).
4. Add a **UI Image** GameObject to the scene and make this **Image** a child of the Dropdown Game-Object.
5. Drag the **hearts.png** image from the **Project** window into the **Source Image** property of **Image** for the **Image** GameObject. Set its size to 25 x 25 in **Rect Transform** and drag it over the letter **H** in **Hearts** in the Label GameObject.
6. In the **Scene** panel, drag the Label GameObject so it appears to the right of the **Hearts** image.

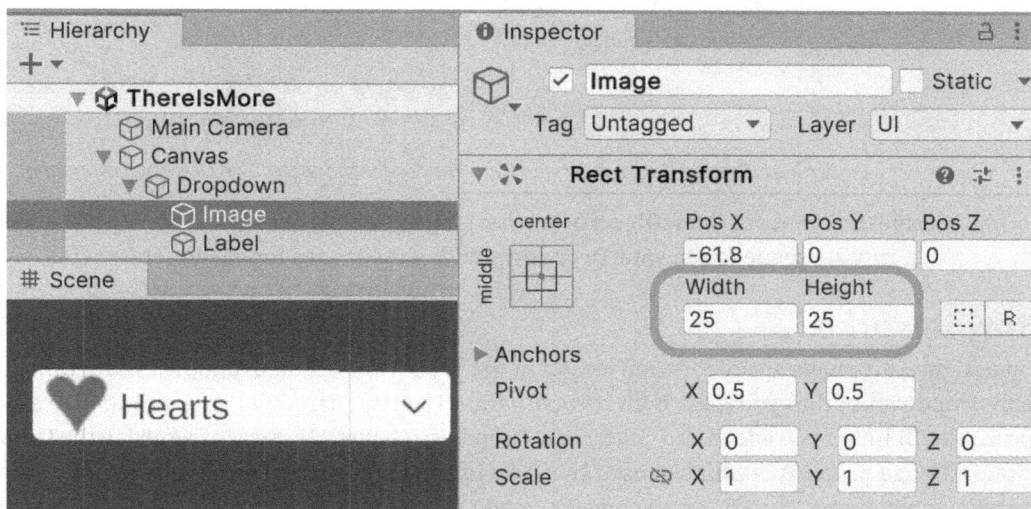

Figure 2.40: Adding a Hearts Image GameObject as a child of the Dropdown GameObject

7. With **Dropdown** selected in the **Hierarchy**, drag the Image GameObject into the **Caption Image** property of the **Dropdown (Script)** component.
8. Enable the Template GameObject (usually, it is disabled). Make it active by checking its **Active** checkbox at the top of the **Inspector**.

9. Duplicate the `Image` GameObject child of **Dropdown** and name the copy `Item Image`. Make this image a child of the `Item` GameObject that is in **Dropdown-Template-Content-Item** (`Item Image` needs to appear below the white `Item Background Image`; otherwise, it will be covered by the background and not be visible). Also, delete the `Item Checkmark` GameObject that is in **Dropdown-Template-Content-Item**.

10. Since items in the dropdown are slightly smaller, resize `Item Image` to be 20 x 20 in its **Rect Transform**.

11. Position `Item Image` over the letter **O** of **Option A** of `Item Text`, and then move `Item Text` to the right so that the icon and text are not on top of each other.

12. With **Dropdown** selected in the **Project** window, drag the **Item Image** GameObject into the **Item Image** property of the **Dropdown (Script)** component:

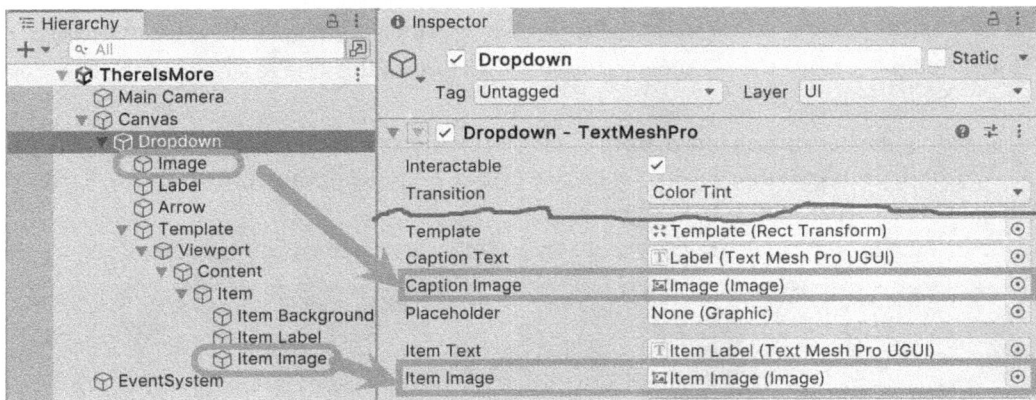

Figure 2.41: Setting the Caption and Item images for the Dropdown UI menu component

13. Disable the `Template` GameObject by unchecking its **Active** checkbox at the top of the **Inspector**. Then run the scene to see your **Dropdown** with icon images for each menu option.

How it works...

You assigned an image sprite for each option in the properties of the Dropdown GameObject. So for each dropdown option you have both its text name (Hearts, Diamonds etc.) and its corresponding sprite. The UI Image you added as a child of the Dropdown GameObject was assigned to the Dropdown's Caption Image property – which is used by Unity to show the current selections image at the top of the dropdown menu. The UI Image copy you made named Item Image was made a child of the Item GameObject, and that is used to display the sprites for each option when the rows of the dropdown menu are being displayed.

Displaying a radar to indicate the relative locations of objects

A radar displays the locations of other objects relative to the player, usually based on a circular display, where the center represents the player and each graphical blip indicates how far away and what relative direction objects are to the player. Sophisticated radar displays will display different categories of objects with different colored or shaped blip icons:

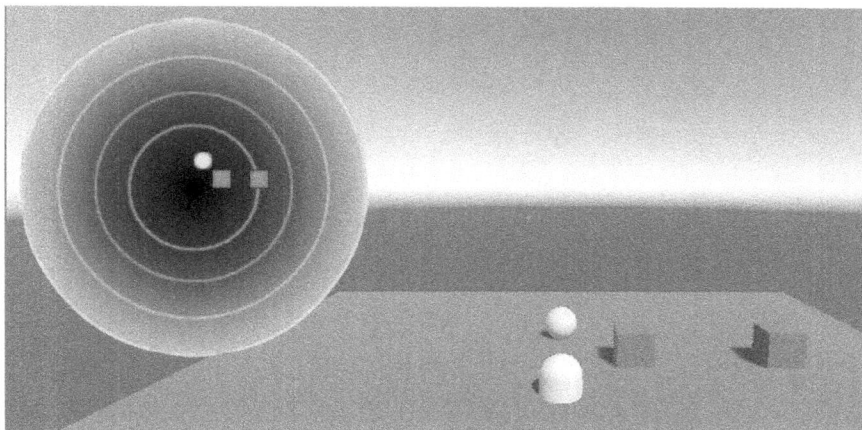

Figure 2.42: Example of a radar

In the preceding screenshot, we can see two red square blips, indicating the relative position of the two red cube GameObjects tagged Cube near the player, and a yellow circle blip indicating the relative position of the yellow sphere GameObject tagged Sphere. The green circle radar background image gives the impression of an aircraft control tower radar or something similar.

Getting ready

For this recipe, we have prepared the radar images and terrain textures that you need in folders named Images and Textures in 02_14.

How to do it...

To create a radar to show the relative positions of the objects, follow these steps:

1. Create a new Unity 3D project.
2. Import the provided Images and Textures folders into the project.
3. Create a terrain by choosing **Create | 3D Object | Terrain**.

4. Change the size of **Terrain** to 20 x 20 by setting the **Terrain Width** and **Terrain Length** properties for the **Terrain** component in the **Inspector**. Also, set its position to (-10, 0, -10) so that its center is at (0, 0, 0):

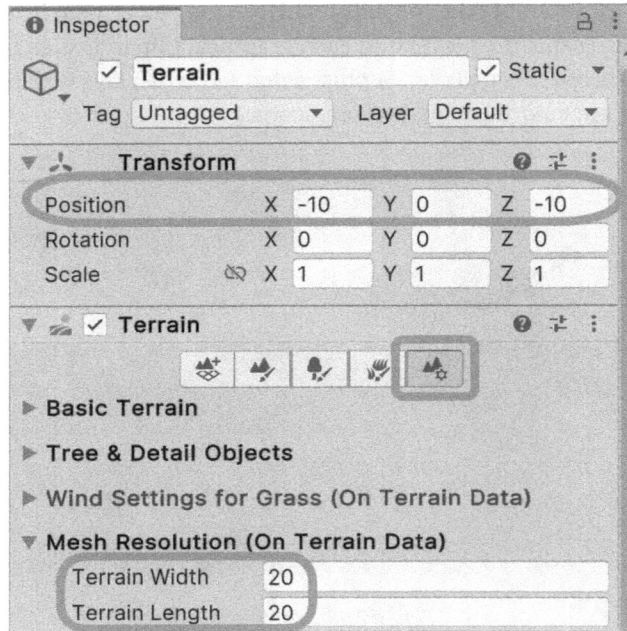

Figure 2.43: Terrain settings for this recipe

> **Note.** We change the size of a **Terrain** through its **Terrain Width** and **Terrain Length** properties in its **Terrain** component. The **Scale** property of a **Terrain's Transform** component does not affect its size.

5. Let's give this whole terrain a sandy look. Select the **Paint Terrain** tool (second from left – mountains and paintbrush icon) for the **Terrain** component in the **Inspector**. Then choose the **Paint Texture** option in the drop-down menu. Create a new **Terrain Layer** by clicking the **Edit Terrain Layers...** button. Finally, you must select the SandAlbedo texture from the imported Textures folder – find this easily by typing sand in the search bar. You should now see a new **Terrain Layer** named **NewLayer**, and the whole terrain should have been textured with the SandAlbedo texture.

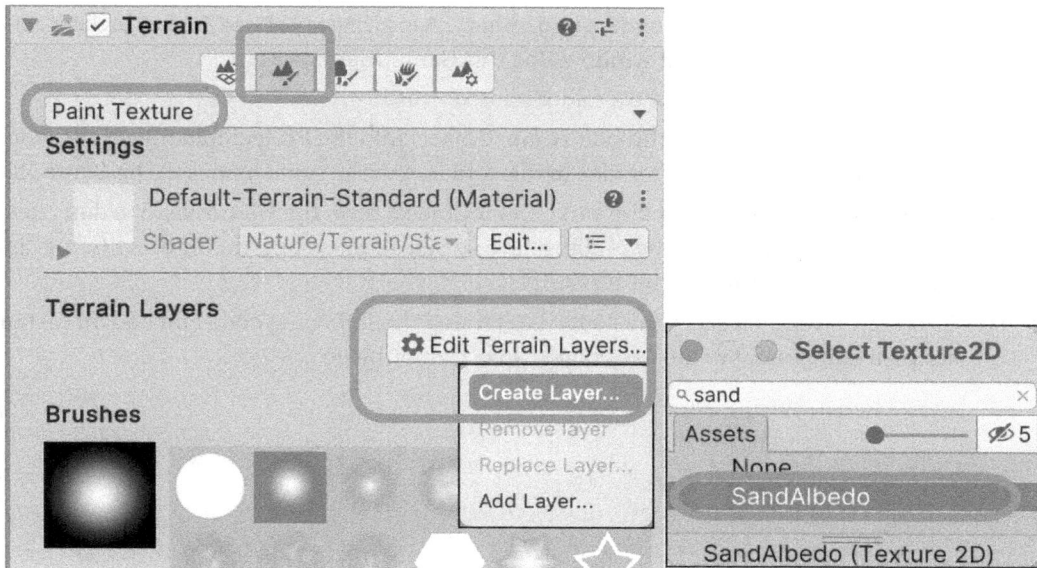

Figure 2.44: Settings for painting the terrain

6. Create a **3D Cube** GameObject at **Position** (2, 0.5, 2). Create a **Cube** tag and tag this GameObject with this new tag. Texture this GameObject with the red image called **icon32_square_red** by dragging the **icon32_square_red** image from the **Project** window over this GameObject in the **Hierarchy** window.

7. Duplicate the cube GameObject and move it to **Position** (6, 0.5, 2).

8. Create a **3D Sphere** GameObject at **Position** (0, 0.5, 4). Create a tag called **Sphere** and tag this GameObject with this new tag. Texture this GameObject with the yellow image called **icon32_square_yellow**.

9. In the **Hierarchy** window, add a **UI RawImage** GameObject to the scene named **RawImage-radar**.

10. Ensure that the RawImage-radar GameObject is selected in the **Hierarchy** window. From the Images folder in the **Project** window, drag the radarBackground image into the **Raw Image (Script)** public property's **Texture**.

11. In **Rect Transform**, position RawImage-radar at the top left using the **Anchor Presets** item. Then, set both **Width** and **Height** to 200 pixels.

12. Create a new Tag named Blip.

13. Create a new **UI RawImage** named blip-cube. Assign it the redSquareBlackBorder texture image file from the **Project** window. Tag this GameObject as Blip.

14. Create a new **UI RawImage** named **blip-sphere**. Assign it the `yellowCircleBlackBorder` texture image file from the **Project** window. Tag this GameObject as `Blip`.

15. In the **Project** window, create a folder named `Prefabs`.

16. Drag the **blip-sphere** and **blip-cube** GameObjects into the **Project** folder `Prefabs`. You should now see two new prefab asset files in this folder with the same names as the GameObjects.

17. Delete the **blip-sphere** and **blip-cube** GameObjects from the **Hierarchy**. We don't need these in the scene initially, and the prefabs have stored all the properties of these GameObjects, to be instantiated at runtime by our scripts.

18. Create a C# script class called `Radar` containing the following code and add an instance as a scripted component to the `RawImage-radar` GameObject:

```csharp
using UnityEngine;
using UnityEngine.UI;

public class Radar : MonoBehaviour {
    public float insideRadarDistance = 20;
    public float blipSizePercentage = 5;
    public GameObject rawImageBlipCube;
    public GameObject rawImageBlipSphere;
    private RawImage rawImageRadarBackground;
    private Transform playerTransform;
    private float radarWidth;
    private float radarHeight;
    private float blipHeight;
    private float blipWidth;

    void Start() {
        rawImageRadarBackground = GetComponent<RawImage>();
        playerTransform =
            GameObject.FindGameObjectWithTag("Player").transform;
        radarWidth = rawImageRadarBackground.rectTransform.rect.width;
        radarHeight = rawImageRadarBackground.rectTransform.rect.height;
        blipHeight = radarHeight * blipSizePercentage / 100;
        blipWidth = radarWidth * blipSizePercentage / 100;
    }

    void Update() {
        RemoveAllBlips();
        FindAndDisplayBlipsForTag("Cube", rawImageBlipCube);
        FindAndDisplayBlipsForTag("Sphere", rawImageBlipSphere);
    }
```

```
    private void FindAndDisplayBlipsForTag(string tag, GameObject
prefabBlip) {
        Vector3 playerPos = playerTransform.position;
        GameObject[] targets = GameObject.FindGameObjectsWithTag(tag);
        foreach (GameObject target in targets) {
            Vector3 targetPos = target.transform.position;
            float distanceToTarget = Vector3.Distance(targetPos,
                playerPos);
            if ((distanceToTarget <= insideRadarDistance))
              CalculateBlipPositionAndDrawBlip (playerPos, targetPos,
                prefabBlip);
        }
    }

    private void CalculateBlipPositionAndDrawBlip (Vector3 playerPos,
Vector3
        targetPos, GameObject prefabBlip) {
        Vector3 normalisedTargetPosition = NormalizedPosition(playerPos,
            targetPos);
        Vector2 blipPosition =
            CalculateBlipPosition(normalisedTargetPosition);
        DrawBlip(blipPosition, prefabBlip);
    }

    private void RemoveAllBlips() {
        GameObject[] blips = GameObject.FindGameObjectsWithTag("Blip");
        foreach (GameObject blip in blips)
            Destroy(blip);
    }

    private Vector3 NormalizedPosition(Vector3 playerPos, Vector3
targetPos) {
        float normalisedyTargetX = (targetPos.x - playerPos.x) /
            insideRadarDistance;
        float normalisedyTargetZ = (targetPos.z - playerPos.z) /
            insideRadarDistance;
        return new Vector3(normalisedyTargetX, 0, normalisedyTargetZ);
    }

    private Vector2 CalculateBlipPosition(Vector3 targetPos) {
```

```
            float angleToTarget = Mathf.Atan2(targetPos.x, targetPos.z) *
                Mathf.Rad2Deg;
            float anglePlayer = playerTransform.eulerAngles.y;
            float angleRadarDegrees = angleToTarget - anglePlayer - 90;
            float normalizedDistanceToTarget = targetPos.magnitude;
            float angleRadians = angleRadarDegrees * Mathf.Deg2Rad;
            float blipX = normalizedDistanceToTarget * Mathf.
Cos(angleRadians);
            float blipY = normalizedDistanceToTarget * Mathf.
Sin(angleRadians);
            blipX *= radarWidth / 2;
            blipY *= radarHeight / 2;
            blipX += radarWidth / 2;
            blipY += radarHeight / 2;
            return new Vector2(blipX, blipY);
    }


    private void DrawBlip(Vector2 pos, GameObject blipPrefab) {
            GameObject blipGO = (GameObject)Instantiate(blipPrefab);
            blipGO.transform.SetParent(transform.parent);
            RectTransform rt = blipGO.GetComponent<RectTransform>();
            rt.SetInsetAndSizeFromParentEdge(RectTransform.Edge.Left, pos.x,
                blipWidth);
            rt.SetInsetAndSizeFromParentEdge(RectTransform.Edge.Top, pos.y,
                blipHeight);
    }
}
```

19. Ensure that the RawImage-radar GameObject is selected in the **Hierarchy**. We now need to populate public variables **Raw Image Blip Cube** and **Raw Image Blip Sphere** the for the Radar scripted component in the **Inspector**.

20. Drag the **blip-sphere** from the Prefab folder in the **Project** panel into the **Raw Image Blip Cube** public variable in the **Inspector**. Then drag the **blip-sphere** Prefab asset file from the Prefab folder into the **Raw Image Blip Sphere** public variable. By doing this, you are setting these public script variables to reference these prefabs, allowing the Radar scripted component to control the display of GameObjects created from these Prefabs at runtime.

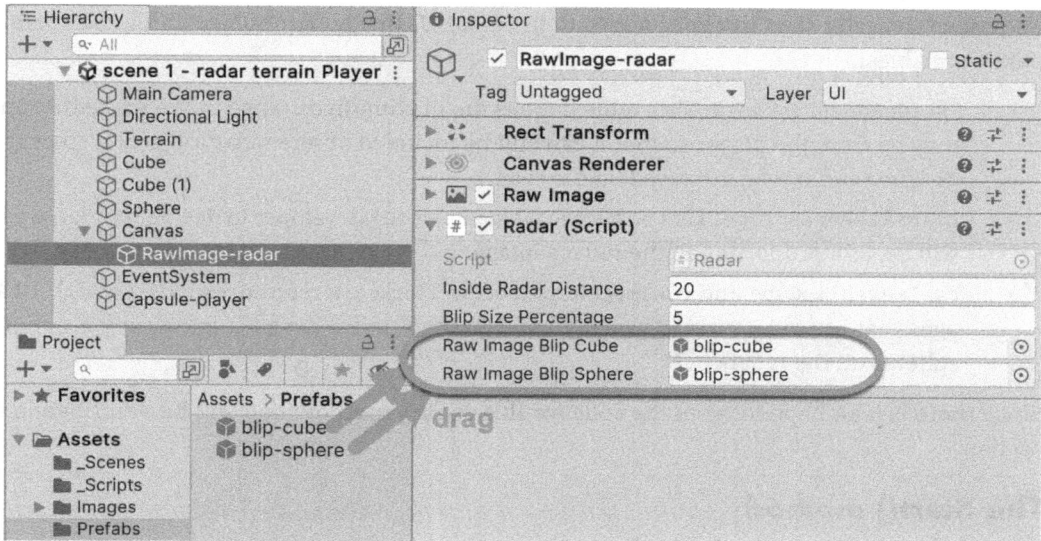

Figure 2.45: Calculation for the blip method

21. In the **Inspector**, set the **Main Camera Transform** properties to have **Position** (0, 5, -10), and **Rotation** (10, 0, 0). This will allow you to see the cubes and sphere GameObjects easily when playing the game.

22. Save and Run your game. We just see an empty radar image at the top left of the screen! The Radar script will draw blips on the radar relative to the position of the GameObject tagged Player.

23. Stop the game and create a new 3D capsule named Capsule-player, positioned at (0, 1, 0), and tag this Player.

24. Save and Run your game. Now we can see a yellow circle blip, and two red square blips on the radar, showing the relative position of the yellow sphere and two red cubes to the capsule tagged Player!

How it works...

A radar background is displayed on the screen. The center of this circular image represents the position of the player's character. In this recipe, you created two prefabs – one for red square images to represent each red cube found within the radar distance, and one for yellow circles to represent yellow sphere GameObjects.

The `Radar` C# script class has been added to the radar **UI Image** GameObject. This class defines four public variables:

- `insideRadarDistance`: This value defines the maximum distance in the scene that an object may be from the player so that it can still be included on the radar (objects further than this distance will not be displayed on the radar).
- `blipSizePercentage`: This public variable allows the developer to decide how large each blip will be, as a proportion of the radar's image.
- `rawImageBlipCube` and `rawImageBlipSphere`: These are references to the prefab UI **RawImages** that are to be used to visually indicate the relative distance and position of cubes and spheres on the radar.

Since there is a lot happening in the code for this recipe, each method will be described in its own section.

The Start() method

The `Start()` method caches a reference to the **RawImage** of the radar background image. Then, it caches a reference to the **Transform** component of the player's character (tagged as **Player**). This allows the scripted object to know about the position of the player's character in each frame. Next, the width and height of the radar image are cached, so that the relative positions for blips can be calculated based on the size of this background radar image. Finally, the size of each blip (`blipWidth` and `blipHeight`) is calculated using the `blipSizePercentage` public variable.

The Update() method

The `Update()` method calls the `RemoveAllBlips()` method, which removes any old UI **RawImage** GameObjects of cubes and spheres that might currently be displayed. If we didn't remove old blips before creating new ones, then you'd see "tails" behind each blip as new ones are created in different positions – which could actually be an interesting effect.

Next, the `FindAndDisplayBlipsForTag(...)` method is called twice. First, for the objects tagged **Cube** to be represented on the radar with the `rawImageBlipCube` prefab, and then again for objects tagged **Sphere** to be represented on the radar with the `rawImageBlipSphere` prefab. As you might expect, most of the hard work of the radar is to be performed by the `FindAndDisplayBlipsForTag(...)` method.

> This code is a simple approach to creating a radar. It is very inefficient to make repeated calls to `FindGameObjectWithTag("Blip")` for every frame from the `Update()` method. In a real game, it would be much better to cache all created blips in something such as a `List` or `ArrayList`, and then simply loop through that list each time.

The FindAndDisplayBlipsForTag(...) method

This method inputs two parameters: the string tag for the objects to be searched for, and a reference to the **RawImage** prefab to be displayed on the radar for any such tagged objects within the range.

First, the current position of the player's character is retrieved from the cached player Transform variable. Next, an array is constructed, referring to all GameObjects in the scene that have the provided tag. This array of GameObjects is looped through, and for each GameObject, the following actions are performed:

1. The position of the target GameObject is retrieved.
2. The distance from this target's position to the player's position is calculated.
3. If this distance is within the range (less than or equal to insideRadarDistance), then the Cal culateBlipPositionAndDrawBlip(...) method is called.

The CalculateBlipPositionAndDrawBlip (...) method

This method inputs three parameters: the position of the player, the position of the target, and a reference to the prefab of the blip to be drawn.

Three steps are now required to get the blip for this object to appear on the radar:

1. The normalized position of the target is calculated by calling NormalizedPosition(...).
2. The position of the blip on the radar is calculated from this normalized position by calling CalculateBlipPosition(...).
3. The **RawImage** blip is displayed by calling DrawBlip(...) and passing the blip's position and the reference to the **RawImage** prefab that is to be created there.

The NormalizedPosition(...) method

The NormalizedPosition(...) method inputs the player's character position and the target GameObject's position. It has the goal of outputting the relative position of the target to the player, returning a Vector3 object (actually, a C# **struct** – but we can think of it as a simple object) with a triplet of *X*, *Y*, and *Z* values. Note that since the radar is only 2D, we ignore the *Y*-value of the target GameObjects, so the *Y*-value of the Vector3 object that's returned by this method will always be 0. So, for example, if a target was at exactly the same location as the player, the *X*, *Y*, and *Z* of the returned Vector3 object would be (0, 0, 0).

Since we know that the target GameObject is no further from the player's character than insideRadarDistance, we can calculate a value in the -1 ... 0 ... +1 range for the *X* and *Z* axes by finding the distance on each axis from the target to the player and then dividing it by insideRadarDistance. An *X*-value of -1 means that the target is fully to the left of the player (at a distance that is equal to insideRadarDistance), while +1 means it is fully to the right. A value of 0 means that the target has the same *X* position as the player's character. Likewise, for -1 ... 0 ... +1 values in the *Z-axis* (this axis represents how far, in front or behind us, an object is located, which will be mapped to the vertical axis in our radar).

Finally, this method constructs and returns a new Vector3 object with the calculated *X* and *Z* normalized values and a *Y*-value of zero.

The CalculateBlipPosition(...) method

First, we calculate angleToTarget, which is the angle from (0, 0, 0) to our normalized target position.

Next, we calculate anglePlayer, which is the angle the player's character is facing. This recipe makes use of the **yaw** angle of the rotation, which is the rotation about the *Y-axis* – that is, the direction that a character controller is facing. This can be found in the Y component of a GameObject's eulerAngles component of its transform. You can imagine looking from above and down at the character controller and seeing what direction they are facing – this is what we are trying to display graphically with the radar.

Our desired radar angle (the angleRadarDegrees variable) is calculated by subtracting the player's direction angle from the angle between the target and the player, since a radar displays the relative angle from the direction that the player is facing to the target object. In mathematics, an angle of zero indicates a direction of east. To correct this, we need to also subtract 90 degrees from the angle.

The angle is then converted into radians since this is required for these Unity trigonometry methods. We then multiply the Sin() and Cos() results by our normalized distances to calculate the X and Y values, respectively (see the following diagram):

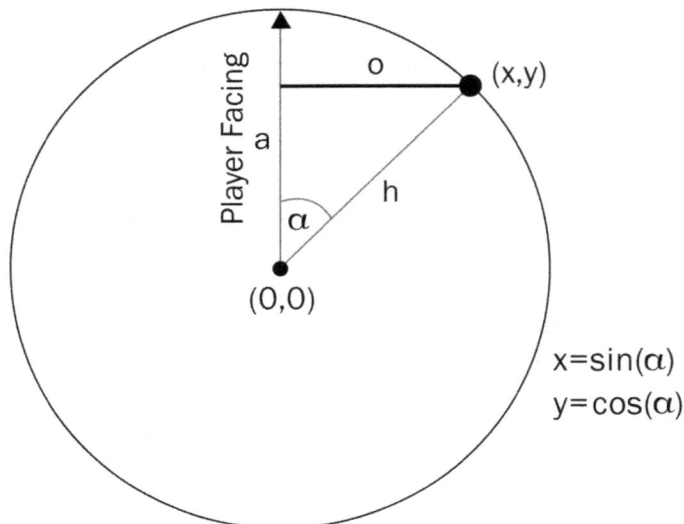

Figure 2.46: Calculation for the blip method

> In the preceding diagram, alpha is the angle between the player and the target object, "a" is the adjacent side, "h" is the hypotenuse, and "o" is the side opposite the angle.

Our final position values need to be expressed as pixel lengths, relative to the center of the radar. So, we multiply our blipX and blipY values by half the width and the height of the radar; note that we only multiply with half the width since these values are relative to the center of the radar. We then add half the width and the height of the radar image to the blipX/Y values so that these values are now positioned relative to the center.

Finally, a new **Vector2** object is created and returned, passing back these final calculated *X* and *Y* pixel values for the position of our blip icon.

The DrawBlip() method

The DrawBlip() method takes the input parameters of the position of the blip (as a Vector2 *X, Y* pair) and the reference to the **RawImage** prefab to be created at that location on the radar.

A new GameObject is created (**instantiated**) from the prefab and is parented to the radar GameObject (of which the scripted object is also a component). A reference is retrieved from the **Rect Transform** component of the new **RawImage** GameObject that has been created for the blip. Calls to the Unity **RectTransform** method, SetInsetAndSizeFromParentEdge(...), result in the blip GameObject being positioned at the provided horizontal and vertical locations over the radar image, regardless of where in the **Game** window the background radar image has been located.

There's more...

This radar script scans 360 degrees all around the player and only considers straight-line distances on the X-Z plane. So, the distances in this radar are not affected by any height difference between the player and target GameObjects. The script can be adapted to ignore targets whose height is more than some threshold different from the player's height.

Also, as presented, this recipe's radar *sees* through *everything*, even if there are obstacles between the player and the target. This recipe can be extended to not show obscured targets by using **raycasting** techniques.

And, of course, you can replace the 3D capsule with a user-controlled animated character, such as those covered in *Chapter 9, Animated Characters*.

Further reading

The following are some useful resources for learning more about working with core UI elements in Unity:

- Learn more about Unity Dropdown UI GameObjects:

    - https://docs.unity3d.com/Manual/script-Dropdown.html.

- Learn more about raycasting:

    - http://docs.unity3d.com/ScriptReference/Physics.Raycast.html.

3

Inventory and Advanced UIs

Many games involve the player collecting or choosing from a selection of items. Examples include collecting keys to open doors, collecting ammo for weapons, and choosing from a collection of spells to cast.

The recipes in this chapter provide a range of solutions for displaying to the player whether they are carrying an item or not, whether they are allowed more than one of a given item, and how many they have.

The two parts of software design for implementing inventories relate to, first, how we choose to represent the data about inventory items (that is, the data types and structures to store the data) and, second, how we choose to display information about inventory items to the player (the UI).

Also, while not strictly inventory items, player properties such as lives left, health, and time remaining can also be designed around the same concepts that we will present in this chapter.

First, we need to think about the nature of different inventory items for any particular game:

- Single items:

  - Examples: The only key for a level or a suit of magic armor.
  - Data type: bool (Boolean – true/false).
  - UI: Nothing (if not carried) or text/image to show being carried. Or perhaps, if we wish to highlight to the player that there is an **option** to carry this item, we could display a text string saying no key/key or two images, one showing an empty key outline and the second showing a full-color key.

- Two or more of the same item:

  - Examples: Lives left; number of arrows or bullets left.
  - Data type: int (integer – whole numbers).
  - UI: Text count or images.

- Collection of related items:

    - Examples: Keys of different colors to open doors of that color; potions of different strengths with different titles.

    - Data structure: A `struct` or `class` for the general item type (for example, the `Key` class's color, stored as an array or `List<>`).

    - UI: Text list or list/grid arrangement of icons.

- Collection of different items:

    - Examples: Keys, potions, weapons, and tools, all in the same inventory system.

    - Data structure: `List<>`, `Dictionary<>`, or an array of objects, which can be instances of different classes for each item type.

Each of the preceding representations and UI display methods will be illustrated in the recipes in this chapter. In addition, we'll learn how to create and use custom **sorting layers** so that we have complete control over which objects appear on top of or below other objects – something that is pretty important when scene content can contain background images, pickups, player characters, and so on.

In this chapter, we will cover the following recipes:

- Creating a simple 2D mini-game – SpaceGirl
- Displaying single object pickups with carrying and not-carrying text
- Displaying single object pickups with carrying and not-carrying icons
- Displaying multiple pickups of the same object with multiple status icons
- Using panels to visually outline the inventory UI area and individual items
- Creating a C# inventory slot UI to display scripted components
- Displaying multiple pickups of different objects as a list of text via a dynamic `List<>` of scripted `PickUp` objects
- Displaying multiple pickups of different objects as text totals via a dynamic `Dictionary<>` of `PickUp` objects and `enum` pickup types

Creating a simple 2D mini-game — SpaceGirl

This recipe will show you how to create the 2D SpaceGirl mini-game that almost all the recipes in this chapter are based on. The following figure shows an example of the mini-game we will be creating:

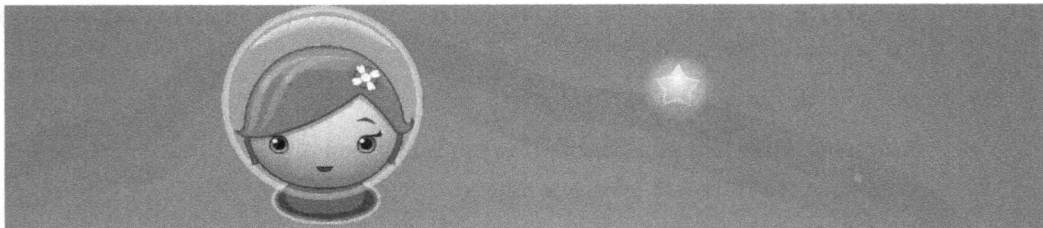

Figure 3.1: Example of the 2D SpaceGirl mini-game

Getting ready

For this recipe, we have prepared the images you need in a folder named `Sprites` in the `03_01` folder. We have also provided the completed game as a Unity package in this folder, named `Simple2DGame_SpaceGirl`: https://github.com/PacktPublishing/Unity-2023-Cookbook-Fifth-Edition.

How to do it...

To create the simple 2D SpaceGirl mini-game, follow these steps:

1. Create a new 2D (Core) project.

2. Import the supplied `Sprites` folder into your project.

3. Since it's a 2D project, each sprite image should be of the **Sprite (2D and UI)** type. Check this by selecting the sprite in the **Project** panel; then, in the **Inspector** panel, check the **Texture Type** property. If you need to change its type, you can change it from the drop-down menu and then click the **Apply** button.

4. Set the Unity Player's screen size to **800 x 600** by choosing it from the drop-down menu on the **Game** panel. If 800 x 600 isn't an offered resolution, then click the plus (+) button and create this as a new resolution for the panel.

5. Display the **Tags and Layers** properties for the current Unity project. Choose **Edit | Project Settings | Tags and Layers**. Alternatively, if you are already editing a GameObject, then you can select the **Add Layer...** menu from the **Layer** drop-down menu at the top of the **Inspector** panel, next to the **Static true/false** toggle.

6. Use the expand/contract triangle tools to contract **Tags and Layers**, as well as to expand **Sorting Layers**. Use the plus (+) button to add two new sorting layers, as shown in the screenshot. First, add one named **Background**, then add one named **Foreground**.

   The sequence is important since Unity will draw items in layers further down this list on top of the items earlier in the list. You can rearrange the layer sequence by clicking and dragging the position control, the wide equals (=) icon to the left of the word **Layer** in each row:

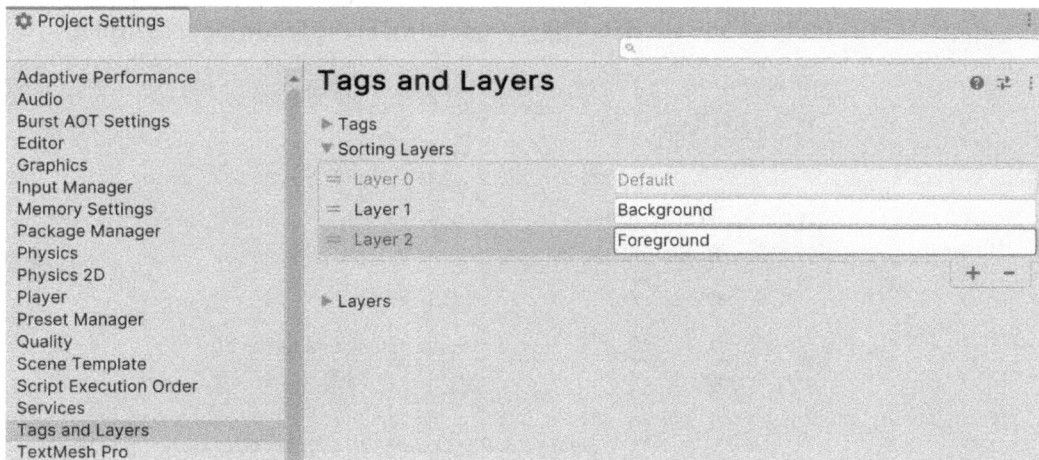

Figure 3.2: Background and Foreground layers

7. Drag the background_blue sprite from the **Project** panel (in the Sprites folder) into either the **Game** or **Hierarchy** window to create a GameObject for the current scene. Set **Position** of this GameObject to (0,0,0). It should completely cover the **Game** panel (at a resolution of **800 x 600**).

8. Set **Sorting Layer** of the background_blue GameObject to **Background** (in the **Sprite Renderer** component):

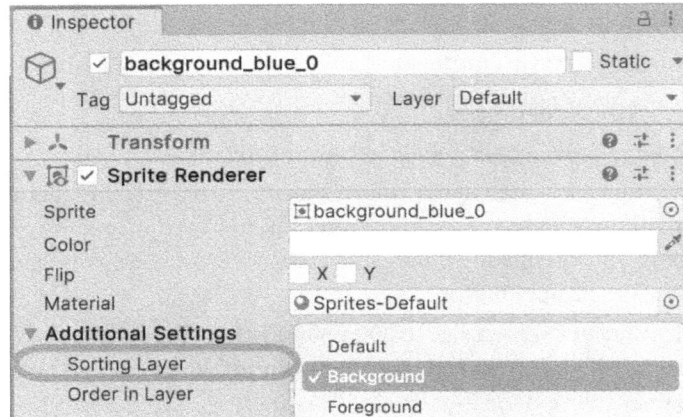

Figure 3.3: Setting the background image GameObject's Sorting Layer

9. Drag the star sprite from the **Project** panel (in the Sprites folder) into either the **Game** or **Hierarchy** window to create a GameObject for the current scene:

    •   Name this GameObject star.

    •   Create a new tag **Star** and assign this tag to the star GameObject (tags are created in the same way as sorting layers earlier in this recipe).

    •   Set **Sorting Layer** of the star GameObject to **Foreground** (in the **Sprite Renderer** component).

    •   Add a **Box Collider 2D** (**Add Component | Physics 2D | Box Collider 2D**) to the star GameObject and check **Is Trigger**, as shown in the following screenshot:

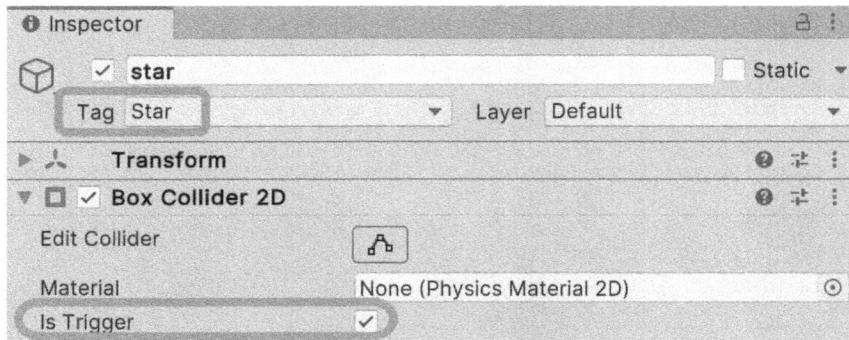

Figure 3.4: Box Collider 2D trigger setting

10. Drag the girl1 sprite from the **Project** panel (in the Sprites folder) into either the **Scene** or **Hierarchy** window to create a GameObject for the player's character in the current scene. Rename this GameObject player-girl1.

11. Set **Sorting Layer** of the player-girl1 GameObject to **Foreground**.

12. Add a **Box Collider 2D** (**Add Component** | **Physics 2D** | **Box Collider 2D**) to the player-girl1 GameObject. Then add a Rigidbody 2D component to this GameObject:

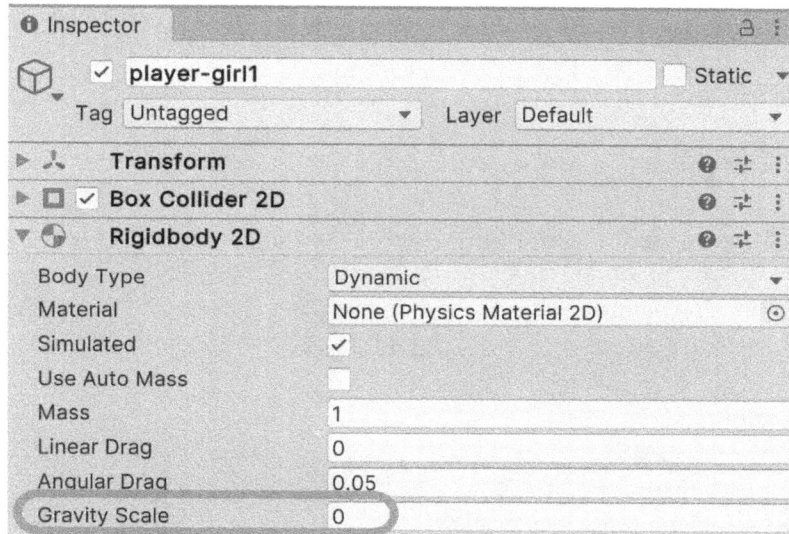

Figure 3.5: Setting Gravity Scale to zero

13. Create a new folder for your scripts named _Scripts.

14. Create a C# script called PlayerMove (in the _Scripts folder):

```
using UnityEngine;

public class PlayerMove : MonoBehaviour {
  public float speed = 10;
  private Rigidbody2D rigidBody2D;
  private Vector2 newVelocity;

void Awake(){
    rigidBody2D = GetComponent<Rigidbody2D>();
}

void Update() {
    float xMove = Input.GetAxis("Horizontal");
    float yMove = Input.GetAxis("Vertical");
```

```
        float xSpeed = xMove * speed;
        float ySpeed = yMove * speed;

        newVelocity = new Vector2(xSpeed, ySpeed);
    }

    void FixedUpdate() {
        rigidBody2D.velocity = newVelocity;
    }


}
```

15. Add an instance of the PlayerMove C# script as a component of the player-girl1 GameObject in the **Hierarchy** window:

16. Save the scene (name it Main Scene and save it in a new folder named _Scenes).

17. Run your game, a simple 2D SpaceGirl mini-game.

How it works...

In this recipe, you created a player character in the scene using the girl1 sprite and added a scripted component instance of the PlayerMove class. You also created a star GameObject (a pickup), a tagged star with a 2D box collider that will trigger a collision when the player's character hits it. When you run the game, you should be able to move the player-girl1 character around using the *W*, *A*, *S*, and *D* keyboard keys, the arrow keys, or a joystick.

Unity maps user inputs such as key presses, arrow keys, and game controller controls to its Input class. Two special properties of the Input class are the Horizontal and Vertical axes, which can be accessed via the Input.GetAxis ("Horizontal") and Input.GetAxis ("Vertical") methods.

> **Managing your input mapping:** You can map from different user input methods (keys, mouse, controllers, and so on) to the axes via **Edit | Project Settings | Input Manager.**

There is a newVelocity variable that is updated each frame in the Update() method based on the inputs. This Vector2 value is then applied to the FixedUpdate() method to become the new velocity for the GameObject.

Currently, nothing will happen if the player-SpaceGirl character hits a star because this has yet to be scripted.

With that, you have added a background (the background_blue GameObject) to the scene, which will be behind everything since it is in the rearmost sorting layer, called **Background.**

Items you want to appear in front of the background (the player character and the star, so far) are placed on the **Foreground** sorting layer.

You can learn more about Unity tags and layers at http://docs.unity3d.com/Manual/class-TagManager.html.

Displaying single object pickups with carrying and not-carrying text

Often, the simplest inventory situation is to display text to tell players whether they are carrying a single item (or not). In this recipe, we'll script the ability to detect collisions with the **star** GameObject and add this to the SpaceGirl mini-game. We will also display an on-screen message stating whether a star has been collected or not:

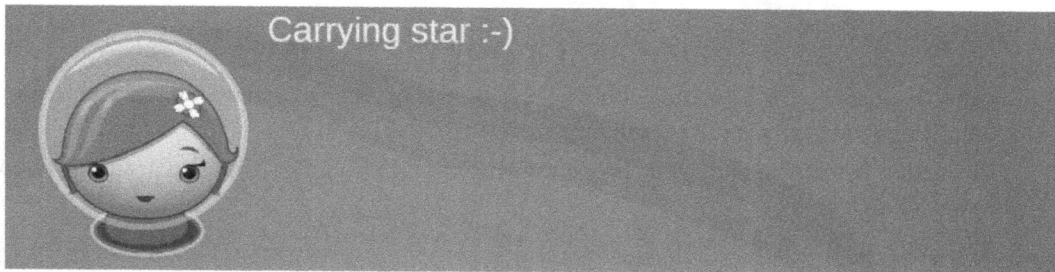

Figure 3.6: Example of text for displaying single-object pickups

At the end of the recipe, in the *There's more...* section, we'll learn how to adapt this recipe to maintain the **Integer** total for how many stars have been collected, for a version of the game with lots of stars to collect.

Getting ready

For this recipe, we have prepared a folder named _Scripts in the 03_02 folder.

This recipe assumes that you are starting with the Simple2Dgame_SpaceGirl project, which we set up in the previous recipe. So, make a copy of that project and work on that.

How to do it...

To display text to inform the user about the status of carrying a single object pickup, follow these steps:

1. Start with a new copy of the Simple2Dgame_SpaceGirl mini-game.

2. Add a **UI Text (TMP)** GameObject by choosing menu **GameObject | UI | Text - TextMeshPro**. When prompted, click the button to import the **TMP Essentials**.

> Unity UI text objects are built with the TextMeshPro package – abbreviated to TMP in the menu.

3. Rename this TMP text GameObject `Text-carrying-star`. In its **TextMeshPro -Text (UI)** property, change its text to `Carrying star: false`.

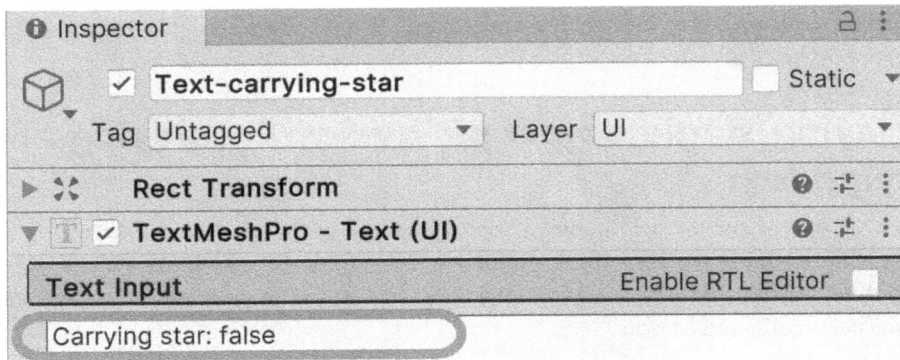

Figure 3.7: Setting the default UI (TMP) text value

4. In the **Inspector** panel, set the color of `Text-carrying-star` to yellow. Center the text horizontally and vertically, and set **Font Size** to 32.

5. Edit its **Rect Transform** and, while holding down *Shift + Alt* (to set its pivot and position), choose the top **stretch** box:

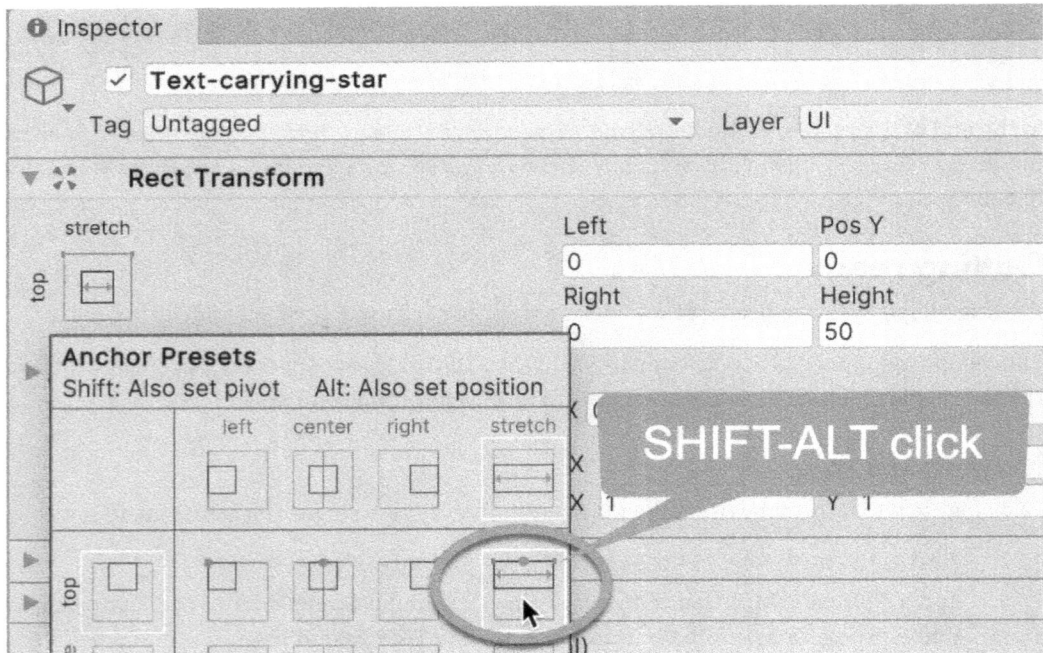

Figure 3.8: Editing Rect Transform

Your text should now be positioned at the top of the **Game** panel, and its width should stretch to match that of the whole panel.

6. Create the following C# script class called `PlayerInventory` in the _Scripts folder:

```
using UnityEngine;

public class PlayerInventory : MonoBehaviour {
    private PlayerInventoryDisplay playerInventoryDisplay;
    private bool carryingStar = false;

    void Awake() {
        playerInventoryDisplay =
            GetComponent<PlayerInventoryDisplay();
    }

    void Start() {
        playerInventoryDisplay.OnChangeCarryingStar( carryingStar);
    }

    void OnTriggerEnter2D(Collider2D hit) {
        if (hit.CompareTag("Star")) {
            carryingStar = true;
            playerInventoryDisplay.OnChangeCarryingStar(
carryingStar);
            Destroy(hit.gameObject);
        }
    }
}
```

7. Create the following C# script class called `PlayerInventoryDisplay` in the _Scripts folder:

```
using UnityEngine;
using TMPro;

[RequireComponent(typeof(PlayerInventory))]
public class PlayerInventoryDisplay : MonoBehaviour
{
    public TextMeshProUGUI starText;

    public void OnChangeCarryingStar(bool carryingStar)
    {
        string starMessage = "no star :-(";
        if(carryingStar){
            starMessage = "Carrying star :-)";
        }
```

```
                    starText.text = starMessage;
        }
    }
```

8. Add an instance of the `PlayerInventoryDisplay` script class to the `player-SpaceGirl` Game-Object in the **Hierarchy** panel.

> Note that since the `PlayerInventoryDisplay` class contains `RequireComponent()`, an instance of the `PlayerInventory` script class will be automatically added to the `player-SpaceGirl` GameObject.

9. From the **Hierarchy** panel, select the `player-SpaceGirl` GameObject. Then, from the **Inspector**, access the **Player Inventory Display (Script)** component and populate the **Star Text** public field with the `Text-carrying-star` GameObject, as shown in the following screenshot:

Figure 3.9: Populating the Star Text public field

10. When you play the scene, after moving the character into the star, the star should disappear, and the on-screen text message should change to **Carrying star :-)**.

How it works...

In this recipe, you created a **UI Text (TMP)** GameObject called `Text-carrying-star` to display a text message stating whether the player is carrying a star. You created two script classes, and an instance of each was added as a component of the player's `player-SpaceGirl` character GameObject.

> The `PlayerInventory` script class detects player-star collisions, updates internal variables stating whether a star is being carried, and asks for the UI display to be updated each time a collision is detected.
>
> The `PlayerInventoryDisplay` script class handles communication with the user by updating the text message that's displayed by the `Text-carrying-star` **UI Text (TMP)** GameObject.

Several software design patterns, such as the **Model-View-Controller** (MVC) pattern, separate the code that updates the UI from the code that changes player and game variables, such as score and inventory item lists. Although this recipe has only one variable and one method for updating the UI, well-structured game architectures scale up to cope with more complex games, so it is often worth the effort of using a little more code and an extra script class, even for this relatively simple game, if we want our final game architecture to be well structured and maintainable.

One additional advantage of this design pattern is that the method that's used to communicate information to the user via the UI can be changed (for example, from text to an icon – see the next recipe!), without any need to change the code in the PlayerInventory script class.

The PlayerInventory script class

The playerInventoryDisplay variable is a reference to an instance object of the PlayerInventoryDisplay class.

The bool variable called carryingStar represents whether the player is carrying the star at any point in time; it is initialized to false.

The Awake() method caches a reference to the playerInventoryDisplay sibling component.

When the scene begins via the Start() method, we call the OnChangeCarryingStar(...) method of the playerInventoryDisplay script component, passing in the initial value of carryingStar (which is false). This ensures that we are not relying on text that's been typed into the Text-carrying-star **UI Text (TMP)** object at **design time**, so that the UI that's seen by the user is always set by our **runtime** methods. This avoids problems where the words to be displayed to the user are changed in code and not in the **Inspector** panel, which leads to a mismatch between the on-screen text when the scene first runs and after it has been updated via a script.

> A golden rule in Unity game design is to avoid duplicating content in more than one place so that we avoid having to maintain two or more copies of the same content. Each duplicate is an opportunity for maintenance issues to occur when some, but not all, copies of a value are changed.
>
> Maximizing the use of Prefabs is another example of this principle in action. This is also known as the **DRY** principle – **Don't Repeat Yourself.**

Each time the player's character collides with any object that has its **Is Trigger** set to true, an OnTriggerEnter2D() event message is sent to both objects involved in the collision. The OnTriggerEnter2D() message is passed as a parameter that is a reference to the Collider2D component inside the object that we just collided with.

Our player's OnTriggerEnter2D() method tests the tag string of the object that was collided with to see whether it has the Star value.

Since the star GameObject we created has its trigger set and has the Star tag, the if statement inside this method will detect a collision with the star GameObject and complete the following three actions:

- The Boolean (flag) carryingStar variable will be set to true.
- The OnChangeCarryingStar(...) method of the playerInventoryDisplay script component will be called, passing in the updated value of carryingStar.
- The GameObject that was just collided with will be destroyed – that is, the **star GameObject**.

> Boolean variables are often referred to as flags. The use of a bool (true/false) variable to represent whether some feature of the game state is true or false is very common. Programmers often refer to these variables as flags. So, programmers might refer to the carryingStar variable as the star-carrying flag.

The PlayerInventoryDisplay script class

The public TextMeshProUGUI variable named starText is a reference to the Text-carrying-star UI **Text (TMP)** GameObject in our game. Its value was set via drag-and-drop at design time.

The OnChangeCarryingStar (carryingStar) method updates the text property of starText with the value of the starMessage string variable. This method takes a bool argument called carryingStar as input. The default value of the starMessage string tells the user that the player is not carrying the star. However, an if statement tests the value of carryingStar, and if that is true, then the message is changed to inform the player that they are carrying the star.

There's more...

Let's look at some details you won't want to miss.

Collecting multiple items and displaying the total number carried

Often, there are pickups that the player can collect more than one of. In such situations, we can use an integer to represent the total number collected and use a UI **Text (TMP) GameObject** object to display this total to the user. Let's modify this recipe to allow SpaceGirl to collect lots of stars!

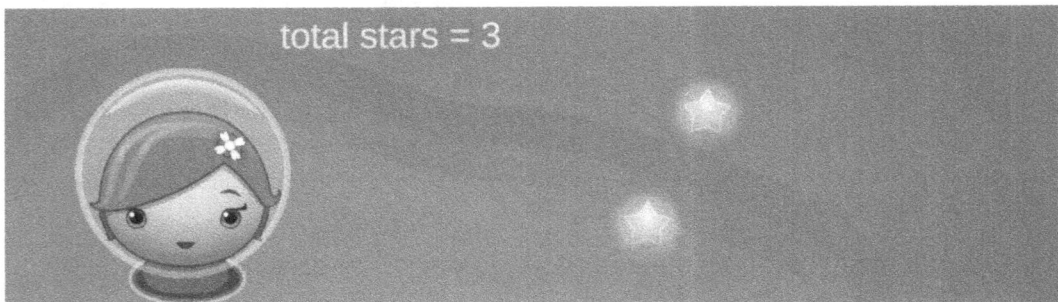

Figure 3.10: Example of collecting and displaying the multiple item total

To convert this recipe into one that shows the total number of stars that have been collected, do the following:

1. Make three or four more copies of the star GameObject and spread them around the scene. This gives the player several stars to collect rather than just one.

> Use the *Ctrl + D* (Windows) or *Cmd + D* (Mac) keyboard shortcut to quickly duplicate GameObjects.

2. Change the contents of the C# PlayerInventory script class so that it contains the following:

```csharp
using UnityEngine;

public class PlayerInventory : MonoBehaviour
{
    private PlayerInventoryDisplay playerInventoryDisplay;
    private int totalStars = 0;
    void Awake()
    {
        playerInventoryDisplay =
GetComponent<PlayerInventoryDisplay>();
    }

    void Start()
    {
        playerInventoryDisplay.OnChangeCarryingStar(totalStars);
    }

    void OnTriggerEnter2D(Collider2D hit)
    {
        if (hit.CompareTag("Star"))
        {
            totalStars++;
            playerInventoryDisplay.
OnChangeCarryingStar(totalStars);
            Destroy(hit.gameObject);
        }
    }
}
```

3. Change the contents of the C# PlayerInventoryDisplay script class so that it contains the
    following:

```
using UnityEngine;
using TMPro;


[RequireComponent(typeof(PlayerInventory))]
public class PlayerInventoryDisplay : MonoBehaviour
{
    public TextMeshProUGUI starText;

    public void OnChangeCarryingStar(int numStars)
    {
        string starMessage = "total stars = " + numStars;
        starText.text = starMessage;
    }
}
```

As you can see, in PlayerInventory, we now increment totalStars by 1 each time a star GameObject
is collided with. In PlayerInventoryDisplay, we display a simple text message stating **total stars =**
on-screen, followed by the integer total that was received by the OnChangeStarTotal(...) method.

Now, when you run the game, you should see the total stars start at 0 and increase by 1 each time the
player's character hits a star.

Alternative – combining all the responsibilities into a single script

The separation of the player inventory (what they are carrying) and how to display the inventory to
the user is an example of a best practice design pattern called **Model-View-Controller** (**MVC**), whereby
we separate the code that updates the UI from the code that changes player and game variables, such
as score and inventory item lists.

However, for *very simple games*, we may choose to display its status in a single script class. For an example of this approach for this recipe, remove the PlayerInventory and PlayerInventoryDisplay
script components, create a C# script class called PlayerInventoryCombined, and add an instance of
the script to the player-SpaceGirl GameObject in the **Hierarchy** panel:

```
using UnityEngine;
using TMPro;


public class PlayerInventoryCombined : MonoBehaviour {
```

```
public TextMeshProUGUI starText;
private bool carryingStar = false;

private void Start() {
    UpdateStarText();
}

void OnTriggerEnter2D(Collider2D hit) {
    if (hit.CompareTag("Star")) {
        carryingStar = true;
        UpdateStarText();
        Destroy(hit.gameObject);
    }
}

private void UpdateStarText() {
    string starMessage = "no star :-(";
    if (carryingStar) {
        starMessage = "Carrying star :-)";
    }

    starText.text = starMessage;
}
}
```

There is no difference in terms of the experience of the player; the change is simply in the architectural structure of our game code.

This recipe demonstrates how to create a simple inventory that includes display text to tell players whether they are carrying a single item (or not). In addition, you can detect collisions with GameObjects and display an on-screen message stating whether the GameObject has been collected. While it's possible to do all this work in a single script, it's a good idea to break different responsibilities into different script classes whenever we can, since it means at any one time, we are working on a small amount of code that does one key action. Smaller components are easier to debug, more likely to be reused, and less likely to have to be changed.

Displaying single-object pickups with carrying and not-carrying icons

Graphic icons are an effective way to inform the player that they are carrying an item. In this recipe, if no star is being carried, a gray-filled icon in a blocked-off circle will be displayed in the top-left corner of the screen:

Figure 3.11: Example of a single-object pickup

Then, once a star has been picked up, a yellow-filled star icon will be displayed. In many cases, icons are clearer (they don't require reading and thinking about) and can also be smaller on-screen than text messages that indicate player status and inventory items.

This recipe will also illustrate the benefits of the MVC design pattern, which we described in the previous recipe – we are changing how to communicate with the user (using the **View** via icons rather than text), but we can use, with no changes required, the PlayerInventory script class (the **Model-Controller**), which detects player-star collisions and maintains the Boolean flag that tells us whether a star is being carried.

Getting ready

For this recipe, we have prepared a folder named _Scripts in the 03_03 folder.

This recipe assumes that you are starting with the Simple2Dgame_SpaceGirl project, which we set up in the first recipe of this chapter.

How to do it...

To toggle carrying and not-carrying icons for a single-object pickup, follow these steps:

1. Start with a new copy of the Simple2Dgame_SpaceGirl mini-game.
2. Import the _Scripts folder from the provided files (this contains a copy of the PlayerInventory script class from the previous recipe, which we can use unchanged for this recipe).
3. Add a **UI Image** object to the scene (**GameObject | UI | Image**). Rename it Image-star-icon.
4. Select the Image-star-icon object in the **Hierarchy** panel. Drag the icon_nostar_100 sprite from the Sprites folder in the **Project** panel to the **Source Image** field in the **Inspector** panel.

5. Click on the **Set Native Size** button for the **Image** component. This will resize the **UI image** so that it fits the physical pixel's width and height of the icon_nostar_100 sprite file:

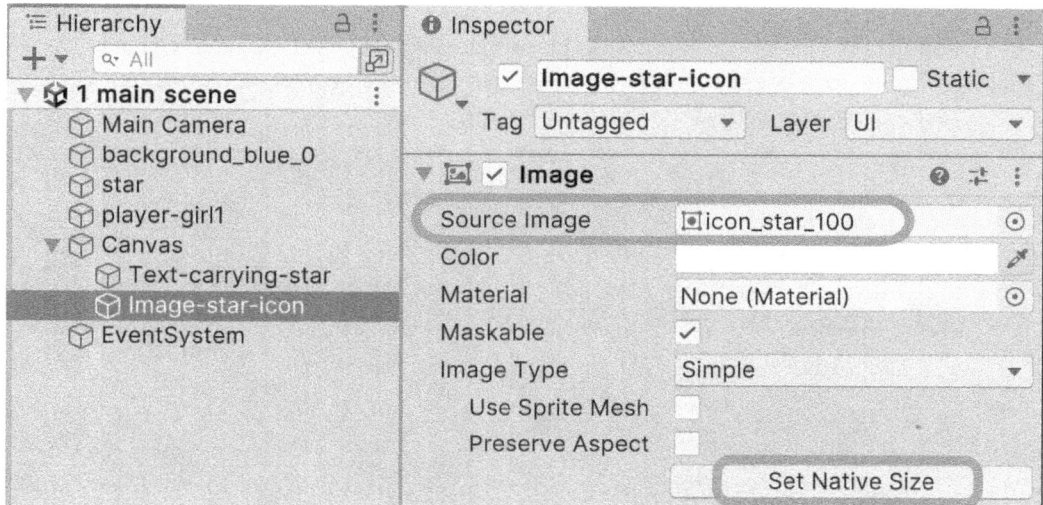

Figure 3.12: Image source and native size settings

6. Position the image icon at the top, to the left of the **Game** panel, in **Rect Transform**. Choose the **top-left** box component while holding down *Shift + Alt* (to set its pivot and position).

7. Create the following C# PlayerInventoryDisplay script class and add an instance of the script to the player-SpaceGirl GameObject in the **Hierarchy** panel:

```
using UnityEngine;
using UnityEngine.UI;


[RequireComponent(typeof(PlayerInventory))]
public class PlayerInventoryDisplay : MonoBehaviour  {
    public Image imageStarGO;
    public Sprite iconNoStar;
    public Sprite iconStar;

    public void OnChangeCarryingStar(bool carryingStar) {
        if (carryingStar)
            imageStarGO.sprite = iconStar;
        else
            imageStarGO.sprite = iconNoStar;
    }
}
```

8. From the **Hierarchy** panel, select the player-girl1 GameObject. Then, from the **Inspector** panel, access the **Player Inventory Display (Script)** component and populate the **Star Image** public field with the Image-star-icon UI image object.

9. Populate the **Icon No Star** public field from the **Project** panel with the icon_nostar_100 sprite, and then populate the **Icon Star** public field from the **Project** panel with the icon_star_100 sprite, as shown in the following screenshot:

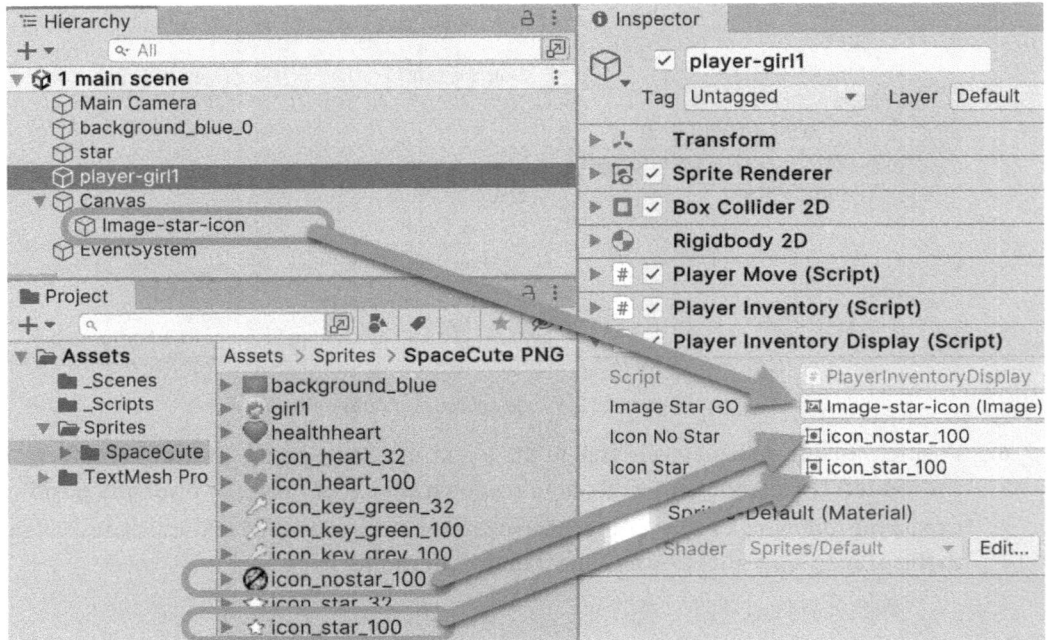

Figure 3.13: Populating the public fields for Player Inventory Display

10. Save and play the scene. You should see the no-star icon (a gray-filled icon in a blocked-off circle) in the top left until you pick up the star, at which point it will change to show the carrying-star icon (yellow-filled star). Note that you might need to reposition either the star or player-girl1 so that they don't collide on play.

How it works...

In the PlayerInventoryDisplay script class, the imageStarGO image variable is a reference to the Image-star-icon UI image object. The iconStar and iconNoStar sprite variables are references to the sprite files in the **Project** panel – the sprites that inform the player of whether or not a star is being carried.

Each time the OnChangeCarryingStar(carryingStar) method is invoked by the PlayerInventory object, this method uses an if statement to set the **UI image** to the sprite that corresponds to the value of the bool argument that's been received.

Displaying multiple pickups of the same object with multiple status icons

If there is a small, fixed total number of an item to be collected rather than text totals, an effective UI approach is to display placeholder icons (empty or grayed-out pictures) to show the user how many of the same item are left to be collected. Each time an item is picked up, a placeholder icon is replaced with a full-color, collected icon.

In this recipe, we will use gray-filled star icons as the placeholders and yellow-filled star icons to indicate each collected star, as shown in the following screenshot:

Figure 3.14: UI showing multiple status icons

Getting ready

This recipe assumes that you are starting with the Simple2Dgame_SpaceGirl project, which we set up in the first recipe of this chapter.

How to do it...

To display multiple inventory icons for multiple pickups of the same type of object, follow these steps:

1. Start with a new copy of the Simple2Dgame_SpaceGirl mini-game.

2. Create a C# script class called PlayerInventory in the _Scripts folder:

```csharp
using UnityEngine;

public class PlayerInventory : MonoBehaviour {
    private PlayerInventoryDisplay playerInventoryDisplay;
    private int totalStars = 0;

    void Awake() {
        playerInventoryDisplay = GetComponent<PlayerInventoryDisplay>
            ();
    }

    void Start() {
        playerInventoryDisplay.OnChangeStarTotal(totalStars);
```

```
        }

    void OnTriggerEnter2D(Collider2D hit) {
        if (hit.CompareTag("Star")) {
            totalStars++;
            playerInventoryDisplay. OnChangeStarTotal(totalStars);
            Destroy(hit.gameObject);
        }
    }
}
```

3. Select the star GameObject in the **Hierarchy** panel and make three more copies of it. There should now be four star GameObjects in the scene. Move these new star GameObjects to different parts of the screen.

4. Add the following C# script class, called PlayerInventoryDisplay, to the player-girl1 GameObject in the **Hierarchy** panel:

```
using UnityEngine;
using System.Collections;
using UnityEngine.UI;

[RequireComponent(typeof(PlayerInventory))]
public class PlayerInventoryDisplay : MonoBehaviour {
    public Image[] starPlaceholders;
    public Sprite iconStarYellow;
    public Sprite iconStarGrey;

    public void OnChangeStarTotal(int starTotal){
        for (int i = 0;i < starPlaceholders.Length; ++i){
            if (i < starTotal)
                starPlaceholders[i].sprite = iconStarYellow;
            else
                starPlaceholders[i].sprite = iconStarGrey;
        }
    }
}
```

5. In the **Project** panel, select the `icon_star_gray_100` sprite (in the `Sprites` folder), and in the **Inspector**, ensure its **Sprite Mode** is set to **Single**. If it's not, then select **Single** and click **Apply**. This will ensure our 100x100 sprites don't lose any size due to transparent pixels around the edge.

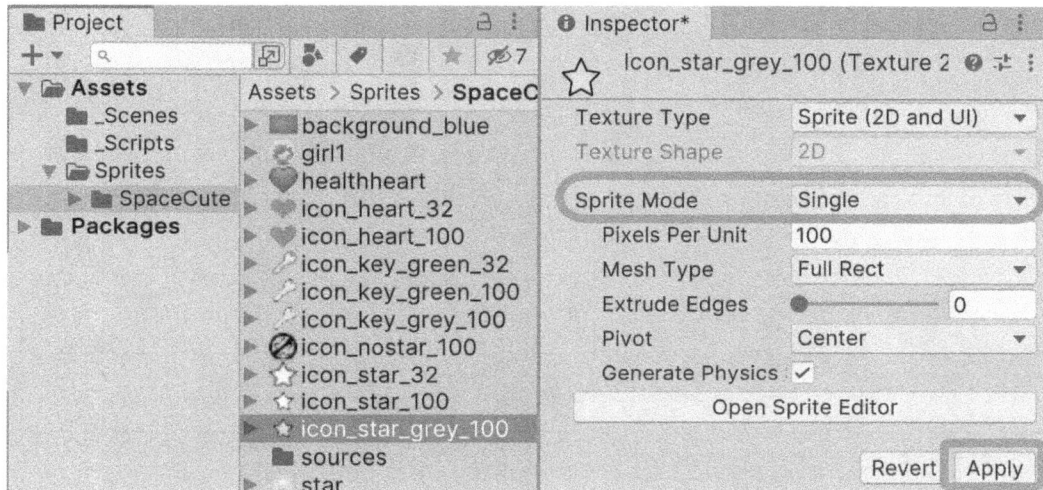

Figure 3.15: Ensuring image is treated full-size since containing a single sprite

6. Repeat this check for the **Sprite Mode** setting of **Single** for the `icon_star_100` sprite.

7. Select **Canvas** in the **Hierarchy** panel and add a new **UI Image** object (**Create | UI | Image**). Rename it `Image-star0`.

8. Select `Image-star0` in the **Hierarchy** panel.

9. From the **Project** panel, drag the `icon_star_grey_100` sprite (in the `Sprites` folder) into the **Source Image** field in the **Inspector** panel for the **Image (Script)** component.

10. Click on the **Set Native Size** button for the **Image (Script)** component. This will resize the UI image so that it fits the physical pixel width and height of the `icon_star_grey_100` sprite file.

11. Now, we will position our icon at the top left of the **Game** panel. Edit the **UI image's Rect Transform** component and, while holding down *Shift + Alt* (to set its pivot and position), choose the **top-left** box. The **UI Image** `Image-star0` should now be positioned at the top left of the **Game** panel.

12. Make three more copies of `Image-star0` in the **Hierarchy** panel, naming them `Image-star1`, `Image-star2`, and `Image-star3`.

13. In the **Inspector** panel, change the **Pos X** position (in the **Rect Transform** component) of Image-star1 to 100, Image-star2 to 200, and Image-star3 to 300.

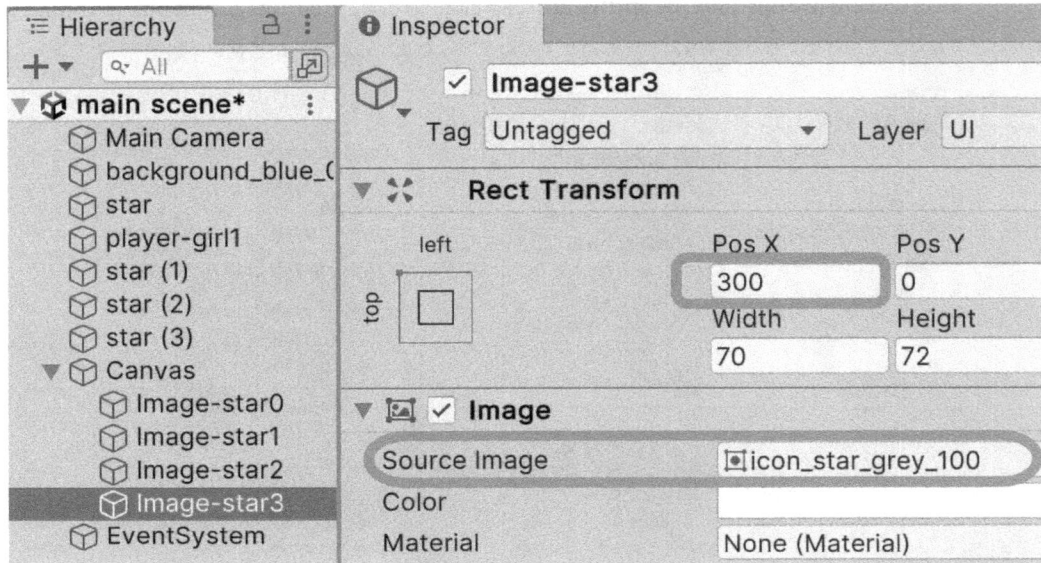

Figure 3.16: Rect Transform settings for Image-star3

14. In the **Hierarchy** panel, select the player-girl1 GameObject. Then, from the **Inspector** panel, access the **Player Inventory Display (Script)** component and set the **Size** property of the **Star Placeholders** public field to 4.

15. Next, populate the **Element 0/1/2/3** array values of the **Star Placeholders** public field with **UI Image** objects Image-star0/1/2/3.

16. Now, populate the **Icon Star Yellow** and **Icon Star Grey** public fields from the **Project** panel with the icon_star_100 and icon_star_grey_100 sprites.

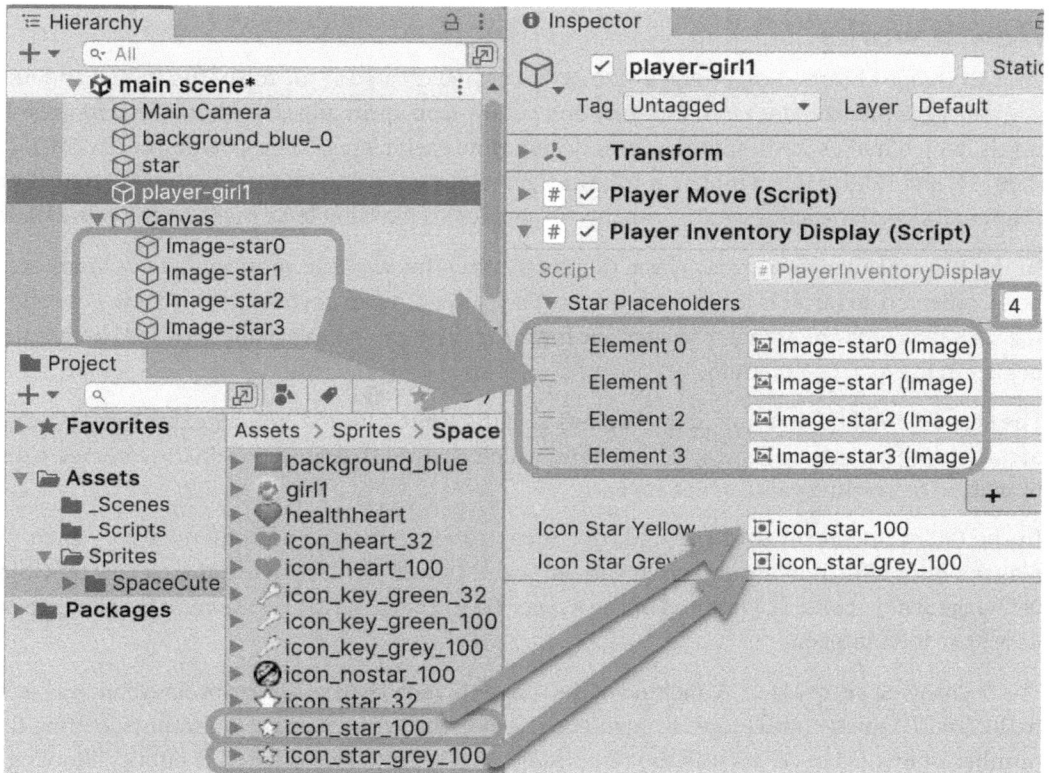

Figure 3.17: Populating the public fields with four UI Image GameObjects and two sprite asset files

17. Now, when you play the scene, you should see the sequence of four gray placeholder star icons. Then, each time you collide with a star, the next icon at the top should turn yellow.

How it works...

Four **UI Image** objects, Image-star0/1/2/3, have been created at the top of the screen and initialized with the gray placeholder icon. The gray and yellow icon sprite files have been resized to be 100x100 pixels, which makes positioning them at **design time** easier, since their positions are (0,0), (100,0), (200,0), and (300,0). In a more complicated game screen or one where screen real estate is precious, the actual size of the icons would probably be smaller – a decision to be made by the game UI designer.

In the PlayerInventory script class, the totalStars int variable represents how many stars have been collected so far; it is initialized to zero. The playerInventoryDisplay variable is a reference to the scripted component that manages our inventory display – this variable is cached before the scene begins in the Awake() method.

The Start() method that runs at the beginning of the scene calls the OnChangeStarTotal(...) method of the PlayerInventoryDisplay component, which ensures that the icons on the screen are displayed to match the starting value of totalStars.

In the OnTriggerEnter2D() method, the totalStars counter is incremented by 1 each time the player's character hits an object tagged as Star. As well as destroying the hit GameObject, the OnChangeStarTotal(...) method of the PlayerInventoryDisplay component is called, passing the new star total integer.

The OnChangeStarTotal(...) method of the PlayerInventoryDisplay script class contains references to the four UI images and loops through each item in the array of image references, setting the given number of images to yellow and the remaining ones to gray. This method is public, allowing it to be called from an instance of the PlayerInventory script class.

Note that when Unity believes there may be multiple sprites "packaged" into a single image, it will ignore any transparent pixels bordering a sprite image file. So while the image file might be 100x100 pixels, Unity will only see the sprite as a smaller size (e.g., 72x72) for the rectangle enclosing non-transparent pixels. Since hiding/revealing icons in this recipe is dependent on precise image sizes on screen, we must fix this by ensuring Unity knows that each sprite asset file contains a single image only, and so will use all pixels regardless of transparency. For this reason, we had to ensure that both icon_star_gray_100 and icon_star_100 were declared as in **Sprite Mode Single**.

There's more...

Let's look at some details you won't want to miss.

Revealing icons for multiple object pickups by changing the size of a tiled image

Another approach that could be taken to show increasing numbers of images is to make use of tiled images. The same visual effect as in the previous recipe can also be achieved by making use of a tiled gray star image with a width of 400 (showing four copies of the gray star icon) behind a tiled yellow star image, whose width is 100 times the number of stars collected.

If the yellow-starred image is less wide than the gray-starred imaged beneath, then we'll see gray stars for any remaining locations. For example, if we are carrying 3 stars, we'll make the width of the yellow-starred image *3 x 100 = 300* pixels wide. This will show 3 yellow stars and reveal 100 pixels; that is, 1 gray star from the gray-starred image beneath it.

To display gray and yellow star icons for multiple object pickups using tiled images, let's adapt our recipe to illustrate this technique:

1. In the **Hierarchy** panel, delete the entire **Canvas** GameObject (and therefore, delete all four UI images).

2. Add a new **UI Image** object to your scene (**Create | UI | Image**). Rename the GameObject `Image-stars-gray`.

3. Ensure **Image-stars-gray** is selected in the **Hierarchy** panel. From the **Project** panel, drag the `icon_star_gray_100` sprite (in the `Sprites` folder) into the **Source Image** field in the **Inspector** panel (in the **Image (Script)** component).

4. Click on the **Set Native Size** button for the **Image (Script)** component. This will resize the UI image so that it fits the physical pixel width and height of the `icon_star_gray_100` sprite file.

5. Now, position the icon at the top left of the screen. Edit the **UI Image's Rect Transform** component and, while holding down *Shift + Alt* (to set its pivot and position), choose the **top-left** box. The **UI image** should now be positioned at the top left of the **Game** panel.

6. In the **Inspector** panel, change **Width** (in the **Rect Transform** component) of **Image-stars-gray** to `400`. Also, set **Image Type** (in the **Image (Script)** component) to **Tiled**, as shown in the following screenshot:

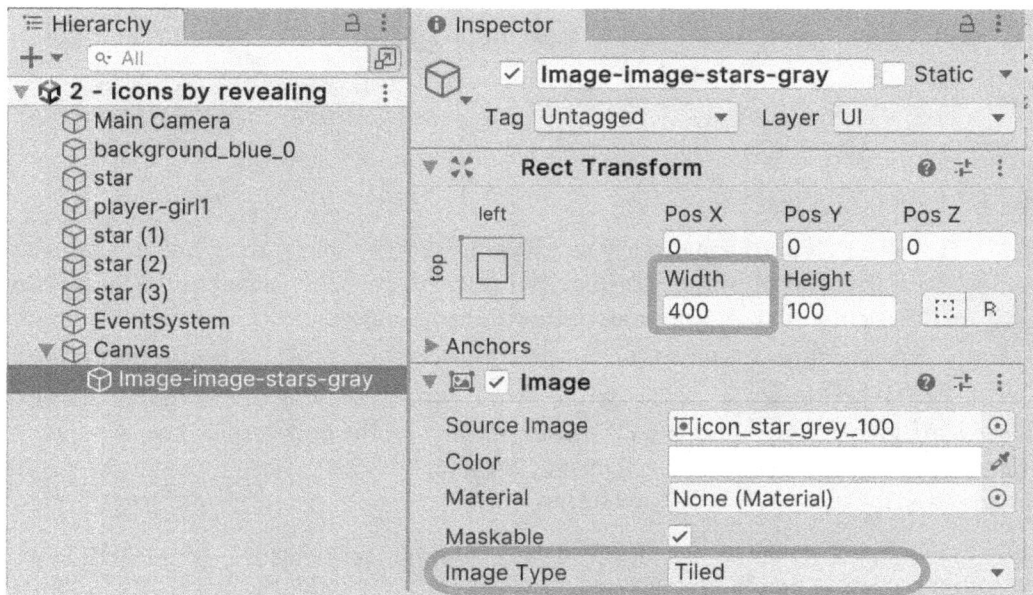

Figure 3.18: Width and Tiled Image settings for UI Image-stars-grey

For a simple game like this, we are choosing simplicity over memory efficiency. You'll see a notice, suggesting that you use an advanced texture with **Wrap** mode repeated and a cleared packing tag. While more memory efficient, it's more complicated to do for small, simple tiling such as in this recipe.

7. Make a copy of **Image-stars-gray** in the **Hierarchy** panel, naming the copy **Image-stars-yellow**.

8. With **Image-stars-yellow** selected in the **Hierarchy** panel, from the **Project** panel, drag the icon_star_100 sprite (in the Sprites folder) into the **Source Image** field in the **Inspector** window (in the **Image (Script)** component).

9. Set the width of **Image-stars-yellow** to 0 (in the **Rect Transform** component). So, now, we have the yellow stars tiled image above the gray stars tiled image, but since its width is zero, we don't see any of the yellow stars yet.

10. Replace the content of the existing C# script, called PlayerInventoryDisplay, with the following code:

```
using UnityEngine;
using UnityEngine.UI;

[RequireComponent(typeof(PlayerInventory))]
public class PlayerInventoryDisplay : MonoBehaviour {
    public Image iconStarsYellow;

    public void OnChangeStarTotal(int starTotal) {
        float newWidth = 100 * starTotal;
        iconStarsYellow.rectTransform.SetSizeWithCurrentAnchors(
            RectTransform.Axis.Horizontal, newWidth );
    }
}
```

11. From the **Hierarchy** panel, select the **player-girl1** GameObject. Then, from the **Inspector** panel, access the **Player Inventory Display (Script)** component and populate the **Icons Stars Yellow** public field with the **Image-stars-yellow** UI image object.

How it works...

Image-stars-gray is a tiled image, wide enough (400px) for the gray sprite, icon_star_gray_100, to be shown four times. **Image-stars-yellow** is a tiled image, above the gray one, initially with its width set to zero so that no yellow stars can be seen.

Each time a star is picked up, a call is made from the PlayerInventory scripted object to the OnChangeStarTotal() method of PlayerInventoryDisplay, passing the new integer number of stars collected. By multiplying this by the width of the yellow sprite image (100 px), we get the correct width to set **Image-stars-yellow** so that the corresponding number of yellow stars will now be seen by the user. Any stars that remain to be collected will still be seen as gray stars.

The actual task of changing the width of **Image-stars-yellow** can be completed by calling the `SetSizeWithCurrentAnchors(...)` method. The first parameter is the axis, so we pass the `RectTransform.Axis.Horizontal` constant so that the width will be changed. The second parameter is the new size for that axis, so we must pass a value that is 100 times the number of stars collected so far (the `newWidth` variable).

This recipe has demonstrated two approaches to displaying multiple pickups of the same object with multiple status icons that are commonly used in games, such as health, key collection, or, in this case, stars. Both approaches use a placeholder image that is replaced when an item is then collected. The first is best used when the number of items to collect is known, while the second is best for when the number of items is either unknown or there's a large quantity of them. The next recipe will explore improving the UI elements of an inventory.

Using panels to visually outline the inventory UI area and individual items

There are four kinds of objects we see when playing a game:

- GameObjects that have some visual elements, such as 2D and 3D objects.
- UI elements located in **World Space,** so they appear next to GameObjects in the scene.
- UI elements located in **Screen Space - Camera,** so they appear at a fixed distance from the camera (but can be obscured by GameObjects closer to the camera than these UI elements).
- UI elements located in **Screen Space - Overlay.** These always appear above the other three kinds of visual elements and are perfect for **Heads-Up Display (HUD)** elements, such as inventories.

Sometimes, we want to visually make it clear which elements are part of the UI's HUD and which are visual objects in the scene. **HUDs are often used to display multiple pieces of information at once, such as the player's health, inventory, and game progress (score or level).** Using Unity **UI Panels,** along with an opaque or translucent background image, is a simple and effective way to achieve this.

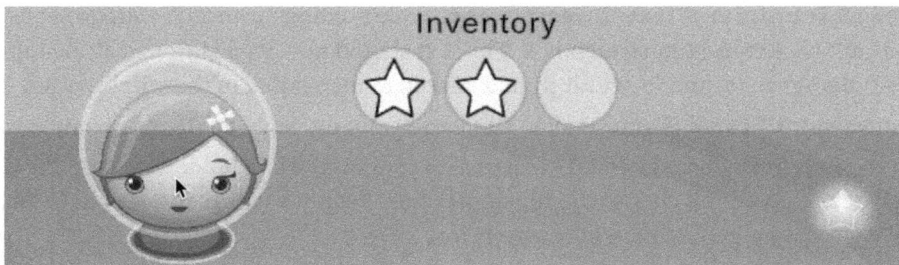

Figure 3.19: Example of using Unity UI Panels

UI Panels can also be used to display locations (slots) with shaped or colored backgrounds indicating where items may be placed, or how many may be collected.

As shown in the preceding screenshot, in this recipe, we'll create a panel with some title text and three inventory slots, two of which will be filled with star icons, communicating to the player that there is one more star that can be collected/carried. Note that this recipe is a preparatory recipe illustrating how to use panels to create an effective inventory UI – the actual display won't change when running this recipe's project. You'll learn how to update the panel slots dynamically in the recipe after this one.

Getting ready

This recipe assumes that you are starting with the Simple2Dgame_SpaceGirl project, which we set up in the first recipe of this chapter.

How to do it...

To use panels to visually outline the inventory area and individual items, follow these steps:

1. Start with a new copy of the Simple2Dgame_SpaceGirl mini-game.

2. In the **Hierarchy** panel, create a **UI Panel** (**GameObject | UI | Panel**) and rename it Panel-background.

3. Now, let's position Panel-background at the top of the **Game** panel, stretching the horizontal width of the canvas. Edit the UI image's **Rect Transform** component and, while holding down *Shift + Alt* (to set its pivot and position), choose the **top-stretch** box.

4. The panel will still be taking up the whole game window. Now, in the **Inspector** panel, change **Height** (in the **Rect Transform** component) of Panel-background to 100.

5. With the Panel-background still selected in the **Hierarchy**, add a **UI Text (TMP)** object (**Create | UI | Text - TextMeshPro**) and name it Text-inventory. When prompted, click the button to import the **TMP Essentials**. For its **TextMeshPro -Text UI** component, change the text to **Inventory**.

6. In the **Inspector**, ensure the Text-inventory GameObject is selected, and for its **Alignment** component vertically and horizontally center its text. Then set its **Font Size** to 23, **Vertex Color** to black, and make the text bold.

7. Edit **Rect Transform** of Text-inventory and, while holding down *Shift + Alt* (to set its pivot and position), choose the **top-stretch** box. The text should now be positioned at the top-center of Panel-background, and its width should stretch to match that of the whole panel.

8. Create a new **UI Panel** (**GameObject | UI | Panel**) and name it Panel-inventory-slot.

9. Edit the **Rect Transform** of Panel-inventory-slot and, while holding down *Shift + Alt* (to set its pivot and position), choose the **top-center** box. Set both **Width** and **Height** to 70 and **Pos Y** to -30, as shown in the following screenshot:

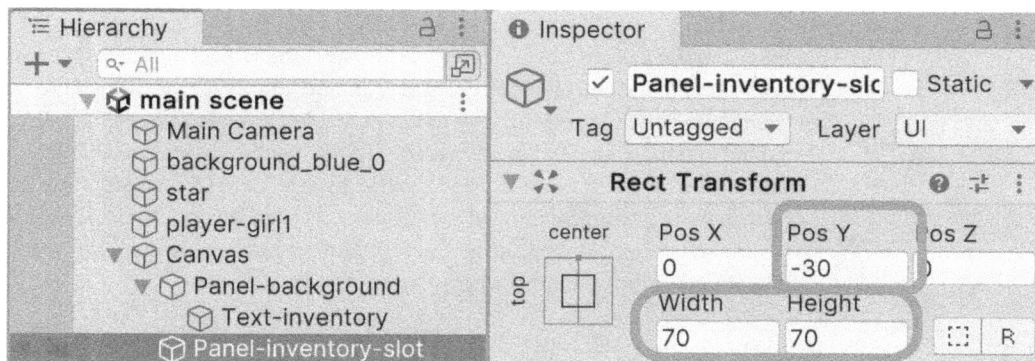

Figure 3.20: Rect Transform settings for Panel-inventory-slot

10. Ensure that the Panel-inventory-slot GameObject is selected in the **Hierarchy**. In the **Image (Script)** component, change **Source Image** from the **UI Panel** default of **Background** to the circular **Knob** image (this is one of the built-in images that comes as part of the Unity UI system). As shown in the following screenshot, you should now see a circle centered below the title text in our inventory HUD rectangle. This circle visually tells the user that there is space in the inventory for an item to be collected:

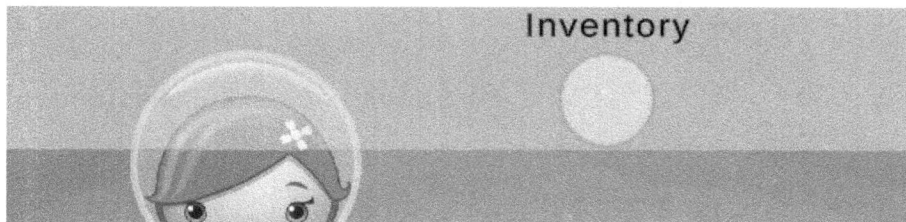

Figure 3.21: Example of a built-in circular Knob image for the UI

11. Imagine that the player has collected a star. Now, inside our inventory slot circle panel, let's add a yellow star icon image. Add a **UI Image** object to the scene (**GameObject | UI | Image**). Rename it Image-icon and drag the Image-icon GameObject to make it a child to Panel-inventory-slot.

Child GameObjects can be hidden, making the GameObject **inactive**. By creating a new **UI Image** GameObject for our star icon and adding it as a child of our `Panel-inventory-slot` GameObject, we can now display the star icon when the image is enabled and hide it by making it inactive. This is a general approach, which means that so long as we have a reference to the **Image** GameObject, we don't have to do extra work swapping images, as we had to do in some of the previous recipes. This means we can start writing more general-purpose code that will work with different inventory panels for keys, stars, money, and so on.

12. With `Image-icon` selected in the **Hierarchy** panel, drag the `icon_star_100` sprite (in the `Sprites` folder) from the **Project** panel into the **Source Image** field in the **Inspector** panel (in the **Image (Script)** component).

13. Edit **Rect Transform** of `Image-icon` and, while holding down *Shift + Alt* (to set its pivot and position), choose the **stretch-stretch** box. Set the **X** and **Y Scale** values to 0.7. The star icon should now be stretched enough that it fits inside the 70x70 parent panel, so that we can see a star nicely arranged inside the circle:

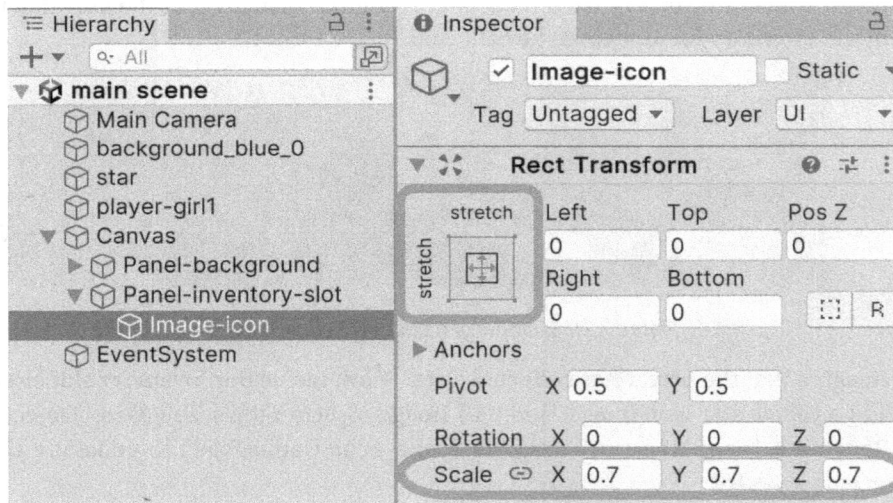

Figure 3.22: Rect Transform settings of the Image-icon child of Panel-inventory-slot

14. Save and run the scene and play the game. You should see a clearly defined rectangle at the top of the screen, with `Inventory` as the title text. Inside the inventory rectangular area, you will see a circular slot, currently showing a star.

15. Let's display three slots to the player. First, change the **Pos X** horizontal position of the `Panel-inventory-slot` panel to `-70`. This moves it left of center, making space for the next one, and allowing us to center the three slots when we've finished.

16. Duplicate the `Panel-inventory-slot` panel, renaming (if necessary) the copy `Panel-inventory-slot (1)`. Set **Pos X** of this copy to 0.

17. Duplicate `Panel-inventory-slot` again, this time renaming (if necessary) the copy `Panel-inventory-slot (2)`. Set **Pos X** of this copy to 70. Now, select the `Image-star-icon` child of this third panel and make it inactive (at the top of the **Inspector** panel, uncheck its **active** checkbox, to the left of the GameObject's name). The star for this panel should now be hidden so that only the circle background of the slot's panel is visible.

18. Save and run the scene, and you should see a nice UI showing two stars collected and an empty third slot.

How it works...

In this recipe, we created a simple panel (**Panel-background**) with the title **UI Text (TMP)** as a child GameObject at the top of the game canvas, which shows a grayish background rectangle and the title text stating **Inventory**. This indicates to the player that this part of the screen is where the inventory HUD will be displayed.

To illustrate how this might be used to indicate a player carrying stars, we added a smaller panel for one slot in the inventory with a circular background image and in that, added a start icon as a child GameObject. We then duplicated the slot panel two more times, positioning them 70 pixels apart. After that, we disabled (made inactive) the star icon of the third slot so that an empty slot circle is shown.

Our scene presents the user with a display, indicating that two out of a possible three stars are being carried. This recipe is a good start to a more general-purpose approach to creating inventory UIs in Unity, and we'll build on this in the next recipe...

> We'll learn how to limit the player's movement to prevent their character from moving into the rectangles of HUD items like this in *Chapter 11, Controlling and Choosing Positions*.

Creating a C# inventory slot UI to display scripted components

In the previous recipe, we started to work with **UI panels** and images to create a more general-purpose GameObject for displaying inventory slots, as well as images to indicate what is stored in them. In this recipe, we will take things a little further with the graphics and also create a C# script class that works with each inventory slot object:

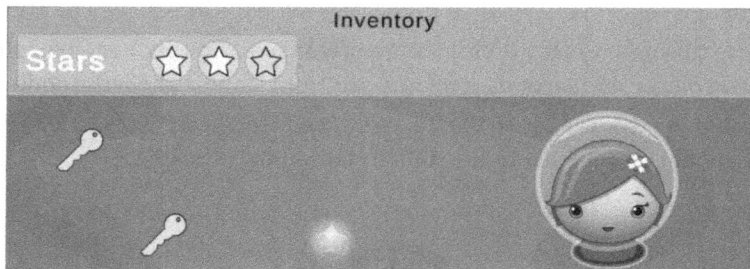

Figure 3.23: Example showing stars inventory panel, with three slots

As shown in the above screenshot, in this recipe, we'll create a UI (and scripts) for an inventory that has three locations for stars, and three more for keys, using colored and gray icons to indicate how many have been collected.

Getting ready

This recipe adapts the previous one. So, make a copy of the project from the previous recipe and work on this copy.

For this recipe, we have prepared a folder named _Scripts in the 03_06 folder.

How to do it...

To create a C# inventory slot display script component, follow these steps:

1. Import the _Scripts folder from the provided files (this contains a copy of the PlayerInventory script class for an integer number of stars, from one of the previous recipes, which we can use unchanged for this recipe).

2. Delete two of the three inventory slot GameObjects – that is, Panel-inventory-slot (1) and Panel-inventory-slot (2) – so that only Panel-inventory-slot remains.

3. First, we'll create a panel for three star slots. In the **Hierarchy** panel, create a **UI Panel** (**Create | UI | Panel**) and rename it Panel-stars.

4. We'll now position Panel-stars at the top-left of the **Game** panel and make it fit within the left-hand side of our general inventory rectangle. Edit the UI **Image's Rect Transform** component and, while holding down *Shift + Alt* (to set its pivot and position), choose the **top-left** box. Now, set **Width** to 300 and **Height** to 60. We'll nudge this away from the top-left corner by setting **Pos X** to 10 and **Pos Y** to -30.

5. Add a **UI Text (TMP)** object (**Create | UI | Text - TextMeshPro**) and rename it Text-title. For its **TextMeshPro -Text UI** component, change the text to Stars. Make Text-title a child of Panel-stars.

6. Edit the **Rect Transform** of Text-title and, while holding down *Shift + Alt* (to set its pivot and position), choose the **left-middle** box. The text should now be positioned to the left of Panel-stars.

7. In the **Inspector** panel, left-align the text horizontally, and center-align the text vertically, set its **Height** to 50, and set **Font Size** to 32. Choose a **yellow** text color.

8. We'll now nudge Text-title away from the very left edge by setting **Pos X** to 10.

9. Make Panel-inventory-slot a child of Panel-stars. Edit its **Rect Transform** and, while holding down *Shift + Alt* (to set its pivot and position), choose the **left-middle** box.

10. Resize Panel-inventory-slot so that its **Width** and **Height** are both 50 x 50 pixels. Set its **Pos X** to 140. It should now appear to the right of the yellow Stars text:

Figure 3.24: Panel-inventory-slot as child inside Panel-stars

11. Rename the `Image-icon` GameObject `Image-icon-gray`. Then, duplicate this GameObject, naming the copy `Image-icon-color`. Both should be child GameObjects of `Panel-inventory-slot`. In the **Hierarchy** panel, the sequence should be that the first child is `Image-icon-gray` and the second child is `Image-icon-color`. If this isn't the order, then swap them around.

12. Select `Image-icon-gray` and drag the `icon_star_grey_100` sprite (in the `Sprites` folder) from the **Project** window into the **Source Image** field of the **Inspector** window (in the **Image (Script)** component). If you make the `Image-icon-color` GameObject inactive, you should see the gray star icon inside the slot panel's circle.

13. Create a C# script called `PickupUI` (a copy can be found in the `_Scripts` folder).

14. Add an instance of the `PickupUI` C# script as a component to the GameObject of `Panel-inventory-slot` in the **Hierarchy** panel:

```csharp
using UnityEngine;

public class PickupUI : MonoBehaviour {
    public GameObject iconColor;
    public GameObject iconGrey;

    void Awake() {
        DisplayEmpty();
    }

    public void DisplayColorIcon() {
        iconColor.SetActive(true);
        iconGrey.SetActive(false);
    }
}
```

```
public void DisplayGreyIcon() {
    iconColor.SetActive(false);
    iconGrey.SetActive(true);
}

public void DisplayEmpty() {
    iconColor.SetActive(false);
    iconGrey.SetActive(false);
}
}
```

15. Select `Panel-inventory-slot` in the **Hierarchy**. In the **Inspector** panel, for the **Pickup UI (Script)** component, populate the **Icon Color** public field by dragging `Image-icon-color` from **Hierarchy**. Likewise, populate the **Icon Gray** public field by dragging `Image-icon-gray` from **Hierarchy**. Now, the scripted **PickupUI** component in `Panel-inventory-slot` has references to the colored and gray icons for this inventory slot GameObject.

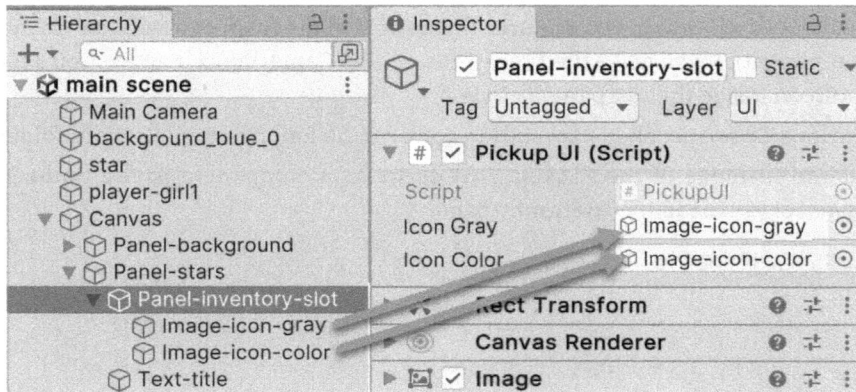

Figure 3.25: Panel-inventory-slot with PickupUI scripted component referencing the gray and color icons

16. Duplicate `Panel-inventory-slot`, rename this copy `Panel-inventory-slot2`, and set its **Pos X** to 190.

17. Duplicate `Panel-inventory-slot` for a second time, rename this copy `Panel-inventory-slot3`, and set its **Pos X** to 240. You should now see all three star inventory icons lined up nicely spaced to the right of the yellow `Stars` title text:

Figure 3.26: Panel-stars showing three star inventory icons lined up

18. Add the following C# script, called `PlayerInventoryDisplay`, to the `player-girl1` GameObject in the **Hierarchy** panel:

```csharp
using UnityEngine;

[RequireComponent(typeof(PlayerInventory))]
public class PlayerInventoryDisplay : MonoBehaviour
{
  public PickupUI[] slots = new PickupUI[1];

  public void OnChangeStarTotal(int starTotal)
  {
    int numInventorySlots = slots.Length;
    for (int i = 0; i < numInventorySlots; i++)
    {
      PickupUI slot = slots[i];
      if (i < starTotal)
        slot.DisplayColorIcon();
      else
        slot.DisplayGreyIcon();
    }
  }
}
```

19. From the **Hierarchy** panel, select the `player-girl1` GameObject. Then, do the following in the **Inspector** panel for the **Player Inventory Display (Script)** component:

   - Set the size of the **Slots** public array to 3.
   - Populate the **Element 0** public field with the `Panel-inventory-slot` GameObject.
   - Populate the **Element 1** public field with the `Panel-inventory-slot2` GameObject.

- Populate the **Element 2** public field with the `Panel-inventory-slot3` GameObject:

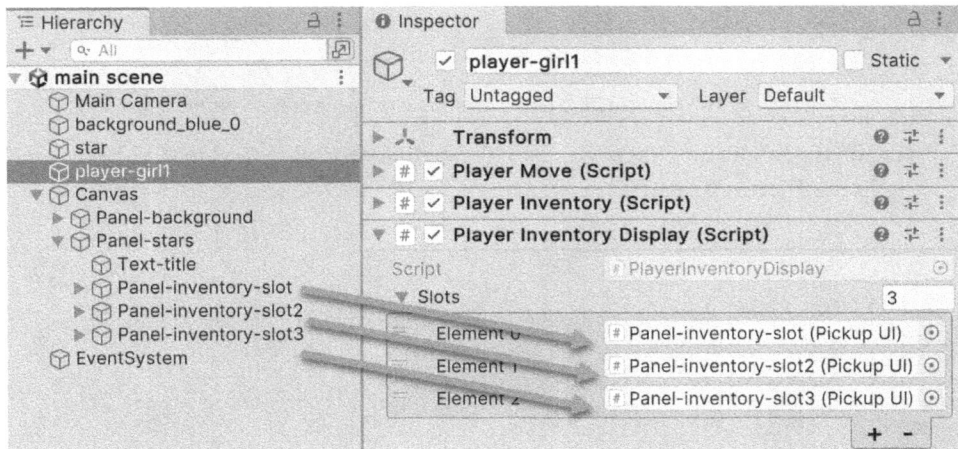

Figure 3.27: Populating the public fields of Player Inventory Display (Script)

20. Finally, make two more copies of the `star` GameObject in the scene and move them around. So, there are now three GameObjects tagged `Star` for the player to collect.

21. When you run the game and the player's character hits each `star` GameObject, it should be removed from the scene, and the next free inventory star icon should change from gray to yellow.

How it works...

In this recipe, we created a panel (`Panel-stars`) in which to display the large **Stars** title text and three inventory slot panels to show how many stars can be collected, as well as how many have been collected at any point in the game. Each star panel slot is a **UI Panel** with a circular **Knob** background image and two children, one showing a gray icon image and a second showing a colored icon image. When the colored icon image GameObject is disabled, it will be hidden, which reveals the gray icon. When both the colored and gray images are disabled, then an empty circle will be shown, which could, perhaps, be used to indicate to the user that a general-purpose location is empty and available in the inventory.

The `PickupUI` script class has two public variables that are references to the gray and colored icons for the GameObject they relate to. Before the scene starts (the `Awake()` method is used), the script hides the gray and colored icons and displays an empty circle. This script class declares three public methods (they are public so that they can be invoked from another scripted object when the game is running). These methods hide/reveal the appropriate icons to display the related inventory panel UI object as either empty, gray, or colored. The methods are clearly named `DisplayEmpty()`, `DisplayGrayIcon()`, and `DisplayColorIcon()`.

The `PlayerInventory` script class maintains an integer total, `starTotal`, of how many stars have been collected (initialized to zero). Each time the player character collides with an object, if that object is tagged as **Star**, the `AddStar()` method is invoked. This method increments the total and sends a message, passing the new total to the `OnChangeStarTotal(...)` method of its sibling scripted component, `PlayerInventoryDisplay`.

The PlayerInventoryDisplay script class contains a public array of references to PickupUI objects, and a single public method called OnChangeStarTotal(...). This method loops through its array of PickupUI scripted objects, setting them to display color icons while the loop counter is less than the number of stars carried, and thereafter, setting them to display gray icons. This results in the color icons being displayed to match the number of stars being carried.

> Note: It might seem that we could make our code simpler by assuming that slots always display gray (no star) and just change one slot to yellow each time a yellow star is picked up. But this would lead to problems if something happens in the game (for example, hitting a black hole or being shot by an alien) that makes us drop one or more stars. The C# PlayerInventoryDisplay script class makes no assumptions about which slots may or may not have been displayed as gray, yellow, or empty previously. Each time it is called, it ensures that an appropriate number of yellow stars are displayed, and that all the other slots are displayed with gray stars.

The **UI Panel** GameObject's slots for the three stars have PickupUI scripted components, with each linked to its gray and colored icons.

Several star GameObjects are added to the scene (all tagged Star). The array of PickupUI object references in the PlayerInventoryDisplay scripted component in the player-SpaceGirl GameObject is populated with references to the PickupUI scripted components in the three **UI Panels** for each star.

There's more...

Here are some details you won't want to miss.

Modifying the game for a second inventory panel for keys

We have created a great display panel for collecting star objects. Now, we can reuse what we've done to create a second panel to display the collection of key objects in the game.

Figure 3.28: Example showing two panels, each with three slots

To modify the game to make a second inventory panel for key collection, do the following:

1. Make a copy of the project and work on the copy (if you wish to retain a version that works just with stars).

2. Duplicate the Panel-stars GameObject, naming the copy Panel-keys.

3. For the Panel-keys GameObject in the **Hierarchy**, do the following:

   - For the Text-title child, change the text of the **TextMeshPro - TextUI (Script)** component from **Stars** to Keys.

   - Ensuring Panel-keys is still selected, for its **Rect Transform**, choose **top-right**, set **Pos X** to -10 (to move away from the right edge), and set **Pos Y** to -30 (to vertically align with Panel-keys).

   - For each Image-icon-gray GameObject that is a child of all three Panel-inventory-slots, change **Image (Script) Source Image** to icon_key_grey_100.

   - For each Image-icon-color GameObject that is a child of all three Panel-inventory-slots, change **Image (Script) Source Image** to icon_key_green_100.

> Note: You can replace the **Source Image** for multiple UI Image GameObjects in a single action, selecting multiple GameObjects and then changing their properties in the **Inspector**. For example, using the *Ctrl* key (Windows) or Command key (macOS) select the three Image-icon-gray child GameObjects of the three panels; then in the **Inspector**, set the **Source Image** to icon_key_grey_100. All three GameObjects should now have been set to the gray key icon.

4. Remove the PlayerInventory and PlayerInventoryDisplay script components from the player-girl1 GameObject.

5. Create the following C# script, called PlayerInventoryKeysAndStars, in the _Scripts folder:

```
using UnityEngine;

public class PlayerInventoryKeysAndStars : MonoBehaviour {
    private int starTotal = 0;
    private int keyTotal = 0;
    private PlayerInventoryDisplayKeysStars playerInventoryDisplay;

    void Awake() {
        playerInventoryDisplay =
            GetComponent<PlayerInventoryDisplayKeysStars>();
    }

    void Start() {
        playerInventoryDisplay.OnChangeStarTotal(starTotal);
        playerInventoryDisplay.OnChangeKeyTotal(keyTotal);
    }
```

```
    void OnTriggerEnter2D(Collider2D hit) {
        if(hit.CompareTag("Star")){
            AddStar();
            Destroy(hit.gameObject);
        }

        if(hit.CompareTag("Key")){
            AddKey();
            Destroy(hit.gameObject);
        }
    }

    private void AddStar() {
        starTotal++;
        playerInventoryDisplay.OnChangeStarTotal(starTotal);
    }

    private void AddKey() {
        keyTotal++;
        playerInventoryDisplay.OnChangeKeyTotal(keyTotal);
    }
}
```

6. Add a component of the following C# script, called `PlayerInventoryDisplayKeysStars`, to the `player-SpaceGirl` GameObject in the **Hierarchy:**

```
using UnityEngine;

[RequireComponent(typeof(PlayerInventoryKeysAndStars))]
public class PlayerInventoryDisplayKeysStars : MonoBehaviour  {
    public PickupUI[] slotsStars = new PickupUI[1];
    public PickupUI[] slotsKeys = new PickupUI[1];

    public void OnChangeStarTotal(int starTotal)
    {
        UpdateSlotDisplays(starTotal, slotsStars);
    }

    public void OnChangeKeyTotal(int keyTotal) {
        UpdateSlotDisplays(keyTotal, slotsKeys);
    }
```

```
    private void UpdateSlotDisplays(int pickupTotal, PickupUI[]
iconSlots)
    {
        int numInventorySlots = iconSlots.Length;
        for(int i = 0; i < numInventorySlots; i++){
            PickupUI slot = iconSlots[i];
            if(i < pickupTotal)
                slot.DisplayColorIcon();
            else
                slot.DisplayGreyIcon();
        }
    }
}
```

7. With player-girl1 selected in the **Hierarchy**, for its **PlayerInventoryDisplayKeysStars** scripted component, set both **slotsKeys** and **slotsStars** to 3 (making the size of each of these arrays 3). Then, drag the corresponding inventory-slot GameObjects from the **Hierarchy** to populate these arrays.

8. Create a new GameObject called key by dragging a copy of the icon-key-green-100 sprite image from the **Project** panel into the scene. Rename this GameObject key.

9. Add to the GameObject key a **Box Collider 2D** component (**Physics 2D**) and tick its **Is Trigger** setting. In its **Sprite Renderer** component, set **Sorting Layer** to **Foreground**.

10. Create a new **Tag** called Key and add it to the GameObject key.

11. Make two duplicates of the key GameObject, moving them to different locations in the scene (so that the player can see all three stars and all three keys).

As you can see, we have duplicated and adjusted the visual **UI Panel** and components of the **star**-carrying inventory to give us a second one for the **key**-carrying inventory. Likewise, we have added code to detect collisions with objects tagged as **Key** and added this to the inventory display script to update the **UI Panel** for keys, when notified that a change has been made in terms of the number of keys being carried.

Displaying multiple pickups of different objects as a list of text via a dynamic List<> of scripted PickUp objects

When working with different kinds of pickups, one approach is to use a C# **List** to maintain a flexible-length data structure containing the items currently in the inventory. In this recipe, we will show you how, each time an item is picked up, a new object is added to such a List collection. An iteration through the List will show how the text display for items is generated each time the inventory changes.

Here, we will introduce a very simple `PickUp` script class, demonstrating how information about a pickup can be stored in a scripted component, extracted upon collision, and stored in our `List`:

Figure 3.29: Example of the UI displaying multiple pickups of different objects

Getting ready

This recipe assumes that you are starting with the `Simple2Dgame_SpaceGirl` project, which we set up in the first recipe of this chapter.

How to do it...

To display the inventory total text for multiple pickups of different object types, follow these steps:

1. Start with a new copy of the `Simple2Dgame_SpaceGirl` mini-game.
2. Create a new (general purpose) tag named `Pickup`.
3. Now change the tag of the **star** GameObject from **Star** to **Pickup**:

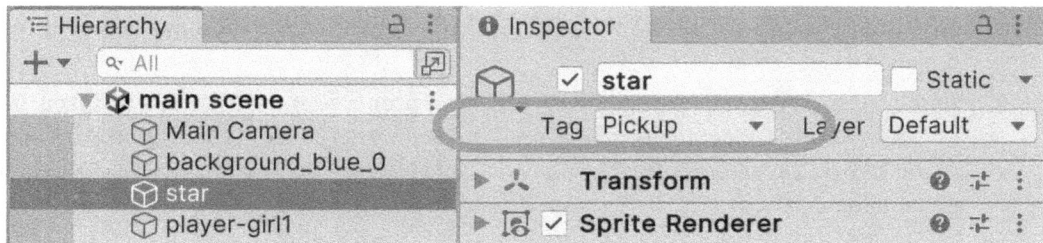

Figure 3.30: Changing Star to Pickup

4. Create a C# `PickUp` script class, and add an instance of this script class as a component of the **star** GameObject:

```csharp
using UnityEngine;

public class PickUp : MonoBehaviour
{
    public string description;
}
```

5. In the **Inspector,** set the **Description** property of the **PickUp (Script)** component of the **star** GameObject to star:

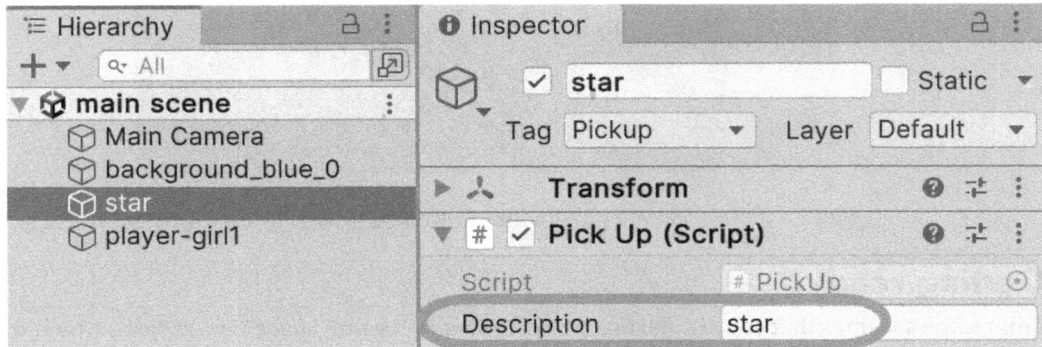

Figure 3.31: Setting the description property of the **PickUp** *(Script) component to star*

6. Select the star GameObject in the **Hierarchy** panel and make a copy of it, renaming the copy heart.

7. In the Inspector panel, change the **Description** property of the **Pick Up (Script)** component of the **heart** GameObject to heart. Also, drag the healthheart.png image from the Sprites folder in the **Project** panel into the **Sprite** property of the heart GameObject. The player should now see the heart image on the screen for this pickup item.

8. Select the star GameObject in the **Hierarchy** panel and make another copy of it, renaming the copy key.

9. In the **Inspector** panel, change the **Description** property of the **Pick Up (Script)** component of the key GameObject to key. Also, drag the icon_key_green_100 image from the Sprites folder into its **Sprite** property. The player should now see the key image on the screen for this pickup item.

10. Make a few more copies of each pickup GameObject and place them around the screen so that there are two or three each of the star, heart, and key pickup GameObjects.

11. Add a **UI Text (TMP)** object (**Create | UI | Text - TextMeshPro**) and rename it Text-inventory-list. For the **TextMeshPro - Text UI** component, change its text to several copies of the text the quick brown fox jumped over the lazy dog, or another long list of nonsense words, to test the overflow settings you'll change in the next step. Set the **Vertex Color** to yellow.

12. In the **TextMeshPro -Text UI** component, ensure that **Wrapping** is set to **Enabled,** and that **Overflow** has the value **Overflow.** This will ensure that the text will wrap onto a second or third line (if needed) and not be hidden if there are lots of pickups.

13. In the **Inspector** for the **Alignment** property, center the text horizontally and top-aligned vertically. Then, set **Font Size** to 28.

14. Edit its **Rect Transform** and set its **Height** to 50. Then, while holding down *Shift + Alt* (to set its pivot and position), choose the **top-stretch** box. The text should now be positioned at the middle-top of the **Game** panel, and its width should stretch to match that of the whole panel.

15. Your text should now appear at the top of the **Game** panel as several lines of overflowing yellow text.

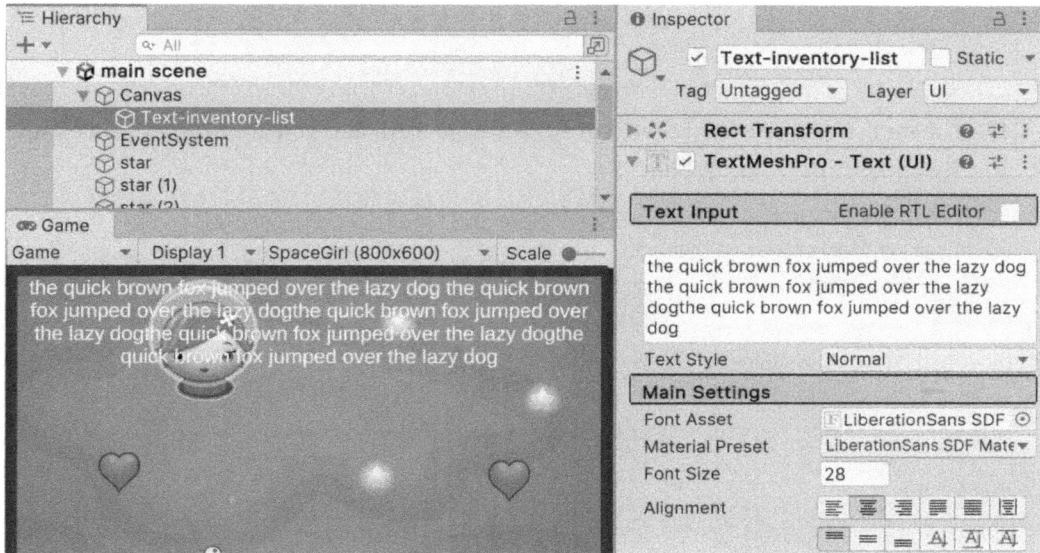

Figure 3.32: Overflowing text at top of Game panel

16. Create the following `PlayerInventory` C# script class in the `_Scripts` folder:

```csharp
using UnityEngine;
using System.Collections.Generic;

public class PlayerInventory : MonoBehaviour {
    private PlayerInventoryDisplay playerInventoryDisplay;
    private List<PickUp> inventory = new List<PickUp>();

    void Awake() {
        playerInventoryDisplay = GetComponent<PlayerInventoryDisplay>
            ();
    }

    void Start() {
        playerInventoryDisplay.OnChangeInventory(inventory);
    }

    void OnTriggerEnter2D(Collider2D hit) {
        if(hit.CompareTag("Pickup")){
            PickUp item = hit.GetComponent<PickUp>();
            inventory.Add( item );
```

```
                    playerInventoryDisplay.OnChangeInventory(inventory);
                    Destroy(hit.gameObject);
                }
            }
        }
```

17. Create the `PlayerInventoryDisplay` C# script class, and add an instance as a component of the `player-girl1` GameObject:

```csharp
using UnityEngine;
using System.Collections.Generic;
using TMPro;

[RequireComponent(typeof(PlayerInventory))]
public class PlayerInventoryDisplay : MonoBehaviour {
    public TextMeshProUGUI inventoryText;

    public void OnChangeInventory(List<PickUp> inventory) {
        // (1) clear existing display
        inventoryText.text = "";

        // (2) build up new set of items
        string newInventoryText = "carrying: ";
        int numItems = inventory.Count;
        for(int i = 0; i < numItems; i++){
            string description = inventory[i].description;
            newInventoryText += " [" + description+ "]";
        }

        // if no items in List then set string to empty message
        if(numItems < 1)
            newInventoryText = "(empty inventory)";

        // (3) update screen display
        inventoryText.text = newInventoryText;
    }
}
```

18. From the **Hierarchy** panel, select the `player-girl1` GameObject. Then, from the **Inspector** window, access the **Player Inventory Display (Script)** component and populate the **Inventory Text** public field with the `Text-inventory-list` UI Text (TMP) object.

19. Play the game. Each time you pick up a star, key, or heart, the updated list of what you are carrying should be displayed in the form carrying [key] [heart].

How it works...

In the `PlayerInventory` script class, the variable inventory is a C# `List<>`. This is a flexible data structure that can be sorted, searched, and dynamically (at runtime, when the game is being played) have items added to and removed from it. `<PickUp>`, which is in pointy brackets, means that the variable inventory will contain a list of `PickUp` objects. For this recipe, our `PickUp` class just has a single field – a string description – but we'll add more sophisticated data items to the `PickUp` classes we'll look at in later recipes. This variable inventory has been initialized to be a new, empty C# `List` of `PickUp` objects.

Before the scene starts, the `Awake()` method of the `Player` script class caches a reference to the `PlayerInventoryDisplay` scripted component.

When the scene starts, the `Start()` method invokes the `OnChangeInventory(...)` method of the `PlayerInventoryDisplay` scripted component. This is so that the text that's displayed to the user at the beginning of the scene corresponds to the initial value of the variable inventory. (This might, for some games, not be empty. For example, a player might start a game with some money, or a basic weapon, or a map.)

When the `OnTriggerEnter2D(...)` method detects collisions with items tagged as `Pickup`, the `PickUp` object component of the item that's been hit is added to our inventory list. A call is also made to the `OnChangeInventory(...)` method of `playerInventoryDisplay` to update our inventory display to the player, passing the updated inventory **List** as a parameter.

The `playerInventoryDisplay` script class has a public variable that's linked to the **Text-inventory-list** UI Text object. First, the `OnChangeInventory(...)` method sets the UI Text object to empty, and then loops through the inventory list, building up a string of each item's description in square brackets (`[key]`, `[heart]`, and so on). If there were no items in the list, then the string is set to the text (`empty inventory`). Finally, the text property of `Text-inventory-list` is set to the value of this string representation; that is, what is inside the variable inventory.

There's more...

Here are some details you won't want to miss.

Ordering items in the inventory list alphabetically

It would be nice to alphabetically sort the words in the inventory list, both for neatness and consistency (so that, in a game, if we pick up a key and a heart, it will look the same, regardless of the order they are picked up in), but also so that items of the same type will be listed together so that we can easily see how many of each item we are carrying.

To implement alphabetical sorting for the items in the inventory list, we need to do the following:

1. Add the following C# code to the beginning of the `OnChangeInventory(...)` method in the `PlayerInventoryDisplay` script class:

    ```
    public void OnChangeInventory(List<PickUp> inventory){
        inventory.Sort(
            delegate(PickUp p1, PickUp p2){
    ```

```
            return p1.description.CompareTo(p2.description);
        }
    );


    // rest of the method as before ...
}
```

2. You should now see all the items listed in alphabetical order.

> This C# code takes advantage of the C# `List.Sort(...)` method, a feature of collections whereby each item can be compared to the next, and they are swapped if they're in the wrong order (if the `CompareTo(...)` methods return `false`). Learn more at https://learn.microsoft.com/en-us/dotnet/api/system.collections.generic.list-1.sort?view=net-8.0.

Displaying multiple pickups of different objects as text totals via a dynamic Dictionary<> of PickUp objects and enum pickup types

While the previous recipe worked fine, any old text could have been typed into the description of a pickup or perhaps mistyped (star, Sstar, starr, and so on). A much better way of restricting game properties to one of a predefined (enumerated) list of possible values is to use C# enums. As well as removing the possibility of mistyping a string, it also means that we can write code to deal with the predefined set of possible values. In this recipe, we will improve our general-purpose `PickUp` class by introducing three possible pickup types (`Star`, `Heart`, and `Key`), and write inventory display code that counts the number of each type of pickup being carried before displaying these totals via a **UI Text** object on the screen. We will also switch from using a `List` to using a `Dictionary`, since the `Dictionary` data structure is designed specifically for key-value pairs, which is perfect for associating a numeric total with an enumerated pickup type:

Figure 3.33: Example of multiple pickups being implemented via a dynamic dictionary

In this recipe, we will also manage any additional complexity by separating the controller (user collection event) logic from the stored inventory data by introducing an inventory manager scripted class. By doing this, our player controller will be simplified to just containing two methods (`Awake`, for getting a reference to the inventory manager, and `OnTriggerEnter2D`, for responding to collisions by communicating with the inventory manager).

Getting ready

This recipe adapts the previous one. So, make a copy of the project from the previous recipe and work on this copy.

How to do it...

To display multiple pickups of different objects as text totals via a dynamic `Dictionary`, follow these steps:

1. Replace the contents of the `PickUp` C# script class with the following code:

```
using UnityEngine;

public class PickUp : MonoBehaviour
{
    public enum PickUpType { Star, Key, Heart }
    public PickUpType type;
}
```

2. Replace the contents of the `PlayerInventory` C# script class with the following code:

```
using UnityEngine;
using System.Collections.Generic;

public class PlayerInventory : MonoBehaviour
{
    private PlayerInventoryDisplay playerInventoryDisplay;
    private Dictionary<PickUp.PickUpType, int> items = new
Dictionary<PickUp.PickUpType, int>();

    void Awake()
    {
        playerInventoryDisplay = GetComponent<PlayerInventoryDisplay>();
        playerInventoryDisplay.OnChangeInventory(items);
    }

    public void Add(PickUp pickup)
    {
        PickUp.PickUpType type = pickup.type;
        int oldTotal = 0;
        if (items.TryGetValue(type, out oldTotal)) {
            items[type] = oldTotal + 1;
```

```
            } else {
                items.Add (type, 1);
            }

            playerInventoryDisplay.OnChangeInventory(items);
        }

        void OnTriggerEnter2D(Collider2D hit) {
            if(hit.CompareTag("Pickup")) {
                PickUp item = hit.GetComponent<PickUp> ();
                Add(item);
                Destroy(hit.gameObject);
            }
        }
    }
```

3. Replace the content of the `PlayerInventoryDisplay` script class with the following code:

```
using UnityEngine;
using System.Collections.Generic;
using TMPro;

[RequireComponent(typeof(PlayerInventory))]
public class PlayerInventoryDisplay : MonoBehaviour
{
    public TextMeshProUGUI inventoryText;

    public void OnChangeInventory(Dictionary<PickUp.PickUpType, int>
inventory)
    {
        inventoryText.text = "";
        string newInventoryText = "carrying: ";

        foreach (var item in inventory)
        {
            int itemTotal = item.Value;
            string description = item.Key.ToString();
            newInventoryText += " [ " + description + " " + itemTotal + "
]";
        }

        int numItems = inventory.Count;
```

```
        if (numItems < 1) {
            newInventoryText = "(empty inventory)";
        }


        inventoryText.text = newInventoryText;
    }
}
```

4. In the **Hierarchy** (or **Scene**) panel, select *each pickup* GameObject in turn, and from their drop-down menus, choose their corresponding **Type** in the **Inspector** panel (so choose **Star** for star GameObjects, **Key** for key GameObjects and so on). As the screenshot shows, public variables that are of the enum type are automatically restricted to the set of possible values as a combo box drop-down menu in the **Inspector** panel:

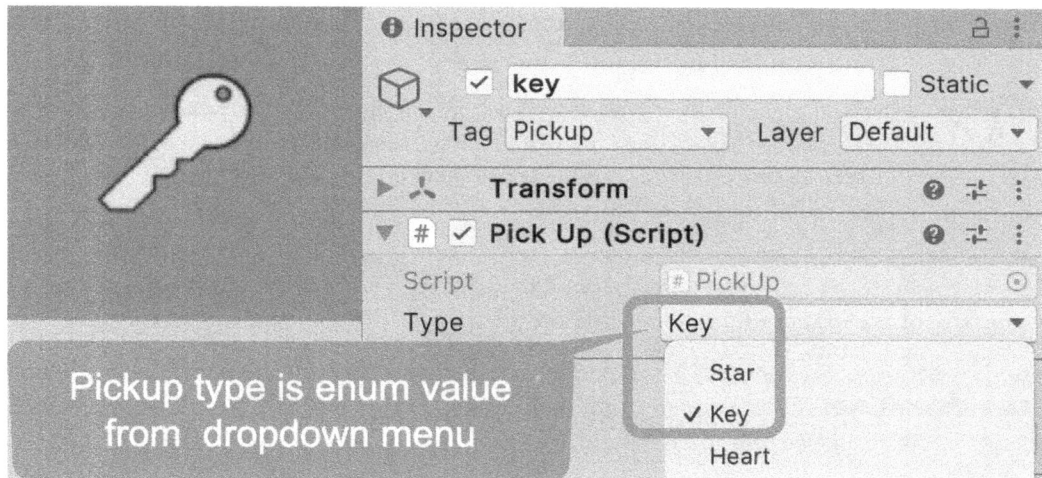

Figure 3.34: Setting the possible values as a combo box drop-down menu

5. Play the game. First, you should see a message on the screen stating that the inventory is empty. Then, as you pick up one or more items of each pickup type, you'll see text totals for each type you have collected.

How it works...

Each PickUp GameObject in the scene has a scripted component of the PickUp class. The PickUp object for each PickUp GameObject has a single property, a PickUp type, which has to be one of the enumerated sets of Star, Key, or Heart. The use of an enumerated type means that the value has to be one of these three listed values, which means no misspelling/mistyping errors that could have happened with a general text string type can happen here, as in the previous recipe.

In the PlayerInventory script class, the inventory being carried by the player is represented by C# Dictionary variable items.

A `Dictionary` is made up of a sequence of *key-value pairs*, where the key is one of the possible `PickUp`. `PickUpType` enumerated `values`, and the `value` is the integer total of how many of that type of pickup is being carried. This `Dictionary` states what type will be used for a key, and then what type (or script class) will be stored as the value for that key. The following statement is used to declare our `Dictionary` variable items:

```
items = new Dictionary<PickUp.PickUpType, int>()
```

C# dictionaries provide a `TryGetValue(...)` method, which receives the parameters of a key and is passed a reference to a variable the same data type as the value for the `Dictionary`. When the `Add(...)` method is called, the type of the `PickUp` object is tested to see if a total for this type is already in the `Dictionary` items. If an item total is found inside the `Dictionary` for the given type, then the value for this item in the `Dictionary` is incremented. If no entry is found for the given type, then a new element is added to the `Dictionary` with a total of 1.

> **TryGetValue call-by-reference parameter**
>
> Note the use of the C# out keyword before the `oldTotal` parameter in the `items.TryGetValue(type, out oldTotal)` statement. This indicates that a reference to the actual variable, `oldTotal`, is being passed to the `TryGetValue(...)` method, not just a copy of its value. This means that the method can change the value of the variable.

The method returns `true` if an entry is found in the `Dictionary` for the given type, and if so, sets the value of `oldTotal` to the value against this key.

The last action of the `Add(...)` method is to call the `OnChangeInventory(...)` method of the `PlayerInventoryDisplay` scripted component of the `player-girl1` GameObject. This will update the text totals displayed on screen.

The `OnChangeInventory(...)` method of the `PlayerInventoryDisplay` script class initializes the `newInventoryText` string variable and then iterates through each item in the `Dictionary`, appending a string of the type name and total for the current item to `newInventoryText`. Finally, the text property of the **UI Text (TMP)** object is updated with the completed text inside `newInventoryText`, showing the pickup totals to the player.

You can learn more about using C# lists and dictionaries in Unity in the Unity Technologies tutorial at `https://unity3d.com/learn/tutorials/modules/intermediate/scripting/lists-and-dictionaries`.

There's more...

Let's look at some details you won't want to miss.

Separating responsibilities with MVC

Our PlayerInventory script class has two sets of responsibilities:

- Maintaining the internal record of items being carried (this is the **Model** – the data about the player)
- Detecting collisions, updating the state, and asking the display class to inform the player visually of the changed items being carried (this is **Controller** work – deciding what should be done based on messages received by the player)

So let's separate these two sets of responsibilities into separate script classes:

- An InventoryManager script class to maintain the internal record of items being carried (and ask the display class to inform the player visually each time a change is made to the items being carried)
- A PlayerController script class to detect collisions and ask InventoryManager to update what is being carried

The addition of this extra software layer separates the player collision detection behavior from how the inventory is internally stored, and it also prevents any single script class from becoming too complex by attempting to handle too many different responsibilities. This recipe is an example of the **low coupling** of the MVC design pattern. We have designed our code to not rely on or make too many assumptions about other parts of the game, which reduces the likelihood of a change in some other part of our game breaking our inventory display code. The display (view) is separated from the logical representation of what we are carrying (inventory manager model), and changes to the model are made by public methods that are called from the player (controller).

To refactor the project separating the inventory model and player controller responsibilities into separate C# script classes, do the following:

1. Remove the **PlayerInventoryDisplay (Script)** component from the player-SpaceGirl GameObject.
2. Next, remove the **PlayerInventory (Script)** component from the player-SpaceGirl GameObject.
3. Create a new C# script class called PlayerController containing the following code, and add an instance as a component of the player-girl1 GameObject:

```
using UnityEngine;

public class PlayerController : MonoBehaviour {
    private InventoryManager inventoryManager;

    void Awake() {
        inventoryManager = GetComponent<InventoryManager>();
    }
```

```csharp
        void OnTriggerEnter2D(Collider2D hit) {
            if(hit.CompareTag("Pickup")){
                PickUp item = hit.GetComponent<PickUp> ();
                inventoryManager.Add(item);
                    Destroy(hit.gameObject);
            }
        }
    }
```

4. Add an instance of the InventoryManager C# script class to the player-player-girl1 Game-Object in the **Hierarchy** panel:

```csharp
using UnityEngine;
using System.Collections.Generic;

public class InventoryManager : MonoBehaviour {
    private PlayerInventoryDisplay playerInventoryDisplay;

    private Dictionary<PickUp.PickUpType, int> items = new
        Dictionary<PickUp.PickUpType, int>();

    void Awake() {
        playerInventoryDisplay = GetComponent<PlayerInventoryDisplay>
            ();
        playerInventoryDisplay.OnChangeInventory(items);
    }

    public void Add(PickUp pickup) {
        PickUp.PickUpType type = pickup.type;
        int oldTotal = 0;

        if(items.TryGetValue(type, out oldTotal))
            items[type] = oldTotal + 1;
        else
            items.Add (type, 1);

        playerInventoryDisplay.OnChangeInventory(items);
    }
}
```

5. Add an instance of the PlayerInventoryDisplay C# script class as a component of the player-girl1 GameObject.

6. Edit the first few lines of the `PlayerInventoryDisplay` C# script class to remove the requirement for a `PlayerInventory` component, and to require the `InventoryManager` and `PlayerController` components:

```
using UnityEngine;
using System.Collections.Generic;
using TMPro;

[RequireComponent(typeof(InventoryManager))]
[RequireComponent(typeof(PlayerController))]
public class PlayerInventoryDisplay : MonoBehaviour
{
    // content of class is unchanged ...
```

We can see that the function methods that were in the `PlayerInventory` C# script class have been distributed between the new `InventoryManager` and `PlayerController` C# script classes:

- `PlayerController` gets a reference to the `InventoryManager` component via its `Awake()` method and responds to collections through its `OnTriggerEnter2D()` function method, and calls the `Add()` method of the inventory manager when something tagged `Pickup` has been hit

- `InventoryManager` responds to calls to the `Add()` method to update the `Dictionary` items, and itself calls the `PlayerInventoryDisplay` component to update the **UI Text (TMP)** display each time the pickup totals have been added to

Further reading

The recipes in this chapter demonstrated a range of C# data representations for inventory items and a range of Unity UI interface components for displaying the status and contents of player inventories at runtime. The `Inventory UI` needs good-quality graphical assets for a high-quality result. Some sources of assets that you might wish to explore include the following:

- The graphics for our SpaceGirl mini-game are from the Space Cute art by Daniel Cook; he generously publishes lots of free 2D art for game developers to use:

    - http://www.lostgarden.com/
    - http://www.lostgarden.com/search?q=planet+cute

- Sethbyrd – lots of fun 2D graphics:

    - http://www.sethbyrd.com/

- Royalty-free art for 2D games:

    - https://www.gameart2d.com/freebies.html

- Unity article on Model-View-Controller and Model-View-Presenter architectures:

    - https://unity.com/how-to/build-modular-codebase-mvc-and-mvp-programming-patterns#mvp-and-unity

Learn more on Discord

To join the Discord community for this book – where you can share feedback, ask questions to the author, and learn about new releases – follow the QR code below:

https://packt.link/unitydev

4

Playing and Manipulating Sounds

Sound is a very important part of the gaming experience. In fact, we can't stress enough how crucial sound is to a player's immersion in a virtual environment. Just think of the engine running in your favorite racing game, the distant urban buzz in a simulator game, or the creeping noises in horror games. Think of how these sounds transport you *into* the game.

Before we get into the recipes, first, let's review how different sound features work in Unity. A project with audio needs one or more audio files. These are called AudioClips in Unity, and they sit in your Project folders. At the time of writing, Unity supports four audio file formats: .wav, .ogg, .mp3, and .aif. Files of these types are re-encoded when Unity builds for a target platform (PC, Mac, Linux Standalone, Android, or WebGL, for example). It also supports tracker modules in four formats: .xm, .mod, .it, and .s3m.

A scene (or prefab) GameObject can have an AudioSource component, which can either be linked to an AudioClip sound file at **design time** or through scripting at **runtime**. For a player to hear sounds when a game scene is playing, there needs to be one GameObject that has an active AudioListener component. When you create a new scene, one is added automatically for you in the main camera. You can think of an AudioListener component as a simulated digital "ear," since the sounds Unity plays are based on the relationship between playing AudioSources and the active AudioListener component. There should be only one active AudioListener component in a scene, since Unity would not know how to calculate 3D sounds if there is more than one digital "ear" – and you'll see a warning in the **Console** panel if there is more than one AudioListener active in a scene.

So the three things we need for sound to work are:

- One GameObject with an active AudioListener component
- One or more GameObjects with AudioSource components, and
- One or more AudioClip asset files, being played by GameObject with AudioSource components

Simple sounds, such as pickup effects and background soundtrack music, can be defined as **2D sound**. However, Unity supports 3D sounds, which means that the location and distance between playing AudioSources and the active AudioListener component determine the way the sound is perceived in terms of loudness and left/right balance.

Additionally, you can engineer synchronized sound playing and scheduling through AudioSettings.dspTime – this is a value based on the samples in the audio system, so it is far more precise than the Time.time value. Also, dspTime will pause/suspend with the scene, so no logic is required for rescheduling when using dspTime.

Some great effects are possible through Unity's 3D sound features, and to allow us to explore 3D, several recipes in this chapter modify Unity's Third-Person Starter assets. So the first recipe in this chapter goes through how to get these free assets from the Unity Asset Store and set up a project using them.

In this chapter, we will cover the following recipes:

- Setting up the Third Person Character Controller project
- Playing sound when a scene begins
- Removing redundant AudioListener components
- Enabling and customizing 3D sound effects
- Adding effects with Audio Reverb Zones
- Playing different one-off sound effects with a single AudioSource component
- Playing and controlling different sounds each with its own AudioSource component
- Creating just-in-time AudioSource components at runtime through C# scripting
- A button to play a sound with no scripting
- Preventing an audio clip from restarting if it is already playing
- Waiting for the audio to finish playing before auto-destructing an object
- Creating audio visualization from sample spectral data
- Synchronizing simultaneous and sequential music to create a simple 140 bpm music-loop manager
- Recording sound clips with the free Audacity application

Setting up the Third Person Character Controller project

Unity provides a great Playground scene with a Third Person robot character set up and ready to use. We'll get those assets and set up a project with this scene in this recipe, making it really easy to work with 3D sounds later this chapter.

How to do it...

To set up the Third Person Character Controller project, perform the following steps:

1. Create a new **Unity 3D** project.
2. Open a web browser tab to Unity Asset Store, by choosing **Menu: Window | Asset Store**.

3. Search for the free **Starter Assets Third Person Character Controller**.

4. Select the assets, and when viewing their details, click the button **Add to My Assets**.

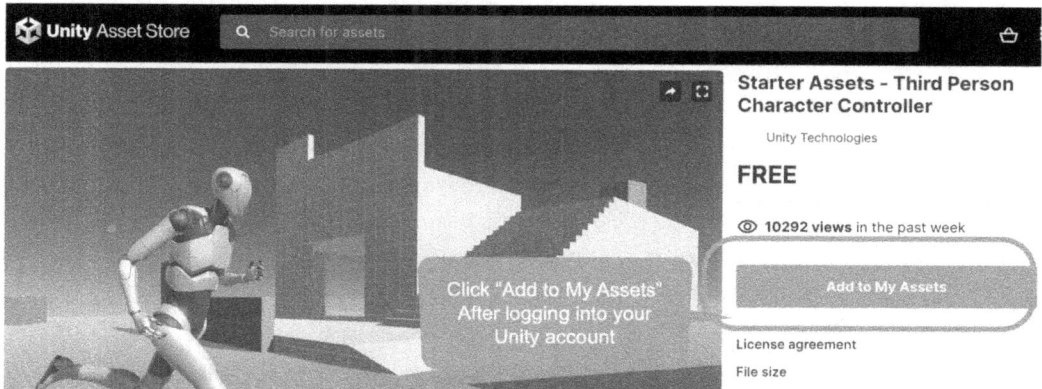

Figure 4.1: Adding Third-Person Starter to your assets on the Unity Asset Store website

5. In the Unity editor, open the **Package Manager** to list your asset store assets by choosing the menu **Window | Package Manager**. Dock the **Package Manager** panel alongside the **Inspector** panel. From the Packages dropdown choose My Assets (see the following screenshot).

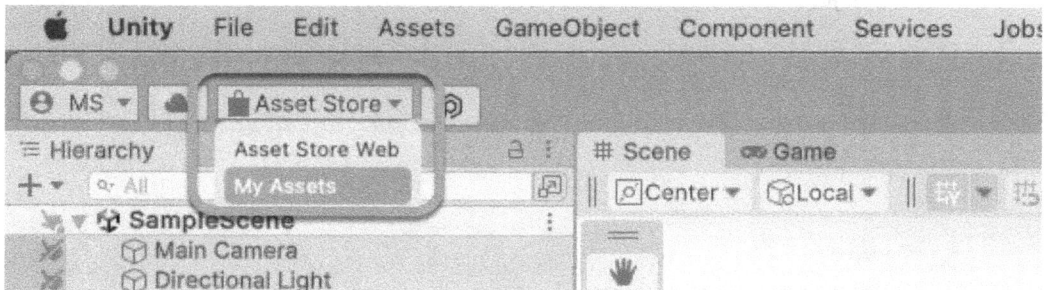

Figure 4.2: Option to view My Assets in the Package Manager in Unity editor

6. In the **Package Manager** panel locate and import the **Starter Assets Third Person Character Controller** package.

> When importing the package, if asked, agree to additional Package Manager dependencies (click **Install/Upgrade**), and agree to enable the new input systems backends and restart the editor (click **Yes**).
>
> **Note:** If some GameObjects in the scene are pink, see the *Fixing pink textures* steps at the end of this recipe.

7. You should now have a folder **Starter Assets** in your **Project** panel.

8. Open the Playground scene, in the **Project** panel folder: **Starter Assets | Third-Person Starter | Scenes**.

9. Run the scene, and move your robot character around the 3D environment using the arrow keys/WASD.

How it works...

You've added the free Unity third-person assets to your account. So, as we did with this new 3D project, you can import these assets into any project you are working with, providing you with a ready-made 3D character controller. We'll make use of these assets for some of the 3D sound recipes in this chapter, and we'll use them again in *Chapter 8, Animated Characters*.

There's more...

There are some details that you don't want to miss.

Fixing pink textures

When importing assets into a project, sometimes there are issues when assets were set up for a different / older render pipeline. If you see some bright pink assets in your scenes, this indicates rendering issues. However, Unity has made it very easy to automatically fix almost all the issues that arise. If you see lots of pink objects when you load the Playground scene, follow these steps to fix the problem:

1. Open the **Rendering Pipeline Converter** by choosing the menu **Window | Rendering | Rendering Pipeline Converter**.

2. Check all four boxes, and click the **Initialize And Convert** button.

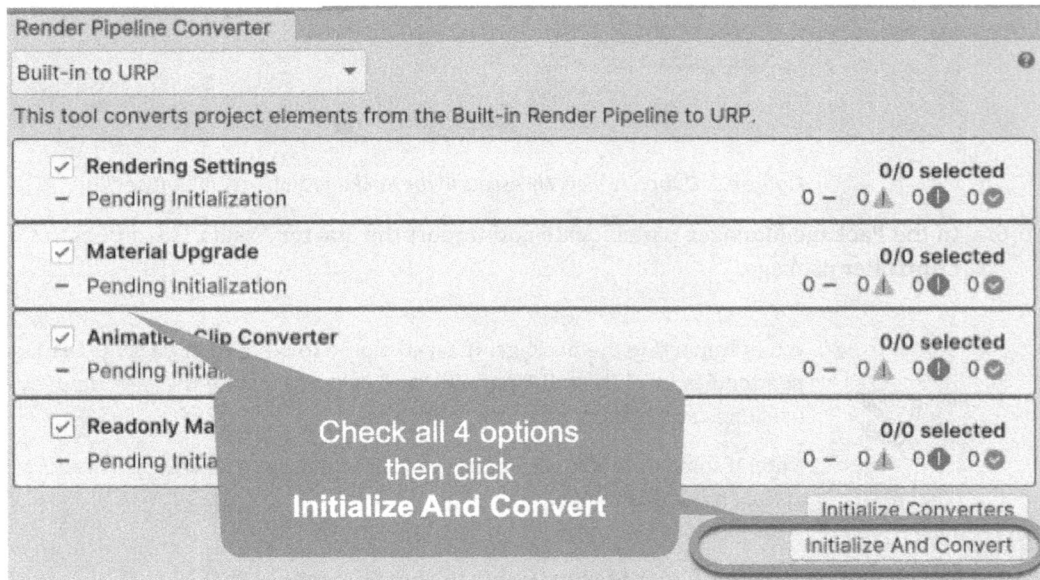

Figure 4.3: Fixing pink textures with the Rendering Pipeline Converter

3. That's it! Save and run the scene, and hopefully, everything looks fine.

If there are any remaining pink texture objects in the scene, you can usually fix the offending materials manually by doing the following:

1. Select the GameObject with the pink texture in the **Hierarchy**.
2. In the **Inspector**, locate its **Material** in the **Mesh Render Component**.
3. With the **Material** selected in the **Project** panel, in the **Inspector**, set its **Shader** to **Universal Render Pipeline | Unit**.

Playing sound when a scene begins

Often, we want a sound to play as soon as a scene is loaded; this may be to indicate to a player that the new scene is ready, or perhaps it's some background music or audio sound effect for the scene. As you'll see in this recipe, Unity makes it very easy for us to play as soon as a scene is loaded.

Before we start using the 3D scene created in the first recipe of this chapter, we'll first explore some core audio features in a 2D project in this recipe.

Getting ready

Try out this recipe using any audio clips you might have on your computer. We have included a number of classic *Pacman* game sound clips inside the 04_02 folder.

How to do it...

To play a sound when the scene is loaded, perform the following steps:

1. Create a new **Unity 2D** project. There is a single sample scene, containing a Main Camera GameObject that has an **Audio Listener** component.
2. Import the sound clip files.
3. Select the Main Camera GameObject in the **Hierarchy**, and then from the **Project** panel, drag the Pacman Opening Song audio clip into the **Inspector**.
4. You should see that Unity has added an **Audio Source** component to the Main Camera, and its **AudioClip** property is linked to the Pacman Opening Song audio clip.

By default, AudioSource components have their **Play On Awake** options checked when created – leave this option checked, since this is the behavior we want for this recipe.

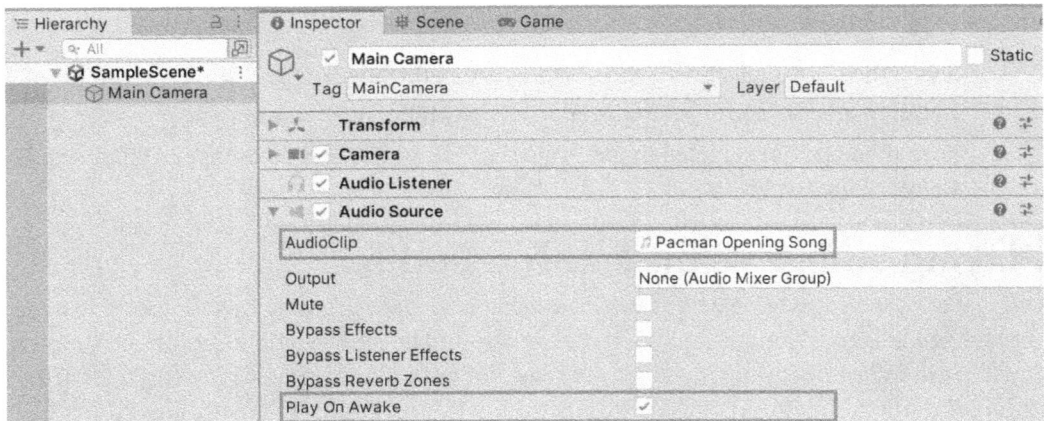

Figure 4.4: Dragging sound clips into the scripted component

5. Save and play the scene. You should hear the Pacman song play once as soon as the scene begins.

How it works...

All three requirements for sound have been met; since the default `Main Camera` has an **Audio Listener** component, it also has an **Audio Source** component, and that component is linked to the `Pacman Opening Song` audio clip. When you drag an **AudioClip** asset file onto a GameObject, Unity will automatically add an **Audio Source** component if the GameObject does not already have one.

> **Note:** You can, of course, manually add an Audio Source component, and then drag an AudioClip asset file from the **Project** panel to populate the AudioClip public variable of the Audio Source component.

The default setting for an **Audio Source** component is to **Play On Awake**. This means it plays when the scene begins, since when a scene begins every GameObject is sent an `Awake()` message.

There's more...

There are some details that you don't want to miss.

Create a GameObject with AudioSource linked to an AudioClip in a single step

If you don't mind having a GameObject in your **Hierarchy** named after an **AudioClip** file, in a single step you can create a new GameObject that has an **AudioSource** component linked to a specific **AudioClip**. Just drag the **AudioClip** from the Project panel into the **Hierarchy** (or **Scene**) panel. A new GameObject will automatically be created, with the same name as the **AudioClip** file!

Make the sound clip keep looping

The default setting for an **AudioSource** component is to play once. However, if you check the **Loop** property, then the clip will start playing from the beginning again after finishing.

Customizing other AudioSource settings

As well as being the component that plays **AudioClips**, there are a range of settings we can customize in **AudioSource** components. These include volume, pitch, and stereo (left-right) balance. When you have several sounds playing at the same time in a scene, some tweaking may be needed so that the player can hear things as desired.

Removing redundant AudioListener components

There are times when we may end up with more than one **AudioListener** in a scene, so need to remove or deactivate one to avoid problems. We'll learn how to do so in this recipe.

Getting ready

You can find several Pacman game sound clips inside the 04_02 folder.

How to do it...

To remove **AudioListener** components, perform the following steps:

1. Create a new **Unity 2D** project. There is a single sample scene, containing a Main Camera GameObject that has an **Audio Listener** component.
2. Import the sound clip files.
3. Select the Main Camera GameObject in the **Hierarchy**, and then from the **Project** panel, drag the Pacman Opening Song audio clip into the **Inspector**.
4. Save and play the scene. You should hear the Pacman song play once as soon as the scene begins.
5. Now, add a second camera to the scene, choosing **GameObject | Camera**. Rename this GameObject Camera2.
6. Save and play the scene. You should hear the Pacman song again, but also look in the **Console**, where you'll see a warning message stating that there are two audio listeners in the scene.

7. Select the `Camera2` GameObject in the **Hierarchy**, and then in the **Inspector**, remove the **AudioListener** component using the three-dots options menu to the right of this component (see the following screenshot).

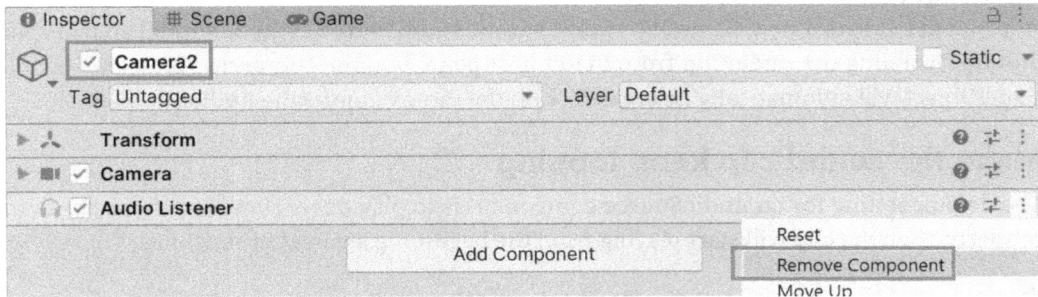

Figure 4.5: Removing Audio Listener component from a camfera GameObject

8. Save and play the scene. You should hear the Pacman song again, but with no warning messages this time.

How it works...

When a camera GameObject is added to a scene, by default, it has three components, a **Transform**, a **Camera**, and an **AudioListener**. In this example, since our `Main Camera` already had an **AudioListener**, we needed to remove the **AudioListener** from `Camera2` to avoid problems.

Enabling and customizing 3D sound effects

The default spatial setting for an **AudioSource** component is to be fully 2D. However, the **Spatial Blend** between 2D and 3D interpretation of a sound playing from a GameObject's **AudioSource** is easily changed. Once the 3D location of an **AudioSource** is brought into the equation, many interesting audio effects are possible to enhance the player's experience. In this recipe, we'll move the spatial blend of an **AudioSource** fully to 3D, and customize the **Volume Rolloff**, which determines how a sound becomes quieter the further away it is located from the active **AudioListener** in a scene.

There are scenes where the best location of an **AudioListener** is not on a camera GameObject. This is often the case for third-person games, where the camera is behind/above the player's character, and we want the player to hear the sounds that their animated character should be hearing (not where the observing camera is located). In this recipe, we'll modify the Playground scene from Unity's Third-Person starter assets to make the only active **AudioListener** the one that is a part of the player's animated character.

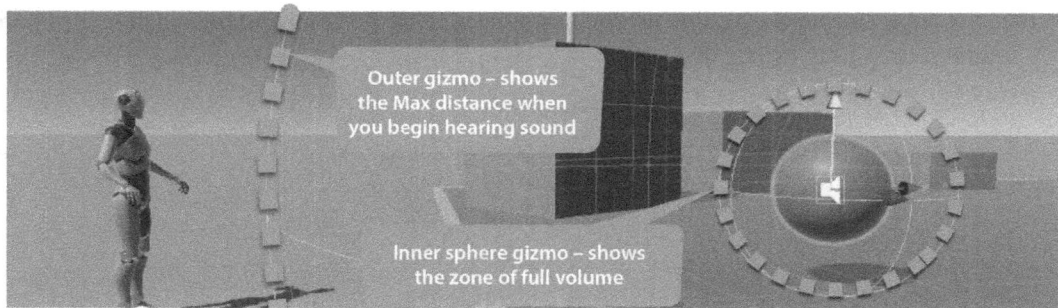

Figure 4.6: Scene panel gizmos indicating 3D sound distances

Getting ready

You can find several Pacman game sound clips inside the 04_02 folder.

This recipe builds on the project created in the first recipe in this chapter, so make a copy of that project and work from it.

How to do it...

To enable 3D sound effects through **Spatial Blend** and **Volume Rolloff** settings, perform the following steps:

1. Open the project created in the first recipe of this chapter, containing the **Starter Assets - Third Person Character Controller**.

2. You will see that the Main Camera (which has an **AudioListener** component) is positioned away from the robot character.

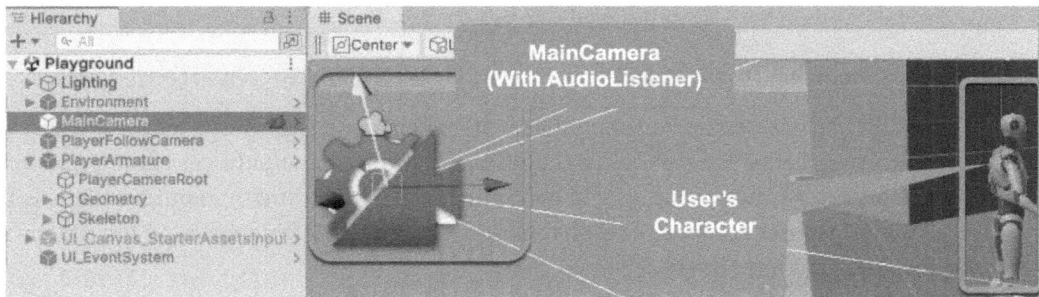

Figure 4.7: The Main Camera (with AudioListener) looking at the robot character

3. Create a new **3D Sphere** GameObject (**GameObject | 3D | Sphere**) and rename this Sphere-music. Position this GameObject just in front of the Third-Person robot character in the scene.

4. With GameObject Sphere-music selected in the **Hierarchy**, drag from the **Project** panel the Pacman Opening Song **AudioClip** into the **Inspector**. This should have added an **AudioSource** component to the GameObejct, linked to that **AudioClip**.

5. In the **Inspector** for the **AudioSource** component, check the **Loop** option, and drag the **Spatial Blend** setting fully to the right (3D).

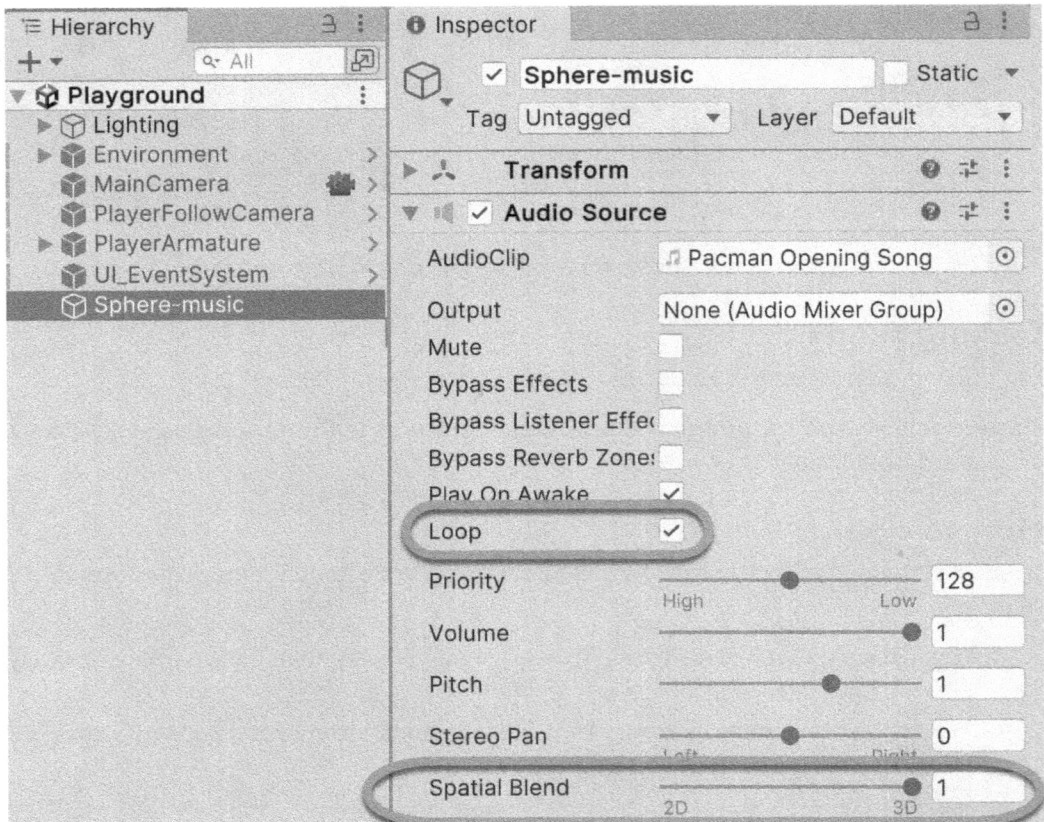

Figure 4.8: Making AudioSource 3D by setting the Spatial Blend property in the Inspector

6. Save and play the scene. You should hear the Pacman song play in a loop. As your character moves towards the Sphere-music GameObject, the sound should get louder, and in fact, as your character walks past it, it will continue to get louder, and then quieter! This is because the **AudioListener** component is on the Main Camera GameObject behind the robot character, not on the character itself.

    Let's fix this issue. Stop the scene and then do the following.

7. In the **Hierarchy** select the Main Camera GameObject and remove its **Audio Listener** component.

8. In the **Hierarchy** select the Player Armature GameObject and add an **AudioListener** component, using the **Add Component** button in the **Inspector**.

9. Save and play the scene. Now, as your character moves toward the Sphere-music GameObject, the sound should get louder, and as soon as your character starts to move away from the Sphere GameObject, it should start to get quieter.

How it works...

You have removed the **AudioListener** component from the Main Camera GameObject, and added a **AudioListener** component to the Player Armature GameObject. This means that for 3D sounds, Unity will calculate what sounds and effects should be heard at the location of the player's character, rather than the separately located Main Camera location.

> Note: For first-person scenes, there is no need to relocate the active AudioListener, since the camera represents the location of the player.

We have set the **Spatial Blend** property of the Sphere-music Sphere's AudioSource component to be fully 3D. Unity uses the distance from the active AudioListener in the scene to the GameObject with this AudioSource component to decide what audio should be heard. The volume of the Pacman music heard by the user is determined by the distance between the robot character (with the **AudioListener** component) and the Sphere-music GameObject (with the **AudioSource** component).

There's more...

There are some details that you don't want to miss.

Changing the way the volume changes with distance

The default setting for the 3D settings of an **AudioSource** component defines a **Logarithmic Rolloff** for **Volume Rolloff**. This means that very rapidly as the **AudioListener** moves away from the **Audio-Source**, the volume gets quiet. Also, the **Max Distance** defaults to **500**, so the sound will still be heard very quietly until the distance from the **AudioListener** to the **AudioSource** is greater than **500** Unity units (virtual meters).

Let's change this a little. Do the following:

1. Select the Sphere-music GameObject in the **Hierarchy**.
2. Click the triangle to reveal the details of the **3D Sound Settings** in the **Inspector**.

3. Change the **Volume Rolloff** from **Logarithmic Rolloff** to **Custom Rolloff**. Set the **Max Distance** to **20**.

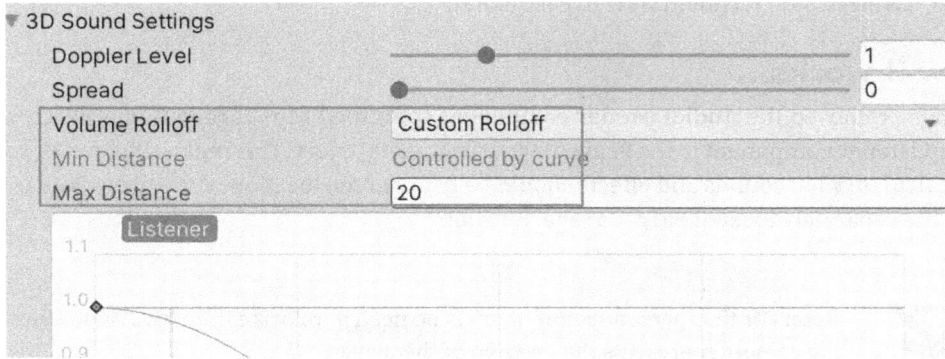

Figure 4.9: Changing AudioSource Volume Rolloff settings

4. Save and play the scene.

When your character is more than 20 units away, you should hear nothing. When the player's character moves closer to the Sphere-music GameObject (within 20 units), the music should gradually get louder.

Adding keys to customize the volume falloff curve

You can customize the volume falloff curve further – by double-clicking on the curve to add key points, which you can drag to change the curve shape. Here's a curve where a key was added so that the maximum volume is reached once the player is within 7 units of the **AudioSource** location:

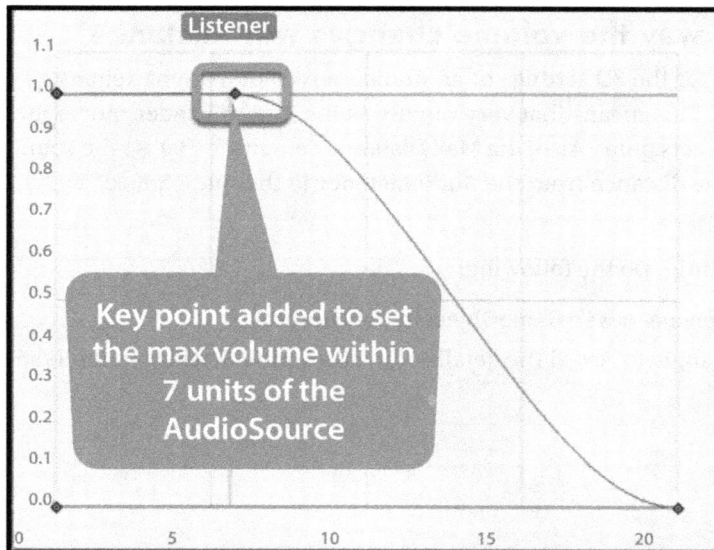

Figure 4.10: Custom volume rolloff curve with a max volume within 7 units

Save and play the scene.

When your character is more than 20 units away, you should hear nothing. When the player's character moves closer to the Sphere-music GameObject (within 20 units), the music should gradually get louder.

Adding effects with Audio Reverb Zones

One way to make the sound in a game more realistic is to apply effects about how sound behaves in different physical environments. The way sounds echo changes when they bounce off surfaces made of different materials, such as in bathrooms, sewers, hallways, or outdoors in a city or quarry, and so on. **Audio Reverb Zones** are a Unity component that can add such effects to **AudioSources** in spherical zones in a scene, which we'll learn to create and customize in this recipe.

Getting ready

This recipe follows from the previous one, so make a copy of that project and work on the copy.

How to do it...

To add effects with **Audio Reverb Zones**, perform the following steps:

1. Open the copy you made of the previous recipe.
2. Open the Playground scene customized in the previous recipe.
3. In the **Hierarchy** panel, double-click the Sphere-music GameObject – this should select the object and align the **Scene** panel to the center of this GameObject.
4. Let's remove any odd Doppler effects. Do this in the Inspector by dragging the **3D Sound Setting Doppler Level** all the way to the left (zero).
5. With the Sphere-music GameObject still selected in the **Hierarchy** panel, add an **Audio Reverb Zone** to the scene by choosing **Menu: GameObject | Audio | Audio Reverb Zone**. A new **Audio Reverb Zone** should be created in the scene, as a child of the Sphere-music GameObject.
6. Select the Audio Reverb Zone GameObject, and in the **Inspector**, set **MinDistance** to 0.5 and **MaxDistance** to 5. Also, change the **Reverb Preset** setting to **Psychotic** (to really emphasize the echo effect).

Figure 4.11: Customizing the Audio Reverb Zone settings in the Inspector

7. Save and play the scene. You should hear the Pacman song play once as you get near the sphere, and when very near the sphere, you should hear the reverb echo effect strongly.

How it works...

By double-clicking on Sphere-music, that GameObject was selected in the **Hierarchy**. So when the new **Audio Reverb Zone** was added to the scene, its center was set to the same position as the center of Sphere-music, and it was automatically made a child of this GameObject in the **Hierarchy**. The **MinDistance** setting is when the reverb echo effect will be at its maximum. This value was set to **0.5**, since this is where the edge of the sphere is from the center of the **Audio Reverb Zone**. The default size for a sphere is **1** unit, so its radius is **0.5** units. The **MaxDistance** setting determines when the reverb echo effect becomes zero, so having it set to **5** means that when the user's character is close to the sphere (**0.5–5**) the reverb effect will be heard, but when a little further away (> **5**) the music will still be heard, but with no echo effect. This kind of "tweaking" of settings is common to get the desired audio effects for different objects and locations in a scene.

There's more...

There are some details that you don't want to miss.

Customize Reverb settings with the User preset

If you change the **Audio Reverb Zone Reverb Preset** setting to **User**, you can customize all the different revert settings for the zone.

Playing different one-off sound effects with a single AudioSource component

As we've seen in the recipes at the beginning of this chapter, the basics of playing a sound are very straightforward in Unity – by adding an AudioSource component to a GameObject and linking it to an AudioClip sound file. For simple sound effects, such as short, one-off plays of pickup confirmation noises, it's useful to have a single AudioSource component that you can then reuse to play different sound effects. That is what we'll do in this recipe.

Getting ready

Try out this recipe using any short audio clip that is less than one second in duration. You can also find several Pacman game sound clips inside the 04_02 folder.

How to do it...

To play multiple sounds using the same AudioSource component, perform the following steps:

1. Create a new **Unity 2D** project and import the sound clip files.
2. Create a C# script class, called SoundPlayer, in a new folder, _Scripts, that contains the following code:

```
using UnityEngine;
```

```
[RequireComponent(typeof(AudioSource))]
public class SoundPlayer : MonoBehaviour
{
    public AudioClip clipEatCherry;
    public AudioClip clipExtraLife;

    private AudioSource audioSource;

    void Awake() {
        audioSource = GetComponent<AudioSource>();
    }

    void Update() {
        if (Input.GetKey(KeyCode.UpArrow)) {
            audioSource.PlayOneShot(clipEatCherry);
        }
        if (Input.GetKey(KeyCode.DownArrow)) {
            audioSource.PlayOneShot(clipExtraLife);
        }
    }
}
```

3. Create a new empty GameObject named SoundManager.

4. Add an instance of the C# script class SoundPlayer as a scripted component to the SoundManager GameObject.

5. Ensure that the SoundManager GameObject is selected in the **Hierarchy** panel.

6. Then, in the **Inspector** panel, drag the Pacman Eating Cherry sound clip from the **Project** panel into the public **Clip Eat Cherry** variable of the scripted **SoundPlayer (Script)** component.

7. Repeat this procedure for the Pacman Extra Life sound clip and the public **Clip Extra Life** variable.

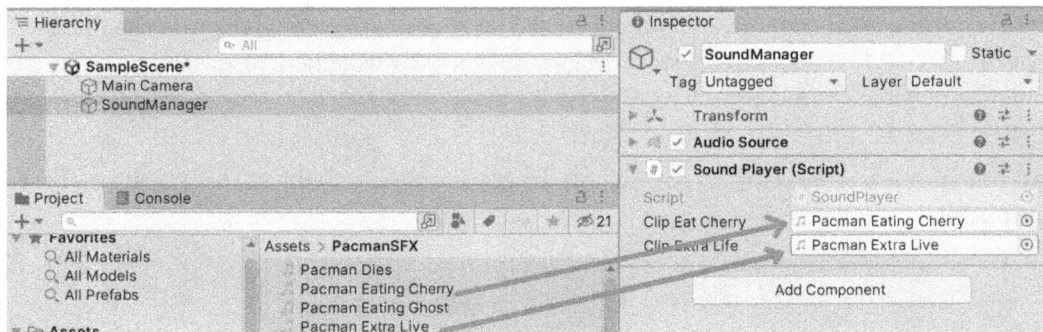

Figure 4.12: Dragging sound clips into the scripted component of the SoundManager GameObject

8. Run the scene, and press the *UP* and *DOWN* arrow keys to play the different sound effects.

How it works...

You have created a C# script class, called `SoundPlayer`. The script class includes a `RequireComponent` attribute. It declares that any GameObject containing a scripted object component of this class must also have an **AudioSource** component (note that one will be added automatically if such a component does not exist when the scripted component is added).

The `SoundPlayer` script class has two public **AudioClip** properties: `Pacman Eating Cherry` and `Pacman Extra Life`. At `design time`, we associated **AudioClip** sound files from the **Project** panel with these public properties.

At `runtime`, the `Update()` method is executed in every frame. This method checks whether the *UP* and *DOWN* array keys are being pressed. If so, it plays the **Eat Cherry or Extra Life** sounds, respectively – sending the **AudioSource** component a `PlayOneShot()` message with the appropriate **AudioClip** sound file link.

> **Note:** We cannot pause sounds played with PlayOneShot.
>
> While this is great for short, one-off sound effects, one limitation of the `PlayOneShot()` method is that you cannot then find out the status of the playing sound (for instance, whether it has finished, or at what point it was playing, and more). Nor can you pause/restart a sound played with `PlayOneShot()`. For such detailed control of sounds, each sound needs its own **AudioSource** component.
>
> You can learn more about the `PlayOneShot()` method in the official Unity documentation at `https://docs.unity3d.com/ScriptReference/AudioSource.PlayOneShot.html`.

There's more...

There are some details that you don't want to miss.

Playing a sound at a static point in 3D world space

Similar to `PlayOneShot()` is the `PlayClipAtPoint()` **AudioSource** method. This allows you to play a sound clip for an `AudioSource` component created at a specific point in 3D world space. Note that this is a static class method – so you don't need an `AudioSource` component to use this method. An `AudioSource` component is created (at the location you provide) and will exist as long as the `AudioClip` sound is playing. The `AudioSource` component will be automatically removed by Unity once the sound has finished playing.

All you need is a Vector3 (*x,y,z*) position object and a reference to the `AudioClip` file that you want to be played:

```
Vector3 location = new Vector3(10, 10, 10);
AudioSource.PlayClipAtPoint(soundClipToPlay, location);
```

Playing and controlling different sounds, each with its own AudioSource component

The approach in the previous recipe (using PlayOneShot(...) with a single AudioSource component) is fine for one-off sound effects. When further control is required over a playing sound, each sound will need to be played in its own AudioSource component. In this recipe, we'll create two separate AudioSource components and pause/resume each of them with different arrow keys.

Getting ready

Try out this recipe with two audio clips that are several seconds long. We have included two free music clips inside the 04_02 folder.

How to do it...

To play different sounds with their own AudioSouce components, perform the following steps:

1. Create a new Unity 2D project and import the sound clip files.
2. Create a new empty GameObject in the scene named Music medieval, containing an **Audio-Source** component that is linked to the 186772__dafawe__medieval **AudioClip**. You can do this in a single step by dragging the music clip from the **Project** panel into either the **Hierarchy** panel or the **Scene** panel.
3. Create a second empty GameObject in the scene named Music arcade, containing an **Audio-Source** component that is linked to the 251461__joshuaempyre__arcade-music-loop **AudioClip**.
4. Uncheck the **Play On Awake** property of the AudioSource **components** for both GameObjects. This is so that the sounds do not begin playing as soon as the scene has been loaded.
5. Create a C# script class, called SourceController, in a new folder, _Scripts, that contains the following code:

```
using UnityEngine;

public class SourceController : MonoBehaviour
{
    public AudioSource audioSourceMedieval;
    public AudioSource audioSourceArcade;

    void Update()
    {
        if (Input.GetKey(KeyCode.RightArrow)) {
            PlayUnpause(audioSourceMedieval);
        }

        if (Input.GetKey(KeyCode.LeftArrow)) {
            audioSourceMedieval.Pause();
```

```
        }

        if (Input.GetKey(KeyCode.UpArrow)) {
            PlayUnpause(audioSourceArcade);
        }

        if (Input.GetKey(KeyCode.DownArrow)) {
            audioSourceArcade.Pause();
        }
    }

    private void PlayUnpause(AudioSource audioSource)
    {
        if (audioSource.time > 0) {
            audioSource.UnPause();
        } else {
            audioSource.Play();
        }
    }
}
```

6. Create an empty GameObject named SoundManager.

7. Add an instance of the C# script class SourceController as a scripted component to the SoundManager GameObject.

8. Ensure that the SoundManager GameObject is selected in the **Hierarchy** panel. In the **Inspector** panel, drag **GameObject Music medieval** into the public Audio Source Medieval variable in the scripted SourceController (Script) component.

9. Then, drag **GameObject Music arcade** into the public Audio Source Arcade variable in the scripted SourceController (Script) component.

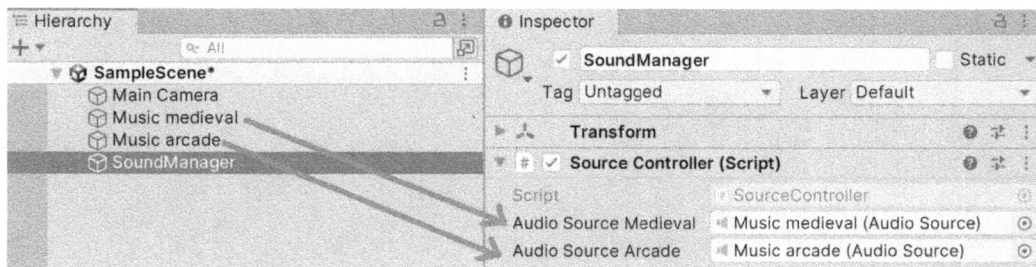

Figure 4.13: Dragging AudioSource components from GameObejcts into the scripted public variables

10. Run the scene, and press the *UP* and *DOWN* arrow keys to start/resume and pause the medieval sound clip. Press the *RIGHT* and *LEFT* arrow keys to start/resume and pause the arcade sound clip.

How it works...

You created a C# script class, called SourceController, and added an instance of this class as a component to the SoundManager GameObject. Additionally, you also created two **GameObjects**, called Music medieval and Music arcade in the scene, with each containing an AudioSource component linked to a different music clip.

The script class has two public AudioSource properties: Audio Source Medieval and Audio Source Arcade. At design time, we associated the AudioSource components of the Music medieval and Music arcade **GameObjects** with these public properties.

At runtime, the Update() method is executed in every frame. This method checks for the *UP/DOWN/RIGHT/LEFT* array keys being pressed. If the *UP* arrow key is detected, the function method PlayUnpause(...) is passed the medieval music audio source variable (audioSourceMedieval). If the *DOWN* arrow key is pressed, the medieval music audio source is sent a Pause() message.

The function method PlayUnpause(...) receives a reference to an **AudioSource** component. The time property of an **AudioSource** object indicates the position of the audio playhead, which is zero if at the beginning of the clip, and greater than zero if it is playing or has been paused while playing. Method PlayUnpause(...)will unpause the **AudioSource's** audio clip if it was paused (time > 0), by sending it an UnPause() message. Otherwise, it will begin playing the clip from the beginning, by sending a Play() message.

We control the playing of the arcade music in a corresponding way through the detection of the *RIGHT/LEFT* arrow keys.

Since each AudioClip sound file is associated with its own AudioSource component, it allows us to simultaneously play and manage each sound independently.

Creating just-in-time AudioSource components at runtime through C# scripting

In the previous recipe, for each sound clip that we wanted to manage in the scene, we had to manually create GameObjects with **AudioSource** components at design time. However, using C# scripting, we can create our own GameObjects that contain **AudioSources** at runtime – just when they are needed.

This method is similar to the built-in **AudioSource** PlayClipAtPoint() method, but the created **AudioSource** component is completely under our programmatic control.

This code was inspired by some of the code posted, in 2011, in the online *Unity Answers* forum, by user Bunny83. Unity has a great online community, with users helping each other and posting interesting ways of adding features to games. You can find out more about that post at `http://answers.unity3d.com/questions/123772/playoneshot-returns-false-for-isplaying.html`.

Getting ready

This recipe adapts the previous one. So, make a copy of the project from the previous recipe, and work on this new copy.

How to do it...

To create just-in-time **AudioSource** components at runtime through C# scripting, perform the following steps:

1. Open the project, which is a copy of the previous recipe.
2. Delete the Music medieval and Music arcade GameObjects from the scene – we'll be creating the **AudioSource** components we need at **runtime** in this recipe!
3. Edit the SourceController C# script class to read as follows:

```csharp
using UnityEngine;

public class SourceController : MonoBehaviour
{
    public AudioClip clipMedieval;
    public AudioClip clipArcade;

    private AudioSource audioSourceMedieval;
    private AudioSource audioSourceArcade;

    void Awake()
    {
        audioSourceMedieval = CreateAudioSource(clipMedieval, true);
        audioSourceArcade = CreateAudioSource(clipArcade, false);
    }

    void Update()
    {
        if (Input.GetKey(KeyCode.RightArrow)) {
            PlayUnpause(audioSourceMedieval);
        }
```

```
            if (Input.GetKey(KeyCode.LeftArrow)) {
                audioSourceMedieval.Pause();
            }

            if (Input.GetKey(KeyCode.UpArrow)) {
                PlayUnpause(audioSourceArcade);
            }

            if (Input.GetKey(KeyCode.DownArrow)) {
                audioSourceArcade.Pause();
            }
        }

        private void PlayUnpause(AudioSource audioSource)
        {
            if (audioSource.time > 0) {
                audioSource.UnPause();
            } else {
                audioSource.Play();
            }
        }

        private AudioSource CreateAudioSource(AudioClip audioClip, bool
startPlayingImmediately)
        {
            GameObject audioSourceGO = new GameObject();
            audioSourceGO.transform.parent = transform;
            audioSourceGO.transform.position = transform.position;
            AudioSource newAudioSource = audioSourceGO.
AddComponent<AudioSource>();
            newAudioSource.clip = audioClip;
            if (startPlayingImmediately) {
                newAudioSource.Play();
            }

            return newAudioSource;
        }
    }
```

4. Ensure that the SoundManager GameObject is selected in the **Hierarchy** panel. In the **Inspector** panel, drag the **AudioClip** 186772__dafawe__medieval sound clip from the **Project** panel into the public **Clip Medieval AudioClip** variable in the scripted SourceController (Script) component. Repeat this procedure with the 251461__joshuaempyre__arcade-music-loop **AudioClip** for the **Clip Arcade** variable.

5. Run the scene, and press the *UP* and *DOWN* arrow keys to start/resume and pause the medieval sound clip. Press the *RIGHT* and *LEFT* arrow keys to start/resume and pause the arcade sound clip.

How it works...

The key feature of this recipe is the new CreateAudioSource(...) method. This method takes input as a reference to a sound clip file and a Boolean true/false value as to whether the sound should start playing immediately. The method does the following:

- It creates a new **GameObject.** (with the same parent and at the same location as the **GameObject** doing the creating).
- It adds a new AudioSource component to the new **GameObject.**
- It sets the audio clip of the new **AudioSource** component to the provided **AudioClip** parameter.
- If the Boolean parameter is true, the **AudioSource** component is immediately sent a Play() message so that it starts playing the sound clip.
- A reference to the AudioSource component is returned.

The remainder of the SoundController script class is very similar to that of the previous recipe. There are two public **AudioClip** variables, clipMedieval and clipArcade, which are set through drag-and-drop at design time to link to the sound clip files inside the Sounds Project folder.

The audioSourceMedieval and audioSourceArcade **AudioSource** variables are now private. These values are set up in the Awake() method by calling and storing values returned by the CreateAudioSource(...) method with the clipMedieval and clipArcade **AudioClip** variables.

To illustrate how the Boolean parameter works, the medieval music **AudioSource** component is created to play immediately, while the arcade music won't start playing until the *UP* arrow key has been pressed. Playing, resuming, and pausing the two audio clips is just the same as in the previous recipe – via the arrow-key detection logic in the (unchanged) Update() and PlayUnpause(...) methods.

There's more...

There are some details that you don't want to miss.

Adding the CreateAudioSource(...) method as an extension to the MonoBehavior class

Since the CreateAudioSource(...) method is a general-purpose method that could be used by many different game script classes, it doesn't naturally sit within the SourceController class.

The best place for general-purpose generative methods such as this is to add them as static (class) methods to the component class they work with. In this case, it would be great if we could add this method to the MonoBehavior class itself – so any scripted component could create **AudioSource** GameObjects on the fly.

All we have to do is create a class (I placed this class in a Unity project folder named ExtensionMethods) with a static method, as follows:

```
using UnityEngine;

public static class ExtensionMethods {
    public static AudioSource CreateAudioSource(this MonoBehaviour parent,
AudioClip audioClip, bool startPlayingImmediately)
    {
        GameObject audioSourceGO = new GameObject("music-player");
        audioSourceGO.transform.parent = parent.transform;
        audioSourceGO.transform.position = parent.transform.position;
        AudioSource newAudioSource =
            audioSourceGO.AddComponent<AudioSource>() as AudioSource;
        newAudioSource.clip = audioClip;

        if (startPlayingImmediately)
            newAudioSource.Play();

        return newAudioSource;
    }
}
```

As you can see, we add a first parameter to the extension method, stating which class we are adding this method to. Since we have added this to the MonoBehavior class, we can now use this method in our scripted classes as though it were built-in. So, our Awake() method in our SourceController class appears as follows:

```
void Awake() {
    audioSourceMedieval = this.CreateAudioSource(clipMedieval, true);
    audioSourceArcade = this.CreateAudioSource(clipArcade, false);
}
```

That's it – we can now remove the method from our SourceController class and use this method in any of our MonoBehavior scripted classes.

A button to play a sound with no scripting

Unity can send messages to GameObjects, including built-in components such as **AudioSource** and components that are instances of C# script classes we write ourselves. In this recipe, we'll create a scene with a button that sends a Play() message to the **AudioSource** component of a GameObject – to illustrate how some messages can be sent with no custom scripting at all. Then, in the following recipe, we'll write a custom C# script class that will refine the scene's behavior, preventing the **AudioSource** from being restarted if it is already playing.

Getting ready

Try out this recipe with any audio clip that is one second or longer in duration. We have included the engineSound audio clip inside the 04_09_engine_sound/Sounds/ folder.

How to do it...

To create a button to play a sound with no scripting, perform the following steps:

1. Create a new Unity 2D project and import the sound clip file.

2. Create a new empty GameObject in the scene named EngineObject, containing an **AudioSource** component that is linked to the engineSound **AudioClip**. You can do this in a single step by dragging the sound clip assert file from the **Project** panel into either the **Hierarchy** panel or the **Scene** panel.

3. Uncheck the **Play On Awake** property for the **AudioSource** component of the EngineObject **GameObject**. This is to ensure the sound does not begin playing as soon as the scene has been loaded.

4. Create a UI button named Button-play, by choosing **Menu: GameObject | UI | Button – Text-MeshPro**. If asked, import the **TextMeshPro Essentials**.

> **Note:** Unity UI button text objects are built with the TextMeshPro package.

5. Change the text of the button Play Sound to the center-left of the screen. Do this by editing its **Rect Transform** and, while holding down *Shift + Alt* (to set its pivot and position), choose the **left-middle** box.

6. With Button-play selected in the **Hierarchy**, in the **Inspector** panel create a new **On-click** event-handler by clicking the plus-sign (+) button.

7. Make this event handler send a Play() message to the EngineObject **GameObject's Audio-Source** component. Do this by dragging the EngineObject into the **Object** slot (which initially says **None (Object)**) and selecting the **AudioSource | Play()** function.

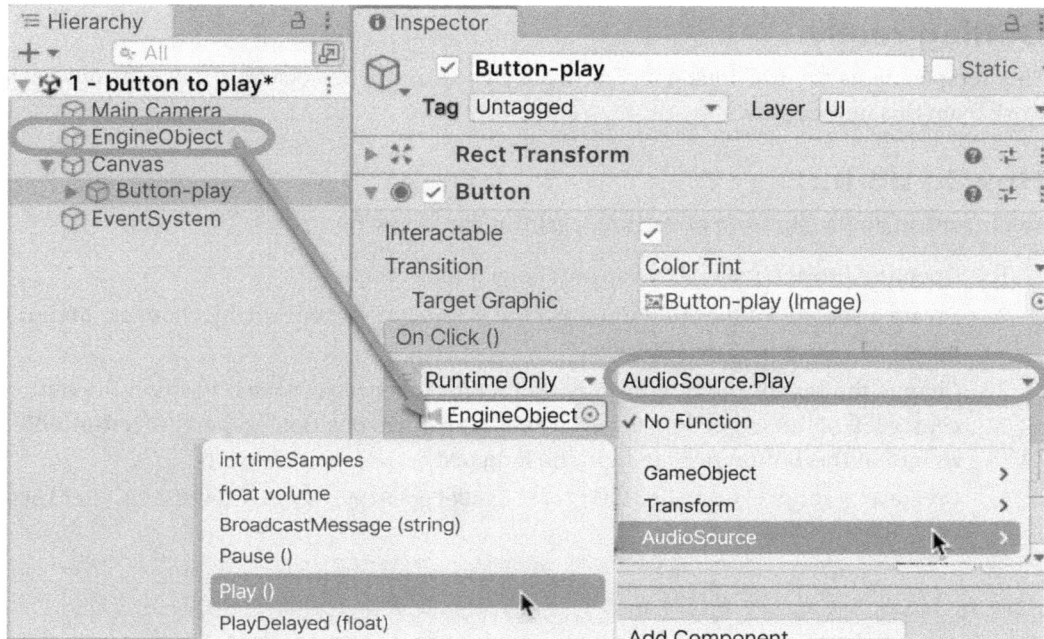

Figure 4.14: Creating the OnClick action for the button to send a Play() message to the engine object AudioSource

8. Save and run the scene. When you click the button, the sound should play.

How it works...

The **AudioSource** component of a GameObject will respond to a number of messages. Those messages may set the value of public variables, or execute function methods such as Play() and Pause(). When we create an **On Click** action for a button, and tell it which GameObject to send a message to, we then see a list of all the components of that GameObject, and for a selected component, all the public variables we can set and all the function methods we can call. In this recipe, we defined our button to send a Play() message to the **AudioSource** component of the EngineObject **GameObject**.

Preventing an audio clip from restarting if it is already playing

In a game, there might be several different events that cause a particular sound effect to start playing. If the sound is already playing, then in almost all cases, we won't want to restart the sound. A limitation of the previous recipe was that if we repeatedly clicked the button, the **AudioClip** would keep being interrupted and restart from the beginning with each button click.

This recipe involves a custom C# script class that includes a test so that an **AudioSource** component is only sent a Play() message if it is currently not playing.

Getting ready

This recipe adapts the previous one. So, make a copy of the project from the previous recipe, and work from this new copy.

How to do it...

To prevent an audio clip from restarting, perform the following steps:

1. Open the project, which is a copy of the previous recipe.

2. Create a second UI button named Button-play-not-interrupt by choosing **Menu: GameObject | UI | Button – TextMeshPro.**

3. Change the text of the button Play But Not Interrupt Sound. Position the button in the center-left of the screen by setting its **Rect Transform** position property to **left-middle.** Set the **Height** of this button to **90** to fit in the long text.

4. Create a C# script class, called WaitToFinishBeforePlaying, in a new folder, _Scripts, which contains the following code:

```
using UnityEngine;
using TMPro;

public class WaitToFinishBeforePlaying : MonoBehaviour  {
    public AudioSource audioSource;
    public TextMeshProUGUI buttonText;

    void Update() {
        string statusMessage = "(Re)Play sound";
        if(audioSource.isPlaying ) {
            statusMessage = "(sound playing)";
        }

        buttonText.text = statusMessage;
    }

    public void ACTION_PlaySoundIfNotPlaying() {
        if( !audioSource.isPlaying ) {
            audioSource.Play();
        }
    }
}
```

> **Note:** Personally, I prefer prefixing method names with a word in CAPS to make them easy to locate when setting as a button handler actions, as is the case above for the method `ACTION_PlaySoundIfNotPlaying()`. However, this method could have equally been named simply `PlaySoundIfNotPlaying()` to follow the usual C# method naming conventions...

5. Create an empty GameObject named `SoundManager`.

6. Add an instance of the C# script class `WaitToFinishBeforePlaying` as a scripted component to the `SoundManager` GameObject.

7. Ensure that the `SoundManager` GameObject is selected in the **Hierarchy** panel. In the **Inspector** panel, drag the GameObject `EngineObject` into the public **Audio Source** variable in the scripted `WaitToFinishBeforePlaying` (`Script`) component.

8. Then, drag the **Text (TMP)** child of `Button-play-not-interrupt` into the public **Button Text** variable.

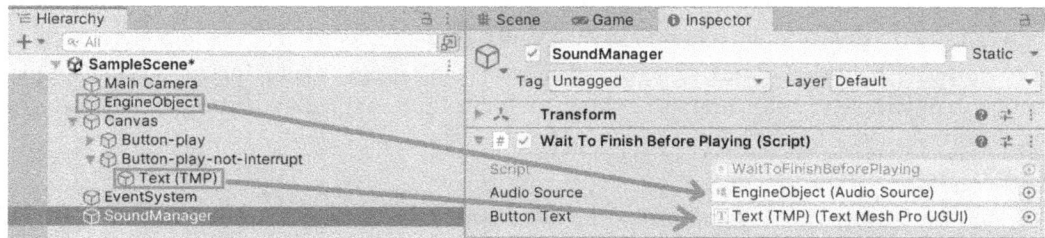

Figure 4.15: Linking the audio source and button text to the scripted component

9. With `Button-play` selected in the **Hierarchy** panel, create a new **On-click** event handler. You can do this by dragging the `SoundManager` into the **Object** slot and selecting the `ACTION_PlaySoundIfNotPlaying()` function.

10. With `Button-play-not-interrupt` selected in the **Hierarchy**, in the **Inspector** panel create a new **On-click** event handler by clicking the plus-sign (+) button. Then, drag the `SoundManager` GameObject into the **Object** slot and select the `WaitToFinishBeforePlaying | ACTION_PlaySoundIfNotPlaying()` function.

11. Save and run the scene. While the `Button-play` button on the left will restart a playing sound, if the sound is playing, the button to the right displays text (sound playing) and will not interrupt the sound if clicked. When the sound is not playing, this button changes its text to (Re)Play Sound, and it will play the sound when clicked.

How it works...

AudioSource components have a public readable property, isPlaying, which is a Boolean true/false flag that indicates whether a sound is currently playing. In this recipe, in every frame the Update() function method ensures that the text of the button is set to display (Re)Play Sound when the sound is not playing and (sound playing) when it is, by testing this isPlaying flag.

When the button is clicked on, the ACTION_PlaySoundIfNotPlaying() function method is executed. This method uses an if statement, ensuring that a Play() message is only sent to the AudioSource component if its isPlaying is false, and it updates the button's text as appropriate. If the isPlaying flag is true, then the sound is playing, and no Play() message will be sent.

Waiting for the audio to finish playing before auto-destructing an object

An event might occur (such as an object pickup or the killing of an enemy) that we wish to notify the player of by playing an audio clip and an associated visual object (such as an explosion particle system or a temporary object in the location of the event). However, as soon as the clip has finished playing, we will want the visual object to be removed from the scene. This recipe provides a simple way in which to link the end of an audio clip playing with the automatic destruction of its containing object.

Getting ready

Try out this recipe with any audio clip that is one second or more in duration. We have included the engineSound audio clip inside the 04_09_engine_sound/Sounds/ folder.

How to do it...

To wait for audio to finish playing before destroying its parent GameObject, perform the following steps:

1. Create a new Unity 2D project, and import the sound clip file.

2. Create a new empty GameObject in the scene, containing an **AudioSource** component that is linked to the engineSound **AudioClip**. You can do this in a single step by dragging the sound clip assert file from the **Project** panel into either the **Hierarchy** panel or the **Scene** panel. Rename this GameObject EngineObject.

3. Uncheck the **Play On Awake** property for the **AudioSource** component of the EngineObject **GameObject.** This is to ensure the sound does not begin playing as soon as the scene has been loaded.

4. Create a C# script class, called AudioDestructBehaviour, in a new folder, _Scripts, that contains the following code:

```
using UnityEngine;

public class AudioDestructBehaviour : MonoBehaviour {
    private AudioSource audioSource;
```

```
        void Awake() {
                audioSource = GetComponent<AudioSource>();
        }

        private void Update() {
                if( !audioSource.isPlaying ) {
                        Destroy(gameObject);
                }
        }
    }
```

5. Add an instance of the C# script class `AudioDestructBehaviour` as a scripted component of the `EngineObject` GameObject:

6. In the **Inspector** panel, disable (uncheck) the scripted `AudioDestructBehaviour` component of `EngineObject` (when needed, it will be re-enabled via C# code):

Figure 4.16: Disabling (unchecking) the scripted component of EngineObject

7. Create a C# script class called `ButtonActions` in the `_Scripts` folder that contains the following code:

```
using UnityEngine;

public class ButtonActions : MonoBehaviour {
    public AudioSource audioSource;

    public AudioDestructBehaviour audioDestructScriptedObject;

    public void ACTION_PlaySound() {
            if( !audioSource.isPlaying ) {
                    audioSource.Play();
            }
    }

    public void ACTION_DestroyAfterSoundStops(){
```

```
                        audioDestructScriptedObject.enabled = true;
            }
        }
```

8.   Create an empty GameObject named GameManager.

9.   Add an instance of the C# script class ButtonActions as a scripted component to the GameManager GameObject.

10.   With GameManager selected in the **Hierarchy** panel, drag EngineObject into the **Inspector** panel for the public Audio Source variable.

11.   With the GameManager selected in the **Hierarchy** panel, drag EngineObject into the **Inspector** panel for the public Audio Destruct Scripted Object variable.

> **Note:** It might seem strange to drag GameObject GameManager twice into both public variables for the scripted component ButtonActions. However, what is actually happening is that each variable is being linked to a component of the GameObject GameManager. The public **Audio Source** variable is linked to the **AudioSource** component inside GameManager, and the **Audio Destruct Scripted Object** variable is linked to the component that is an instance of the C# script class AudioDestructBehaviour inside GameManager.

12.   Create a UI button named Button-play-sound by choosing **Menu: GameObject | UI | Button – TextMeshPro**. If asked, import the **TextMeshPro Essentials**.

> **Note:** Unity UI button text objects are built with the TextMeshPro package.

13.   Change the text of the button Play Sound. Position the button in the center-left of the screen by setting its **Rect Transform** position property to **middle-left**.

14.   With Button-play-sound selected in the **Hierarchy**, in the **Inspector** panel, create a new **On-click** event handler by clicking the plus-sign (+) button. Then, drag the GameManager GameObject into the object slot, and select the GameManager | ButtonActions | ACTION_PlaySound() function.

15.   Create a second UI button named Button-destroy-when-finished-playing, and change its text to Destroy When Sound Finished. Position the button in the center-right of the screen by setting its **Rect Transform** property to the **middle-left**. Set the **Width** of this button to **400**.

16. With `Button-destroy-when-finished-playing` selected in the **Hierarchy,** in the **Inspector** panel, create a new **On-click** event handler by clicking the plus-sign (+) button. Then, drag the `GameManager` GameObject into the object slot, and select the `GameManager` | `BttonActions` | `ACTION_DestroyAfterSoundStops()` function.

17. Run the scene. Clicking the **Play Sound** button will play the engine sound each time. However, if you click the **Play Sound** button and then quickly click the **Destroy When Sound Finished** button, as soon as the `engineSound` audio clip has finished playing, you'll see the `EngineObject` GameObject disappear from the **Hierarchy** panel. This is because the GameObject has destroyed itself.

How it works...

In this recipe, you created a `ButtonActions` script class and added an instance as a component to the `GameManager` **GameObject.** This has two public variables, one to an **AudioSource** component and one to an instance of the scripted `AudioDestructBehaviour` component.

The GameObject named `EngineObject` contains an **AudioSource** component, which stores and manages the playing of the audio clip. `EngineObject` also contains a scripted component, which is an instance of the `AudioDestructBehaviour` class. This scripted component is initially disabled. When enabled, every frame in this object (via its `Update()` method) tests whether the audio source is playing (`!audio.isPlaying`). As soon as the audio is found to be not playing, the GameObject is destroyed.

Two UI (TMP) buttons are created. The `Button-play-sound` button calls the `ACTION_PlaySound()` method of the scripted component in the `GameManager` GameObject. This method will start playing the audio clip if it is not already playing.

The second button, `Button-destroy-when-finished-playing`, calls the `ACTION_DestoryAfterSoundStops()` method of the scripted component in the `GameManager` GameObject. This method enables the scripted `AudioDestructBehaviour` component in the `AudioObject` GameObject. This is so that the `EngineObject` GameObject will be destroyed once its **AudioSource** sound has finished playing.

See also

Please refer to the *Preventing an audio clip from restarting if it is already playing* recipe in this chapter.

Creating audio visualization from sample spectral data

The Unity audio systems allow us to access music data via the `AudioSource.GetSpectrumData(...)` method. This gives us the opportunity to use that data to present a `runtime` visualization of the overall sound being heard (from the **AudioListener**) or the individual sound being played by individual **AudioSource** components.

The following screenshot shows lines drawn using a sample script provided by Unity at `https://docs.unity3d.com/ScriptReference/AudioSource.GetSpectrumData.html`:

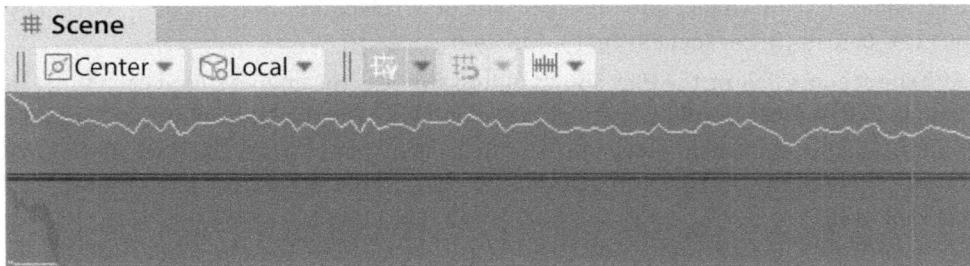

Figure 4.17: Debug `DrawLines` *audio visualization in the Scene panel*

Note that, in the preceding sample code, the use of `Debug.DrawLine()` only appears in the **Scene** panel when running the game in the Unity editor (not for final builds). Therefore, it cannot be seen by the game player. In this recipe, we'll take that same spectral data and use it to create a runtime audio spectral visualization in the **Game** panel. We'll do this by creating a row of 512 small cubes and then changing the heights of each frame, based on 512 audio data samples for the playing **AudioSource** component.

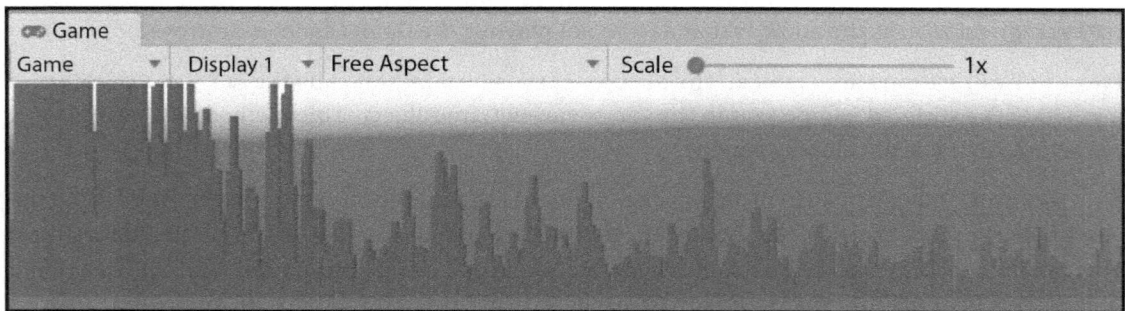

Figure 4.18: The audio spectrum visualization in the Game panel from this recipe

Getting ready

For this recipe, we have provided several free 140 bpm music clips inside the `04_12` folder.

How to do it...

To create audio visualization from the sample spectral data, perform the following steps:

1. Create a new 3D project, using the build-in render pipeline, and import the provided sound clip files.

2. In the **Inspector** panel, set the **background** of the **Main Camera** to black.

3. Set the `Main Camera` **Transform Position** to (224, 50, -200).

4. Set the `Main Camera` **Camera** component to have the following settings:

   • **Projection = Perspective**

- Field of View = 60
- Clipping Planes: Near = 0.3
- Clipping Planes: Far = 300

5. If there is not already a **Directional Light** in the scene, then add one. You can do this by navigating to **GameObject | Light | Directional Light**.

6. Add a new empty **GameObject**, named visualizer, to the scene. Add an **AudioSource** component to this **GameObject**, and set its **AudioClip** to one of the 140 bpm loops provided. Check the **Loop** option.

7. Create a C# script class, called SpectrumCubes, in a new folder, _Scripts, that contains the following code:

```csharp
using UnityEngine;

public class SpectrumCubes : MonoBehaviour
{
    const int NUM_SAMPLES = 512;
    public Color displayColor;
    public float multiplier = 5000;
    public float startY;
    public float maxHeight = 50;
    private AudioSource audioSource;
    private float[] spectrum = new float[NUM_SAMPLES];
    private GameObject[] cubes = new GameObject[NUM_SAMPLES];

    void Awake() {
        audioSource = GetComponent<AudioSource>();
        CreateCubes();
    }

    void Update() {
        audioSource.GetSpectrumData(spectrum, 0, FFTWindow.
BlackmanHarris);
        UpdateCubeHeights();
    }

    private void UpdateCubeHeights() {
        for (int i = 0; i < NUM_SAMPLES; i++)
        {
            Vector3 oldScale = cubes[i].transform.localScale;
            Vector3 scaler = new Vector3(oldScale.x,
                HeightFromSample(spectrum[i]), oldScale.z);
```

```
        cubes[i].transform.localScale = scaler;
        Vector3 oldPosition = cubes[i].transform.position;
        float newY = startY + cubes[i].transform.localScale.y / 2;
        Vector3 newPosition = new Vector3(oldPosition.x, newY,
            oldPosition.z);
        cubes[i].transform.position = newPosition;
    }
}

private float HeightFromSample(float sample) {
    float height = 2 + (sample * multiplier);
    return Mathf.Clamp(height, 0, maxHeight);
}

private void CreateCubes() {
    for (int i = 0; i < NUM_SAMPLES; i++) {
        GameObject cube = GameObject.CreatePrimitive(PrimitiveType.
Cube);
        cube.transform.parent = transform;
        cube.name = "SampleCube" + i;

        Renderer cubeRenderer = cube.GetComponent<Renderer>();
        cubeRenderer.material = new Material(Shader.
Find("Specular"));
        cubeRenderer.sharedMaterial.color = displayColor;

        float x = 0.9f * i;
        float y = startY;
        float z = 0;
        cube.transform.position = new Vector3(x, y, z);

        cubes[i] = cube;
    }
}

}
```

8. Add an instance of the C# script class SpectrumCubes as a scripted component to the visualizer GameObject:

9. With the visualizer GameObject selected in the **Hierarchy** panel, click to choose a visualization color from the public **Display Color** variable for the SpectrumCubes (**Script**) component in the **Inspector** panel.

10. Run the scene. You should see the cubes jump up and down, presenting a runtime visualization of the sound data spectrum for the playing sound.

How it works...

In this recipe, you created a C# script class, called SpectrumCubes. You created a **GameObject** with an **AudioSource** component and an instance of your scripted class. All the work is done by the methods of the SpectrumCubes C# script class. Here is an explanation of each of these methods:

- The void Awake() method: This method caches references to the sibling AudioSource component and then invokes the CreateCubes() method.

- The void CreateCubes() method: This method loops for the number of samples (the default is 512) to create a **3D Cube** GameObject, in a row along the x axis. Each cube is created with the name of SampleCube<i> (where "i" is from 0 to 511) and then parented to the visualizer GameObject (since the scripted method is running in this GameObject). Then, each cube has the color of its renderer set to the value of the public displayColor parameter. The cube is then positioned on the x axis according to the loop number, the value of the public startY parameter (so that multiple visualizations can be viewed in different parts of the screen), and Z = 0. Finally, a reference to the new cube GameObject is stored in the cubes[] array.

- The void Update() method: Each frame in this method updates the values inside the spectrum[] array through a call to GetSpectrumData(...). In our example, the FFTWindow.BlackmanHarris frequency window technique is used. Then, the UpdateCubeHeights() method is invoked.

- The void UpdateCubeHeights() method: This method loops for each cube to set its height to a scaled value of its corresponding audio data value in the spectrum[] array. The cube has its y value scaled by the value returned by the HeightFromSample(spectrum[i]) method. Then, the cube is moved up (that is, its **Transform Position** is set) from the value of startY by half its height so that all the scaling appears upward (rather than up and down). This is to ensure there is a flat line along the base of our spectrum of cubes.

- The float HeightFromSample(float) method: The HeightFromSample(float) method does a simple calculation (the sample value times the public parameter multiplier) with a minimum value of two added to it. The value returned from the function is this result, limited to the maxHeight public parameter (via the Mathf.Clamp(...) method).

There's more...

There are some details that you don't want to miss.

Adding visualizations to a second AudioSource component

The script has been written so that it is easy to have multiple visualizations in a scene. So, to create a second visualizer for a second audio clip in the scene, perform the following steps:

1. Duplicate the visualizer GameObject.

2. Drag a different **audio clip** from the **Project** panel into the AudioSource component of your new **GameObject**.

3. Set the `startY` public parameter in the **Inspector** panel to 60. (so that the new row of cubes will be above the original row).

4. In the **Inspector** panel, choose a different **Display Color** public variable for the **SpectrumCubes (Script)** component:

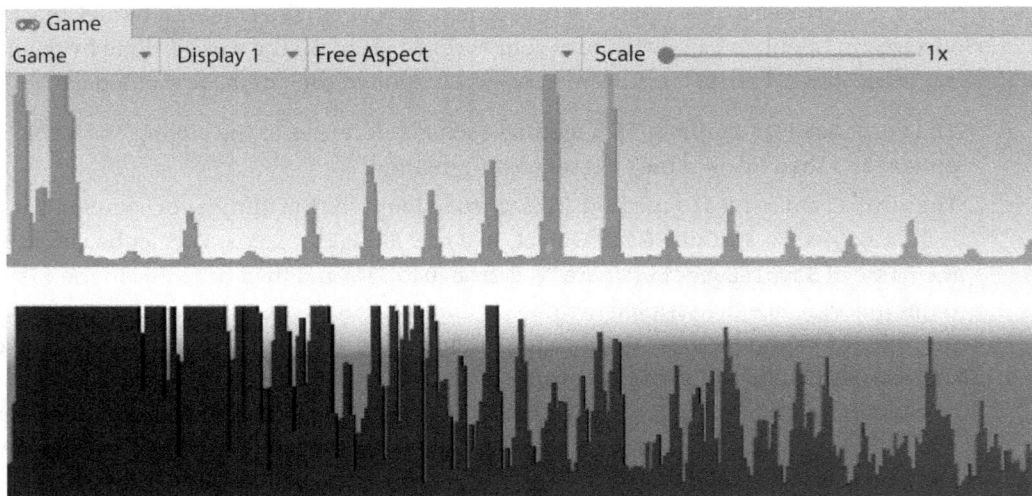

Figure 4.19: Two visualizer GameObjects in play mode

Trying out different Fast Fourier Transform (FFT) window types

There are several different approaches to the frequency analysis of audio data; our recipe currently uses the `FFTWindow.BlackmanHarris` version. You can learn about (and try out!) some of the others from the Unity FFTWindow documentation page at `https://docs.unity3d.com/ScriptReference/FFTWindow.html`.

Synchronizing simultaneous and sequential music to create a simple 140 bpm music-loop manager

There are times when we need to precisely schedule audio start times to ensure a smooth transition from one music track to another, or to ensure simultaneous music tracks play in time together.

In this recipe, we'll create a simple 4-track 140 **bpm** (**beats-per-minute**) music manager that starts playing a new sound after a fixed time – the result of which is that the tracks fit together perfectly, and those that do overlap do so in synchronicity.

Getting ready

For this recipe, we have provided several free 140 bpm music clips inside the `04_12` folder.

How to do it...

To create a music-loop manager, perform the following steps:

1. Create a new Unity 3D project, and import the provided sound clip files.

2. Create four GameObjects in the scene that contain an AudioSource component, linked to a different **AudioClip** loop from the 140 bpm files provided. You can do this in a single step by dragging the music clip from the **Project** panel into either the **Hierarchy** panel or the **Scene** panel.

3. In the **Inspector** panel, uncheck the **Play On Awake** parameter for all four AudioSource components (so that they don't start playing until we tell them to).

4. Add a new empty **GameObject** named musicScheduler to the scene.

5. Create a C# script class, called LoopScheduler, in a new folder, _Scripts, that contains the following code:

```
using UnityEngine;

public class LoopScheduler : MonoBehaviour {
    public float bpm = 140.0F;
    public int numBeatsPerSegment = 16;
    public AudioSource[] audioSources = new AudioSource[4];
    private double nextEventTime;
    private int nextLoopIndex = 0;
    private int numLoops;
    private float numSecondsPerMinute = 60F;
    private float timeBetweenPlays;

    void Start() {
        numLoops = audioSources.Length;
        timeBetweenPlays = numSecondsPerMinute / bpm *
numBeatsPerSegment;
        nextEventTime = AudioSettings.dspTime;
    }

    void Update() {
        double lookAhead = AudioSettings.dspTime + 1.0F;
        if (lookAhead > nextEventTime)
            StartNextLoop();

        PrintLoopPlayingStatus();
    }
```

```csharp
private void StartNextLoop() {
    audioSources[nextLoopIndex].PlayScheduled(nextEventTime);
    nextEventTime += timeBetweenPlays;

    nextLoopIndex++;
    if (nextLoopIndex >= numLoops)
        nextLoopIndex = 0;
}

private void PrintLoopPlayingStatus(){
    string statusMessage = "Sounds playing: ";
    int i = 0;

    while (i < numLoops) {
        statusMessage += audioSources[i].isPlaying + " ";
        i++;
    }

    print(statusMessage);
}
```

6. Add an instance of the C# script class `LoopScheduler` as a scripted component to the `musicScheduler` GameObject.

7. With the `musicScheduler` GameObject selected in the **Hierarchy** panel, drag each of the music-loop GameObjects into the four available slots for the **AudioSources** public array variable inside the `Loop Scheduler` **(Script)** component:

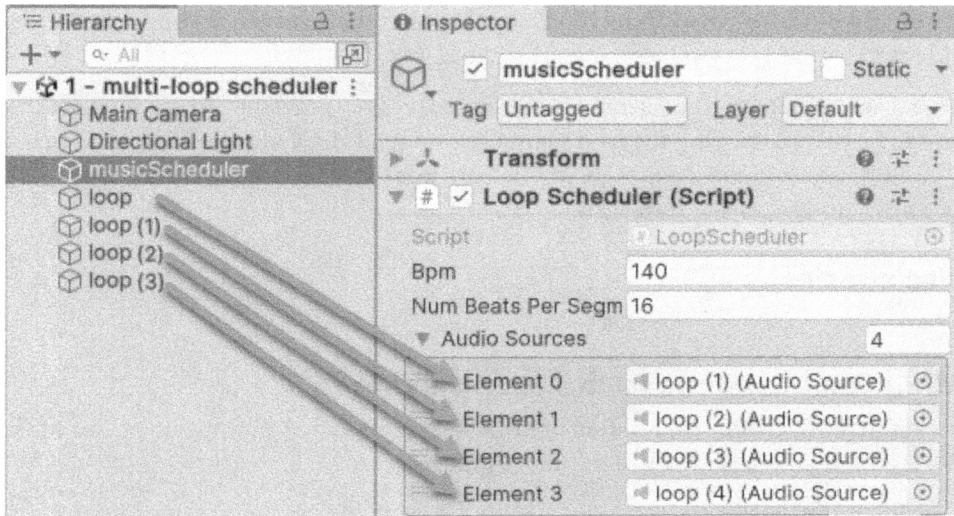

Figure 4.20: Linking AudioSource GameObjects to the public variables of the LoopScheduler (Script) component

8. Run the scene. Each clip should start, in turn, after the same time delay. If you choose one or two longer clips, they will continue playing while the next clip begins – all overlapping perfectly, since they are all 140 bpm sound clips.

How it works...

In this recipe, you added four **GameObjects to the scene**, each containing **AudioSources** linked to 140 bpm music clips. You created a C# script class, called LoopScheduler, and added an instance to an empty **GameObject**. You associated the four **AudioSources** in your **GameObjects** with the four slots in the public AudioSources array variable in your scripted component.

The number of music clips you use can easily be changed by changing the size of the public array variable in the C# script class LoopScheduler.

The Start() method counts the length of the array to set the numLoops variable. Then, it calculates the number of seconds to delay before starting each clip (this is fixed according to the BPM and beats-per-measure). Finally, it sets the current time to be the time to start the first loop.

The Update() method decides whether it's time to schedule the next loop. It does this by testing whether the current time, plus one-second into the future, is past the time to start the next loop. If so, the StartNextLoop() method is invoked. Regardless of whether we have started the next loop, the PrintLoopPlayingStatus() method is printed to the console, displaying to the user which loops are playing or not.

The `PrintLoopPlayingStatus()` method loops for each `AudioSource` reference in the array, creating a string of `true` and `false` values to be printed out.

The `StartNextLoop()` method sends a `PlayScheduled(...)` message to the next `AudioSource` component to be played, passing the `nextEventTime` value. It then adds the time between plays for the next event time. Then, the next value of the loop index is calculated (add one, if past the end of the array, and then reset this to zero again).

There's more...

There are some details that you don't want to miss.

Adding visualizations to the four playing loops

It's great fun to watch the visualization of the loop sounds as they play together. To add visualizations to the four AudioSources, perform the following steps:

1. Import the `SpectrumCubes.cs` C# script file from the previous recipe into this project.
2. Select the Main Camera in the **Hierarchy** panel. Then, in the **Inspector** panel, set the **Transform Position** to (224, 50, -200).
3. With the Main Camera still selected, set the **Camera** in the **Inspector** panel to have the following settings:
   - **Projection = Perspective**
   - **Field of View = 60**
   - **Clipping Planes = 0.3 - 300**
4. Add a `DirectionalLight` **GameObject** to the scene.
5. For each of the four **GameObjects** containing your **AudioSources**, add an instance of the `SpectrumCubes` **script class**.
6. In the **Inspector** panel for the Spectrum Cubes (**Script**) component, change the `displayColors` variable for each `AudioSource` **GameObject**.
7. Set the `startY` values of the Spectrum Cubes (**Script**) components for the four `GameObjects` to -50, 0, 50, and 100. For most screen sizes, this should allow you to see all four visualization spectrums.

Recording sound clips with the free Audacity application

When I quickly want a "placeholder" sound clip, and don't have a file on my computer, or can't find one quickly on the web, then often I'll just record a sound file using the free, open-source, multi-platform Audacity application. I would also use this application to record the final version of dialogue, such as a narrative voice-over for introduction scenes, or spoken words when a player interacts with a non-player character in a game.

This recipe shows you how to record and save audio clips easily with this application.

Figure 4.21: The free, open-source Audacity audio editor application

Getting ready

Download and install the Audacity application for your computer system (Windows/Mac/Linux). You can find it at https://www.audacityteam.org/download/.

How to do it...

To record sound clips with the free Audacity application, perform the following steps:

1. Open the Audacity application.
2. Click the red-circle record button and speak/play sound into your computer's microphone. When finished, click the black-square stop button.

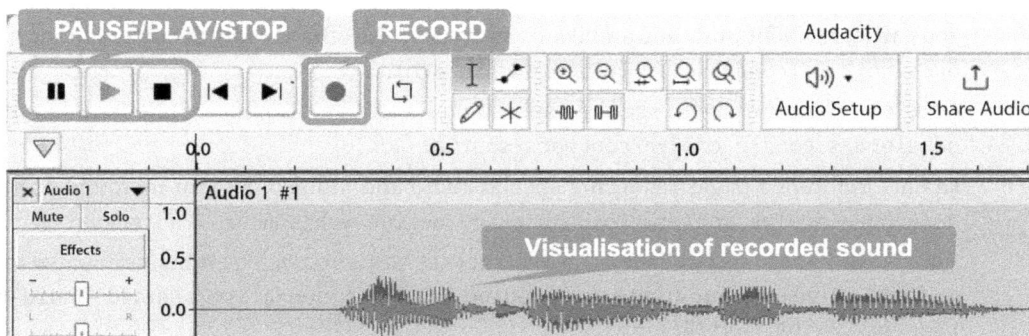

Figure 4.22: The Audacity application interface

3. You can select all of some of the recorded sound, and apply effects like echo from the **Application** menu.

4. When finished, you can export the sound files as MP3, WAV, or OGG, using the **File | Export** menu.

5. Once exported, you can import the sound file into your Unity projects.

How it works...

Audacity is a simple yet powerful sound editing application that is perfect for quickly recording and editing sound files for use with your Unity projects.

Further reading

The recipes in this chapter demonstrate both scripted and Unity audio system approaches to help you manage audio and introduce dynamic effects at runtime. Games can become far more engaging when the audio environment of effects and music subtly changes, based on the context of what is happening in the game. You may wish to explore further with the Unity Audio Mixer and Audio Reverb Zones, for example. Please refer to the following:

- `https://docs.unity3d.com/2023.1/Documentation/Manual/AudioMixer.html`
- `https://docs.unity3d.com/Manual/class-AudioReverbZone.html`

What is possible with special audio is now becoming even more interesting with the introduction of ambisonic audio when playing 3D VR games. This allows listeners to enjoy rich audio experiences based on whether sounds are above or below the listener, as well as their distance from an audio source. To find out more about ambisonic audio, please view the following references:

- Rode offers some information on the history of ambisonics: `https://www.rode.com/blog/all/what-is-ambisonics`.

- You can learn about Unity and ambisonic audio from the official Unity documentation: `https://docs.unity3d.com/Manual/AmbisonicAudio.html`.

- Google's reference pages regarding special audio and ambisonics can be found at `https://developers.google.com/vr/concepts/spatial-audio`.

- Oculus' reference pages regarding special audio and ambisonics can be found at `https://developer.oculus.com/downloads/package/oculus-ambisonics-starter-pack/`.

- Robert Hernadez has published a great article at *Medium.com*, disambiguating how to record and edit ambisonic audio: `https://medium.com/@webjournalist/spatial-audio-how-to-hear-in-vr-10914a41f4ca`.

5

Textures, Materials, and 3D Objects

3D games need 3D objects in their scenes! In this chapter, we will explore different ways of adding 3D objects to our scenes, including the following:

- Using and adapting the built-in 3D primitives that Unity offers, such as cubes and cylinders
- Importing 3D models from third parties, and converting other modeling formats into the .fbx format, the easiest format to use in Unity

We also want to customize how 3D objects look when seen by the user (via a Unity Camera). An understanding of **textures** and **materials** is needed to customize the look of 3D objects. A **texture** is a 2D image file that is used by the game engine as a source of pixels (colored dots), which can be used to influence the display of a 3D object. A **material** is an asset file that defines how the rendering engine should treat a 3D object, which includes textures for colors (albedo), textures for height maps, transparency, how metallic (shiny) the surface of the object should be treated, how to offset and tile (duplicate) textures, whether the 3D object should be emitting light, and so on.

In this chapter, we will also explore the use of image textures and materials to customize the look of 3D objects, including examples of how to dynamically change GameObject materials at runtime, such as responding to when the user clicks a GameObject with the mouse.

In this chapter, we will cover the following recipes:

- Creating a scene with 3D primitives and a texture
- Creating material asset files and setting Albedo textures
- Exporting Blender files as **Filmbox** (**FBX**) files for use in Unity
- Importing FBX models into a Unity project
- Highlighting GameObject materials on mouseover
- Fading the transparency of a material

Creating a scene with 3D primitives and a texture

In this recipe, we'll create a simple signpost using an image containing text, as well as a combination of 3D cubes, cylinders, and plane primitives.

While in some cases we'll use complex 3D models that have been imported from modeling apps or third parties, there are cases where 3D primitives are quick, simple, and sufficient for a game task. Examples of 3D primitives in games include invisible objects with trigger colliders (for example, to open doors or signal a checkpoint), the use of spheres as projectiles, the use of scaled cubes and planes for signposts, and so on. The speed and simplicity of using 3D primitives also make them perfect for fast prototyping, where the objects act as placeholders that can be replaced with more sophisticated models at a later stage in the production of a game. Materials can be quickly created that reference images to textured 3D primitives.

Figure 5.1: A signpost created with 3D primitives

Getting ready

For this recipe, you'll need an image. Either use one of your own or the beware.png image that we've provided in the 05_01 folder.

How to do it...

To create elements in a scene using 3D primitives, follow these steps:

1. Create a new 3D project.
2. Import the beware.png image (or your own) into the newly created 3D project.
3. Add a 3D cube named Cube-signpost to the scene using menu: **GameObject | 3D Object | Cube**.
4. Set the **Position** of the Cube-signpost GameObject to (0,0,0) and its **Scale** to (2, 1, 0.1). This should make it a portrait-style rectangle shape that only has a little depth (like a big, flat piece of wood for a signpost).

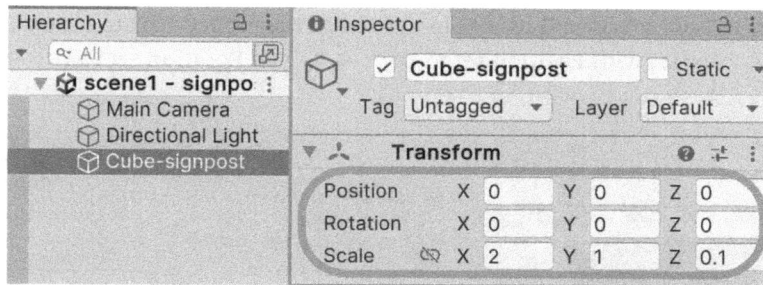

Figure 5.2: Setting the properties of the 3D cube

5. Create a 3D plane named Plane-sign using the menu **GameObject | 3D Object | Plane**. Then, in the **Hierarchy** panel, make it a child of Cube-signpost (drag Plane-sign onto Cube-signpost). This means its properties will be relative to Cube-signpost. Don't worry about its size, orientation, and so on – we'll fix that soon.

6. Drag the beware texture asset file from the **Project** panel onto the Plane-sign GameObject in the **Hierarchy** panel. You'll now see the image painted on the panel (don't worry that it's in mirror writing – we'll fix that soon) and that a new folder has been created, named Materials, in the **Project** panel. This folder contains the new **material** asset file with the same name as the image asset file – that is, beware:

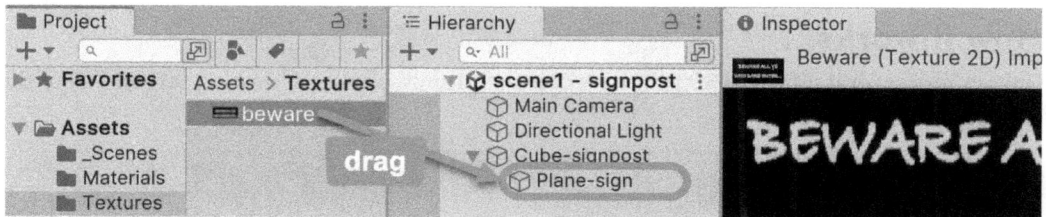

Figure 5.3: Dragging a texture asset file onto a 3D plane GameObject in the Hierarchy

7. In the **Inspector** panel, set the **Position** of the Plane-sign GameObject to (0, 0, -0.51), its **Rotation** to (-90, 0, 0), and its **Scale** to (-0.08, 1, -0.08). This should now display the textured panel on the front of our cube, fixing the mirror-writing issue and showing a nice border for the cube around the textured panel:

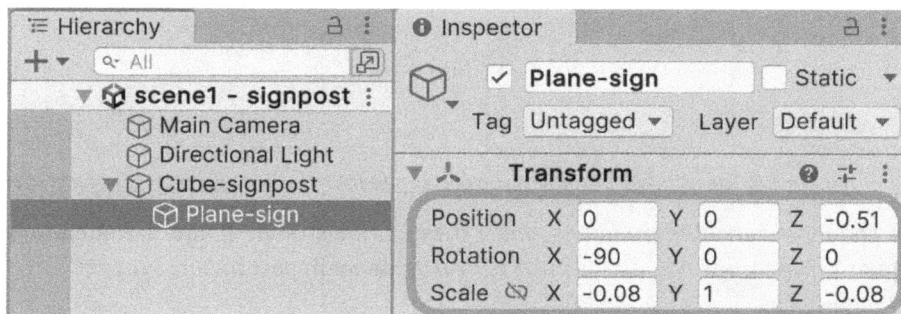

Figure 5.4: Setting the properties of the 3D plane primitive as a child of the cube

8. A cylinder will work well as a pole to hold our signpost. Create a new 3D cylinder, named Cylinder-pole, with **Position** set to (0, -0.4, 0), **Rotation** set to (0, 0, 0), and **Scale** set to (0.1, 1, 0.1):

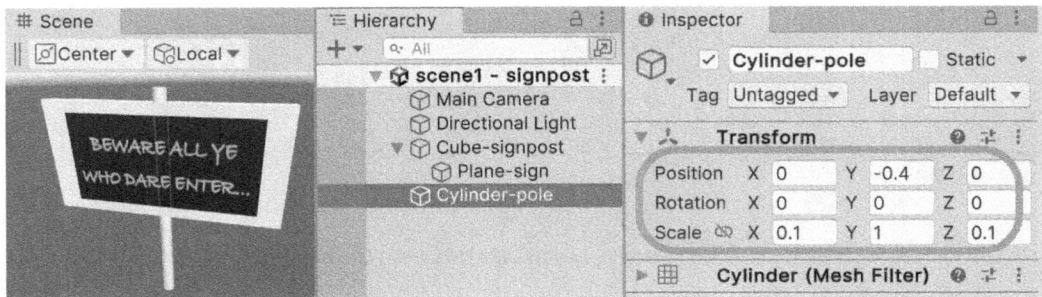

Figure 5.5: Setting the properties of the 3D Cylinder for the pole of our signpost

9. Finally, let's add the ground. Again, we can use a scaled and positioned 3D cube for this. Create a new **cube**, named Cube-ground, with its **Position** set to (0, -2, 0) and its **Scale** set to (20, 1, 20).

10. Save and run the scene. Note that we'll see how to add a spotlight to highlight the signpost at the end of this recipe, making your scene match the original recipe screenshot.

How it works...

Unity offers several different 3D primitive objects, as shown in the following screenshot of the **Game-Object | 3D Object** menu:

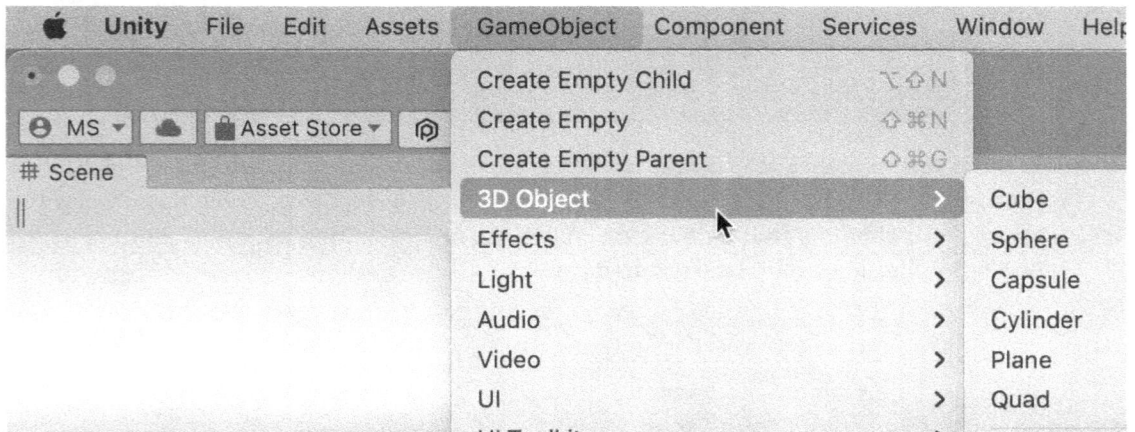

Figure 5.6: The list of 3D primitives you can create from the GameObject | 3D Object menu

Perhaps the most versatile are **cubes,** since they can be stretched and flattened into many useful rectangular polyhedrons. We used a cube to create the main part of our signpost.

However, when a cube is textured with an image, the image will appear, in different orientations, on all six sides. For the text image of our signpost, we used a **plane**, since a plane has only two sides (front and back) and only displays a textured image on its front face. In fact, we cannot even see the plane at all from the side or behind.

As we found in this recipe when we created the pole for our signpost with a **cylinder**, cylinders are straight, round objects with flat circle ends, which makes them useful to approximate metal and wooden poles, or when making very shallow, circular objects such as plates or discs.

Some game developers have a set of prefabs to quickly prototype a scene. However, as we've seen in this recipe, Unity's 3D primitives can be quite effective to test or provide placeholders with recognizable approximations of objects in a scene.

There's more...

Here are some ways to enhance this recipe.

Enhancing the scene — adding a spotlight

Let's enhance the scene a little more by creating a long shadow effect. We can do this by changing the **Directional Light** settings, adding a spotlight, and turning off the default **Environmental Lighting Skybox**:

1. Change the **Rotation** of **Directional Light** in the scene to (15, 20, 0). This should make the scene a little darker, and we should see a shadow behind the sign on the ground.

2. Add a new **Spot Light** to the scene (**Create | Light | Spot Light**), and set its **Position** to (0, 0, -2), its **Rotation** to (0, 0, 0), its **Range** to 5, its **Spot Angle** to 50, and its **Intensity** to 5:

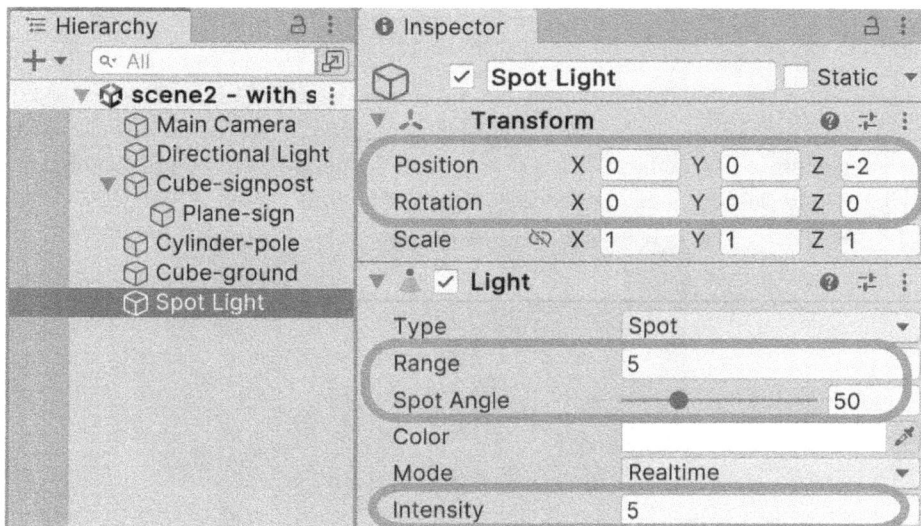

Figure 5.7: Spot Light settings to add atmosphere to our signpost scene

3. Display the **Lighting** panel by choosing the menu **Window | Rendering | Lighting**.

4. Select the **Environment** tab. Then, set the **Environment Skybox Material** to **None,** by clicking the circle target button and choosing **None** from the top of the list.

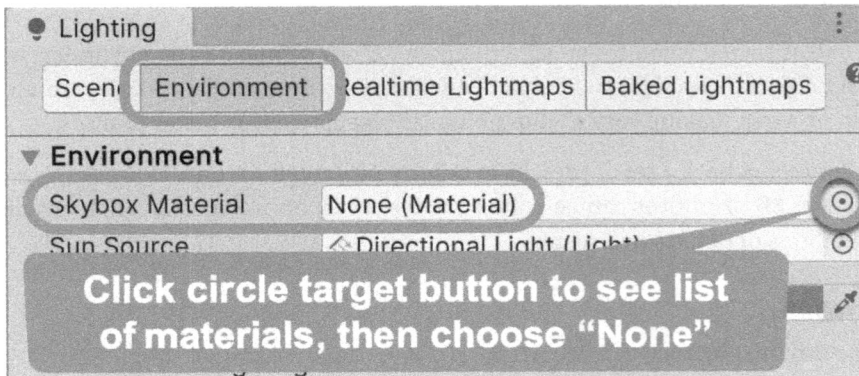

Figure 5.8: Removing the Environmental Skybox Material

You've now added a spotlight, removed the default skybox, and rotated the directional light to enhance how your signpost (and its shadow) appear when the scene is played.

Creating material asset files and setting Albedo textures

Unity 3D GameObjects have a **mesh renderer** component, and that component needs a **material** in order to display a **texture** image asset file. In one of the steps of the previous recipe, you dragged the beware image texture asset file from the **Project** panel onto the Plane-sign GameObject. What happened was that Unity created a material asset file, named beware, linked to the texture image, and that material was set for the Plane-sign mesh renderer component.

When using a text image in a simple way, an automated material created like this is fine. However, sometimes we may wish to create a new material from scratch, not linked to an image texture, or we may wish to customize the material properties. In this recipe, we'll create and customize two materials and apply them to GameObjects, improving the scene further.

Figure 5.9: The signpost and pole rendered with materials we've created

Getting ready

This recipe follows from the previous one, so make a copy of that project and work from the copy.

For this recipe, you'll also need an image to texture the sign to look as if it's made of wooden planks. Either use one of your own or the wood_plank_weathered.jpg image that we've provided in the 05_02 folder (many thanks to FreeStockTextures.com for publishing this texture with a Creative Commons Zero license).

How to do it...

To create material asset files and set **Albedo** textures, perform the following steps:

1. Open the copy you have made of the previous recipe.
2. Import the wood_plank_weathered.jpg image (or your own) into the newly created 3D project.
3. Navigate to the Materials folder in the **Project** panel.
4. Create a new material asset file named m_brown, by clicking the plus (+) sign (the **Create** button) of the **Project** panel, and choose **Material**:

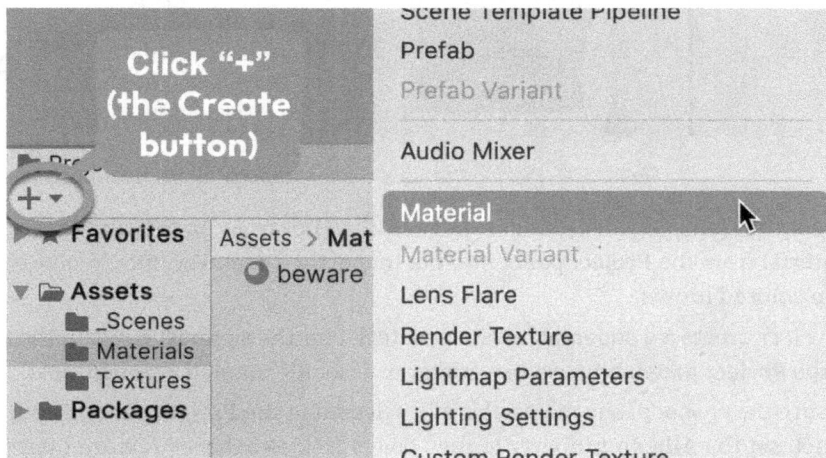

Figure 5.10: Creating a new material asset file in the Project panel

5. Select the new material in the **Project** panel. Then, in the **Inspector** panel, set the **Albedo** property for material m_brown to a brown color:

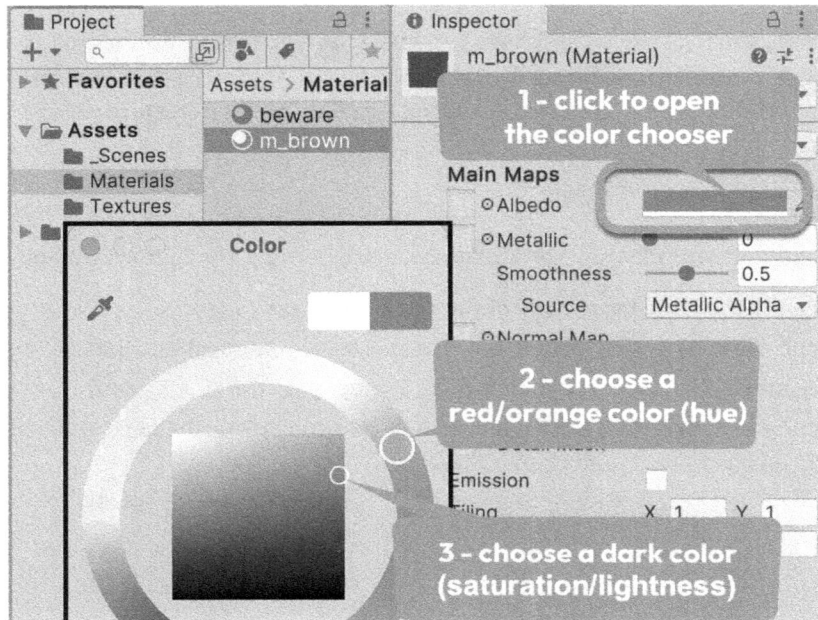

Figure 5.11: Picking the Albedo color property for the new material

6. Select the GameObject Cylinder-pole in the **Hierarchy** panel, and then drag the m_brown material from the **Project** panel into the **Inspector** panel. You should now see the pole of the sign colored brown.

7. Now let's create a wooden plank-style material for the signpost cube. In the Materials folder in the **Project** panel, create a new material asset file, named m_woodplank.

8. Ensure the m_woodplank material is still selected in the **Project** panel. Then, in the **Inspector** panel, set the **Albedo** property to the wood_plank_weathered texture image. One method is to drag the wood_plank_weathered texture image into the square to the left of the **Albedo** property in the **Inspector** panel.

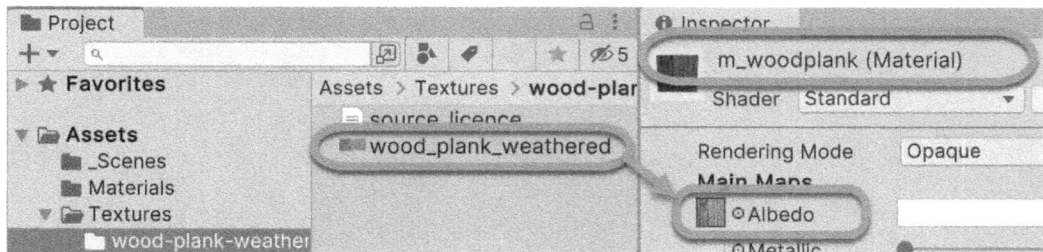

Figure 5.12: Setting the beware image as the material's Albedo property

> **Note:** An alternative method to set an **Albedo** texture image is to click the target circle asset chooser (left of the **Albedo** property), filter with the text wood, and then select the wood_plank_weathered texture image.
>
> In the **Inspector** panel, wherever you see a target circle button next to a property, you can open an asset file search dialog for the property type.

9. Drag the m_woodplank material asset file from the **Project** panel onto the Cube-signpost GameObject in the **Hierarchy** panel, making that appear to be made of wood.

10. While this works, let's make the planks smaller. Ensure the m_woodplank material is still selected in the **Project** panel. In the **Inspector** panel, explore the different tiling settings. For example, try setting its tiling to **X = 5** and **Y = 5**.

How it works...

In this recipe, you have learned how to create materials from scratch and adjust their properties. The m_brown material was created just by choosing a color and lightness setting. The m_woodplank material was more sophisticated, using a texture image file and then setting it to be duplicated through the *tiling* property.

Creating materials with colors and textures is a very useful skill for quickly mocking up the environment in a scene. The textures and properties of materials can be improved later on, and those changes will be applied to all GameObjects in the scene using those materials.

Exporting Blender files as FBX for use in Unity

There are times where, to get the visual effect we want, we need a complex 3D object. One option is to create our own using the ProBuilder modeling tools built into Unity, which we will explore later in *Chapter 7, Creating 3D Geometry with ProBuilder*. However, another option is to use a 3D model that's been created by a third party. The Unity Asset Store is one source for models, and they are already prepared for importing into Unity as FBX models or even Unity prefabs. However, there are thousands of free 3D models available online that have been made with third-party 3D modeling applications, and all we need to do is open them in their applications and export them as FBX models, ready to be imported into Unity.

In this recipe, we'll convert a free 3D model of some pumpkin jack-o'-lanterns and other items into a
.fbx file, which we can then import and use in a Unity project:

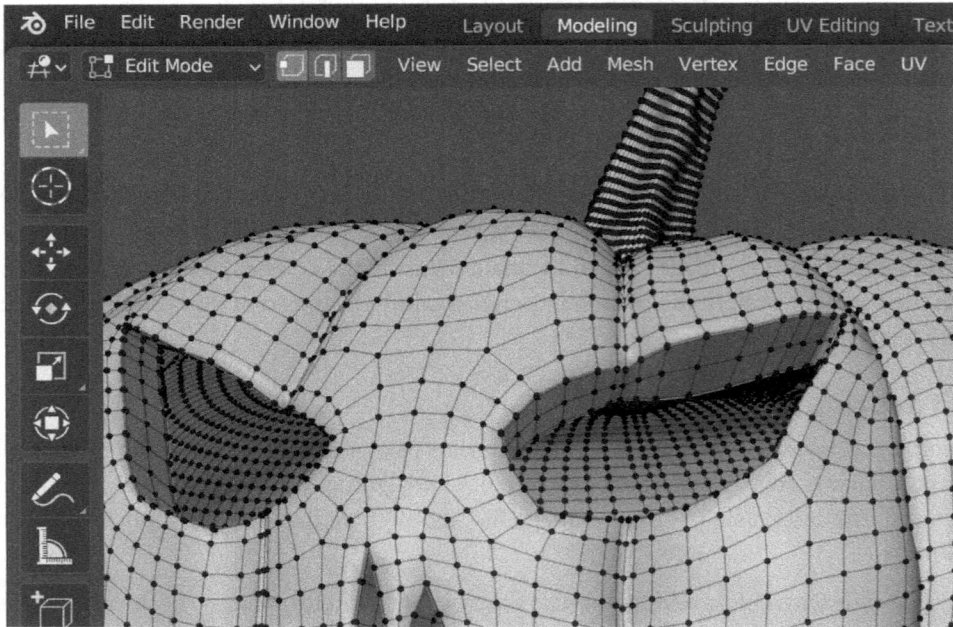

Figure 5.13: Closeup of the pumpkin model in the free Blender 3D modeling application

Getting ready

In this recipe, we will use the free Blender 3D modeling application to convert a .blend 3D scene file
into a .fbx format model that can be imported into Unity. We have provided a file called pumpkin.
blend in the 05_03 folder. Thanks to Oltsch for posting this model free on the TurboSquid website at
https://www.turbosquid.com/3d-models/3d-pumpkins-candles-model-1484325.

You can download the installer (for Windows, Mac, or Linux) for Blender from the following URL:
https://www.blender.org/download/.

However, if you can't install or use the Blender application, in the 05_04 folder, we have also provided
the .fbx format model created in this recipe, pumpkin.fbx.

How to do it...

To convert a Blender 3D model into an FBX model for use in Unity, follow these steps:

1. Open the Blender file in the Blender application.
2. You'll see that there are four main objects: some pumpkins, flames, candles, and a glass jar.
3. For this recipe, we only want the pumpkin with the face and its lid. Keep Plane.003 and
   Plane.011, and then right-click to delete the other objects. To delete an object (and all its children) in Blender, select the object, and then choose **Delete Hierarchy** from the right-click menu.

4. Rename `Plane.003` to `Pumpkin` by right-clicking the menu and going to **ID Data | Rename**. Then, rename `Plane.011` to `Pumpkin-lid`:

Figure 5.14: Deleting the Bottle.001 object in Blender

5. Centralize the pivot point to the `Pumpkin` object, by selecting this object and then choosing the menu **Object | Set Origin | Origin to Centre of Mass (Surface)**.

6. Now, go to **File**, choose **Export**, and select the **FBX (.fbx)** format. Save the exported model as `pumpkin`.

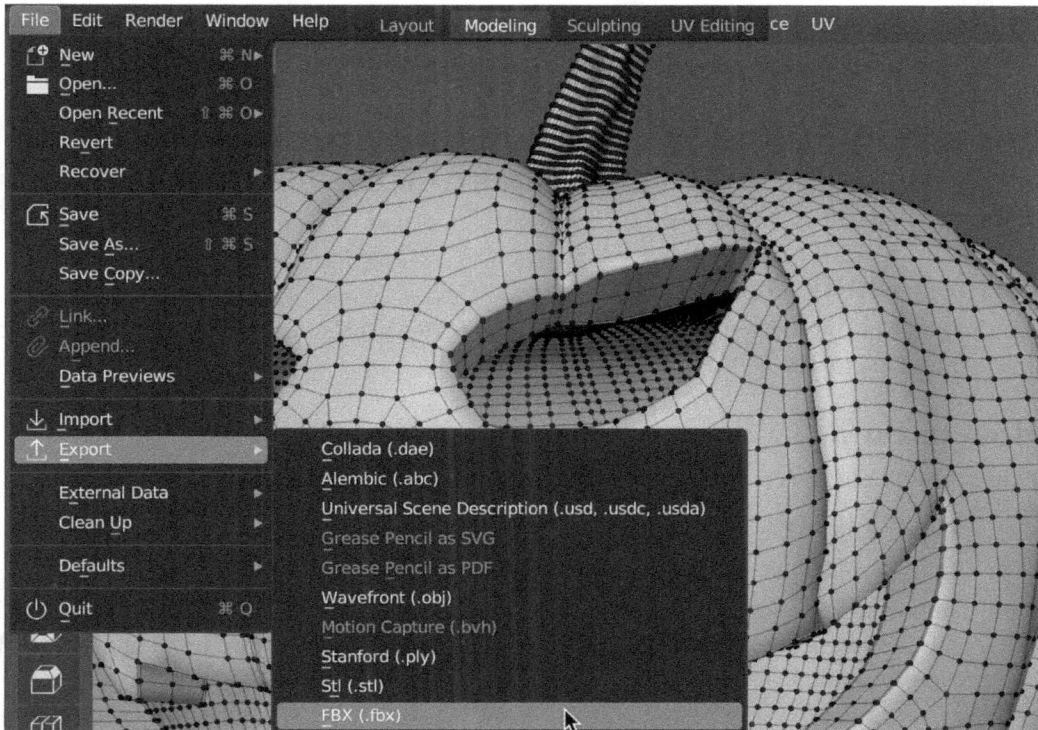

Figure 5.15: Exporting the edited model in Blender as an FBX file

7. You can now quit the Blender application.

8. You should now have a file, `pumpkin.fbx`, which is the 3D model in the FBX format ready to be imported by Unity.

How it works...

For third-party 3D models already in the FBX format, we can just import them directly into our Unity project. For models in other formats, if we have the same application that we used to create the model, then we can do what we did in this recipe – that is, we can open the model in the modeling application, remove any unwanted elements, and then export it in the FBX format.

3D modeling applications are complex, sophisticated applications. However, as demonstrated in this recipe, all we really need to know is how to view, select, and delete elements, as well as how to export the model in FBX format.

In this recipe, we removed the unwanted elements of the model in the Blender application. It is also possible to remove some elements after a model has been imported into Unity – some elements will appear as child GameObjects when the model has been added to a scene.

If the model we wish to use requires no editing before being converted into the FBX format, then there are some online converters that will do this for us, such as Aspose's *X-to-FBX* and Green Token's *onlineconv*. These two online converters can be found at the following links:

- **Free online X to FBX Converter:** `https://products.aspose.app/3d/conversion/x-to-fbx`
- **Online 3D Model Converter:** `https://www.greentoken.de/onlineconv/`

Importing FBX models into a Unity project

In the previous recipe, you created an FBX model file. In this recipe, we'll import that model into Unity and work with its materials.

Getting ready

Use the FBX file you created in the previous recipe. We have also provided an FBX file for your use inside the `05_04` folder.

How to do it...

To import FBX models into a Unity project, perform the following steps:

1. Create a new 3D project.
2. In the **Project** panel, create a folder named `Models`.
3. Import the `pumpkin.fbx` model into the `Models` folder. You do this either by dragging the FBX file into the **Project** panel from the file explorer, or by choosing the menu **Assets | Import New Asset...**.
4. Create a GameObject by dragging the `pumpkin` asset file from the **Project** panel into the scene.
5. Create a new 3D cube in the scene, by choosing the menu **GameObject | 3D Object | Cube**. You can see the pumpkin object is quite a bit smaller than the 1 x 1 x 1 cube.
6. Delete the cube GameObject (it was just useful to get an idea of the size of the pumpkin GameObject).

7. Select the pumpkin asset file in the Models folder in the **Project** panel.

8. In the **Inspector** panel, ensure the **Model** tab is selected. Then, change **Scale Factor** to 5 and click **Apply**. Now, the model is five times larger:

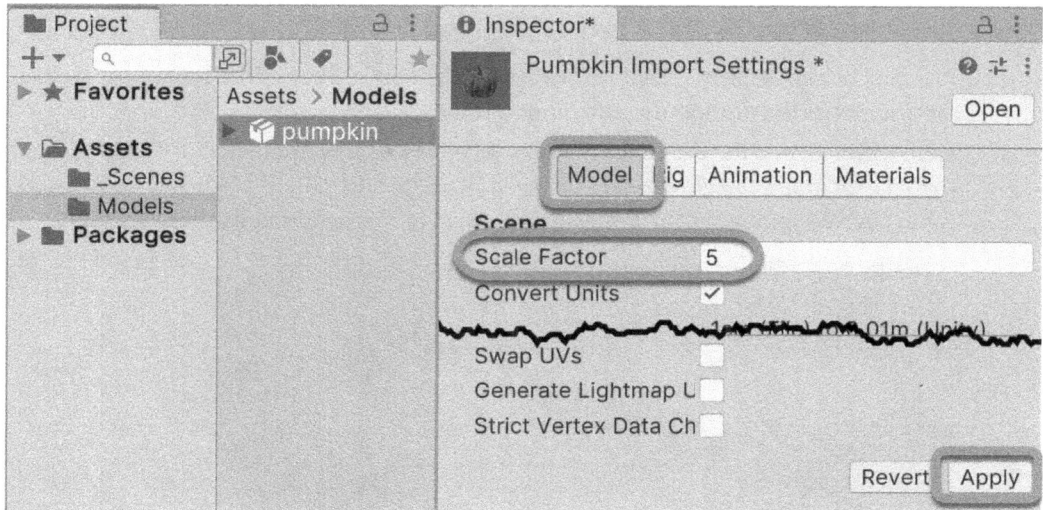

Figure 5.16: Increasing the Scale Factor property of our pumpkin mode in the Inspector panel

Since we've changed the core import settings for this model asset file, any GameObject in the scene already created from this asset will also be scaled by the new scale factor.

9. Select the pumpkin asset file in the Models folder in the **Project** panel.

10. In the **Inspector** panel, ensure the **Materials** tab is selected. Then, click the **Extract Materials...** button and select your Models folder – so that the materials for the pumpkin model are in the same folder as the pumpkin model.

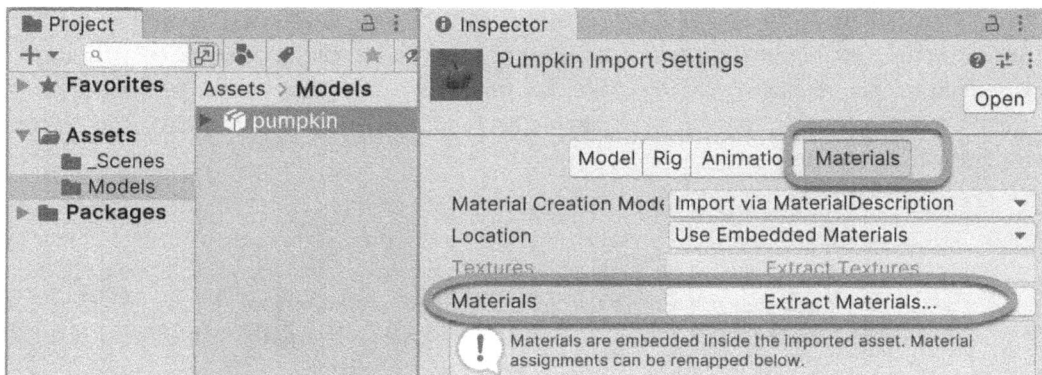

Figure 5.17: Extracting the materials of the pumpkin model

11. You'll now find three new materials in the `Models` folder:

    • `Pumpkin`: This is for the inside of the pumpkin.

    • `Pumpkin_skin`: This is for the outside of the pumpkin.

    • `Pumpkin_stem`: This is for the lid of the pumpkin.

12. You can change these materials to customize the appearance of the pumpkin, such as making the gray lid of the pumpkin green, and so on.

13. A fun effect is to make the `Pumpkin` material emissive so that the inside of the pumpkin glows. To make the pumpkin glow, select the `Pumpkin` asset file in the **Project** panel and select an HDR color, such as yellow for the **Emission** property of the material, as shown here:

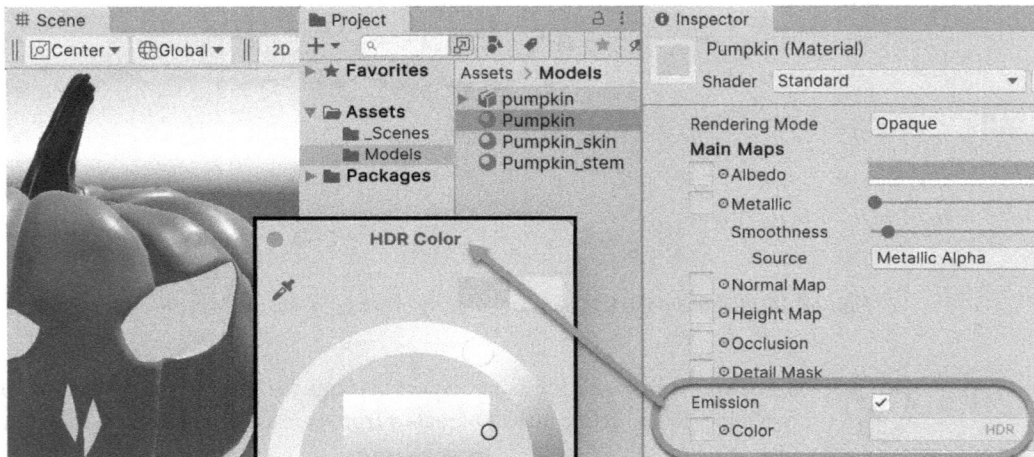

Figure 5.18: Making the inside of the pumpkin an emissive glowing material

How it works...

As we've seen, once you have a model in the FBX format, it can be easily imported into a Unity project. You can extract materials, to customize the object further, and also change model import settings such as the scaling factor.

Different modeling applications work at different scales, so Unity has made it easy for us to adjust a **scale factor** for models we have imported. For this recipe, the pumpkin seemed about one-fifth the size of a 1 x 1 x 1 cube, so we applied a **Scale Factor** of 5 to make the pumpkin a size that was easier to work with in our project.

> When working with several models, it's a good idea to create a separate sub-folder (inside the folder `Models`) for each model's FBX and materials. For example, we could have created a `Models` or `Pumpkin` folder, and inside that folder would be the Pumpkin FBX and the three extracted materials. Keeping asset files well organized in a hierarchy of folders will save lots of time later when searching for resources.

There's more...

There are some details that you don't want to miss.

Using Blender files directly in a Unity project

If you have the Blender application installed on the same computer that you are using Unity, then you don't actually have to perform the FBX export/import procedure – Unity will do this for you automatically, each time it spots that the Blender .blend file has been changed, since it was last automatically converted to an FBX.

To try this out, do the following:

1. Create a new 3D project.
2. In the **Project** panel, create a folder named Models.
3. Import the pumpkin.blend model into the Models folder. You do this either by dragging the Blender file into the **Project** panel from the file explorer, or by choosing the menu **Assets | Import New Asset....**
4. Create a GameObject by dragging the pumpkin asset file from the **Project** panel into the scene.

You will find that the pumpkin model in the **Project** panel works just like our exported FBX – since Unity has (in the background) used the Blender application to create an FBX from the Blender project file.

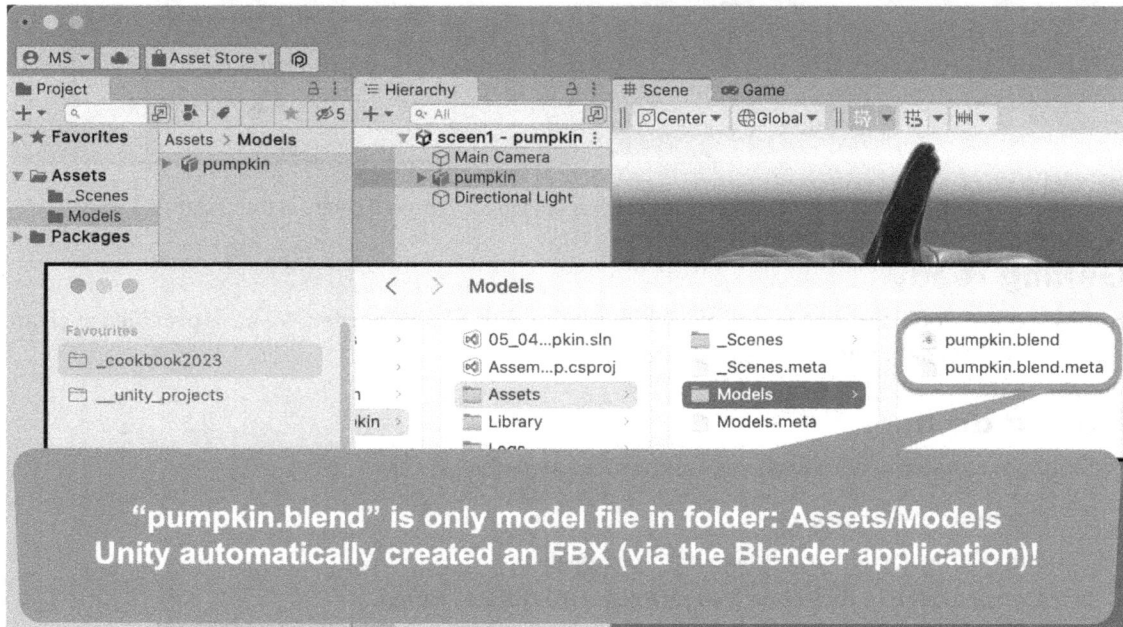

Figure 5.19: Using a native Blender project file in our Unity Assets folder (for automatic FBX creation by Unity)

Highlighting GameObject materials on mouseover

Changing the look or color of an object at runtime can be a very effective way of letting players know that they can interact with it. This is very useful in a number of game genres, such as puzzles and point-and-click adventures, and it can also be used to create 3D user interfaces.

In this recipe, we'll swap materials when the mouse is over an object by using a second material, showing a border:

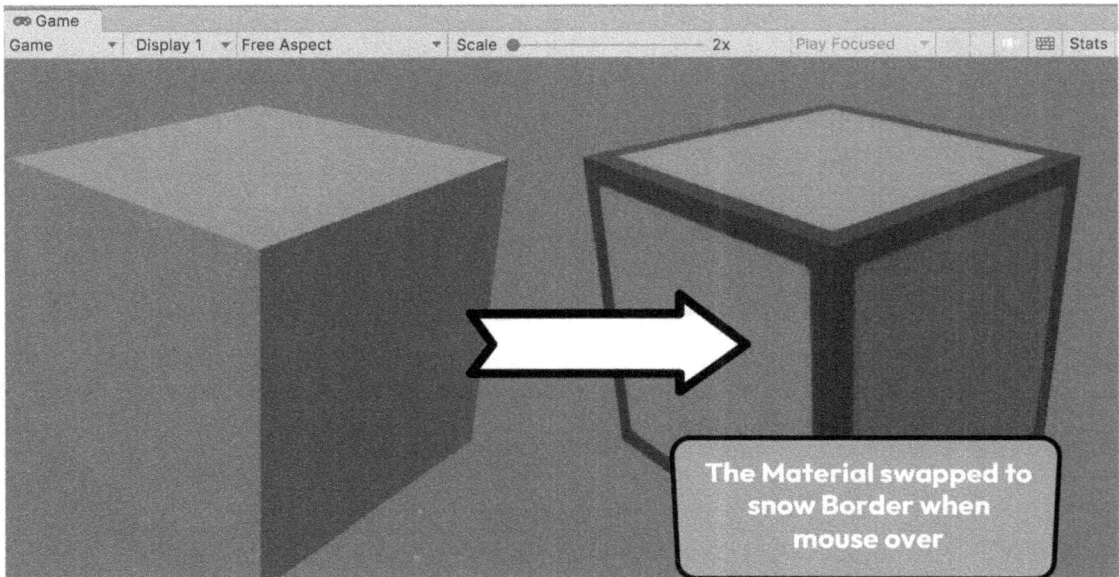

Figure 5.20: The interactive cube rollover highlight effect we'll create in this recipe

Getting ready

For this recipe, you'll need an image of a white square with a black border. We have provided such an image, called border.png, in the 05_05 folder.

How to do it...

To highlight a material at mouseover, follow these steps:

1. Create a new 3D project.
2. Import the border.png image into your project.
3. Create a **cube** in the scene (**GameObject | 3D Object | Cube**).
4. In the **Project** panel, create a folder named Materials, and inside this, create a new **Material** asset, named m_cube. Set its **Albedo** color to red.

5. Duplicate your material, naming the copy m_cubeHighlighted. Set the **Albedo** image for this new material to the **border** option:

Figure 5.21: Assigning the border image to the Albedo texture for the Material m_cubeHighlighted

6. In the **Hierarchy** panel, select the **cube** GameObject and assign it the m_cube material. To do so, drag the asset file from the **Project** panel onto the GameObject.

7. Create a new C# script class called MouseOverSwap and add an instance object as a component to the cube:

```csharp
using UnityEngine;

public class MouseOverSwap : MonoBehaviour
{
    public Material mouseOverMaterial;
    private Material _originalMaterial;
    private MeshRenderer _meshRenderer;

    void Awake()
    {
        _meshRenderer = GetComponent<MeshRenderer>();
        _originalMaterial = _meshRenderer.sharedMaterial;
    }

    void OnMouseOver()
    {
        _meshRenderer.sharedMaterial = mouseOverMaterial;
    }
}
```

```
void OnMouseExit()
{
  _meshRenderer.sharedMaterial = _originalMaterial;
}
}
```

8. Ensure that **Cube** is selected in the **Hierarchy** panel. Then, drag the m_cubeHighlighted asset file from the **Project** panel into the **Mouse Over Material** variable slot, in the **Inspector** panel of the **Mouse Over Swap (Script)** component:

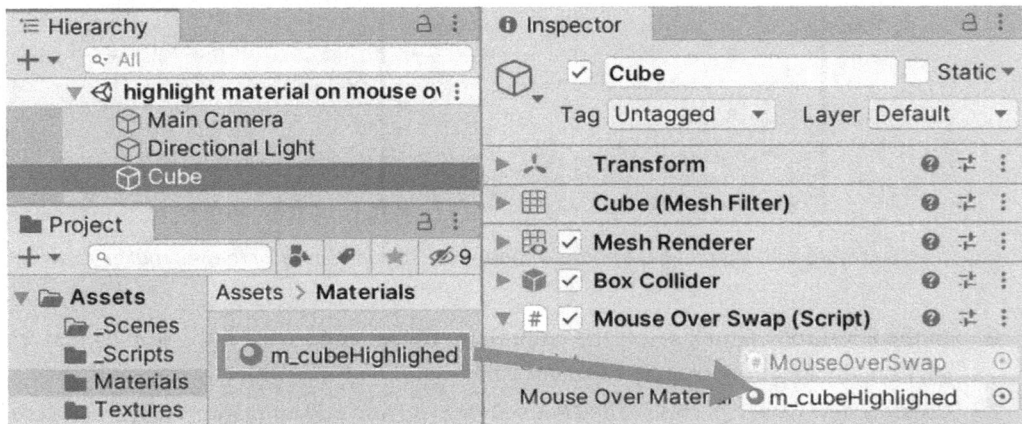

Figure 5.22: Assigning the highlighted material to the public scripted component variable of the cube

9. Run the scene. **The Cube** will be red initially; then, when the mouse hovers over it, you'll see a black border on each edge.

> If the main camera isn't showing much of the cube, try this trick. Arrange the **Scene** panel so that you can see the cube clearly, then single-click the main camera in the **Hierarchy** to select the GameObject, and click **GameObject | Align With View**. The main camera should now show the cube just as you see it in the **Scene** panel!

How it works...

The Awake() method performs two actions:

*   Stores a reference to the mesh renderer component in the _meshRenderer variable
*   Stores a reference to the original material of the GameObject in the _originalMaterial variable

The cube is automatically sent the mouse enter/exit events as the user moves the mouse pointer over and away from the part of the screen where the cube is visible. Our code has added a behavior to the cube to change the material when these events are detected.

We can change the material for the cube by changing which material is referred to by the `sharedMaterial` property of the mesh renderer GameObject.

When the `OnMouseOver` message is received, the method with that name is invoked, and the GameObject's material is set to `mouseOverMaterial`. When the `OnMouseExit` message is received, the GameObject's material is returned to `_originalMaterial`.

> **Note:** If the material of a GameObject is shared by several objects, we must be careful when changing the material properties so that we only change those we want to. If we wish to only change the values of a particular GameObject, we can use the `.material` property of `Renderer`, since a separate clone is created if there is more than one object that uses the same material. If we want all GameObjects using the same material to be affected by changes, we should use the `.sharedMaterial` property of `Renderer`. Since there was only one GameObject in this recipe, either could have been used.
>
> Read more at `https://docs.unity3d.com/ScriptReference/Renderer-material.html`.

There's more...

Here are some ways to enhance this recipe.

A collider needed for custom meshes

In this recipe, we created a primitive cube, which automatically has a **box collider** component. If you were to use the preceding script with a custom 3D mesh object, ensure the GameObject has a **Physics | Collider** component so that it will respond to mouse events.

Changing the material's color in response to mouse events

Another way to indicate that a GameObject can be interacted with by using the mouse is to just change the **Albedo** color of the material, rather than swapping to a new material. To illustrate this, we can have one color for a mouseover and a second color for a mouse click.

Do the following:

1. Remove the scripted `MouseOverSwap` from the cube GameObject.

2. Create a new C# script class, called `MouseOverDownHighlighter`, and add an instance object as a component of the cube:

```csharp
using UnityEngine;

public class MouseOverDownHighlighter : MonoBehaviour {
  public Color mouseOverColor = Color.yellow;
  public Color mouseDownColor = Color.green;
```

```csharp
private Material _originalMaterial;
private Material _mouseOverMaterial;
private Material _mouseDownMaterial;
private MeshRenderer _meshRenderer;

private bool _mouseOver = false;

void Awake()
{
  _meshRenderer = GetComponent<MeshRenderer>();
  _originalMaterial = _meshRenderer.sharedMaterial;
  _mouseOverMaterial = NewMaterialWithColor(mouseOverColor);
  _mouseDownMaterial = NewMaterialWithColor(mouseDownColor);

}

void OnMouseEnter()
{
  _mouseOver = true;
  _meshRenderer.sharedMaterial = _mouseOverMaterial;
}

void OnMouseDown()
{
  _meshRenderer.sharedMaterial = _mouseDownMaterial;
}

void OnMouseUp()
{
    if (_mouseOver) {
       OnMouseEnter();
    } else {
       OnMouseExit();
    }
}


void OnMouseExit()
{
  _mouseOver = false;
```

```
        _meshRenderer.sharedMaterial = _originalMaterial;
    }

    private Material NewMaterialWithColor(Color newColor)
    {
      Material material = new Material(_meshRenderer.sharedMaterial);
      material.color = newColor;

      return material;
    }
}
```

There are two public **colors:** one for a mouseover and one for mouse-click highlighting.

3. Run the **scene.** You should now see different highlight colors when the mouse pointer is over
    the cube **GameObject,** and when you click the mouse button when the mouse pointer is over
    the cube **GameObject.**

Since we are creating two new **materials,** the reusable NewMaterialWithColor() C# method is included here to simplify the content of the Start() method. A boolean (true/false) variable has
been introduced so that the correct behavior occurs once the mouse button is released, depending
on whether the mouse pointer is still over the object (mouseOver = true) or has moved away from
the object (mouseOver = false).

Fading the transparency of a material

A feature of many games is for objects to fade away so that they're invisible, or so that they appear gradually until they're fully visible. Unity provides a special **rendering mode** called **Fade** for this purpose.

In this recipe, we will create an object that, once clicked, fades out and disappears. We'll also look at
how to enhance the code so that it takes the GameObject's own initial alpha value into account. This
will make it self-destruct and destroy the GameObject completely when the object has faded out.

Getting ready

For this recipe, we have provided the script in the 05_06 folder.

How to do it...

Follow these steps:

1. Create a new Unity 3D project.
2. Create a sphere by choosing the menu **GameObject | 3D Object | Sphere.**
3. In the **Hierarchy** panel, you can select the **Sphere** GameObject and see its collider component
    in the **Inspector** panel. All the 3D primitives such as cubes and spheres are created with ap-
    propriate colliders.

> If you are using a custom 3D object in this recipe, you'll have to add a collider. To do this, in the **Inspector** panel, go to **Add Component | Physics | Mesh Collider.**

4. In the **Project** panel, create a folder named Materials, and in this folder, create a new material named m_fade.

5. With the m_fade asset file selected in the **Project** panel, change its rendering mode to **Fade in the Inspector panel:**

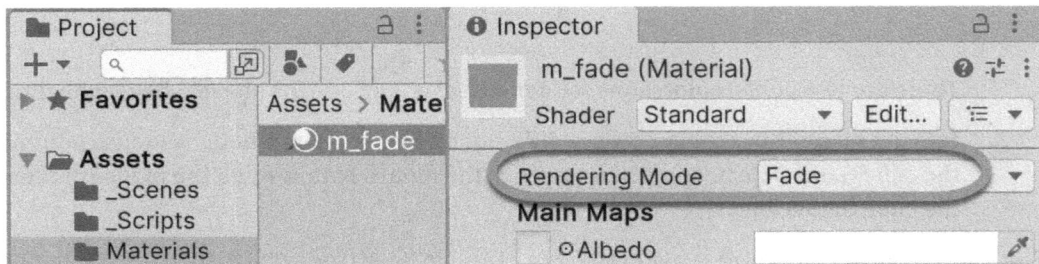

Figure 5.23: Setting the Rendering Mode of a material to Fade

> The **Fade** rendering mode is specifically designed for situations such as this recipe.
>
> Other rendering modes, such as **Transparent**, will turn the **Albedo** color transparent, but not the specular highlights or the reflections, in which case the object will still be visible.

6. Apply the m_fade material asset file to the sphere GameObject by dragging it from the **Project** panel into the sphere GameObject in the **Hierarchy** panel.

7. Create a new C# script class called FadeObject.cs, containing the following:

```
using UnityEngine;

public class FadeObject: MonoBehaviour {
  public float fadeDurationSeconds = 1.0f;
  public float alphaStart = 1.0f;
  public float alphaEnd = 0.0f;
  private float startTime;
  private MeshRenderer meshRenderer;
  private Color fadeColor;
  private bool isFading = false;

  void Awake () {
```

```
    meshRenderer = GetComponent<MeshRenderer>();
    fadeColor = meshRenderer.material.color;
    UpdateMaterialAlpha(alphaStart);
  }

  void Update() {
    if (isFading)
      FadeAlpha();
  }

  void OnMouseUp() {
    StartFading();
  }

  private void StartFading() {
    startTime = Time.time;
    isFading = true;
  }

  private void FadeAlpha() {
    float timeFading = Time.time - startTime;
    float fadePercentage = timeFading / fadeDurationSeconds;
    float alpha = Mathf.Lerp(alphaStart, alphaEnd, fadePercentage);
    UpdateMaterialAlpha(alpha);

    if (fadePercentage >= 1) {
      isFading = false;
    }
  }

  private void UpdateMaterialAlpha(float newAlpha) {
    fadeColor.a = newAlpha;
    meshRenderer.material.color = fadeColor;
  }
}
```

There are many parts to this code. A fully commented listing of this code is available in the folder for this chapter. See the *Technical requirements* section in the preface.

8. Add an instance of the C# script class `FadeObject.cs` as a component to the sphere GameObject.

9. Save and play your **scene**. When you mouse-click the sphere GameObject, it will start to fade away.

How it works...

The opaqueness of the material that's using a **Fade Shader** is determined by the alpha value of its main **color**. This recipe is based on changing the `Alpha` value of the color of the mesh renderer.

There are three public variables:

- `fadeDurationSeconds`: The time we want our fading to take, in seconds
- `alphaStart`: The initial `Alpha` (transparency) value with which we want the GameObject to start (1 = fully visible and 0 = invisible)
- `alphaEnd`: The `Alpha` value into which we want to fade the GameObject

The `UpdateMaterialAlpha(...)` method updates the `Alpha` value of the GameObject's `color` object with the given value by updating the `Alpha` value of the `fadeColor` `color` variable and then forcing the mesh renderer material to update its **color** value to match those in `fadeColor`.

When the scene begins, the `Awake()` method caches a reference to the Mesh renderer component (the `meshRenderer` variable), and also the `Color` object of the material of the mesh renderer (the `fadeColor` variable). Finally, the GameObject's `Alpha` variable is set to match the value of the `alphaStart` variable, which it does by invoking the `UpdateMaterialAlpha(...)` method.

The `OnMouseUp()` method is invoked when the user clicks the GameObject with their mouse. This invokes the `StartFading()` method.

> The actions to start fading weren't put in this method, since we may also wish to start fading due to some other events (such as keyboard clicks, a timer hitting some value, or an NPC going into a mode such as dying). So, we separated the logic that detects that the event we are interested in has taken place from the logic for the actions we wish to perform – in this case, to start the fading process.

The `StartFading()` method records the current time, since we need that to know when to finish fading (the time when we started fading, plus `fadeDurationSeconds`). Also, it sets the `isFading` boolean flag to `true` so that logic elsewhere related to fading will know it's time to do things.

The `Update()` method, which is called each frame, tests whether the `isFading` flag is `true`. If it is, the `FadeAlpha()` method is invoked for each frame.

The `FadeAlpha()` method is where the majority of our alpha-fading logic is based:

- `timeFading` is calculated: The time since we started fading.
- `fadePercentage` is calculated: How far we are from the start (0) to the finish (1) of our fading.
- `Alpha` is calculated: The appropriate `Alpha` value for our fade percentage, using the `Lerp(...)` method to choose an `intermedia` value based on a `0..1` percentage.

- The UpdateMaterialAlpha(...) method with the new Alpha value.
- If fading has finished (fadePercentage >= 1), we set the isFading boolean flag to false to indicate this.

There's more...

Here are some ways to enhance our fading features.

Destroying objects when fading is complete

If fading a GameObject to invisible is how we communicates to the player that it is leaving the scene (for example, something is completed or dying), then we may want that GameObject to be destroyed after the fading process is completed. Let's add this feature to our code.

Do the following:

1. Add a new public boolean variable to our script (default to false):

```
public bool destroyWhenFadingComplete = true;
```

2. Add a new EndFade() method that sets isFading to false and then tests whether the public destroyWhenFadingComplete variable was set to true and, if so, destroys the GameObject:

```
private void EndFade() {
    isFading = false;

    if(destroyWhenFadingComplete) {
        Destroy (gameObject);
    }
}
```

3. Refactor the FadeAlpha() method so that it invokes EndFade() when the fading is complete (fadeProgress >= fadeDurationSeconds):

```
private void FadeAlpha()
{
    float fadeProgress = Time.time - startTime;
    float alpha = Mathf.Lerp(alphaStart, alphaEnd, fadeProgress
        / fadeDurationSeconds);
    UpdateMaterialAlpha(alpha);

    if (fadeProgress >= fadeDurationSeconds) {
        EndFade();
    }
}
```

Using the GameObject's alpha as our starting alpha value

It may be that the game designer has set the Alpha value of a GameObject in the **Inspector** panel to the initial value they want. So, let's enhance our code to allow this to be indicated, by checking a public boolean flag variable in the **Inspector** panel and adding code to read and use the GameObject's alpha if that option is chosen.

Do the following:

1. In the **Inspector** panel, click the **color picker** for the **Albedo** material and set the Alpha value to something other than 255 (for example, set it to 43, which is almost transparent):

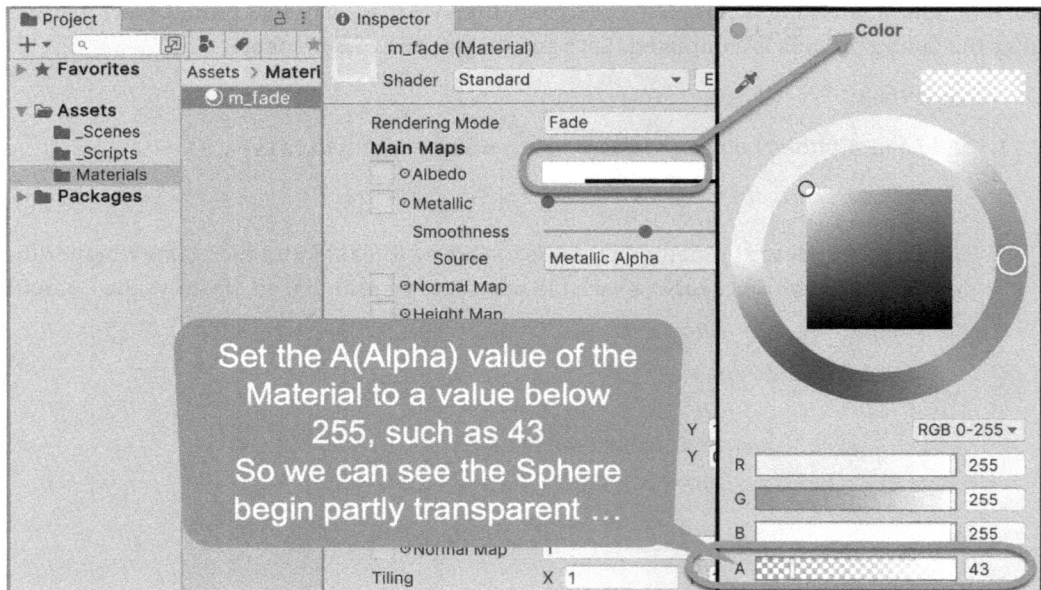

Figure 5.24: Setting the Alpha transparency value of a material

2. Add a new public boolean variable to our script (default to false):

```
public bool useMaterialAlpha = false;
```

3. Add logic to the Awake() method so that if this flag is true, we use the Alpha value of the **color** that was read from the GameObject's **material** as the **scene** begins (fadeColor.a):

```
void Awake () {
    meshRenderer = GetComponent<MeshRenderer>();

    // set object material's original color as fadeColor
    fadeColor = meshRenderer.material.color;

    // IF using material's original alpha value, THEN use
```

```
        //material's alpha value for alphaStart
    if (useMaterialAlpha) {
        alphaStart = fadeColor.a;
    }

    // start object's alpha at our alphaStart value
    UpdateMaterialAlpha(alphaStart);
}
```

Using a coroutine for our fading loop

Where possible, we should avoid adding code to the Update() method since this is invoked every frame, which means it can reduce the performance of our games, especially if many objects have scripted components with Update() methods, all testing flags every frame.

One very effective solution is to invoke a **coroutine** when we want some actions to be performed over several frames. This is because a coroutine can perform some actions, then **yield** control back to the rest of the **scene**, and then resume its actions from where it left off, and so on, until its logic is completed.

Do the following:

1. Remove the Update() method.

2. Add a new using statement at the top of the script class, since coroutines return an IEnumerator value, which is part of the System.Collections package:

    ```
    using System.Collections;
    ```

3. Add a new method:

    ```
    private IEnumerator FadeFunction() {
        while (isFading)
        {
            yield return new WaitForEndOfFrame();
            FadeAlpha();
        }
    }
    ```

4. Refactor the StartFading() method so that it starts our coroutine:

    ```
    private void StartFading() {
        startTime = Time.time;
        isFading = true;
        StartCoroutine(FadeFunction());
    }
    ```

That's it – once the coroutine has been started, it will be called each frame until it completes its logic, temporarily suspending its execution each time a yield statement is executed. See the *Further reading* section for more information about Unity coroutines.

Further reading

This section contains some sources that provide more information about the topics that were covered in this chapter.

Double Prism offers a nice introduction to the differences between textures and materials:

* `https://2xp-studio.com/materials-and-textures-in-3d-graphics-differences/`

Learn more about the different types of asset files Unity uses here:

* `https://docs.unity3d.com/Manual/AssetTypes.html`

Every game needs textures. Here are some sources of free textures that are suitable for many games:

* textures.com: `https://www.textures.com/`
* GAMETEXTURES: `https://gametextures.com/`
* Free Stock Textures: `https://freestocktextures.com/`

Take a look at the following links to import models into Unity:

* Importing models: `https://docs.unity3d.com/Manual/ImportingModelFiles.html`
* Limitations of importing models: `https://docs.unity3d.com/Manual/HOWTO-ImportObjectsFrom3DApps.html`
* Maxon's Free Cinema3D Importer at the Asset Store: `https://assetstore.unity.com/packages/tools/integration/cineware-by-maxon-158381`

Learn more about using coroutines in Unity:

* `https://docs.unity3d.com/Manual/Coroutines.html`
* `https://learn.unity.com/tutorial/coroutines`

Learn more on Discord

To join the Discord community for this book – where you can share feedback, ask questions to the author, and learn about new releases – follow the QR code below:

`https://packt.link/unitydev`

6

Creating 3D Environments with Terrains

3D games need 3D objects in their scenes! In this chapter, we will explore the powerful Unity terrain tools to create landscapes with a range of characteristics. Through texture and height painting, and the addition of trees and details such as grass and plants, complete environments can be modeled in a Unity scene. We'll also explore how to use the new visually realistic water modeling features available for **High Definition Render Pipeline** (HDRP) projects.

In this chapter, we will cover the following recipes:

- Creating and texture-painting terrains
- Unity terrain tools and samples for powerful height-painting
- Adding terrain holes
- Adding trees and vegetation
- Realistic water features for HDRP projects

Creating and texture-painting terrains

In this recipe, we will introduce terrains by creating, sizing, positioning, and **texture-painting** (with an image) a flat **terrain**. Later recipes will enhance this terrain with heights, trees, and other details.

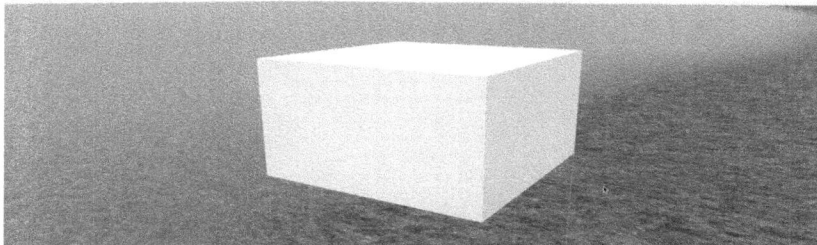

Figure 6.1: Our flat terrain texture painted with sand and grass around a cube

Getting ready

You'll need at least two different images to texture-paint the terrain in this recipe. In the folder 06_01, we have provided two free terrain textures (**SandAlbedo** and **GrassHillAlbedo** .PSD Photoshop asset files) from the legacy Unity sample assets.

How to do it...

To create a **terrain**, follow these steps:

1. Create a new 3D project, and import the two terrain textures.

2. Add a **cube** to the scene by choosing the menu: **GameObject | 3D Object | Cube**. In the **Inspector** panel, set its **Position** to (0, 0, 0).

3. Add a terrain to the scene by choosing the menu: **GameObject | 3D Object | Terrain**. In the **Inspector** panel, set its **Position** to (0, 0, 0).

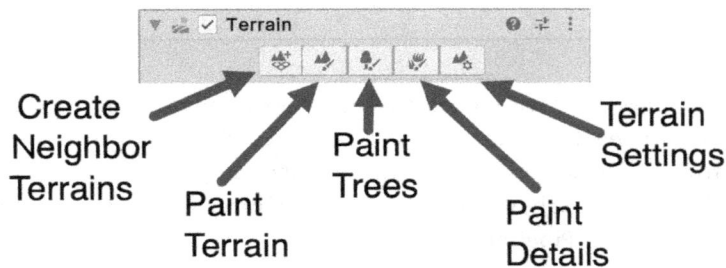

Figure 6.2: The five terrain tools in the Inspector

4. When a terrain is selected in the **Hierarchy** window, you'll see the icons for the five terrain tools in the **Terrain** component in the **Inspector** panel.

5. You'll notice that the cube appears at one of the corners of the terrain! This is because Unity terrains are positioned by a corner and not their center, as with 3D primitives such as cubes.

6. We'll be setting the size of this **terrain** to **100 x 100**. Then, to make the center of the terrain (0, 0, 0), we need to position it at (-width/2, 0, -length/2).

7. Click the **Terrain Settings** icon (on the right, as shown in Figure 6.2). Then, for the **Mesh Resolution** properties of this component, set both **Terrain Width** and **Terrain Length** to 100:

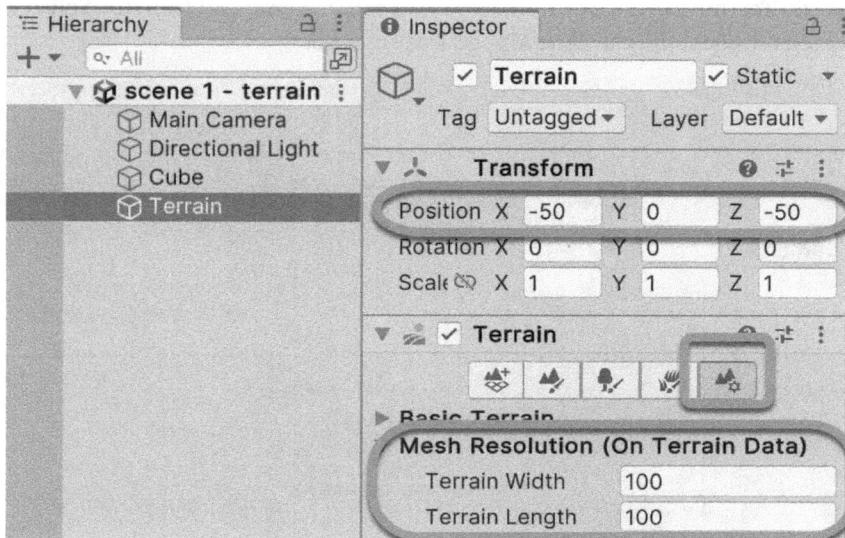

Figure 6.3: Terrain settings for both a width and length of 100

8. Since the terrain's **Transform Position** is (-50, 0, -50), you should see that our cube at (0, 0, 0) sits in the center of our **terrain**.

9. One of the most powerful ways to make terrains feel like natural outdoor surfaces is by texturing different parts of them with environmental images such as sand, grass, stones, moss, and so on. The next few steps will do this with some of the **Standard Assets** textures we imported at the beginning of this recipe.

10. In the **Inspector** panel, select the **Paint Terrain** tool (the second icon, which looks like mountains and a paintbrush). Ensure **Paint Texture** is selected in the drop-down menu. Then, click the **Edit Terrain Layers...** button, choose **Create Layer...**, and from the **Select Texture2D** panel, locate and select the the **SandAlbedo** texture asset file.

Figure 6.4: Adding SandAlbedo as the first terrain layer

11. When there is only a single terrain layer, this texture will be painted over the entire terrain. Follow the same steps you did previously to add a second terrain layer, this time choosing GrassHillAlbedo.

12. Ensure your GrassHillAlbedo layer is selected, and select the second brush (white-centered but with a fuzzy outline). Set **Brush Size** to 20. Now, use your brush to paint half of the terrain with grass. You may wish to increase the brush size to speed up the painting of large areas.

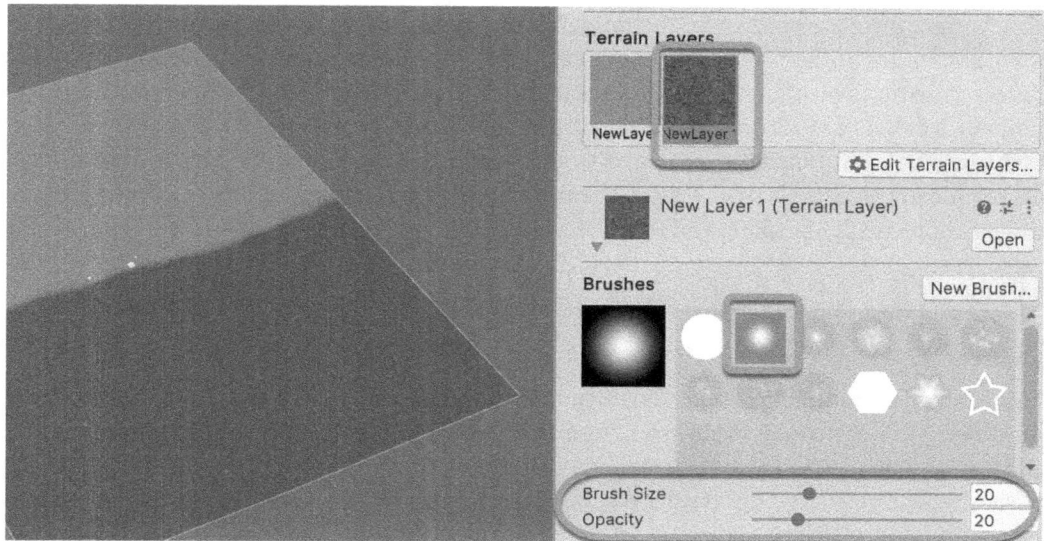

Figure 6.5: Setting the gradient-edged grass brush's size to 20

13. Save your scene.

14. Your scene should now contain a large, flat terrain that is half covered in sand and half in grass, meeting near the cube at the center.

How it works...

Terrains are sophisticated 3D object meshes in Unity. Since the position of terrains is based on a corner, if we want the terrain to be centered on a particular point such as (0, 0, 0), we have to subtract half the width from the X-coordinate and half the length from the Z-coordinate. So, for our 100 x 100 terrain, we set its corner position to (-50, 0, -50) so that its center is (0, 0, 0), which is where our **cube** was located.

Textures are painted onto terrains through terrain layers. When there is only one terrain layer, the texture is used to paint the entire terrain, as it did when we set the first terrain layer to the SandAlbedo image. Once there are two or more terrain layers, we can use the brushes to paint different parts of the terrain with different textures.

The "softer" the edge of a brush, the more mixing there is between the texture that was present and the new texture of the brush being applied. The softer outline brushes create soft edges where two different textures meet on a painted terrain. In addition, reducing the **Opacity** of a brush below 100 percent will also reduce the weighting of the new brush texture compared to the existing terrain texture. When painting with vegetation such as grass and moss, it is effective to simulate where the vegetation at the edges of the rocky areas and fully green areas begin. The opacity of a brush affects the *mix* of textures all over the **brush** area.

Unity terrain tools and samples for powerful height-painting

Using a combination of textures and terrains with variations in height enables sophisticated and realistic physical environments to be created. In this recipe, we'll build on the previous texture-painted terrain by varying the height to create an interesting landscape in our scene. To make things a bit easier, we'll import the terrain samples, which provide a range of useful textures and height brushes for common terrain features.

Figure 6.6: We'll create a circle hill, bordered by mountains, painted with sand and grass, with a canyon entrance

Getting ready

This recipe builds on the previous one, so make a copy of that and work on this copy.

How to do it...

To use Unity terrain tools and samples for powerful height-painting, follow these steps:

1. Open your copy of the Unity project from the previous recipe.

2. Open the **Asset Store**, ensure you are logged in to your Unity account, and locate the *Terrain Sample Asset Pack*.

> You can find the *Terrain Sample Asset Pack* at `https://assetstore.unity.com/packages/3d/environments/landscapes/terrain-sample-asset-pack-145808`

3. In the Asset Store, click the **Add to My Assets** button to link these assets to your Unity account.

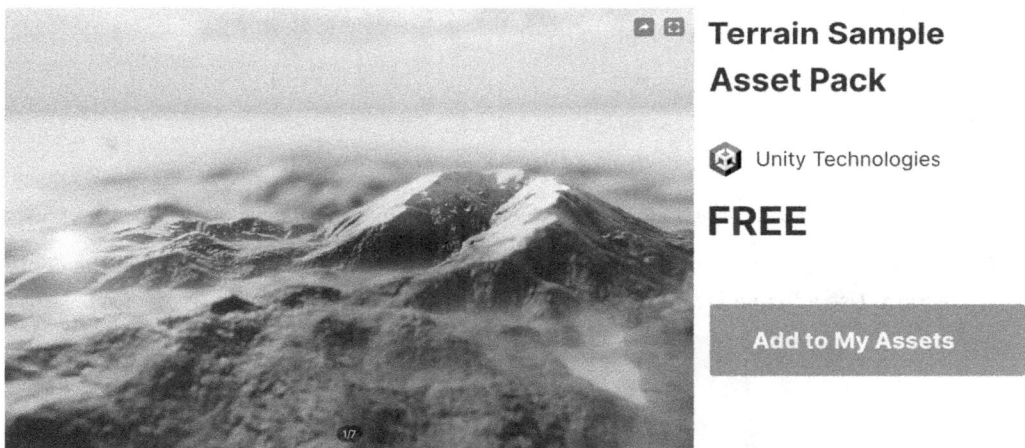

Figure 6.7: Adding the Terrain Sample Asset Pack to My Assets in the Asset Store

4. In Unity, open the **Package Manager,** selecting **My Assets,** In Unity, open the Package Manager and select My Assets from the Package dropdown at the top-left of the panel.

5. In the **Package Manager** panel, locate the *Terrain Sample Asset Pack.* **Download** and **Import** these assets. Note: you may have to agree to install/update dependent packages when importing.

6. Next, still in the **Package Manager**, switch from listing My Assets to listing the packages from the Unity Registry instead, then install the Terrain Tools package list the packages from the **Unity Registry,** and install the **Terrain Tools** package.

7. Ensure the Terrain GameObject is selected in the **Hierarchy** panel. Then, in the **Inspector** panel, select the **Paint Terrain** tool (the second icon, which looks like mountains and a paintbrush).

8. Choose **Set Height** from the drop-down menu. Then, in the **Set Height Controls**, set the **Height** to 2.

9. Choose the first brush (the large white circle with a well-defined border).

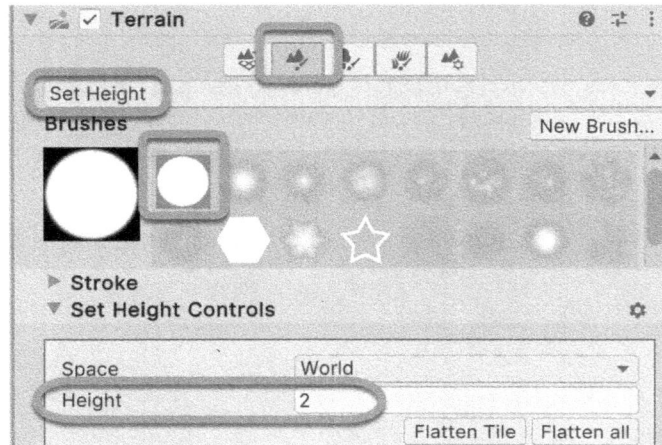

Figure 6.8: Choosing the Set Height option from the terrain tools in the Inspector panel

10. In the **Scene** panel, paint a raised area with a **Height** of 2 around the cube at the center of the terrain. Use a **Brush Size** of around 10 (in the **Stroke** options in the **Inspector**).

Figure 6.9: The raised area around our cube

11. Although it isn't a very natural terrain shape, it nicely illustrates how we can use a brush to raise the height to a given limit.

> If we want to remove raised areas, we can either hold down the *Ctrl* key when clicking the mouse to lower them, or we can set the height to 0 and then height-paint with the brush.

12. Let's smooth those shape-raised edges. From the drop-down menu in the **Inspector** panel, immediately below the five terrain tool icons, select **Smooth Height**. Now, paint with the same brush – the steep edges of our raised circle around the cube should be smoothed into more natural slopes.

Figure 6.10: Choosing the Smooth Height option for terrain painting

13. Texture-paint (the **Paint Texture** drop-down) all the raised areas with yellow sand.

14. Next, texture-paint all the low areas inside the circle with Grass_Moss_TerrainLayer. To do this, add a new **Terrain Layer**, and select the texture Grass_Moss_TerrainLayer. This is one of the samples we added from the **Package Manager**. We have made an interesting location for our cube, which might be a special pickup or location in a game.

15. Finally, let's reduce the chance of our player falling off the edge of the world by creating a range of spiky high mountains all around the edge of our terrain. From the drop-down menu in the **Inspector** panel, choose the **Raise or Lower Terrain** option.

16. Choose an interesting brush and add one or two mountains to the edge of the terrain. The "canyon" ones are fun to work with. When you move the mouse over the terrain, you'll see a "ghostly" preview of how the terrain will be raised if you click the mouse. You can modify the brush in the stroke settings in the Inspector: **Stroke Strength** (opacity), **Stroke Size**, and **Stroke Rotation**.

> One of the usability benefits of the **Terrain Tools** package you added earlier is that there are three keyboard shortcuts that, combined with the mouse, allow you to change these three properties without having to drag slides in the **Inspector** panel. Press *A* and mouse drag to change the stroke strength (you'll see the height of the preview rise and lower). Press *S* and mouse drag to change the stroke size. Press *D* and mouse drag to change the stroke rotation. Combining different values of these three properties means a single brush can have quite different effects on a terrain.

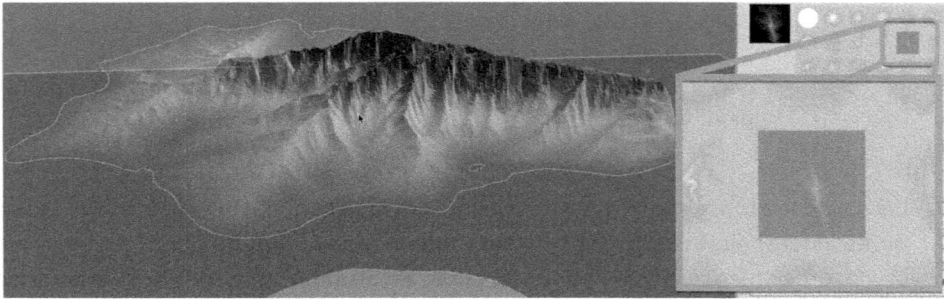

Figure 6.11: Two very different effects from the same brush with different stroke properties

17. Just using one or two brushes adds an interesting mountain range around the edges of the terrain, by varying the properties as you use the brushes.

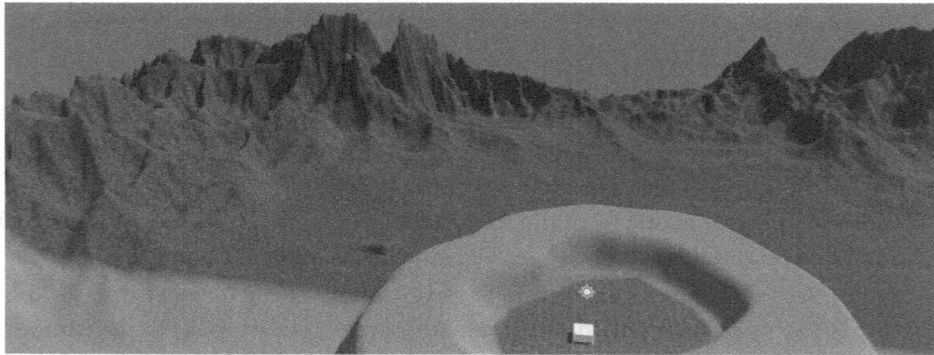

Figure 6.12: Our Cube in the middle of a terrain bordered by spiky mountains

18. Finally, let's add a way to walk from the main terrain flat area into the circle hill around our cube. Hold down the *Ctrl* key and mouse-click over one part of the circle hill, lowering the terrain for the canyon shape. You should now have an entrance into the circle around the cube.

Figure 6.13: A lowered canyon entrance into the circle around our cube

How it works...

In this recipe, we explored three height painting tools. The first areas were raised straight up to a height of 2 using the **Set Height** tool. The **Set Height** tool is useful to quickly raise or lower areas to a specific height. However, it leaves very steep walls between areas of different heights. It's also great for returning areas of the terrain back to zero.

The second tool we used was **Smooth Height**. This tool is used to finesse the final quality of the areas of the terrain where the height changes are too steep, so it is often used in combination with the **Set Height** tool.

Finally, we used the **Raise or Lower Terrain** height painting tool. This incrementally adds (or, with *Ctrl* pressed, subtracts) height to the terrain areas being painted upon. In combination with complex brushes, and varying brush properties, it allows us to quickly add complex details to areas of the terrain.

The terrain tools provide more control over the brushes and, perhaps most usefully, allow us to stay working in the **Scene** panel through the keyboard + mouse combinations to vary brush strength, size, and rotation. The **Terrain Sample Assets** provide a great starting point, with a range of useful environment textures and brushes.

There's more...

The terrain tools have also added some sophisticated ways to finesse your terrains by applying functions such as erosion, noise, and smudge transformations. Explore these through the drop-down menu of the **Paint Terrain** tool.

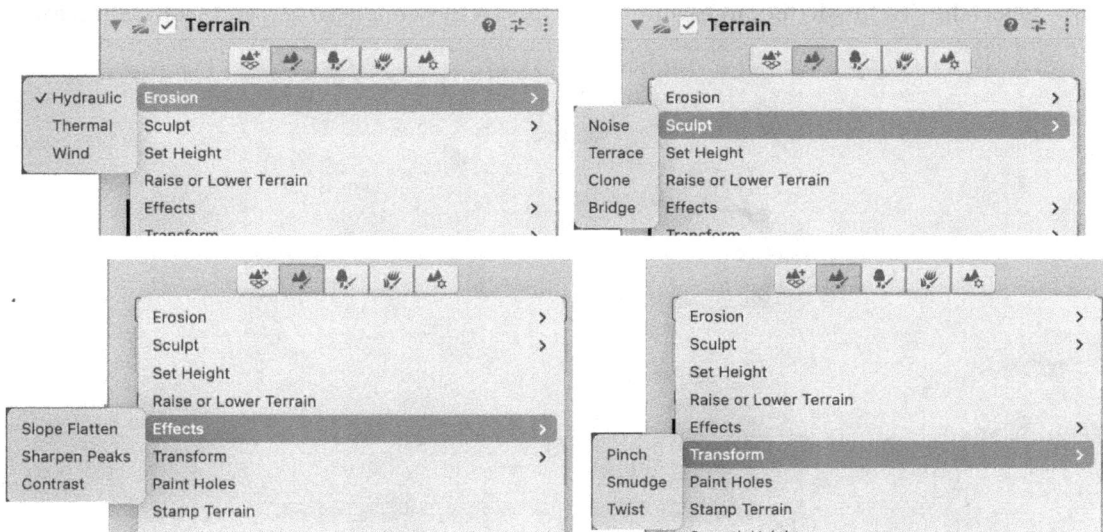

Figure 6.14: Sophisticated additional paint terrain transformations

Adding terrain holes

A limitation of terrains is that they are a single mesh, with points raised or lowered. Sometimes, we want more complex, concave geometry for effects such as caves, tunnels, or just simple holes in the terrain. In this recipe, we'll build on the previous one to add a terrain hole that leads to another piece of terrain.

Getting ready

This recipe builds on the previous one, so make a copy of that and work from it.

How to do it...

To paint and make use of a hole in a **terrain**, follow these steps:

1. Open your copy of the Unity project from the previous recipe.
2. Add a spotlight that shines on where the cube is sitting by choosing **menu: GameObject**. In the **Inspector** panel, set its **Position** to (0, 0.5, 0), **Range** to 50, **Spot Angle** to 125, and **Intensity** to 2. This will allow this light to light up a bonus level that we'll create beneath our **terrain**.
3. Delete the cube from the scene – in a full game, this cube might be removed when a goal has been achieved, revealing the entrance to the bonus level we are going to create.
4. Ensure the Terrain GameObject is selected in the **Hierarchy** panel. Then, in the **Inspector** panel, select the **Paint Terrain** tool (the second icon, which looks like mountains and a paintbrush).
5. From the drop-down menu, select **Paint Holes**.
6. Select the first brush (the large white circle with a well-defined border) and set **Brush Size** to 1:

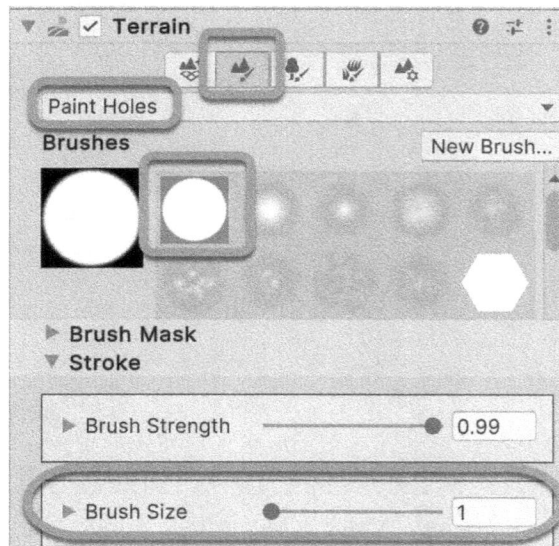

Figure 6.15: Paint Holes with a circle brush size of 1

7. Now, paint a hole in the **terrain** in the area where we see the white circle lit by our spotlight. This is the area that would have been beneath the cube we just deleted. Although the brush is a circle, Unity will remove rectangular faces from the terrain when making terrain holes. So you should end up with a jagged, straight-line-edged hole in the terrain mesh.

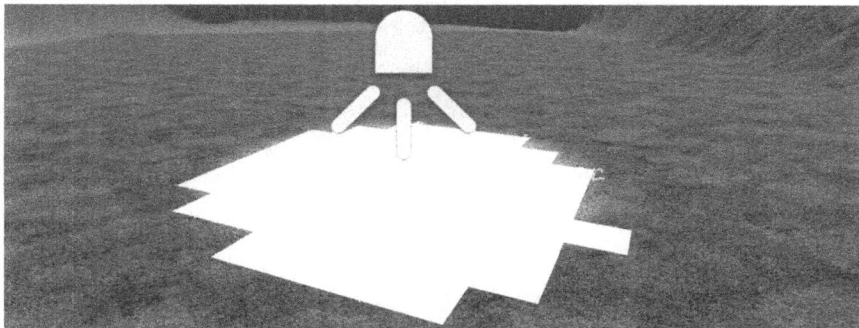

Figure 6.16: The ragged-edged hole in our terrain

8. Let's tidy up the edges of this hole. For speed, we'll use four squashed cubes. Create a **cube** named Cube-wall and scale it to (1.7, 1.7. 0.25). Make three more copies (Cube-wall2/3/4), rotating two of them by 90 degrees about the Y-axis (0, 90, 0). Move these squashed cubes to make a neat edge for the hole:

Figure 6.17: Four cubes squashed and positioned to make an entrance near the terrain hole

9. Now, let's create a simple, flat level beneath this hole. Create a **plane** named Cube-bonus-level (**GameObject | 3D Object | Plane**) and set its **Position** to (0, -10, 0) and **Scale** to (2, 2, 2).

10. With that, we have created a scene containing a **terrain**, and a neat hole in that **terrain** leading to a subterranean lit level where we can place some bonus items or a bonus challenge.

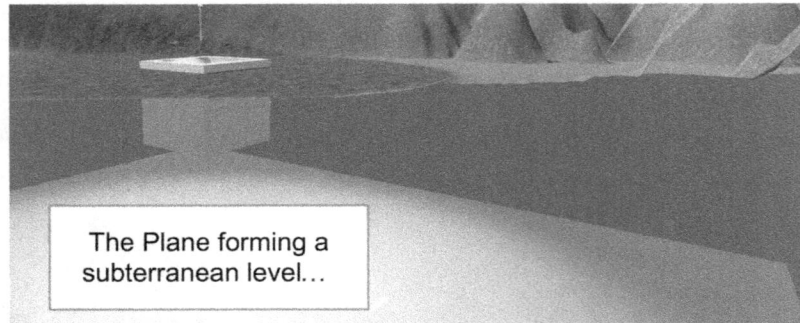

The Plane forming a subterranean level...

Figure 6.18: The bonus level cube and drop through the terrain hole, as seen from below

How it works...

As we have seen, holes can be painted in **terrain** meshes, as simply as adjusting height and texture-paint terrains. Adding holes opens up a wide range of features we can add to scenes, allowing us to make complex combinations of one or more terrains with other scene objects and geometry.

When creating holes in terrains, the edges may be unsightly, and in most games, they will need tidying up. In this recipe, we used four cubes positioned and scaled to form a neat entrance and channel from the surface of the terrain, down to the bonus Cube level. For a more natural entrance to a **terrain** hole, you could use a pre-existing 3D model or a custom mesh that's been textured using the **ProBuilder** package (see *Chapter 7, Creating 3D Geometry with ProBuilder*).

Adding trees and vegetation

As well as textures and heights, Unity provides tools to make it easy to "paint" trees and vegetation models onto a terrain. While, theoretically, you could just create tens, hundreds, or thousands of GameObjects from a tree or grass 3D model, animating them using a grass shader, it's much easier and faster to use the tree and vegetation detail painting tools for terrains. In this recipe, we'll work with two models; one is a model for a heather bush (since, for some reason, Unity does not provide any tree models with the **Terrain Sample Assets**), and the second model is of a kind of grass.

Figure 6.19: The flat entrance to our circle hill detail, painted with grass and plants

Getting ready

This recipe builds on the previous one, so make a copy of that and work with this copy.

How to do it...

To add trees and vegetation, follow these steps:

1. Open your copy of the Unity project from the previous recipe.
2. Ensure the Terrain GameObject is selected in the **Hierarchy** panel. Then, in the **Inspector** panel, select the **Paint Trees** tool (the middle icon, which looks like a tree and a paintbrush).
3. Click the **Edit Trees...** button, and then click **Add Tree**. In the dialog box, locate the Heather_A plant Prefab, and then click the **Add** button.

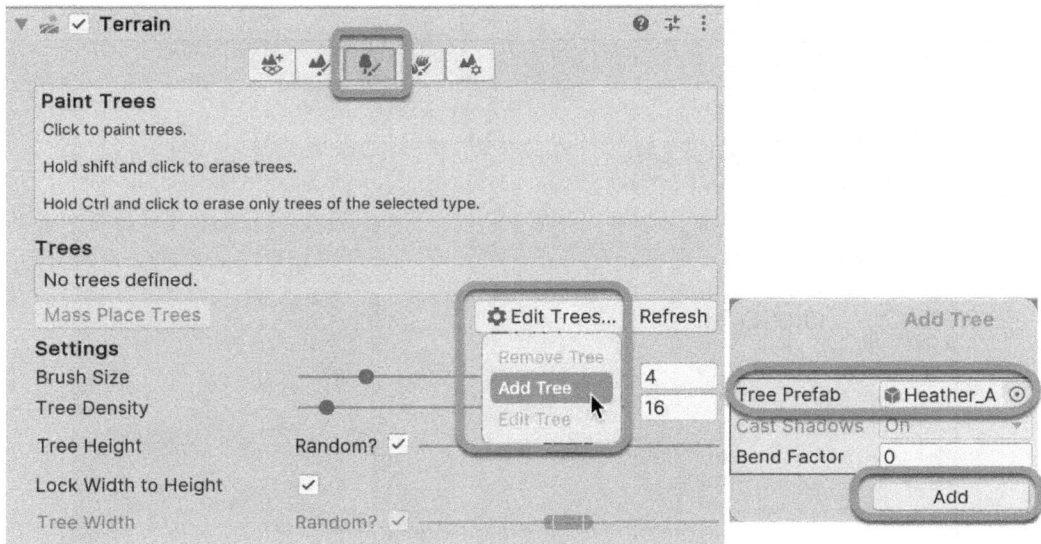

Figure 6.20: Adding the Heather_A prefab model as a tree for painting on the terrain

4. Now we have a tree added, we can use a brush to add trees to our terrain. Set the **Brush Size** to 4 and the **Tree Density** to 16. Use the brush to paint heather trees (bushes!) on the top of our circle-shaped hill. You can use the *Ctrl* key to delete painted trees.

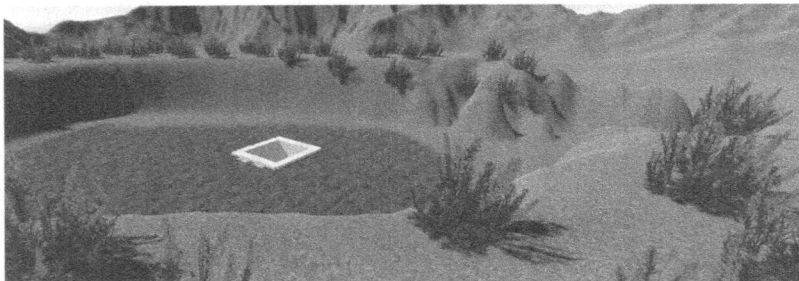

Figure 6.21: The top of our circle hill painted with heather "trees"

5. Now, select the **Paint Details** tool (the second-from-right icon that looks like grass and a paint-brush).

6. Click the + button to add new grass details. In the dialog box, locate the Grass_A plant prefab, and then click the **Add** button.

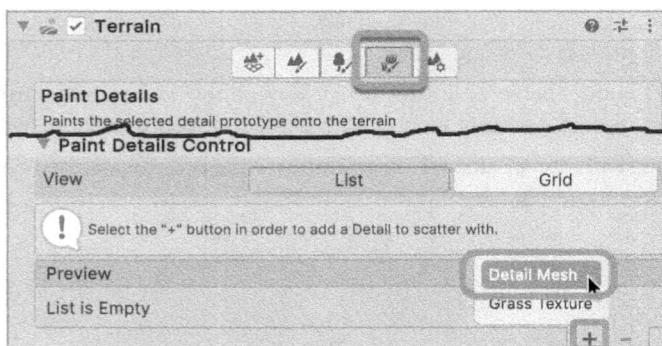

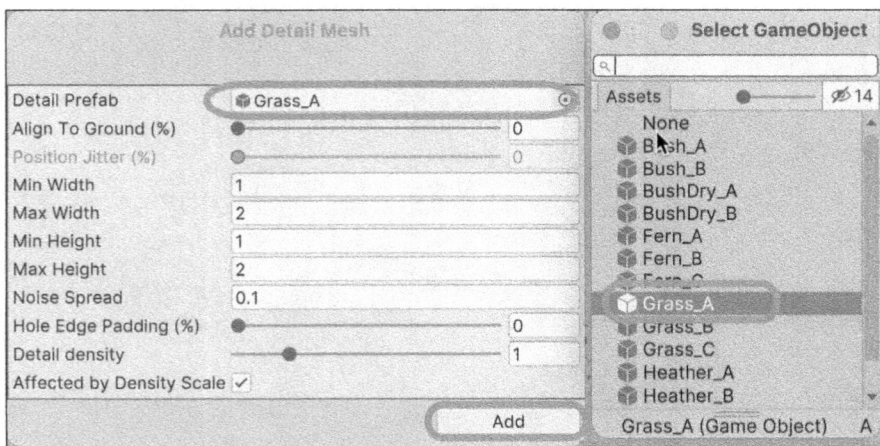

Figure 6.22: Adding the grass detail prefab for detail painting

7. Repeat the previous step to add the `Plant_A` plant prefab.

8. With both details' prefabs checked in the **Inspector** panel, paint this mixture of grass and plants in the flat entrance to the circle hill around our terrain hole.

How it works...

With prefabs, we can use the **Paint Trees** feature to make Unity randomly paint trees in areas covered by a brush. Likewise, rather than just having flat texture images of grass, moss, and so on, we can use the **Paint Details** feature to add one or more detail meshes (we added two: grass and a dried plant), and then randomly paint these models over areas of our terrain. With grass shaders making branches/fronds sway in the wind, trees and vegetation details add much realism and life to the outdoor scenes of our Unity games.

Realistic water features for HDRP projects

Some photorealistic and real-time animated visual effects are only possible when working with **HDRP** projects. Recent versions of the Unity Editor have added some fantastic features for animated, realistic water effects, which we'll explore with this HDRP project recipe.

Figure 6.23: Our swimming pool scene with real-time reflected sunlight

How to do it...

To create realistic water features for HDRP projects, perform the following steps:

1. Create a new 3D (HDRP) project.

2. Open the **OutdoorsScene** scene.

3. Open the **Project Settings** panel by choosing the menu **Edit | Project Settings**. Select the **Quality | HDRP** tab, and locate the **Water** section under **Rendering** in the settings. Check the **Enable** box and wait a few seconds. You have now enabled water rendering for the HDRP quality settings.

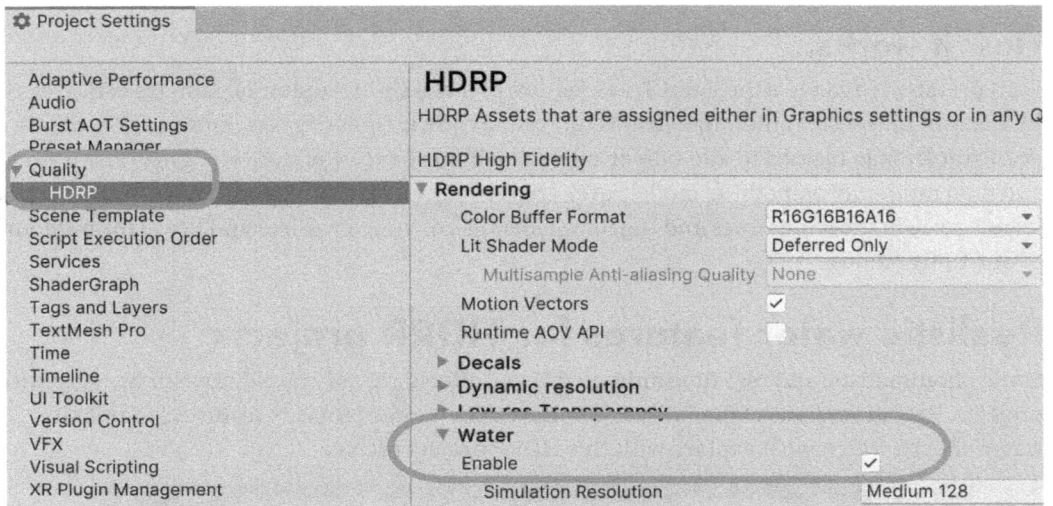

Figure 6.24: Enabling water rendering in the Quality | HDRP section of Project Settings

4. Next, in three places (!) in the **Project Settings** panel, you need to enable water rendering for the **Graphics | HDRP Global Settings**: for **Camera Rendering**, **Realtime Reflection Rendering**, and **Baked or Custom Reflection Rendering**. You have now enabled water rendering for the HDRP graphics settings.

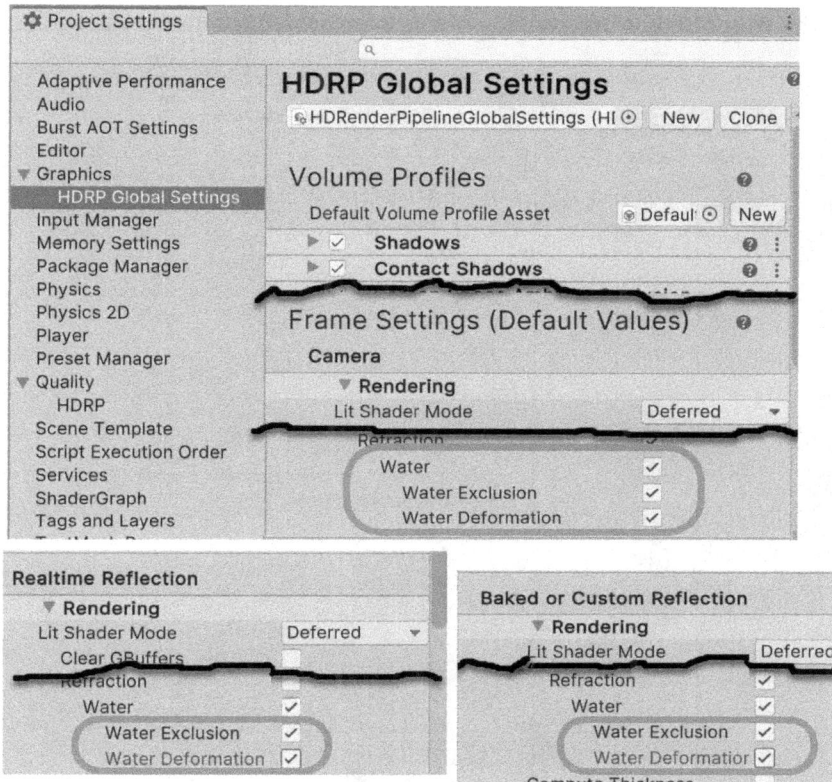

Figure 6.25: Enabling water rendering in the Graphics | HDRP Global Settings section of Project Settings

5. Select the **Sky and Fog Volume** GameObject in the **Hierarchy** panel.

6. In the **Inspector** panel, set the **Volume Ambient Mode** to **Dynamic**.

7. Click the **Add Override** button, and add a **Water Rendering** override.

8. For the **Water Rendering** override, set its **State** drop-down menu value to **Enabled.**

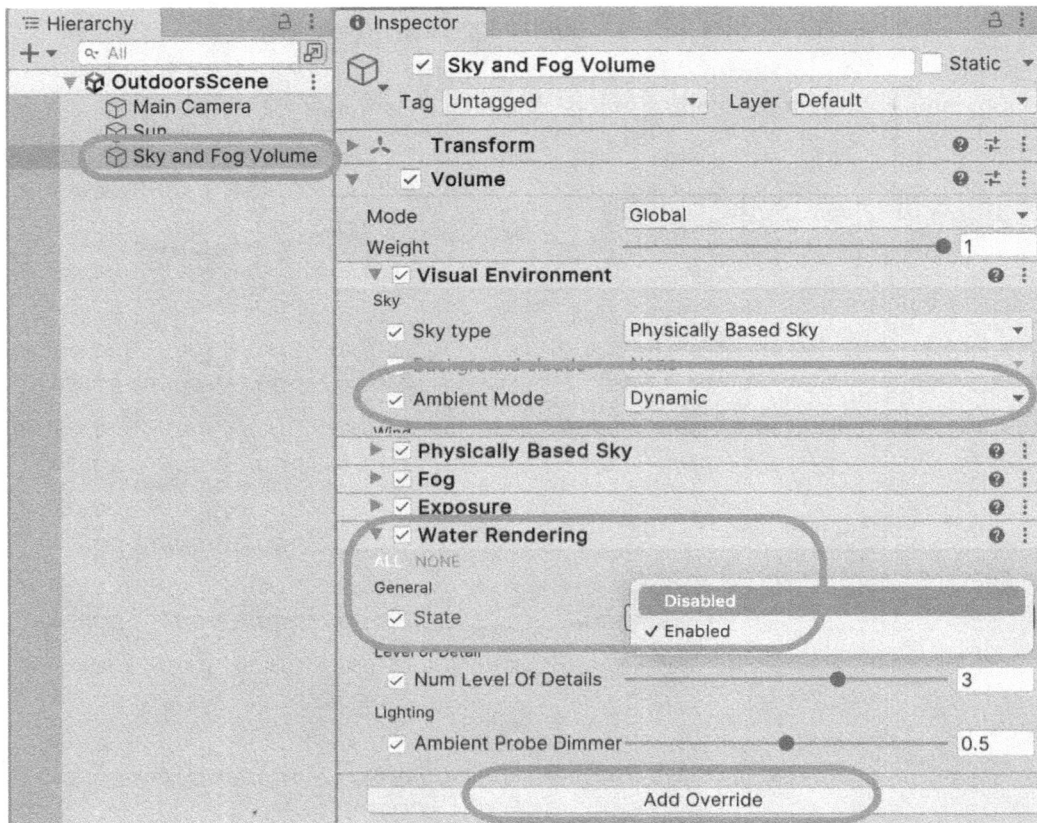

Figure 6.26: Enabling water rendering for the Volume component of the Sky and Fog Volume GameObject

9. Now that the rendering is all enabled, let's add an ocean to the scene! Add a new water Game-Object to the scene by choosing the menu **Create | Water | Surface | Ocean Sea or Lake.**

10. Add a cube to the scene, and scale it to (10, 10, 10).

11. Orient the **Scene** panel so that you can see the ocean water lapping at the sides of the cube, then single-click the main camera, and make this camera show the same view by choosing the menu **GameObject | Align With View.**

12. Save and run the scene. You should see realistic, moving ocean ripples, lapping up to the cube, including real-time shadows.

Figure 6.27: Simulated dynamic ocean surface, with real-time lighting and shadows

13. Save the scene with a new name, such as `OutdoorsScene-Pool`.

14. Delete the ocean water GameObject and the cube.

15. Let's add a swimming pool to the scene! Add a new water GameObject to the scene by choosing the menu **Create | Water | Surface | Pool**. Set the **Position** of this **Pool** GameObject to (0, 0.3, 0), and ensure its **Scale** is (10, 1, 10).

16. Add a sphere to the scene, and set its **Position** to (-1, 0, -1).

17. We can't just have some water and a floating ball – we need to add a bottom and sides to make this a swimming pool. Create a new cube named `Cube-bottom`. Set the **Position** to (0, 0, 0) and **Scale** to (10, 0.1, 10).

18. Create a new light-blue material named `m_blue`, and assign this to `Cube-bottom`.

19. Create a new cube named `Cube-side`, with the **Position** set to (5, 0, 0) and the **Scale** to (1, 1, 12).

20. Duplicate GameObject `Cube-side` three times so that you now also have the GameObjects `Cube-side (1)`, `Cube-side (2)`, and `Cube-side (3)`.

21. Select **Cube-side (1)**, and set its **Position** to (-5, 0, 0).

22. Select **Cube-side (2)**, and set its **Position** to (0, 0, 5) and **Rotation** to (0, -90, 0).

23. Select **Cube-side (3)**, and set its **Position** to (0, 0, -5) and **Rotation** to (0, -90, 0).

24. Save and run the scene. You should see realistic, moving swimming pool surface ripples, lapping up to the sphere, including real-time reflections from the sun.

How it works...

Once you have created an HDRP 3D scene, in order to have your scene benefit from water-based rendering effects, you need to enable several rendering components to identify and include water-based GameObjects in the rendering system. Unity provides several GameObjects to model different types of water, including **Pool** and **Ocean**, which were explored in this recipe. Another very useful water GameObject is **River**.

Most of the work is done for you by the powerful HDRP rendering pipeline. For example, we can see real-time shadows on the ocean surface where the water is obscured from the sun by the cube. Another example is where the blue of the pool bottom cube means that the surface of the pool water is light blue, and if viewing the pool from a right angle, reflections from the simulated sun in the scene can be seen.

Real-time, dynamic surfaces, shadows, and reflections for environments with water-based features now allow you to create highly realistic scenes, using the Unity game engine.

Further reading

This section contains some sources that provide more information about the topics that were covered in this chapter.

Every game needs textures and materials. Here are some sources of free textures that are suitable for many games:

- Free, seamless outdoor textures suitable for tiling: `https://seamless-pixels.blogspot.com/2012/09/free-seamless-ground-textures.html`
- textures.com: `https://www.textures.com/`
- GAMETEXTURES: `https://gametextures.com/`

A great collection of free-to-use (Creative Commons-licensed) materials can be found at AmbientCG: `https://ambientcg.com/`.

To learn more about Unity terrains, take a look at the following links (although a couple are a few years old, they're still very useful):

- Creating and editing terrains: `https://docs.unity3d.com/Manual/terrain-UsingTerrains.html`
- A Unity blog post outlining the terrain tools: `https://blogs.unity3d.com/2019/05/28/speed-up-your-work-with-the-new-terrain-tools-package/`
- A Unity blog post about using terrain tools to sculpt a volcano: `https://blogs.unity3d.com/2019/08/15/accelerate-your-terrain-material-painting-with-the-2019-2-terrain-tools-update/`
- A Unity blog post with tips about adding features by adding terrain holes: `https://blogs.unity3d.com/2020/01/31/digging-into-terrain-paint-holes-in-unity-2019-3/`

To learn more about trees and vegetation, try the following resources:

- Unity SpeedTree 3D vegetation modeling – the industry-standard vegetation toolkit for projects: `https://unity.com/products/speedtree`
- Building your own trees, using the menu **Create** | **3D Object** | **Tree**: `https://docs.unity3d.com/Manual/tree-FirstTree.html`

Some great tutorials for working with terrains and vegetation tools can be found at the following (although a couple of years old, they're still very useful):

- A tutorial demonstrating the creation of 3D landscapes: `https://www.youtube.com/watch?v=pZuXDZkBMow`
- A tutorial demonstrating using terrain tools for textures and trees: `https://medium.com/nerd-for-tech/getting-started-with-terrain-tools-in-unity-part-2-textures-and-trees-5f6b2f57393c`

Unity has published two exemplary scenes illustrating the power of the terrain modeling and vegetation features. One is for URP and the second is for the HDRP:

- A **URP** (**Universal Render Pipeline**) demo: `https://assetstore.unity.com/packages/3d/environments/unity-terrain-urp-demo-scene-213197`
- An **HDRP** demo: `https://assetstore.unity.com/packages/3d/environments/unity-terrain-hdrp-demo-scene-213198`

Height maps for terrains can be created outside of Unity and imported to customize terrains. The "basic" edition of World Machine is now free for non-commercial use. World Machine is a powerful terrain builder and creates assets that you can directly import into your Unity projects. Note that World Machine is only available for the Windows operating systems: `https://www.world-machine.com/`.

We have only dipped our toes (sorry!) into the features of the Unity HDRP water system. Learn more with these resources:

- Unity water package documentation: `https://docs.unity3d.com/Packages/com.unity.render-pipelines.high-definition@16.0/manual/WaterSystem-use.html`
- A Unity 2022 video tutorial about the water system that is still very relevant: `https://www.youtube.com/watch?v=Iu2_aKpo7bM`

7

Creating 3D Geometry with ProBuilder

There are two main ways to get 3D objects for use in your Unity games. One way is to import 3D models created outside of Unity, such as with dedicated 3D modeling applications like **Blender**, **3D Studio Max**, and **Maya**. Many 3D models are available from the Unity Asset Store and other online websites such as *TurboSquid.com*. A second way is to create your own 3D objects inside the Unity Editor, using the ProBuilder package. **ProBuilder** is a Unity package that can be added to any project. It allows you to create and manipulate geometry inside the Unity editor. While Terrains are great for creating terrains, **ProBuilder** allows you to create 3D objects and then manipulate them, such as by extruding or moving **vertices**, **edges**, or **faces**. You can also paint them with colors or texture them with **materials**.

Whether working with dedicated 3D modeling applications or ProBuilder, there are some fundamental concepts that you need to understand to get the most out of these tools. A 3D object is a mesh of connected polygons (2D shapes, such as triangles and rectangles). Each polygon of a 3D object is called a **face**. A face is defined by a boundary of straight line edges. Each edge is a straight line between two vertices. So, to be more specific, when working with 3D modeling tools, including ProBuilder, we talk about the following four levels of detail/editing modes:

- **Object**: This is the whole object, which can be moved, rotated, and scaled.
- **Face**: The flat 2D surfaces (rectangles or triangles). Faces can be painted with colors or texture images. Faces can be moved (deforming the object, as connected faces are affected by the move). Faces can also be extruded – where the face becomes like an opening from which an elongated object with the same outline is created.
- **Edge**: A straight line between two vertices. When we move an edge, we change the faces that share the edge. We can color objects from edges.
- **Vertex**: The two endpoints of each edge are vertices. So when we move a vertex, we change connected edges. We can color objects from vertices.

ProBuilder has a floating menu for us to select one of these four modes at any time.

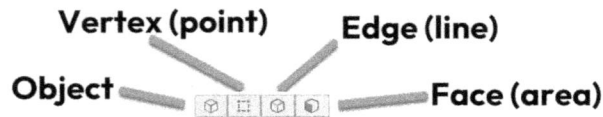

Figure 7.1: The ProBuilder Scene panel selection modes

In this chapter, you'll find the following ProBuilder recipes:

- Getting started with ProBuilder
- Transforming an object through scaling and coloring
- Creating a house with ProBuilder
- Exploring ProBuilder Boolean operations to add a window to our house
- Organizing level geometry as empty GameObject children

Getting started with ProBuilder

In this recipe, we'll set up the ProBuilder tools and create several 3D game assets.

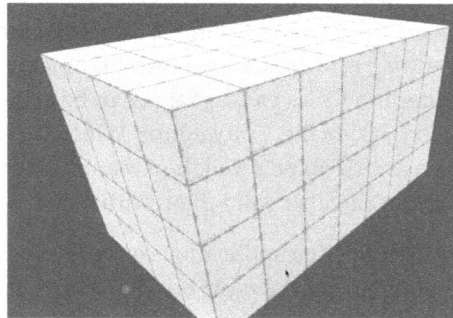

Figure 7.2: The 4 x 4 x 4 cube-based object we'll create in this recipe

Getting ready

For this recipe, you need a brick-like texture. We've provided the file `Yellow_brick_wall.jpg` in folder `07_01`.

How to do it...

To create geometry with ProBuilder, follow these steps:

1. Create a new Unity 3D project.
2. Open the **Package Manager** panel by choosing **Windows | Package Manager**. Set the **Packages** list to **Unity Registry**. Locate the **ProBuilder** package and install it for this project. You should now have a new Unity menu, **Tools**, with a single option, **ProBuilder**.

3. Open the **ProBuilder** tools panel by choosing **Menu: Tools | ProBuilder | ProBuilder Window**. Dock this panel somewhere convenient so you can always see it, for example, as a column between the **Hierarchy** and **Scene** panels.

4. Use the right-mouse-button menu to choose your preference of Icon or Text mode.

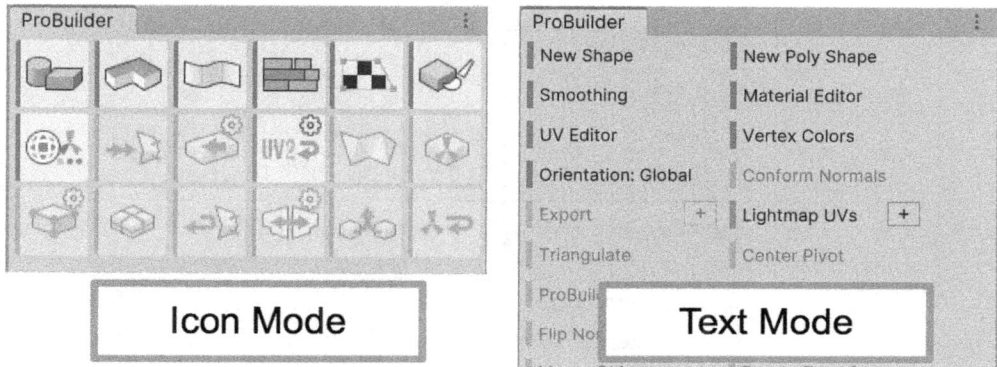

Figure 7.3: ProBuilder Text mode and Icon mode

5. If you've not used ProBuilder much before, take a few minutes to explore the seven main tools. The first tool, **New Shape**, is perhaps the one you'll use the most.

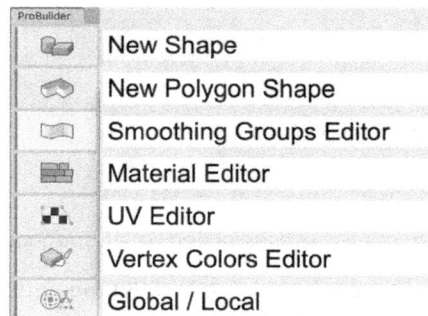

Figure 7.4: The seven main ProBuilder tool icons

By default, ProBuilder will apply a grid-box texture (like a schoolbook's squared graph paper) for objects. However, you can change that in the **ProBuilder** tab of the Unity **Preferences**. ProBuilder works best when you have a visible and active snapping grid setup.

6. First, click the **Grid Visual** button in the Scene panel, ensure **Grid Plane** is set to **Y**, and drag **Opacity** all the way to the right (maximum visibility).

7. Click **Grid Snapping**, and ensure the grid size is 1 and locked for all three axes (**X**, **Y**, and **Z**).

8. Now, click the **Incremental Snapping** button and set its Move value to 1, again locked for all three axes. Ensure the **Grid Visual** and **Grid Snapping** buttons are highlighted in blue, indicating that they are active.

You should now have a horizontal (Y-plan) visible snapping grid that will constrain your Pro-Builder edits to increments of 1 Unity unit. This will make it much easier to resize and line up different objects, faces, edges, and vertices.

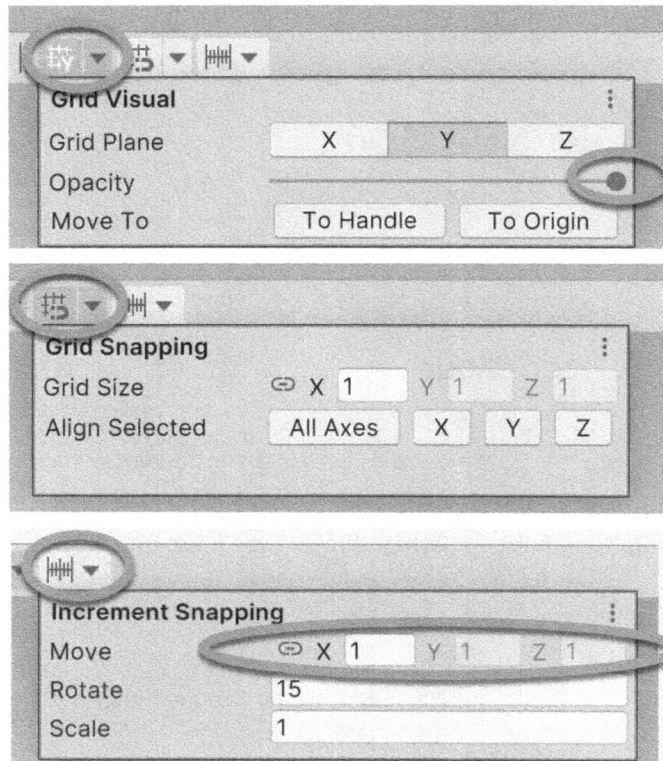

Figure 7.5: Recommended settings for Grid Visual, Grid Snapping, and Increment Snapping

9. Let's create a **ProBuilder cube.** Click the **New Shape** button/icon, and in the floating **Create Shape** panel that appears in the **Scene** panel, select the **Cube** shape (the bottom-left option).

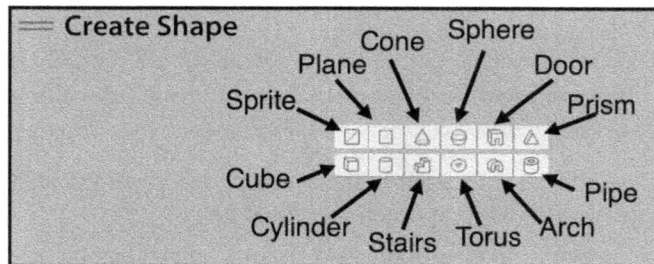

Figure 7.6: ProBuilder new shape icons

10. Now, in the **Scene** panel, use the mouse (click and drag) to draw a 2 x 4 rectangle to define the "base" of our new 3D object.

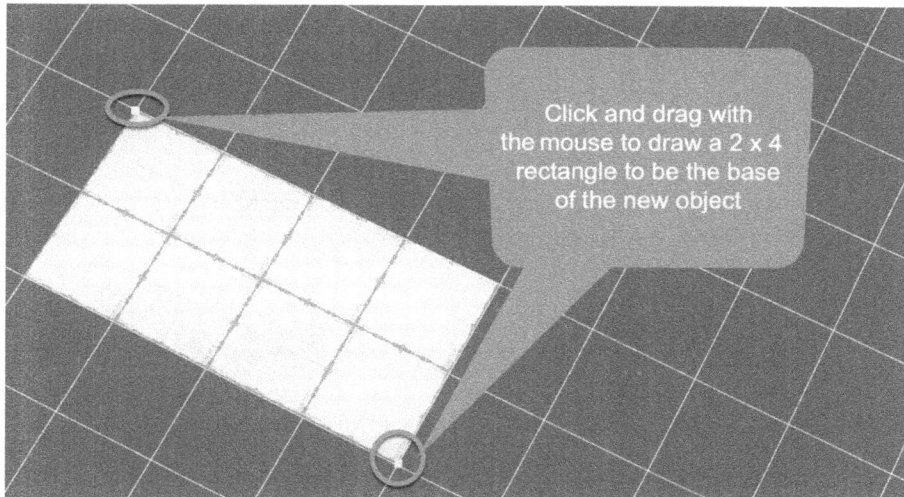

Figure 7.7: Using the mouse to define the 2x4-unit rectangle base for the new 3D ProBuilder object

11. Now, move the mouse pointer upward until the base rectangle has been "extruded" upward to create an object 4 units high. Click the left mouse button to confirm object creation.

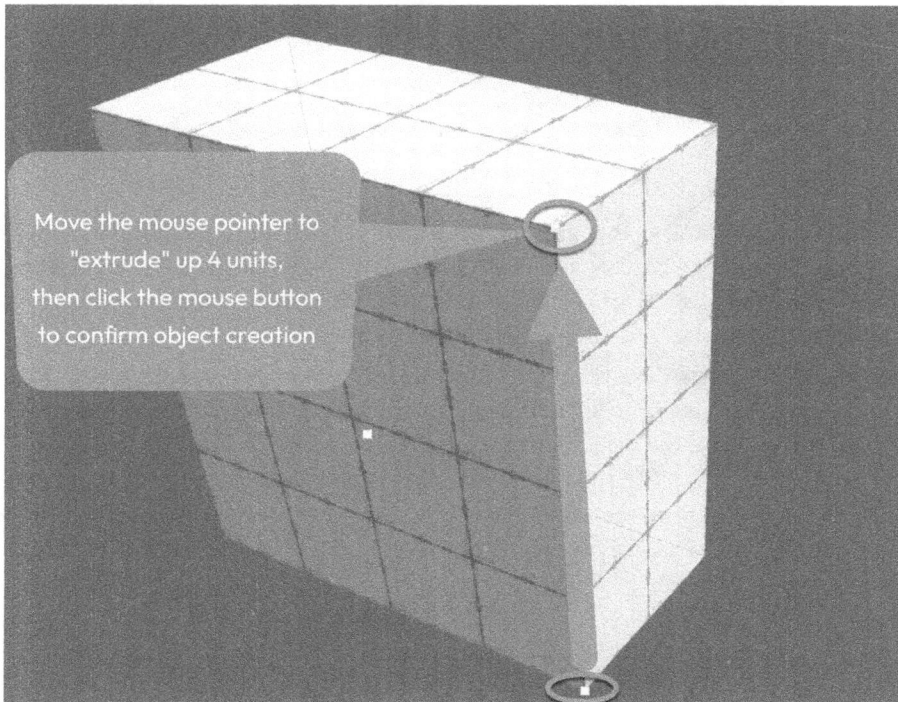

Figure 7.8: Using the mouse to extrude the base to create a 4-unit-high 3D ProBuilder object

12. In the ProBuilder tools panel, click the **New Shape** button/icon to go out of **Object Creation** mode. You should see the floating **Create Shape** panel removed from the Scene panel.

13. When we've just created a new object, ProBuilder will be in **Object Selection** mode. We are now going to change this object by working with faces, so click the **Face** mode icon in the Scene panel (it's on the right).

14. Select the 4 x 4 sized face, and use the red (X-axis) movement arrow to drag the face, resizing the object to be 7 units wide. When we **move** a face, we resize the object in the direction of that face.

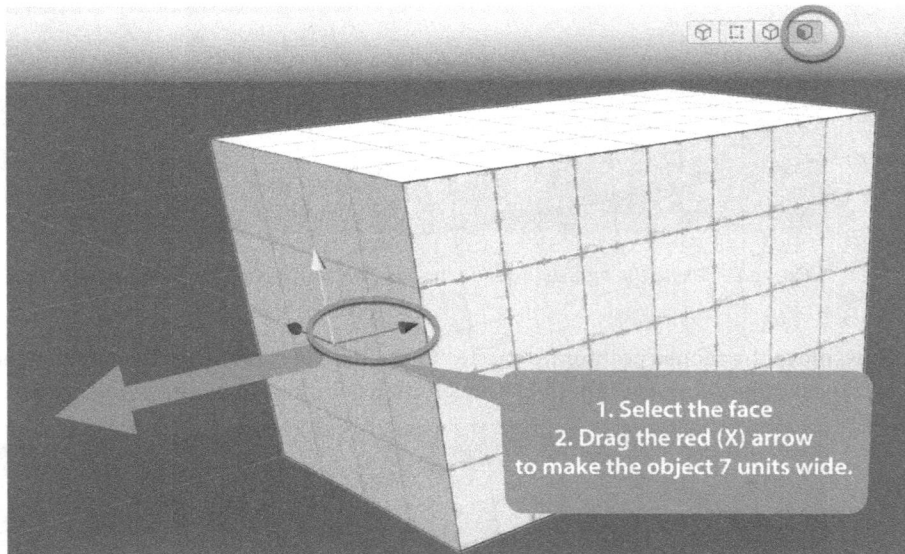

Figure 7.9: Resizing the object by moving a face

15. Create a large plane as the "ground" on which our cube-based object sits, by clicking the Pro-Builder **New Shape** button, and then choosing the **Plane** icon (the second from left in the top row), and dragging a large rectangle. You do this in just the same way you dragged a rectangle to create the base of the Cube-based shape earlier in this recipe.

16. Next, create a new Cone shape, by clicking the ProBuilder **New Shape** button, choosing the **Cone** icon (the third from left in the top row), dragging the circle for its base, then releasing the mouse to confirm the base size, moving the mouse upward for the height, and then clicking to confirm its height. In the **Edit Shape** dialog, you can customize the properties of this new shape, such as setting the number of sides to greater than 20 for a smooth **Cone**, or down to 3 for a triangular-based pyramid. Click theProBuilder **New Shape** button again to confirm the shape object creation.

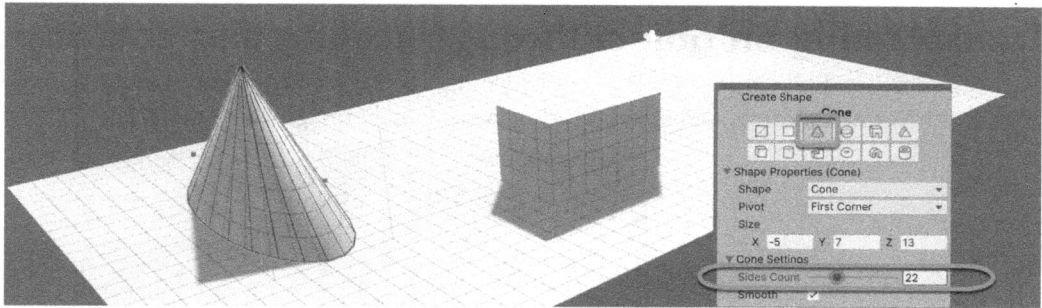

Figure 7.10: Setting the number of sides for a smooth Cone shape

17. Note that you can return to the **Edit Shape** properties dialog, by selecting the ProBuilder **GameObject** in the **Hierarchy,** and then clicking the **ProBuilder (PB)** properties tool in the **Scene** panel.

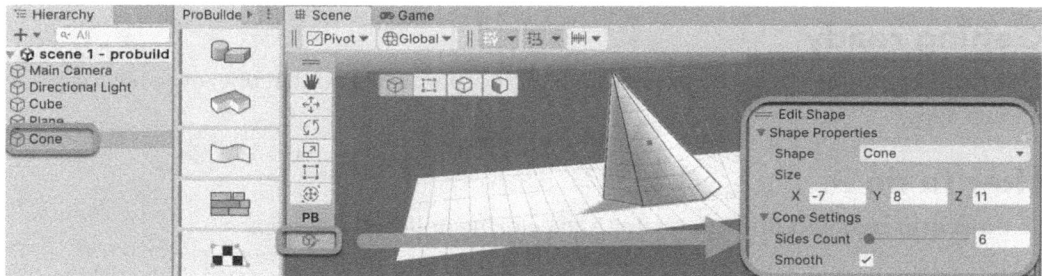

Figure 7.11: Returning to the ProBuilder Edit Shape dialog

18. Save your scene. You've now a scene containing some ProBuilder objects that you have created and customized.

How it works...

By adding the **ProBuilder** package to a new 3D project, we can enable ProBuilder's features.

ProBuilder allows mesh objects to be added to the **scene.** Dragging faces using the ProBuilder **Face-selection** tool allows you to select and then move some of the faces to resize the object.

When you create a new ProBuilder object, you edit its properties in the **Edit Shape** dialog. You can return to edit the properties of a previously created shape, by selecting that object and clicking the ProBuilder **Edit Shape** tool in the **Scene** panel.

Transforming an object through scaling and coloring

In this recipe, we'll take things further, transforming the object with some scaling, adding simple colors with Vertex Coloring, and then texturing faces with an image-based material.

Figure 7.12: The swimming-pool-topped pyramid object we'll create in this recipe

Getting ready

This recipe continues from the previous one. So keep working on that one, or make a copy and work on that copy.

How to do it...

To transform an object through scaling and coloring, follow these steps:

1. In the ProBuilder tools panel, display the **Vertex Colors Editor** panel by clicking the **Vertex Colors** icon/button.

2. Select the top face of your **Cube** object, and click the **Apply** button next to the yellow vertex color. Your object should now have a yellow top face.

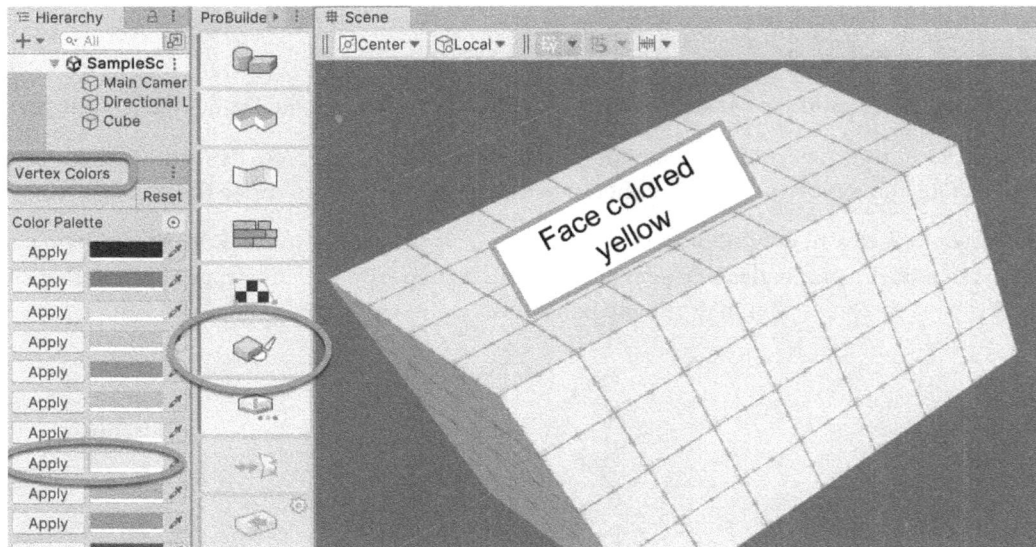

Figure 7.13: Setting the color of a face using Vertex Colors

3. Select the yellow top face, and scale this face a little smaller in all 3 axes (X, Y, and Z). Your object should begin to look a little like a pyramid.

4. With the yellow top face selected, and while holding down the *Shift* key, again scale this face a little smaller in all three axes (X, Y, and Z). You should see a new face created, within the top of the object. You have extruded the face.

5. Select this inner face, and color it blue by clicking the **Apply** button next to the blue color in the **Vertex Colors** panel.

6. Now, move this blue face downward. This should make an indentation on the top of your object.

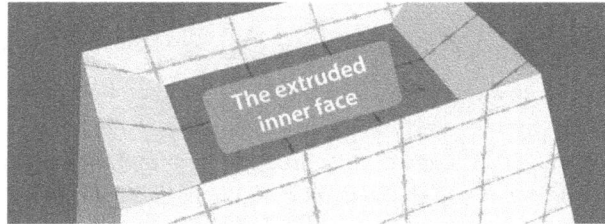

Figure 7.14: Indenting an object by moving the extruded face downward

7. Create a new **Material** named **m_yellowBrick**. Drag the provided texture into the **Albedo** property of your new material. Set its X and Y tiling to 0.3.

8. In the ProBuilder tools panel, display the **Material Editor** panel by clicking the **Material Editor** icon/button. For the second material, choose m_yellowBrick.

9. Select the side faces of your object, and click the **Alt+2** button next to the second material. Your object should now have yellow brick-textured sides.

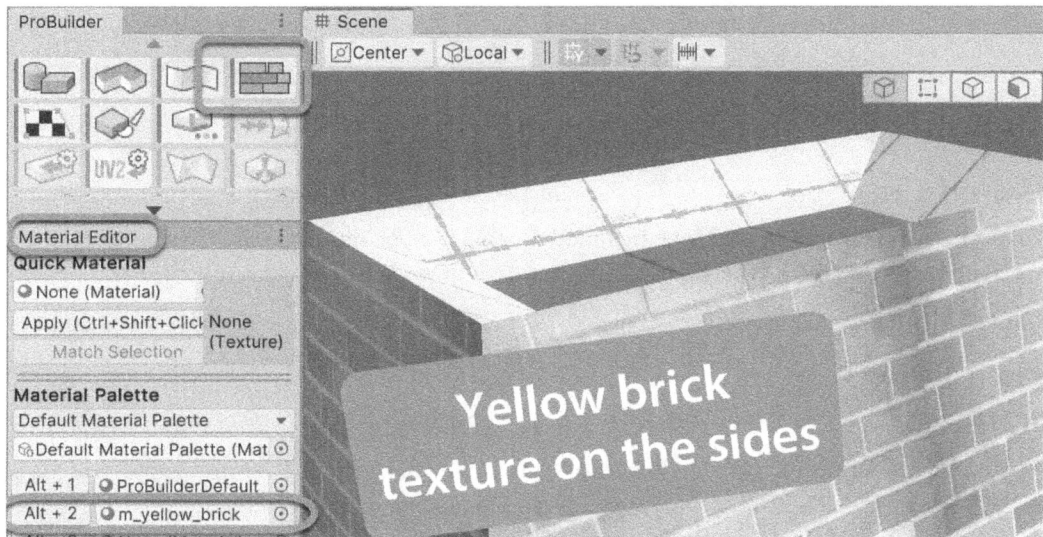

Figure 7.15: Adding a yellow brick texture to the sides of the object

10. Save your scene. You should now have a yellow brick-style pyramid, with what looks like a blue swimming pool at the top!

How it works...

Scaling a face deforms the shape, since connected faces move too. Using **Shift** when dragging or scaling results in a new, extruded face being created. When we move an extruded face, we can create an indentation or a convex (sticking-out) feature.

The **Vertex Color Editor** and **Material Editor** allow faces to be colored and textured.

Creating a house with ProBuilder

If you quickly want to create a more detailed prototype, maybe based on an actual building, we can combine the techniques that we used in the previous recipe to extrude a complex building. In this recipe, we will demonstrate how to make a square room that includes a window and a door. Following the same process, a building of any shape can be made using ProBuilder:

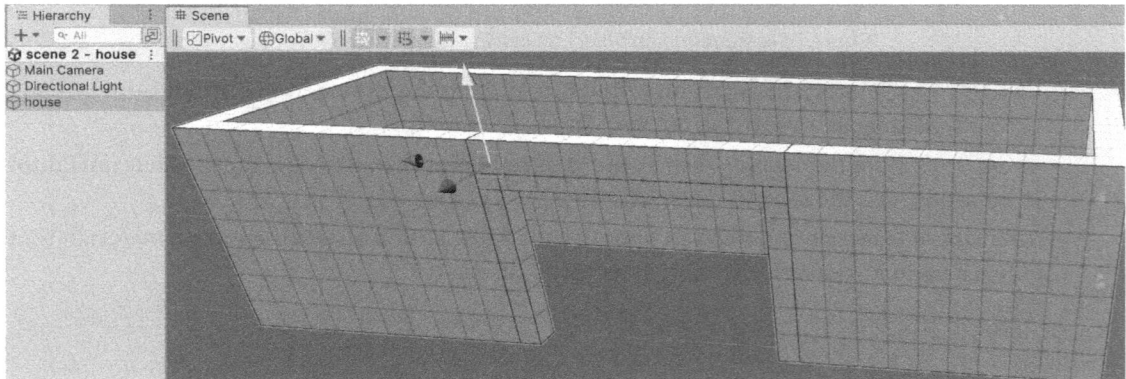

Figure 7.16: The one-room house created in this recipe

Getting ready

You'll need the ProBuilder package installed for this recipe, and also the grid snapping setup. Either work on a copy of the previous recipe, or create a new 3D project and follow *steps 1—8* of the first recipe in this chapter, *Getting Started with ProBuilder*.

How to do it...

To create a house with ProBuilder, follow these steps:

1. Create a new 3D project. Add the **ProBuilder** package, and set up visible grid snapping in increments of 1 Unity unit (as shown in the first recipe in this chapter).

2. Choose the **New Poly Shape** tool (the second item) from the **ProBuilder** window, and draw 12 points in a kind of *C* shape, representing the walls and a space for a doorway for a small, one-room building:

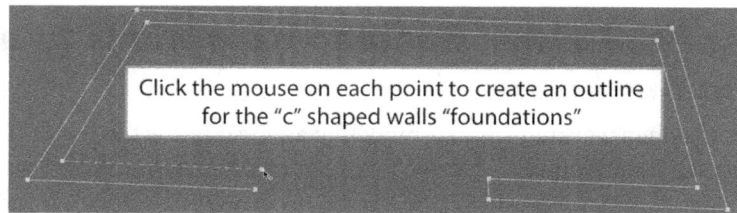

Figure 7.17: the Poly Shape tool creating a C-shaped wall footprint for our room

3. Once you've joined the first point to the last point, use the mouse to extrude up from your 2D shape for walls that are 7 units high.

4. Now, choose the **New Shape** tool, and choose the **Door** shape (second from last in the top row).

5. Drag a rectangle 1 unit wide in the gap you left in the walls, and extrude upward by 6 units. You should now have a mesh for a doorway.

Figure 7.18: Adding a Door shape to our room

6. Again, with the **New Shape** tool, create a **Cube** shape, that is one unit by the door width in size. Then, complete our doorway by moving this shape above the door shape you just created.

7. Select all three objects and join them together using the **Merge Objects** ProBuilder tool.

8. Save the scene. You should now have a one-room house object.

How it works...

The **New Poly Shape** tool is perfect for laying out the floorplan of a building, and then it allows us to extrude upward to create several walls at the same time. If you get the height wrong, you can choose the top face, and move it up or down to change the height.

The **Door Shape** is designed to avoid hidden faces – that is, faces that would never be seen when we combine several objects to make shapes like doorways.

If you have created a complex object with several different shapes, the **Merge Objects** tool turns them into a single GameObject, keeping your **Hierarchy** tidy and allowing you to move/scale/duplicate the object easily.

Exploring ProBuilder Boolean operations to add a window to our house

A powerful feature when working with 3D-modeling applications is the use of Boolean operations to create or change shapes by combining shapes together in additive/subtractive ways. ProBuilder offers three Boolean operations (currently in a beta-style "experimental" mode). In this recipe, we'll learn how to subtract one shape from another to create window spaces in the walls of our house.

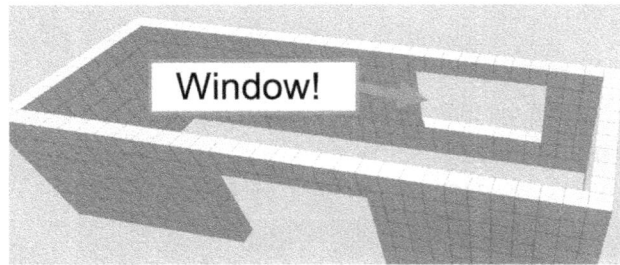

Figure 7.19: The window added to the back wall of our house using a Boolean shape subtraction

Getting ready

This recipe builds on the previous one, so make a copy of that project and work on that copy.

How to do it...

To explore ProBuilder Boolean operations to add a window to our house, perform the following steps:

1. Enable the **ProBuilder** experimental tools by opening the **Preferences** panel, selecting the **ProBuilder** tab, and checking the **Experimental Features Enabled** option. Wait a few seconds, since Unity will need to refresh the ProBuilder package.

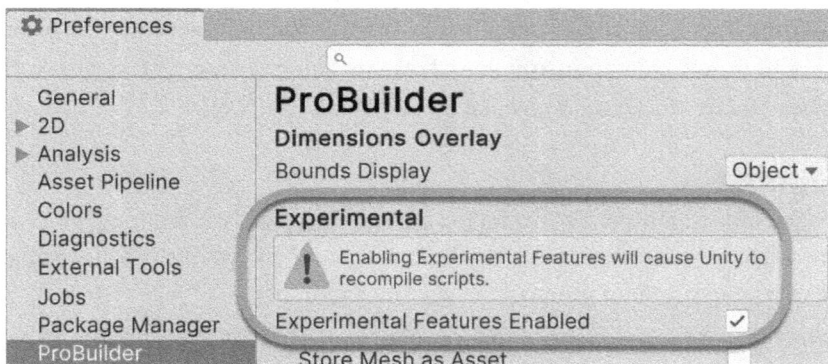

Figure 7.20: Enabling the ProBuilder Experimental Features

2. Once enabled, you should now see a new option on the ProBuilder menu: **Tools | ProBuilder | Experimental | Boolean (CSG) Tool**.

3. We need two objects for these Boolean tools, so use the **New Shape** tool, and create a cube that is 1 unit deep, and a rectangular shape of the size of a window space for the back wall of our house, for example, 8 units wide by 4 units high.

4. Move your new cube onto the back wall, where you want the window space.

5. Select both the house and cube in the **Hierarchy**, and then choose **Menu: Tools | ProBuilder | Experimental | Boolean (CSG) Tool**.

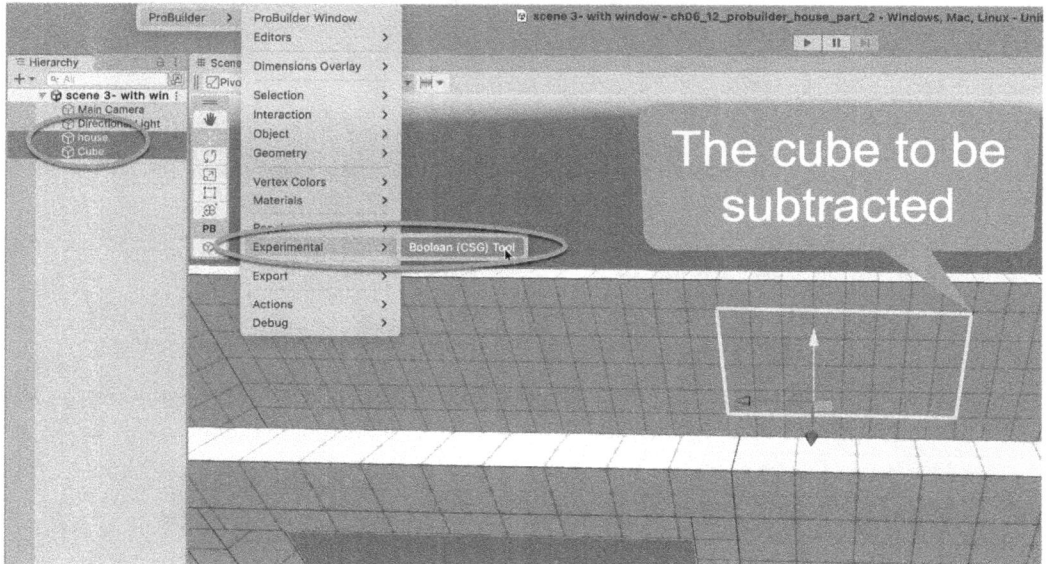

Figure 7.21: Using the experimental Boolean ProBuild tool

6. The **Boolean (Experimental)** window should appear, already populated with the house on the left side and the cube on the right. (If they are the other way around, click the double arrow to swap them.)

7. Select the **Subtraction** operation from the drop-down menu. You should see the text **house Subtracts Cube** on the left-hand side, above the image of the house. Now, click the **Apply** button.

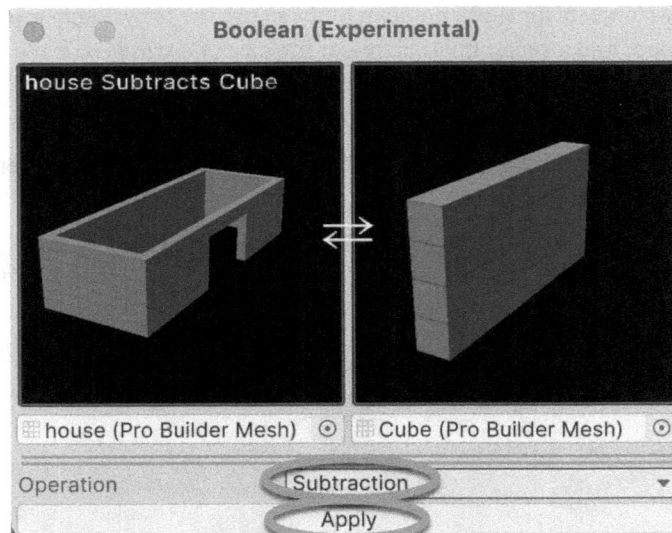

Figure 7.22: Subtracting the Cube from the House to create a window space

8. A new GameObject, named **New Game Object**, should have been created in the **Hierarchy**. Delete the house and cube GameObjects, and rename **New Game Object** as house-with-window.

9. Save your scene. You should now have a single-shape GameObject in your scene, which is the house with a window gap in the back wall.

How it works...

Once enabled, there are several additional experimental ProBuilder features. Perhaps the most exciting feature is the application of Boolean operations to combine/subtract/intersect between two ProBuilder objects. In the previous recipe, we used the *Door* shape to create a doorway. As we've seen in this recipe, if we want a tidy hole in something, the subtraction operation is a very easy way to achieve this – we just create an object the shape of the hole we want, move it into place, and then subtract it from the shape needing a hole.

Between the range of different shapes possible, their manipulation with face/edge/vertex changes, and then adding in extrusions and Boolean operations, ProBuilder is a powerful tool for creating and manipulating 3D objects right inside the Unity editor. If your game is to have high-quality visuals, you'll still need to create or import assets created by an experienced 3D modeler, but combined with vertex coloring and texturing, ProBuilder offers a way to very quickly create scene objects (geometry) to try out ideas or prototype game levels.

Organizing level geometry as empty GameObject children

Many beginners working with 3D modeling tools have trouble navigating and orienting the scene being edited once it contains many objects of different sizes. A simple yet effective approach is to make all related 3D objects children to an empty GameObject, positioned at the "center" of those objects. You can use this method if you want to quickly create a more detailed prototype, maybe based on an actual building. We can combine the techniques that we used in the previous recipe to extrude a complex building. In this recipe, we will demonstrate how to parent **GameObject** whose children are a yellow floor plane, a blue staircase, and a green cone. Following the same process, a building of any shape can be made using ProBuilder:

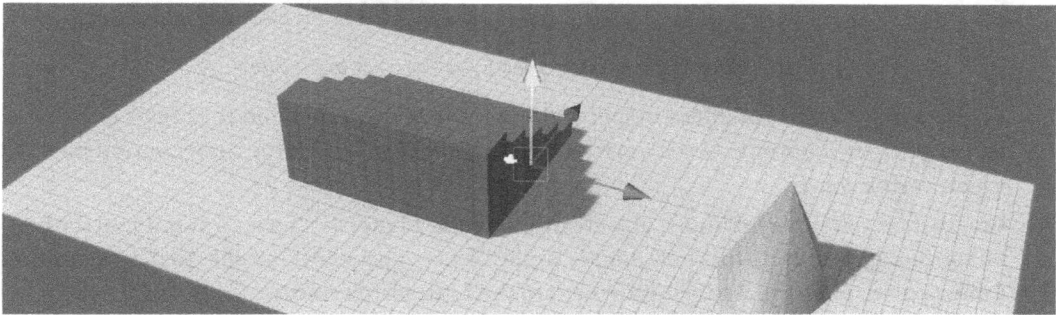

Figure 7.23: Collection of ProBuilder objects as children of an empty GameObject

Getting ready

You'll need the ProBuilder package installed for this recipe, and also grid snapping set up. Either work on a copy of one of the previous recipes, or create a new 3D project and follow *steps 1–5* of the first recipe to set things up to work with ProBuilder.

How to do it...

To organize level geometry as empty GameObject children, follow these steps:

1. Create a new 3D project. Add the ProBuilder package, and set up visible grid snapping in increments of 1 Unity unit.

2. Create a new empty GameObject named geometry_parent by choosing **Menu: Create | Create Empty**. Set the **Transform** position of this to the center of the collection of objects you are going to work with. Often, setting this to (0, 0, 0) is simplest – and then later (if required) you can always move this (and all its children) elsewhere in the scene.

3. Create a new 10 x 10 ProBuilding Plane, whose center is in approximately the same position as the **GameObject** geometry_parent. Child this new Plane **GameObject** to geometry_parent.

4. In the ProBuilder tools panel, display the **Vertex Colors Editor** panel by clicking the **Vertex Colors** icon/button.

5. Select the **Plane** GameObject, and in the **Vertex Colors Editor** panel, click the **Apply** button next to the **yellow** vertex color.

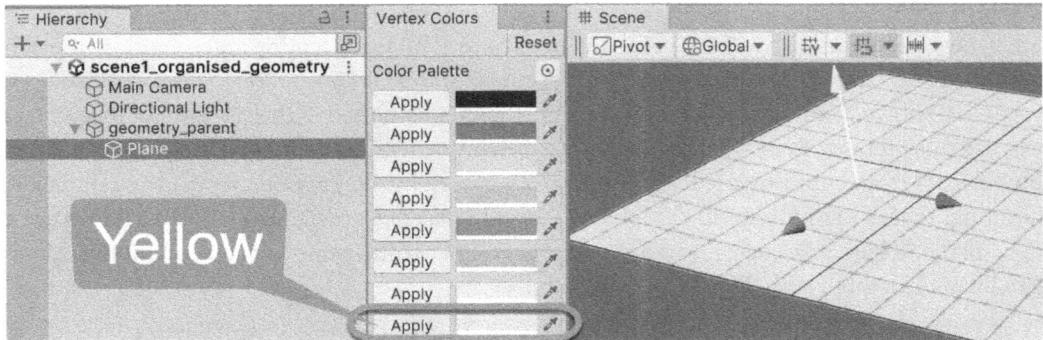

Figure 7.24: Appling a yellow vertex color to the plane

6. Create a new **Cone** ProBuilder object sitting on the **Plane,** and child this new Cone GameObject to geometry_parent. In the **Vertex Colors Editor** panel, click the **Apply** button next to the **green** vertex color.

7. Create a new **Stairs** ProBuilder object sitting on the **Plane,** set the step count to 5 in the **Edit Shape** dialog, and child this new **Stairs** GameObject to geometry_parent. In the **Vertex Colors Editor** panel, click the **Apply** button next to the **blue** vertex color.

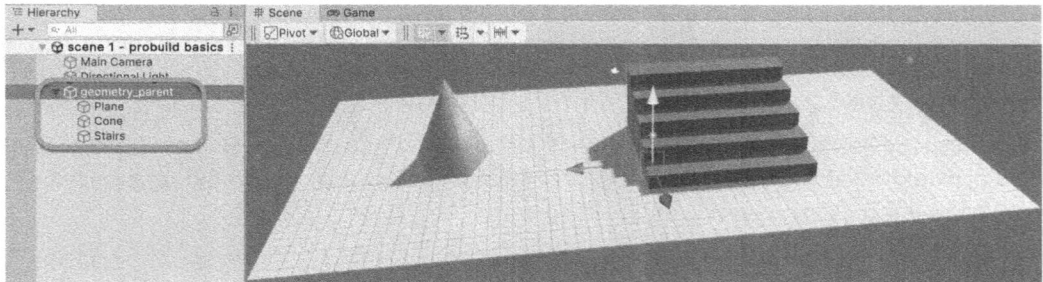

Figure 7.25: ProBuiler objects centered on an empty geometry_parent GameObject

8. Double-click the **GameObject** geometry_parent in the **Hierarchy**, to focus the **Scene** panel on the **Transform** position of this GameObject.

9. While holding *Alt*, drag the mouse to rotate the view, and use the mouse wheel to zoom in and out of the view.

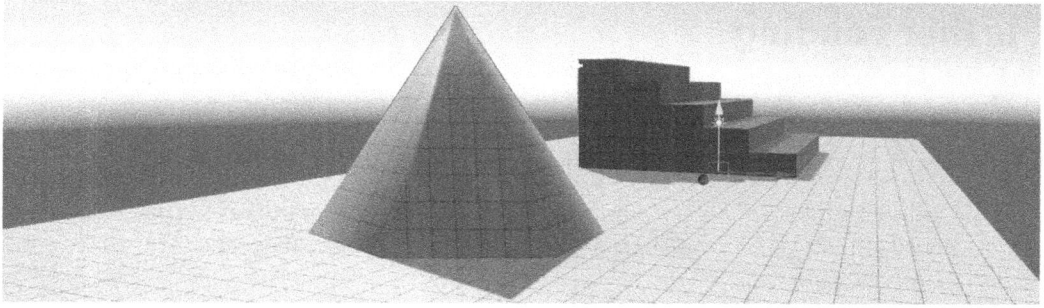

Figure 7.26: Another angle of the view for the zoomed and rotated ProBuilder collection of objects

10. Save the scene.

How it works...

The Unity **Scene** panel was designed to allow it to focus (center) upon the **Transform** of a selected GameObject. By having an empty parent GameObject whose **Transform** is the center of a collection of children that are ProBuilder (and other) 3D objects that we are interested in viewing/editing, we have made it very easy to focus on that collection by double-clicking the parent object in the **Hierarchy**.

> **Note:** you can use this technique hierarchically (it's the **Hierarchy** panel for a reason!). For example, you could start with an empty GameObject for a house (named, say, geometry_house), and its contents (the roof, chimney, door, tables, chairs, and so on). Then, when that's all been modeled, create a new empty GameObject called geometry_scene, and child geometry_house to this GameObject. You could then duplicate the house, and add planes for road, grass, etc., all childed to geometry_scene. To focus on the scene, double-click geometry_scene in the **Hierarchy**, and then zoom/rotate as desired. To focus on the house, double-click geometry_house in the **Hierarchy**, and so on.
>
> Don't waste time "flying" through Scene panel contents with *WASD* and panning to try to center the panel's view on the center of the collection you are working with. Take a little time to organize things with a parent GameObject whose **Transform** is the center point, which will make your life much easier!

Further reading

This section contains some sources that provide more information about the topics that were covered in this chapter.

You can learn more about modeling in Unity with ProBuilder:

- Unity blog post in 2018 outlining the core ProBuilder features: `https://blogs.unity3d.com/2018/02/15/ProBuilder-joins-unity-offering-integrated-in-editor-advanced-level-design/`
- Unity Technology ProBuilder documentation manual: `https://docs.unity3d.com/Packages/com.unity.probuilder@5.0/manual/index.html`
- Unity Technology ProBuilder training videos: `https://www.youtube.com/user/Unity3D/search?query=ProBuilder`
- Kodeco (previously Ray Wenderlich) ProBuilder prototyping tutorial: `https://www.kodeco.com/12008376-probuilder-tutorial-rapid-prototyping-in-unity`

Learn more on Discord

To join the Discord community for this book – where you can share feedback, ask questions to the author, and learn about new releases – follow the QR code below:

`https://packt.link/unitydev`

8

2D Animation and Physics

Since Unity 4.6 in 2014, Unity has shipped with dedicated 2D features, and in this chapter, we will present a range of recipes that introduce the basics of 2D animation to help you understand the relationships between the different animation elements.

In Unity 2D, animations can be created in several different ways – one way is to create many images, each slightly different, which give the appearance of movement frame by frame.

A second way to create animations is by defining keyframe positions for individual parts of an object (for example, the arms, legs, feet, head, and eyes) and getting Unity to calculate all the in-between positions when the game is running:

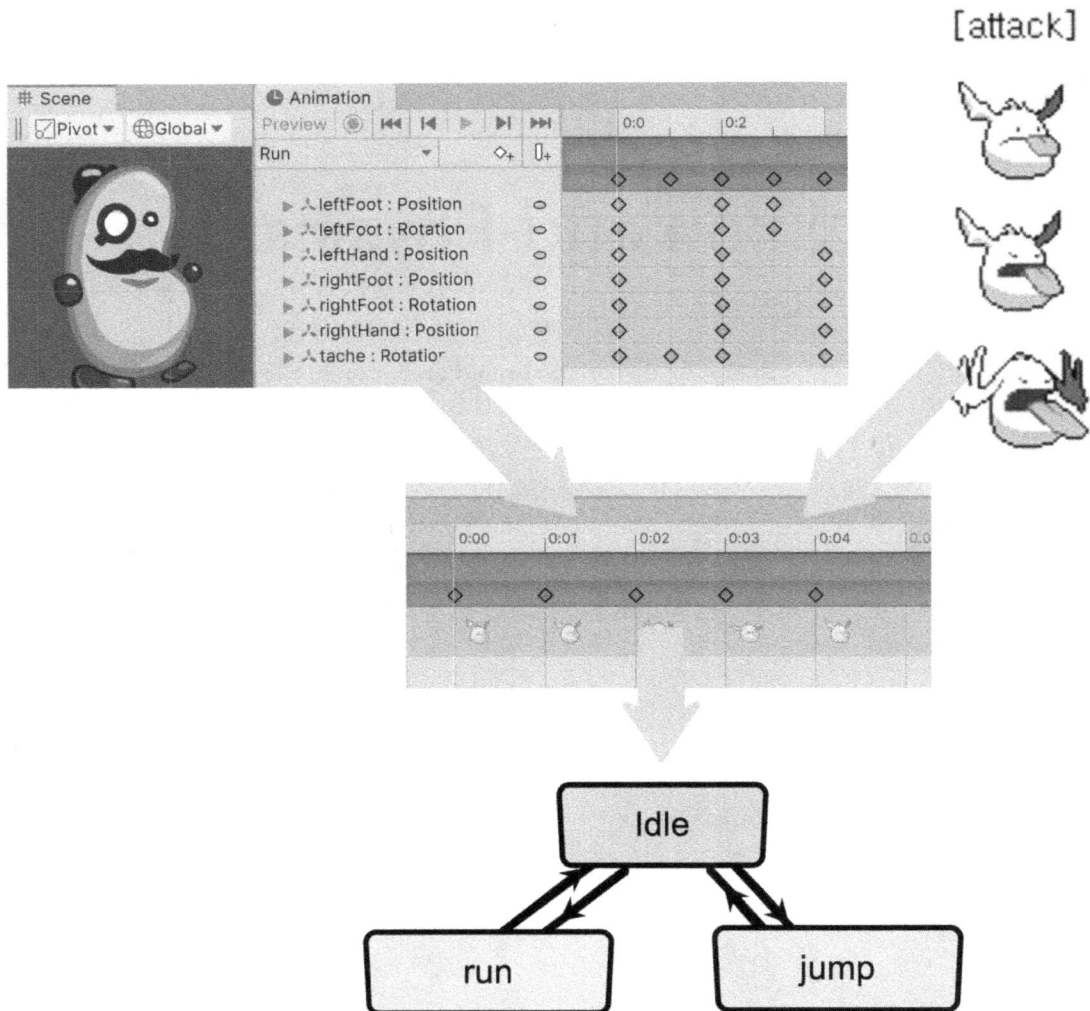

Figure 8.1: Overview of animation in Unity

Both sources of animations become animation clips in the **Animation** panel. Each animation clip then becomes a **state** in an **Animator Controller State Machine**. We can also duplicate **states** based on animation clips, or create new state and add scripted behaviors. We can also define sophisticated conditions, under which the GameObject will **transition** from one animation **state** to another.

In this chapter, we will look at the animation system for 2D game elements. The `PotatoMan` character is from Unity's 2021 **2D Platformer**. We have provided the character as a package in the downloadable assets for this chapter. The following projects are a good place to see lots more examples of 2D game design and animation techniques.

In this chapter, we will cover the following recipes:

- Flipping a sprite horizontally – the DIY approach
- Flipping a sprite horizontally – using Animator State chart and transitions
- Animating body parts for character movement events
- Creating a three-frame animation clip to make a platform continually animate
- Making a platform start falling once stepped on using a Trigger to move the animation from one state to another
- Creating animation clips from sprite sheet sequences
- Creating a platform game with Tiles and Tilemaps
- Using sprite placeholders to create a simple physics scene
- Editing polygon Colliders for more realistic 2D physics
- Creating an `explosionForce` method for 2D physics objects
- Clipping via Sprite Masking

Flipping a sprite horizontally – the DIY approach

Perhaps the simplest 2D animation is a simple flip, from facing left to facing right, facing up to facing down, and so on. In this recipe, we'll add a cute bug sprite to the scene and write a short script to flip its horizontal direction when the *Left* and *Right* arrow keys are pressed:

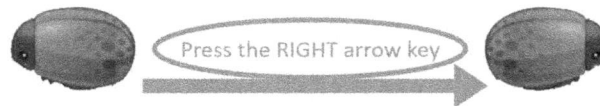

Figure 8.2: Example of flipping a sprite horizontally

Getting ready

For this recipe, we have prepared the image you need in a folder named `Sprites` in the `08_01` folder.

How to do it...

To flip an object horizontally with arrow key presses, follow these steps:

1. Create a new Unity 2D project.

> If you are working on a project that was originally created in 3D, you can change the default project behavior (for example, new **Sprite Texture** additions and **Scene** mode) to 2D by going to **Edit | Project Settings | Editor** and then choosing 2D for **Default Behavior Mode** in the **Inspector** panel:

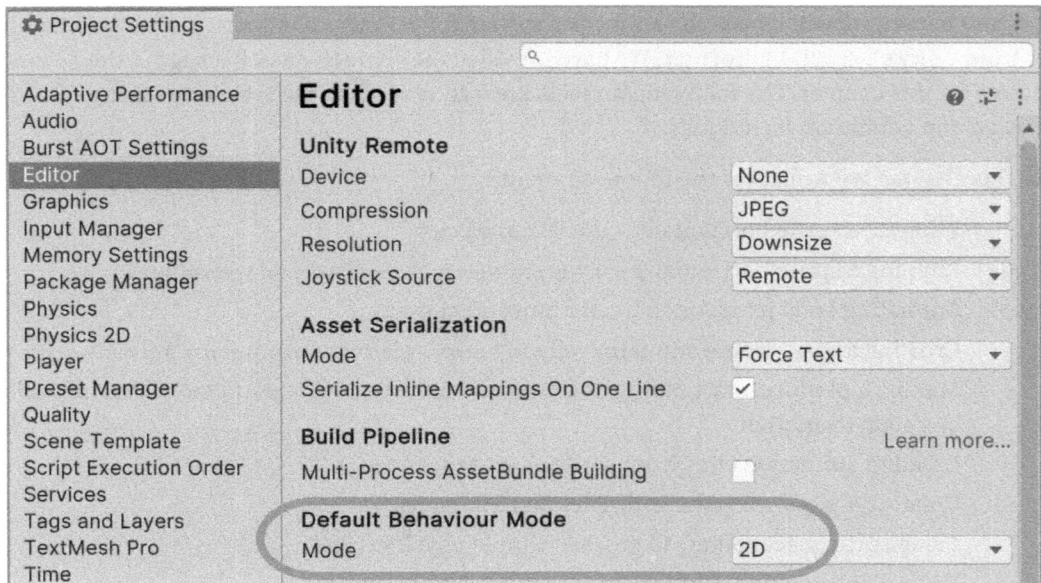

Figure 8.3: Setting Default Behavior Mode to 2D

2. Import the provided image; that is, `Enemy Bug.png`.

3. Drag an instance of the red `Enemy Bug` image from the **Project | Sprites** folder into the scene. Position this GameObject at (`0, 0, 0`) and scale it to (`2, 2, 2`).

4. Create a C# script class called `BugFlip` and add an instance object as a component of the `Enemy Bug`:

```csharp
using UnityEngine;
using System.Collections;

public class BugFlip : MonoBehaviour {
    private bool facingRight = true;

    void Update() {
        if (Input.GetKeyDown(KeyCode.LeftArrow) && facingRight)
            Flip ();
        if (Input.GetKeyDown(KeyCode.RightArrow) && !facingRight)
            Flip();
    }

    void Flip (){
        // Switch the way the player is labelled as facing.
        facingRight = !facingRight;
```

```
            // Multiply the player's x local scale by -1.
            Vector3 theScale = transform.localScale;
            theScale.x *= -1;
            transform.localScale = theScale;
        }
    }
```

5. When you run your scene, pressing the *Left* and *Right* arrow keys should make the bug face left or right.

How it works...

The C# class defines a Boolean variable, facingRight, that stores a true/false value that corresponds to whether the bug is facing right. Since our bug sprite was initially facing right, we set the initial value of facingRight to true to match this.

Every frame, the Update() method checks whether the *Left* or *Right* arrow key has been pressed. If the *Left* arrow key is pressed and the bug is facing right, then the Flip() method is called; likewise, if the *Right* arrow key is pressed and the bug is facing left (that is, facing right is false), the Flip() method is called.

The Flip() method performs two actions; the first simply reverses the true/false value in the facingRight variable. The second action changes the +/- sign of the X value of the localScale property of the transform. Reversing the sign of localScale results in the 2D flip that we desire. Look inside the PlayerControl script for the PotatoMan character in the next recipe – you'll see that the same Flip() method is used.

There's more...

As an alternative to changing the X component of the local scale, we could change the flipX Boolean property of the GameObject's SpriteRenderer component by changing the contents of our BugFlip script to the following:

```
using UnityEngine;
using System.Collections;

public class BugFlip : MonoBehaviour {
    private bool facingRight = true;
    private SpriteRenderer _spriteRenderer;

    void Awake() {
        _spriteRenderer = GetComponent<SpriteRenderer>();
    }
```

```
    void Update() {
        if (Input.GetKeyDown(KeyCode.LeftArrow) && facingRight) {
            _spriteRenderer.flipX = facingRight;
            facingRight = false;
        }

        if (Input.GetKeyDown(KeyCode.RightArrow) && !facingRight){
            _spriteRenderer.flipX = facingRight;
            facingRight = true;
        }
    }
}
```

Once again, we are basing our code on the assumption that the sprite begins by facing right.

> **Note:** Using the **FlipX** property of the `SpriteRenderer` component only affects how the sprite is displayed (rendered). If there are other aspects of the sprite that are not vertically symmetrical, such as its polygon Collider, then you should avoid the **FlipX** property and always use the X-scale approach in the main recipe above, since that approach horizontally flips all parts of the GameObject.

Flipping a sprite horizontally — using Animator State chart and transitions

In this recipe, we'll use the Unity animation system to create two states corresponding to two animation clips.

Unity **Animator Controllers** manage a **State Machine** to control how and when an animated GameObject should change. At any point in time, a GameObject with an **Animator Controller** component is either in one state or transitioning between states.

We'll have one script class that changes `localScale` according to which animation state is active. There will also be a second script class, which will map the arrow keys to the **Horizontal** input axis values as a **parameter** in the **State Machine**, which will drive the transition from one state to the other.

While this may seem like a lot of work, compared to the previous recipe, such an approach illustrates how we can map from input events (such as key presses or touch inputs) to parameters and triggers in a **State Machine**.

Getting ready

For this recipe, we have prepared the image you need in a folder named Sprites in the 08_02 folder.

How to do it...

To flip an object horizontally using **Animator State** chart and transitions, follow these steps:

1. Create a new Unity 2D project.

2. Import the provided image; that is, Enemy Bug.png.

3. Drag an instance of the red Enemy Bug image from the **Project | Sprites** folder into the scene. **Position** this GameObject at (0, 0, 0) and **Scale** it to (2, 2, 2).

4. With the Enemy Bug GameObject selected in the **Hierarchy** panel, open the **Animation** panel (**Window | Animation | Animation**) and click the **Create** button to create a new **Animation Clip** asset. Save the new **Animation Clip** asset as beetle-right. You will also see that an Animator component has been added to the Enemy Bug GameObject:

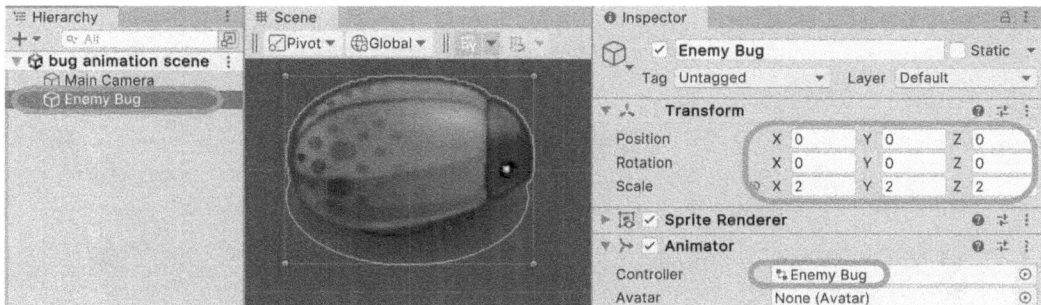

Figure 8.4: Animator component added to the Enemy Bug GameObject

5. If you look in the **Project** panel, you'll see that two new asset files have been created: an **Animation Clip** called beetle-right and an **Animator Controller** named Enemy Bug:

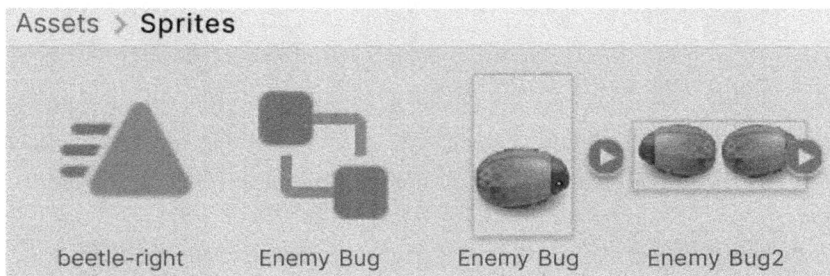

Figure 8.5: Two new asset files included as part of the project

6. Close the **Animation** panel and double-click **Enemy Bug Animator Controller** to start editing it – it should appear in a new **Animator** panel. You should see four states; the **Any State** and **Exit** states should be unlinked, and **Entry** should have a **Transition** arrow connecting to the beetle-right state.

This means that as soon as **Animator Controller** starts to play, it will enter the beetle-right state. beetle-right is tinted orange to indicate that it is in the **Default** state:

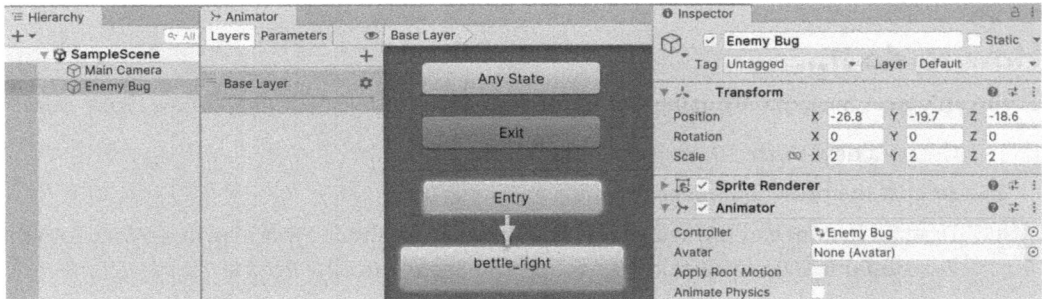

Figure 8.6: Animator Controller for Enemy Bug

> If there is only one **Animation Clip** state, that will be the **Default** state automatically. Once you have added other states to the state chart, you can right-click a different state and use the context menu to change which state is entered first.

7. Select the beetle-right state and make a copy of it, renaming the copy beetle-left (you can right-click and use the menu that appears or the *CTRL/CMD + D* keyboard shortcut). It makes sense to position beetle-left to the **left** of beetle-right:

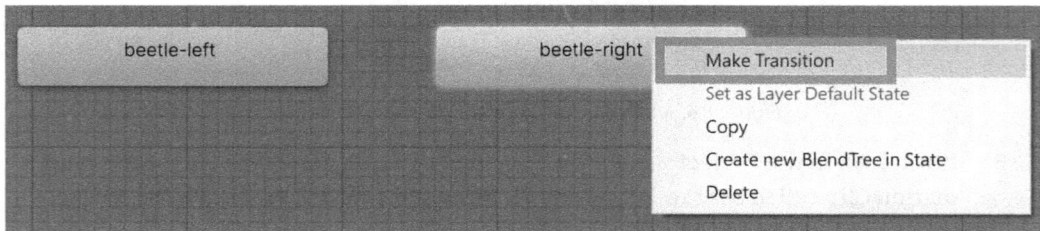

Figure 8.7: Adding a Transition to beetle-right

8. Move your mouse pointer over the beetle-right state. Then, in the right-click context menu, choose **Make Transition** and drag the white arrow that appears into the beetle-left state.

9. Repeat this step with beetle-left to create a **transition** back from beetle-left to beetle-right:

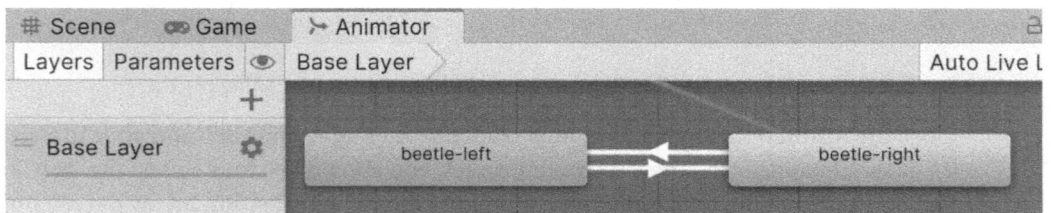

Figure 8.8: Creating the beetle-left to beetle-right Transition

10. We want an instant **Transition** between the left- and right-facing beetles. So, for each **Transition**, uncheck the **Has Exit Time** option. Click the **Transition** arrow to select it (it should turn blue) and then uncheck this option in the **Inspector** panel.

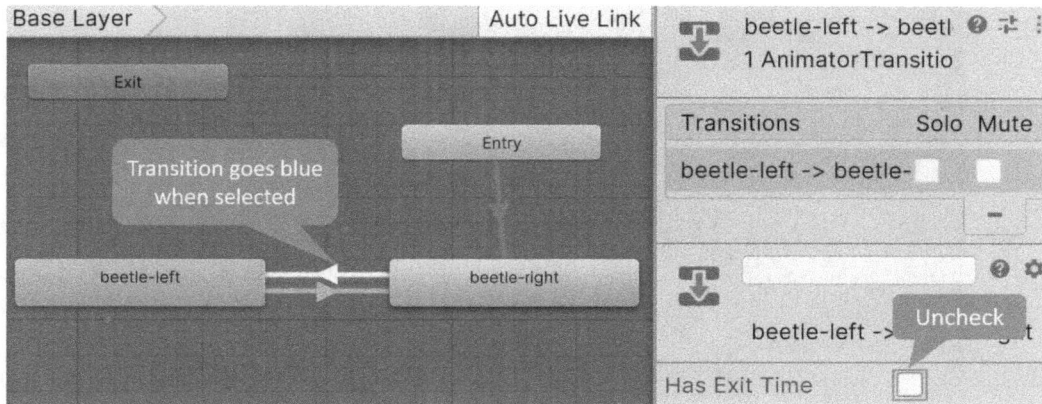

Figure 8.9: Unchecking the Has Exit Time option for an instant transition

> To delete a **Transition**, select it and then use the *Delete* key (**Windows**) or press *Fn + Backspace* (**macOS**).

11. To decide when to change the active state, we need to create a parameter indicating whether the *Left/Right* arrow keys have been pressed. *Left/Right* key presses are indicated by the Unity input system's **horizontal** axis value. Create a state chart float parameter named axisHorizontal by selecting **Parameters** (rather than **Layers**) from the top left of the **Animator** pane, clicking the plus (+) button, and choosing **Float**. Name it axisHorizontal:

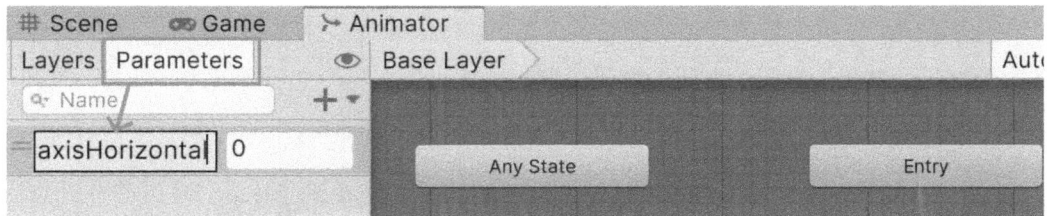

Figure 8.10: Adding a Float as a parameter and naming it axisHorizontal

12. With our parameter, we can define the conditions for changing between the left- and right-facing states. When the *Left* arrow key is pressed, the Unity input system's **Horizontal** axis value is negative, so select the **Transition** from beetle-right to beetle-left and in the **Inspector** panel, click the plus (+) symbol in the **Conditions** section of the **Transition** properties. Since there is only one parameter, this will automatically be suggested, with defaults of **Greater** than **zero**.

Use the drop-down menu to change from **Greater** to **Less** so that we get our desired condition.

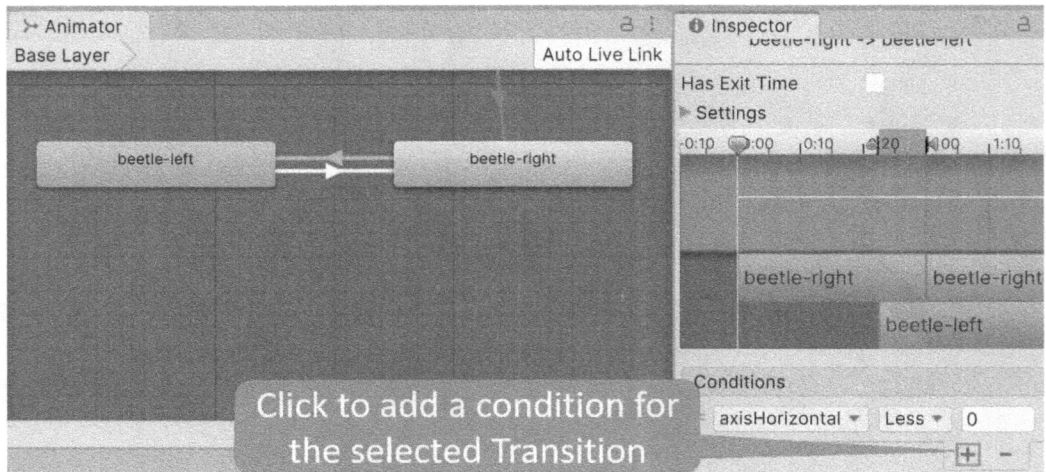

Figure 8.11: Changing the condition of a Transition to Less

13. Now, select the **Transition** from beetle-left to beetle-right and add a **condition.** In this case, the default for axisHorizontal is **Greater** than **0**, which is just what we want (since a positive value is returned by the Unity input system's **Horizontal** axis when the *Right* arrow key is pressed).

14. We need a method to actually map from the Unity input system's **Horizontal** axis value (from the *Left/Right* arrow keys) to our **Animator** state chart parameter, called axisHorizontal. We can do this with a short script class, which we'll create shortly.

15. Create a C# script class named InputMapper and add an instance object as a component to the Enemy Bug GameObject:

```
using UnityEngine;

public class InputMapper : MonoBehaviour {
    Animator animator;

    void Start() {
        animator = GetComponent<Animator>();
    }

    void Update() {
        animator.SetFloat("axisHorizontal", Input.
GetAxisRaw("Horizontal"));
    }
}
```

16. Now, we need to change the local scale property of the GameObject when we switch to the left- or right-facing state. Create a C# script class named `LocalScaleSetter`:

```
using UnityEngine;

public class LocalScaleSetter : StateMachineBehaviour {
    public Vector3 scale = Vector3.one;

    override public void OnStateEnter(Animator animator,
AnimatorStateInfo stateInfo, int layerIndex) {
        animator.transform.localScale = scale;
    }


}
```

17. In the **Animator** panel, select the `beetle-right` state. In the **Inspector** panel, click the **Add Behaviour** button and select **LocalScaleSetter**. The default public `Vector3` value of (`1, 1, 1`) is fine for this state.

18. In the **Animator** panel, select the `beetle-left` state. In the **Inspector** panel, click the **Add Behaviour** button and select **LocalScaleSetter**. Change the public `Vector3` scale to a value of (`-1, 1, 1`) – that is, we need to swap the X scaling to make our **sprite** face to the left:

Figure 8.12: Setting the public Vector3 scale to a value of (-1,1,1) for beetle-left

> Adding instance objects of C# script classes to **Animator** states is a great way to link the logic for actions when entering into/exiting a state with the **Animator** states themselves.

19. When you run your scene, pressing the *Left* and *Right* arrow keys should make the bug face left or right.

How it works...

Each frame of the `Update()` method of the `InputMapper` C# script class reads the Unity input system's **Horizontal** axis value and sets the **Animator State** chart's `axisHorizontal` parameter to this value.

If the value is less than (left arrow) or greater than (right arrow) 0, if appropriate, the **Animator** state system will switch to the other state.

The LocalScaleSetter C# script class actually changes the localScale property (with an initial value of 1, 1, 1, or reflected horizontally to make it face left at -1, 1, 1). For each state, the public Vector3 variable can be customized to the appropriate values.

The OnStateEnter(...) method is involved each time you enter the state that an instance object of this C# class is attached to. You can read about the various event messages for the StateMachineBehaviour class at https://docs.unity3d.com/ScriptReference/StateMachineBehaviour.html.

When we press the *Left* arrow key, the value of the Unity input system's **Horizontal** axis value is negative, and this is mapped to the **Animator** state chart's axisHorizontal parameter, causing the system to **Transition** to the beetle-left state and OnStateEnter(...) of the LocalScaleSetter script class instance to be executed. This sets the local scale to (-1, 1, 1), making the texture flip horizontally so that the beetle faces left.

There's more...

Here is a suggestion for enhancing this recipe.

Instantaneous swapping

You may have noticed a delay, even though we set **Exit Time** to 0. This is because there is a default blending when transitioning from one state to another. However, this can be set to 0 so that the **State Machine** switches instantaneously from one state to the next.

Do the following:

1. Select each **Transition** in the **Animator** panel.
2. Expand the **Settings** properties.
3. Set both **Transition Duration** and **Transition Offset** to 0:

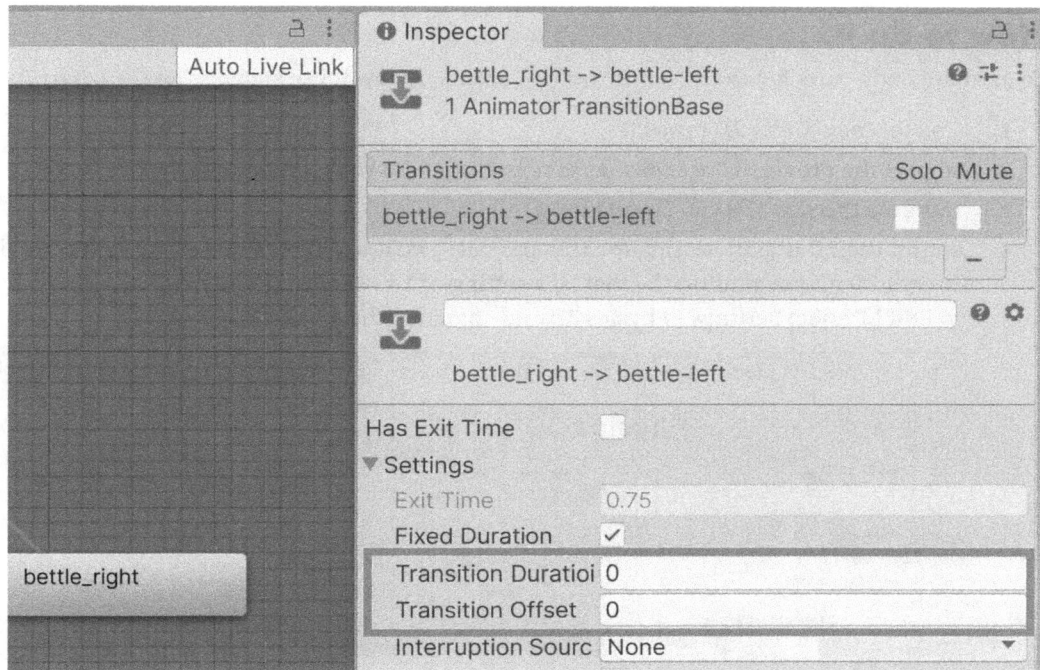

Figure 8.13: Setting both Transition Duration and Transition Offset to 0

Now, when you run the scene, the bug should immediately turn left and right as you press the corresponding arrow keys.

Animating body parts for character movement events

In the previous recipe, we used the Unity animation tool to alter the transition of a sprite based on input. In this recipe, we'll learn how to animate the hat of the Unity PotatoMan character in response to a jumping event using a variety of animation techniques, including keyframes and transforms.

Getting ready

For this recipe, we have prepared the files you need in the 08_03 folder.

How to do it...

To animate body parts for character movement events, follow these steps:

1. Create a new Unity 2D project.

2. Import the provided PotatoManAssets package into your project.

3. Increase the size of Main Camera to 10.

4. Set up the 2D gravity setting for this project – we'll use the same setting that's provided in Unity's 2D platform tutorial – that is, a setting of Y= -30. Set 2D gravity to this value by going to **Edit** | **Project Settings** | **Physics** 2D and changing the **Y** value to -30:

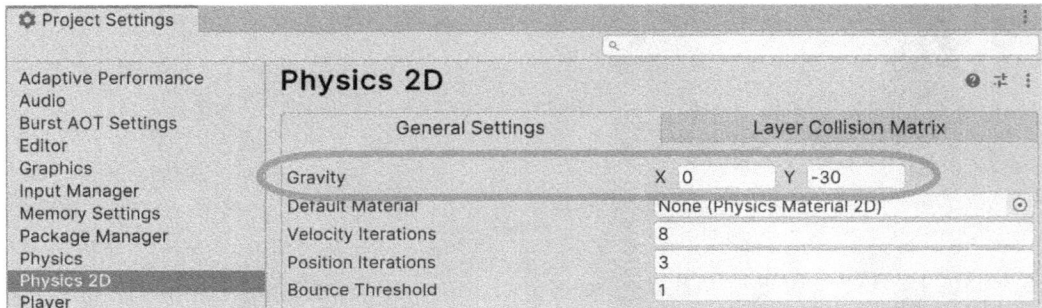

Figure 8.14: Setting up 2D gravity for this project

5. Drag an instance of the character2D **hero** character from the **Project** | Prefabs folder into the scene. Position this GameObject at (0, 3, 0). Rename this GameObject hero.

6. Drag an instance of the platformWallBlocks sprite from the **Project** | **Sprites** folder into the scene. Position this GameObject at (0, -4, 0).

7. Add a **Box Collider 2D** component to the platformWallBlocks GameObject by going to **Component** | **Physics 2D** | **Box Collider 2D**.

8. We now have a stationary platform that the player can land on and walk left and right on. Create a new **layer** named Ground and assign the platformWallBlocks GameObject to this new layer, as shown in the following screenshot. Pressing the *Spacebar* when the character is on the platform will now make him jump:

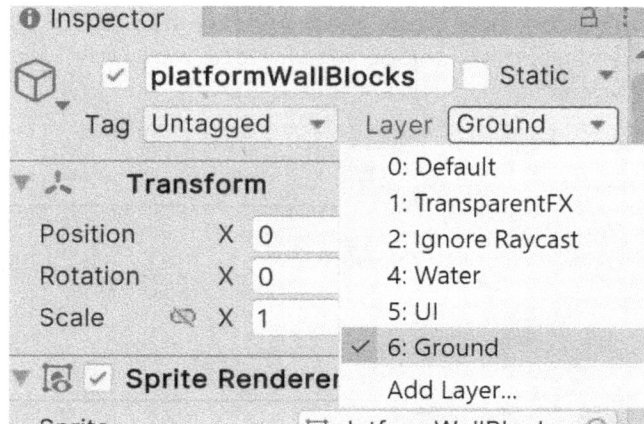

Figure 8.15: Adding a new layer called Ground

9. Currently, the hero character is animated (arms and legs moving) when we make him jump. Let's remove the **Animation Clip** components and **Animator Controller** and create our own from scratch. Delete the **Clips** and **Controllers** folders from the **Project** panel folder in **Assets | PotatoMan2DAssets | Character2D | Animation**, as shown here:

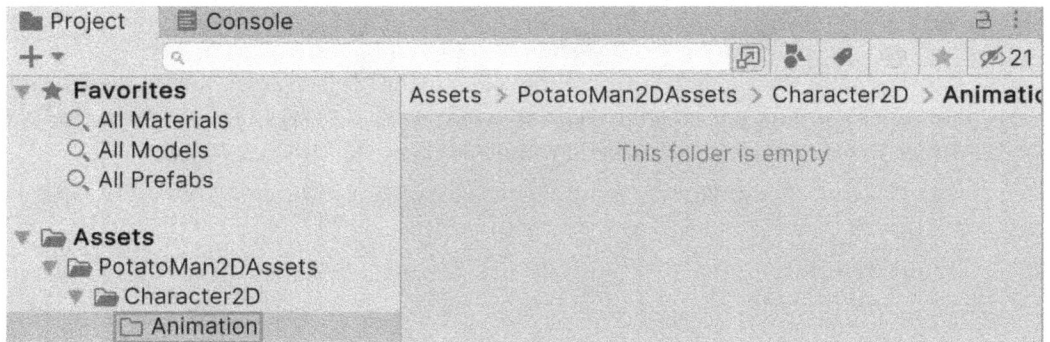

Figure 8.16: Animation folder with Clips and Controllers removed

10. Let's create an **Animation Clip** (and its associated **Animator Controller**) for our hero character. In the **Hierarchy** panel, select the hero GameObject. Ensuring that the hero GameObject is selected in the **Hierarchy** panel, open the **Animation** panel and ensure it is in **Dopesheet** view (this is the default).

11. Click the **Animation** panel's **Create** button and save the new clip in the **Character2D | Animation** folder, naming it `character-potatoman-idle`. You've now created an **Animation Clip** for the **Idle** character state (which is not animated):

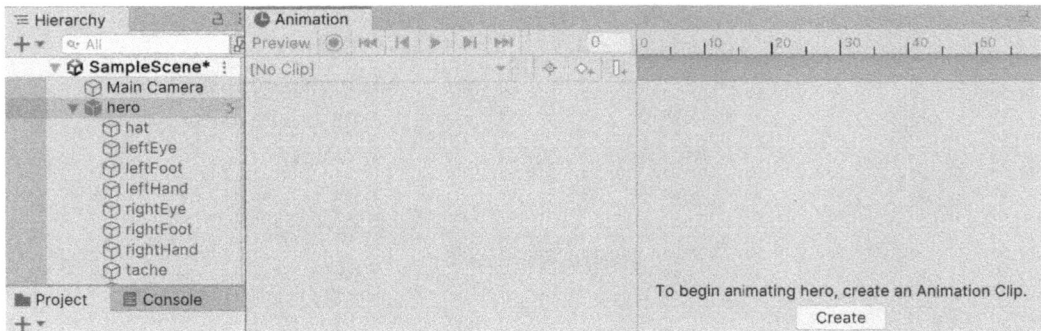

Figure 8.17: Animation panel for the PotatoMan character

> Your final game may end up with dozens, or even hundreds, of animation clips. Make things easy to search for by prefixing the names of the clips with the object's type, name, and then a description of the animation clip.

12. Looking at the **Character2D | Animation** folder in the **Project** panel, you should see both animation clips you have just created (`character-potatoman-idle`) and a new **Animator Controller**, which has defaulted to the name of your hero Character2D GameObject:

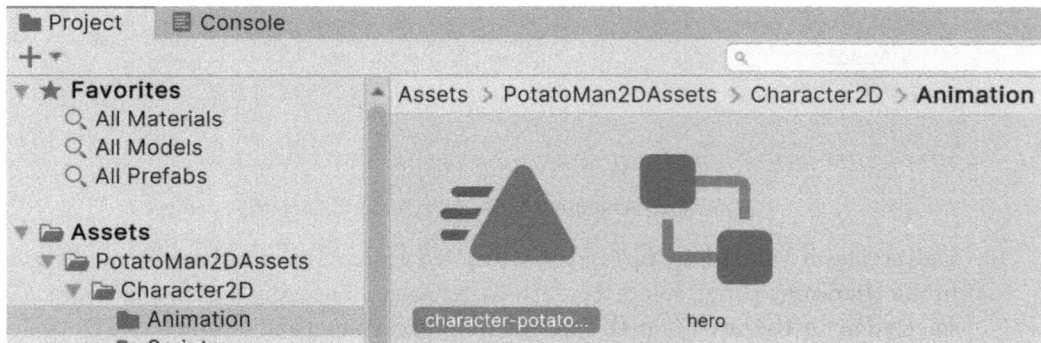

Figure 8.18: Animation clips and Animator Controller for the PotatoMan character

13. Ensuring the hero GameObject is selected in the **Hierarchy** panel, open the **Animator** panel. You'll see a **State Machine** for controlling the animation of our character. Since we only have one **Animation Clip** (`character-potatoman-idle`), upon entry, the **State Machine** immediately enters this state:

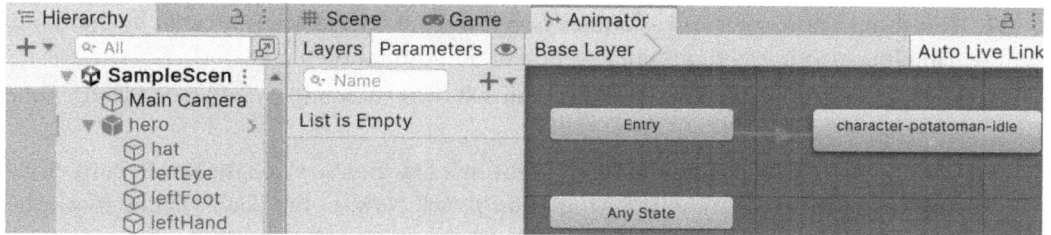

Figure 8.19: State Machine for controlling the animation of the character

14. Run your scene. Since the character is always in the "idle" state, we see no animation yet when we make it jump.

15. Create a jump **Animation Clip** that animates **hat**. Ensure that the hero GameObject is still selected in the **Hierarchy** panel. Click the drop-down menu in the **Animation** panel and create a new clip in your **Animation** folder, naming it character-potatoman-jump:

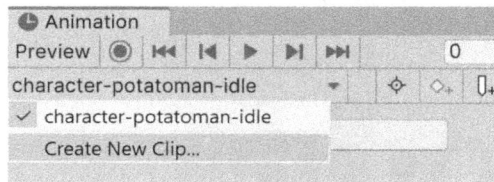

Figure 8.20: Creating a new clip in your Animation folder

16. Click the **Add Property** button and choose **Transform | Position** for the **hat** child object, by clicking its plus (+) button. We are now ready to record changes that are made to the (X, Y, Z) position of the **hat** GameObject in this Animation Clip:

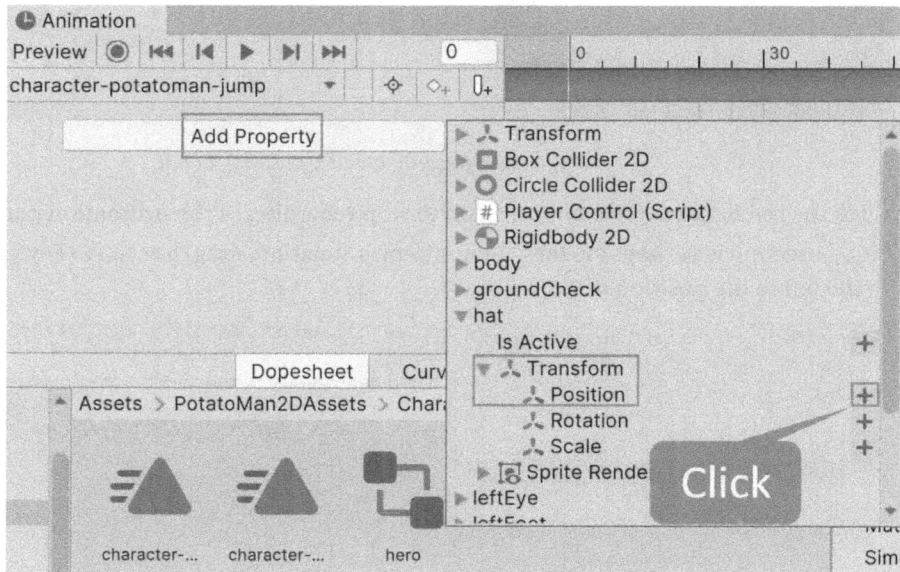

Figure 8.21: Adding the Position property

17. You should now see two keyframes at 0.0 and 1.0. These are indicated by diamonds in the **Timeline** area in the right-hand section of the **Animation panel**.

18. Click to select the first keyframe (at time 0.0) – the diamond should turn blue to indicate it is selected.

19. Let's record a new position for the **hat** for this first frame. Click the red **Record** circle button once to start recording in the **Animation** panel. Now, in the **Scene** panel, move the **hat** up and left a little, away from the head. You should see that all three X, Y, and Z values have a red background in the **Inspector** panel – this is to inform you that the values of the **Transform** component are being recorded in an **Animation Clip**:

Figure 8.22: Recording a new position for hat

20. Click the red **Record** circle button again to stop recording in the **Animation** panel.

21. Since 1 second is perhaps too long for our jump animation, drag the second keyframe diamond to the left to the position of 0.5:

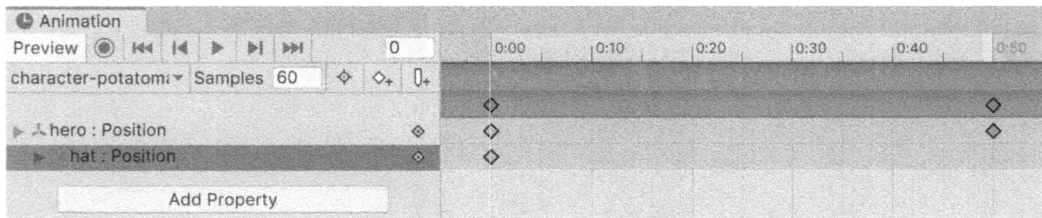

Figure 8.23: Adjusting the second keyframe to a position of 0.5

22. We need to define when the character should **Transition** from the **Idle** state to the **Jump** state. In the **Animator** panel, select the `character-potatoman-idle` state and create a **Transition** to the `character-potatoman-jump` state by right-clicking and choosing the **Make Transition** option. Then, drag the **Transition** arrow to the `character-potatoman-jump` state, as shown in the following screenshot:

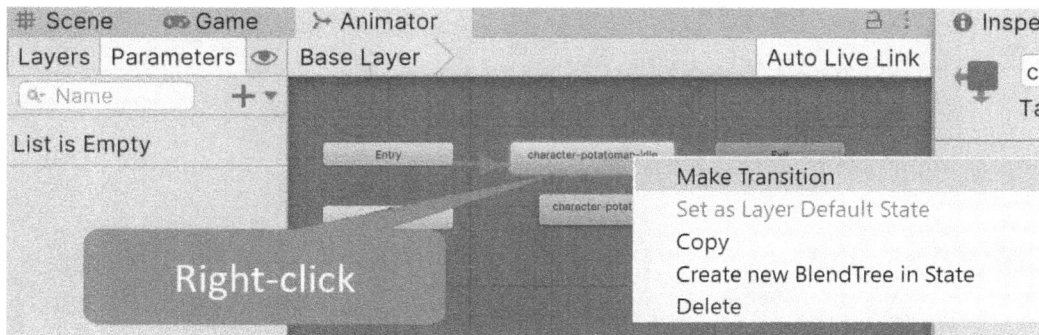

Figure 8.24: Making a Transition for character-potatoman-idle

23. Let's add a **Trigger** parameter named **Jump** by clicking on the plus (+) button at the top left of the **Animator** panel, choosing **Trigger**, and typing in `Jump`:

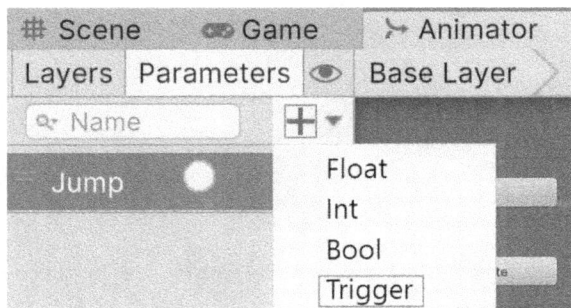

Figure 8.25: Adding a Trigger parameter named Jump

24. We can now define the properties for when our character should **Transition** from **Idle** to **Jump**. Click the **Transition** arrow to select it, set the following two properties, and add one condition to the **Inspector** panel:

- **Has Exit Time:** Uncheck this option.
- **Transition Duration(s):** Set to `0.01`.

- **Conditions:** Add **Jump** (click the plus (+) button at the bottom):

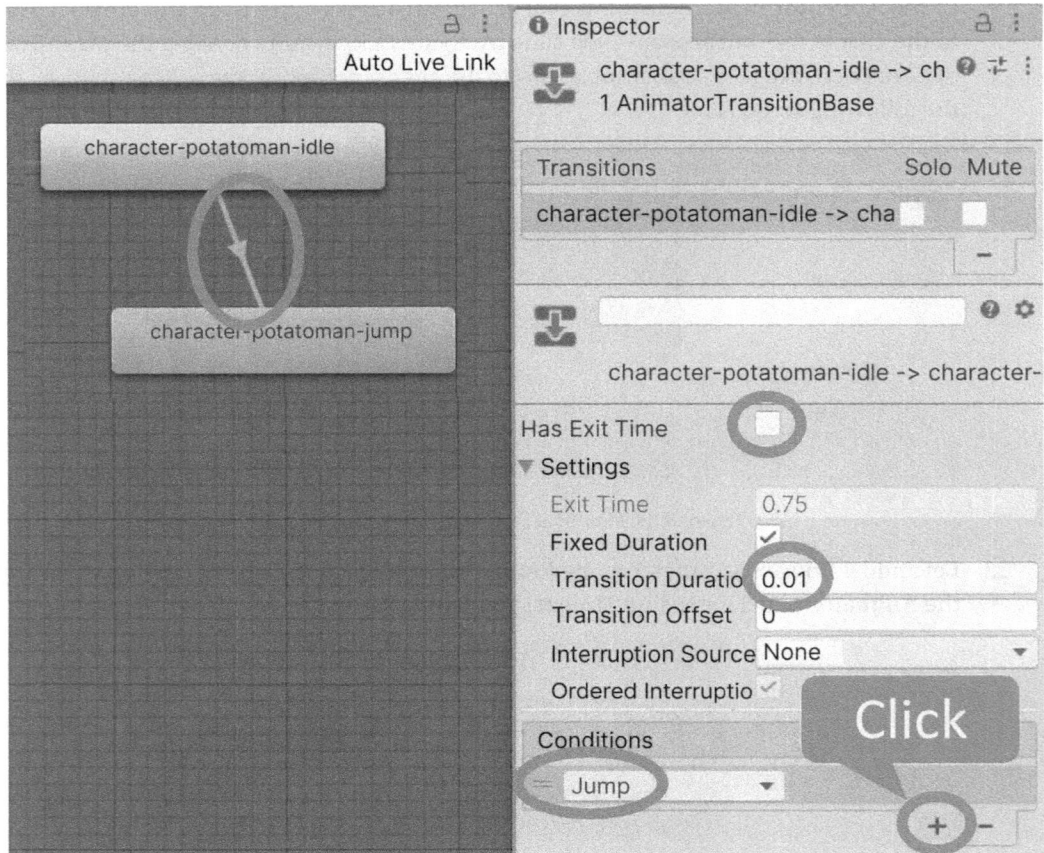

Figure 8.26: Inspector panel for character Transition properties

25. Save and run your scene. Once the character has landed on the platform and you press the *Spacebar* to **jump**, you'll see the character's hat jump away from his head and slowly move back. Since we haven't added a **Transition** to leave the **Jump** state, this **Animation Clip** will loop so the **hat** keeps on moving, even when the jump is completed.

26. In the **Animator** panel, select the character-potatoman-jump state and add a new **Transition** to the character-potatoman-idle state. Select this **Transition** arrow and in the **Inspector** panel, set its properties as follows:

- **Has Exit Time:** Leave checked.
- **Exit Time:** 0.5 (this needs to have the same time value as the second keyframe of our **Jump** animation clip):

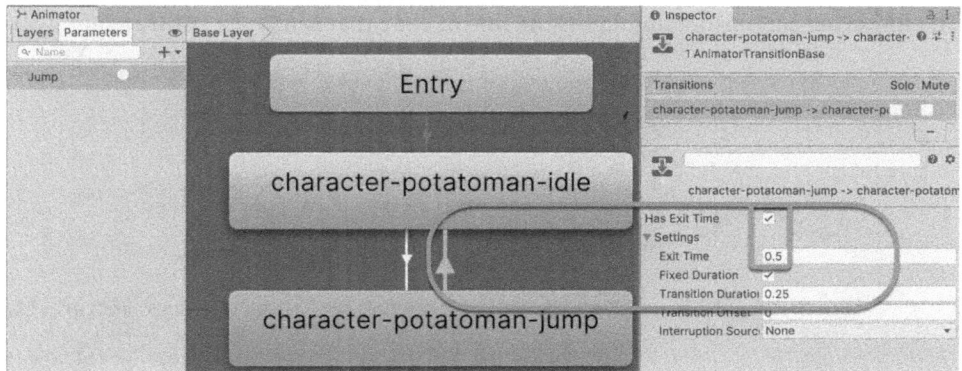

Figure 8.27: The character-potatoman-jump state and character-potatoman-idle
state Transition properties

27. Save and run your scene. Now, when you jump, the **hat** should play its animation once, after which the character will immediately return to its **Idle** state.

How it works...

In this recipe, you added an **Animator Controller State Machine** to the hero GameObject. The two animation clips you created (**idle** and **jump**) appear as states in the **Animator** panel. You created a **Transition** from **Idle** to **Jump** when the **JumpTrigger** parameter is received by the **State Machine**. You then created a second **Transition**, which transitions back to the **Idle** state after waiting 0.5 seconds (the same duration between the two keyframes in our **Jump Animation Clip**).

The player makes the character jump by pressing the *Spacebar*. This causes the code in the PlayerControl C# scripted component of the hero GameObject to be invoked, which makes the sprite move upward on the screen and also sends a SetTrigger(...) message to the **Animator Controller** component for the **Jump** trigger.

The difference between a **Boolean parameter** and a **Trigger** is that a **Trigger** is temporarily set to true. Once the SetTrigger(...) event has been consumed by a state transition, it automatically returns to false. So, triggers are useful for actions we wish to complete once and then revert to a previous state. A **Boolean parameter** is a variable that can have its value set to **True** or **False** at different times during the game. So, different **Transitions** can be created to fire, depending on the value of the variable at any time. Note that Boolean parameters have to have their values explicitly set back to **False** with SetBool(...); otherwise, they will remain on the **True** value.

The following screenshot highlights the line of code that sends the `SetTrigger(...)` message.

```
                                       PlayerControl.cs — Scripts
68
69        // If the player should jump...
70        if(jump)
71        {
72            // Set the Jump animator trigger parameter.
73            anim.SetTrigger("Jump");
74
75            // Add a vertical force to the player.
76            GetComponent<Rigidbody2D>().AddForce(new Vector2(0f, jur
77
78            // Make sure the player can't jump again until the jump
    Update are satisfied.
79            jump = false;
80        }
```

Figure 8.28: Example of the code that sends the Trigger message

The **State Machines** of animations with a range of motions (running/walking/jumping/falling/dying) will have more states and transitions. The Unity-provided `PotatoMan` **hero** character has a more complex **State Machine**, and more complex animations (for its hands and feet, eyes and hat, and so on for each **Animation Clip**), which you may wish to explore.

You can learn more about the **Animation** view at `http://docs.unity3d.com/Manual/AnimationEditorGuide.html`.

Creating a three-frame animation clip to make a platform continually animate

In this recipe, we'll make a wooden-looking platform continually animate, moving upward and downward. This can be achieved with a single three-frame **Animation Clip** (starting at the top, then positioned at the bottom, and finally back at the top position). Note that each frame is a static position and that we will employ the animation technique known as **in-betweening** to create the necessary movement:

Figure 8.29: Example of a moving platform game

Getting ready

This recipe builds on the previous one, so make a copy of that project and work on the copy for this recipe.

How to do it...

To create a continually moving animated platform, follow these steps:

1. Drag an instance of the platformWoodBlocks sprite from the **Project | Sprites** folder into the scene. Position this GameObject at (-4, -5, 0) so that these wood blocks are neatly to the left and slightly below the wall blocks platform.

2. Add a **Box Collider 2D** component to the platformWoodBlocks GameObject so that the player's character can stand on this platform too. To do this, go to **Add Component | Physics 2D | Box Collider 2D**.

3. Create a new folder named Animations to store the controller and **Animation Clip** we'll create next.

4. Ensuring the platformWoodBlocks GameObject is still selected in the **Hierarchy** panel, open an **Animation** panel and ensure it is in **Dopesheet** view (this is the default).

5. Click the **Animation** panel's **Create** button and save the new clip in your new Animation folder, naming it platform-wood-moving-up-down.

6. Click the **Add Property** button, choose **Transform,** and then click the plus (+) button by **Position.** We are now ready to record changes to the (X, Y, Z) position of the platformWoodBlocks GameObject in this **Animation Clip**:

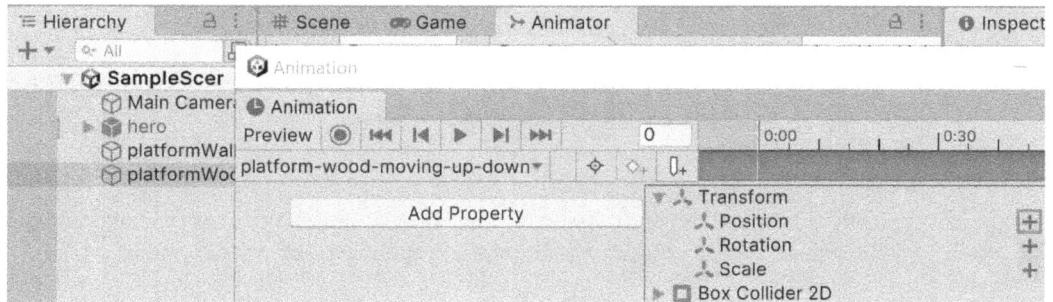

Figure 8.30: Animation setting for Transform | Position of platformWoodBlocks

7. You should now see two keyframes at **0.0** and at **1.0**. These are indicated by diamonds in the **Timeline** area in the right-hand section of the **Animation** panel.

8. We need three keyframes, with the new one at **2:00** seconds. Click at **2:00** in the **Timeline** area, along the top of the **Animation** panel, so that the red line for the current playhead time moves to **2:00**.

Then, click the diamond + button to create a new keyframe at the current playhead position:

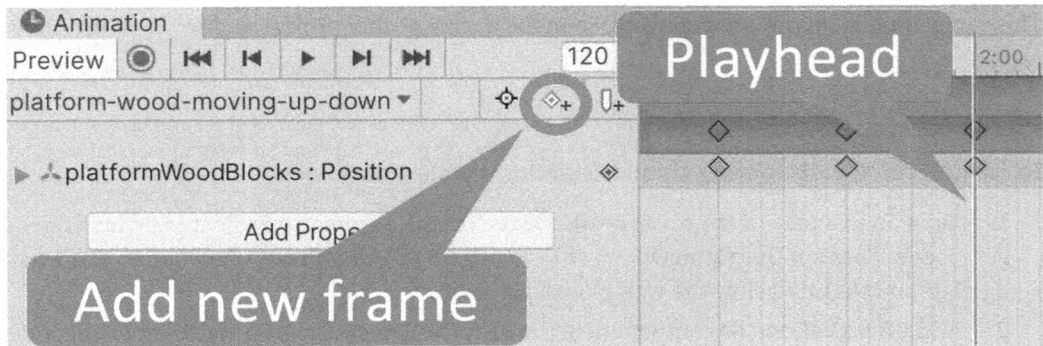

Figure 8.31: Adding a new keyframe at 2:00 seconds

9. The first and third keyframes are fine – they record the current height of the wood platform at **Y** = -5. We need to make the middle keyframe record the height of the platform at the top of its motion, and Unity **in-betweening** will do the rest of the animation work for us.

10. Select the middle keyframe (at the **1:00 position**) by clicking on the diamond at **1:00** (they should both turn blue, and the red playhead vertical line should move to **1:00** to indicate that the middle keyframe is being edited).

11. Click the red **Record** circle button to start recording changes.

12. In the **Inspector** panel, change the Y position of the platform to 0. You should see that all three X, Y, and Z values have a red background in the **Inspector** panel – this is to inform you that the values of the **Transform** component are being recorded in the **Animation Clip**.

13. Click the red **Record** circle button again to finish recording your changes.

14. Save and run your scene. The wooden platform should now be animating continuously, moving smoothly up and down between the positions we set up.

> If you want the PotatoMan character to be able to jump when on the moving wooden block, you'll need to select the `block` GameObject and set its layer to **Ground**.

How it works...

In this recipe, you added an animation to the `platformWoodBlocks` GameObject. This animation contains three keyframes. A keyframe represents the values of the properties of the object at a point in time. The first keyframe stores a Y value of -4, the second keyframe a Y value of 0, and the final keyframe -4 again. Unity calculates all the in-between values for us, and the result is a smooth animation of the Y position of the platform.

In-betweening is the automated calculation of intermediate values to provide smoothly animated changes between GameObject properties.

There's more...

Here's a suggestion for enhancing this recipe.

Copying the animation relative to a new parent GameObject

If we wanted to duplicate the moving platform, simply duplicating the platformWoodBlocks Game-Object in the **Hierarchy** panel and moving the copy would n't work. When you run the scene, each duplicate would be animated back to the location of the original animation frames (that is, all the copies would be positioned and moving from the original location).

The solution is to create a new, empty GameObject named movingBlockParent, and then a platformWoodBlocks child for this GameObject. Once we've done this, we can duplicate the movingBlockParent GameObject (and its platformWoodBlocks child) to create more moving blocks in our scene, all of which move relative to where the parent GameObject is located at design time.

Making a platform start falling once stepped on using a Trigger to move the animation from one state to another

In many cases, we don't want an animation to begin until some condition has been met, or some event has occurred. In these cases, a good way to organize an **Animator Controller** is to have two animation states (clips) and a **Trigger** on the transition between the clips. We can use code to detect when we want the animation to start playing, and at that time, we send the **Trigger** message to the **Animator Controller,** causing a **Transition** to start.

In this recipe, we'll create a water platform block in our 2D platform game. Such blocks will begin to slowly fall down the screen as soon as they have been stepped on, and so the player must keep on moving; otherwise, they'll fall down the screen with the blocks too!

Figure 8.32: Example of a falling platform

Getting ready

This recipe builds on the previous one, so make a copy of that project and work on the copy for this recipe.

How to do it...

To construct an animation that only plays once a **Trigger** has been received, follow these steps:

1. In the **Hierarchy** panel, create an empty GameObject named water-block-container, positioned at (2.5, -4, 0). This empty GameObject will allow us to make duplicates of animated water blocks that will animate relative to their parent GameObject's position.

2. Drag an instance of the Water Block sprite from the **Project | Sprites** folder into the scene and make it a child of the water-block-container GameObject. Ensure the position of your new child Water Block GameObject is (0, 0, 0) so that it appears neatly to the right of the wall blocks platform:

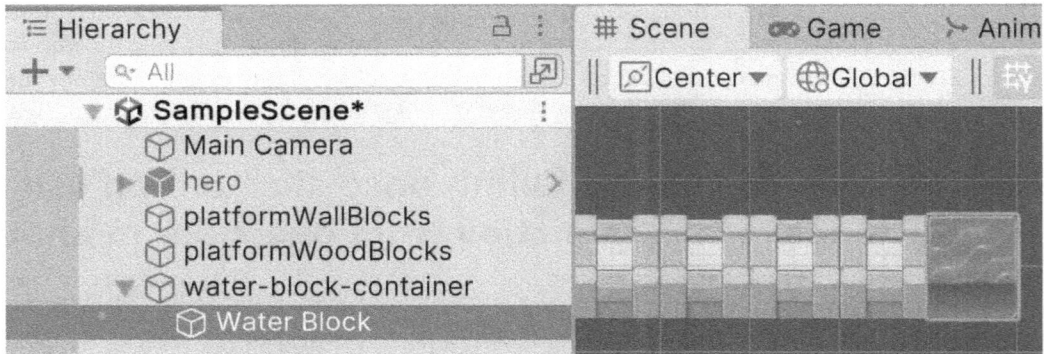

Figure 8.33: Dragging the Water Block sprite to the scene and ensuring its position is (0,0,0)

3. Add a **Box Collider 2D** component to the child Water Block GameObject. Set the layer of this GameObject to **Ground** so that the player's character can stand and jump on this water block platform.

4. Ensuring the child Water Block GameObject is selected in the **Hierarchy** panel, open an **Animation** panel and create a new clip named platform-water-up, saving it in your Animations folder. Click the **Add Property** button, choose **Transform** and **Position**, and delete the second keyframe at **1:00**.

5. Create a second **Animation Clip** named platform-water-down. Again, click the **Add Property** button, choose **Transform** and **Position**, and delete the second keyframe at **1:00**.

6. With the first keyframe at **0:00** selected, click the red **Record** button once to start recording changes and set the Y value of the GameObject's **transform position** to -5. Press the red **Record** button again to stop recording changes. You have now created the water-block-down **Animation Clip**.

7. You may have noticed that, as well as the up/down animation clips that you created, another file was created in your Animations folder – an **Animator Controller** named Water Block. Select this file and open the **Animator** panel to view and edit the **State Machine** diagram:

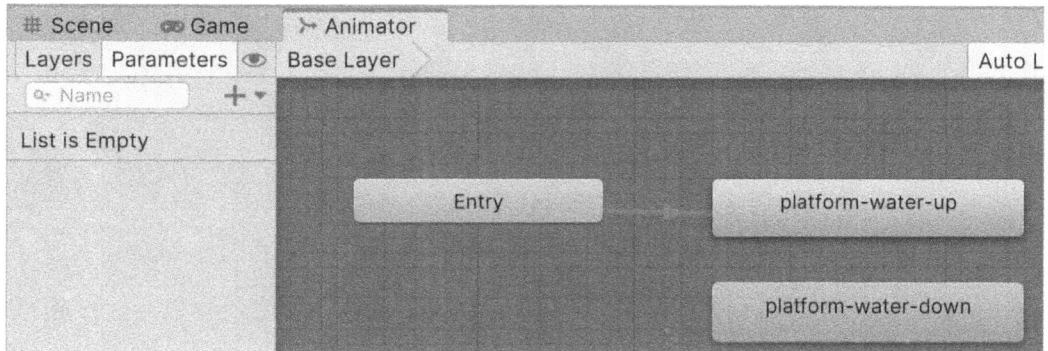

Figure 8.34: Animator Controller for the Water Block GameObject

8. Currently, although we've created two animation clips (states), only the **Up** state is ever active. This is because when the scene begins (**Entry**), the object will immediately go into the platform-water-up state, but since there are no **Transition** arrows from this state to platform-water-down, at present, the Water Block GameObject will always be in its **Up** state.

9. Ensure the platform-water-up state is selected (it will have a blue border around it) and create a **Transition** (arrow) to the platform-water-down state by choosing **Make Transition** from the right-click menu.

10. If you run the scene now, the default **Transition** settings will be provided after 0.75 seconds (default **Exit Time value**), and Water Block will **Transition** into its **Down** state. We don't want this – we only want them to animate downward after the player has walked onto them.

11. Create a **Trigger** named Fall by choosing the **Parameters** tab in the **Animator** panel, clicking the + button and selecting **Trigger**, and then selecting Fall.

12. Do the following to create the transition to wait for our **Trigger**:

    • In the **Animator** panel, select **Transition**.

    • In the **Inspector** panel, uncheck the **Has Exit Time** option.

    • Set **Transition Duration** to 3.0 (so that Water Block slowly transitions to its **Down** state over a period of 2 seconds).

- In the **Inspector** panel, click the + button to add a **condition**. This should automatically suggest the only possible **condition parameter**, which is our `Fall` **Trigger**:

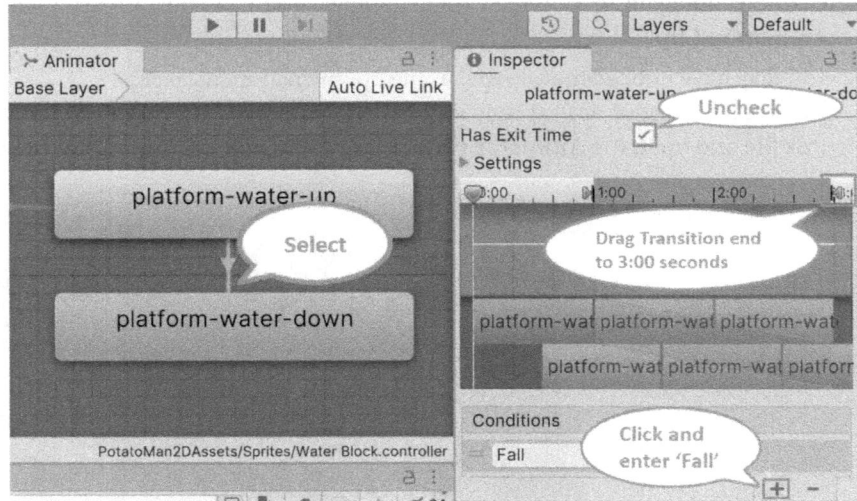

Figure 8.35: Fall Trigger settings for the Down state of the platform

> An alternative to setting **the transition duration** numerically is to drag the **Transition** end time to **3:00** seconds in the **Animation Timeline**, under **Transition Settings** in the **Inspector** panel.

13. Ensure the child `Water Block` GameObject is selected, add a (second) **2D Box Collider** with a **Y Offset** of 1, and tick its **Is Trigger** checkbox:

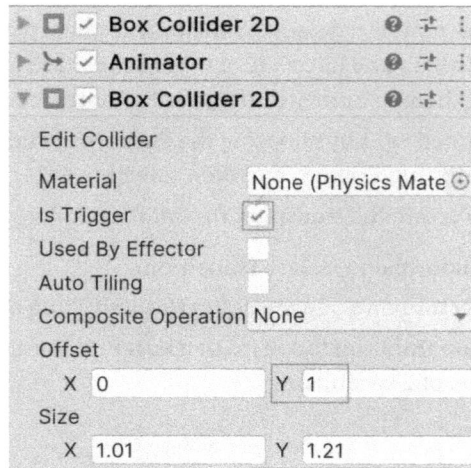

Figure 8.36: Box Collider 2D settings for Water Block

14. Create a C# script class called WaterBlock and add an instance object as a component to the child Water Block GameObject:

```csharp
using UnityEngine;
using System.Collections;

public class WaterBlock : MonoBehaviour {
    const string TAG_PLAYER = "Player";
    const string ANIMATION_TRIGGER_FALL = "Fall";

    private Animator animatorController;

    void Start(){
        animatorController = GetComponent<Animator>();
    }

    void OnTriggerEnter2D(Collider2D hit){
        if(hit.CompareTag(TAG_PLAYER)){
            animatorController.SetTrigger(ANIMATION_TRIGGER_FALL);
        }
    }
}
```

15. Make six more copies of the water-block-container GameObject, with their X positions increasing by 1 each time; that is, 3.5, 4.5, 5.5, and so on.

16. Run the scene. As the player's character runs across each water block, they will start falling down, so they had better keep running!

How it works...

In this recipe, you created an empty GameObject called water-block-container to act as a container for a WaterBlock. By adding a WaterBlock to this parent GameObject, you made it possible to make copies and move them in the scene so that the animations were **relative** to the location of each parent (container) GameObject.

By adding a **Box Collider 2D** and setting the layer of WaterBlock to **Ground**, you enabled the player's character to walk on these blocks.

You created a two-state **Animator Controller State Machine**. Each state was an **Animation Clip**. The **Up** state is for WaterBlock at normal height (Y = 0), while the **Down** state is for WaterBlock further down the screen (Y = -5). You created a **Transition** from the **Water Block Up** state to its **Down** state that will take place when the **Animator Controller** receives a Fall **Trigger** message.

After that, you added a second **Box Collider 2D** with a **Trigger** to WaterBlock so that our script could detect when the player (tagged **Player**) enters its Collider. When the player triggers the Collider, the Fall **Trigger** message is set, which makes the WaterBlock GameObject start gently transitioning into its **Down** state further down the screen.

You can learn more about Animator Controllers at http://docs.unity3d.com/Manual/class-AnimatorController.html.

Creating animation clips from sprite sheet sequences

The traditional method of animation involved hand-drawing many images, each slightly different, which were displayed quickly frame by frame to give the appearance of movement. For computer game animation, the term **sprite sheet** is given to an image file that contains one or more sequences of sprite frames. Unity provides tools to break up individual sprite images into large sprite sheet files so that individual frames, or sub-sequences of frames, can be used to create animation clips that can become states in **Animator Controller** State Machines. In this recipe, we'll import and break up an open source monster sprite sheet into three animation clips for **Idle**, **Attack**, and **Death**, as follows:

[attack]

Figure 8.37: An example of a sprite sheet

Getting ready

For all the recipes in this chapter, we have prepared the sprite images you need in the 08_06 folder. Many thanks to Rosswet Mobile for making these sprites available as open source.

How to do it...

To create a frame-by-frame animation using a sprite sheet, follow these steps:

1. Create a new Unity 2D project.
2. Import the provided image; that is, monster1.
3. With the monster1 image selected in the **Project** panel, change its sprite mode to **Multiple** in the **Inspector** panel. Then, click the **Apply** button at the bottom of the panel:

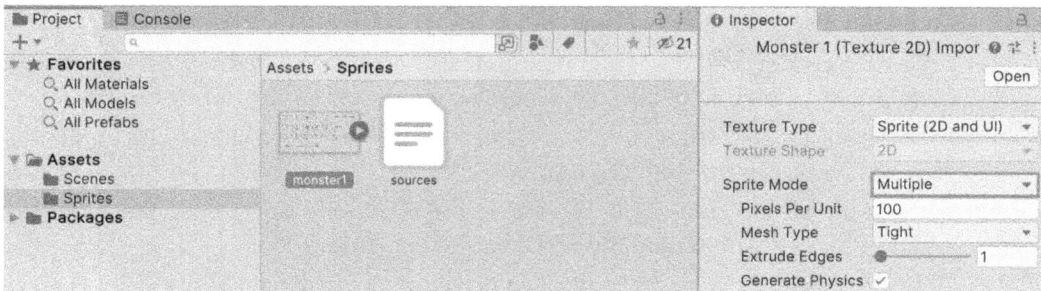

Figure 8.38: Setting for changing Sprite mode to Multiple

4. In the **Inspector** panel, open the **Sprite Editor** panel by clicking the **Sprite Editor** button.

5. In **Sprite Editor**, open the **Slice** drop-down dialog. For **Type**, choose the **Grid by Cell Size** drop-down option and set **X** and **Y** to 64. Click the **Slice** button, and then the **Apply** button in the bar at the top right of the **Sprite Editor** panel:

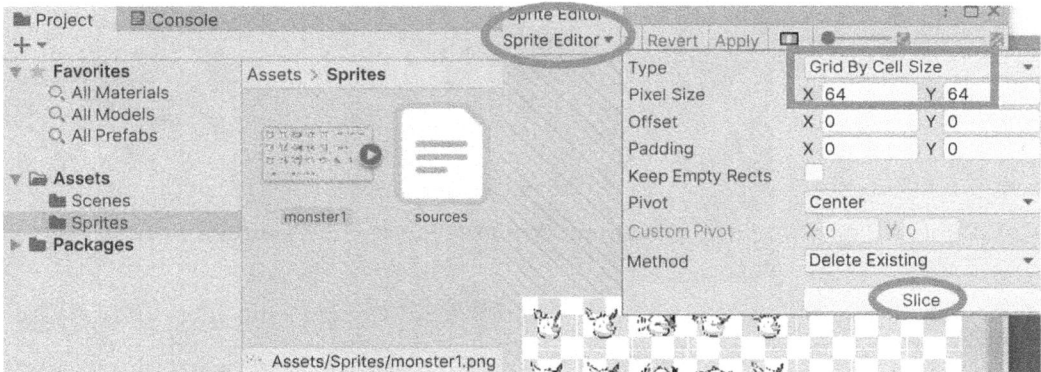

Figure 8.39: Sprite Editor settings for splicing the sprite sheet

6. In the **Project** panel, you can now click the triangle button on the right-hand side of the sprite. You'll see all the different child frames of this sprite (as highlighted in the following screenshot):

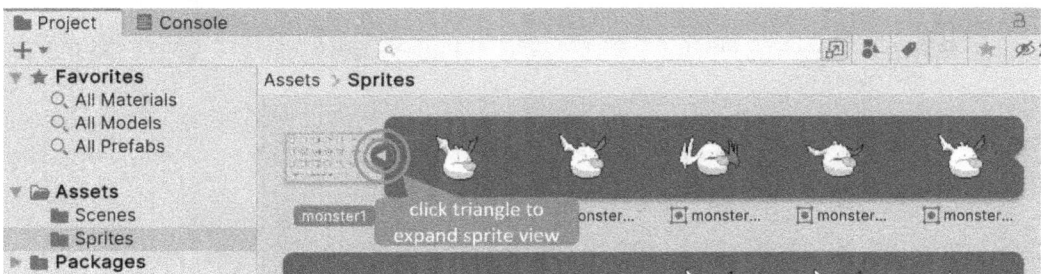

Figure 8.40: Clicking the triangle to expand the sprite view to access each individual sprite

7. Create a folder named `Animations`.

8. In your new folder, create an **Animator Controller** asset file named `monster-animator` by going to the **Project** panel, right-clicking, and selecting **Create | Animator Controller**.

9. In the scene, create a new empty GameObject named `monster1` (at position `0, 0, 0`) and drag your `monster-animator` into this GameObject.

10. With the `monster1` GameObject selected in the **Hierarchy** panel, open the **Animation** panel and create a new **Animation Clip** named `monster1-idle`.

11. Select the `monster1` image in the **Project** panel (in its expanded view) and select and drag the first five frames (`monster1_0` to `monster1_4`) into the **Animation** panel.

12. Display the **Sample** properties by clicking the **More Options** (three dots) button in the right-hand corner of the **Animation** panel and select **Show Sample Rate**.

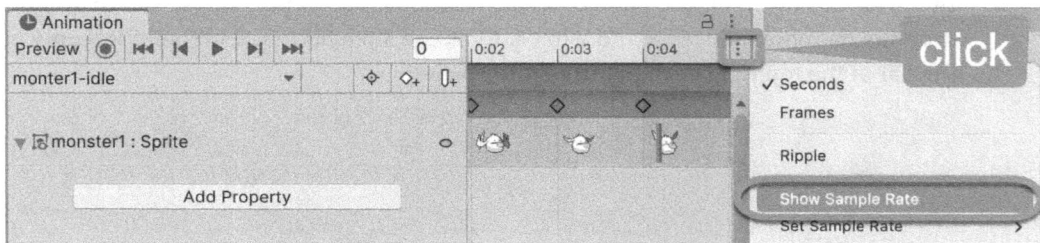

Figure 8.41: Displaying the sample properties

13. You will need to change the sample rate to `12` (since this animation was created to run at 12 frames per second).

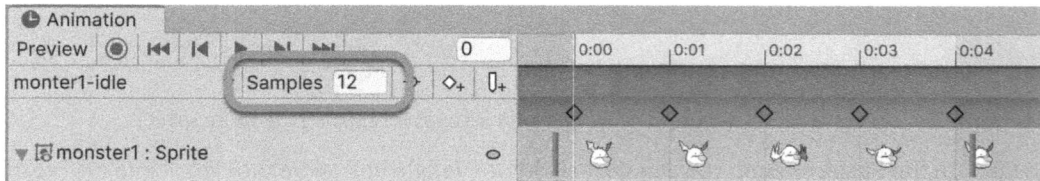

Figure 8.42: Setting up the Animation panel with the sample rate set to 12

14. If you look at **State Machine** for `monster-animator`, you'll see that it has a default state (clip) named `monster-idle`.

15. When you run your scene, you should see the `monster1` GameObject animating in its `monster-idle` state. You may wish to make the size of **Main Camera** a bit smaller (size 1) since these are quite small sprites:

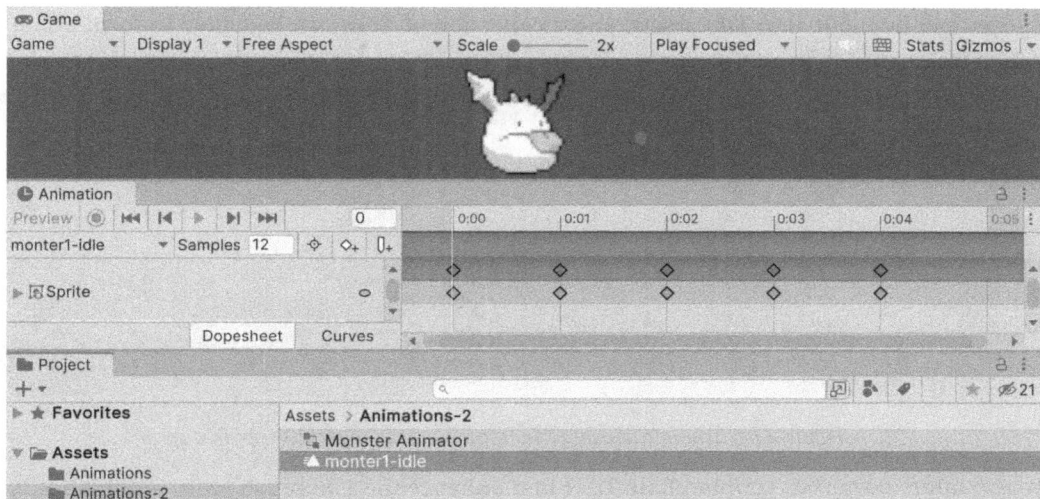

Figure 8.43: Animation panel for the sprite animation of monster1

How it works...

Unity's **Sprite Editor** knows about sprite sheets, and once the correct grid size has been entered, it treats the items in each grid square inside the sprite sheet image as an individual image, or frame, of the animation. In this recipe, you selected sub-sequences of sprite animation frames and added them to several animation clips. You added an **Animator Controller** to your GameObject so that each **Animation Clip** appears as a state in **Animator Controller State Machine**.

You can now repeat this process, creating an **Animation Clip** called monster-attack that uses frames 8-12, and a third clip called monster-death that uses frames 15-21. You can then create triggers and transitions to make the monster 1 GameObject transition into the appropriate states as the game is played.

You can learn more about the Unity Sprite Editor by looking at the Unity video tutorials at https://learn.unity.com/tutorial/introduction-to-sprite-editor-and-sheets.

You can learn more about 2D animation with sprite sheets by reading the following article by John Horton on GameCodeSchool.com: http://gamecodeschool.com/unity/simple-2d-sprite-sheet-animations-in-unity/.

Creating a platform game with Tiles and Tilemaps

Unity has introduced a set of **Tile** features that makes creating tile-based scenes quick and easy. A Tile Grid GameObject acts as the parent to Tilemaps. These are the GameObjects that Tiles are painted on, from the **Tile Palette** panel.

Sprites can be made into **Tile** assets, and a collection of Tiles can be added to form a **Tile Palette**, which we can use to paint a scene:

Figure 8.44: Example of using Tilemapper and GameArt2D-supplied sprites

It also offers powerful, scripted Rule Tiles that enhance the Tile brush tools, automatically adding the top, left, right, and bottom edge Tiles as more Grid elements are painted with Tiles. Rule Tiles can even randomly choose from a selection of Tiles under defined conditions. You can learn more at https://unity3d.com/learn/tutorials/topics/2d-game-creation/using-rule-tiles-tilemap.

In this recipe, we'll create a simple 2D platformer by building a Grid-based scene using some free Tile sprite images.

Getting ready

For this recipe, we have prepared the Unity package and images you need in the 08_07 folder.

Special thanks to GameArt2D.com for publishing the desert image sprites under the **Creative Commons Zero** license: https://www.gameart2d.com/free-desert-platformer-tileset.html.

How to do it...

To create a platform game with Tiles and Tilemaps, follow these steps:

1. Create a new Unity 2D project.
2. Import the provided images. An easy option to import the images is to drag them from your file explorer onto the **Project** panel.
3. The tile sprites we're using for this recipe are 128 x 128 pixels in size. It's important to ensure that we set the pixels per unit to 128 so that our **sprite** images will map to a grid of 1 x 1 Unity units. Select all the sprites in the **Project | DesertTilePack | Tile** folder and in the **Inspector** panel, set **Pixels Per Unit** to 128. Click on the **Apply** button at the bottom right of the **Inspector** panel.

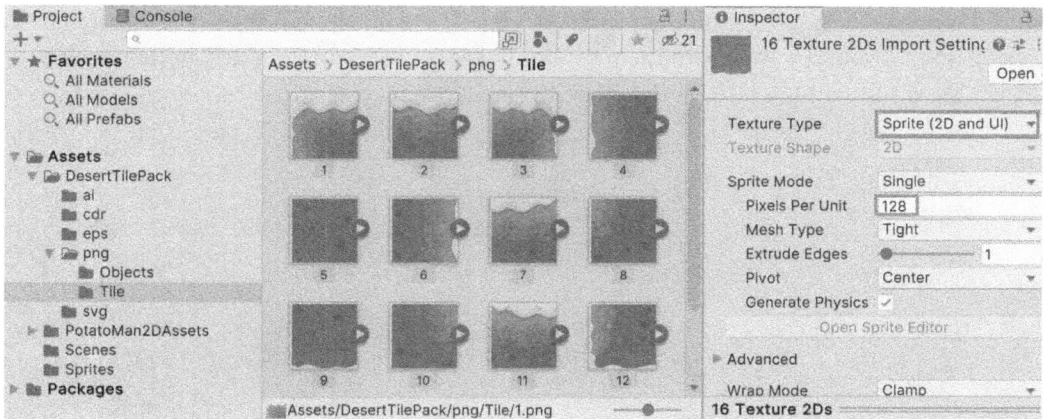

Figure 8.45: Import settings for DesertTilePack

4. Display the **Tile Palette** window by going to **Window | 2D | Tile Palette**.

5. In the **Project** panel, create a new folder named `Palettes` (this is where you'll save your `TilePalette` assets).

6. Click the **Create New Palette** button in the **Tile Palette** window and create a new **Tile Palette** named `DesertPalette`:

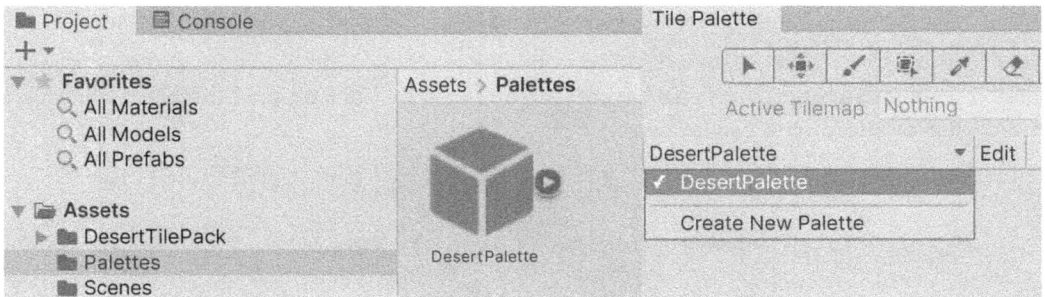

Figure 8.46: Creating a new Tile Palette named DesertPalette

7. In the **Project** panel, create a new folder named `Tiles` (this is where you'll save your **Tile** assets).

8. Ensure that `DesertPalette` is selected in the **Tile Palette** panel. Then, select all the sprites in the **Project / DesertTilePack / Tile** folder and drag them into the **Tile Palette** panel. When asked where to save these new `Tile` asset files, select your new **Assets / Tiles** folder.

You should now have 16 `Tile` assets in your `Tiles` folder, and these Tiles should be available so that you can work with them in your `DesertPalette` in the **Tile Palette** panel:

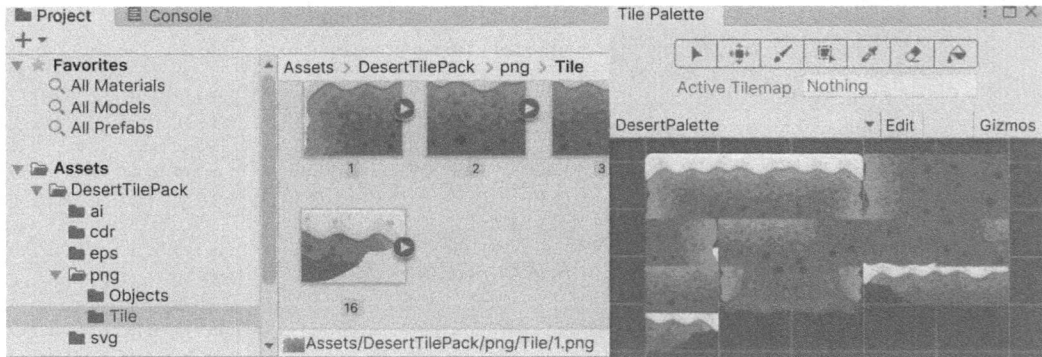

Figure 8.47: DesertPalette created, which includes 16 tile assets

9. Drag the asset file `Sprite BG` (included in `DesertTilePack`) into the scene. Resize `Main Camera` (it should be **Orthographic** since this is a 2D project) so that the desert background fills the entire **Game** panel.

10. Add a `Tilemap` GameObject to the scene by going to **2D Object | Tilemap | Rectangular**. You'll see a `Grid` GameObject added, and as a child of that, you'll see a `Tilemap` GameObject. Rename the `Tilemap` GameObject `Tilemap-platforms`.

11. You will need to create a new layer called `Background` for the Tilemap platform. Go to **Edit | Project Settings | Tags and Layers | Sorting Layers**. Click on the little plus (+) button and add **Layer 1**, calling it `Background`.

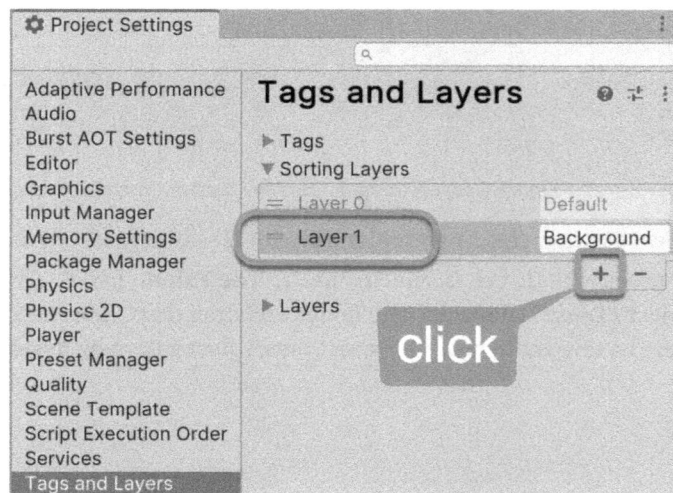

Figure 8.48: Creating a sorting layer

12. Select the `Tilemap-platforms` GameObject, and in the **Inspector** for the **TileMap renderer** component, ensure that the **Additional Settings Sorting Layers** property is set to **Background**.

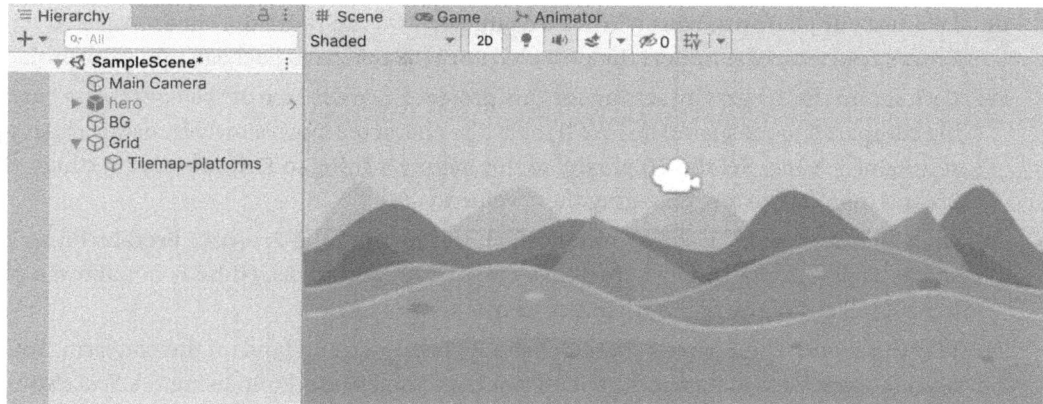

Figure 8.49: Adding a rectangular Tilemap Grid named Tilemap-platforms

> Just as **UI** GameObjects are children of a **Canvas**, **Tilemap** GameObjects are children of a **Grid**.

13. We can now start *painting* Tiles onto our Tilemap. Ensure Tilemap-platforms is selected in the **Hierarchy** panel and that you can see the **Tile Palette** panel. In the **Tile Palette** panel, select the **Paint with active brush** tool (the *paintbrush* icon). Now, click on a **Tile** in the **Tile Palette** panel. Then, in the **Scene** panel, each time you click the mouse button, you'll add a Tile to Tilemap-platforms that's automatically aligned with the Grid.

Figure 8.50: Painting tiles onto an active Tilemap by selecting the active brush tool

14. If you want to delete a Tile, *Shift* + click over that Grid position.
15. Use the **Tile Palette** brush to paint two or three platforms.
16. Add a suitable Collider to the Tilemap-platforms GameObject. Select the Tilemap-platforms GameObject in the **Hierarchy** panel and, in the **Inspector** panel, select to add a **Tilemap Collider 2D** instance. Click **Add Component** and then choose **Tilemap | Tilemap Collider 2D**.
17. Create a new **layer** named Ground and set the Tilemap-platforms GameObject to be on this **layer** (this will allow characters to jump when standing on a platform).

18. Let's test our platform scene with a 2D character – we can reuse the `PotatoMan` character from Unity's free tutorials. Import the provided `PotatoManAssets` package into your project.

19. Let's set up the 2D gravity setting for this project since the size of the `PotatoMan` character is big compared to the platforms. We'll make the character move slowly by using a heavy gravity setting of **Y** = -60. Set the 2D gravity to this value by going to **Edit | Project Settings | Physics 2D** and then, at the top, changing the **Y** value to -60.

20. Drag an instance of the `PotatoMan` **hero** character from the **Project / Prefabs** folder into the scene. Position him somewhere above one of your platforms. (If he is not above a platform, he will fall out of the scene when you play.)

21. Play the scene. The 2D **hero** character should fall down and land on the platform. You should be able to move the character left and right, and make him jump using the *Spacebar*.

22. You may wish to decorate the scene by dragging some of the object sprites onto the scene (in the **Project** panel, go to the **Project | DesertTilePack | Object** folder).

How it works...

By having a set of platform sprites that are all a regular size (128 x 128), it is straightforward to create a **Tile Palette** from those sprites, and then to add a `Grid` and Tilemap to the scene, allowing the **Tile Palette** brush to paint Tiles into the scene. By doing this, we added platforms to this scene that are all well aligned with each other, both horizontally and vertically.

You had to set the **sprite** pixels per unit to 128, matching the size of these sprites, so that each **Tile** maps to a 1 x 1 Unity Grid unit. If we were to use different size sprites (say, 256 x 256), then the pixels per unit must be set to that size, again to achieve a 1 x 1 Unity Grid.

You added a **Tilemap Collider 2D** to the `Tilemap` GameObject so that characters (such as the `PotatoMan`) can interact with the platforms. Without a Collider, these Tiles would have seemed just part of the background graphics. By adding a **Ground** layer and setting the `Tilemap` GameObject to this **layer**, the jump code in the `PotatoMan` `character-controller` script can test the layer of the object being stood on so that the jump action will only be possible when standing on a platform **Tile**.

Sorting layers determine how GameObjects are rendered – what they appear below and what they appear above. The top-to-bottom sequence (Layer 0, Layer 1...) in which sorting layers are arranged in the **Settings** panel determines their rendering sequence. Content rendered earlier will appear below content rendered later. For example, in a project with foreground and background graphics, we'd want the background layer to appear above the foreground sorting layer in our **Settings** panel, so that it would be rendered before (and so behind) the foreground sorting layer content.

When there are two or more GameObjects in the same sorting layer, we can use the integer (whole number) sprite renderer **Order in Layer** property to control the rendering sequence. GameObjects for the same sorting layer will be rendered in the numeric sequence of their **Order in Layer** value, so 0 before 1, 1 before 2, and so on.

There's more...

Here's a suggestion for enhancing this recipe.

Tile palettes for objects and walls

The sprite objects in the Desert free pack are all different sizes, and certainly not consistent with the 128 x 128 sprite size for the platform Tiles.

However, if the sprites for the objects and walls in your game *are* the same size as your platform sprites, you can create a **Tile Palette** for your objects and paint them into the scene using the **Tile Palette** brush.

Using sprite placeholders to create a simple physics scene

Unity offers physics for 2D scenes by providing Colliders, Rigidbodies, gravity, and so on, just as it does for 3D scenes. In this recipe, we'll create a 2D mountain-style landscape made up of some colored triangles, and then have some square blocks fall down due to gravity into one of the low dips where two triangles overlap. Rather than using images for our sprites, we'll learn how to use 2D geometric **sprite placeholders**, a feature in Unity allowing for quick prototyping and scene layouts, where the placeholder sprites can easily be replaced with texture images later. The following screenshot illustrates the starting and ending locations of the square blocks as they fall down and collide with the triangular landscape sprites:

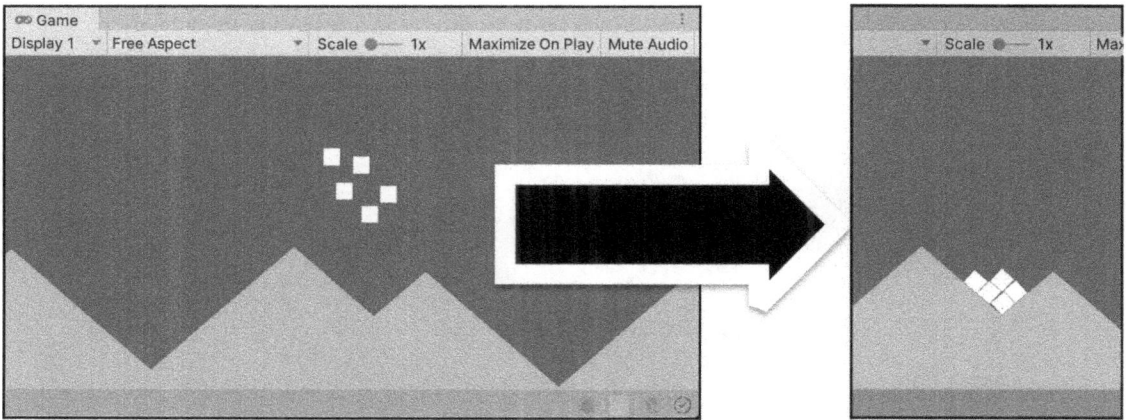

Figure 8.51: A physics scene where 2D squares fall into a dip in the landscape

Getting ready

We'll create everything from scratch in Unity for this recipe, so no preparation is required.

How to do it...

To use sprite placeholders to create a simple physics scene, follow these steps:

1. Create a new Unity 2D project.
2. Create a new Sprite Placeholder asset file named Triangle by going to **Assets | Create | 2D | Sprites | Triangle**. You should now see an asset file in the **Project** panel named Triangle.

3. Create a folder in the **Project** panel named `Sprites` and move your `Triangle` sprite file into it:

Figure 8.52: The new Triangle sprite asset file in the Project panel

4. Drag the `Triangle` sprite file into the scene to create a GameObject named `Triangle`.

5. Ensure the `Triangle` GameObject is selected in the **Hierarchy** panel. Then, in the **Inspector** panel, set the **Color** property of the **Sprite Renderer** component to a red-pink color.

6. In the **Inspector** panel, add a **Polygon Collider 2D** component by clicking the **Add Component** button, then choose **Physics 2D**.

7. Then, add a **Rigidbody 2D** component.

8. We don't want our triangles to move at all (they are mountains!), so we will freeze their X, Y, and Z positions. In the **Inspector** panel, for the **Rigidbody 2D** component, check **Freeze Position** for **X** and **Y** and **Freeze Rotation** for **Z**:

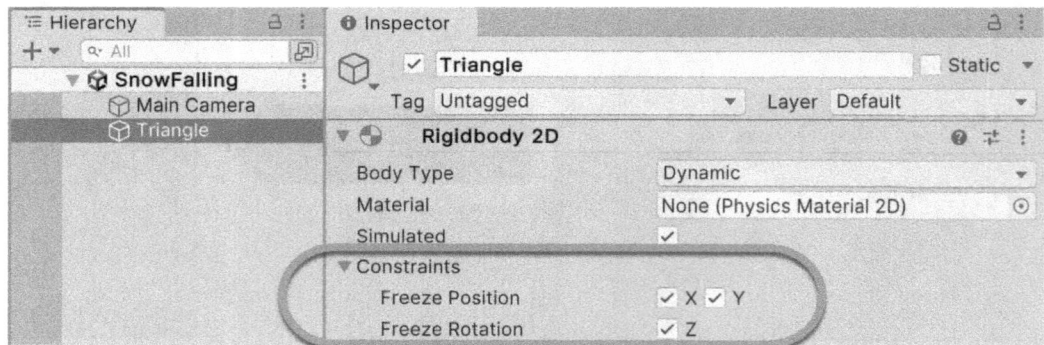

Figure 8.53: Freeze Position and Freeze Rotation for our Triangle GameObject landscape

9. Now make the `Triangle` sprite larger, and make 4 or 5 copies, which will automatically be named `Triangle 1`, `Triangle 2`, and so on.

10. Arrange them at the lower part of the **Scene** panel, so they look like pointy mountains at the bottom of the **Game** panel. See *Figure 8.53* for the effect you are trying to achieve. The aim is to have no *gap* at the bottom of the screen, so our square sprites will fall down onto this *landscape* of triangle mountains. See *Figure 8.54*:

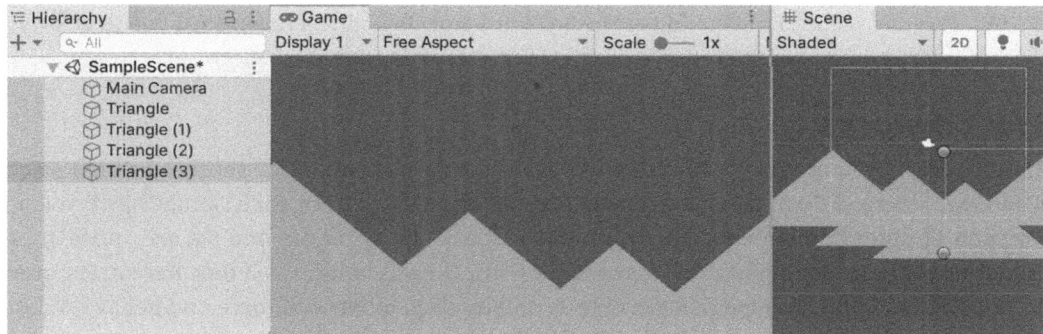

Figure 8.54: Our landscape of Triangle GameObjects

11. Create a new Square **sprite placeholder** in the **Project** panel and move it into the Sprites folder.

12. Drag the Square sprite asset file from the **Project** panel into the scene to create a GameObject.

13. Ensure the Square GameObject is selected in the **Hierarchy** panel. Then, in the **Inspector** panel, add a **Polygon Collider 2D** component and a **Rigidbody 2D** component.

14. Duplicate the Square GameObject four times so that you have five GameObjects named Square, Square 1, Square 2, Square 3, and Square 4.

15. Then, arrange these Square GameObjects so that they are now overlapping in the middle of the scene. This will ensure that when they fall down due to gravity, they'll end up rolling down into one of the dips in our Triangle landscape:

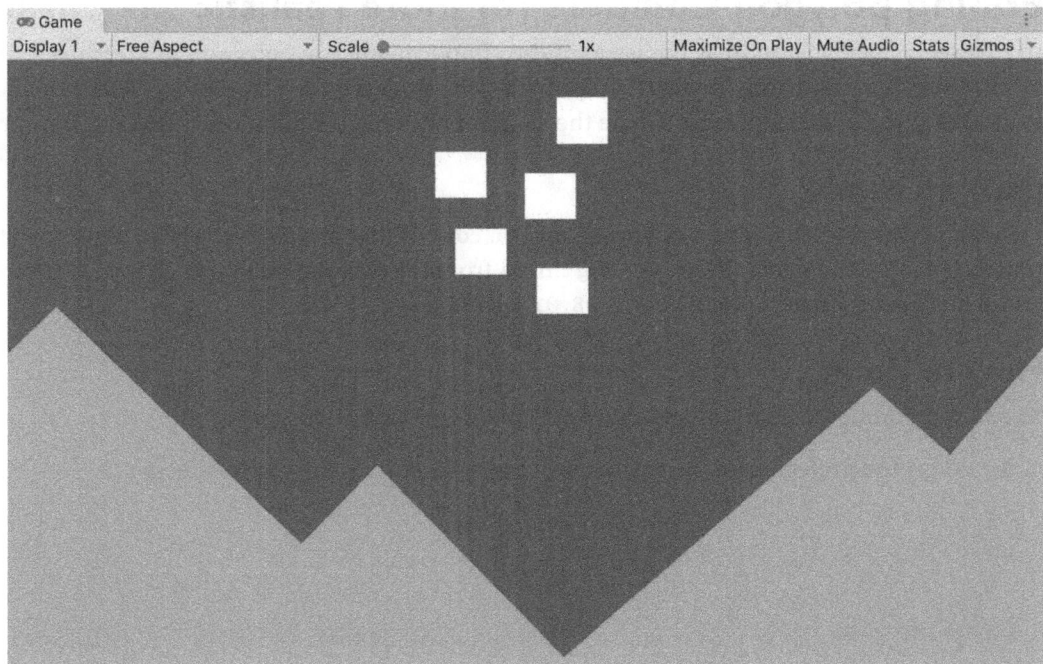

Figure 8.55: Our collection of Square GameObjects, ready to fall down into the Triangle landscape

16. Play the scene. You should see the Square GameObjects fall downward due to gravity and roll down into one of the low points where two triangles overlap.

How it works...

In this recipe, you created Triangle and Square **sprite placeholder** asset files in the **Project** panel. Then, you dragged them into the scene to create GameObjects. For each GameObject, you added 2D **Polygon Collider** components (following the shape of the Triangle and Square sprites), as well as **Rigidbody** 2D components, which make these GameObjects behave as if they have mass, gravity, and so on. The Colliders indicate that the objects will hit each other with force and behave accordingly.

Then, you froze the position and rotation of the Triangle GameObjects to make them solid and immobile, as if they were mountains of solid rock. The Square GameObjects were positioned in the middle of the **Scene** panel – as if they were starting in the air, above the ground of Triangle. So, when the scene is played, gravity is applied by the Unity 2D physics system, making the Square GameObjects fall downward until they hit each of the immobile Triangle mountains. Then, they roll around until they end up being collected in the low points (or valleys) where the two triangles overlap.

Once a scene has been successfully prototyped with GameObjects based on sprite placeholders, it's easy to replace the 2D geometric shape with an actual image. For example, we could select all the Square GameObjects and, in their **Sprite Renderer** components, select the Enemy Bug sprite to replace the Square GameObjects – we'll do exactly this in the next recipe...

Editing polygon Colliders for more realistic 2D physics

In this recipe, we will address the issue of the simple geometric polygon Colliders used as sprite placeholders not matching the final image sprite that replaces the placeholder. This will allow you to use polygon Colliders to approximate the shape of any image outline you might use in your game.

Getting ready

This recipe builds on the previous one, so make a copy of that project and work on the copy for this recipe. We'll also be using the Enemy Bug image from the first recipe in this chapter, which can be found in a folder named Sprites in the 08_01 folder.

How to do it...

To edit polygon Colliders for more realistic 2D physics, follow these steps:

1. Copy the project from the previous recipe and use this copy for this recipe.

2. If you haven't done so already, import the Enemy Bug image from the provided assets. Then, in the **Inspector** panel, for each Square GameObject, change the sprite in the **Sprite Renderer** component to the Enemy Bug sprite (replacing Square).

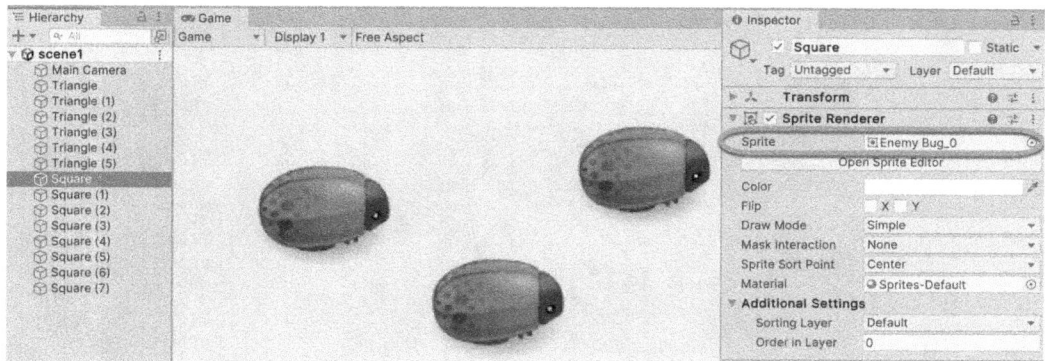

Figure 8.56: Enemy Bug sprite replacing the Square sprite placeholder

3. Run the scene. You should see all the Enemy Bug GameObjects fall down into a pile at a low point where two triangles meet:

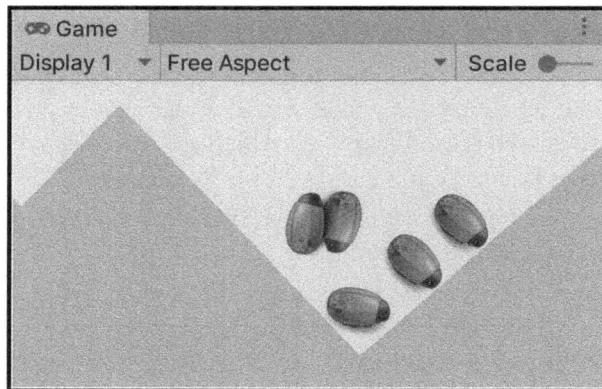

Figure 8.57: Bugs falling into a pile. There's space in between them due to the Square polygon Collider

4. However, looking at the screen, we can see there is some space between the Enemy Bug Game-Objects. When we look at the Colliders in the **Scene** panel, we can see that there are Square Colliders with lots of space around each oval Enemy Bug image:

Figure 8.58: Square Colliders around oval Enemy Bug images

5. Select the Square GameObject in the **Hierarchy** panel. Then, in the **Inspector** panel, for the **Polygon Collider 2D** component, click the **Edit Collider** icon. In the **Scene** panel, you'll now have a small square on the Collider lines, which you can drag to deform the Collider polygon so that it matches the oval outline of the Enemy Bug image.

6. When your polygon Collider approximates the outline of Enemy Bug, stop editing by clicking the **Edit Collider** icon again.

7. Delete the Square (1), Square (2), Square (3), and Square (4) GameObjects. Now, duplicate Square four times and arrange them so that they are not overlapping. You now have five GameObjects with polygon Colliders that match what the user can see.

8. Run the scene. You should see all the Enemy Bug GameObjects fall down into a pile at a low point where two triangles meet.

9. This time, they should be almost touching the Collider and each other based on the polygon Collider that matches the oval shape of the **Enemy Bug** icon:

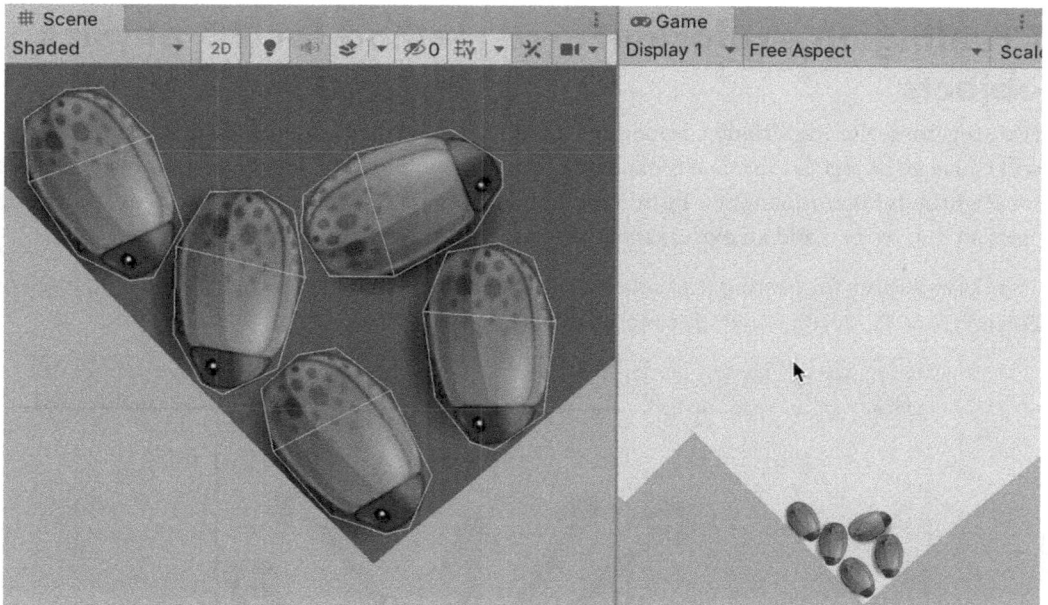

Figure 8.59: Enemy bugs falling and colliding based on a more realistic bug polygon shape

How it works...

We created GameObjects with the Square sprite placeholders. However, for flexibility, these have a polygon Collider added to them. When first created, the polygon Colliders match the geometric **square** shape of the Square sprite placeholders – however, these polygon Colliders can be edited to make the Collider a different shape.

In this recipe, you replaced the Square sprite with the Enemy Bug sprite image, then edited the **Polygon Collider 2D** component so that it matches the oval shape of the outline of the Enemy Bug image. By removing the GameObjects with the Square polygon Colliders, and then duplicating the one containing the oval bug-shaped polygon, all the falling GameObjects collided to match the visible shape of the Enemy Bug images.

This approach can be used to create polygon Colliders to approximate the shape of any image outline you might wish to use in your game.

Creating an explosionForce method for 2D physics objects

For 3D games, the **Rigidbody** component has a useful method called `AddExplosionForce(...)` that will cause an object to look as if it explodes from a given position. However, there is no such method for **Rigidbody2D** components. In this recipe, we'll write an extension method for the `Rigidbody2D` class so that we can add an `explosionForce` method for 2D objects.

Thanks to Swamy for posting the code for the `Rigidbody2D` extension class's explosion force on Unity Forum (Nov. 21, 2013), which this recipe is based on:

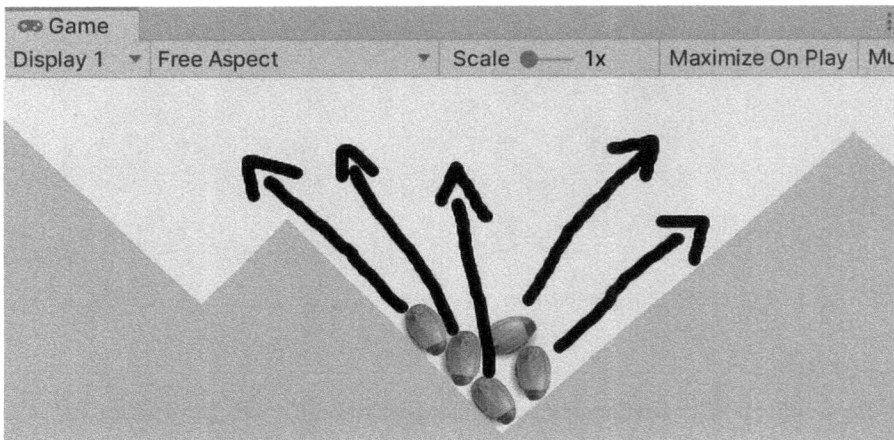

Figure 8.60: Bugs being violently forced away from an invisible explosion point

Getting ready

This recipe builds on the previous one, so make a copy of that project and work on the copy for this recipe.

How to do it...

To create an `explosionForce` method for 2D physics objects, follow these steps:

1. Copy the project from the previous recipe, and use this copy for this recipe.

2. We can add a method to the `Rigidbody2D` script class by creating the following C# script class called `Rigidbody2DExtension`:

```
using UnityEngine;

public static class Rigidbody2DExtension{
    public static void AddExplosionForce(this Rigidbody2D body, float
explosionForce, Vector3 explosionPosition, float explosionRadius){
        Vector3 forceVector = (body.transform.position -
explosionPosition);
```

```
                    float wearoff = 1 - (forceVector.magnitude / explosionRadius);
                    body.AddForce(forceVector.normalized * explosionForce * wearoff);
            }
    }
```

3. Create a new empty **GameObject** named explosion.

4. Create a C# script class called ExplodeCircle and add an instance object as a component to the child explosion GameObject. Choose **Add Component | New Script** in the **Inspector** panel:

```csharp
using System.Collections;
using System.Collections.Generic;
using UnityEngine;

public class ExplodeCircle : MonoBehaviour{
    public float power = 800f;
    public float radius = 3f;

    void Update(){
        if (Input.GetKeyUp(KeyCode.Space)){
            print("Exploding ...");
            Explode();
        }
    }

    void Explode(){
        Vector2 explosionPos = transform.position;
        Collider2D[] Colliders = Physics2D.OverlapCircleAll(explosionPos,
    radius);
        foreach (Collider2D hit in Colliders){
            Rigidbody2D rigidbody = hit.GetComponent<Rigidbody2D>();

            if (rigidbody != null)
                rigidbody.AddExplosionForce(power, explosionPos, radius);
        }
    }

    void OnDrawGizmosSelected(){
        // Draw a red circle to show range of explosion radius
        Gizmos.color = Color.red;
        Gizmos.DrawWireSphere(transform.position, radius);
    }
}
```

5. Move the explosion GameObject just below the low point where the bugs will fall to when the scene runs.

6. In the **Inspector** panel, you can change the radius of the explosion. In the **Scene** panel, you can resize the red circle to indicate the extent of the explosion's radius:

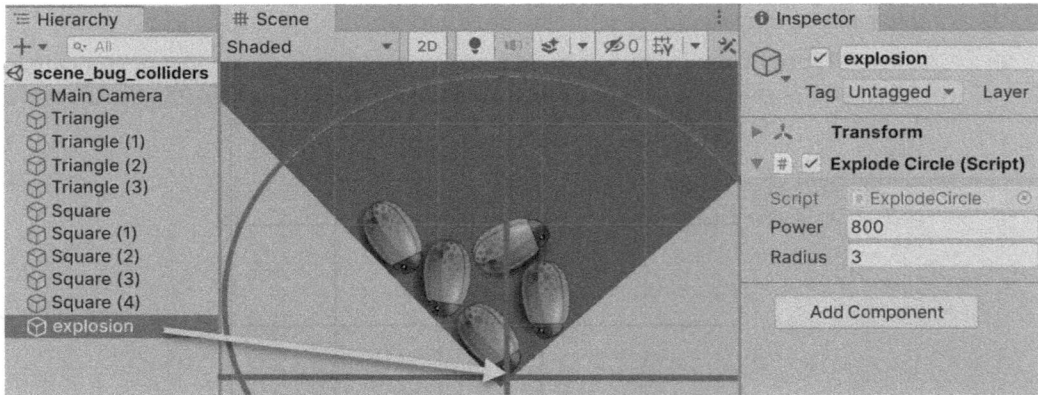

Figure 8.61: Circle in the Scene window showing the radius of the explosion's force

7. Run the scene. The bugs should fall toward the low point inside the explosion radius. Each time you press the *Spacebar*, an explosive force will be applied to all the bugs inside the explosion radius.

How it works...

The C# Rigidbody2DExtension script class adds a new method called AddExlosionForce(...) to all **Rigidbody2D** components in the project. The advantage of using an extension class method is that this method now becomes available to use by all other scripts in the project, with no need to refer to this extension class explicitly – it's as if Unity provided this extra method as part of its core code library.

This AddExlosionForce(...) method takes three parameters when called:

- float explosionForce: The magnitude of the force to be applied (the larger the value, the more force will be applied)

- Vector3 explosionPosition: The center (origin) point of where the explosion is to take place (the further a GameObject is from this position, the less force it will receive)

- float explosionRadius: The furthest point beyond which no force will be applied by the simulated explosion

This method will apply a force to **Rigidbody2D** based on the direction and position it is in relative to explosionPosition, applying the force to move the object away from explosionPosition. The wearoff variable reduces the force that's applied based on how far away **Rigidbody2D** is from explosionPosition – objects closer to this position will have more force applied to them, simulating how an explosion works in the real world.

The C# ExplodeCircle script class has two public variables called power and radius that will be passed to AddExlosionForce(...) when applied to a **Rigidbody2D**. The Update() method checks each frame to see whether the user has pressed the *Spacebar*, and when detected, the Explosion() method is invoked. However, you could have some other way to decide when to apply an explosion, such as when a timer has finished, when a Collider has been triggered, and so on.

The OnDrawGizmosSelected() method of the ExplodeCircle class draws a red circle in the **Scene** panel, which indicates the extent of the radius value inside which the explosion force will be applied. It uses the **Gizmos** wire sphere drawing command.

The Explosion() method of the ExplodeCircle class uses the Physics2D OverlapCircleAll() method to return an array of all **Rigidbody2D** components of GameObjects within a circle of the value of the radius from the position of the GameObject. Each of these Colliders is looped through, and the AddExplosionForce(...) method is invoked on them so that all the GameObjects within the circle for the current radius value will have the explosion force applied to them.

There's more...

Rather than using a Physics2D circle to decide which GameObjects are to be affected by the explosion, another game situation might be to apply an explosive force to all the objects of a particular tag. In this case, you would replace the ExplodeCircle scripted component of the explosion GameObject with an instance of the following script class, setting the tagName string to the desired tag in the **Inspector** panel:

1. You first need to create a new **tag** by going to **Edit | Project Settings | Tags and Layers | Tag** and adding a new tag called Bug.
2. This new tag needs to be assigned individually to some of the square GameObjects in the **Inspector** panel.
3. Remove the ExplodeCircle script class component from the explosion GameObject.
4. Create a C# script class called ExplodeTag and add an instance object as a component to the child explosion GameObject:

```csharp
using UnityEngine;

public class ExplodeTagged : MonoBehaviour {
    public string tagName = "Bug";
    public float force = 800f;
    public float radius = 3f;

    private GameObject[] _gameObjects;

    private void Awake(){
        _gameObjects = GameObject.FindGameObjectsWithTag(tagName);
    }

    void Update(){
```

```
        if (Input.GetKeyUp(KeyCode.Space))
        {
            Explode();
        }
    }


    private void Explode(){
        foreach(var gameObject in _gameObjects){
            Rigidbody2D rigidbody2D = gameObject.
GetComponent<Rigidbody2D>();
            rigidbody2D.AddExplosionForce(force, transform.position,
radius);
        }
    }
}
```

Clipping via Sprite Masking

Clipping is the computer graphics term for choosing which parts of a graphical object to display or hide. In 2D graphics, it's very common to use one image to define parts of a screen to either only show other images or to never show other images. Such an image is known as an **image mask**. In this recipe, we'll use two related images. One is an image of the inside of a room, showing a panel out to a night skyscape. The second image is transparent, except for the rectangle where the panel is. We'll use the second image to only allow other sprites to be seen when they pass by the rectangle panel. The following figure shows how we can only see the parts of the moving blue bird sprite when it is overlapping with the rectangle panel:

Figure 8.62: Image mask being used to only show the bird when it flies through the window rectangle area of the screen

Getting ready

For this recipe, we have prepared the files you need in the 08_10 folder.

How to do it...

To clip images via **Sprite Masking**, follow these steps:

1. Create a new Unity 2D project.
2. Create a folder named Images and import the three provided images into it; that is, GAME_ROOM. png, GAME_ROOM_windowMask.png, and blueBird.png.
3. Drag the GAME_ROOM image into the scene and resize it so that it completely fills the **Game** panel. For its **Sprite Renderer** component, set **Order in Layer** to -1 (so that this room image will be behind other images in the scene).
4. With the GAME_ROOM GameObject selected, from the **Create** menu in the **Hierarchy** panel, go to **2D Object | Sprite Mask**. A GameObject named Sprite Mask should now appear in the **Hierarchy** panel as a child of the GAME_ROOM GameObject:

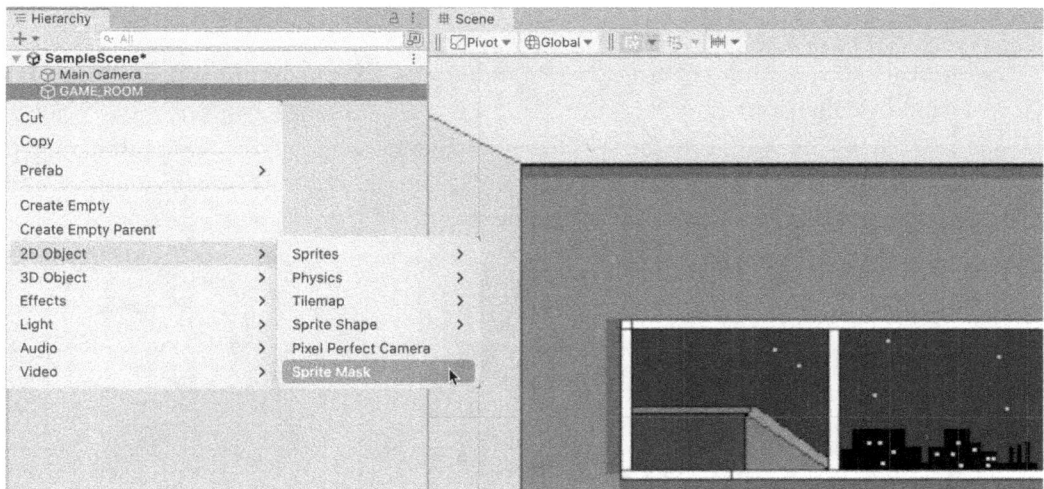

Figure 8.63: Adding a Sprite Mask child to the GAME_ROOM GameObject

5. Select the Sprite Mask GameObject in the **Hierarchy** panel. Then, in the **Inspector** panel, for its **Sprite Mask** component, click the **Sprite** selection circle icon. Then, from the list of sprites, choose GAME_ROOM_windowMask.

If you look in the **Scene** panel carefully, you'll see an orange border around the rectangle of the window of the GAME_ROOM image. You may need to resize the orange **Sprite Mask** rectangle to fully outline the night time window.

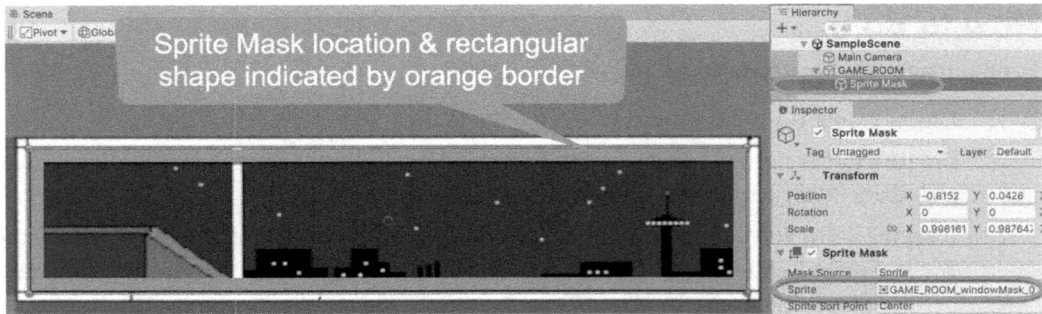

Figure 8.64: Selecting the image to use as the mask for the Sprite Mask child of GAME_ROOM

6. Drag the blueBird image from the **Project** panel into the scene to create a blueBird GameObject. Make the newly created blueBird GameObject a child of the GAME_ROOM GameObject in the **Hierarchy** panel.

7. Resize and reposition the blueBird GameObject so that its right-hand side is just inside the window's night skyscape.

8. In the **Inspector** panel, for the **Sprite Renderer** component, set the **Mask Interaction** property to **Visible Inside Mask**. You should now only see the parts of the blueBird image that are inside the rectangle of the night skyscape window:

Figure 8.65: Setting blueBird to only be visible inside the window rectangle mask

9. Run the scene to confirm that the blueBird sprite is masked (unseen) except for the parts inside the night skyscape window rectangle.

How it works...

In this recipe, you added the GAME_ROOM sprite to the scene, then added a GameObject containing a **Sprite Mask** component as a child of the GAME_ROOM GameObject.

You set the sprite of **Sprite Mask** to GAME_ROOM_windowMask. This was a special image that only contained the rectangular area of the night skyscape:

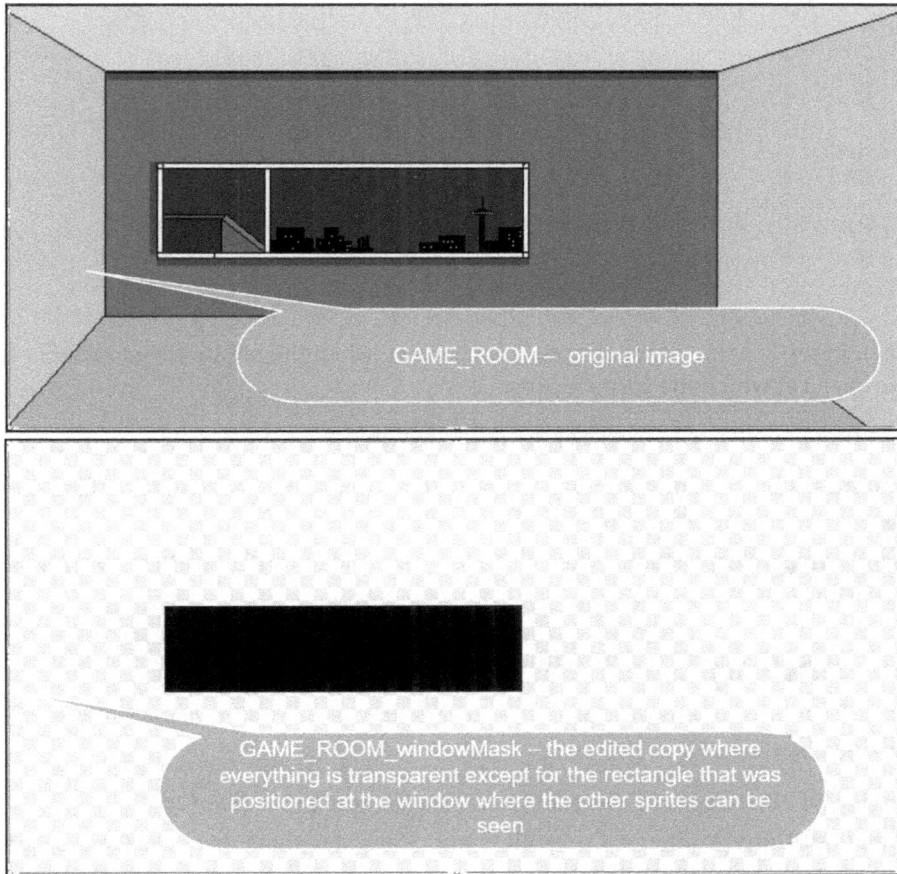

Figure 8.66: Original GAME_ROOM and mask images

The black rectangle in the GAME_ROOM_window **mask** image defines the only places where a masked sprite will be displayed – that is, the rectangle where the window appears in the GAME_ROOM image.

By adding the blueBird GameObject to the scene as a child of GAME_ROOM and setting its relationship to **Sprite Mask,** only the parts of the blueBird image that appear inside the rectangular mask image of GAME_ROOM_windowMask will be seen by the user.

You can think of a sprite mask as a piece of cardboard with a hole cut out of it – we can only see images behind the mask that pass by the cut-out hole.

There's more...

If you wanted to have the `blueBird` GameObject fly from left to right, you could add a scripted component to `blueBird` based on a new C# script class called `MoveBird` containing the following code:

```
using UnityEngine;

public class MoveBird : MonoBehaviour
{
    public float speed = 1f;

    void Update()
    {
        transform.Translate(speed * Vector3.right * Time.deltaTime);
    }
}
```

As can be seen in the last line of the scripted component, when the public `speed` variable is changed, it adjusts the speed at which the sprite moves.

You can learn more about sprite masks at the following links:

*   Sprite masks in the Unity manual: `https://docs.unity3d.com/Manual/class-SpriteMask.html`
*   The following inScope Studios video (it's a few years old but still useful): `https://www.youtube.com/watch?v=1QktsHJwXCQ`

Further reading

Take a look at the following links for useful resources and sources of information regarding the 2D features provided by Unity:

*   Overview of 2D features in Unity: `https://unity.com/solutions/2d`
*   More about Unity Animator Controllers and State Machines: `https://docs.unity3d.com/Manual/class-AnimatorController.html`
*   Unity's 2D rogue-like tutorial series: `https://unity3d.com/learn/tutorials/s/2d-roguelike-tutorial`
*   Platform sprites from Daniel Cook's Planet Cute game resources: `https://lostgarden.home.blog/2007/05/12/dancs-miraculously-flexible-game-prototyping-tiles/`
*   Creating a basic 2D platformer game: `https://learn.unity.com/project/2d-platformer-template`
*   A fantastic set of modular 2D characters released under the free Creative Commons license from Kenney. These assets would be perfect for animating body parts in a similar way to the `PotatoMan` example in this chapter and in the Unity 2D platformer demo: `http://kenney.nl/assets/modular-characters`

- Joe Strout's illuminating *Game Developer* article on three approaches to 2D character animation with Unity's scripting and animation states: `https://www.gamedeveloper.com/programming/2d-animation-methods-in-unity`

Here are some learning resources on using Tilemap:

- Unity Tilemap tutorial: `https://learn.unity.com/tutorial/introduction-to-tilemaps`
- Lots of 2D extra resources, free from Unity Technologies: `https://github.com/Unity-Technologies/2d-extras`
- Sean Duffy's great tutorial on using Tilemap on the Ray Wenderlich (now `Kodeco.com`) site: `https://www.kodeco.com/23-introduction-to-the-new-unity-2d-tilemap-system`

9

Animated Characters

In this chapter, you'll learn how to import characters and animations from Mixamo and also how to change the character for the Third Person Starter Assets. The recipes in this chapter build on the Unity Technologies Third Person Starter Assets packages – these assets provide a great starting point to try out content and characters with first- and third-person cameras.

In this chapter, we'll learn how to do the following:

- Unity's Third Person Character Controller assets
- Adding a Third Person Character Controller to a scene
- Adding a clothing accessory pickup for a character
- Swapping the Third Person Armature for a different character
- Creating a 3D character with Autodesk Character Generator
- Selecting and downloading a character from Mixamo
- Selecting and downloading an animation clip from Mixamo
- Creating an animated NPC in Unity using a character and animation clip
- Using scripts to control 3D animations (old input system)
- Using scripts to control 3D animations (new input system)

Unity's Third Person Character Controller assets

Many of the recipes in this chapter make use of Unity's Third Person Character Controller starter project. We met this briefly in *Chapter 4*, *Playing and Manipulating Sounds*, and we'll repeat those steps and take things a little further here.

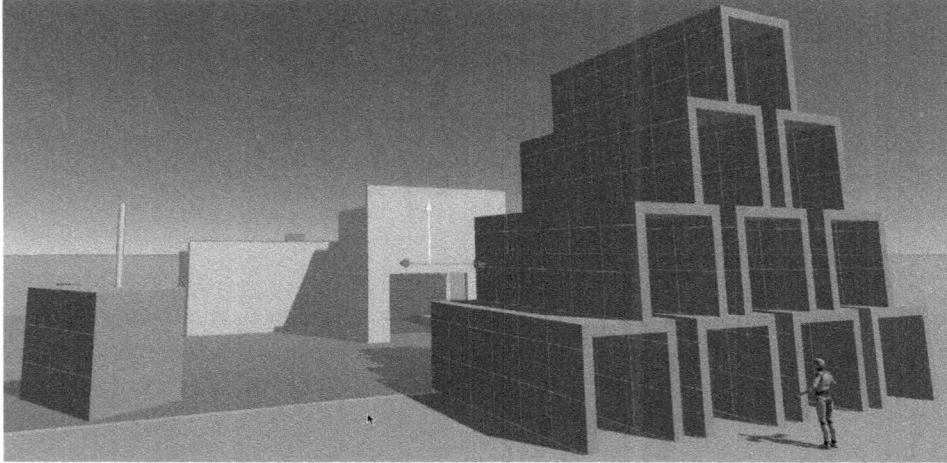

Figure 9.1: The pyramid of prefab tunnels that we'll make in this recipe

How to do it...

To work with Unity's Third Person Character Controller project, perform the following steps:

1. Create a new **Unity 3D** project.

2. Open a web browser tab to the Unity Asset Store, by choosing **Asset menu: Window | Asset Store**.

3. Search for the free **Starter Assets Third Person Character Controller**.

4. Select the assets, and when viewing their details, click the **Add to My Assets** button.

5. In the Unity Editor, open **Package Manager** to list your **Asset Store** assets by choosing **menu: Window | Package Manager** From the Packages dropdown choose My Assets (see the following screenshot). Dock the **Package Manager** panel alongside the **Inspector** panel.

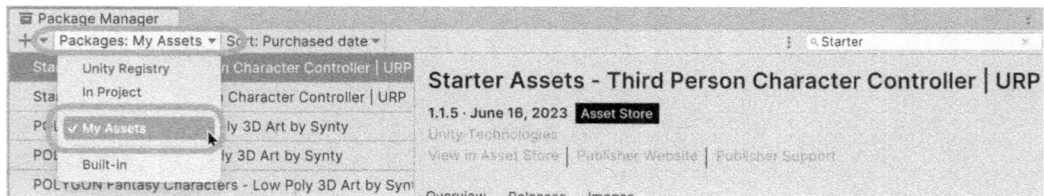

Figure 9.2: Option to view My Assets in Package Manager in Unity Editor

6. In the **Package Manager** panel, locate and import the **Starter Assets Third Person Character Controller** package.

> **Note:** When importing the package, if asked, agree to additional Package Manager dependencies (click **Install/Upgrade**), and agree to enable the new input systems backends and restart the Editor (click **Yes**).

7. You should now have a **Starter Assets folder** in your **Project** panel.

8. Open the Playground scene, in the **Project** panel folder: **Starter Assets | Third-Person Starter | Scenes**.

> **Note:** If you have any pink textures, see the recipe in *Chapter 4* for steps to fix this.

9. Run the scene, and move your robot character around the 3D environment using the arrow keys/WASD.

10. Let's create a pyramid of tunnels. You'll find the tunnel Prefab in the **Project** panel:

    Assets | StarterAssets | Environment | Prefabs | Tunnel_prefab.

11. Drag four Prefabs to form the base, then three more sitting on top of them, then two, and then the final one.

12. Let's make one of the **Tunnel_prefab** GameObjects blue by changing its material. Select a **Tunnel_prefab** object and, in the **Inspector** window, drag the **Blue_Mat material** into its **Mesh Renderer** component. You will find this material in the **Project** panel folder: **Assets | Starter-Assets | Environment | Art | Materials.**

13. Run the scene, and move your robot character around and through the tunnel pyramid!

How it works...

You've added the free Unity Third Person Character Controller starter project assets to your **Asset Store** account. The assets provide a demo scene, and a ready-made third-person character controller and camera. We added to the scene by creating GameObjects from the provided Environment Prefabs. We were able to change the color of one of the tunnel objects by using one of the Art Material asset files provided.

Adding a Third Person Character Controller to a scene

Once we've added the Unity Third Person Character Controller starter assets to a project, we can use them for our own scenes. In this recipe, we'll create an empty scene, add a squashed Cube as the floor, and then add a controllable character with a linked camera by using a Prefab from our imported assets.

Figure 9.3: The scene we'll build in this recipe, with a third-person character added to the scene

Getting ready

This project builds on the previous one – so either work on a copy of that one, or add the Unity Third Person Character Controller starter assets to another project using the steps from that recipe.

How to do it...

To add a Third Person Character Controller to a scene, perform the following steps:

1. Create a new scene.

2. Add a **Cube** to the scene named Cube-ground and scale it to be large and thin by setting **Transform Scale** to (20, 0.01, 20) in the **Inspector**.

3. Add some texture to Cube-ground by dragging onto this GameObject the GridBlue_01_Mat material asset file, which can be found in the **Project** panel folder: **StarterAssets | Environment | Art | Materials**.

4. Delete the **MainCamera** GameObject – since the Third Person prefab we are going to use already has a **Main Camera** GameObject.

5. Drag into the scene the NestedParentArmature_Unpack prefab asset file, which can be found in the **Project** panel folder: **StarterAssets | ThirdPersonController | Prefabs**.

6. A GameObject linked to a Prefab will appear blue in the **Hierarchy**. We want to customize this asset, so let's break the link to the Prefab. Right-click the NestedParentArmature_Unpack GameObject and choose **Prefab | Unpack Completely**. The GameObject should no longer be blue, so we can now change it from its original Prefab asset file settings.

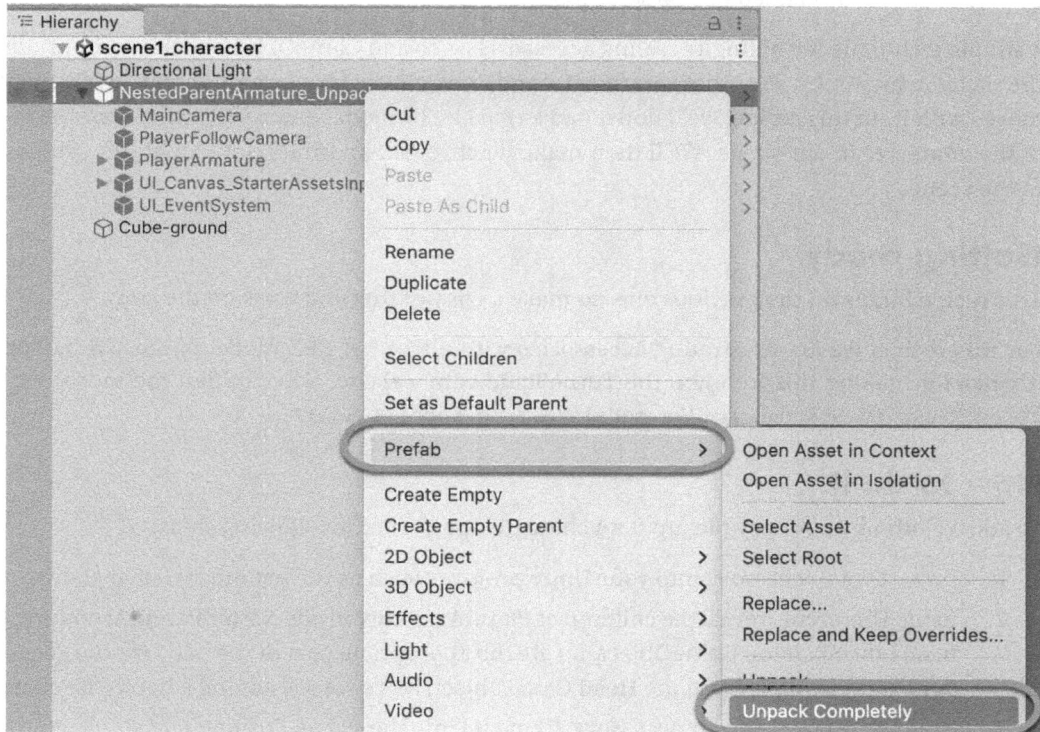

Figure 9.4: Unpacking the Third Person Armature Prefab

7. In the **Inspector**, select the PlayerArmature GameObject (a child of **NestedParentArmature_Unpack**), and set **Transform Position** to (0, 0, 0). This will stop it from falling through the ground when we run the scene.

8. Delete the **UI_Canvas_StarterAssetsInputs_Joysticks** GameObject since we won't be creating a mobile application build from this project.

9. Save and play the scene. You should be able to control the camera with your mouse, and move the character with the *WASD-SPACE* keys.

How it works...

The Unity Third Person Character Controller assets include a hierarchical prefab, containing a Main Camera, player character, and various scripted components for making the character controllable by the user for a 3D game. In this recipe, you created a new GameObject from the provided Prefab, and then "unpacked" the Prefab so it could be customized. The Prefab comes with additional UI Game-Objects for mobile application builds, which we removed since they weren't needed for this recipe.

Adding a clothing accessory pickup for a character

A great feature to add to Third Person Controllers in games is the ability to accessorize them! Whether it's clothing like hats and gloves, jewelry, or weapons, players engage with their characters more if they can see pickups added to character models.

A simple technique for hiding/revealing accessories is to add GameObjects to the appropriate part of the skeleton hierarchy, so as that part (head, hand, foot, etc.) of the skeleton is animated the accessory moves with it. In this recipe, we'll download a free hat 3D model and add it to the skeleton hierarchy of the character in our scene. We'll then make it active (so, invisible) until an object collides with it in the scene.

Getting ready

This recipe builds on the previous one, so make a copy of that and work on the copy.

For this recipe, we are illustrating accessories with a free hat FBX model published by **1pudge101** (thanks for making this free!) on the TurboSquid.com website. You can find the model we used at `https://www.turbosquid.com/3d-models/hat-3d-model-1956617`.

How to do it...

To add a clothing accessory pickup for a character, perform the following steps:

1. Import your accessory into your Unity project – such as the hat model listed above.
2. In the **Hierarchy**, reveal the children of **PlayerArmature** inside **NestedParentArmature_Unpack**. Inside the **Skeleton** GameObject, locate the appropriate part of the body for the accessory. In our example, we located the **Head** GameObject, since we are adding a hat to the character.
3. Drag an instance of your accessory 3D model into the scene. Position and size it appropriately, and then child the GameObject to the part of the **Skeleton** GameObject it should move with. We added a hat GameObject to the scene, moved it to sit on the robot character's head, and made this hat GameObject a child of the **Head** GameObject, inside **Skeleton**.

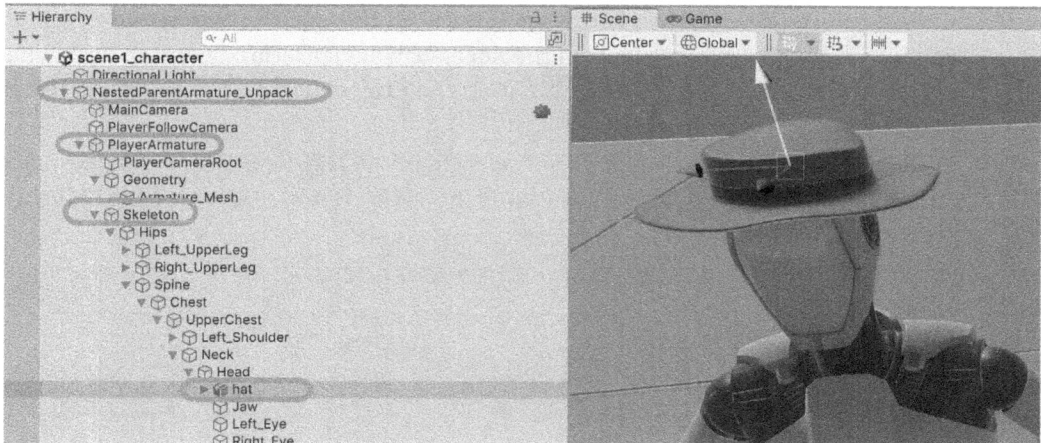

Figure 9.5: Positioning the hat GameObject as a child of Head in the PlayerArmature's Skeleton

4. With your accessory GameObject selected in the **Hierarchy**, make this whole GameObject inactive by unchecking its active option at the top of the **Inspector** panel.

5. Next, add a 3D Cube to the scene, named `Cube-pickup`. In the **Inspector**, check **Is Trigger** for this GameObject.

6. In the **Inspector**, create a new `Hat` **Tag** and assign this tag to the `Cube-pickup` GameObject. (Use a different, appropriate Tag if you are adding a watch, glove, or weapon...)

7. In the **Project** panel, create a new C# script class named `AccessoryPickup.cs` containing the following:

```
using UnityEngine;

public class AccessoryPickup : MonoBehaviour {
    public GameObject hatInPlayer;

    private void OnTriggerEnter(Collider hit) {
        if (hit.CompareTag("Hat")) {
            Destroy(hit.gameObject);
            hatInPlayer.SetActive(true);
        }
    }
}
```

8. In the **Hierarchy**, select the **PlayerArmature** GameObject inside **NestedParentArmature_Unpack**, and add an instance object of the **AccessoryPickup** script class to the character in the scene by dragging the script asset file from the **Project** panel onto the **PlayerArmature** GameObject.

9. With the **PlayerArmature** GameObject selected in the **Hieararchy**, drag the **hat** GameObject that is a child of the **Head** GameObject into the public **Hat In Player** variable in the **Inspector** for the scripted component.

Figure 9.6: Linking the hat GameObject to the public scripted variable

10. Save and run your scene. When the player walks into Cube-pickup, the Cube should disappear, and the hat should appear on the head of the robot character the player is controlling.

How it works...

Because you placed the **hat** GameObject model as a child of the player's **Skeleton** GameObject hierarchy, whenever that part of the character's body is animated, the hat accessory will be animated with it – just as if the accessory were being worn/held. Having an inactive GameObject made active after a collision with an object with a particular tag is a common and effective technique for many game features.

The script has a public GameObject variable so that once a collection has been detected, it has a reference to the accessory GameObject to be made active.

Swapping the Third Person Armature for a different character

For quick prototyping, the provided robot Third Person Character model is fine. But when making your own games, you'll want to replace that model with a character in keeping with your game genre, such as magicians, elves, soldiers, merchants, zombies, and so on. In this recipe, we'll import a free 3D character model from the **Asset Store**, and customize our scene to animate this new character rather than the default robot.

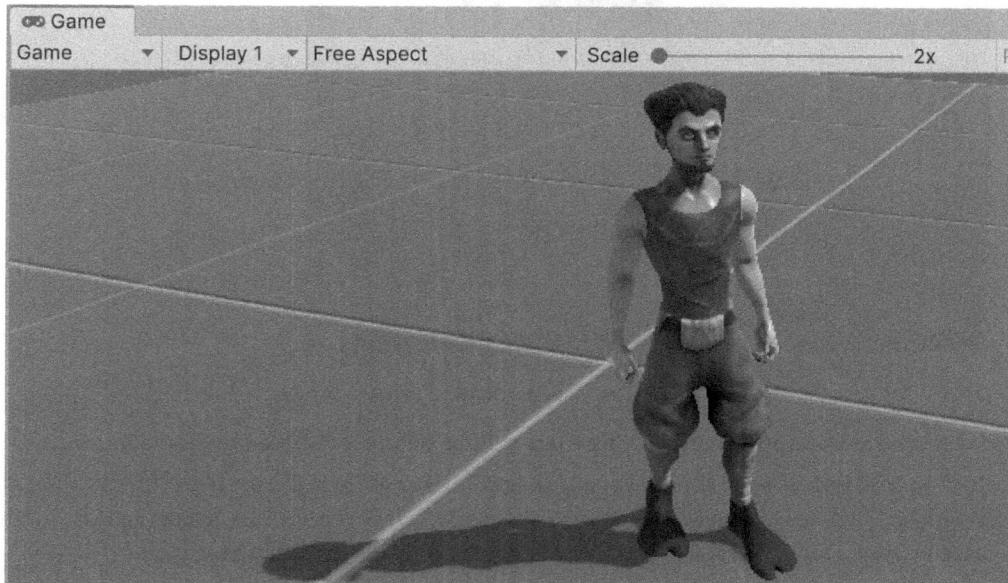

Figure 9.7: Our downloaded Elf character replacing the default robot

Getting ready

This recipe works on any scene using the Unity Third Person Character assets. So you could create a new project or work on a copy of a project from any of the earlier recipes in this chapter.

Note that you can find the Character Elf free character from the **Asset Store** at this URL (thanks to Maksim Bugrimov for publishing this as a free asset): `https://assetstore.unity.com/packages/3d/characters/humanoids/character-elf-114445`

How to do it...

To swap the Third Person Armature for a different character, perform the following steps:

1. Open the scene containing the Third Person Character.

2. Open a web browser tab to the Unity Asset Store by choosing **Asset Store | Asset Store Web** from the **Asset Store** menu at the top left of the Unity Editor application window.

3. Search for the free **Character Elf** character.

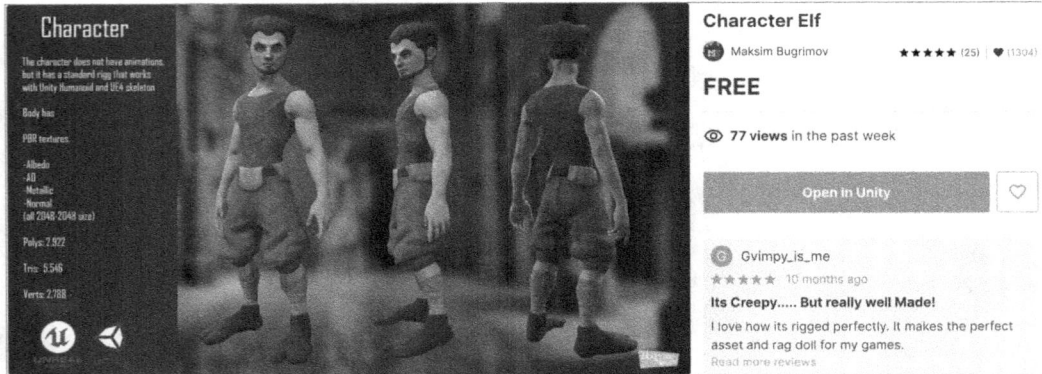

Figure 9.8: The free Character Elf asset in the Asset Store

4. Select the assets and, when viewing their details, click the **Add to My Assets** button.

5. In the Unity Editor, open the **Package Manager** to list your **Asset Store** assets by choosing **menu: Window | Package Manager** From the Packages dropdown choose My Assets (see the following screenshot). Dock the **Package Manager** panel alongside the **Inspector** panel.

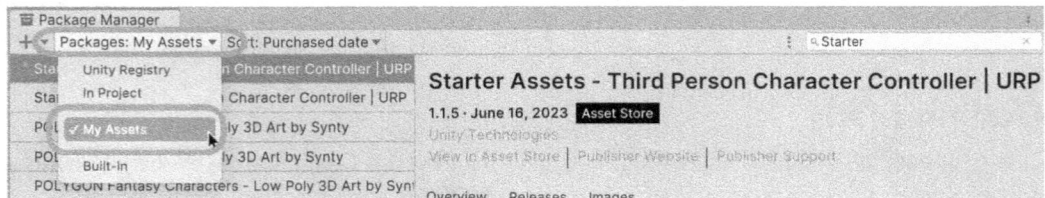

Figure 9.9: Option to view My Assets in the Package Manager in Unity Editor

6. In the **Package Manager** panel, locate and import the **Character Elf** package. Click **Download** and then **Import**.

7. You should now have a Character_Elf folder in your **Project** panel.

> **Note:** You may need to fix pink textures. You can do this by choosing **Window | Rendering | Render Pipeline Converter,** selecting all options, and clicking **Initialize and Convert.** After a minute or so, all textures should have been upgraded to work with the render pipeline for your project.

8. Select the Character_Elf asset file in the **Project** folder: Character_Elf | Prefabs. Look at the model's properties in the **Inspector,** and see its preview at the bottom of the **Inspector** panel.

9. In the **Inspector,** click the **Rig** tab and ensure **Animation Type** is set to **Humanoid** (and click the **Apply** button if you had to change it to **Humanoid**).

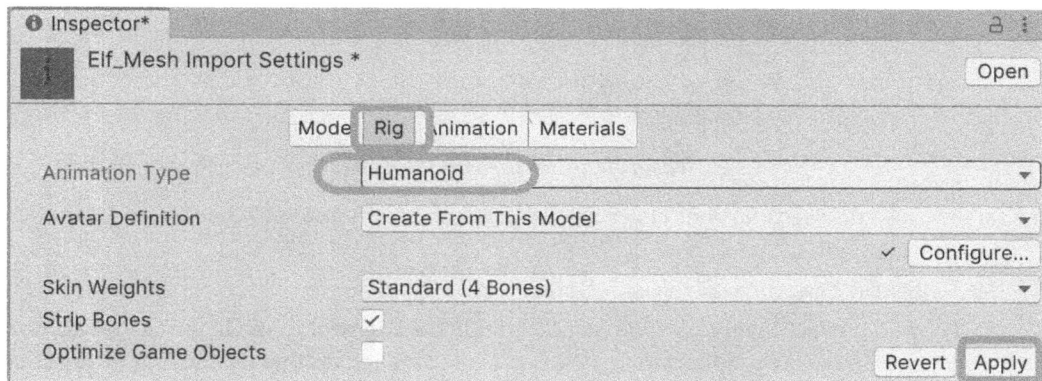

Figure 9.10: Ensuring the character's Animation Type is Humanoid in the Inspector panel

10. Add an elf character GameObject to the scene, by dragging the Character_Elf Prefab asset file into the scene from the **Project** panel folder: **Character_Elf | Prefabs.**

11. Now focus the **Scene** panel on the Armature_Mesh GameObject – do this by double-clicking this GameObject with the mouse. This is in the **Hierarchy** as a child: **NestedParentArmature_Unpack | PlayerArmature | Geometry.**

12. Now move the Character_Elf GameObject to the same location as the **Armature_Mesh** GameObject. Do this by single-clicking the Character_Elf GameObject and then choosing **GameObject | Move To View.**

13. Now make the `Character_Elf` GameObject a sibling of the `Armature_Mesh` GameObject by dragging it into the **Hierarchy** to also be a child of **PlayerArmature | Geometry**.

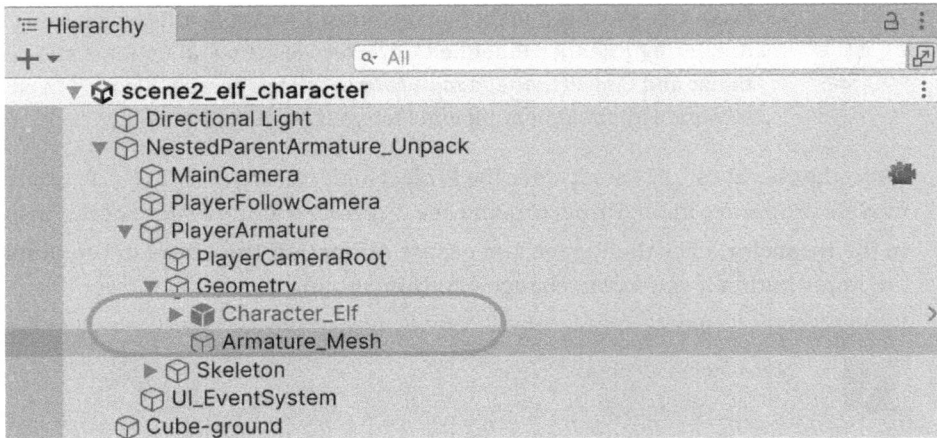

Figure 9.11: Childing Character Elf to the PlayerArmature Geometry GameObject

14. You should see the robot character (**Armature_Mesh**) and elf (`Character_Elf`) at the same position in the scene.

15. Delete the `Armature_Mesh` GameObject.

16. Now select the `PlayerArmature` GameObject in the **Hierarchy**, and in the **Inspector,** change the **Animator Avatar** property to `Elf_MeshAvatar`. You can do this by clicking the circular target button to open an asset chooser dialog, which will list all the **Avatar** asset files for this project.

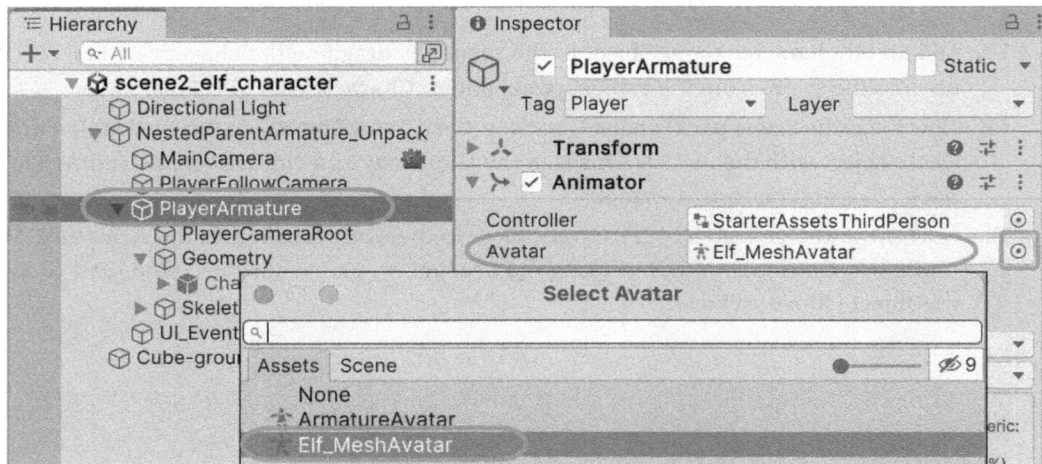

Figure 9.12: Setting the PlayerArmature GameObject Animator to the Elf_MeshAvatar

17. Save and play the scene. You should now be controlling an animated elf character instead of the default robot character.

How it works...

Whatever GameObjects are children of the Geometry child of the PlayerArmature GameObject will be moved by the Character Controller scripts. By bringing a new model into the scene, and positioning it just where the robot Armature_Mesh character was, you made sure the elf character would be positioned where the Character Controller scripts and camera expected. The **Animator** component of the PlayerArmature GameObject controls the animation clips being played by the character the user is controlling. So, you had to update this component to make the animation clips move elf characters instead of the robot.

Of course, instead of the **Character Elf** asset from the **Asset Store**, you could use any character model you have, from third-party sources or ones you've made yourself – as long as the character is "rigged" for animation. Rigs are virtual skeletons that are recognized by game engines such as Unity. See the next recipe to learn one way to create your own rigged character.

There's more...

There are some details that you won't want to miss.

Extracting textures when colors are not showing

For this **Character Elf** model, the textures had already been extracted, so we could see the correct images/colors on the character's clothes. However, sometimes when everything is packed into a single FBX file, Unity is not able to immediately find the textures for rendering the model.

> **Note:** It's a good idea to have a separate folder for each character model so that any additional files such as textures, animation clips, and controllers can all be stored in the same place. One common way to organize character models and associated asset files is to have a main Models folder and then a subfolder for each model and its files.

When you have selected the character asset file in the **Project** panel and are previewing how it looks at the bottom of the **Inspector** panel, you can follow these steps to extract textures if the model is initially previewing as gray all over:

1. Select your character asset file in the **Project** panel. View its properties in the **Inspector** panel.
2. In the **Inspector**, click the **Materials** tab, and click the **Extract Textures** button. A file save dialog will open for you to choose the location for the Material asset files to be saved. We usually save them in the same folder as the model itself.

3. Once the Materials have been extracted, you should see the character preview in the **Inspector** panel is fully textured/colored and ready to use in the scene.

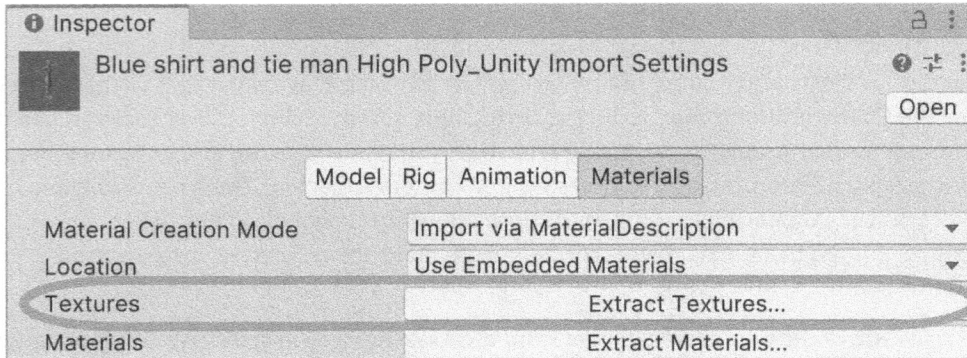

Figure 9.13: Extracting model textures

> **Note:** When extracting textures, you may get a pop-up dialog suggesting that you fix any issues – Unity is pretty good at fixing model textures so clicking **Fix** will usually make the character appear correctly.

Creating a 3D character with Autodesk Character Generator

In the last recipe, we used a character downloaded from the Unity Asset Store. While it is possible to create and rig a 3D character from scratch using tools like Blender, many hours (perhaps hundreds of hours) are needed to become proficient in a 3D modeling tool to get high-quality results. Another way to create 3D models is to use an application that offers many templates and customizable properties. In this recipe, we'll use the **Autodesk Character Generator** web application to create a ready-for-Unity, rigged 3D character model.

Figure 9.14: Using our generated character in a Unity project

Getting ready

You'll need an Autodesk account and access to the **Character Generator** web application – this is free for many students through Autodesk's Education Community program at https://www.autodesk.com/education/edu-software

> **Note:** At the time of writing, access to Autodesk Character Generator is erratic. Many users are able to make an account, but some are not. Hopefully Autodesk will resolve this issue to make this powerful character generator available to everyone in the near future.

We've provided the character created in this recipe as an FBX asset file, ready for you to use in your Unity projects inside the `09_05` folder.

How to do it...

To create a 3D character with Autodesk Character Generator, perform the following steps:

1. Open a web browser and log in to the Autodesk Character Generator web application site at https://charactergenerator.autodesk.com/.

2. Click the **+New** button to start creating a new character.

3. Choose an artistic style (**Standard**, **Bulk** (muscles!), or **Gorn** (Star Trek alien!)), and then a base character template, and then click the **Customize** button. For example, choose **Standard** and **Carter**.

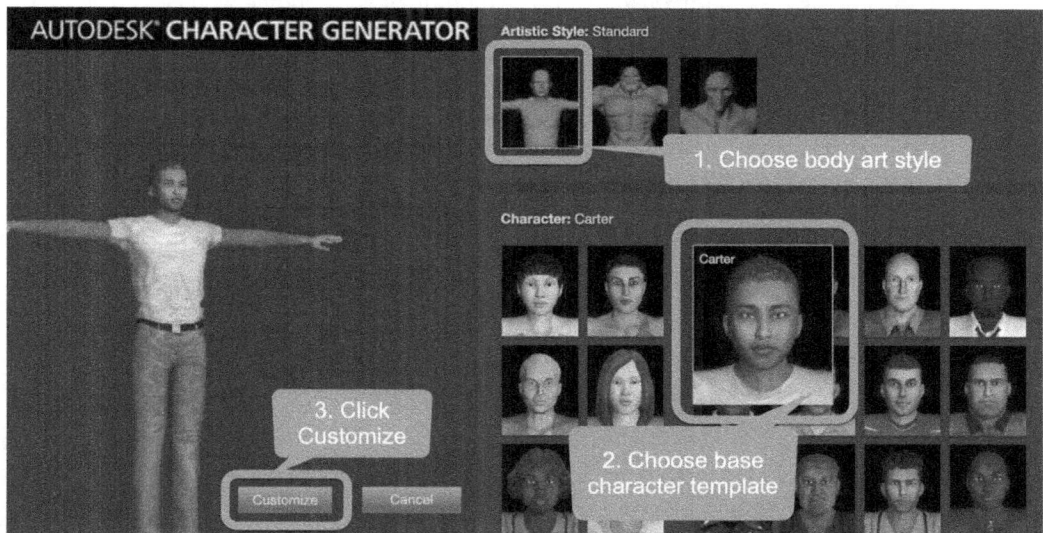

Figure 9.15: Choosing the initial artistic style and character in Character Generator

4. The customization screen allows you to edit properties for **Face**, **Skin**, **Eyes**, **Hair**, **Body**, and **Clothing**. At the top left of the screen, there are arrows to change the head orientation, and a drop-down menu to zoom into different parts of the face.

5. The **Face** and **Body** tabs allow you to choose two examples and then "tween" between them using the slider. The **Skin**, **Eyes**, **Hair**, and **Clothes** tabs let you select one of a range of examples.

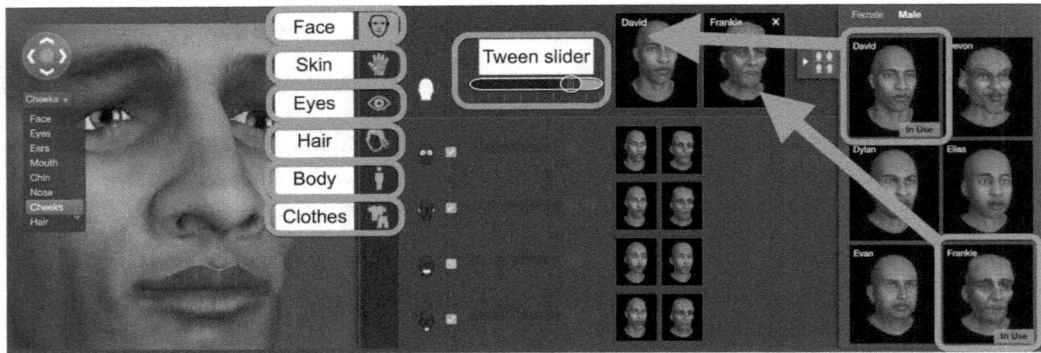

Figure 9.16: Using the slider for property values in between two face templates

6. **Face** lets you individually customize the eye sockets, ears, mouth, nose, and cheeks. You can drag two different examples and tween between them for each part of the face.

7. **Clothes** lets you select **Top**, **Bottom**, and **Shoes** (one of the shoe options is simply two bare feet). Each item of clothing has a range of color variations to choose from.

8. When you are satisfied with your character modeling, name your new character and click the **Finish** button. We named ours blue shirt and tie man.

9. After a few seconds, your new character will appear in your **Character Designs** list.

10. When viewing **Character Designs** for the module of your choice, you can click the download button (rectangle with an arrow, at the bottom left of the character) to open the **Generate Character** dialog choices.

11. In the **Generate Character** dialog, you can choose different quality and export settings. We recommend you choose the higher-quality settings initially (and you can generate lower-quality modules also if you wish):

   • **Poly Resolutions:** High

   • **Geometry:** Triangles

   • **Facial Expression:** Facial bone rig

   • **Textures:** (all options): Clothes + Specular Map + Normal Map

   • **Skeleton Resolution:** High

12. Leave **Character Orientation** as the default **Y-up**. You may change the name. We chose blue shirt and tie man HighPoly.

13. Finally, choose to generate a **Unity (.fbx)** file, then click the **Generate** button.

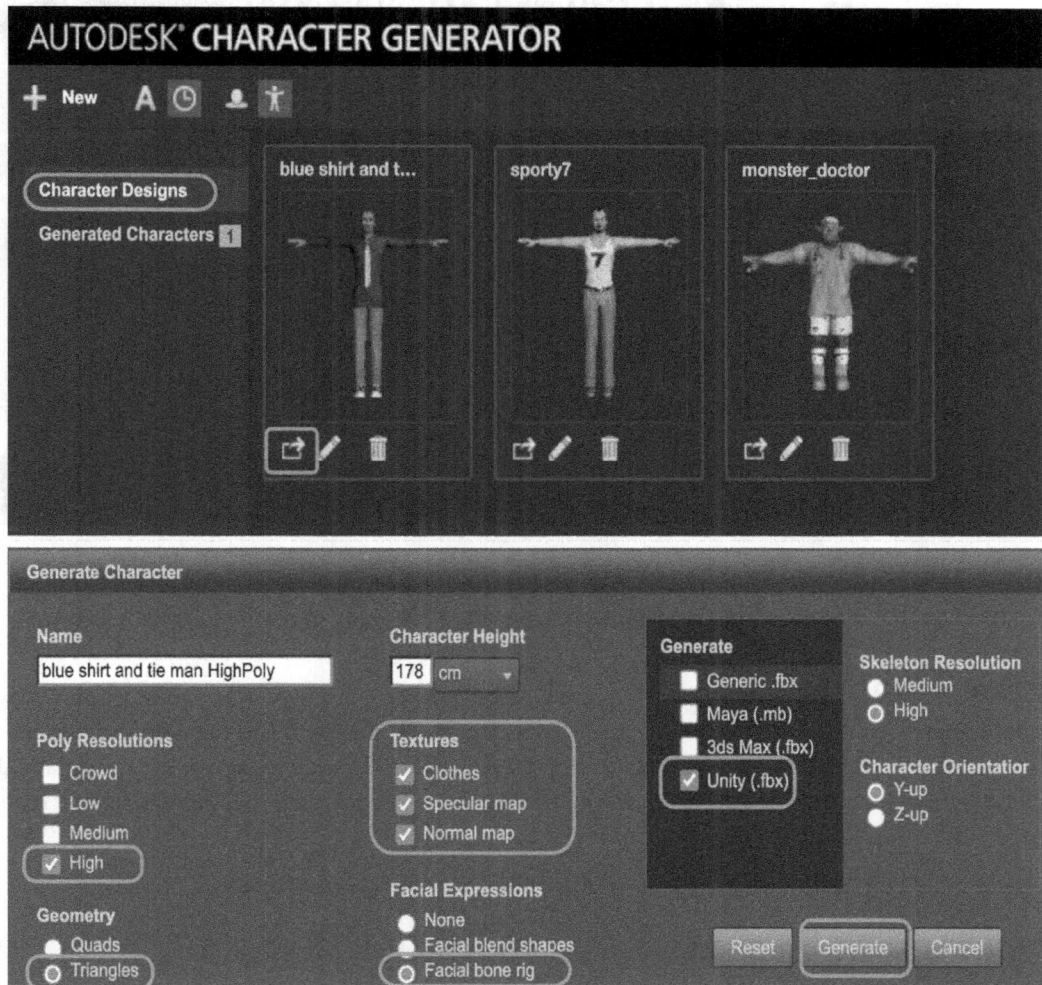

Figure 9.17: The Generate Character dialog

14. After a minute or so, your FBX will be listed in **Generated Characters** on the web page. You can now download the FBX file ready for use in Unity.

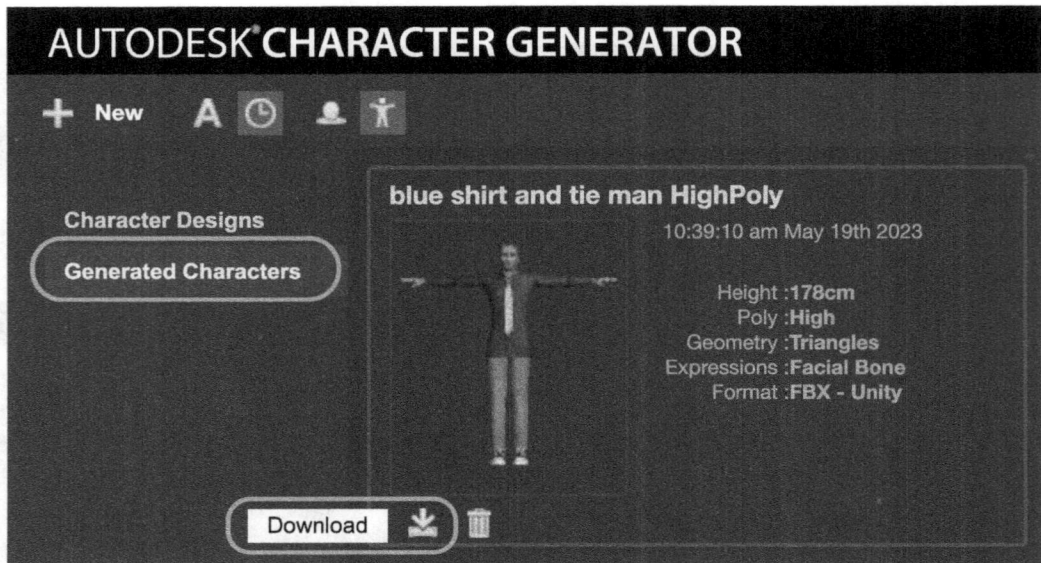

Figure 9.18: The Generated Characters dialog

15. You'll download a ZIP file, which contains the FBX ready to use in your own games.

How it works...

Applications such as Autodesk's Character Generator make it easy to create/customize 3D characters by choosing options for body shape properties and clothes. These applications can automatically "rig" characters for export to common 3D model file standards, including the FBX model format recognized by Unity.

The created FBX files can then easily be imported into your game projects.

There's more...

There are some details that you won't want to miss.

Replace the robot character with your created character

Use your new character, rather than **Character Elf**, and follow the steps for the previous recipe. You will soon be playing with your own created character as a Third Person Controller!

Remove the Light child GameObject from the character Hierarchy

When using characters from Autodesk's Character Generator in a Unity project, you may find that there is an unnecessary child GameObject named **Light** in the **Hierarchy**. Delete this GameObject, since it isn't a Unity light, and its size and position may mess up how the character is treated for animation purposes.

Selecting and downloading a character from Mixamo

For some years now, Adobe has been offering a free, powerful web application for animating characters called **Mixamo**. This web application provides a range of free, ready-made characters and also allows you to upload your own characters. It also allows you to choose from 100s of free animation clips. You can export just an animation clip or a combined character model and an animation clip, and one of the export options is for Unity!

In this recipe, we'll select a character and export it as an FBX asset file suitable for use in Unity. This recipe, plus the following two recipes, together allow you to create an animated NPC character in a Unity project from Mixamo resources.

Getting ready

This recipe uses the free **Adobe Mixamo** web application, so you'll need to sign up for an account with them if you don't have one already.

How to do it...

To select and download a character from Mixamo, follow these steps:

1. Open a web browser and visit Mixamo.com.
2. Sign up/log in with your **Mixamo/Adobe** account.
3. Select the **Characters** tab (navigation bar – top left of web page).
4. Select your character, such as **Remy**. There is a search box, allowing you to type in letters or words to filter the items listed. After clicking your chosen character, you should see the character appear in the right-hand preview panel.

Figure 9.19: Selecting the "Remy" character in Mixamo

5. Download your character by clicking the large orange **DOWNLOAD** button, ensuring that you choose the **FBX Binary (.fbx)** format and **T-pose** for **Pose**.

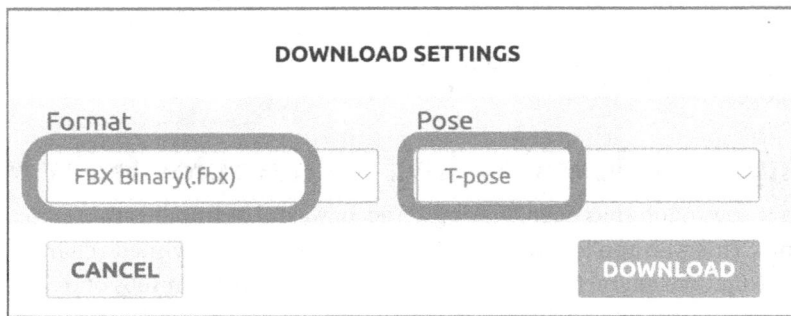

Figure 9.20: Download settings for a Mixamo character for use in Unity

6. You now have an FBX asset file ready for use in a Unity project.

How it works...

Mixamo exports the selected 3D character models, including the virtual skeleton "rig," in the FBX format, which is recognized by Unity.

There's more...

There are some details that you won't want to miss.

Using Mixamo to convert and fix character models

Mixamo provides a range of 3D character models to choose from. However, you may have other character models you have made yourself, or downloaded from other sources. Mixamo is able to load FBX and OBJ model files, and rig and export characters suitable for use in Unity. So if you have a model that isn't in the right format or is not rigged suitably, you can upload your model to Mixamo, fix it, and then download the character FBX ready for Unity.

The example used here is the Toon Girl free 3D model, which wasn't set up correctly for Unity, and so we used Mixamo to fix the rigging issues. The free Toon Girl model was created by DuDeHTM (thanks for making it free!) on the TurboSquid.com website. DuDeHTM's free models can be found at this URL: https://www.turbosquid.com/3d-models/girl-model-1637866?

To upload and rig your own model with Mixamo, do the following:

1. Open a web browser and visit Mixamo.com.
2. Sign up/log in with your **Mixamo/Adobe** account.

3. Click the **UPLOAD CHARACTER** button, and upload your character model.

4. Mixamo will present a percentage progress bar as it imports the model – it should take less than a minute to upload most models.

5. Mixamo asks you to orient your model – in case it's rotated the wrong way. Mixamo looks for the model to be facing you, and standing in a standard T-pose (legs together and arms stretched out to the sides). Once it looks fine, click **NEXT**.

6. Next, Mixamo asks you to place markers at key locations on the character: CHIN, WRISTS, ELBOWS, KNEES, AND GROIN. Most models are symmetrical, so with this option checked, you place the left-hand wrist/elbow/knee marker, and the right-hand one is automatically mirrored for you.

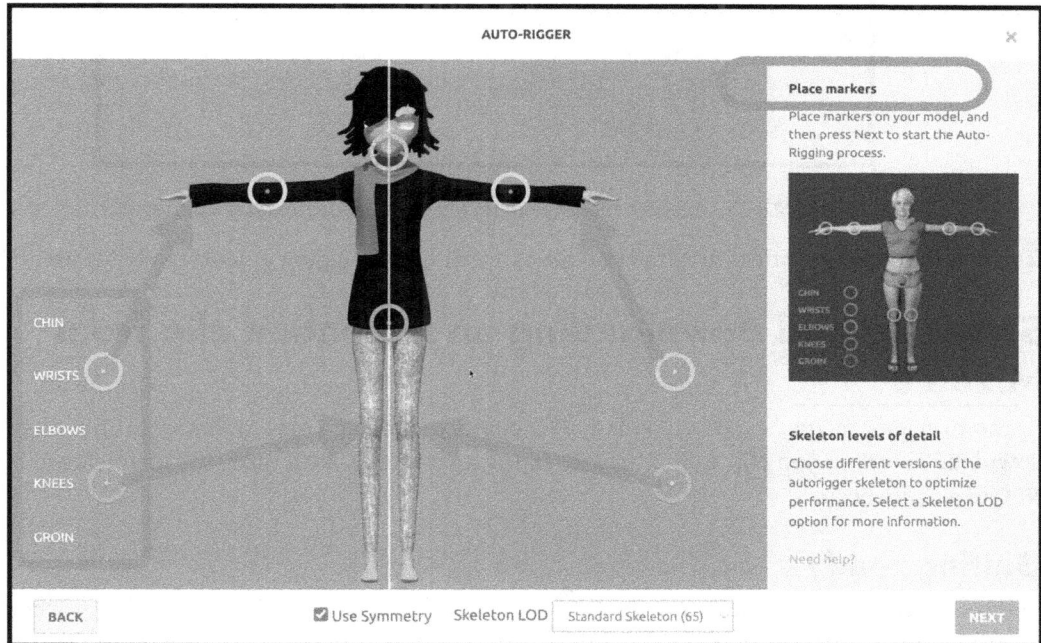

Figure 9.21: Placing the joint markers for the Mixamo AUTO-RIGGER on the Toon Girl model

7. After positioning all the markers, click **NEXT** and then wait for a minute or two for the AUTO-RIGGER to do its magic! If Mixamo isn't happy with any marker placements, you'll return to the **Place markers** page.

8. Once the AUTO-RIGGER has successfully completed its work, you can review the character as it's animated, and then continue to use this character model on the main Mixamo page.

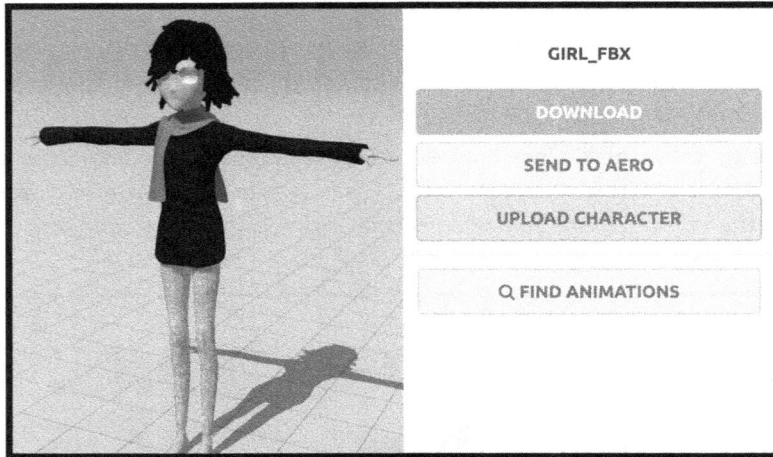

Figure 9.22: The Toon Girl model, with rig fixed and ready to export as Unity FBX

It's great that Mixamo offers this free tool for converting and rigging models ready for use in Unity!

Selecting and downloading an animation clip from Mixamo

In the previous recipe, we downloaded a 3D character model from Mixamo. In this recipe, we'll download an animation clip from Mixamo. In the next recipe, we'll bring the character to life as an NPC in a Unity project.

Getting ready

This recipe uses the free **Adobe Mixamo** web application, so you'll need to sign up for an account with them if you don't have one already.

How to do it...

To select and download an animation clip from Mixamo, perform the following steps:

1. Open a web browser and visit `Mixamo.com`.

2. Sign up/log in with your **Mixamo/Adobe** account.

3. Click the **Animations** tab, and search for an appropriate "idle" animation. We like the **Breathing Idle** animation for NPCs that are animated as they stand in place in a scene. Once you click an animation, you'll see it previewed in the main character window, for whatever character you have selected.

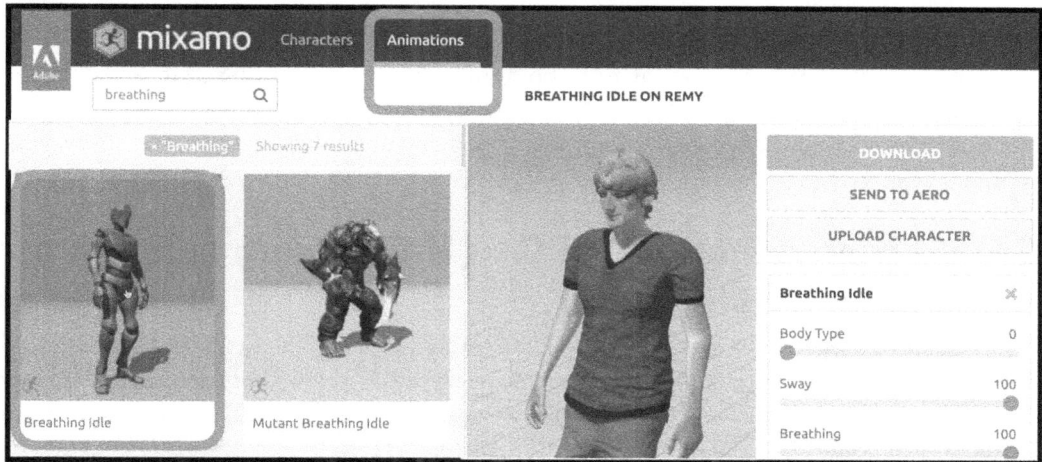

Figure 9.23: Previewing the Breathing Idle animation clip on the Remy character model

4. Click the **DOWNLOAD** button to open the **DOWNLOAD SETTINGS** dialog:

    *   For **Format**, choose **FBX For Unity(.fbx)**.
    *   If you just want the animation clip (with no character model), then for **Skin**, choose **Without Skin**. However, if you want to download both the character model and animation clip, then choose **With Skin**.
    *   Accept the defaults of **30** and **none** for **Frames per Second** and the **Keyframe Reduction** settings.

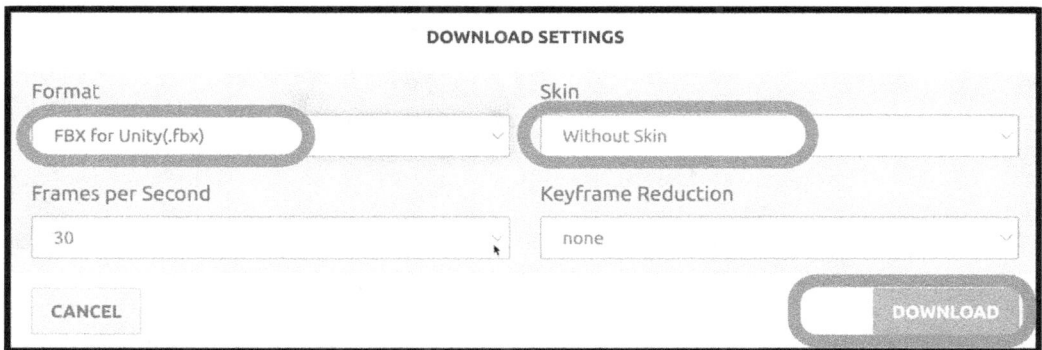

Figure 9.24: Previewing the Breathing Idle animation clip on the Remy character model

5. Once you've chosen your settings, click the **DOWNLOAD** button.
6. You now have a character animation clip FBX file ready to use in Unity.

How it works...

Mixamo allows you to select an animation clip from its library, and download it as an FBX file formatted for use with Unity's character animation system. Once you have the combination of a character model and some animation clips, you are ready to create animated characters in Unity, which we'll do in the next recipe.

Creating an animated NPC in Unity using a character and animation clip

Once you have a rigged character model and an animation clip, you are ready to add an animated NPC character to a Unity project. For this recipe, we'll create a new Unity project and import a character and animation clip, then add an Animator Controller asset that will allow Unity to animate the character in a scene.

Figure 9.25: Our animated NPC character

Getting ready

For this recipe, you can use your own FBX animation clips and 3D characters, or use the ones from previous recipes in this chapter.

We have provided the character (GIRL_unity.fbx) and animation clip (@Breathing Idle.fbx) from the previous two recipes for you to use in the 09_08 folder.

How to do it...

To create an animated NPC in Unity using a character and animation clip, perform the following steps:

1. Create a new 3D Unity project.

2. Import your character model FBX into the **Project** panel's **Assets** folder. In the **Inspector** panel, you may need to set the model's **Rig Animation Type** settings to **Humanoid**.

> **Note:** If colors/textures aren't displaying, you may need to extract textures. The steps for how to do this are shown at the end of the *Swapping the Third Person Armature for a different character* recipe.

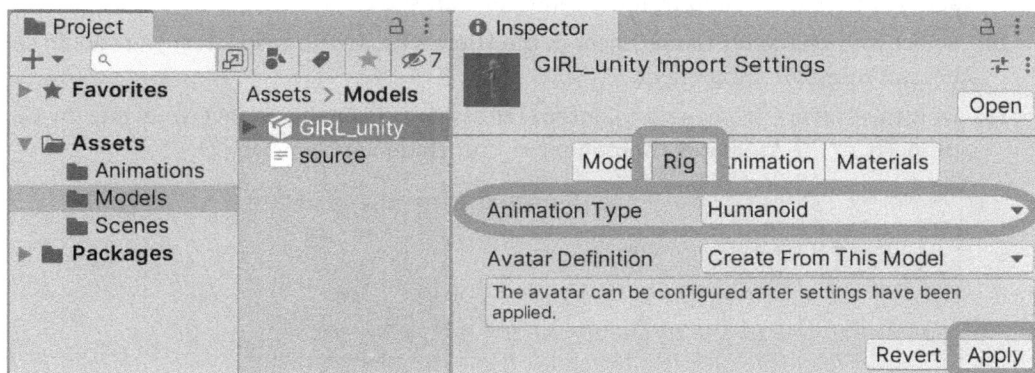

Figure 9.26: Setting the Rig Animation Type of the imported character to Humanoid

3. Select your character's folder in the **Project** panel, and create a new **Animator Controller** asset file named Character Animator Controller, by choosing the **Project** panel menu **Create | Animator Controller**.

4. Double-click the Character Animator Controller asset file to open the **Animator** panel to graphically edit this state chart animator controller asset.

5. Import the @Breathing Idle.fbx animation asset file into your project, then drag @Breathing Idle.fbx from the Project panel into the Animator panel @Breathing Idle.fbx from the **Project** panel into the **Animator** panel. A new animation state named **Breathing Idle** should be added to the animation state chart.

6. Since this is the only animation state, it should become orange and be automatically linked from the **Entry** state.

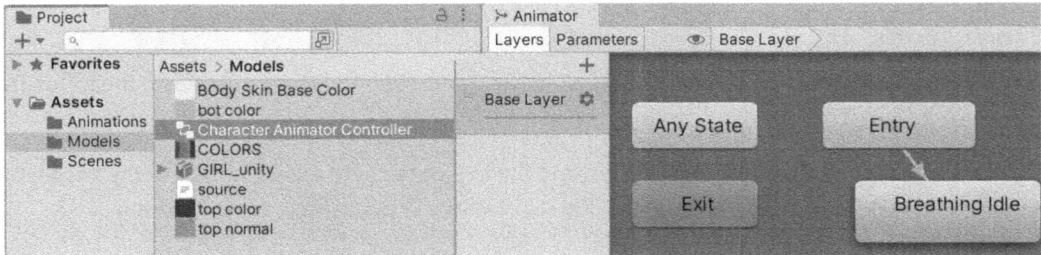

Figure 9.27: The Breathing Idle animation state in the Animator state chart

7. Create a new scene for your project.

8. Add a 3D Cube to the scene, named Cube-ground, and set **Transform Scale** to **(20, 0.01, 20)**.

9. Drag your character into the scene and set **Transform Position** to **(0, 0, 0)**. Your character should be located standing at the center of your squashed Cube.

10. Select your character GameObject in the **Hierarchy**. In the **Inspector**, you should see that it has an **Animator** component.

11. Drag the Character Animator Controller asset file from the **Project** panel into the **Controller** property of the **Animator** component of your character GameObject.

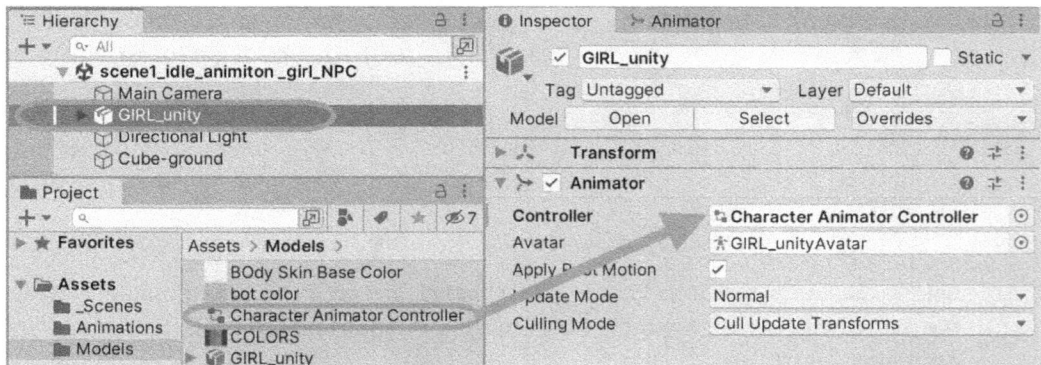

Figure 9.28: Linking your Animator Controller asset file to the Animator component of your NPC character in the scene

12. Save and play the scene. You should see your character animated on the spot with the animation clip that was imported.

How it works...

A character needs an **Animator** component to be animated in a Unity scene. In this recipe, you imported both a character and an animation clip. You created a new **Animator Controller** asset file and added your animation clip as the default animation state.

By creating a scene GameObject from your character model, and linking its **Animator** component to your **Animator Controller** asset file, you made Unity apply the animation clip to the character model, which resulted in the animated 3D character in this scene.

Using scripts to control 3D animations (old input system)

In the previous recipe, we created a simple transition in our **Animator Controller** from **Entry** to the **Idle** animation. However, when we control a character, we will want it to run different animations depending on our inputs – idle, walking running, jumping, and so on. Different events will determine which animation clip we want our character to execute – such as pressing the *WASD-SPACE* keys or changing from a walking animation to a running one depending on our character's speed of movement, and so on.

Unity has two input systems. The old input system works fine, however, the new input system adds more abstraction between devices generating input events (such as keyboards, mice, game controllers, etc.) and the code that responds to actions indicated by those input events. In this recipe, we'll use the old input system, and in the recipe that follows, we'll explore how to use the new input system. Therefore, between the two recipes, you'll be able to choose which is most appropriate for your own game projects.

We need to be able to trigger animation transitions and parameters through our scripted code. In this recipe, we'll create a very simple controller script for our human model and its **Animator Controller**. We will use the *WS-Up-Down-ARROW* keys to make our character walk backwards/forwards, and change from an idle animation to a walking one and back again as appropriate to its speed.

After learning the basics in this recipe, you'll have the knowledge and skills to understand third-party character controllers, such as those from the Unity Starter Assets and **Asset Store** assets.

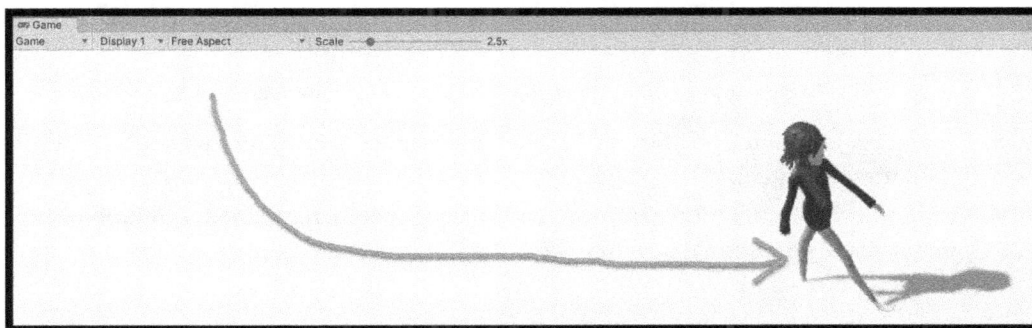

Figure 9.29: What we'll create in this recipe – a user-controlled walking animated character

Getting ready

This recipe builds upon the previous one, so make a copy of that and use that copy.

How to do it...

To use scripts to control 3D animations, perform the following steps:

1. Open the copy of the project from the previous recipe.

2. Open a web browser and visit Mixamo.com.

3. Sign up/log in with your **Mixamo/Adobe** account.

4. Click the **Animations** tab, and search for an appropriate "walk" animation. We like the **Standard Walk** animation. Download this animation clip, formatted for Unity FBX and **Without Skin**.

5. Import your walk FBX animation clip into your project (into the **Project** panel folder named **Animations**).

6. In the **Project** panel, select your walk animation clip, and in the **Inspector**, click the **Rig** tab and set **Animation Type** to **Humanoid**.

7. Now click the **Animation** tab, check the **Loop Time** check box, and click the **Apply** button – since we want this walking animation to loop.

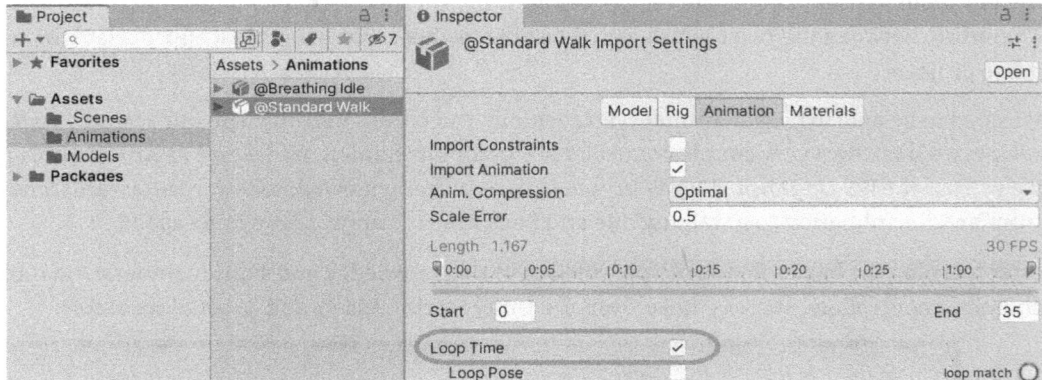

Figure 9.30: Checking the Loop Time property of the Animation Clip asset file

8. In the **Project** panel, locate your **Character Animator Controller** and double-click it to open the **Animator** panel.

9. Drag your FBX animation clip from the **Project** panel folder named **Animations** into the **Animator** panel. You should now have both the **Idle** and **Walk** animation clips in your animation state chart.

10. Select the **Parameters** tab at the top left of the **Animator** panel, and then click the plus (+) button to add a new **Float** parameter named **Speed**:

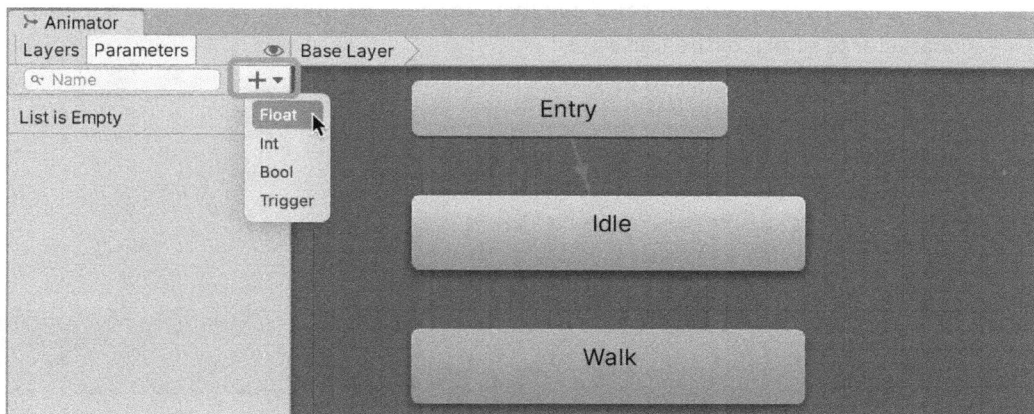

Figure 9.31: Creating the Speed Float parameter in the Animator Controller

11. Click the **Idle** animation clip state, and then right-click and choose **Make Transition**. This should give you an array that you can then link to the **Walk** animation clip state.

12. Now, select the **Transitions** arrow from **Idle** to **Walk**. In the **Inspector** panel, uncheck the **Has Exit Time** option.

13. Then, at the bottom of this component, add a new **Condition** by clicking the plus (+) button. Since we only have one parameter, **Speed**, this is automatically chosen for the Condition variable. The default values of **Greater** for **operator** and **zero** for **value** are also what we want, so we have no more work to do here. We now have a conditional transition from **Idle** to **Walk** that will only fire when the **Speed** parameter is greater than zero.

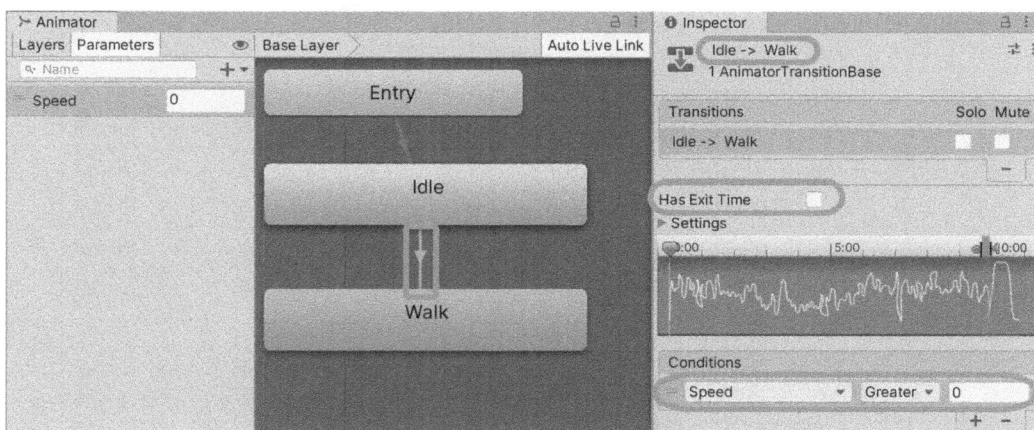

Figure 9.32: The conditional transition from Idle to Walk when Speed is greater than 0

14. Add a second conditional transition to our **Animator Controller** state chart. This transition should be from the **Walk** state back to the **Idle** state, again with no exit time (**Has Exit Time** option unchecked), and the condition that the **Speed** parameter has to be less than 0.1 for the character to return to the **Idle** state.

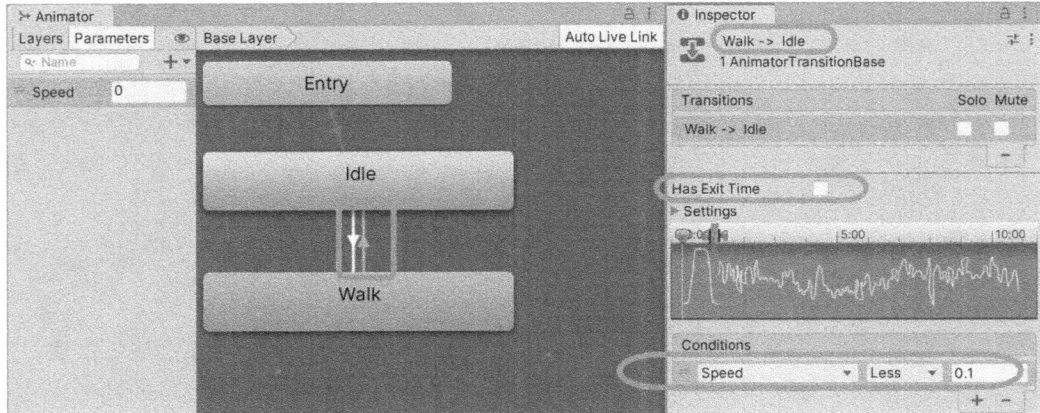

Figure 9.33: The conditional transition from Walk to Idle when Speed is less than 0.1

15. Let's now write a script that allows the *Up/Down* arrows (and *W/S* keys) to change the **Speed** parameter in our **Animator Controller**. In the **Project** panel, create a new C# script class named PlayerMove.cs containing the following code:

```
using UnityEngine;


public class PlayerMove : MonoBehaviour {
    public float rotateSpeed = 1;

    private float vertical = 0;
    private float horizontal = 0;
    private Animator animator;

    private void Awake() {
        animator = GetComponent<Animator>();
    }

    private void Update() {
        vertical = Input.GetAxis("Vertical");
        horizontal = Input.GetAxis("Horizontal");
    }
```

```
void FixedUpdate()
{
    animator.SetFloat("Speed", vertical);

    // turn
    transform.Rotate(0, horizontal * rotateSpeed, 0);
}
}
```

16. Add an instance object of this script class to the character in the scene by dragging the PlayerMove script asset file from the **Project** panel onto the **character** GameObject in the **Hierarchy**.

17. Save and run your scene. The character should begin in the **Idle** animation state. If you press the *Up arrow* or *W* keys, you should see the character start walking. If you arrange your panels so you can see the **Animator** panel when the **Scene** is running, then you'll see the value of the **Speed** parameter being changed. However, if you release the keys, while the value of the **Speed** parameter returns to zero, the character stays in the **Walk** state.

How it works...

By making conditional transitions between animation states, we enable the whole animation of a character to be driven through in-game events and scripting. Unity **Animation Controllers** provide a range of different types of parameters, from simple integer and float numeric values to true/false **Boolean** variables, and consumable Boolean "triggers."

While sometimes we do want an automatic, timed transition from one animation state to another, in many cases, we need to uncheck the **Exit Time** property for **Transitions** and define the appropriate condition for when we wish the **Animation State** change to take place.

In this recipe, we created a parameter named **Speed**, whose value we set in our script based on the **vertical** input axis. Unity's default input setup means that the *W/S* and *Up/Down* keyboard keys affect the vertical axis values in the range -1 ... 0 .. +1. Our scripted variable, vertical, is set based on this Unity input axis: vertical = Input.GetAxis("Vertical"). Therefore, our code has two functions:

- Get values from the user/game events.
- Set the corresponding parameter in the Animator Controller.

It's best to process user input in Update() methods (each frame), and to communicate with **Animator Controllers** in FixedUpdate() (in sync with the physics and animation engines). So, we declare variables in our script that can be accessed by any method.

Using scripts to control 3D animations (new input system)

At present, new Unity 3D projects default to the old input system (this may change in later versions of the Unity Editor). However, many assets and projects now use the new input system, so let's learn how to control our players using the new system. In this recipe, we'll switch the project from the previous recipe to the new input system and then set up the input controls, event handlers, and scripts to work with the new input system.

Getting ready

This recipe is based on the previous one, so make a copy of that and work on that copy.

How to do it...

To use scripts to control animations with the new input system instead of the old `Input.GetAxis(<>)`, perform the following steps:

1. Open the **Package Manager** panel, by going to **Window | Package Manager**. Choose **Unity Registry** and locate the **Input System** package. Install this package.

2. Open the **Project Settings** panel by choosing **Edit | Project Settings...**.

3. Select the **Player** tab, and open the **Other Settings** section. Then, for **Configuration Active Input Handling**, choose **Input System Package (New)** from the drop-down menu.

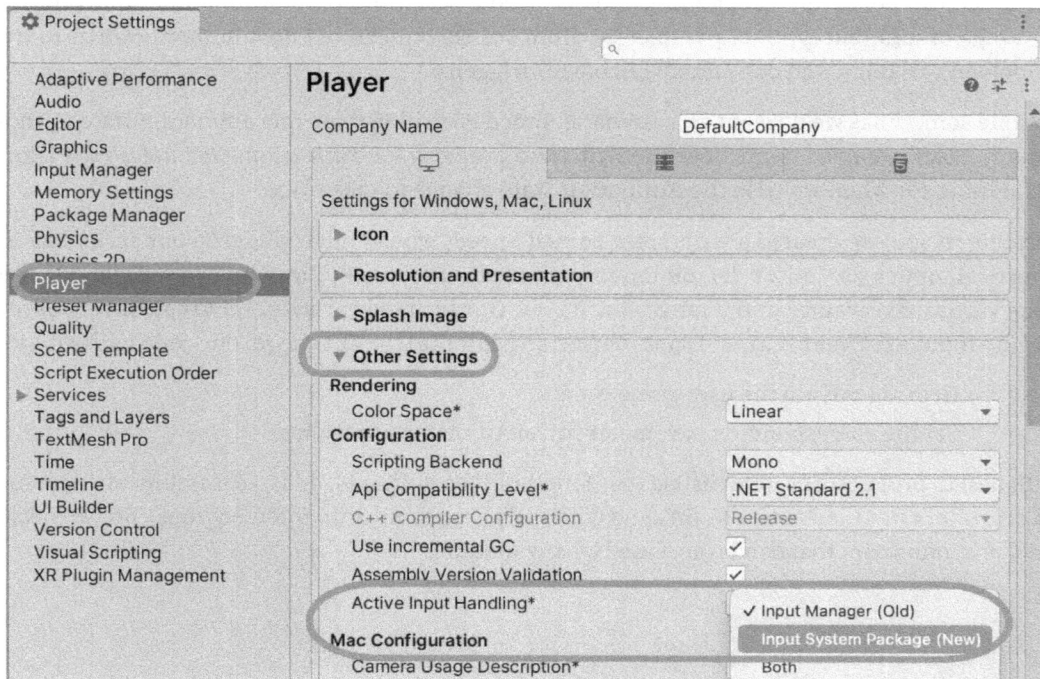

Figure 9.34: Switching the project Player settings to the new input system

4. You will be prompted to confirm the change and the Unity Editor will restart itself to load the new settings.

5. Select your character GameObject in the **Hierarchy**, and in the **Inspector**, click the **Add Component** button and enter Player Input. This will add a **Player Input** component to your character GameObject.

6. Click the **Create Actions…** button for this new component and select a project folder and name for the new **Input Actions Asset** file. For example, create an **InputActions** folder and name the new asset file Controls.

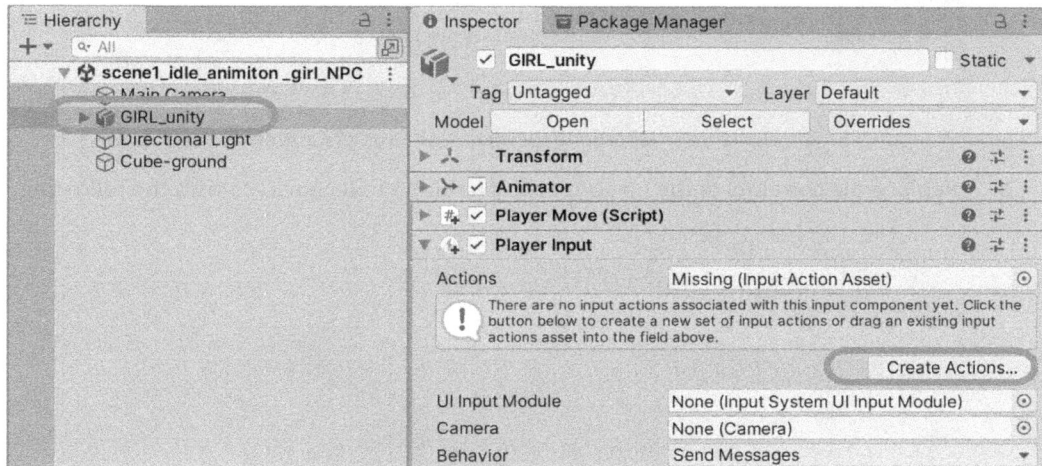

Figure 9.35: Creating a new Input Action Asset file for the player inputs

7. The default contents of a new **Input Action Asset** file already have a WASD set of player inputs, so the new asset doesn't need to be edited.

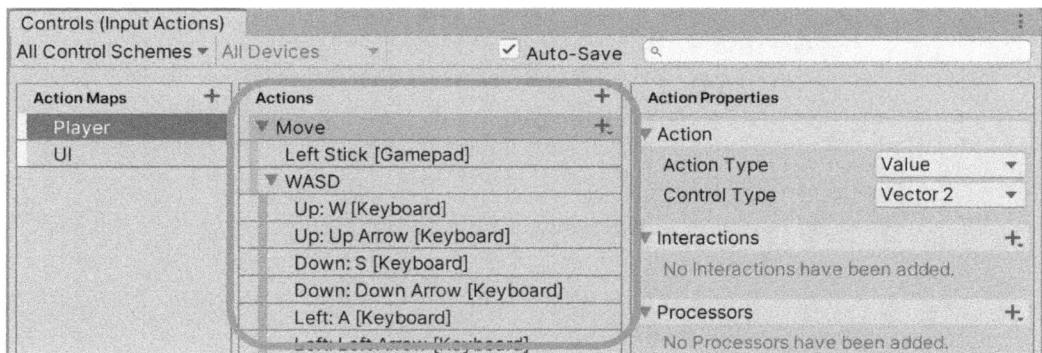

Figure 9.36: The Input Action Player Map with its default WASD player inputs

8. In the **Project** panel, select the Controls **InputActions** asset file. Then, in the **Inspector** check the **Generate C# Class** option, then click the **Apply** button.

You will see a new C# script file added to the contents of the **InputActions** folder in the **Project** panel.

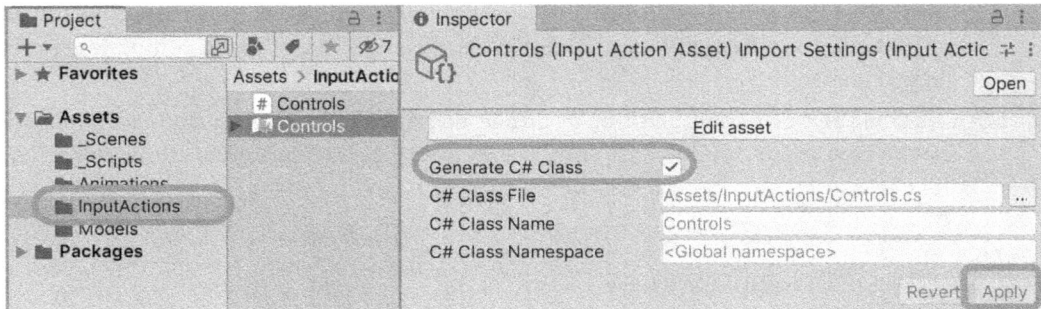

Figure 9.37: Generating the InputActions asset Controls C# script class

9. Replace the contents of the C# script class named PlayerMove.cs with the following:

```
using UnityEngine;
using UnityEngine.InputSystem;

public class PlayerMove : MonoBehaviour, Controls.IPlayerActions {
    private Vector2 _direction;
    private float vertical = 0;
    private float horizontal = 0;
    private Animator animator;
    public float rotateSpeed = 1;

    public void OnMove(InputAction.CallbackContext context) {
        _direction = context.ReadValue<Vector2>();
        vertical = _direction.y;
        horizontal = _direction.x;
    }

    private void Awake() {
        animator = GetComponent<Animator>();
    }

    void FixedUpdate() {
        animator.SetFloat("Speed", vertical);
        transform.Rotate(0, horizontal * rotateSpeed, 0);
        transform.Rotate(0, horizontal * rotateSpeed, 0);
    }
}
```

10. Select your character GameObject in the **Hierarchy**, and in the **Inspector**, for the **Player Input** component, set **Behavior** to **Invoke Unity Events** from the drop-down menu.

11. Next, expand the **Events | Player** part of this component and click the add event (+plus sign) button. Where a GameObject is needed (below the **Runtime Only** dropdown), drag your player GameObject.

12. Next, choose the **PlayerMove.OnMove** method from the drop-down menu.

Figure 9.38: Changing Player Input to Invoke Unity Events

13. Save and run the scene. Just as before, you can use the *WASD* and *Arrow* keys to make the character walk forward and turn.

How it works...

You converted the project to use the new input system. This requires **Input Actions Asset** files to map input devices to input actions and their values. By default, new **Input Actions Asset** files include a **Move** property, which is a **Vector2** (X and Y value pair).

You changed the PlayerMove C# script class to listen for and respond to **OnMove(...)** events. The **Player Input** component added to the player's character GameObject in the **Hierarchy** was linked to the **Input Actions Asset** file, and set to **Invoke Unity Events** when input actions (such as *WASD* and *Arrow* keys) are detected. You registered the OnMove() method of the PlayerMove C# script class to respond to movement input actions. So each time *WASD* or *Arrow* keys are detected as being input, the Vector2 (X and Y value pair) is changed accordingly, and the new values are sent as a message to the OnMove() method of the PlayerMove C# script class object in the character's GameObject. This OnMove() method extracted the Y value and set it to the vertical variable and the X value and set it to the horizontal value. Every time the FixedUpdate() method is invoked (many times a second), the value of the vertical variable is used to set the speed inside the Animator Controller, and the value of the horizontal variable is used for any turning.

While the new input system may take some time to get used to, its event-based approach is very powerful. Also, were you to use a game controller, its properties could very easily be mapped to the **Input Actions** asset files, with no change needed in the code. The old input system would have needed changes to the code when swapping a project from keyboard input to a game controller.

Further reading

Here are some links to further reading/resources regarding the topics covered in the chapter.

Learn more about 3D models and animation importing from the following sources:

- Unity docs on importing **3D Models**: `https://docs.unity3d.com/Manual/HOWTO-importObject.html`
- Unity docs about the **Model Import Settings** window: `https://docs.unity3d.com/Manual/class-FBXImporter.html`
- Unity docs about the **Model** tab: `https://docs.unity3d.com/Manual/FBXImporter-Model.html`
- Unity docs about **Model** file formats: `https://docs.unity3d.com/Manual/3D-formats.html`
- Samples of **Mixamo** free assets in the **Asset Store**:
  - `https://assetstore.unity.com/packages/3d/animations/melee-axe-pack-35320`
  - `https://assetstore.unity.com/packages/3d/animations/magic-pack-36269`

Learn more about Unity Animator Controllers and scripting animations from the following sources:

- Unity video tutorial about Animator Controllers: `https://www.youtube.com/watch?v=JeZkctmoBPw`
- Unity video tutorial about using scripting to influence Animator Controllers: `https://www.youtube.com/watch?v=s7EIp-OqVyk`
- Invector offers a free version of its powerful human player controller in the **Asset Store**. It is well worth checking out if you are writing a third-person controller game: `https://assetstore.unity.com/packages/tools/utilities/third-person-controller-basic-locomotion-free-82048`.

If you want to create and rig your own characters from scratch, Blender is a fantastic, free, open-source modeling application. Learn more about Blender:

- Perhaps the best introduction to Blender is the official online manual:
  - `https://docs.blender.org/manual`
- This site lists several tutorials introducing 3D character modeling with Blender:
  - `https://cgian.com/2022/06/blender-character-modeling-tutorial`
- Rigging characters with the free Rigify Blender plug-in:
  - `https://www.pluralsight.com/blog/tutorials/rigging-minutes-blenders-rigify-addon`

A nice guide to the new input system by Alex Somerville can be found here:

- `https://mrlovelies.medium.com/unity-dev-player-movement-new-input-system-6291f257a77d`

Learn more on Discord

To join the Discord community for this book – where you can share feedback, ask questions to the author, and learn about new releases – follow the QR code below:

```
https://packt.link/unitydev
```

10

Saving and Loading Data

Times when we wish to load data include retrieving high scores from previous plays of a game, or perhaps remembering data values between scenes. Another time to load data is when a level layout is stored in a data file (such as text characters, or a data format like XML or JSON), and when a scene begins, that data is loaded and used to dynamically create GameObjects for the scene. And of course, we must have saved the data previously in order to be able to load it at a later time.

Some of the saving/loading is ephemeral – just while a game is playing – and everything is reset the next time the game is run. Other times, data can be stored that is remembered between game plays, either as part of the build application's private data, or to shared folders that can be changed from outside the application, as data on a web server, or via a web communication.

Sometimes, data to be saved are as simple as individual numbers or text strings, and other times they can be hundreds or thousands of lines of text or name/value pairs. When communicated by the web interface, or to local storage using Unity's PlayerPrefs class, we are restricted in the types of data that we can work with and limited to saving and loading integers, floats, and strings. In this chapter, we will provide several recipes illustrating ways to save and load data at runtime, including the use of static variables and the PlayerPrefs class.

> PlayerPrefs offers a great, multi-platform way to store persistent data locally for Unity games. However, it is also very easy to hack into, so sensitive or confidential data should not be stored using this technique. To securely store data online in a hashed/encrypted format, data storage methods should be used. However, for data such as lives left, checkpoints reached, scores and time remaining, and so on, it is an easy and simple way to implement memory after a scene has been exited.

Text-based external data is very common and very useful, as it is both computer- and human-readable. Text files can be used to allow non-technical team members to edit written content or to record game performance data during development and testing.

Text-based formats also permit serialization – the encoding of live object data suitable for transmission, storing, and later retrieval. Unity treats all of the following file types as Text Assets:

- `.txt`: Plain text file
- `.html`, `.htm`: HTML (HyperText Markup Language)
- `.xml`: XML (eXtensible Markup Language) data
- `.json`: JSON (JavaScript Object Notation)
- `.csv`: CSV (Comma Separate Values)
- `.yaml`: YAML (YAML Ain't Markup Language)
- `.fnt`: Bitmap font data (with the associated image texture file)
- `.bytes`: Binary data (accessed through the bytes property)

Three of the file formats listed above are widely used text interchange standards: XML, JSON, and CSV. XML is a meta-language, that is, a set of rules that allows markup languages to be created to encode specific kinds of data. Some examples of data-description language formats using the XML syntax include the following:

- SVG: Scalable Vector Graphics – an open standard method of describing graphics, supported by the Worldwide Web consortium.
- SOAP: Simple Object Access Protocol for the exchange of messages between computer programs and web services.
- X3D: XML 3D – an ISO standard to represent 3D objects.

JSON is sometimes referred to as the "fat-free" alternative to XML – offering similar data interchange strengths but being smaller, and simpler, both by not offering extensibility and using just three characters for formatting:

- `property : value`: The colon character separates a property name from its value
- `{ }`: Braces are for an object
- `[ ]`: Square brackets are for an array of values/objects

CSV is one of the oldest and simplest text data exchange formats:

- `value1, value2, value3, etc.`: comma characters (,) separate each piece of data

Network communications to databases and web servers are another strategy to save and load game data. A **server** waits for messages requesting something for a **client**, and when one is received, it attempts to interpret and act upon the message, sending back an appropriate response to the **client**. A **client** is a computer program that can communicate with other **clients** and/or **servers**. **Clients** send **requests** and receive **responses** in return. It is useful to keep the following four concepts in mind when thinking about and working with client-server architectures:

- **Client**: an application that sends messages to servers, for example, our game!
- **Server**: an application that responds to received requests from clients, for example, a web server or database server

- **Request:** the message our game sends, asking for data or sending data to be stored
- **Response:** the message the server sends back to the client, containing either requested data or a status/success message

The world is networked, which involves many different **clients** communicating with other **clients**, and also with **servers**. Each of the Unity deployment platforms illustrates an example of a **client**:

- WebGL (running in a web browser)
- Windows and Mac applications
- Nintendo Switch
- Microsoft Xbox
- Sony Playstation
- Mobile devices, such as tablets and smartphones

The **servers** that these games can communicate with include dedicated multiplayer **game servers**, regular web servers, and online database servers. Multiplayer game development is a topic for a whole book of its own. Web and database servers can play many roles in game development and runtime interaction. One form of Unity game interaction with web servers involves a game communicating with an online **server** for data, such as high scores, inventories, player profiles, and chat forums, and we'll explore an example of this in this chapter's recipe for a database-support leaderboard.

The recipes in this chapter explore a range of data saving and loading strategies, from internal Play-erPrefs and static variables to local text files and online web and database server communications.

In this chapter, we will cover the following recipes:

- Saving data between scenes using static properties
- Saving data between scenes and games using PlayerPrefs
- Reading data from a text file
- Loading game data from a text file map
- Writing data to a file
- Logging player actions and game events to a file
- Reading data from the web
- Setting up a leaderboard using PHP and a database
- Unity game communication with a web server leaderboard

Since some of the recipes in this chapter make use of web servers and a database; for those recipes, you will require either the PHP 8 language (which comes with its own web server and SQLite database features) or an AMP package.

If you are installing the PHP language, refer to the installation guide and download links:

- `https://www.php.net/manual/en/install.php`
- `https://www.php.net/downloads`

If you do want to install a web server and database server application, a great choice is XAMPP. It is a free, cross-platform collection of everything you need to set up a database and web server on your local computer. The download page also contains FAQs and installation instructions for Windows, Mac, and Linux: `https://www.apachefriends.org/download.html`.

Saving data between scenes using static properties

Keeping track of the player's progress and user settings during a game is vital to give your game a greater feeling of depth and content. In this recipe, we will learn how to make our game remember the player's score between the different levels (scenes).

Note that this example game is rigged! In this game, higher will always win, and lower will always lose, but we can build on this simple game to count, store, and retrieve the number of wins and games that have been played.

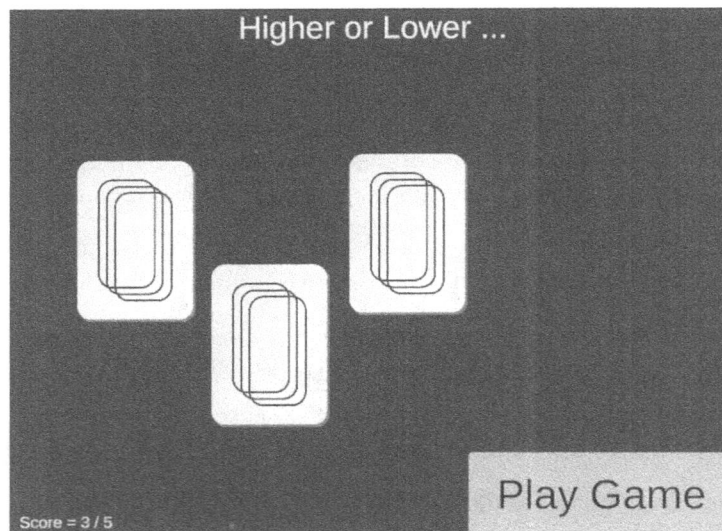

Figure 10.1: Our Higher or Lower game, with the score being remembered between scenes

Getting ready

We have included a complete project in a Unity package named game_HigherOrLower in the 10_01 folder. To follow this recipe, we will import this package as the starting point.

How to do it...

To save and load player data, follow these steps:

1. Create a new **Unity 2D project,** and import the game_HigherOrLower package.

If asked, agree to import the **TMP Essentials** so that the text and buttons will appear correctly in the imported scenes.

2. Open the Build Settings panel with menu **Edit | Build Settings....**

3. Remove the default scene from the build, and add each of the scenes from the imported package to the build in their name sequence (scene0_mainMenu, then scene1_gamePlaying, and so on).

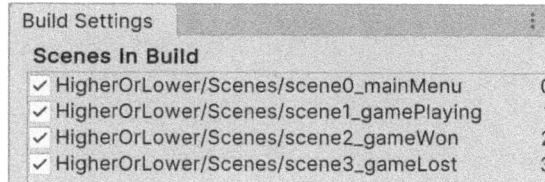

Build Settings	⋮
Scenes In Build	
✓ HigherOrLower/Scenes/scene0_mainMenu	0
✓ HigherOrLower/Scenes/scene1_gamePlaying	1
✓ HigherOrLower/Scenes/scene2_gameWon	2
✓ HigherOrLower/Scenes/scene3_gameLost	3

Figure 10.2: Adding scenes to the build in the scene name sequence

4. Make yourself familiar with the game by playing it a few times and examining the content of the scenes. The game starts on the scene0_mainMenu scene, inside the **Scenes** folder.

5. Ensure you are editing scene scene0_mainMenu.

6. Let's create a class to store the number of correct and incorrect guesses made by the user. Create a new C# script class, PlayerData, with the following code:

```
using UnityEngine;

public class PlayerData : MonoBehaviour {
    public static int scoreCorrect = 0;
    public static int scoreIncorrect = 0;
}
```

7. In the lower-left corner of the scene0_mainMenu scene, create a **UI Text (TMP)** GameObject named Text - score, containing the placeholder text Score: 99 / 99. Change the font color to light gray.

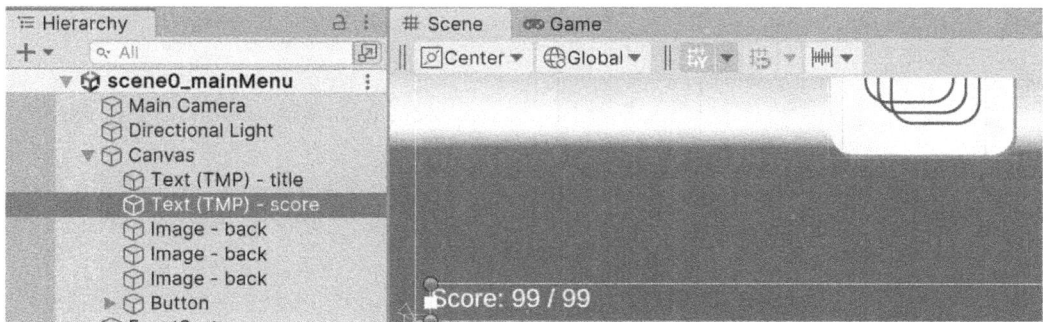

Figure 10.3: Creating a UI Text GameObject to display the score

8. Create a new C# script class, **UpdateScoreText**, with the following code.

```
using UnityEngine;
using TMPro;

public class UpdateScoreText : MonoBehaviour {
    private TextMeshProUGUI _scoreText;

    private void Awake() {
        TextMeshProUGUI _scoreText =
GetComponent<TextMeshProUGUI>();
    }

    void Start() {}
        int totalAttempts = PlayerData.scoreCorrect + PlayerData.
scoreIncorrect;
        string scoreMessage = "Score = ";
        scoreMessage += PlayerData.scoreCorrect + " / " +
totalAttempts;

        _scoreText.text = scoreMessage;
    }
}
```

9. Add an object of C# script class UpdateScoreText as a component of GameObject Text - score, by dragging the script from the **Project** panel onto this GameObject.

10. Create a new C# script class, IncrementCorrectScore, with the following code.

```
using UnityEngine;

public class IncrementCorrectScore : MonoBehaviour {
  void Start () {
    PlayerData.scoreCorrect++;
  }
}
```

11. Create a new C# script class, IncrementIncorrectScore, with the following code:

```
using UnityEngine;

public class IncrementIncorrectScore : MonoBehaviour {
  void Start () {
```

```
            PlayerData.scoreIncorrect++;
        }
    }
```

12. In the `scene2_gameWon` scene, create a new Empty GameObject named `score-updater`, and add an object of the C# script class `IncrementCorrectScore` as a component of this new GameObject.

13. In the `scene3_gameLost` scene, create a new Empty GameObject named `score-updater`, and add an object of the C# script class `IncrementIncorrectScore` as a component of this new GameObject.

14. Save your scripts and play the game. As you progress from level (scene) to level, you will find that the score and the player's name are remembered between scenes.

15. Quit the Unity application, and you'll find that the numbers reset to zero.

How it works...

The `PlayerData` class uses static (class) properties called `scoreCorrect` and `scoreIncorrect` to store the current total number of correct and incorrect guesses, respectively. Since these are public static properties, any object from any scene can access (set or get) these values, since the static properties are remembered from scene to scene. This class also provides the public static method called `ZeroTotals()`, which resets both values to zero.

When the `scene0_mainMenu` scene is loaded, all the GameObjects with scripts will have their `Start()` methods executed. The **UI Text** GameObject `Text-score` has an instance of the `UpdateScoreText` class as a script component so that the script's `Start()` method will be executed, which retrieves the correct and incorrect totals from the `Player` class, creates a `scoreMessage` string about the current score, and updates the **UI Text** GameObject `Text-score` text property so that the user sees the current score.

When the game is running and the user guesses correctly (higher), then the `scene2_gameWon` scene is loaded. When the `scene2_gameWon` scene is loaded, the `Start()` method of the `IncrementCorrectScore` script component of GameObject `score-updater` will be invoked, which adds 1 to the `scoreCorrect` variable of the `PlayerData` class.

When the game is running and the user guesses incorrectly (lower), then the `scene3_gameLost` scene is loaded. When the `scene3_gameLost` scene is loaded, the `Start()` method of the `IncrementIncorrectScore` script component of GameObject `score-updater` will be invoked, which adds 1 to the `scoreIncorrect` variable of the `PlayerData` class.

The next time the user visits the main menu scene, the new values of the correct and incorrect totals will be read from the `PlayerData` class, and the UI Text on the screen will inform the user of their updated total score for the game.

There's more...

There are some details that you don't want to miss.

Hiding the score before the first attempt is completed

Showing a score of zero out of zero isn't very professional. Let's add some logic so that the score is only displayed (a non-empty string) if the total number of attempts is greater than zero:

```
void Start()
    {
    int totalAttempts = PlayerData.scoreCorrect + PlayerData.scoreIncorrect;

    string scoreMessage = "";
    if (totalAttempts > 0)
    {
        scoreMessage = "Score = ";
        scoreMessage += PlayerData.scoreCorrect + " / " + totalAttempts;
    }
    _scoreText.text = scoreMessage;
}
```

See also

Refer to the *Saving data between scenes and games using PlayerPrefs* recipe in this chapter for more information about alternative ways to save and load player data.

Saving data between scenes and games using PlayerPrefs

While the previous recipe illustrates how the static properties allow a game to remember values between different scenes, these values are forgotten once the game application is exited. Unity provides the **PlayerPrefs** feature to allow a game to store and retrieve data between the different game-playing sessions.

Getting ready

This recipe builds upon the previous recipe, so make a copy of that project and work from it.

How to do it...

To save and load the player data using `PlayerPrefs`, follow these steps:

1. Delete the C# script `PlayerData`.
2. Edit the C# script called `UpdateScoreText` by replacing the `Start()` method with the following code:

    ```
    void Start(){
    ```

```
int scoreCorrect = PlayerPrefs.GetInt("scoreCorrect");
int scoreIncorrect = PlayerPrefs.GetInt("scoreIncorrect");

int totalAttempts = scoreCorrect + scoreIncorrect;
string scoreMessage = "Score = ";
scoreMessage += scoreCorrect + " / " + totalAttempts;

_scoreText.text = scoreMessage;
}
```

3. Now, edit the C# script called `IncrementCorrectScore` by replacing the `Start()` method with the following code:

```
void Start () {
    int newScoreCorrect = 1 + PlayerPrefs.GetInt("scoreCorrect");
    PlayerPrefs.SetInt("scoreCorrect", newScoreCorrect);
}
```

4. Now, edit the C# script called `IncrementIncorrectScore` by replacing the `Start()` method with the following code:

```
void Start () {
    int newScoreIncorrect = 1 + PlayerPrefs.GetInt("scoreIncorrect");
    PlayerPrefs.SetInt("scoreIncorrect", newScoreIncorrect);
}
```

5. Save your scripts and play the game. Quit Unity and then restart the application. You will find that the player's name, level, and score are now remembered between the game sessions.

How it works...

We had no need for the `PlayerData` class, since this recipe uses the built-in runtime class called `PlayerPrefs`, which is provided by Unity.

Unity's `PlayerPrefs` runtime class is capable of storing and accessing information (the `string`, `int`, and `float` variables) in the user's machine. Values are stored in a `plist` file (Mac) or the registry (Windows), in a similar way to web browser cookies, which means they are remembered between game application sessions.

Values for the total correct and incorrect scores are stored by the `Start()` methods in the `IncrementCorrectScore` and `IncrementIncorrectScore` classes. These methods use the `PlayerPrefs.GetInt("")` method to retrieve the old total, add 1 to it, and then store the incremented total using the `PlayerPrefs.SetInt("")` method.

These correct and incorrect totals are then read each time the `scene0_mainMenu` scene is loaded, and the score totals are displayed via the **UI Text** object on the screen.

For more information on **PlayerPrefs**, see Unity's online documentation at `http://docs.unity3d.`
`com/ScriptReference/PlayerPrefs.html`.

See also

Refer to the *Saving data between scenes using static properties* recipe in this chapter for more information
about alternative ways to save and load player data.

Reading data from a text file

Text files are used to store and communicate data for many purposes. Unity makes it easy for you to
read in the contents of a text file via the built-in `TextAsset` script class. In this recipe, you'll add a UI
Text item to a scene, and then write a script that at runtime will read the contents from the text file
and display them to the player.

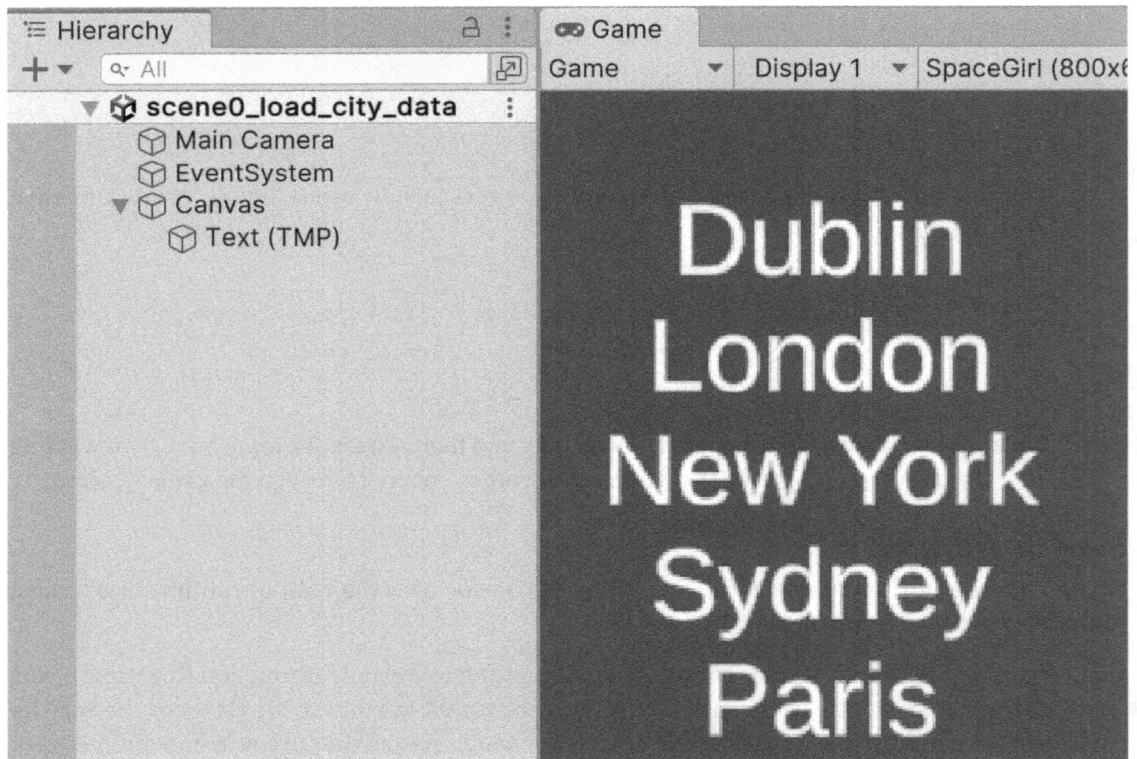

Figure 10.4: Showing the text file contents on screen at runtime

Getting ready

You can use any text file as the data for this recipe. We've provided a small text file containing a list of
cities, in a `cities.txt` file in the **10_03** folder.

cities.txt	
1	Dublin
2	London
3	New York
4	Sydney
5	Paris

Figure 10.5: Contents of the text file `cities.txt`

How to do it...

To read data from a text file, perform the following steps:

1. Create a new **Unity 2D** project.

2. Import the provided **cities.txt** text file into the project (or use some other text file you have on your computer).

3. Load the TextMeshPro essentials resources, by choosing the menu **Window | TextMeshPro | Import TMP Essential Resources**.

4. In the **Hierarchy** panel, add a **Text** TMP GameObject to the scene by choosing the menu **GameObject | UI | Text - TextMeshPro**. Rename this GameObject `Text-file-contents`.

5. Make this UI Text GameObject cover the whole screen in the **RectTransform** component by holding down *Shift + Alt* and clicking the bottom-right option of **stretch-stretch**. Also, set its text to be vertically and horizontally centered.

6. Create a new C# script class, `TextFileReader`, with the following code:

```
using UnityEngine;
using TMPro;
public class TextFileReader : MonoBehaviour {
    public TextAsset textFile;
    private TextMeshProUGUI _textComponent;

    private void Awake() {
        _textComponent = GetComponent<TextMeshProUGUI>();
    }

    void Start() {
        string citiesText = textFile.text;
        _textComponent.text = citiesText;
    }
}
```

7. Select the GameObject `Text-file-contents`, and add an object of the C# script class `TextFileReader` as a component of this new GameObject.

8. With the GameObject Text-file-contents selected in the Hierarchy, drag the text file cities.txt from the **Project** panel into the public variable **Text File** for the **Text File Reader (Script)** property.

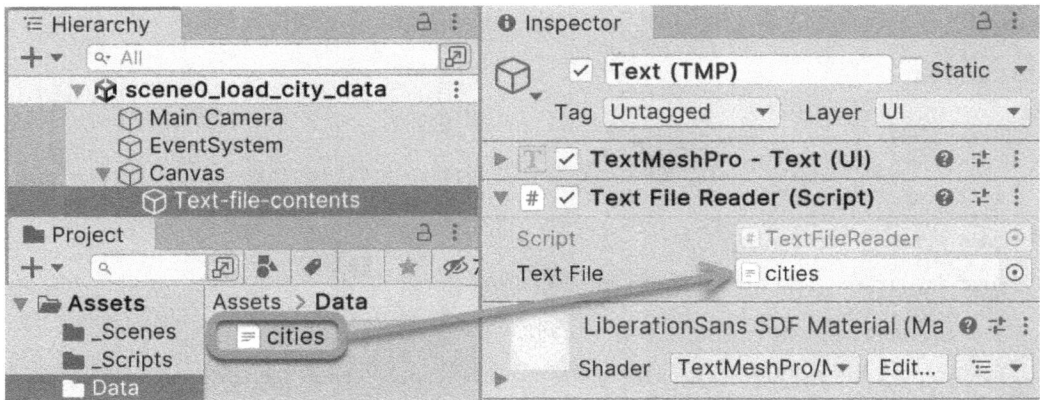

Figure 10.6: Linking the cities.txt text file to the public variable of the TextFileReader script component

9. Save and Play the scene. You should see the contents of the text file displayed on the screen in the **UI text** GameObject.

How it works...

The public **TextAsset** variable **textFile** in the C# script class TextFileReader allows us to link a text file asset in the Project panel with our scripted component. The text from the file cities.txt is then set to be the text of the UI Text object in the scene.

> **Note:** This TextAsset variable technique is only appropriate when there will be no change to the data file after game compilation, since the text file data is serialized into the general build resources, and so it cannot be changed after the build has been created.

Loading game data from a text file map

Rather than having to create and place every GameObject on the screen by hand for every level of a game, a better approach can be to create the text files of rows and columns of characters, where each character corresponds to the type of GameObject that is to be created in the corresponding location.

In this recipe, we'll use a text file and a set of prefab sprites to display a graphical version of a text data file, for a screen from the classic game *NetHack*:

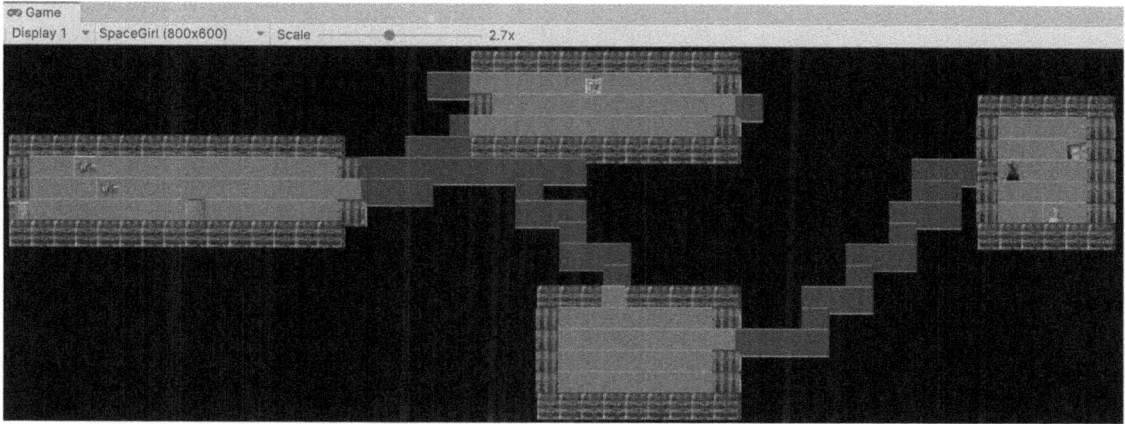

Figure 10.7: The level we've created from a text file level description

In the 10_04 folder, we have provided the following two files for this recipe:

- level1.txt: A text file representing a level
- absurd128.png: A 128 x 128 sprite sheet for NetHack

The level data came from the NetHack Wikipedia page, while the sprite sheet came from SourceForge:

- http://en.wikipedia.org/wiki/NetHack
- http://sourceforge.net/projects/noegnud/files/tilesets_nethack-3.4.1/absurd%20 128x128/

Note that we also included a Unity package with all the prefabs setup, since this can be a time-consuming task.

How to do it...

To load game data from a text file map, do the following:

1. Create a new **Unity 2D project**.
2. Import the level1.txt text file and the absurd128.png image file.
3. Select absurd128.png in the **Inspector** panel, and set **Texture Type** to **Sprite (2D and UI)** and **Sprite Mode** to **Multiple**. Then, click the **Sprite Editor** button to open **Sprite Editor**.
4. In **Sprite Editor**, click the **Slice** drop-down option, then for **Type**, choose **Grid by Cell Size**, and set **Pixel Size** to 128 x 128. Then, click the **Slice** button, followed by the **Apply** button to apply these settings and slice up the image into many 128 x 128 sprites that we can work with.

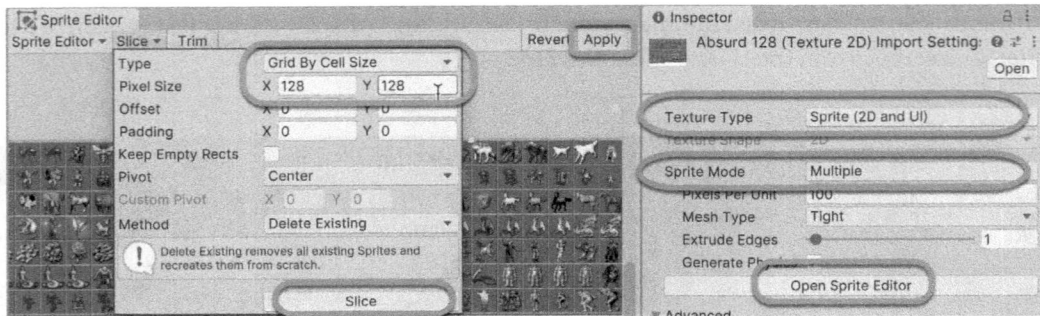

Figure 10.8: Slicing the sprite sheet into a grid of 128 x 128-pixel images

5. In the **Project** panel, click on the right-facing white triangle to explode the icon to show all the sprites in this sprite sheet individually:

Figure 10.9: Many individual 128 x 128 sprites viewed in the Project panel

6. Drag the sprite called absurd128_175 onto the scene. Rename this GameObject corpse_175. This way, the **sprite** asset file named absurd128_175 gives us a **Scene** GameObject named corpse_175.

7. In the **Project** panel, create a folder named Prefabs, and then create a new prefab by dragging the corpse_175 GameObject into this Prefabs folder. Finally, delete corpse_175. You have now created a prefab containing corpse_175.

8. Repeat this process for the following sprites (that is, create GameObjects, rename them, and then create prefabs for each one). The default name for each sprite will be absurd128_<n>, where <n> is the number for each sprite in the list.

    *   alter_583
    *   chest_586
    *   corridor_849
    *   door_844
    *   floor_848
    *   horiz_1034
    *   potion_675
    *   stairs_up_994

- stairs_down_993
- vert_1025
- wizard_287

> **Note:** You can quickly find an asset file in the Project panel by typing part of its name in the search box.

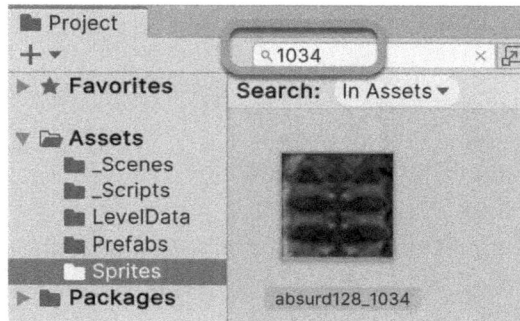

Figure 10.10: Using the Project panel asset name search tool to find absurd128_1034

9. Select **Main Camera** in the **Inspector** panel, and ensure that it is set to an **orthographic** camera, sized **20**, with **Clear Flags** set to **Solid Color** and **Background** set to a dark **Black** color.

10. Create a new C# script class, called `LoadMapFromTextfile`, containing the following:

```
using UnityEngine;
using System.Collections.Generic;

public class LoadMapFromTextfile : MonoBehaviour
{
  public TextAsset levelDataTextFile;

  public GameObject alter_583;
  public GameObject chest_586;
  public GameObject corpse_175;
  public GameObject corridor_849;
  public GameObject door_844;
  public GameObject floor_848;
  public GameObject horiz_1034;
  public GameObject potion_675;
  public GameObject stairs_down_993;
  public GameObject stairs_up_994;
```

```
    public GameObject vert_1025;
    public GameObject wizard_287;

    public Dictionary<char, GameObject> dictionary = new Dictionary<char,
GameObject>();

    void Awake(){
      char newlineChar = '\n';

      dictionary['_'] = alter_583;
      dictionary['('] = chest_586;
      dictionary['%'] = corpse_175;
      dictionary['#'] = corridor_849;
      dictionary['+'] = door_844;
      dictionary['.'] = floor_848;
      dictionary['-'] = horiz_1034;
      dictionary['!'] = potion_675;
      dictionary['>'] = stairs_down_993;
      dictionary['<'] = stairs_up_994;
      dictionary['|'] = vert_1025;
      dictionary['@'] = wizard_287;

      string[] stringArray = levelDataTextFile.text.Split(newlineChar);
      BuildMaze( stringArray );
    }

    private void BuildMaze(string[] stringArray){
      int numRows = stringArray.Length;

      float yOffset = numRows / 2;

      for(int row=0; row < numRows; row++){
        string currentRowString = stringArray[row];
```

```
            float y = -1 * (row - yOffset);
            CreateRow(currentRowString, y);
        }
    }

    private void CreateRow(string currentRowString, float y) {
        int numChars = currentRowString.Length;
        float xOffset = numChars / 2;

        for(int charPos = 0; charPos < numChars; charPos++){
            float x = charPos - xOffset;
            char prefabCharacter = currentRowString[charPos];

            if (dictionary.ContainsKey(prefabCharacter)){
                CreatePrefabInstance( dictionary[prefabCharacter], x, y);
            }
        }
    }

    private void CreatePrefabInstance(GameObject objectPrefab, float x,
float y){
        float z = 0;
        Vector3 position = new Vector3(x, y, z);
        Quaternion noRotation = Quaternion.identity;
        Instantiate (objectPrefab, position, noRotation);
    }
}
```

11. Create a new Empty GameObject named **level loader**, and attach an instance of a C# script class, called LoadMapFromTextfile, as a component of this new GameObject.

12. With the GameObject level loader in the **Hierarchy**, drag level1.txt text file from the **Project** panel into the **Level Data Text File** public variable in the **Inspector** for the **Load Map From Textfile (Script)** component.

13. Then, drag the appropriate prefabs into the prefab slots in the **Inspector** for the **Load Map From Textfile (Script)** component.

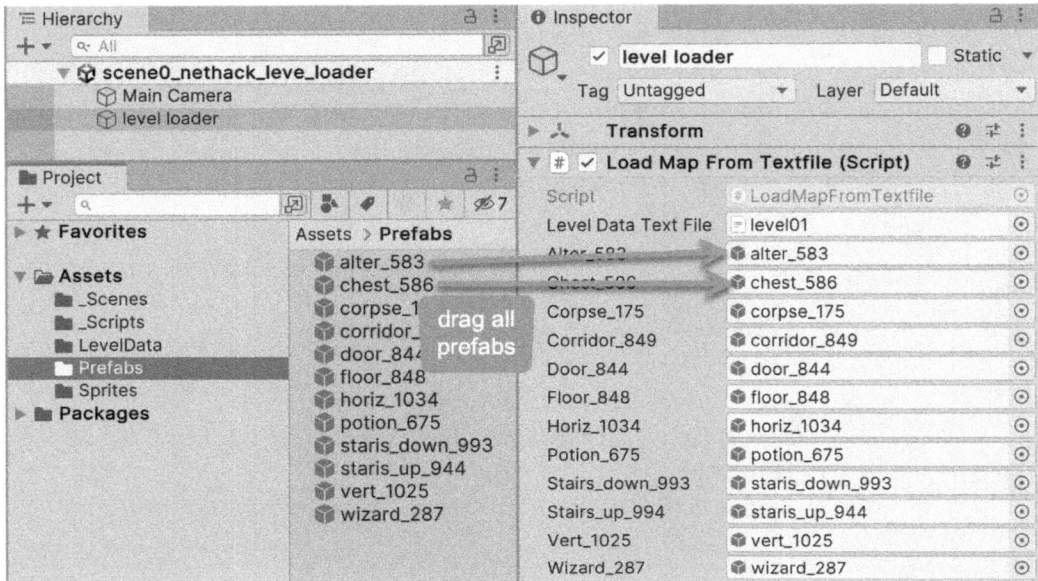

Figure 10.11: Assigning all the public prefab variables of the scripted level loader component

14. When you run the scene, you will see that a sprite-based NetHack map will appear, using your prefabs.

If you want to have a bit of fun, you could replace the image prefabs with 3D mesh objects and adapt this recipe, creating a 3D scene from a 2D map file.

How it works...

The sprite sheet was automatically sliced up into hundreds of 128 x 128-pixel **sprite** squares. We created the prefab objects from some of these sprites so that the copies can be created at runtime when needed.

The text file called level1.txt contains the lines of text characters. Each non-space character represents where a sprite prefab should be instantiated (column = x and row = y). These characters have been chosen so that the text file is human-readable. As shown in the following screenshot, when displayed in a text editor with a fixed-width (typewriter) font, the text file gives us a good idea of how the actual sprite-based graphical 2D map will look:

```
 level01.txt  ×

1                                    - - - - - - - - - - -
2                          ##. . . . ._ . . . . .|
3                           |. . . . . . . . . . .#                  - - - - - -
4                           #. . . . . . . . . .|                 |. . . .|
5        - - - - - - - - - - - - -   ###- - - - - - - - - - -        |. . . .(|
6        |. .%. . . . . . . . . .|##########              ###-@. . .|
7        |. . .%. . . . . . . . .###    #              ## |. . . .|
8        +. . . . . . .<. . . . . .|       ###          ### |. .!.|
9        - - - - - - - - - - - - - -          #              #    - - - - - -
10                                     ###          ###
11                              |    #              #
12                              - - - .- - - - -   ###
13                              |. . . . . . .|   #
14                              |. . . . . . . .####
15                              |. . . . . . .|
16                              |. . . . . . .|
17                              - - - - - - - - -
```

Figure 10.12: The text-level data viewed in a code editor

In the LoadMapFromTextfile script class, a C# dictionary variable named dictionary is declared and initialized in the Start() method, associating the specific prefab GameObjects with some particular characters in the text file.

The Awake() method splits the string into an array using the newline character as a separator. So, now, we have stringArray with an entry for each row of the text data. The BuildMaze(...) method is called with stringArray.

The BuildMaze(...) method interrogates the array to find its length (the number of rows of data for this level) and sets yOffSet to half this value. This is done to allow the prefabs to be placed half above *y = 0* and half below, so (0, 0, 0) is the center of the level map. A for loop is used to read each row's string from the array. The loop passes the string to the CreateRow(...) method, along with the *y* value corresponding to the current row.

The CreateRow(...) method extracts the length of the string and sets xOffSet to half this value. This is done to allow half of the prefab to be placed to the left of *x = 0* and the other half to the right, so (0, 0, 0) is the center of the level map. A for loop is used to read each character from the current row's string, and (if there is an entry in our dictionary for that character) then the CreatePrefabInstance (...) method is called, passing the prefab reference in the dictionary for that character, as well as the *x* and *y* values.

The CreatePrefabInstance(...) method instantiates the given prefab at a position of *x, y, z*, where *z* is always zero, and there is no rotation (Quarternion.identity).

Writing data to a file

Just as we may wish to read data from a file, there are times when it is useful to save data to a file. For example, we may want to save data from our game into our local computer, whether for debugging, recording game testing experiments, and so on.

In this recipe, we'll declare a data structure for player's names and scores, and save the data for a player as a JSON text file.

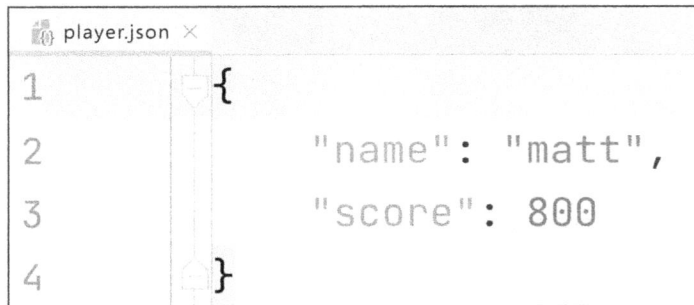

```
player.json ×
1    {
2            "name": "matt",
3            "score": 800
4    }
```

Figure 10.13: Player name and score data saved as a JSON text file

How to do it...

To write data to a file, perform the following steps:

1. Create a new **Unity 2D** project.
2. Create a new, empty folder in the **Project** panel, named **Data**. We'll write our JSON text file into this folder.
3. Create a new C# script class, PlayerScore, with the following code:

```
using UnityEngine;
using System;

[Serializable]
public class PlayerScore {
```

```csharp
    public string name;
    public int score;

    public string ToJson() {
        bool prettyPrintJson = true;
        return JsonUtility.ToJson(this, prettyPrintJson);
    }
}
```

4. Create a new C# script class, MyJsonFileWriter, with the following code:

```csharp
using UnityEngine;
using System.IO;

public class MyJsonFileWriter : MonoBehaviour
{
    public string folderName = "Data";
    public string fileName = "player.json";
    private PlayerScore player1Score;

    private void Awake() {
        player1Score = new PlayerScore();
        player1Score.name = "matt";
        player1Score.score = 800;
    }

    void Start() {
        string stringData = player1Score.ToJson();
        string folderAndFileName = Path.Combine(folderName, fileName);
        string filePath = Path.Combine(Application.dataPath,
folderAndFileName);
        WriteTextFile(filePath, stringData);
    }

    public void WriteTextFile(string pathAndName, string stringData) {
        FileInfo textFile = new FileInfo(pathAndName);
        StreamWriter writer = textFile.CreateText();
        writer.Write(stringData);
        writer.Close();
    }
}
```

5. Create a new Empty GameObject named file-manager. Then, add an object of the C# script class, MyJsonFileWriter, as a component of this new GameObject.

6. Save and play the scene.

7. When you explore the Data folder, you should now find a new JSON file named player.json. containing the player's name and score as JSON text.

How it works...

The C# script class PlayerScore declares the names and types of the data items we wish to store as JSON text, that is, a name that is a string, and a score that is an integer. This script class also declares the ToJson() method, which uses the built-in JsonUtility library to create a JSON-formatted string of the data for each PlayerScore object in our game. We don't add a component of this script class to any GameObject – you'll notice it does not "extend" the Unity Monobehaviour class, so it cannot be added as a GameObject component.

Note the compiler directive [Serializable] immediately before the C# script class name PlayerScore. This is an indication that the data (state) of objects of this class can be serialized. Serialization is when a computer program stores the data of objects, for example, to text or binary files, or sends it to a web server for storage or transmission..

The C# script class MyJsonFileWriter has been designed to be added as a component to a GameObject in the scene – we added an instance of this script class to the GameObject file-manager. This script class declares two public variables, allowing you to customize the filename and folder name of the text file to be created. This script class also declares a PlayerScore variable, named player1Score. When the scene begins, the Awake() function method is executed, which creates a new PlayerScore object with the name matt and score **800**. Then, the Start() function method is executed, which gets the JSON text representation of the contents of the variable player1Score, stored in the variable stringData. The filename and folder name variables are used to create the variable filePath, and this file path variable and the variable stringData are passed to the method function WriteTextFile(…).

The method WriteTextFile(…) uses the C# StreamWriter library to create a new text file and save the player's JSON text data to this file.

There's more...

There are some details that you don't want to miss.

Writing lists of objects to a JSON text file

Unity and C# make it very easy to write a collection of data objects to a single JSON file. To do this for a list of PlayerScore objects, do the following:

1. Create a new C# script class, PlayerScoreList, with the following code:

```
using UnityEngine;
using System.Collections.Generic;
```

```
using System;

[Serializable]
public class PlayerScoreList
{
    public List<PlayerScore> list = new List<PlayerScore>();

    public string ToJson()
    {
        return JsonUtility.ToJson(this, true);
    }
}
```

2. Create a new C# script class, MyJsonFileWriter, with the following code:

```
using UnityEngine;
using System.IO;

public class MyJsonListFileWriter : MonoBehaviour
{
    public string folderName = "Data";
    public string fileName = "playerList.json";
    private PlayerScore playerScore1 = new PlayerScore();
    private PlayerScore playerScore2 = new PlayerScore();
    private PlayerScoreList playerScoreList = new PlayerScoreList();

    void Awake() {
        playerScore1.name = "matt";
        playerScore1.score = 800;
        playerScore2.name = "joelle";
        playerScore2.score = 901;
        playerScoreList.list.Add(playerScore1);
        playerScoreList.list.Add(playerScore2);
    }

    void Start() {
        string folderAndFileName = Path.Combine(folderName, fileName);
        string filePath = Path.Combine(Application.dataPath,
folderAndFileName);
        string stringData = playerScoreList.ToJson();
        WriteTextFile(filePath, stringData);
    }
```

```
public void WriteTextFile(string pathAndName, string stringData) {
    FileInfo textFile = new FileInfo(pathAndName);
    StreamWriter writer = textFile.CreateText();
    writer.Write(stringData);
    writer.Close();
    }
}
```

3. Add an object of the C# script class `MyJsonFileListWriter` as a component of the GameObject `file-manager`.

4. Save and Play the scene.

5. When you explore the `Data` folder, you should now find a new JSON file named `playerList.json`, containing the names and scores for the two players (`matt` and `joelle`) in the list as JSON text.

Logging player actions and game events to a file

The **comma separated values** (CSV) text file format is a good format to save data in, since it is easily read into rows and columns by almost all spreadsheet applications.

In this recipe, we'll write a script that offers a method that can be called from anywhere in our game, from any scene, which will add a row of text to a log file. The following is an example of the saved data when viewed in Microsoft Excel.

	A	B	C	D
1	05/05/2023	18:05:04	created	
2	05/05/2023	18:05:04	scene1 log some events	Scene has started
3	05/05/2023	18:05:05	scene1 log some events	SPACE key was pressed
4	05/05/2023	18:05:05	scene1 log some events	SPACE key was pressed
5	05/05/2023	18:05:06	scene1 log some events	SPACE key was pressed

Figure 10.14: CSV log data viewed in a spreadsheet application

How to do it...

To log player actions and game events to a file, follow these steps:

1. Create a new 2D project.

2. Create a new `AddToLogFile` C# script class containing the following code:

```
using UnityEngine;
using System.IO;
using System;
using UnityEngine.SceneManagement;

public class AddToLogFile : MonoBehaviour {
```

```
      private static string _fileName = "";
      private static string _folderName = "Logs";
      private static string _filePath = "(no file path yet)";

      private static void CreateNewLogFile() {
        _fileName = DateTime.Now.ToString("yyy_MM_dd--HH:mm:ss") + ".csv";
        _filePath = Path.Combine(Application.dataPath, _folderName);
        _filePath = Path.Combine(_filePath, _fileName);

        // Create a file to write to
        using (StreamWriter sw = File.CreateText(_filePath)) {
          sw.WriteLine(TimeStamp() + ",created");
        }
      }

      private static string TimeStamp() {
        return DateTime.Now.ToString("yyyy/MM/dd") + "," + DateTime.Now.
ToString("HH:mm:ss");
      }

      public static void LogLine(string textLine) {
        // the first time we try to log a line, we need to create the file
        if (!File.Exists(_filePath)) {
          CreateNewLogFile();
        }

        string sceneName = SceneManager.GetActiveScene().name;
        textLine = TimeStamp() + "," + sceneName + "," + textLine;

        using (StreamWriter sw = File.AppendText(_filePath)) {
          sw.WriteLine(textLine);
        }
      }
    }
```

3. Create a new `SimpleLogging` C# script class, containing the following code:

```
using UnityEngine;

public class SimpleLogging : MonoBehaviour
{
    void Start() {
```

```
        AddToLogFile.LogLine("Scene has started");
    }

    void Update() {
        if (Input.GetKeyUp(KeyCode.Space)) {
            AddToLogFile.LogLine("SPACE key was pressed");
        }
    }
}
```

4. Create a new Empty GameObject named `log-manager`. Then, add an object of the C# script class `SimpleLogging` as a component of this new GameObject.

5. Run the scene. Press the *spacebar* key a few times, and then stop running the project.

6. You should now find a new timestamped CSV log file in the `Logs` folder:

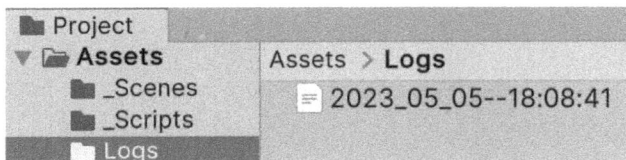

Figure 10.15: The CSV log file in the Logs folder

How it works...

In this recipe, you created a C# script class called `AddToLogFile`. This contains a public method called `LogLine(...)`, which takes in a string message and adds a new line at the end of the CSV file. This method is **static**, which means it can be called from any script in any scene through a single line of code, since it is not a component of a GameObject. This method creates a new text file if one does not already exist (by executing `CreateNewLogFile()`). Then, it creates a timestamp of the current date and time (with `TimeStamp()`), adds the name of the scene that is currently running, and appends this long, comma-separated string to the text file.

To illustrate how to use our file-logging script, we created a second C# script class called `SimpleLogging`. An instance of this class was added as a component to the empty GameObject `log-manager`. This script adds a line to the log file when its `Start()` method is invoked (when the scene begins), and also each time the user presses and releases the *spacebar*.

In this example logged events such as the scene starting and when the *spacebar* was pressed. However, the events resulting in data being appended to the log file could be anything, such as the user passing a checkpoint or losing a life, completing a level, and so on. This type of live game logging can be useful to record data about whatever aspects of gameplay we might want to record and analyze for debugging, gameplay testing, and so on.

There's more...

There are some details that you don't want to miss.

Automatically log the username of the system user

C# offers a handy variable, which is the username of the currently logged-in system user: Environment. UserName. So you can update the string created for the UserName variable in a method called LogLine(...), as follows, to add the username to the logged data:

```
    textLine = TimeStamp() + "," + Environment.UserName + "," + sceneName + ","
 + textLine;
```

This could be useful to compare actions and timings between different users/testers of your game...

See also

This concept can also be used to log data to a live web database rather than a text file. For an example of this, see this open-source Unity user action logging project, making use of the UnityWebRequest library: https://github.com/dr-matt-smith/unity-user-action-logger-web

Reading data from the web

Much data and authentication is now available via web communications. In this recipe, we'll create a simple Unity project using a C# script class to retrieve a small sample JSON text file from the web.

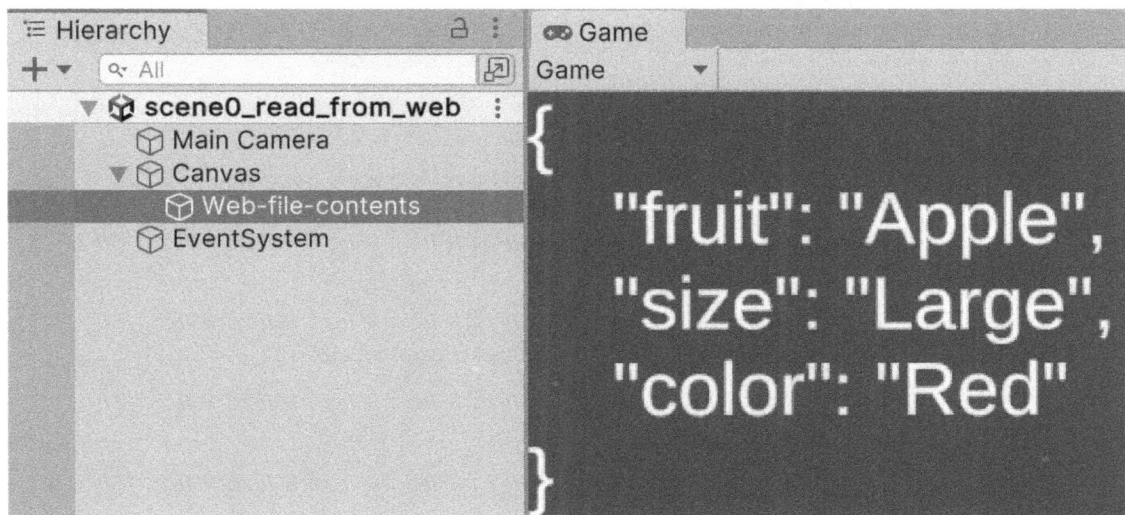

Figure 10.16: The JSON data downloaded from the web

Getting ready

The example of the web-published text file for this recipe can be found at the following URL (thanks to filesamples.com for publishing a range of test files on their website): https://filesamples.com/samples/code/json/sample1.json

However, you can replace this web address with any URL that returns a text file of some kind.

How to do it...

To read data from the web, perform the following steps:

1. Create a new **Unity 2D** project.

2. Load the TextMeshPro essentials resources, by choosing the menu **Window | TextMeshPro | Import TMP Essential Resources**.

3. In the **Hierarchy** panel, add a **Text TMP** GameObject to the scene by choosing the menu **GameObject | UI | Text - TextMeshPro**. Rename this GameObject **Web-file-contents**.

4. Make this UI Text GameObject cover the whole screen in the **RectTransform** component by holding down *Shift + Alt* and clicking the bottom-right option of **stretch-stretch**. Also, set its text to be left and top-aligned.

5. Create a new C# script class, WebTextFileReader, with the following code:

```csharp
using System.Collections;
using UnityEngine;
using UnityEngine.Networking;
using TMPro;

public class WebTextFileReader : MonoBehaviour {
    public string url = "https://filesamples.com/samples/code/json/sample1.json";
    private TextMeshProUGUI _textComponent;

    private void Awake() {
        _textComponent = GetComponent<TextMeshProUGUI>();
    }

    void Start() {
        StartCoroutine( LoadWWW() );
    }

    private IEnumerator LoadWWW() {
        using (UnityWebRequest webRequest = UnityWebRequest.Get(url)) {
```

```
                    webRequest.certificateHandler = new CertHandler();

                    yield return webRequest.SendWebRequest();

                    switch (webRequest.result) {
                        case UnityWebRequest.Result.ConnectionError:
                        case UnityWebRequest.Result.DataProcessingError:
                            Debug.LogError(": Error: " + webRequest.error);
                            _textComponent.text = "ERROR: " + webRequest.error;
                            break;
                        case UnityWebRequest.Result.ProtocolError:
                            Debug.LogError(": HTTP Error: " + webRequest.error);
                            _textComponent.text = "ERROR: " + webRequest.error;
                            break;
                        case UnityWebRequest.Result.Success:
                            Debug.Log(":\nReceived: " + webRequest.
downloadHandler.text);
                            _textComponent.text = webRequest.downloadHandler.
text;
                            break;
                    }
                }

            }
        }
```

6. Select the GameObject Web-file-contents, and add an object of the C# script class WebTextFileReader as a component of this new GameObject.

7. Save and Play the scene. You should see the contents of the text file downloaded from the web, displayed on screen in the UI text GameObject.

How it works...

All the work is being done by the WebTextFileReader C# script class. There is a public **url** variable, which allows you to change the web address for the text file, whose content will be downloaded and displayed.

There is a private variable, _textComponent, which is a reference to the TextMeshProUGUI **UI Text** component of the Web-file-contents GameObject. This is set when the scene begins and the Awake() method is executed.

Then, the Start() method is executed, which runs the method LoadWWW() as a co-routine. A co-routine is a method function that can paused while it is being executed and then continued later from the same point.

This is perfect for working with web requests, since there may be a delay between the request being made to some web server and the response being received. The **url** variable is used to create a `UnityWebRequest` object variable named `webRequest`.

Once a response has been received, the value of the `result` property of the **webRequest** variable is tested in a `switch` statement. If it is some form of error, then an error message is printed to the console, and also, the UI Text is set as an error message. Otherwise, if the web request was successful, the UI Text is set as the text received from the web – which is the JSON data about large, red apples for the default URL we've declared.

> **Note:** We have provided a `CertHandler.cs` script class to solve the HTTPS authentication for this recipe. For real-world games, you should not use self-signed certificates like this, since connected applications will require full security to protect your connections at runtime.

Setting up a leaderboard using PHP and a database

Games are more fun when there is a leaderboard of high scores that the players have achieved. Even single-player games can communicate with a shared web-based leaderboard. This recipe creates the web **server-side** (PHP) scripts to set and get player scores from a SQL database. The recipe after this one then sees us creating a Unity game **client** that can communicate with this web leaderboard's **server**.

Getting ready

This recipe assumes that you either have your own web hosting or are running a local web server. You could use the built-in PHP web **server** or a **web server**, such as Apache or Nginx. For the database, you could use a SQL database **server** such as MySQL or MariaDB. However, we've tried to keep things simple using SQLite – a file-based database system. So all you actually need on your computer is PHP 8, since it has a built-in web **server** and can talk to SQLite databases, which is the setup on which this recipe was tested.

Install PHP locally as follows:

- Installing PHP on Windows 10 or 11:

    - `https://www.sitepoint.com/how-to-install-php-on-windows/`

- Installing PHP on MacOS

    - `https://www.geeksforgeeks.org/how-to-install-php-on-macos/`

- Installing PHP on Linux

    - `https://www.zend.com/blog/installing-php-linux`

All the PHP scripts for this recipe, along with the SQLite database file, can be found in the `10_08` folder.

How to do it...

To set up a leaderboard using PHP and a database, perform the following steps:

1. Copy the PHP project provided to where you will be running your web server:

    - **Live website hosting:** Copy the files to the live web folder on your server (often `www` or `htdocs`).

    - **Running on a local machine:** At the command line, you can use the Composer script shortcut to run the PHP built-in web server, by typing `composer serve` **or** `php -S localhost:8000 -t ./public`.

```
Terminal:   Local ×   + ∨
% composer serve
> php -S localhost:8000 -t ./public
[Fri May  5 20:50:23 2023] PHP 8.1.13 Development Server (http://localhost:8000)
```

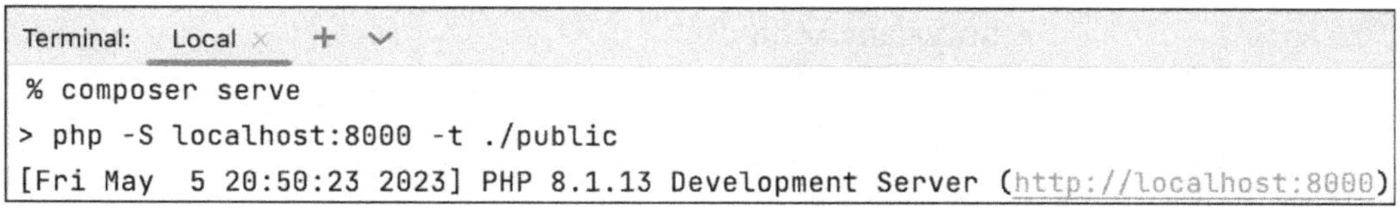

Figure 10.17: Running a web server in a macOS command-line Terminal using the shortcut composer serve

2. Open a web browser at your website location:

    - **Live website hosting:** Visit the URL for your hosted domain.

    - **Running on a local machine:** Visit the URL `localhost:8000`.

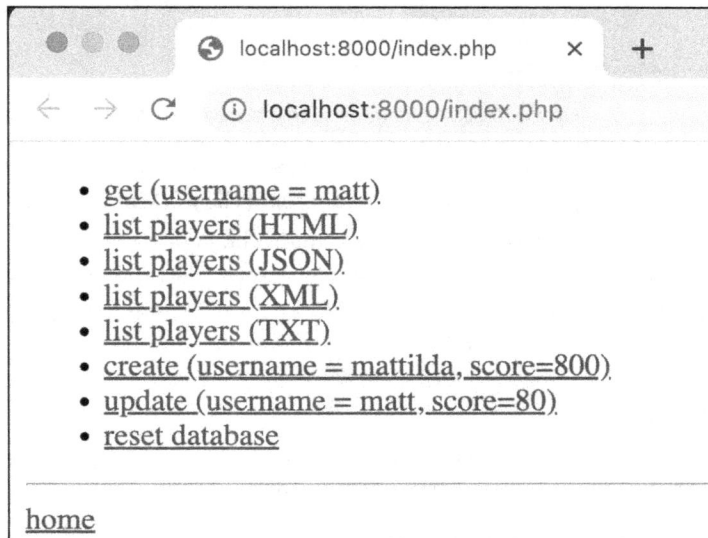

Figure 10.18: Viewing the high score website home page in a web browser

3. Create/reset the database to a set of default players and scores by clicking the last bulleted link: **reset database**. You should see a page with the message **database has been reset**, and a link back to the **home** page (click that link).

4. To view the leaderboard scores as a web page in your web browser, click the second link: **list players (HTML)**.

Figure 10.19: Viewing scores as HTML

5. Try the fifth link – **list players (TXT)** – to retrieve the leaderboard data as a text file. Note how it looks different when viewed in the web browser (which ignores line breaks), compared to how it looks when you view the actual source file returned from the server:

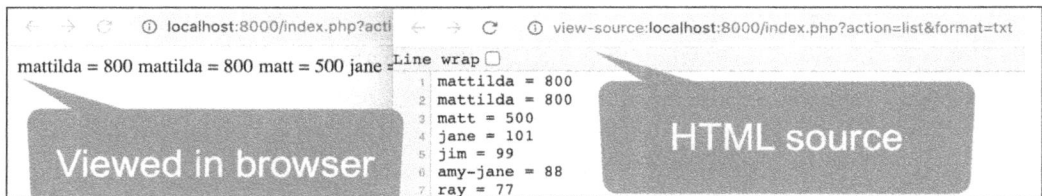

Figure 10.20: Viewing the source page to see the raw text data

6. Do the same with the JSON and XML options, seeing how our **server** can return the contents of the database wrapped up as HTML, plain text (TXT), XML, or JSON.

7. Click the sixth link – **create (username = mattilda, score=800)**. When you next retrieve the contents, you'll see that there is a new database record for the player **mattilda**, with a score of **800**. This shows that our server can receive data and change the contents of the database, as well as just return values from it.

How it works...

The player's scores are stored in an SQLite database. Access to the database is facilitated through the PHP scripts provided. In our example, all the PHP scripts were placed in a folder on our local machine, from which we'll run the **server** (it can be anywhere when using the PHP built-in server). So, the scripts are accessed via http://localhost:8000.

All access is facilitated through a PHP file called index.php. This is called a **front controller**, acting like a receptionist in a building, interpreting **requests** and asking the appropriate function to execute some actions and return a result in **response** to the **request**.

There are five actions implemented, and each is indicated by adding the `action` name at the end of the URL – these are known as URL parameters or query strings. (This is the `GET HTTP` method, which is sometimes used for web forms. Take a look at the address bar of your browser next time you search Google, for example.) The actions and their parameters (if any) are as follows:

- **action = list & format = HTML / TXT / XML / JSON**: This action asks for a listing of all player scores to be returned. Depending on the value of the second variable format (`html/txt/xml/json`), the list of users and their scores are returned in different text file formats.

- **action = reset**: This action asks for a set of default player names and score values to replace the current contents of the database table. This action takes no argument. It returns some HTML, stating that the database has been reset, with a link to the home page.

- **action = get & username = & format = HTML / TXT**: This action asks for the integer score of the named player that is to be found. It returns the score integer. There are two formats: HTML, for a web page giving the player's score, and TXT, where the numerical value is the only content in the HTTP message returned.

- **action = update & username = <username> & score = <score>**: This action asks for the provided score of the named player to be stored in the database (but only if this new score is greater than the currently stored score). It returns the word *success* (if the database update was successful); otherwise, it returns -1 (to indicate that no update took place).

There's more...

Here are some ways to go further with this recipe.

SQLite, PHP, and database servers

The PHP code in this recipe used the PDO data object functions to communicate with a SQLite local file-based database. Learn more about PHP and SQLite at http://www.sqlitetutorial.net/sqlite-php/.

When SQLite isn't a solution (not supported by a web-hosting package), you may need to develop locally with a SQL **Server,** such as MySQL Community Edition or MariaDB, and then deploy with a live database **server** from your hosting company.

A good solution to try things out on your local machine can be a combined web application collection, such as XAMP/WAMP/MAMP. Your web **server** needs to support PHP, and you also need to be able to create the MySQL databases:

- **XAMP:** https://www.apachefriends.org/index.html
- **WAMP:** https://www.wampserver.com/
- **MAMP:** https://www.mamp.info/

phpLiteAdmin

When writing code that talks to database files and database **servers,** it can be frustrating when things don't work and you can't see *inside* the database. Therefore, database **clients** exist to allow you to interact with database **servers** without having to use code.

A lightweight (single file!) solution when using PHP and SQLite is phpLiteAdmin, which is free to use (although you may consider donating if you use it a lot). It is included in the **phpLiteAdmin** folder with this recipe's PHP scripts. It can be run using the Composer script shortcut command – `composer dbadmin` – and will run locally at `localhost:8001`. Once it is running, just click on the link for the player table to see the data for each player's score in the database file.

Note that you'll need to edit the sample configuration file as **phpliteadmin.config.php**; the settings to get this working are as follows:

- Set your preferred password to access this admin page
- Set the directory (**../data** in the provided code):

  - `$directory = '../data';`

- The `$databases` array names the SQLite database file being used:

```
$databases = array(
  array(
    'path'=> 'leaderboard.sqlite',
      'name'=> 'Database 1'
    )
);
```

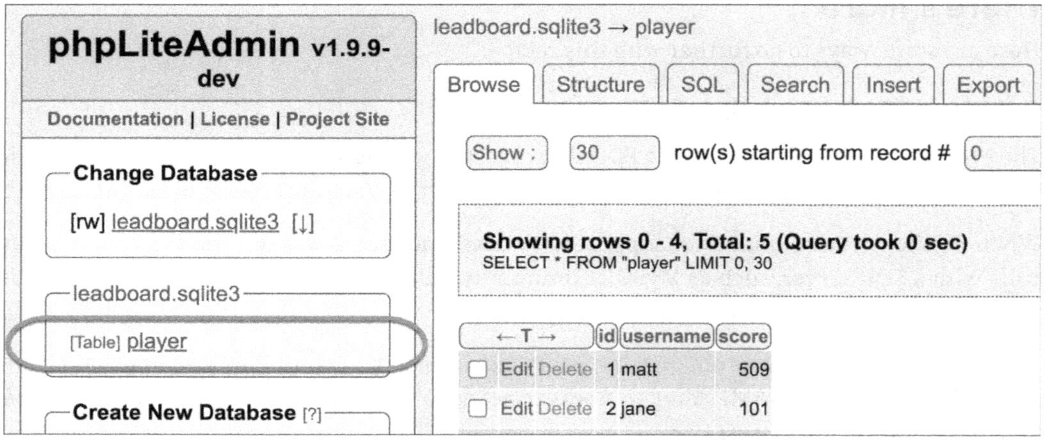

Figure 10.21: Using phpLiteAdmin in a web browser

See also

Learn more about phpLiteAdmin at the project's GitHub repository and website:

- `https://github.com/phpLiteAdmin/pla`
- `https://www.phpliteadmin.org/`

Unity game communication with a web server leaderboard

In this recipe, we create a Unity game client that can communicate, via UI buttons, with our web server leaderboard from the previous recipe:

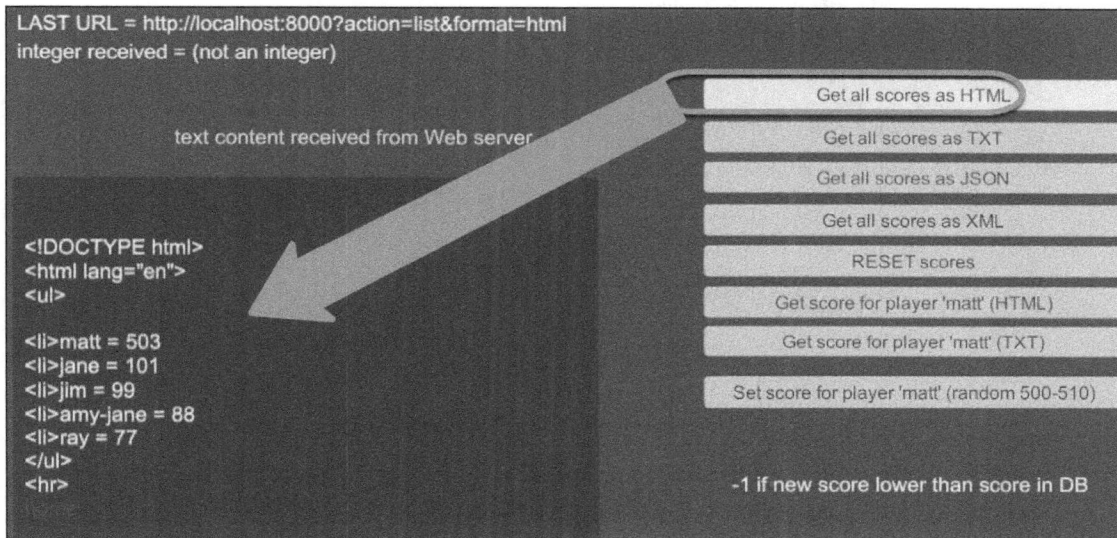

LAST URL = http://localhost:8000?action=list&format=html
integer received = (not an integer)

text content received from Web server

Get all scores as HTML
Get all scores as TXT
Get all scores as JSON
Get all scores as XML
RESET scores
Get score for player 'matt' (HTML)
Get score for player 'matt' (TXT)
Set score for player 'matt' (random 500-510)

<!DOCTYPE html>
<html lang="en">
<ul>

<li>matt = 503
<li>jane = 101
<li>jim = 99
<li>amy-jane = 88
<li>ray = 77
</ul>
<hr>

-1 if new score lower than score in DB

Figure 10.22: Screenshot showing a Unity game retrieving scores from a web server

Getting ready

Since this scene contains several UI elements and the code of the recipe is the communication with the PHP scripts and SQL database, in the 10_09 folder, we have provided a Unity package called **UnityLeaderboardClient**, containing a scene with everything set up for the Unity project.

How to do it...

To create a Unity game that communicates with the web server leaderboard, perform the following steps:

1. Create a new 2D Unity project.
2. Import the Unity package provided, UnityLeaderboardClient.
3. Run the scene provided.
4. Ensure that your PHP leaderboard is up and running.

5. If you are not running locally (localhost:8000), you'll need to update the URL by selecting Main Camera in the **Hierarchy** panel, and then editing the **Leader Board URL** text for the **Web Leader Board (Script)** component in the **Inspector** panel:

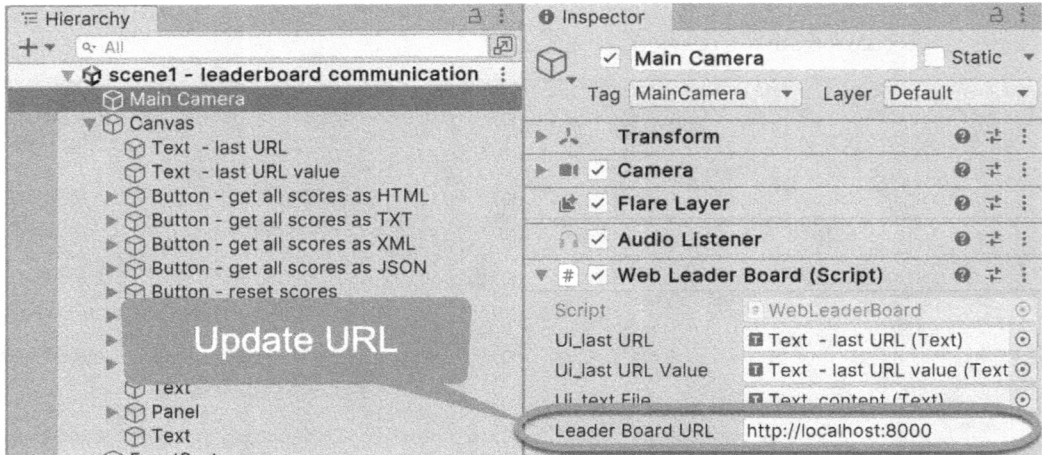

Figure 10.23: Setting a website URL public variable in the Inspector panel

6. Click the buttons on the right-hand side of the screen, making Unity communicate with the PHP scripts that have access to the high-score database.

How it works...

The player's scores are stored in a SQL database. Access to the database is facilitated through the PHP scripts provided by the web server project that was set up in the previous recipe.

In our example, all the PHP scripts were placed in a web **server** folder for a local web server. So the scripts are accessed via http://localhost:8000/. However, since the URL is a public string variable, this can be set to the location of your **server** and site code before running the scene.

There are buttons in the Unity scene (corresponding to the actions the web leaderboard understands) that set up the corresponding action and the parameters to be added to the URL, for the next call to the web **server**, via the LoadWWW() method. The OnClick() actions have been set up for each button to call the corresponding methods of the WebLeaderBoard C# script of Main Camera.

There are also several **UI Text** objects. One displays the most recent URL string sent to the server. Another displays the integer value that was extracted from the response message that was received from the server (or a **not an integer** message if some other data was received).

The third **UI Text** object is inside a **UI panel**, and it has been made large enough to display a full, multiline text string received from the **server** (which is stored inside the textFileContents variable).

We can see that the contents of the HTTP text **Response** message are simply an integer when a random score is set for player *Matt*, when the **Get score for player 'matt' (TXT)** button is clicked, and a text file containing 509 is returned:

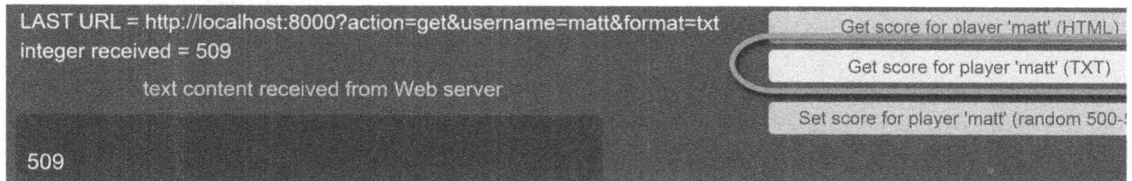

LAST URL = http://localhost:8000?action=get&username=matt&format=txt
integer received = 509
text content received from Web server

Get score for player 'matt' (HTML)
Get score for player 'matt' (TXT)
Set score for player 'matt' (random 500-

509

Figure 10.24: Score of 509 retrieved for player Matt from the web server

The **UI Text** objects have been assigned to public variables of the WebLeaderBoard C# script for the **main camera**. When any of the **UI buttons** are clicked, the corresponding method of the WebLeaderBoard method is called, which builds the URL string with parameters, and then calls the LoadWWW() method. This method sends the **request** to the URL and waits (by virtue of being a coroutine) until a **response** is received. It then stores the content, received in the textFileContents variable, and calls the UpdateUI() method. There is a prettification of the text received, inserting newline characters to make the JSON, HTML, and XML easier to read.

> **Note:** We have provided a **CertHandler.cs** script class to solve the HTTPS authentication for this recipe. For real-world games, you should not use self-signed certificates like this, since connected applications will require full security to protect your connections at runtime.

There's more...

Here are some ways to go further with this recipe.

Extracting the full leaderboard data for display within Unity

The XML/JSON text that can be retrieved from the PHP web server provides a useful method to allow a Unity game to retrieve the full set of the leaderboard data from the database. Then, the leaderboard can be displayed to the user in the Unity game (perhaps, in a nice 3D fashion, or through a game-consistent UI).

Using secret game codes to secure your leaderboard scripts

The Unity and PHP code that is presented illustrates a simple, unsecured web-based leaderboard. To prevent players from hacking into the board with false scores, we should encode some form of secret game code (or key) into the communications. Only update requests that include the correct code will actually cause a change to the database.

> **Note:** The example here is an illustration using the MD5 hashing algorithm. To protect a personal game, this is fine, but it's quite old and would not be suitable for sensitive or valuable data, such as financial or personal details. Some links to articles on more secure approaches can be found at the end of this topic.

The Unity code will combine the secret key (in this example, the **harrypotter string**) with something related to the communication – for example, the same MySQL/PHP leaderboard may have different database records for different games that are identified with a game ID:

```csharp
// Unity Csharp code
string key = "harrypotter";
string gameId = 21;
string gameCode = Utility.Md5Sum(key + gameId);
```

The server-side PHP code will receive both the encrypted game code, and the piece of game data that is used to create that encrypted code. In this example, it is the game ID and MD5 hashing function, which is available in both Unity and PHP.

The secret key (**harrypotter**) is used with the game ID to create an encrypted code that can be compared with the code received from the Unity game (or whatever user agent or browser is attempting to communicate with the leaderboard **Server** scripts). The database actions will only be executed if the game code created on the **Server** matches that sent along with the request for a database action:

```php
// PHP - security code
$key = "harrypotter";
$game_id =  filter_input(['game_id'], INPUT_GET);
$provided_game_code =  filter_input(['game_ code'], INPUT_GET);
$server_game_code = md5($key.$game_id);

if( $server_game_code == $provided_game_code ) {
  // codes match - do processing here
}
```

Further reading

The following are some useful resources to learn more about saving and loading data with Unity:

- Learn more about Unity Text Assets in the manual pages:

    - https://docs.unity3d.com/Manual/class-TextAsset.html

If you really want to get serious about game and web server security, then a proper **API key strategy** is probably the way to go. Here are some articles on sophisticated approaches to securing game data:

- *Securing Unity Games with DexGuard and iXGuard* from GuardSquare:

  - `https://www.guardsquare.com/en/blog/securing-unity-games-dexguard-and-ixguard-how-it-works`

- *Creating API keys using Okta in PHP,* published by Krasimir Hristozov:

  - `https://dev.to/oktadev/build-a-simple-rest-api-in-php-2k0k`

- *How to Secure a PHP Based API Using JWT by Mauro Chojrin:*

  - `https://www.securecoding.com/blog/secure-php-api-using-jwt/`

11

Controlling and Choosing Positions

In this chapter, we will introduce a few recipes and demonstrate a selection of approaches to character control, spawn points, and checkpoints. In the next chapter, we'll look at waypoints for AI-controlled characters traversing Unity navigation meshes.

Many **GameObjects** in games move! Movement can be controlled by the player, by the (simulated) laws of physics in the environment, or by **Non-Player Character** (NPC) logic. For example, objects follow the path of a waypoint, or seek (move toward) or flee (away) from the current position of a character. Unity provides several controllers for first- and third-person characters, and vehicles such as cars and airplanes. **GameObject** movement can also be controlled through the state machines of the Unity **Mecanim** animation system.

However, there may be times when you wish to tweak the player character controllers from Unity or write your own. You might wish to write directional logic – simple or sophisticated **Artificial Intelligence** (AI) to control a game's NPC and enemy characters – which we'll explore in *Chapter 12*, *Navigation Meshes and Agents*. Such AI might involve your computer program making objects orient and move toward or away from characters or other game objects.

This chapter (and the chapter that follows) presents a range of such directional recipes, from which many games can benefit in terms of a richer and more exciting user experience.

Unity provides sophisticated classes and components, including the Vector3 class and rigid body physics, to model realistic movements, forces, and collisions in games. We will make use of these game engine features to implement some sophisticated NPC and enemy character movements in the recipes of this chapter.

For 3D games (and to some extent, 2D games as well), a fundamental data object (struct) is the Vector3 class, whose objects store and manipulate the X, Y, *and* Z values, representing locations in a 3D space. Suppose we draw an imaginary arrow from the origin (0, 0, 0) to a point on the space.

In that case, the direction and length of this arrow (vector) can represent a velocity or force – that is, a certain amount of magnitude in a certain direction. In most cases, we can think of 1 Unity unit (X, Y, or Z) modeling 1 meter in the real world.

Suppose we ignore all the character controller components, colliders, and the physics system in Unity. In that case, we can write code that **teleports** objects directly to a particular **(X, Y, Z)** location in our scene. Sometimes, this is just what we want to do; for example, we may wish to spawn or teleport an object at a location. However, in most cases, if we want objects to move in more physically realistic ways, then we either apply a force to the object's **RigidBody** or change its velocity component. Alternatively, if it has a **Character Controller** component, then we can send it a Move( ) message. Teleporting objects directly to a new location increases the chance that a collision with another object is missed, which can lead to frustrating game behavior for players.

Some important concepts we must understand for NPC object movement and creation (instantiation) include the following:

- **Spawn points**: Specific locations in the scene where objects are to be created or moved to.
- **Checkpoints**: Locations (or colliders) that, once passed through, change what happens later in the game (for example, extra time, or if a player's character gets killed, they respawn to the last checkpoint they crossed, and so on).
- **Waypoints**: A sequence of locations to define a path for NPCs or, perhaps, the player's character to follow.

In this chapter, we will cover the following recipes:

- Using a rectangle to constrain 2D Player object movement
- Player control of a 3D GameObject (and limiting the movement within a rectangle)
- Choosing destinations – finding a random spawn point
- Choosing destinations – finding the nearest spawn point
- Choosing destinations – respawning to the most recently passed checkpoint
- Moving objects by clicking on them
- Firing projectiles in the direction of movement

Using a rectangle to constrain 2D Player object movement

Basic character movement in 2D (within a bounding rectangle) is a core skill for many 2D games, so this first recipe will illustrate how to achieve these features for a 2D game. The remaining recipes will then build on this approach for 3D games.

Since in *Chapter 3, Inventory and Advanced UIs*, we created a basic 2D game template, we'll adapt this game to restrict the movement to a bounding rectangle:

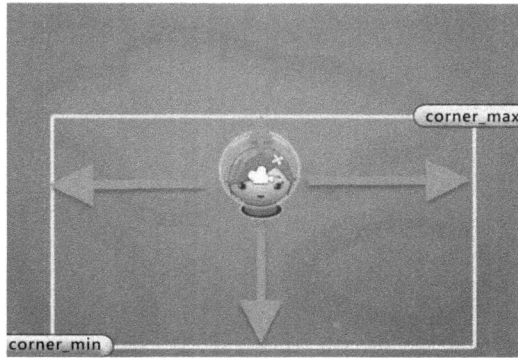

Figure 11.1: Movement of a character within the rectangular area

Getting ready

This recipe builds on the simple 2D mini game called Simple2DGame_SpaceGirl from the first recipe of *Chapter 3, Inventory and Advanced UIs*. Start with a copy of this game, or use the provided completed recipe project as the basis for this recipe.

How to do it...

To create a 2D sprite controlled by the user with movement that is limited to within a rectangle, follow these steps:

1. Create a new, empty GameObject named corner_max, and position it somewhere above and to the right of player_spaceGirl. With this GameObject selected in the Hierarchy panel, choose the large yellow oblong icon in the Inspector panel:

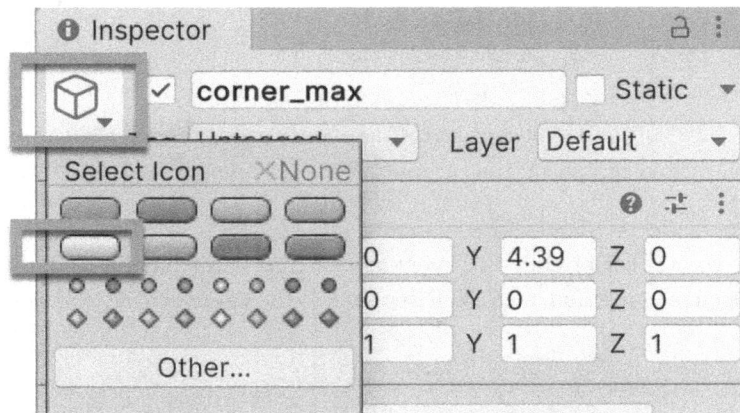

Figure 11.2: Setting a colored design time icon for a GameObject

2. Duplicate the corner_max GameObject by naming the copy corner_min, and position this clone somewhere below and to the left of the player-spaceGirl GameObject. The coordinates of these two GameObjects will determine the maximum and minimum bounds of movement that are permitted for the player's character.

3. Modify the C# script `PlayerMove` to declare some new variables at the beginning of the class:

    ```
    public Transform corner_max;
    public Transform corner_min;

    private float x_min;
    private float y_min;
    private float x_max;
    private float y_max;
    ```

4. Modify the C# script `PlayerMove` so that the `Awake()` method uses the properties of the corner_max and corner_min GameObjects to help set up the maximum and minimum X and Y movement limits:

    ```
    void Awake(){
        rigidBody2D = GetComponent<Rigidbody2D>();
        x_max = corner_max.position.x;
        x_min = corner_min.position.x;
        y_max = corner_max.position.y;
        y_min = corner_min.position.y;
    }
    ```

5. Modify the C# script called `PlayerMove` to declare a new method called `KeepWithinMinMaxRectangle()`:

    ```
    private void KeepWithinMinMaxRectangle(){
        float x = transform.position.x;
        float y = transform.position.y;
        float z = transform.position.z;
        float clampedX = Mathf.Clamp(x, x_min, x_max);
        float clampedY = Mathf.Clamp(y, y_min, y_max);
        transform.position = new Vector3(clampedX, clampedY, z);
    }
    ```

6. Modify the C# script called `PlayerMove` so that, after having updated the velocity in the `FixedUpdate()` method, a call will be made to the `KeepWithinMinMaxRectangle()` method:

    ```
    void FixedUpdate(){
        rigidBody2D.velocity = newVelocity;

        // restrict player movement
        KeepWithinMinMaxRectangle();
    }
    ```

7. With the player-spaceGirl GameObject selected in the **Hierarchy** panel, drag the corner_max and corner_min GameObjects into the **Corner_max** and **Corner_min** public variables in the **Inspector** panel:

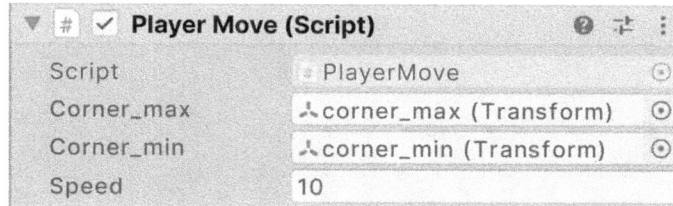

Figure 11.3: Populating the public Corner_max and Corner_min variables with our Game-Objects

Before running the scene in the **Scene** panel, try repositioning the corner_max and corner_min Game-Objects. When you run the scene, the positions of these two GameObjects (**max** and **min**, and **X** and **Y**) will be used as the limits of movement for the player's player-spaceGirl character.

How it works...

In this recipe, you added the empty GameObjects called corner_max and corner_min to the scene. The X and Y coordinates of these GameObjects will be used to determine the bounds of movement that we will permit for the player-controlled character – that is, player-spaceGirl. Since these are empty GameObjects, they will not be seen by the player when in **play mode**. However, we can see and move them in the **Scene** window, and since we added the yellow oblong icons, we can see their positions and names very easily.

When the Awake() method is executed for the PlayerMoveWithLimits object inside the player-spaceGirl GameObject, the maximum and minimum X and Y values of the corner_max and corner_min GameObjects are recorded. Each time the physics system is called via the FixedUpdate() method, the velocity of the player-spaceGirl character is updated to the value set in the Update() method, which is based on the horizontal and vertical keyboard/joystick inputs. However, the final action of the FixedUpdate() method is to call the KeepWithinMinMaxRectangle() method, which uses the Math.Clamp(...) function to move the character back inside the **X** and **Y** limits. This happens so that the player's character is not permitted to move outside the area defined by the corner_max and corner_min GameObjects.

There's more...

There are some details that you don't want to miss out on.

Drawing a gizmo yellow rectangle to visually show a bounding rectangle

As developers, it is useful to *see* elements such as bounding rectangles when **run-testing** our game. Let's make the rectangular bounds of the movement visually explicit in yellow lines in the **Scene** panel by drawing a yellow "gizmo" rectangle.

Add the following method to the C# script class called `PlayerMove`:

```
void OnDrawGizmos(){
    Vector3 top_right = Vector3.zero;
    Vector3 bottom_right = Vector3.zero;
    Vector3 bottom_left = Vector3.zero;
    Vector3 top_left = Vector3.zero;

    if(corner_max && corner_min){
        top_right = corner_max.position;
        bottom_left = corner_min.position;

        bottom_right = top_right;
        bottom_right.y = bottom_left.y;

        top_left = top_right;
        top_left.x = bottom_left.x;
    }

    //Set the following gizmo colors to YELLOW
    Gizmos.color = Color.yellow;

    //Draw 4 lines making a rectangle
    Gizmos.DrawLine(top_right, bottom_right);
    Gizmos.DrawLine(bottom_right, bottom_left);
    Gizmos.DrawLine(bottom_left, top_left);
    Gizmos.DrawLine(top_left, top_right);
}
```

The OnDrawGizmos() method tests that the references to the corner_max and corner_min GameObjects are not null, and then it sets the positions of the four Vector3 objects, representing the four corners defined by the rectangle, with corner_max and corner_min at opposite corners:

Figure 11.4: Use of design time gizmos to show the game creator the rectangular boundaries of player movement

As we can see, the gizmo's color is set to yellow and draws lines, connecting the four corners in the **Scene** window to show the bounding rectangle.

An alternative approach to defining movement bounds

Although the approach in this recipe works fine and is a good introduction to understanding the need to define movement bounds, there are some disadvantages to GameObjects as boundary limits. Positioning two objects to determine the bounds is an extra manual step that can easily get out of sync with the visual representation of the boundaries. Professional game developers would more likely use a collider or sprite, and check against the bounds. Such GameObjects have a Bound component, which is described in the Unity documentation: https://docs.unity3d.com/ScriptReference/Bounds.html.

Player control of a 3D GameObject (and limiting movement within a rectangle)

Many of the 3D recipes in this chapter are built on this basic project, which constructs a scene with a textured terrain, a `Main Camera`, and a red cube that can be moved around by the user with the four directional arrow keys:

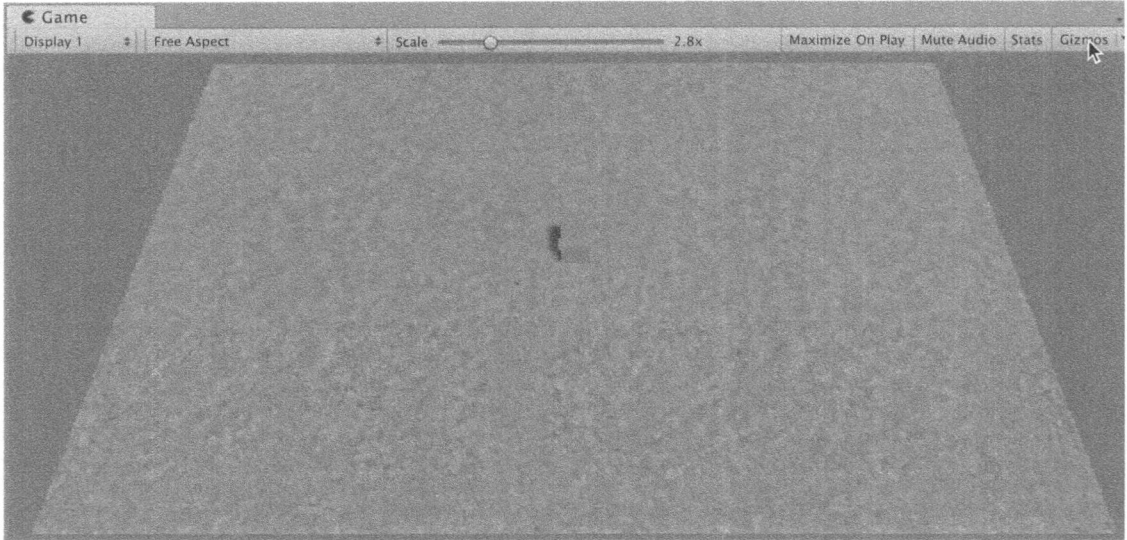

Figure 11.5: Basic 3D scene with a player-controlled cube

The bounds of movement of the cube will be constrained using the same technique that we used in the previous 2D recipe.

Getting ready

For this recipe, we have prepared the image you need in a folder named `Assets`, in the `10_02` folder.

How to do it...

To create a basic 3D cube-controlled game, follow these steps:

1. Create a new, empty **Unity 3D project**.

2. Import the single **Terrain Texture** named `SandAlbedo` from the `10_02` provided folder, in the `Assets` folder.

3. Create a new terrain by going to **GameObject | 3D Object | Terrain**.

4. With this new **Terrain** GameObject selected in the **Hierarchy** panel, in its **Inspector** panel, set its **Transform Position** to (-15, 0, -10).

5. In the **Inspector** panel for the **Terrain | Mesh Resolution** settings, set the terrain's **Width** and **Length** to 30 x 20:

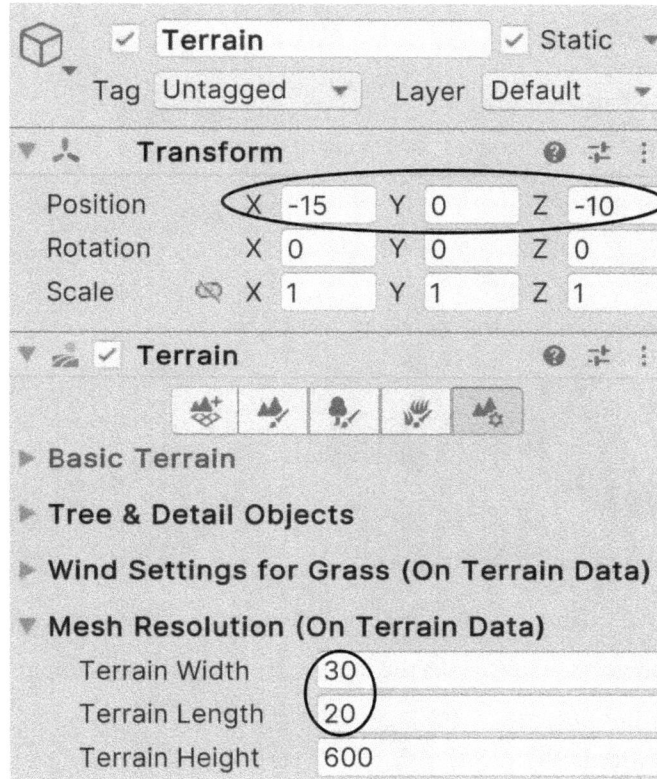

Figure 11.6: Setting the position and size of the terrain

The transform position for terrains relates to the near-left **corner**, not their **center**.

Since the **Transform** position of terrains relates to the corner of the object, we center such objects at (0, 0, 0) by setting the X coordinate equal to (*-1 times width/2*) and the Z-coordinate equal to (*-1 times length/2*). In other words, we slide the object by half its width and half its height to ensure that its center is just where we want it.

In this case, the width is 30 and the length is 20, so we get -15 for X (-1 * 30/2) and -10 for Z (-1 * 20/2).

6. Texture-paint this terrain with the texture we imported earlier – that is, **SandAlbedo**. First, click the paintbrush tool for **terrain texture painting**, choose **Paint Texture** from the drop-down menu, click the **Edit Terrain Textures...** button, click **Create Layer...**, and choose **SandAlbedo** from the pop-up **Select Texture2D** window:

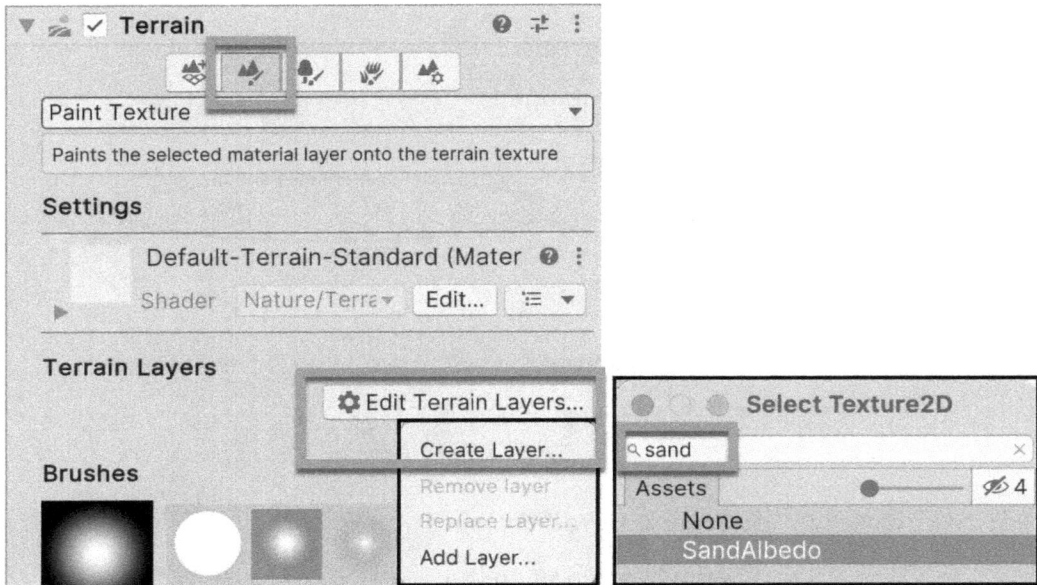

Figure 11.7: Texture-painting the terrain with the SandAlbedo texture

7. In the **Inspector** window, make the following changes to the properties of Main Camera:

   • **Position:** (0, 20, -15)
   • **Rotation:** (60, 0, 0)

8. Change **the aspect ratio** of the **Game** panel from **free aspect** to 16:9. You will now see all the terrain in the **Game** window.

9. Create a new empty GameObject named corner_max, and position it at (14, 0, 9). With this GameObject selected in the **Hierarchy** panel, choose the large, yellow oblong icon highlighted in the **Inspector** window.

10. Duplicate the corner_max GameObject, naming the clone corner_min, and position it at (-14, 0, -9). The coordinates of these two GameObjects will determine the maximum and minimum bounds of the movement permitted for the player's character.

11. Create a new cube GameObject by going to **Create | 3D Object | Cube**. Name this Cube-player, and set its position to (0, 0.5, 0) and its size to (1, 1, 1).

12. With Cube-player selected in the **Hierarchy** panel, in the **Inspector** panel, for the Box Collider component, check the **Is Trigger** option.

13. Add a `RigidBody` component to the `Cube-player` GameObject (**Physics | RigidBody**) and uncheck the **RigidBody** property called **Use Gravity**.

14. Create a red **Material** named `m_red`, and apply it to `Cube-player`.

15. Create a C# script class called `PlayerControl`, and add an instance object as a component of the `Cube-player` GameObject:

```csharp
using UnityEngine;

public class PlayerControl : MonoBehaviour {
  public Transform corner_max;
  public Transform corner_min;
  public float speed = 40;
  private Rigidbody rigidBody;
  private float x_min;
  private float x_max;
  private float z_min;
  private float z_max;
  private Vector3 newVelocity;

  void Awake() {
  rigidBody = GetComponent<Rigidbody>();
  x_max = corner_max.position.x;
  x_min = corner_min.position.x;
  z_max = corner_max.position.z;
  z_min = corner_min.position.z;
  }

  private void Update() {
   float xMove = Input.GetAxis("Horizontal") * speed * Time.deltaTime;
   float zMove = Input.GetAxis("Vertical") * speed * Time.deltaTime;
   float xSpeed = xMove * speed;
   float zSpeed = zMove * speed;
   newVelocity = new Vector3(xSpeed, 0, zSpeed);
  }

  void FixedUpdate() {
   rigidBody.velocity = newVelocity;
   KeepWithinMinMaxRectangle();
  }
```

```
private void KeepWithinMinMaxRectangle() {
float x = transform.position.x;
float y = transform.position.y;
float z = transform.position.z;
float clampedX = Mathf.Clamp(x, x_min, x_max);
float clampedZ = Mathf.Clamp(z, z_min, z_max);
transform.position = new Vector3(clampedX, y, clampedZ);
  }
 }
```

16. With the Cube-player GameObject selected in the **Hierarchy** panel, drag the corner_max and corner_min GameObjects over the corner_max and corner_min public variables in the **Inspector** panel.

When you run the scene, the positions of the corner_max and corner_min GameObjects will define the bounds of movement for the player's Cube-player character.

How it works...

The scene contains a positioned terrain so that its center is at (0, 0, 0). The red cube is controlled by the user's arrow keys through the PlayerControl script.

Just as with the previous 2D recipe, a reference to the (3D) **RigidBody** component is stored when the Awake() method executes, and the maximum and minimum X and Z values are retrieved from the two corner GameObjects and are stored in the x_min, x_max, z_min, and z_max variables. Note that for this basic 3D game, we won't allow any Y-movement, although such movement (and bounding limits by adding a third max-height corner GameObject) can easily be added by extending the code in this recipe.

The KeyboardMovement() method reads the horizontal and vertical input values (which the Unity default settings read from the four directional arrow keys). Based on these left-right and up-down values, the velocity of the cube is updated. The distance it will move depends on the speed variable.

The KeepWithinMinMaxRectangle() method uses the Math.Clamp(...) function to move the character back inside the X and Z limits, ensuring that the player's character is not permitted to move outside the area defined by the corner_max and corner_min GameObjects.

There's more...

There are some details that you don't want to miss out on.

Drawing a thick gizmo yellow rectangle to visually show a bounding rectangle

The Unity gizmos DrawLine(...) method only draws lines that are a single pixel in width. For 3D projects, this can be hard to see. In the Unity forums, in March 2019, **Jozard** kindly suggested the following method to draw multiple single-pixel lines, creating thicker lines that are easier to see.

Here is the updated code to do this:

```
void OnDrawGizmos (){
    Vector3 top_right = Vector3.zero;
    Vector3 bottom_right = Vector3.zero;
    Vector3 bottom_left = Vector3.zero;
    Vector3 top_left = Vector3.zero;

    if(corner_max && corner_min){
        top_right = corner_max.position;
        bottom_left = corner_min.position;

        bottom_right = top_right;
        bottom_right.z = bottom_left.z;

        top_left = bottom_left;
        top_left.z = top_right.z;
    }

    //Set the following gizmo colors to YELLOW
    Gizmos.color = Color.yellow;

    //Draw 4 lines making a rectangle
    Gizmos.DrawLine(top_right, bottom_right);
    Gizmos.DrawLine(bottom_right, bottom_left);
    Gizmos.DrawLine(bottom_left, top_left);
    Gizmos.DrawLine(top_left, top_right);

    // draw thick lines ...
    DrawThickLine(top_right, bottom_right, 5);
    DrawThickLine(bottom_right, bottom_left, 5);
    DrawThickLine(bottom_left, top_left, 5);
    DrawThickLine(top_left, top_right, 5);

}

// from
// https://answers.unity.com/questions/1139985/gizmosdrawline-thickens.html
public static void DrawThickLine(Vector3 p1, Vector3 p2, float width)
{
    int count = 1 + Mathf.CeilToInt(width); // how many lines are needed.
```

```
if (count == 1)
{
    Gizmos.DrawLine(p1, p2);
}
else
{
    Camera c = Camera.current;
    if (c == null)
    {
        Debug.LogError("Camera.current is null");
        return;
    }
    var scp1 = c.WorldToScreenPoint(p1);
    var scp2 = c.WorldToScreenPoint(p2);

    Vector3 v1 = (scp2 - scp1).normalized; // line direction
    Vector3 n = Vector3.Cross(v1, Vector3.forward); // normal vector

    for (int i = 0; i < count; i++)
    {
        Vector3 o = 0.99f * n * width * ((float)i / (count - 1) - 0.5f);
        Vector3 origin = c.ScreenToWorldPoint(scp1 + o);
        Vector3 destiny = c.ScreenToWorldPoint(scp2 + o);
        Gizmos.DrawLine(origin, destiny);
    }
}
}
```

The code works by looping to draw multiple lines next to each other, giving the appearance of a thick line.

Choosing destinations — finding a random spawn point

Many games make use of spawn points and waypoints. This recipe will show you how to choose a random spawn point, and then the instantiation of an object at that chosen point:

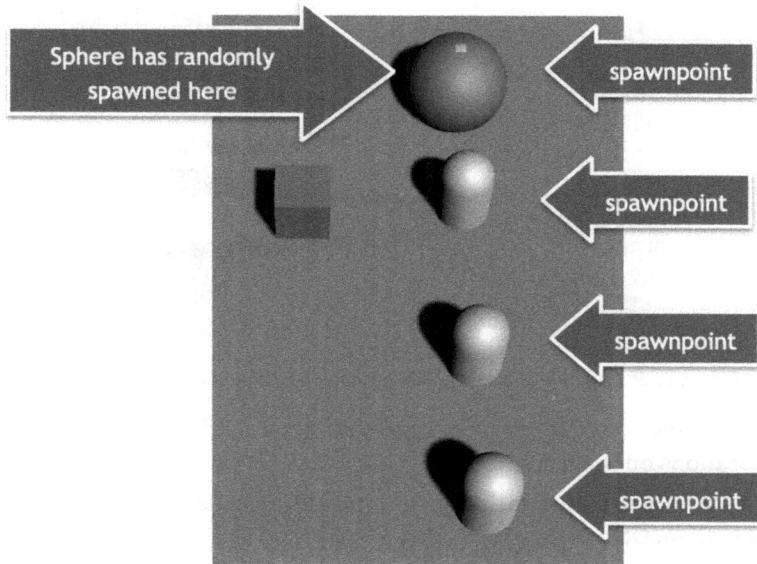

Figure 11.8: Sphere randomly spawned at one of the spawn point capsules

As shown in the preceding figure, the sphere has been spawned at one of the four capsule spawn points.

Getting ready

This recipe builds upon the previous recipe. So make a copy of this project, open it, and then follow the steps in the next section.

How to do it...

To find a random spawn point, follow these steps:

1. In the **Scene** panel, create a sphere (by navigating to **GameObject | 3D Object | Sphere**) scaled to as (2, 2, 2) at position (2, 2, 2) and apply the m_red material.

2. In the **Project** panel, create a new prefab based on your sphere by dragging the Sphere Game-Object from the **Hierarchy** panel into the **Project** panel. Rename this new prefab Prefab-ball.

3. Delete the sphere from the **Hierarchy** panel (we don't need it now we've made our prefab).

4. In the **Scene** panel, create a new capsule (by navigating to **GameObject | 3D Object | Capsule**) named Capsule-spawnPoint at (3, 0.5, 3), and give it a tag of **Respawn** (this is one of the default tags that Unity provides):

> For testing, we'll leave these **Respawn** points visible. For the final game, we would usually uncheck the Mesh Rendered property of each Respawn GameObject so that they are not visible to the player.

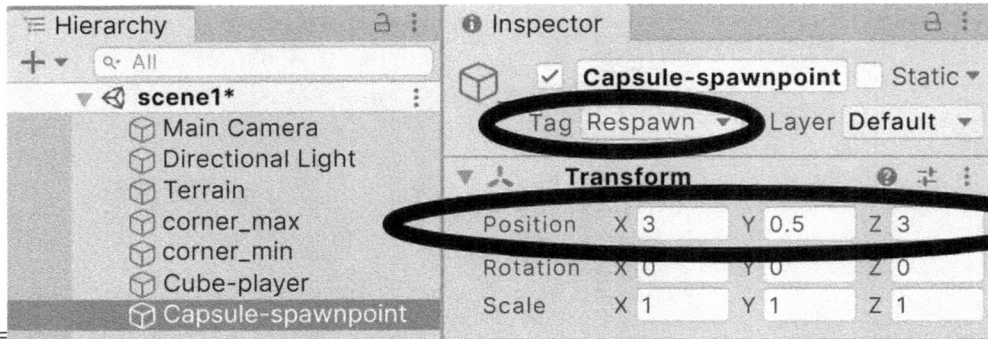

Figure 11.9: Capsule GameObject tagged as Respawn and positioned at (3, 0.5, 3)

5. Make several copies of your Capsule-spawnPoint (use *Ctrl + D/Cmd + D*) and move them to different locations on the terrain.

6. Create a C# script class called BallSpawner, and add an instance object as a component to the Cube-player GameObject:

```csharp
using UnityEngine;

public class BallSpawner : MonoBehaviour {
    public GameObject prefabBall;
    private SpawnPointManager spawnPointManager;
    private float timeBetweenSpawns = 1;

    void Start () {
        spawnPointManager = GetComponent<SpawnPointManager> ();
        InvokeRepeating("CreateSphere", 0, timeBetweenSpawns);
    }

    private void CreateSphere() {
        GameObject spawnPoint =
        spawnPointManager.RandomSpawnPoint();

        GameObject newBall = (GameObject)Instantiate(
            prefabBall, spawnPoint.transform.position,
                Quaternion.identity);
        Destroy(newBall, timeBetweenSpawns/2);
    }
}
```

7. Create a C# script class called SpawnPointManager, and add an instance object as a component to the Cube-player GameObject:

```csharp
using UnityEngine;

public class SpawnPointManager : MonoBehaviour {
    private GameObject[] spawnPoints;

    void Start() {
        spawnPoints = GameObject.FindGameObjectsWithTag("Respawn");
    }

    public GameObject RandomSpawnPoint() {
        int r = Random.Range(0, spawnPoints.Length);
        return spawnPoints[r];
    }
}
```

8. Ensure that Cube-player is selected. Then, in the **Inspector** panel of the BallSpawner scripted component, drag Prefab-ball over the Prefab Ball public variable.

9. Now, run your game. Once per second, a red ball should be spawned and disappear after half a second. The location where each ball is spawned should be randomly chosen from the four capsules tagged **Respawn**.

How it works...

The Capsule-spawnPoint objects represent candidate locations, where we might wish to create an instance of our ball prefab. When our SpawnPointManager object, inside the Cube-player GameObject, receives the Start() message, it creates an array called spawnPoints of all the GameObjects in the scene that have the **Respawn** tag. This can be achieved through a call to the built-in FindGameObjectsWithTag("Respawn") method. This creates an array of all the objects in the scene with the **Respawn** tag – that is, the four Capsule-spawnPoint objects we created in our scene.

When our BallSpawner GameObject, Cube-player, receives the Start() message, it sets the spawnPointManager variable to be a reference to its sibling SpawnPointManager script component. Then, we use the InvokeRepeating(...) method to schedule the CreateSphere() method to be called every **1 second**.

> **Note:** That InvokeRepeating(...) can be a dangerous action to perform in a scene, since it means the named method will be continually invoked at a set interval, so long as the scene is running. So be very careful when using the InvokeRepeating(...) statement.

The SpawnPointManager method, RandomSpawnpoint(...), chooses and returns a reference to a randomly chosen member of spawnPoints – the array of GameObjects in the scene with the **Respawn** tag. This method was declared as public so that our BallSpawner component instance can invoke this method of the SpawnPointManager instance-object component. Note that the second parameter for the Random.Range(...) method is the exclusive limit – so if it's set to 10, we'll get numbers up to, but not including, 10. This makes it perfect for getting random array indexes, since they start at zero and will go up to, but not include, the number of items in the array (.Length).

The BallSpawner method, CreateSphere(), assigns the spawnPoint variable to the GameObject that's returned by a call to the RandomSpawnpoint(...) method of our spawnPointManager object. Then, it creates a new instance of prefab_ball (via the public variable) at the same position as the spawnPoint GameObject. Finally, the built-in Destroy(...) method is used to tell Unity to remove the newly created GameObject after half a second (timeBetweenSpawns / 2).

See also

The same techniques and code can be used for selecting waypoints. Please refer to the *NPC NavMeshAgent to follow waypoints in a sequence* recipe in *Chapter 12, Navigation Meshes and Agents*, for more information about waypoints.

Choosing destinations – finding the nearest spawn point

Rather than just choosing a random spawn point or waypoint, sometimes, we want to select the one that's closest to an object (such as the player's GameObject). In this recipe, we will modify the previous recipe to find the nearest spawn point to the player's cube, and then use that location to spawn a new red ball prefab:

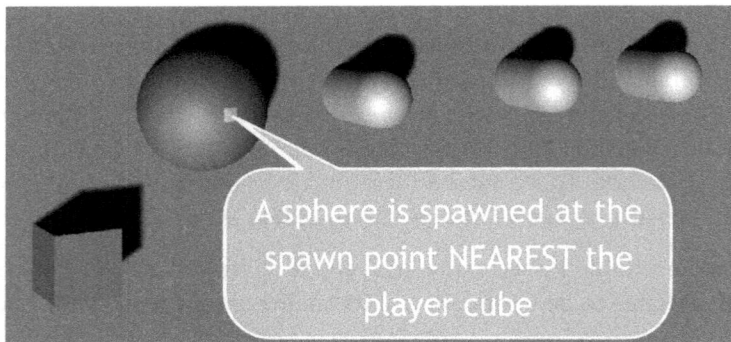

Figure 11.10: Spawning a sphere at the spawn point capsule that is nearest the player cube

Getting ready

This recipe builds upon the previous recipe. So make a copy of that project, open it, and then follow the steps in the next section.

How to do it...

To find the **nearest** spawn point, follow these steps:

1. Add the following method to the C# script class called `SpawnPointManager`:

```
public GameObject GetNearestSpawnpoint (Vector3 source){
    GameObject nearestSpawnPoint = spawnPoints[0];
    Vector3 spawnPointPos = spawnPoints[0].transform.position;
    float shortestDistance = Vector3.Distance(source, spawnPointPos);

    for (int i = 1; i < spawnPoints.Length; i++){
      spawnPointPos = spawnPoints[i].transform.position;
      float newDist = Vector3.Distance(source, spawnPointPos);
      if (newDist < shortestDistance){
        shortestDistance = newDist;
        nearestSpawnPoint = spawnPoints[i];
      }
    }

    return nearestSpawnPoint;
}
```

2. We now need to change the first line in the C# `BallSpawner` class so that the `spawnPoint` variable is set by a call to our new method – that is, `NearestSpawnpoint(...)`:

```
private void CreateSphere(){
    GameObject spawnPoint =
    spawnPointManager.GetNearestSpawnpoint(transform.position);

    GameObject newBall = (GameObject)Instantiate (prefabBall,
    spawnPoint.transform.position, Quaternion.identity);
    Destroy(newBall, timeBetweenSpawns/2);
}
```

3. Now, run your game. Every second, a red ball should be spawned and disappear after half a second. Use the arrow keys to move the player's red cube around the terrain.

4. Each time a new ball is spawned, it should be at the spawn point **closest** to the player.

How it works...

In the NearestSpawnpoint(...) method, we set nearestSpawnPoint to the first (array index 0) GameObject in the array as our default. We then looped through the rest of the array (the array index 1 up to spawnPoints.Length). For each GameObject in the array, we checked that the distance from the current spawnpoint position (Vector3 spawnPointPos) to the provided Vector3 source parameter position is less than the shortest distance so far, and if it is, we update the shortest distance and also set nearestSpawnPoint to the current element. Once we'd searched the array, we returned the GameObject that the nearestSpawnPoint variable refers to.

> **Note:** That to keep the code simple, we are not performing a test for the nearestSpawnPoint(...) method – production code would require that a test be made in case the received parameter was null, in which case the code would exit the method without taking any action.

There's more...

There are some details that you don't want to miss out on.

Avoiding errors due to an empty array

Let's make our code a little more robust so that it can cope with the issue of an empty spawnPoints array – that is, when no objects are tagged as Respawn in the scene.

To cope with no objects being tagged as Respawn, we need to do the following:

1. Improve our Start() method in the C# script class called SpawnPointManager, ensuring that an *error* is logged if the array of the objects tagged as Respawn is empty:

    ```csharp
    void Start() {
        spawnPoints = GameObject.FindGameObjectsWithTag("Respawn");

        // LogError if array empty
        if(spawnPoints.Length < 1)
            Debug.LogError ("SpawnPointManagaer - cannot find any objects
    tagged 'Respawn'");
        }
    ```

2. Improve the RandomSpawnPoint() and NearestSpawnpoint() methods in the C# script class called SpawnPointManager, ensuring that they still return a value (that is, null), even if the array is empty:

    ```csharp
    public GameObject RandomSpawnPoint (){
        // return current GameObject if array empty
        if(spawnPoints.Length < 1)
    ```

```
                    return null;

        // the rest as before ...
```

3. Improve the CreateSphere() method in the C# class called BallSpawner, ensuring that we only attempt to instantiate a new GameObject if the RandomSpawnPoint() and NearestSpawnpoint() methods have returned a non-null object reference:

```
private void CreateSphere(){
    GameObject spawnPoint = spawnPointManager.RandomSpawnPoint ();

    if(spawnPoint){
        GameObject newBall = (GameObject)Instantiate (prefabBall,
            spawnPoint.transform.position, Quaternion.identity);
        Destroy(newBall, timeBetweenSpawns/2);
    }
}
```

These null tests ensure we avoid runtime errors, since no attempt will be made to instantiate a new GameObject if the spawnPoint variable is null in the CreateSphere() method in *step 3* above.

See also

The same techniques and code can be used to select waypoints. Please refer to the *NPC NavMeshAgent to follow waypoints in a sequence* recipe in *Chapter 12, Navigation Meshes and Agents*.

Choosing destinations — respawning to the most recently passed checkpoint

A checkpoint usually represents a certain distance throughout the game (or perhaps a race track) in which an agent (user or NPC) has succeeded in reaching. Reaching (or passing) checkpoints often results in bonus awards, such as extra time, points, ammo, and so on. Also, if a player has multiple lives, then a player will often only be respawned back as far as the most recently passed checkpoint, rather than right to the beginning of the level.

This recipe will demonstrate a simple approach to checkpoints, whereby once the player's character has passed a checkpoint if they die, they are moved back to the most recently passed checkpoint:

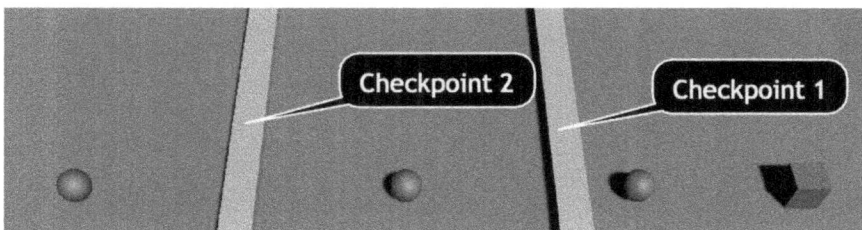

Figure 11.11: The three kill spheres, separated by two checkpoints

In the preceding screenshot, we can see a player-controlled cube on the right. The game area contains three spheres that will kill the player when hit. The game area is divided into three areas by the two elongated cubes specified as checkpoint 1 and checkpoint 2. The player starts on the right, and if they hit the sphere in the right-hand area, they will die and be respawned to where they started. Once the player moves through checkpoint 1 into the middle section, if they hit the middle sphere, they will die but be respawned into the middle section. Finally, if the player moves through checkpoint 2 into the left-hand side of the game area, if they hit the leftmost sphere, they will die but be respawned into the right-hand section of the game area.

Getting ready

This recipe builds upon the player-controlled 3D cube Unity project that you created at the beginning of this chapter: *Player control of a 3D GameObject (and limiting the movement within a rectangle)*. So make a copy of that project, open it, and then follow the steps for this recipe.

How to do it...

To have the respawn position change upon losing a life, depending on the checkpoints that have been passed, follow these steps:

1. Move the Cube-player GameObject to **position** (12, 0.5, 0).
2. Select Cube-player in the **Inspector** panel, and add a **Character Controller** component by clicking on **Component | Physics | Character Controller** (this is to enable the **OnTriggerEnter** collision messages to be received).
3. Create a cube named Cube-checkpoint-1 at (5, 0, 0), scaled to (1, 1, 20).
4. Create a CheckPoint tag and assign this tag to Cube-checkpoint-1.
5. Duplicate Cube-checkpoint-1, name the clone Cube-checkpoint-2, and position it at (-5, 0, 0).
6. Create a sphere named Sphere-Death at (7, 0.5, 0). Assign the m_red material to this sphere to make it red.
7. Create a Death tag, and assign it to Sphere-Death.
8. Duplicate Sphere-Death, and position the clone at (0, 0.5, 0).
9. Duplicate Sphere-Death a second time, and position the second clone at (-10, 0.5, 0).
10. Add an instance of the following C# script class, called CheckPoint, to the Cube-player GameObject:

```
using UnityEngine;

public class CheckPoint : MonoBehaviour {
    private Vector3 respawnPosition;
```

```
            void Start () {
                respawnPosition = transform.position;
            }

            void OnTriggerEnter (Collider hit) {
                if(hit.CompareTag("Checkpoint")) {
                    respawnPosition = transform.position;
                }

              if(hit.CompareTag("Death")){
                    transform.position = respawnPosition;
              }
          }
      }
```

Run the scene. If the cube runs into a red sphere before crossing a checkpoint, it will be respawned back to its starting position. Once the red cube has passed a checkpoint, if a red sphere is hit, then the cube will be moved back to the location of the most recent checkpoint that it passed through.

How it works...

The C# script class called CheckPoint has one variable called respawnPosition, which is a Vector3 object that refers to the position the player's cube is to be moved to (respawned) if it collides with a Death-tagged object. The default setting for this is the position of the player's cube when the scene begins. This default position is set in the Start() method; we set respawnPosition to the player's position before the scene started running.

Each time an object tagged as Checkpoint is collided with, the value of respawnPosition is updated to the current position of the player's red cube at this point in time (that is, where it is when it touches a stretched cube tagged as CheckPoint). The next time an object tagged as Death is hit, the cube will be respawned back to where it last touched the object tagged as CheckPoint.

Moving objects by clicking on them

Sometimes, we want to allow the user to interact with objects through mouse pointer clicks. In this recipe, we will allow the user to move an object in a random direction by clicking on it.

Getting ready

This recipe builds upon the player-controlled 3D cube Unity project that you created at the beginning of this chapter: *Player control of a 3D GameObject*. So make a copy of that project, open it, and then follow the steps for this recipe.

The result of this recipe should look as follows:

Figure 11.12: A pyramid of cubes falling down after being clicked on

How to do it...

To move objects by clicking on them, follow these steps:

1. Delete the Cube-player GameObject.
2. Set the position of Main Camera to (0, 3, -5) and its rotation to (25, 0, 0).
3. Create a C# script class called ClickMove:

```csharp
using UnityEngine;
[RequireComponent(typeof(Rigidbody))]
public class ClickMove : MonoBehaviour {

  public float multiplier = 500f;
  private Rigidbody rigidBody;

  private void Awake() {
    rigidBody = GetComponent<Rigidbody>();
  }

  void OnMouseDown() {
    float x = RandomDirectionComponent();
    float y = RandomDirectionComponent();
    float z = RandomDirectionComponent();
    Vector3 randomDirection = new Vector3(x,y,z);
    rigidBody.AddForce(randomDirection);
  }

  private float RandomDirectionComponent() {
    return (Random.value - 0.5f) * multiplier;
  }
}
```

4. Create a cube GameObject, and add an instance object of the ClickMove script class as a component. You should see that a RigidBody component is automatically added to the new cube, since the script class has the RequireComponent(typeof(Rigidbody)) directive. This only works if the directive is in the code before the script class is added to a GameObject.

5. Make five more duplicates of the cube, and arrange the six cubes into a pyramid by setting their positions like so:

```
(0, 2.5, 0)
  (-0.75, 1.5, 0), (0.75, 1.5, 0)
  (-1.5, 0.5, 0), (0, 0.5, 0), (1.5, 0.5, 0)
```

6. Run the scene. Each time you use the mouse pointer to click on a cube, that cube will have a random directional force applied to it. So with a few clicks, you can knock down the pyramid!

How it works...

The public **float** variable multiplier allows you to change the maximum magnitude of a force by changing the value in the ClickMove scripted component of each cube.

The ClickMove script class has a private variable, called rigidBody, set as a reference to the **RigidBody** component in the Awake() method.

Each time a cube receives a MouseDown() message (such as when it has been clicked with the user's mouse pointer), this method creates a random directional Vector3 object and applies this as a force to the object's rigidBody reference.

The RandomDirectionComponent() method returns a random value between -multiplier and +multiplier.

Firing projectiles in the direction of movement

Another common use of force is to apply it to a newly instantiated object, making it a projectile traveling in the direction that the player's GameObject faces. That's what we'll create in this recipe. The result of this recipe should look as follows:

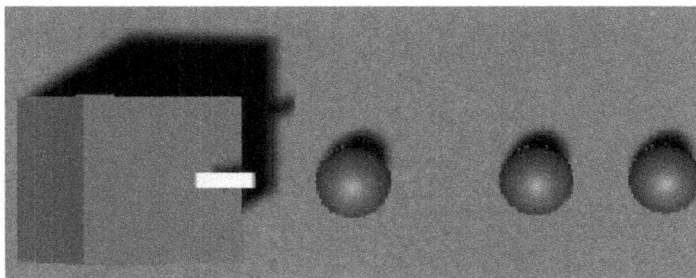

Figure 11.13: Projectiles being fired by the player's object

In the preceding screenshot, on the left, we can see a player-controlled tank, while on the right, we can see three sphere projectiles that have been fired from the player's character.

Getting ready

This recipe builds upon the player-controlled 3D cube Unity project that you created at the beginning of this chapter. So make a copy of that project, open it, and then follow the steps for this recipe.

How to do it...

To fire projectiles in the direction of movement, follow these steps:

1. Create a new Sphere GameObject (by navigating to **Create | 3D Object | Sphere**). Set its size as (0.5, 0.5, 0.5).

2. In the **Inspector** panel, add a **RigidBody** component to Sphere (go to **Physics | RigidBody**).

3. In the **Project** panel, create a new blue material named m_blue (go to **Create | Material**).

4. Apply the m_blue material to your sphere.

5. Use your Sphere GameObject to create a new prefab in the **Project** panel, named prefab_ projectile. First, create a new folder in the **Project** panel, named Prefabs, and then drag the Sphere GameObject from the **Hierarchy** panel into the folder in the **Project** panel. Rename this new prefab file prefab_projectile.

6. Now, delete the Sphere GameObject from the scene (that is, delete it from the **Hierarchy** panel).

7. Ensure to set the position of the Cube-player GameObject to (0, 0.5, 0).

8. Create a new cube named Cube-launcher. Disable its **Box Collider** component and set its transform as follows:

   - **Position:** (0, 1, 0.3)
   - **Rotation:** (330, 0, 0)
   - **Scale:** (0.1, 0.1, 0.5)

9. In the **Hierarchy** panel, make Cube-launcher a child of Cube-player by dragging Cube-launcher onto Cube-player. This means that both objects will move together when the user presses the arrow keys:

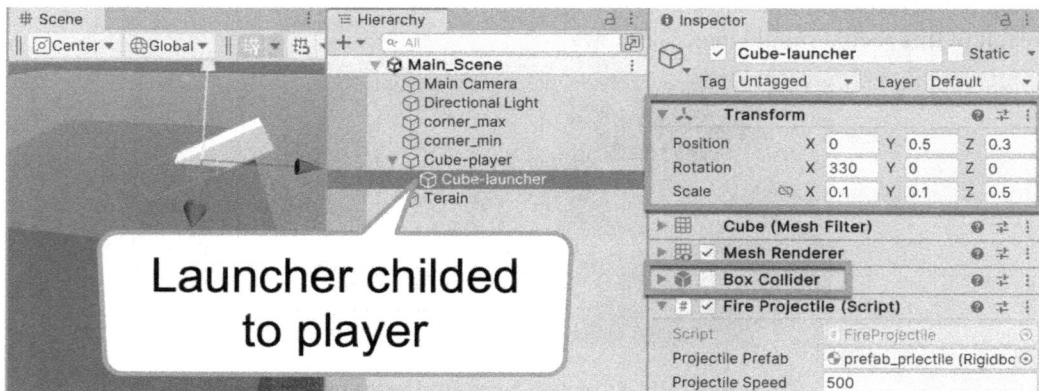

Figure 11.14: The Cube-launcher GameObject's Transform settings (childed to Cube-player)

10. Create a C# script class called `FireProjectile`, and add an instance object as a component to `Cube-launcher`:

```csharp
using UnityEngine;

public class FireProjectile : MonoBehaviour {
 const float FIRE_DELAY = 0.25f;
 const float PROJECTILE_LIFE = 1.5f;


 public Rigidbody projectilePrefab;
 public float projectileSpeed = 500f;

 private float nextFireTime = 0;

 void Update() {
   if (Time.time > nextFireTime)
     CheckFireKey();
 }

 private void CheckFireKey() {
   if(Input.GetButton("Fire1")) {
     CreateProjectile();
     nextFireTime = Time.time + FIRE_DELAY;
   }
 }

 private void CreateProjectile() {
   Vector3 position = transform.position;
   Quaternion rotation = transform.rotation;

   Rigidbody projectileRigidBody = Instantiate(projectilePrefab,
position,
       rotation);
   Vector3 projectileVelocity = transform.TransformDirection(
       Vector3.forward * projectileSpeed);

   projectileRigidBody.AddForce(projectileVelocity);

   GameObject projectileGO = projectileRigidBody.gameObject;
```

```
        Destroy(projectileGO, PROJECTILE_LIFE);
    }
}
```

11. With Cube-launcher selected in the **Inspector** panel, from the **Project** panel, drag prefab_ projectile into the projectile prefab public variable in the fire projectile (Script) component in the **Inspector** panel.

12. Run the scene. You can move around the terrain with the arrow keys, and each time you click the mouse button, you should see a blue sphere projectile launched in the direction that the player's cube faces.

How it works...

In this recipe, you created a blue sphere as a prefab (containing a **RigidBody**). You then created a scaled and rotated cube for the projectile launcher, called Cube-launcher, and then made this object a child of Cube-player.

A common issue is for projectiles to collide with their launcher object, so for this reason, we disabled the **Box Collider** component of the Cube-launcher GameObject. An alternative solution would be to create separate layers and remove the physics interaction between the layer of the launcher object and the layer of the projectiles.

The FireProjectile script class contains a constant called FIRE_DELAY – this is the minimum time between firing new projectiles, set to 0.25 seconds. There is also a second constant called PROJECTILE_ LIFE – this is how long each projectile will "live" until it is automatically destroyed; otherwise, the scene and memory would fill up quickly with lots of old projectiles!

There are also two public variables. The first is for the reference to the sphere prefab, while the second is for the initial speed of newly instantiated prefabs.

There is also a private variable called nextFireTime – this is used to decide whether enough time has passed to allow a new projectile to be fired.

The Update() method tests the current time against the value of nextFireTime. If enough time has passed, then it will invoke the CheckFireKey() method.

The CheckFireKey() method tests to see if the Fire1 button has been clicked. This is usually mapped to the left mouse button, but it can be mapped to other input events via **Project Settings** (**Edit** | **Project Settings** | **Input Manager**). If the Fire1 event is detected, then the next fire time is reset to be FIRE_DELAY seconds in the future, and a new projectile is created by invoking the CreateProjectile() method.

The CreateProjectile() method gets the current position and rotation of the parent GameObject. Remember that the instance object of this class has been added to Cube-launcher, so our scripted object can use the position and rotation of this launcher as the initial settings for each new projectile. A new instance of projectilePrefab is created with these position and rotation settings.

Then, a Vector3 object called projectileVelocity is created by multiplying the projectileSpeed variable by the standard forward vector **(0, 0, 1)**. In Unity, for 3D objects, the Z-axis is usually the direction in which the object faces.

The special TransformDirection(...) method is used to turn the local-space forward direction into a world-space direction, ensuring that we have a vector representing a forward motion relative to the Cube-launcher GameObject. This world-space directional vector is then used to add force to the projectile's **RigidBody**.

Finally, a reference is made to the parent GameObject of the projectile, and the Destroy(...) method is used so that the projectile will be destroyed after 1.5 seconds – the value of PROJECTILE_LIFE.

Further reading

The following are some useful resources to learn more about controlling Unity object positions and movements:

- Learn more about 2D and 3D character controllers at https://docs.unity3d.com/Manual/character-control-section.html.
- Learn more about the local-to-world-space direction TransformDirection(...) method at https://docs.unity3d.com/ScriptReference/Transform.TransformDirection.html.

Throughout this chapter, we have kept to a good rule of thumb:

"Always listen for **input** *in* Update()*.*

Always apply **physics** *in* FixedUpdate()*."*

You can learn more about why we should not check for input in FixedUpdate() in the Unity Answers thread (which is also the source for the preceding quote from user Tanoshimi) at https://answers.unity.com/questions/1279847/getaxis-being-missed-in-fixedupdate-work-around.html.

Learn more on Discord

To join the Discord community for this book – where you can share feedback, ask questions to the author, and learn about new releases – follow the QR code below:

https://packt.link/unitydev

12

Navigation Meshes and Agents

Unity provides **navigation meshes (NavMeshes)** and **artificial intelligence (AI)** agents that can plan pathways and move objects along those calculated paths. **Pathfinding** is a classic AI task, and Unity has provided game developers with fast and efficient pathfinding components that work out of the box.

Having objects that can automatically plot and follow paths from their current location to the desired destination point (or reoriented to a moving target object) provides the components for many different kinds of interactive game characters and mechanics. For example, we can create point-and-click games by clicking on a location or object toward which we wish one or more characters to travel. Or, we can have enemies that "wake up" when our player's character is nearby and move toward (seek) our player, then perhaps go into combat or dialogue mode once they are within a short distance of our player's character.

This chapter will explore ways to exploit Unity's navigation-based AI components to control game character pathfinding and movement.

At the core of Unity's navigation system are four concepts/components:

- Navigation meshes
- Navigation mesh agents
- NavMeshLinks
- NavMeshSurfaces

A navigation **mesh** defines the areas of the world that are navigable. It is usually represented as a set of polygons (2D shapes) so that a path to a destination is plotted as the most efficient sequence of adjacent polygons to follow, taking into account the need to avoid non-navigable obstacles.

The **agent** is the object that needs to calculate (plot) a path through the mesh from its current position to its desired destination position. **NavMesh agents** have properties such as a stopping distance so that they aim to arrive at a point that's a certain distance from the target coordinates, and auto braking so that they gradually slow down as they get close to their destination.

Links allow an agent to travel from one **NavMesh** to another (and back again).

NavMeshSurfaces are what the Unity AI system uses to identify faces of GameObjects in the scene that are candidates to be added to a NavMesh.

A **navigation mesh** can be made up of **areas** that have different "costs." The default cost for an area is **1**. However, to make a more realistic path calculation by AI agent-controlled characters, we might want to model the additional effort it takes to travel through water, mud, or up a steep slope. Therefore, Unity allows us to define custom areas with names that we choose (such as **Water** or **Mud**) and associated costs, such as **2** (that is, water is twice as tiring to travel through).

The most efficient way for games to work with NavMeshes is to pre-calculate the costs of an area in the game world; this is known as **baking** and is performed at **design time** before we run the game.

However, sometimes, there will be features in the game that we wish to use to influence navigation decisions and route planning differently at different times in the game – that is, dynamic **runtime** navigation obstacles. Unity provides a **NavMesh Obstacle** component that can be added to GameObjects and has features such as "carving out" (temporarily removing) areas of a **NavMesh** to force AI agents to recalculate paths that avoid areas blocked by GameObjects with **NavMesh Obstacle** components.

In this chapter, you'll learn how to add **NavMesh Agents** to control characters and how to work with your game environment to specify and bake NavMeshesfor a scene. Some recipes will explore how to create point-and-click style games, where you indicate where you want a character to navigate by clicking on an object or point in the game world.

You'll learn how to add **NavMesh Obstacle** components to moving GameObjects, forcing AI agents to dynamically recalculate their paths at runtime due to objects moving in their way. Finally, you'll learn how to create **NavMeshes** on non-horizontal surfaces and add links between different **NavMesh** objects for your agents to traverse.

In this chapter, we will cover the following recipes:

- NPC to travel to destination while avoiding obstacles
- NPC to seek or flee from a moving object
- Point-and-click move to object
- Point-and-click move to tile
- Point-and-click raycast with user-defined, higher-cost navigation areas
- NPC to follow waypoints in sequence
- Creating a movable NavMesh Obstacle
- Several NavMeshes with a single NavMeshSurface
- Non-horizontal NavMeshes with multiple NavMeshSurfaces and NavMeshLinks

NPC to travel to destination while avoiding obstacles

The introduction of Unity's **NavMeshAgent** has greatly simplified the coding for **Non-Player Character** (**NPC**) and enemy agent behaviors. In this recipe, we'll add some wall obstacles (scaled cubes) and a **NavMeshSurface** so that Unity knows not to try to walk through walls.

We'll then add a **NavMeshAgent** component to our NPC GameObject and tell it to head to a stated destination location by intelligently planning and following a path while avoiding the wall obstacles.

When the **NavMeshSurface** is added, the **Scene** panel displays the blue-shaded walkable areas, as well as unshaded, non-walkable areas at the edge of the terrain and around each of the two wall objects:

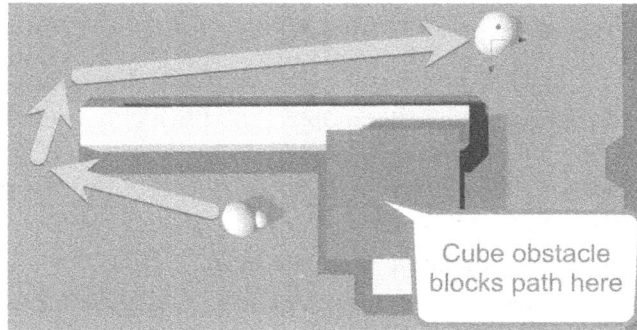

Figure 12.1: Example of an NPC avoiding obstacles

Getting ready

The required SandAlbedo texture terrain can be found in the 12_01 folder.

You will also need to install the **AI Navigation** package, but first, make sure you are signed in with your Unity ID. To install the **AI Navigation** package:

1. Go to **Window | Package Manager**. Then select **Unity Register** from the **Packages** drop-down list, and in the search box, type in **Nav** (short for navigation):

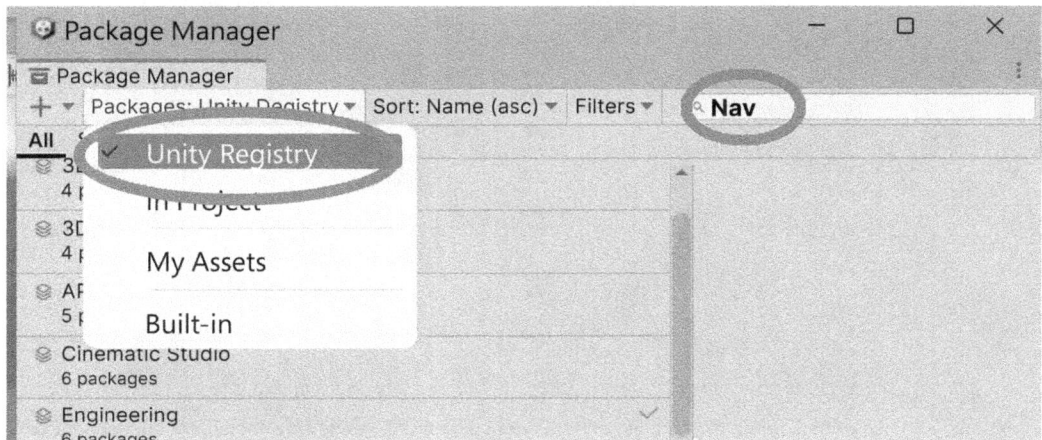

Figure 12.2: Installing the Navigation package

2. Choose **AI Navigation** from the list and then click **Install**.

3. The **Navigation** panel is now accessible through **Window AI | Navigation**. You will see an option called **Navigation (Obsolete)**; this is for backward compatibility with older projects and will not be used in this project.

How to do it...

To make an NPC travel to a destination while avoiding obstacles, follow these steps:

1. Create a new, empty Unity 3D project and add the **AI Navigation** package.

2. Create a new 3D terrain by going to **GameObject | 3D Object | Terrain**. With this new `Terrain` GameObject selected in the **Hierarchy** panel, in the **Inspector** panel's properties, set the **Terrain Width, Terrain Length,** and **Terrain Height** to `30 x 20 x 600` and its position to (`-15, 0, -10`) so we have this GameObject centered at (`0,0,0`).

3. Texture paint this terrain with the `SandAlbedo` texture.

4. Create a 3D capsule named `Capsule-destination` at (`-12, 0, 8`). This will be the target destination for our NPC self-navigating GameObject.

5. Create a sphere named `Sphere-arrow` that is positioned at (`2, 0.5, 2`). Scale it to (`1,1,1`).

6. Create a second sphere named `Sphere-small`. Scale it to (`0.5, 0.5, 0.5`).

7. In the **Hierarchy**, make `Sphere-small` a child of `Sphere-arrow` and position it at (`0, 0, 0.5`).

8. In the **Inspector** panel, add a new **NavMeshAgent** to `Sphere-arrow`. Do this by going to **Add Component | Navigation | Nav Mesh Agent**.

9. Set the **Stopping Distance** property of the **NavMeshAgent** component to 2:

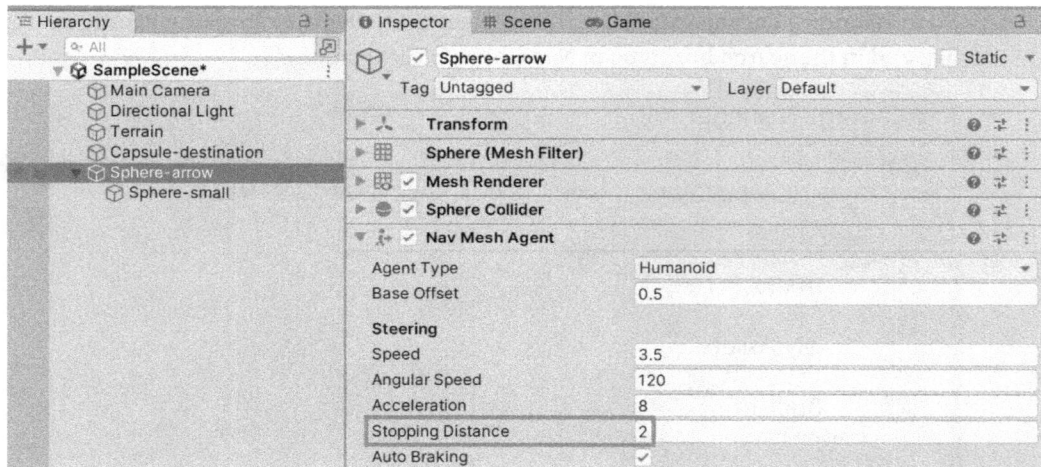

Figure 12.3: Setting the Stopping Distance property of the NavMeshAgent component to 2

10. Create the `ArrowNPCMovement` C# script class, and add an instance object to the `Sphere-arrow` GameObject:

```
using UnityEngine;
```

```
using UnityEngine.AI;

public class ArrowNPCMovement : MonoBehaviour {
    public GameObject targetGo;
    private NavMeshAgent navMeshAgent;

    void Start() {
        navMeshAgent = GetComponent<NavMeshAgent>();
        HeadForDestintation();
    }

    private void HeadForDestintation () {
        Vector3 destination = targetGo.transform.position;
        navMeshAgent.SetDestination (destination);
    }
}
```

11. Ensure that `Sphere-arrow` is selected in the **Inspector** panel. For the **ArrowNPCMovement** scripted component, drag `Capsule-destination` over the **Target Go** variable.

12. Create a 3D cube named `Cube-wall` at (-6, 0, 0) and scale it to (1, 2, 10).

13. Create another 3D cube named `Cube-wall2` at (-2, 0, 6) and scale it to (1, 2, 7).

14. Select the `Cube-wall` object in the **Inspector** panel, and tick the **Static** checkbox. You will find the checkbox in the top-right corner. Do the same for the second `Cube` object:

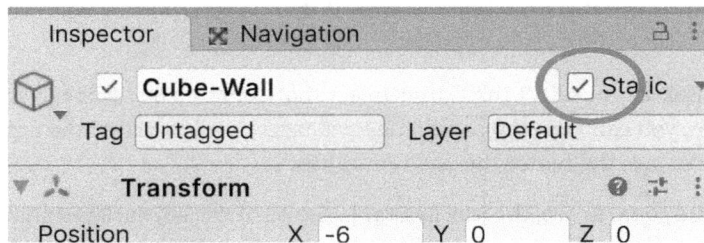

Figure 12.4: Ticking the Static checkbox for both Cube-wall objects

15. To add a `NavMeshSurface` to the terrain, select the terrain in the **Hierarchy** panel, then in the **Inspector** panel, click the '**Add Component**' button and select **NavMeshSurface**. Then click **Bake** on the **NavMeshSurface** component.

16. You will now see a blue tint on the parts of the scene that are areas for a **NavMeshAgent** to consider for its navigation paths. (You may need to change the angle of your main camera to get a better view of the game running.)

17. Now, run your game. You will see that the `Sphere-arrow` GameObject automatically moves toward the `Capsule-destination` GameObject, following a path that avoids the two wall objects.

How it works...

The NavMeshAgent component that we added to the `Sphere-arrow` GameObject does most of the work for us. NavMeshAgents need two things:

- A destination location to head toward
- A **NavMeshSurface** component on the terrain with walkable/non-walkable areas so that it can plan a path by avoiding obstacles

We created two obstacles (the `Cube-wall` objects), and we made these static in the **Inspector** panel. We can see walkable areas forming a blue navigation mesh on the terrain when the **NavMeshSurface** is added and baked.

The location for our NPC object to travel toward is the position of the `Capsule-destination` GameObject at (`-12, 0, 8`); but, of course, we could just move this object in the **Scene** panel at design time, and its new position would be the destination when we run the game.

The `ArrowNPCMovement` C# script class has two variables: one is a reference to the destination GameObject, while the second is a reference to the `NavMeshAgent` component of the GameObject, in which our instance of the `ArrowNPCMovement` class is also a component. When the scene starts, the `NavMeshAgent` sibling component is found via the `Start()` method, and the `HeadForDestination()` method is called, which sets the destination of `NavMeshAgent` to the position of the destination GameObject.

Once `NavMeshAgent` has a target to head toward, it will plan a path there and will keep moving until it arrives (or gets within the stopping distance, if that parameter has been set to a distance greater than zero).

In the **Scene** panel, if you select the GameObject that contains `NavMeshAgent` and choose the **Show Avoidance** gizmo, you can view the candidate local target positions that the agent is considering. The lighter the squares are, the better the position ranks.

The darker red the squares are, the less desirable the position; so, dark red squares indicate positions to avoid since they might, for instance, cause the agent to collide with a **NavMesh** static obstacle:

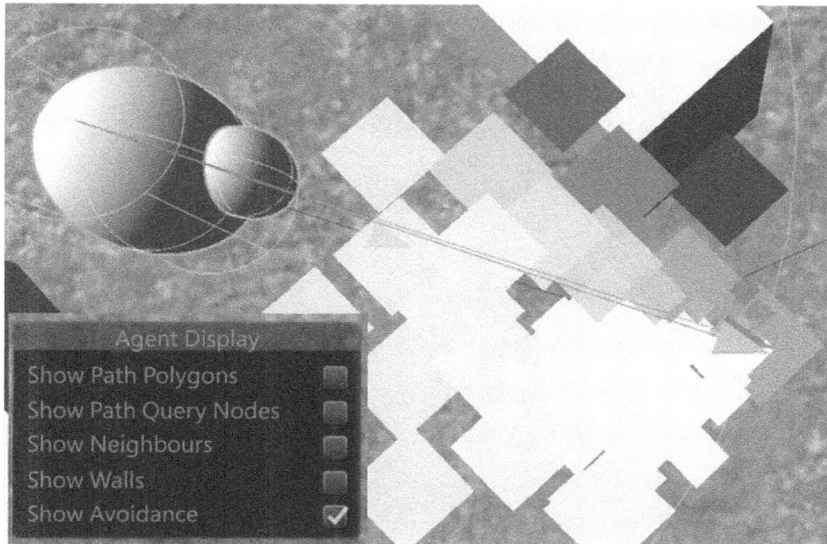

Figure 12.5: Show Avoidance gizmo showing candidate local target positions the agent is considering

Ensure that the object with the **NavMeshAgent** component is selected in the **Hierarchy** panel at run-time to be able to see this navigation data in the **Scene** panel.

NPC to seek or flee from a moving object

Rather than a destination that is fixed when the scene starts, let's allow the Capsule-destination object to be moved by the player while the scene is running. In every frame, we'll get our NPC arrow to reset the destination of **NavMeshAgent** to wherever Capsule-destination has been moved to.

Getting ready

This recipe adds to the previous one, so make a copy of that project folder and do your work for this recipe with that copy.

How to do it...

To make an NPC seek or flee from a moving object, follow these steps:

1. With the Capsule-destination GameObject selected in the **Hierarchy** panel, add a **Rigid Body Physics** component to it by going to **Component | Physics | Rigidbody**.

2. In the **Inspector** panel for the Capsule-destination GameObject, check the **Freeze Position** constraint for the Y axis in the **Constraints** options of the **RigidBody** component. This will prevent the object from moving in the Y axis due to collisions when being moved.

3.   Also, for the `Capsule-destination` GameObject, check the **Freeze Rotation** constraint for the X, Y, and Z axes in the **Constraints** options of the **RigidBody** component. This will prevent the object from falling when being moved.

4.   Create the `SimplePlayerControl` C# script class and add an instance object as a component to the `Capsule-destination` GameObject:

```csharp
using UnityEngine;

public class SimplePlayerControl : MonoBehaviour {
    public float speed = 1000;
    private Rigidbody rigidBody;
    private Vector3 newVelocity;

    private void Start() {
        rigidBody = GetComponent<Rigidbody>();
    }

    void Update() {
        float xMove = Input.GetAxis("Horizontal") * speed * Time.deltaTime;
        float zMove = Input.GetAxis("Vertical") * speed * Time.deltaTime;
        newVelocity = new Vector3(xMove, 0, zMove);
    }

    void FixedUpdate() {
        rigidBody.velocity = newVelocity;
    }
}
```

5.   Update the `ArrowNPCMovement` C# script class so that we call the `HeadForDestintation()` method at every frame – that is, from `Update()` rather than just once in `Start()`:

```csharp
Using UnityEngine;
using UnityEngine.AI;

public class ArrowNPCMovement : MonoBehaviour {
    public GameObject targetGo;
    private NavMeshAgent navMeshAgent;

    void Start() {
        navMeshAgent = GetComponent<NavMeshAgent>();
    }

    private void Update() {
```

```
      HeadForDestintation();
    }

    private void HeadForDestintation() {
      Vector3 destination = targetGo.transform.position;
      navMeshAgent.SetDestination(destination);
    }
  }
```

6. Reposition the Camera to position (-5, 1.6, -8) in the **Inspector** panel to get a better view of the game in action.

How it works...

The SimplePlayerControl script class detects arrow key presses and translates them into a force to apply to move the Capsule-destination GameObject in the desired direction.

The Update() method of the ArrowNPCMovement script class makes NavMeshAgent update its path every frame, based on the current position of the Capsule-destination GameObject. As the user moves Capsule-destination, NavMeshAgent calculates a new path to the object.

There's more...

Here are some details that you don't want to miss.

Using a Debug Ray to show a source-to-destination line

It's useful to use a visual Debug Ray to show us the straight line from the NPC with NavMeshAgent to the current destination it is trying to navigate toward. Since this is a common thing we may wish to do for many games, it's useful to create a static method in a general-purpose class; then, the ray can be drawn with a single statement.

> **Note:** Debug Rays are part of the Debug class provided by Unity. Being able to see lines drawn to indicate the path from an object to its destination is very useful when debugging and tweaking AI navigation during game development. Learn more at https://unity. com/how-to/improve-qa-testing-debugging-debug-class#debugdrawray.

To use a Debug Ray to draw a source-to-destination line, follow these steps:

1. Create a UsefulFunctions.cs C# script class containing the following code:

```
using UnityEngine;

public class UsefulFunctions : MonoBehaviour {
    public static void DebugRay(Vector3 origin, Vector3 destination, Color
c) {
        Vector3 direction = destination - origin;
```

```
        Debug.DrawRay(origin, direction, c);
    }
}
```

2. Now, add a statement at the end of the HeadForDestination() method in the ArrowNPCMovement
    C# script class:

```csharp
using UnityEngine;
using UnityEngine.AI;

public class ArrowNPCMovement : MonoBehaviour {
    public GameObject targetGo;
    private NavMeshAgent navMeshAgent;

    void Start() {
        navMeshAgent = GetComponent<NavMeshAgent>();
    }

    private void Update() {
        HeadForDestintation();
    }

    private void HeadForDestintation() {
        Vector3 destination = targetGo.transform.position;
        navMeshAgent.SetDestination(destination);
        // show yellow line from source to target
        UsefulFunctions.DebugRay(transform.position, destination, Color.
yellow);
    }
}
```

We can now see a yellow line in the **Scene** panel when the scene is running. We can also see this in
the **Game** panel if the **Gizmos** option is selected (at the top right of the **Game** panel's title bar):

Figure 12.6: Yellow line illustrating the Debug Ray source to destination

Constantly updating the NavMeshAgent's destination to flee from the player's current location

There are times when we want an AI-controlled NPC character to move **away** from another character, rather than go toward it. For example, an enemy with very low health might run away, and so allow time to regain its health before fighting again. Or, a wild animal might flee from any other character moving near it.

To instruct our **NavMeshAgent** component to flee from the player's location, we need to replace the ArrowNPCMovement C# script class with the following:

```csharp
using UnityEngine;
using UnityEngine.AI;

public class ArrowNPCMovement : MonoBehaviour {
    public float runAwayDistance = 10;
    public GameObject targetGO;
    private NavMeshAgent navMeshAgent;

    void Start() {
        navMeshAgent = GetComponent<NavMeshAgent>();
    }

    void Update() {
        Vector3 targetPosition = targetGO.transform.position;
        float distanceToTarget = Vector3.Distance(transform.position,
targetPosition);
        if (distanceToTarget < runAwayDistance) {
            FleeFromTarget(targetPosition);
        }
    }

    private void FleeFromTarget(Vector3 targetPosition) {
        Vector3 destination = PositionToFleeTowards(targetPosition);
        HeadForDestintation(destination);
    }

    private void HeadForDestintation (Vector3 destinationPosition) {
        navMeshAgent.SetDestination (destinationPosition);
    }

    private Vector3 PositionToFleeTowards(Vector3 targetPosition) {
```

```
    transform.rotation = Quaternion.LookRotation(transform.position -
targetPosition);
    Vector3 runToPosition = targetPosition + (transform.forward *
runAwayDistance);
    return runToPosition;
  }
}
```

There is a public variable here called runAwayDistance. When the distance to the enemy is less than the value of this runAwayDistance variable, we'll instruct the computer-controlled object to flee in the opposite direction.

The Start() method caches a reference to the **NavMeshAgent** component.

The Update() method calculates whether the distance to the enemy is within runAwayDistance and, if so, it calls the FleeFromTarget(...) method, which passes the location of the enemy as a parameter.

The FleeFromTarget(...) method calculates a point that is runAwayDistance in Unity units away from the player's cube, in a direction that is directly away from the computer-controlled object. This is achieved by subtracting the enemy's position vector from the current transform's position.

Finally, the HeadForDestintation(...) method is called, passing the flee-to position, which results in **NavMeshAgent** being told to set the location as its new destination.

Unity units are arbitrary since they are just numbers on a computer. However, in most cases, it simplifies things to think of distances in terms of meters (1 Unity unit = 1 meter) and mass in terms of kilograms (1 Unity unit = 1 kilogram). Of course, if your game is based on a microscopic world or pan-galactic space travel, then you need to decide what each Unity unit corresponds to for your game context. For more information on units in Unity, check out this post about Unity measurements: http://forum.unity3d.com/threads/best-units-of-measurement-in-unity.284133/#post-1875487.

Debug Rays show the point the NPC is aiming for, whether it is to flee from the player's character or to catch up and maintain a constant distance from it.

> You might need to ensure that the ArrowNPCMovement scripted component is still connected to Sphere-arrow by dragging Capsule-destination over the Target Go variable.

Maintaining a constant distance from the target ("lurking" mode!)

It is simple to adapt the previous code to have an NPC try to maintain a constant distance from a target object. It involves always moving toward a point that is runAwayDistance away from the target, regardless of whether this point is toward or away from the target.

Just remove the IF statement in the Update() method:

```
void Update() {
    Vector3 targetPosition = targetGO.transform.position;
```

```
    float distanceToTarget = Vector3.Distance(transform.position,
targetPosition);
    FleeFromTarget(targetPosition);
}
```

However, with this variation, it might be better to have the method named something like MoveTow ardsConstantDistancePoint() rather than FleeFromTarget(), since our NPC is sometimes fleeing and sometimes following.

Point-and-click move to object

Another way to choose the destination for our Sphere-arrow GameObject is by the user clicking on an object on the screen, and then the Sphere-arrow GameObject moving to the location of the clicked object:

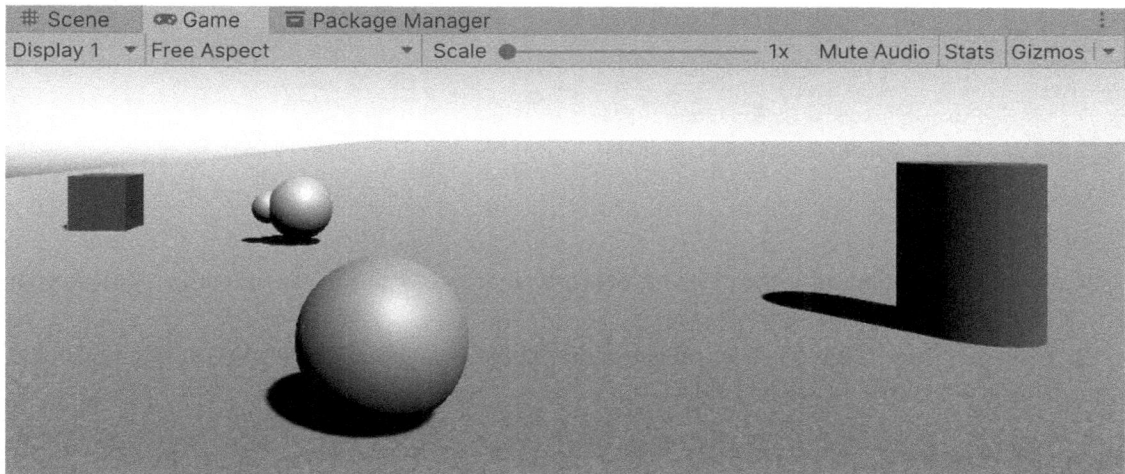

Figure 12.7: Example of point-and-click move to object

Getting ready

This recipe adds to the first recipe in this chapter, so make a copy of that project folder and do your work for this recipe with that copy.

How to do it...

To create an object-based point-and-click mini-game, do the following:

1. In the **Inspector** panel, add the Player tag to the Sphere-arrow GameObject.
2. Delete the two 3D cubes and the 3D Capsule-destination from the scene.
3. Remove the ArrowNCPMovement C# script as a component from the Sphere-arrow Game Object.
4. Create three GameObjects – a 3D cube, a 3D sphere, and a 3D cylinder – and place them in different positions within the **terrain**.

5. On the `NavMeshSurface` component on the terrain, click on **Bake**, to regenerate the `NavMeshSurface`.

6. Create a `ClickMeToSetDestination` C# script class containing the following code:

```
using UnityEngine;
using UnityEngine.AI;

public class ClickMeToSetDestination : MonoBehaviour {
    private NavMeshAgent playerNavMeshAgent;

    void Start() {
        GameObject playerGO = GameObject.FindGameObjectWithTag("Player");
        playerNavMeshAgent = playerGO.GetComponent<UnityEngine.
AI.NavMeshAgent>();
    }

    private void OnMouseDown() {
        playerNavMeshAgent.SetDestination(transform.position);
    }
}
```

7. Add instance objects of the `ClickMeToSetDestination` C# script class as components to your 3D cube, sphere, and cylinder.

8. Run the scene. When you click on one of the 3D objects, the `Sphere-arrow` GameObject should navigate toward the clicked object.

How it works...

The `OnMouseDown()` method of the `ClickMeToSetDestination` C# script class changes the destination of `NavMeshAgent` in the `Sphere-arrow` GameObject to be the position of the clicked 3D object.

The `Start()` method of the `ClickMeToSetDestination` C# script class gets a reference to the `NavMeshAgent` component of the GameObject tagged as `Player` (that is, the `Sphere-arrow` GameObject).

Each time a different object is clicked, the **NavMeshAgent** component inside the `Sphere-arrow` GameObject is updated to make the GameObject move toward the position of the clicked object.

There's more...

There are some details that you don't want to miss.

Creating a mouseover yellow highlight

A good **User Experience (UX)** feedback technique is to visually indicate to the user when an object can be interacted with via the mouse. A common way to do this is to present an audio or visual effect when the mouse is moved over an interactable object.

We can create a `Material` object with a yellow color, which can make an object appear yellow while the mouse is over it, and then make it return to its original material when the mouse is moved away.

Create the `MouseOverHighlighter` C# script class with the following contents. Then, add an instance object as a component to each of the three 3D GameObjects:

```
using UnityEngine;

public class MouseOverHighlighter : MonoBehaviour {
  private MeshRenderer meshRenderer;
  private Material originalMaterial;

  void Start() {
    meshRenderer = GetComponent<MeshRenderer>();
    originalMaterial = meshRenderer.sharedMaterial;
  }

  void OnMouseOver() {
    meshRenderer.sharedMaterial = NewMaterialWithColor(Color.yellow);
  }

  void OnMouseExit() {
    meshRenderer.sharedMaterial = originalMaterial;
  }

  private Material NewMaterialWithColor(Color newColor) {
    Shader shaderSpecular = Shader.Find("Specular");
    Material material = new Material(shaderSpecular);
    material.color = newColor;

    return material;
  }
}
```

Now, when running the game, when your mouse is over one of the three objects, that object will be highlighted in yellow. If you click on the mouse button when the object is highlighted, the `Sphere-arrow` GameObject will make its way up to (but stop just before) the clicked object.

Point-and-click move to tile

Rather than clicking specific objects to indicate the target for our AI-controlled agent, we can create a grid of 3D plane (tile) objects to allow the player to click any tile to indicate a destination for the AI controller character.

So, any location can be clicked, rather than only one of a few specific objects:

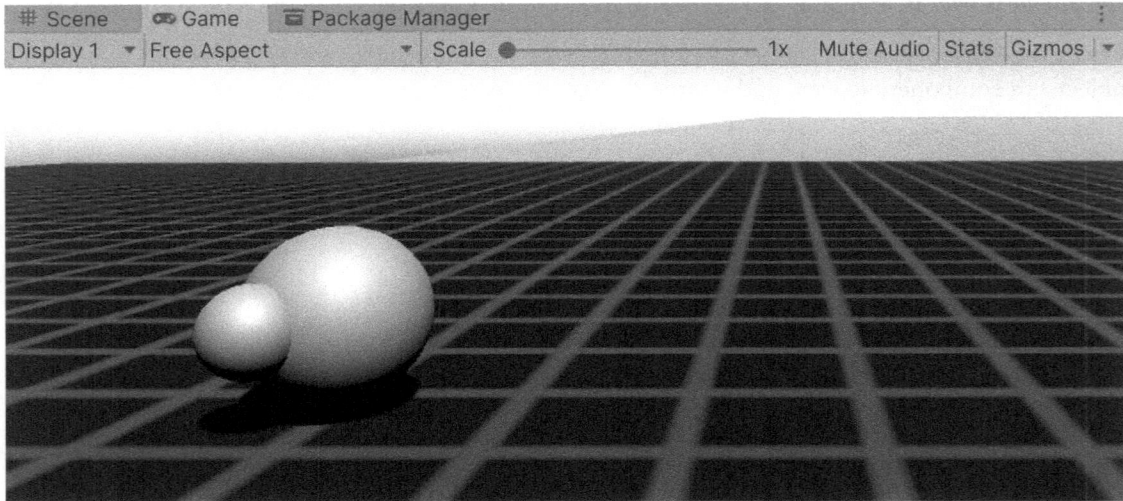

Figure 12.8: Example of point-and-click move to tile with a 3D plane grid

Getting ready

This recipe adds to the previous one, so make a copy of that project folder and do your work for this recipe with that copy.

For this recipe, we have prepared a red-outlined black square texture image named square_outline. png in a folder named Textures in the 12_04 folder.

How to do it...

To create a point-and-click game by setting GameObjects to a selected tile, do the following:

1. Delete your 3D cube, sphere, and cylinder GameObjects from the scene.

2. Remove the script component ArrowNCPMovment from the Sphere-arrow GameObject.

3. Create a new 3D Plane object scaled to (0.1, 0.1, 0.1).

4. Create a new **Material** object with the **Texture** image provided – that is, square_outline.png (a black square with a red outline). Apply this **Material** object to your 3D plane.

5. Add an instance object of the ClickMeToSetDestination script class as a component to your 3D plane.

6. In the **Project** panel, create a new, empty **Prefab** named tile.

7. Populate your tile Prefab with the properties of your 3D Plane GameObject by dragging the Plane GameObject over your tile Prefab (it should change from white to blue to indicate that the Prefab now has the properties of your GameObject).

8. Delete your 3D Plane GameObject from the scene.

9. Create a new TileManager C# script class containing the following code, and add an instance object as a component to the MainCamera GameObject:

```
using UnityEngine;

public class TileManager : MonoBehaviour {
    public int rows = 50;
    public int cols = 50;
    public GameObject prefabClickableTile;

    void Start () {
        for (int r = 0; r < rows; r++) {
            for (int c = 0; c < cols; c++) {
                float y = 0.01f;
                Vector3 pos = new Vector3(r - rows/2, y, c - cols/2);
                Instantiate(prefabClickableTile, pos, Quaternion.identity);
            }
        }
    }
}
```

10. Select MainCamera in the **Hierarchy** panel and, in the **Inspector** panel for the Tile Manager (Script) component, populate the Prefab clickable tile public property with your tile Prefab from the **Project** panel.

11. Run the scene. You should now be able to click on any of the small square tiles to set the destination of the NavMeshAgent-controlled Sphere-arrow GameObject.

How it works...

In this recipe, you created a Prefab containing the properties of a 3D plane named tile, which contained a component instance object of the ClickMeToSetDestination C# script class.

The TileManager script class loops to create 50 x 50 instances of this tile GameObject in the scene.

When you run the game, if you click on the mouse button when the mouse pointer is over a tile, the **NavMeshAgent** component inside the Sphere-arrow GameObject is set to that tile's position. So, the Sphere-arrow GameObject will move toward, but stop just before reaching, the clicked tile position.

The Y value of 0.01 means the plane will be just above the terrain, so we avoid any kind of interference pattern due to meshes being at the same location. By subtracting rows/2 and cols/2 from the X and Z positions, we center our grid of tiles at (0, Y, 0).

Moire patterns or large-scale interference patterns occur when two similar patterns are overlaid in an offset, rotated, or altered pitch position. In Unity 3D, Moire patterns can be commonly identified by texture flicker.

There's more...

There are some details that you don't want to miss.

Using a yellow Debug Ray to show the destination of the AI agent

We can show a Debug Ray from a moving object to its destination tile by creating the MouseOverHighlighter
C# script class with the following contents. We then add an instance object as a component to the
NavMeshAgent component's controlled Sphere-arrow GameObject:

```
using UnityEngine;
using UnityEngine.AI;

public class DebugRaySourceDestination : MonoBehaviour {
    void Update() {
        Vector3 origin = transform.position;
        Vector3 destination = GetComponent<NavMeshAgent>().destination;
        Vector3 direction = destination - origin;
        Debug.DrawRay(origin, direction, Color.yellow);
    }
}
```

The preceding code uses the current position of the character (transform.position – our moment
origin) and the destination point (GetComponent<NavMeshAgent>().destination) as the two endpoints
to display a yellow debug ray.

Point-and-click raycast with user-defined, higher-cost navigation areas

Rather than indicating a desired destination by clicking an object or tile, we can use Unity's built-
in Physics.Raycast(...) method to identify which Vector3 (x,y,z) position relates to the object
surface in the game.

This involves translating from the 2D (x,y) screen position to an imagined 3D "ray" from the user's point
of view, through the screen, into the game world, and identifying which object (polygon) it **hits** first.

This recipe will use Physics.Raycast to set the position of the location that's clicked on as the new
destination for a **NavMeshAgent** controller object. The actual route that's followed can be influenced
by defining navigation mesh areas of different costs. For example, walking through mud or swimming
through water can have a higher cost, since they would take longer, so the AI **NavMeshAgent** can
calculate the lowest-cost route, which may not be the shortest-distance route in the scene:

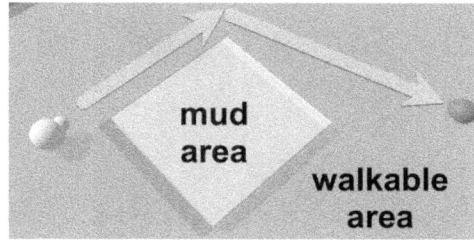

Figure 12.9: Higher cost mud area, inside lower cost walkable area

Getting ready

This recipe adds to the previous one, so make a copy of that project folder and do your work for this recipe with that copy.

How to do it...

To create a point-and-click game using a raycast, do the following:

1. Remove the Tile Manager (Script) component from the MainCamera GameObject.
2. Create a new 3D sphere, named Sphere-destination, scaled to (0.5, 0.5, 0.5).
3. Create a new **Material** object that's red and assign it to the Sphere-destination GameObject.
4. Create a new MoveToClickPoint C# script class containing the following and add an instance object as a component to the Sphere-arrow GameObject:

```
using UnityEngine;
using UnityEngine.AI;

public class MoveToClickPoint : MonoBehaviour {
    public GameObject sphereDestination;
    private NavMeshAgent navMeshAgent;
    private RaycastHit hit;

    void Start() {
        navMeshAgent = GetComponent<NavMeshAgent>();
        sphereDestination.transform.position = transform.position;
    }

    void Update() {
        Ray rayFromMouseClick = Camera.main.ScreenPointToRay(Input.
mousePosition);
```

```
            if (FireRayCast(rayFromMouseClick)){
                Vector3 rayPoint = hit.point;
                ProcessRayHit(rayPoint);
            }
        }

        private void ProcessRayHit(Vector3 rayPoint) {
            if(Input.GetMouseButtonDown(0)) {
                navMeshAgent.destination = rayPoint;
                sphereDestination.transform.position = rayPoint;
            }
        }

        private bool FireRayCast(Ray rayFromMouseClick) {
            return Physics.Raycast(rayFromMouseClick, out hit, 100);
        }
    }
```

5. Select the Sphere-arrow GameObject in the **Hierarchy** panel and, in the **Inspector** panel for the MoveToClickPoint (Script) component, populate the Sphere-destination public property with your red Sphere-destination GameObject.

6. Run the scene. You should now be able to click anywhere on the terrain to set the destination of the NavMeshAgent-controlled Sphere-arrow GameObject. As you click, the red Sphere-destination GameObject should be positioned at this new destination point, toward which the Sphere-arrow GameObject will navigate.

How it works...

In this recipe, you created a small red 3D sphere named Sphere-destination.

There is one public variable for the MoveToClickPoint scripted component of the Sphere-arrow GameObject. This public sphereDestination variable has been linked to the red Sphere-destination GameObject in the scene.

There are two private variables:

*   navMeshAgent: This will be set to refer to the NavMeshAgent component of the Sphere-arrow GameObject so that its destination can be reset when appropriate.

*   hit: This is a RaycastHit object that is passed in as the object to be set by Physics.Raycast(...). Various properties of this object are set after a raycast has been created, including the position in the scene where the raycast hits the surface of an object.

The Start() method caches a reference to the NavMesh component of the Sphere-arrow GameObject and also moves the Sphere-destination GameObject to the current object's location.

For each frame, in the `Update()` method, a ray is created based on the `MainCamera` and the `(z,y)` point that's clicked on the screen. This ray is passed as a parameter to the `FireRayCast(...)` method. If that method returns `true`, then the position of the object that's hit is extracted and passed to the `ProcessRayHit(...)` method.

The `FireRayCast(...)` method receives a `Ray` object. It uses `Phyics.Raycast(...)` to determine whether the raycast collides with part of an object in the scene. If the raycast hits something, the properties of the `RaycastHit hit` object are updated. A `true`/`false` output for whether `Physics.Raycast(...)` hit a surface is returned by this method.

Each time the user clicks on the screen, the corresponding object in the scene is identified with the raycast, the red sphere is moved there, and the **NavMeshAgent** component begins to navigate toward that location.

You can learn more about the Unity `RaycastHit` C# script class at `https://docs.unity3d.com/ScriptReference/RaycastHit.html`.

There's more...

Here are some details that you won't want to miss.

More intelligent pathfinding by setting different costs for custom-defined navigation areas such as mud and water

We can create objects whose meshes are defined as more expensive for NavMeshAgents to travel across, helping AI agent behavior be more realistic in terms of choosing faster paths that avoid water, mud, and so on.

To create a custom **NavMeshArea** (we'll pretend it's mud) with a higher traveling cost, do the following:

1. In the **Navigation** panel, reveal the areas by clicking the **Areas** button. Then, define a new area named Mud with a cost of 2:

Figure 12.10: Defining Mud as a NavMesharea

2. Create a new 3D cube named `Cube-mud` positioned at `(14,-1,10)` and scaled to `(5,5,5)`.
3. Add a **NavMeshModifierVolume** component to the `Cube-mud` GameObject.

4. Resize the **NavMeshModiferVolume** component to $(1,1,1)$ and change the **Area Type** to Mud:

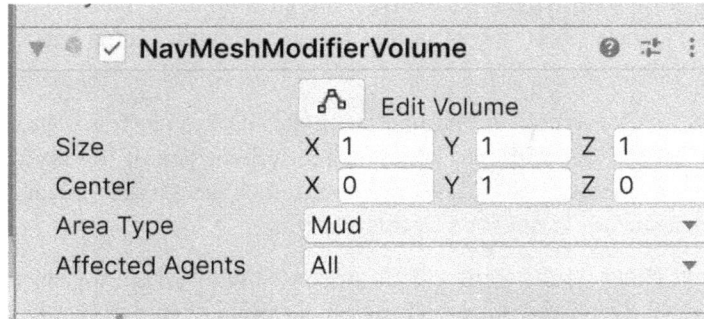

Figure 12.11: Setting Mud as our navigation area in the Navigation panel

5. Now, go to the **NavMeshSurface** component in the Terrain object and click on **Bake** to regenerate the **NavMeshSurface**.

Run the game and click to move the Sphere-arrow GameObject near the edge of the Cube-mud area. Now, click on the opposite side; you will see NavMeshAgent make the Sphere-arrow GameObject follow the lowest-cost path around the edge of the Cube-mud area. We are doing this rather than following a direct-line (as the crow flies) path as the mud area (higher cost) would take longer to traverse:

Figure 12.12: Path of the Sphere-arrow GameObject bypassing the high-cost mud area

Improving the UX by updating a "gaze" cursor in each frame

It's nice to know what our destination will be set to **before** we click the mouse. So, let's add a yellow sphere to show the "candidate" destination for where our raycast is hitting a surface, updating each frame as we move the mouse.

So, we need to create a second yellow sphere. We also need to create a layer for our raycast to ignore. Without such a layer, if we move the yellow sphere to the point where a raycast hits a surface, then in the next frame, our raycast will hit the surface of our yellow sphere – moving it closer and closer to us with each frame!

To improve the UX by updating a "gaze" cursor in each frame, do the following:

1. Create a new yellow **Material** object named m_yellow.

2. Create a second 3D sphere named `Sphere-destination-candidate` that's textured with `m_yellow`.

3. Create a new layer called `UISpheres`.

> It is very important to precisely spell the layer name `UISpheres`. This is because our script will be telling Unity's raycast to ignore objects on the layer named `UISpheres` – so if the spelling (and capitalization) don't exactly match, our spheres won't be ignored.

4. Set this layer for both the `Sphere-destination` and `Sphere-destination-candidate` GameObjects as `LayerUISpheres`.

5. Modify the `MoveToClickPoint` C# script class as follows to add a new public variable called `sphereDestinationCandidate`:

```
public class MoveToClickPoint : MonoBehaviour {
public GameObject sphereDestination;
public GameObject sphereDestinationCandidate;
```

6. Modify the `MoveToClickPoint` C# script class as follows to add an `else` clause to the logic in the `ProcessRayHit(...)` method so that if the mouse is not clicked, the yellow `sphereDestinationCandidate` object is moved to where the raycast hits a surface:

```
private void ProcessRayHit(Vector3 rayPoint) {
   if(Input.GetMouseButtonDown(0)) {
      navMeshAgent.destination = rayPoint;
      sphereDestination.transform.position = rayPoint;
   } else {
      sphereDestinationCandidate.transform.position = rayPoint;
   }
}
```

7. Modify the `MoveToClickPoint` C# script class as follows so that a `LayerMask` is created to ignore the `UISpheres` layer and to pass it as a parameter when `Physics.Raycast(...)` is invoked:

```
private bool FireRayCast(Ray rayFromMousePoint) {
   LayerMask layerMask = ~LayerMask.GetMask("UISpheres");
   return Physics.Raycast(rayFromMousePoint, out hit, 100, layerMask.
value);
}
```

8. Select the `Sphere-arrow` GameObject in the **Hierarchy** panel and, in the **Inspector** panel for the `MoveToClickPoint` (Script) component, populate the `Sphere Destination Candidate` public property with your yellow `Sphere-destination-candidate` GameObject.

9. Run the scene. You should now be able to click anywhere on the terrain to set the destination of the NavMeshAgent-controlled `Sphere-arrow` GameObject. As you click, the red `Sphere-destination` GameObject should be positioned at this new destination point, toward which the `Sphere-arrow` GameObject will navigate.

> We have set a `LayerMask` using the `~LayerMask.GetMask("UISpheres")` statement, which means every layer apart from the named one. This is passed to the `Raycast(...)` method so that our red and yellow spheres are ignored when casting the ray and looking to see which surface the ray hits first.

NPC NavMeshAgent to follow waypoints in sequence

Waypoints are often used as guides to help autonomously moving NPCs and enemies follow a path in a general way but still be able to respond with other directional behaviors, such as fleeing or sensing if friends/predators/prey are nearby. The waypoints are arranged in a sequence so that when the character reaches or gets close to a waypoint, it will select the next waypoint in the sequence as the target location to move toward. This recipe will demonstrate an arrow object moving toward a waypoint. Then, when it gets close enough, it will choose the next waypoint in the sequence as the new target destination. When the last waypoint has been reached, it will start heading toward the first waypoint once more.

Since Unity's NavMeshAgent has simplified coding NPC behavior, our work in this recipe basically becomes finding the position of the next waypoint and then telling NavMeshAgent that this waypoint is its new destination:

Figure 12.13: Example with six waypoints along a "yellow brick road"

Getting ready

This recipe adds to the first recipe in this chapter, so make a copy of that project folder and do your work for this recipe with that copy.

For this recipe, we have prepared the yellow brick texture image that you need in a folder named Textures in the 12_06 folder.

How to do it...

To instruct an object to follow a sequence of waypoints, follow these steps:

1. Replace the contents of the ArrowNPCMovement C# script class with the following:

    ```
    using UnityEngine;
    using UnityEngine.AI;

    public class ArrowNPCMovement : MonoBehaviour {
      private GameObject targetGo = null;
      private WaypointManager waypointManager;
      private NavMeshAgent navMeshAgent;

      void Start () {
        navMeshAgent = GetComponent<NavMeshAgent>();
        waypointManager = GetComponent<WaypointManager>();
        HeadForNextWayPoint();
      }

      void Update () {
        float closeToDestinaton = navMeshAgent.stoppingDistance * 2;
        if (navMeshAgent.remainingDistance < closeToDestinaton) {
          HeadForNextWayPoint ();
        }
      }

      private void HeadForNextWayPoint () {
        targetGo = waypointManager.NextWaypoint (targetGo);
        navMeshAgent.SetDestination (targetGo.transform.position);
      }
    }
    ```

2. Delete the 3D capsule Capsule-destination and the Cube-wall objects and regenerate the **NavMeshSurface** in the terrain.

3. Create a new 3D Capsule object named Capsule-waypoint-0 at (-12, 0, 8).

4. Copy Capsule-waypoint-0, name the copy Capsule-waypoint-3, and position this copy at (8, 0, -8).

> We are going to add some intermediate waypoints numbered 1 and 2 later on. This is why our second waypoint here is numbered 3, in case you were wondering.

5. Create the WaypointManager C# script class with the following contents and add an instance object as a component to the Sphere-arrow GameObject:

```
using UnityEngine;

public class WaypointManager : MonoBehaviour {
  public GameObject wayPoint0;
  public GameObject wayPoint3;

  public GameObject NextWaypoint(GameObject current) {
    if(current == wayPoint0)
        return wayPoint3;

    return wayPoint0;
  }
}
```

6. Ensure that Sphere-arrow is selected in the **Inspector** for the WaypointManager scripted component. Drag Capsule-waypoint-0 and Capsule-waypoint-3 over the public variable projectiles called WayPoint 0 and WayPoint 3, respectively.

7. Now, run your game. The arrow object will move toward one of the waypoint capsules. Then, when it gets close to it, it will slow down, turn around, head toward the other waypoint capsule, and keep doing that continuously.

How it works...

The NavMeshAgent component that we added to the Sphere-arrow GameObject does most of the work for us. NavMeshAgent needs two things:

*   A destination location to head toward
*   A NavMeshSurface so that it can plan a path and avoid obstacles

We created two possible waypoints as the locations for our NPC to move toward – that is, Capsule-waypoint-0 and Capsule-waypoint-3.

The C# script class called WaypointManager has one job: to return a reference to the next waypoint that our NPC should head toward. There are two variables, wayPoint0 and wayPoint3, that reference the two waypoint GameObjects in our scene. The NextWaypoint(...) method takes a single parameter named current, which is a reference to the current waypoint that the object is moving toward (or null).

This method's task is to return a reference to the next waypoint that the NPC should travel toward. The logic for this method is simple: if current refers to waypoint0, then we'll return waypoint3; otherwise, we'll return waypoint0. Note that if we pass this method as null, then we'll get waypoint0 back (this is our default first waypoint).

The ArrowNPCMovement C# script class has three variables. One is a reference to the destination GameObject named Target Go. The second is a reference to the NavMeshAgent component of the GameObject in which our instance of the class called ArrowNPCMovement is also a component. The third variable, called waypointManager, is a reference to the sibling scripted component, which is an instance of our WaypointManager script class.

When the scene starts via the Start() method, the NavMeshAgent and WaypointManager sibling components are found, and the HeadForDestination() method is called.

The HeadForDestination() method sets the variable called targetGO to refer to the GameObject that is returned by a call to NextWaypoint(...) of the scripted component called WaypointManager (that is, targetGo is set to refer to either Capsule-waypoint-0 or Capsule-waypoint-3). Next, it instructs NavMeshAgent to make its destination the position of the targetGO GameObject.

Each frame method called Update() is called. A test is performed to see whether the distance from the NPC arrow object is close to the destination waypoint. If the distance is smaller than twice the stopping distance that was set in our NavMeshAgent, then a call is made to WaypointManager.NextWaypoint(...) to update our target destination to be the next waypoint in the sequence.

There's more...

Here are some details that you won't want to miss.

Working with arrays of waypoints

Having a separate WaypointManager C# script class to simply swap between Capsule-waypoint-0 and Capsule-waypoint-3 may have seemed to be a bit heavy-duty and a case of over-engineering, but this was actually a very good move. An instance object of the WaypointManager script class has the job of returning the next waypoint. It is now very straightforward to add the more sophisticated approach of having an array of waypoints, without having to change any code in the ArrowNPCMovement C# script class. We can choose a random waypoint to be the next destination; for example, see the *Choosing destinations – finding the nearest/a random spawn point* recipe in *Chapter 11, Controlling and Choosing Positions*. Or, we can have an array of waypoints and choose the next one in the sequence.

To improve our game so that it works with an array of waypoints to be followed in sequence, we need to do the following:

1. Make four more copies of Capsule-waypoint-0 (named Capsule-waypoint-1, 2, 4, and 5) and position them as follows:

   - Capsule-waypoint-1: **Position** = (-2, 0, 8)
   - Capsule-waypoint-2: **Position** = (8, 0, 8)

- Capsule-waypoint-4: **Position** = (-2, 0, -8)
- Capsule-waypoint-5: **Position** = (-12, 0, -8)

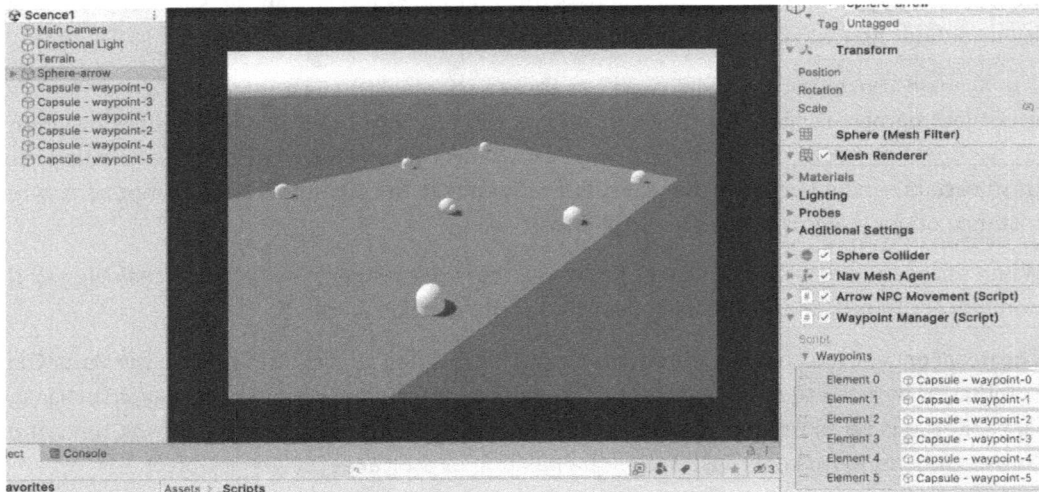

Figure 12.14: Position of the waypoints to create a rectangular path

2. Replace the WaypointManager C# script class with the following code:

```
using UnityEngine;
using System;

public class WaypointManager : MonoBehaviour {
    public GameObject[] waypoints;

    public GameObject NextWaypoint (GameObject current) {
        if( waypoints.Length < 1)
            Debug.LogError ("WaypointManager:: ERROR - no waypoints have
been added to array!");

        int currentIndex = Array.IndexOf(waypoints, current);
        int nextIndex = (currentIndex + 1) % waypoints.Length;

        return waypoints[nextIndex];
    }
}
```

3. Ensure that Sphere-arrow is selected. In the **Inspector** panel for the WaypointManager script-ed component, set the size of the Waypoints array to 6. Now, drag in all six capsule waypoint objects called Capsule-waypoint-0/1/2/3/4/5.

4. Run the game. Now, the `Sphere-arrow` GameObject will move toward waypoint 0 (top left), and then follow the sequence around the terrain.

5. Finally, you can make it look as if the `Sphere` GameObject is following a yellow brick road. Import the provided yellow brick texture, add this to your terrain, and paint the texture to create an oval-shaped path between the waypoints. You may also uncheck the **Mesh Renderer** component for each waypoint capsule so that the user does not see any of the waypoints, just the arrow object following the yellow brick road.

In the `NextWaypoint(...)` method, we check whether the array is empty, in which case an error is logged. Next, the array index for the current waypoint GameObject is found (if present in the array). Finally, the array index for the next waypoint is calculated using a modulus operator to support a cyclic sequence, returning to the beginning of the array after the last element has been visited.

Increased flexibility with the WayPoint class

Rather than forcing a GameObject to follow a single rigid sequence of locations, we can make things more flexible by defining a `WayPoint` class where each `waypoint` GameObject has an array of possible destinations, and each of these has its own array. In this way, a **directed graph (digraph)** can be implemented, of which a linear sequence is just one possible instance. Such a graph means that at each location, there are one or more links to other points – so starting from one position, many different paths can be followed – each being a sequence of connected locations.

To improve our game and make it work with a digraph of waypoints, do the following:

1. Remove the scripted `WayPointManager` component from the `Sphere-arrow` GameObject.

2. Replace the `ArrowNPCMovement` C# script class with the following code:

```
using UnityEngine;
using System.Collections;

public class ArrowNPCMovement : MonoBehaviour {
    public MyWayPoint myWaypoint;
    private bool firstWayPoint = true;
    private UnityEngine.AI.NavMeshAgent navMeshAgent;

    void Start (){
        navMeshAgent = GetComponent<UnityEngine.AI.NavMeshAgent>();
        HeadForNextWayPoint();
    }

    void Update () {
        float closeToDestinaton = navMeshAgent.stoppingDistance * 2;
```

```
              if (navMeshAgent.remainingDistance < closeToDestinaton){
                  HeadForNextWayPoint ();
              }
         }

         private void HeadForNextWayPoint (){
             if(firstWayPoint)
                 firstWayPoint = false;
             else
                 myWaypoint = myWaypoint.GetNextWaypoint();

             Vector3 target = myWaypoint.transform.position;
             navMeshAgent.SetDestination (target);
         }
    }
```

3. Create a new `MyWayPoint` C# script class containing the following code:

```
using UnityEngine;
using System.Collections;

public class MyWayPoint: MonoBehaviour {
    public MyWayPoint[] waypoints;

    public MyWayPoint GetNextWaypoint () {
        return waypoints[ Random.Range(0, waypoints.Length) ];
    }
}
```

4. Select all six GameObjects called `Capsule-waypoint -0/1/2/3/4/5` and add an instance object component of the `MyWayPoint` C# class to them.

5. Select the `Sphere-arrow` GameObject and add to it an instance object component of the `MyWayPoint` C# class.

6. Ensure that the `Sphere-arrow` GameObject is selected. In the **Inspector** panel for the `ArrowNPCMovement` scripted component, drag `Capsule-waypoint-0` into the `MyWayPoint` public variable slot.

7. Now, to make the `Sphere-arrow` object go to a random waypoint after visiting `Capsule-waypoint-0`, increase the size of the `Waypoints` array to 5 in the `Sphere-arrow` object.

Then drag the remaining `Capsule-waypoint` to the array elements.

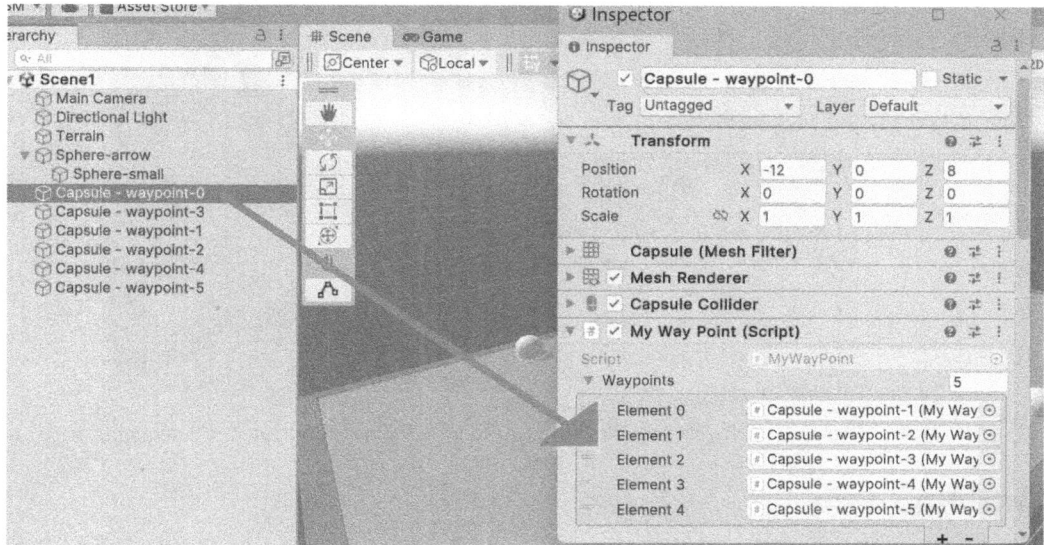

Figure 12.15: Increasing the size of the Waypoints array and adding elements

8. For the remaining `Capsule-waypoint` objects, increase the array size to 1. So, select `Capsule-waypoint-1`, increase the `Waypoints` array size to 1, and drag in `Capsule-waypoint-2`. Next, select `Capsule-waypoint-2`, set its `Waypoints` array size to 1, and drag in `Capsule-waypoint-3`. Continue in this way until you finally link `Capsule-waypoint-5` back to `Capsule-waypoint-0`.

You now have a much more flexible game architecture, allowing GameObjects to randomly select one of several different paths at each waypoint that's reached. In this recipe variation, we implemented a waypoint sequence, since each waypoint has an array of just one linked waypoint.

Creating a movable NavMesh Obstacle

Sometimes, we want a moving object to slow down or to prevent an AI NavMeshAgent-controlled character from passing through an area of our game. Or, perhaps we want something such as a door or drawbridge to sometimes permit travel, and not at other times. We can't "bake" these objects into the NavMesh at design time since we want to change them during runtime.

While computationally more expensive (that is, they slow down your game more than static, non-navigable objects), NavMesh Obstacles are components that can be added to GameObjects, and these components can be enabled and disabled like any other component.

A special property of NavMesh Obstacles is that they can be set to "carve out" areas of the NavMesh, causing NavMeshAgents to then recalculate routes that avoid these carved-out parts of the mesh.

In this recipe, you'll create a player-controlled red cube that you can move to obstruct an AI NavMeshAgent-controlled character.

Also, if your cube stays in one place for half a second or longer, it will carve out part of the NavMesh around it, causing the NavMeshAgent to stop bumping into the obstacle and calculate and follow a path that avoids it:

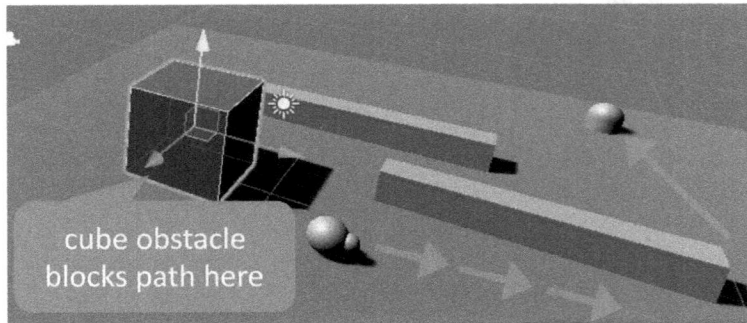

Figure 12.16: Example of the red cube blocking the path of the AI NavMeshAgent

Getting ready

This recipe adds to the first recipe in this chapter, so make a copy of that project folder and do your work for this recipe with that copy.

The required script to control the movement of the red cube (PlayerControl.cs) is provided in the 12_07 folder.

How to do it...

To create a movable NavMesh Obstacle, please follow these steps:

1. Create a **Material** object in the **Project** panel and name it m_red, with the **Color** set to red.

2. Create a red 3D cube for the player to control called Cube-player, making it red by adding the m_red material to it and making it large by setting its scale to (-4,1,3).

3. Add an instance of the provided PlayerControl C# script class as a component to the Cube-player GameObject.

4. In the **Inspector** panel, add a **Navigation | NavMesh Obstacle** component to Cube-player and check its **Carve** property.

   Run the game. You can move the player-controlled red cube so that it gets in the way of the moving Sphere-arrow GameObject.

5. When 0.5 seconds have elapsed since the obstacle (Cube-player) has become stationary, a portion of the NavMesh encompassing the area occupied by the Cube-player and its immediate surroundings is removed. Then, the Sphere-arrow GameObject will recalculate a new route, avoiding the carved-out area where Cube-player is located.

How it works...

At runtime, the AI NavMeshAgent-controlled Sphere-arrow GameObject heads toward the destination point but stops when the player-controlled red cube is in its way. Once the cube is stationary for 0.5 seconds or more, the NavMesh is carved out so that the AI NavMeshAgent-controlled Sphere-arrow GameObject no longer even attempts to plan a path through the space occupied by the cube. It then calculates a new path to completely avoid the obstacle, even if it means backtracking and heading away from the target for part of its path.

Joining several NavMeshes with a single NavMeshSurface

We have a single **NavMeshSurface** component in our terrain. However, what a **NavMeshSurface** does is look for navigable meshes in the scene with a similar orientation – that means close to horizontal for the orientation of the GameObject that has a **NavMeshSurface** component. In this recipe, we'll first add a large cube to the scene, and then rebake the NavMesh to see how the top of the cube gets its own NavMesh. We'll then tilt the cube (a little) to see how NavMeshes can join up if the gap and slope between them are not too great:

Figure 12.17: Joined NavMeshes allowing travel up the slope

Getting ready

This recipe adds to the first recipe in this chapter, so make a copy of that project folder and do your work for this recipe with that copy.

How to do it...

To work with non-horizontal NavMeshes, follow these steps:

1. Add a new 3D cube to the scene, named Cube-slope. Scale this to (12, 12, 12), and position it at (0, 0, -17). Check the **Static** option for this GameObject in the **Inspector**.

2. Move the position of the Capsule-destination GameObject to (0, 10, -17). You should see the destination capsule floating above our new cube.

3. Select the **Terrain** in the **Hierarchy**, and click the **Bake** button for its **Nav Mesh Surface** component. You should now see a second NavMesh on top of the cube.

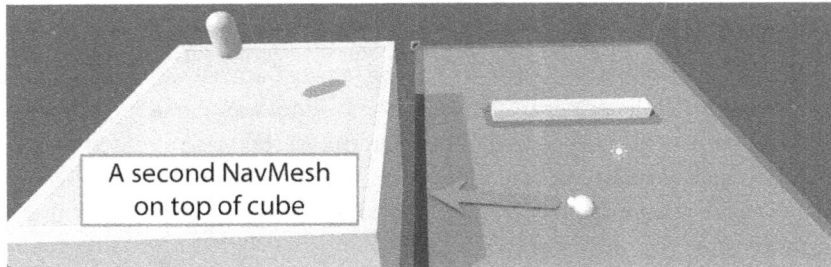

A second NavMesh
on top of cube

Figure 12.18: Second NavMesh on top of the new cube

4. Save and run the scene. You'll see the `Sphere-arrow` GameObject move to the edge of the terrain to get as close as it can to the `Capsule-destination` GameObject.

5. Now set the **Rotation** of `Cube-slope` to (40, 0, 0).

6. Select the **Terrain** in the **Hierarchy**, and click the **Bake** button for its **Nav Mesh Surface** component. You should now see that the two NavMeshes are joined.

7. Save and run the scene. You'll see the `Sphere-arrow` GameObject move to the edge of the terrain and then travel up the slope to arrive directly underneath the `Capsule-destination` GameObject.

How it works...

Before rotating the cube, the top of the cube was horizontal, just like our terrain. So the NavMesh-Surface component in the terrain generated (baked) two NavMeshes – one on the terrain and one on the top of the cube. However, there was no way for our agent to travel from the `Terrain` NavMesh to the one on the top of the cube, even though that one was closer to the destination capsule. When we rotated the cube, its NavMesh was close enough to the Terrain NavMesh to be automatically joined – allowing our agent to travel across the terrain, and then up the slope of the cube to arrive very close to the destination capsule. This was possible since the angle of the slope of the top of the cube was closer to the horizontal of the terrain than any other side of the cube, and the NavMesh of the cube ended up very close to the terrain.

Sometimes, the slopes of the surfaces that we want to have NavMeshes on can be very different, in which case, we need to add additional NavMeshSurface components to objects in the scene. Other times, the gaps are too far between NavMeshes for them to be automatically joined, but we can use NavMeshLinks to create a pathway for an agent to travel from one NavMesh to another – we'll explore both of these approaches in the next recipe.

Non-horizontal NavMeshes with multiple NavMeshSurfaces and NavMeshLinks

NavMeshes don't have to be horizontal (or a slope close to horizontal). Let's investigate this with NavMeshes on vertical walls and ceilings! We'll also link different NavMeshes with NavMeshLinks, so our agents can travel through all NavMeshes in a scene.

Getting ready

This recipe adds to the previous one, so make a copy of that project folder and do your work for this recipe with that copy.

How to do it...

To work with non-horizontal NavMeshes, follow these steps:

1. Add a new 3D cube to the scene, named Cube-vertical, at **Position** (0, 0, 16) and **Scale** (12, 12, 12). Check the **Static** option for this GameObject in the **Inspector**.

2. Move the position of the Capsule-destination GameObject to (0,10,15). You should see the destination capsule floating above our new cube.

3. Select the Terrain in the **Hierarchy**, and click the **Bake** button for its **Nav Mesh Surface** component. You should now see a third NavMesh on top of the Cube-vertical GameObject. However, we wish to have a NavMesh vertically up the side of this new cube Game Object.

4. Select the Cube-vertical GameObject and, in the **Inspector**, use the **Add Component** button to add a **NavMeshSurface** component to this GameObject. Now set the **Rotation** of this Game-Object to (-90, 0, 0). Then click the **Bake** button for its **NavMeshSurface** component.

5. You should now see that Cube-vertical has a NavMesh on top of it and also a new NavMesh vertically up its side:

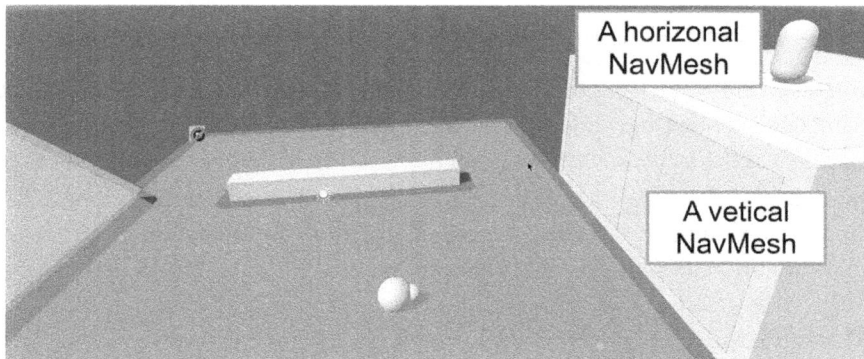

Figure 12.19: The vertical NavMesh as well as the horizontal one on top of the cube

6. Save and run the scene. You'll see the Sphere-arrow GameObject move to the edge of the terrain to get as close as it can to the Capsule-destination GameObject. However, at present, it cannot move from the horizontal NavMesh on the terrain to the vertical NavMesh on the new cube.

7. Let's fix this problem by adding a **NavMesh Link** to the scene. In the **Hierarchy** panel, create a new GameObject containing a **NavMesh Link** by choosing the menu options **GameObject | AI | NavMesh Link**. This will add a new GameObject in the **Hierarchy** named NavMeshLink.

8. A **NavMesh Link** GameObject has two small yellow cubes, representing the endpoints of the link. Position one of these near the edge of the horizontal NavMesh on the terrain, and the second one near the edge of the vertical NavMesh on the Cube-vertical GameObject.

When a NavMesh Link endpoint cube is close enough to a NavMesh to be used by an agent, a line will be displayed from the endpoint cube to the NavMesh (as well as a circle around the cube). Note – it can be a bit fiddly when you first start trying to position the NavMesh Link endpoint:

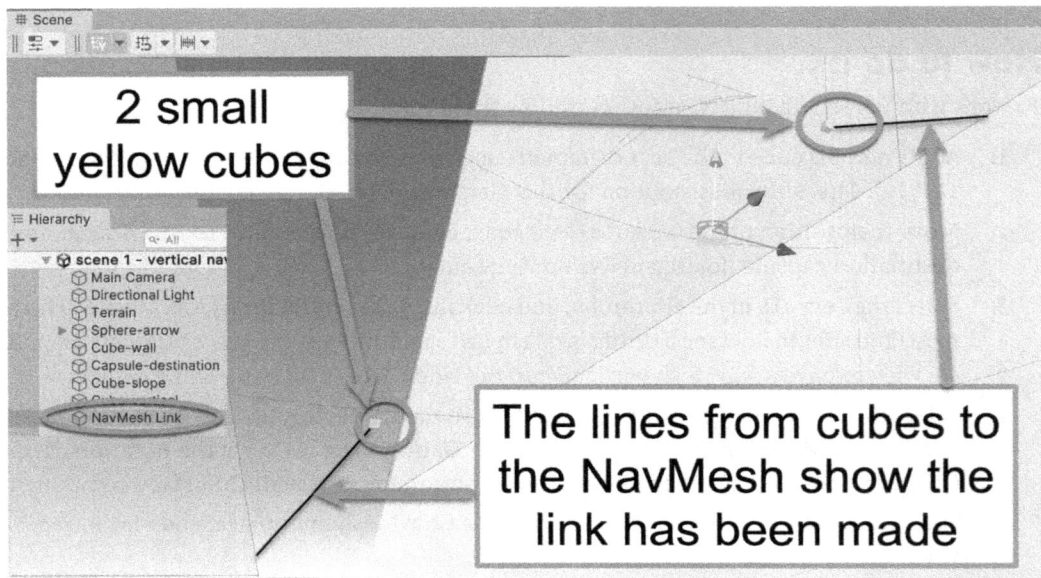

Figure 12.20: NavMesh Link endpoints joining two NavMeshes

9. Add a second **NavMesh Link** to the scene to join the two NavMeshes on the `Cube-vertical` GameObject. Move the endpoints of this second **NavMesh Link** to create a path from the top of the vertical NavMesh to the horizontal NavMesh on the top of the `Cube-vertical` GameObject.

10. Save and run the scene. You should see the `Sphere-arrow` GameObject move across the terrain, then up the side of the vertical NavMesh, and then on to the top of the `Sphere-arrow` Game-Object to arrive at `Capsule-destination`.

How it works...

Each **NavMeshSurface** component enables the Unity AI system to create NavMeshes from static objects in the scene whose UP direction is similar to the GameObject containing the **NavMeshSurface** component. So when we added cubes to the scene and rebaked the **NavMeshSurface** component on the terrain, we saw additional NavMeshes on the horizontal (or near-horizontal) surfaces of the cube. A single **NavMeshSurface** component cannot be used to create NavMeshes on faces that have very different relative orientations, such as vertical, or a ceiling pointing downward. To do this we need to add **NavMeshSurface** components on objects with orientations similar to those where we want the NavMesh – in many cases, we can simply add a **NavMeshSurface** component to the object that has been oriented for the desired NavMesh, such as our `Cube-vertical` GameObject.

When two NavMeshes have a similar orientation and overlap or have very close edges, the NavMeshes are joined automatically so an agent can travel from one NavMesh to another. We saw this when we positioned the `Capsule-destination` GameObject above `Cube-slope` in the previous recipe.

However, when there is some distance between NavMeshes, or their orientations are very different (such as our `Terrain` and `Cube-vertical` GameObject) we must explicitly add NavMeshLinks to the scene, to enable a NavMesh Agent to be able to travel from one NavMesh to another.

Further reading

The following are sources that provide further information about Unity and AI navigation. Some NavMesh features (such as NavMesh Links and dynamic mesh baking at runtime) are not part of the standard Unity installation and require additional installation. You can learn more about these components, their APIs, and how to install them here: `https://docs.unity3d.com/Packages/com.unity.ai.navigation@2.0/manual/`.

The Unity Technologies NavMesh tutorial can be found here: `https://learn.unity.com/tutorial/from-waypoints-to-navmesh`.

While the Unity development community has been asking for 2D NavMeshes for some years now, they've not yet been released as a core feature. There is a lot of online information about how to write your own pathfinding system that would work in 2D. A working project (last checked in November 2023) that extends the Unity navigation components can be found here:

- `https://github.com/h8man/RedHotSweetPepper`

13

Cameras, Lighting, and Visual Effects

Whether you're trying to make a better-looking game or you want to add interesting features, visual effects can add enjoyment and a more professional look and feel to many games:

Figure 13.1: Multiple light sources in Coco VR by Pixar and Magnopus (featured on the Unity website Made with Unity)

At the end of the day, what we can see is the color properties of the rectangular array of pixels (picture elements) that make up the screen (or screens, for VR) that we are looking at. The Unity game engine must use the GameObjects in a scene and their properties to decide what to **render** (draw) in the **Game** window. Some GameObjects model 3D objects, with properties relating to physical **materials**, and **renderer** components for configuring how they should be displayed. Other GameObjects simulate **light** sources.

Modern computers with powerful CPUs and GPUs are beginning to support **real-time raytracing**. Raytracing involves calculating how light from light-emitting GameObjects (lights, **emissive materials**, and so on) bounces from surface to surface and eventually ends up arriving at the location in the scene where the Camera is located.

These different lights can be customized to affect the type of light and directions where light is generated in the scene. The different materials will affect in what directions the light will bounce, and how much of the light hitting a surface of a simulated 3D object will bounce off it – plus, some GameObjects can emit light themselves with **emissive materials**. The different rendering features affect how GameObjects and their materials are drawn for the user to see. There are several techniques that have been developed to reduce the amount of computation required while still producing high-quality, if not quite photorealistic, rendering for 3D scenes. One of these techniques is the **pre-baking** (pre-calculation) of light sources and surface interactions. A second technique is the use of simulated **particle systems**, which can be customized through parameter settings and image textures to give the impression of a wide range of visual effects, including flames, smoke, rain, lightning, and so on.

Unity offers four types of lights (all of which can be added as components of GameObjects in a scene):

- **Directional Light:** For simulating a distance light source, such as sunlight.
- **SpotLight:** A simulated spotlight for projecting light in one direction, increasing in size the further it is from the light, in a cone shape.
- **Point Light:** Similar to a real-world light bulb, which means it emits light in all directions.
- **Area Light:** A processor-intensive, subtle lighting source for rectangular areas of a Scene (pre-baking only, not at **runtime**).

Each light can be customized in many ways, such as its range and color. The intensity of the lights can also be customized. For the **SpotLight** and **Point Light** components, the intensity is reduced based on their distance from the location of the GameObject.

Lights can have a cookie **Texture** applied to them. Cookies are textures that are used to cast shadows or silhouettes in a scene. They are produced using the cookie **Texture** as a mask between the light source and the surfaces being rendered. Their name, and usage, comes from the use of physical devices called cucoloris (nicknamed cookies), which are used in theatre and movie production to produce shadow effects for environments, such as moving clouds, the bars of a prison window, or sunlight being broken up by a jungle leaf canopy.

In addition to the **Light** components of GameObjects, a second source of local lighting in a scene can be from stationary objects that have **emissive materials**. We'll explore this in the final recipe in this chapter:

Figure 13.2: Emissive material in Osiris: New Dawn by Fenix Fire (featured on the Unity website Made with Unity)

The use of processing-intensive **real-time** lighting can be reduced by pre-computing lighting for each scene. This is known as **lightmap baking**. Static – immovable – parts of the scene (lights and other objects) can have their lighting "baked" (pre-computed) into a **Texture Map**, based on the light sources in the scene. Then, at runtime, game performance is improved and the pre-calculated lightmaps can be used, avoiding the lighting of each frame having to be calculated at runtime. As always with computing, there is a memory-versus-speed trade-off, so more memory is required for scenes to store the pre-computations as lightmaps. Unity currently offers the **Progressive CPU** lightmapper (and at the time of writing, a preview **Progressive GPU** lightmapper).

Another source of scene lighting is **ambient lighting** (global environmental lighting). This doesn't come from any locational source as it exists evenly throughout the scene. Ambient light can be used to influence the overall brightness of a scene and is achieved typically through the use of **skybox** materials and simulated sunlight generated from a **Directional Light**. Skyboxes can be defined in the **Lighting settings** window. Such lighting can be generated at runtime or (more efficiently) pre-calculated as a baked **lightmap** before the scene is run:

Figure 13.3: Ambient lightning in D.R.O.N.E. by Five Studios Interactive (featured on the Unity website Made with Unity)

Unity offers two windows for managing lighting in a scene: the **Lighting settings** window and the **Light Explorer** window. The **Lighting** window (accessed via the **Window | Rendering | Lighting Settings** menu) is the hub for setting and adjusting the scene's illumination features, such as **Lightmaps, Global Illumination, Fog,** and much more:

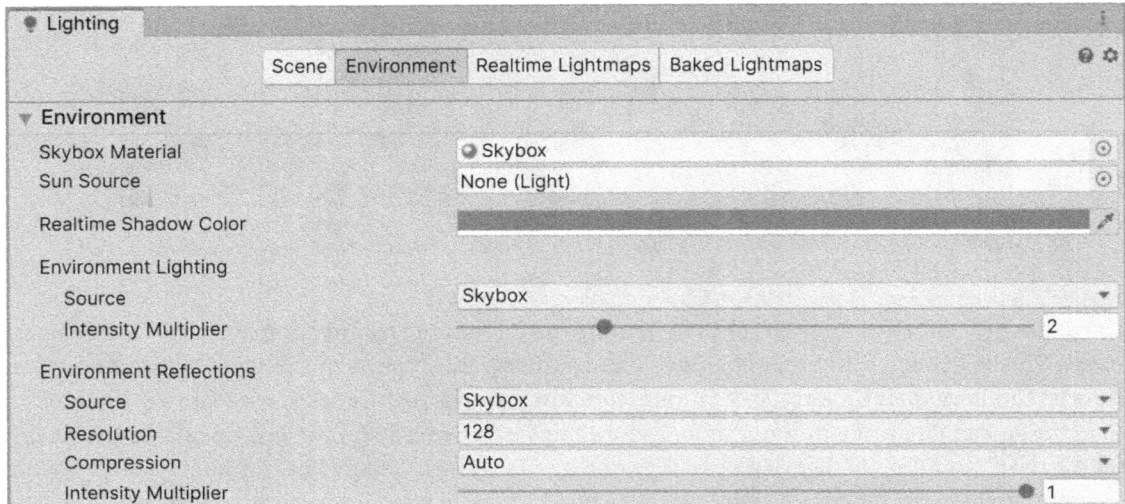

Figure 13.4: The Lighting window for adjusting the Scene's illumination

Then, there's the **Light Explorer** window, used for working with **Lights** in a scene. This panel allows you to edit and view the properties of **all** of the Lights in the current scene. The **Light Explorer** window lists all **Lights** in a single panel, making it easy to work with each individually or change the settings of several at the same time. It can be a great time-saving tool when working with **scenes** involving lots of **Light** GameObjects. To display the **Light Explorer** window, go to **Window | Rendering | Light Explorer**:

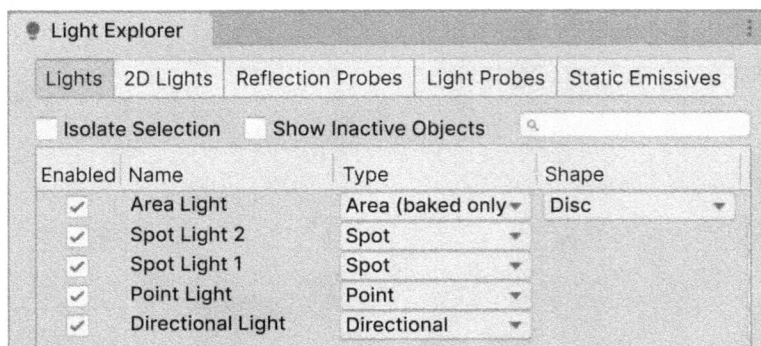

Figure 13.5: The Light Explorer window

We should always pay attention to **Cameras**. They are the windows through which our players see our games. In this chapter, we will explore a range of methods for using **Cameras** to enhance a player's experience.

A scene can contain multiple cameras. Often, we have one **Main Camera** (we're often given one by default with a new **scene**). For **first-person** viewpoint games, we control the position and rotation of the **Camera** directly, since it acts as our eyes. In **third-person** viewpoint games, our main camera follows an animated 3D character (usually from above, behind, or over the shoulder). It can slowly and smoothly change its position and rotation as if a person were holding the camera and moving to keep us in view.

Perspective Cameras have a pyramid-shaped volume of space in front of them, called a *frustrum*. Objects inside this space are projected onto a plane, which determines what we see from the **Camera**. We can control this volume of space by specifying the clipping planes and the **Field of View**. The clipping planes define the minimum and maximum distances that objects have to be between in order to be considered viewable. The Field of View is decided by how wide or narrow the pyramid shape is, as shown in the following figure.

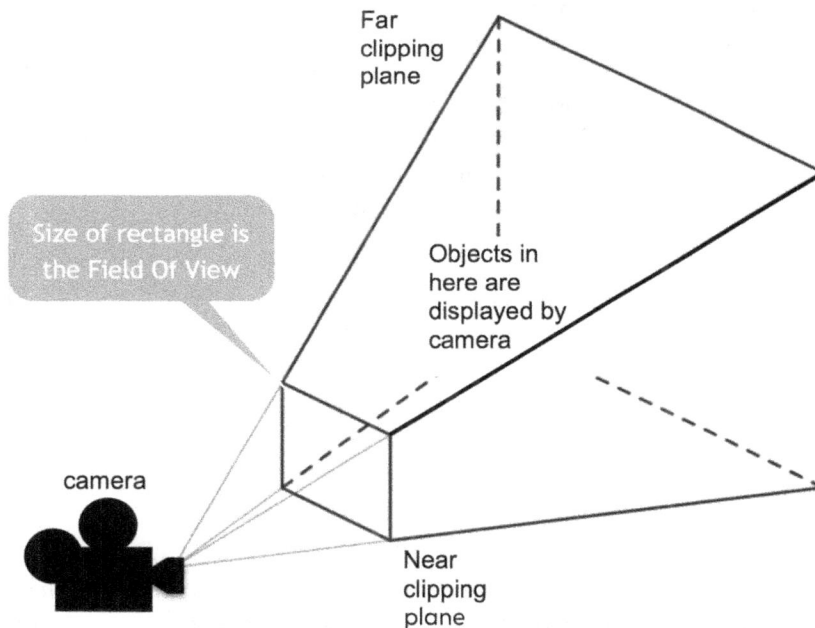

Figure 13.6: A diagram depicting the camera Field of View as a frustrum pyramid-shape space

Cameras can be customized in many ways, including the following:

- They can be set to render in **Orthographic** mode (that is, without **Perspective**).
- They have their **Field of View** manipulated to simulate a wide- or narrow-angle lens.
- They can be rendered on top of other Cameras or within specific rectangular sub-areas of the screen (viewports).
- They send their output to **RenderTexture** asset files (this output can potentially be used to texture other objects in the scene).
- They include/exclude objects on specific layers for/from rendering (using **Culling Masks**).

Cameras have a depth property. This is used by Unity to determine in what sequence **Cameras** are rendered. Unity renders **Cameras** starting with the lowest depth and working up to the highest. This is necessary to ensure that **Cameras** that are not rendered to fill the whole screen are rendered after **Cameras** that are.

Cinemachine is a powerful automated system for camera control. It was developed by Adam Myhill and is now a free Unity package. It offers much to Unity developers, both for **runtime** in-game camera control and in cinematic creation for cutscenes or creating fully animated films. In this chapter, we will present examples of how to add some runtime camera controls to your games using Cinemachine.

At the core of Cinemachine is the concept of a set of virtual cameras in a scene and a Cinemachine Brain component, which decides what virtual camera's properties should be used to control the scene's **Main Camera**.

Unity 2023 offers two main **render pipelines**. These are related to how Unity computes what we see, based on the visible objects in the scene, their materials, the lighting system, and more. For most games, especially those we wish to run efficiently on mobile devices, the default **Universal Render Pipeline (URP)** is the best choice. However, there is also a **High Definition Render Pipeline (HDRP)**, for projects that push the limits of what is now graphically possible on computers with powerful graphics cards, leaning towards photo-realistic games. The **HDRP** uses Cubemap assets that can be used for Skyboxes and other types of environmental visual scene effects.

Additionally, Unity offers many powerful postprocessing visual effects, such as *film grain*, *depth of field blurring*, *wet surface style reflections*, and more. A postprocessing **Volume** can be added to a scene, which will affect how the output of a camera will be rendered and seen by the user.

In this chapter, we will cover the following recipes:

- Creating the basic scene for this chapter
- Working with a fixed **Main Camera**
- Changing how much of the screen a Camera renders
- Adding a top-down orthographic minimap Camera
- Creating an in-game mirror using a RenderTexture to send Camera output to a Plane
- Saving screenshots and RenderTextures to files
- Using Cinemachine ClearShot to switch cameras to keep the player in shot
- A Camera to always look at and follow the Third Person Controller
- Adding film grain and vignette effects with URP postprocessing
- Creating an **HDRP** project with a **High Dynamic Range Imaging (HDRI)** skybox
- Creating and applying a cookie texture to a spotlight
- Baking light from an emissive material onto other scene **GameObjects**

Creating the basic scene for this chapter

All of the recipes in this chapter start off with the same basic scene, featuring a modern house layout (made with ProBuilder meshes), some objects, and a keyboard-controllable 3D character. In this recipe, you'll create a project with this scene that can be duplicated and adapted for each recipe that follows.

> **Note:** For the other recipes in this chapter we provide a completed version of this recipe named `BasicScene_completed.unitypackage`, which can be imported into a new project to save working through this recipe each time.

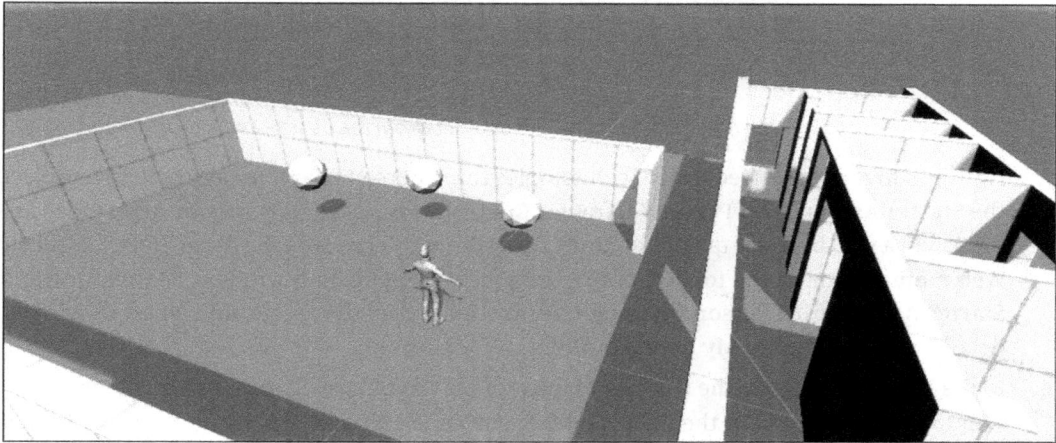

Figure 13.7: An example of the basic scene for this chapter

Getting ready

We have prepared a Unity package named `BasicSceneSetup.unitypackage` that contains all the resources for this recipe. The package can be found in the `13_01` folder.

How to do it...

To create the basic scene for this chapter, perform the following steps:

1. Create a new 3D (Core) project.
2. First, install the ProBuilder package. To do this, navigate to **Window | Package Manager**, set **Packages** to **Unity Registry**, search for `ProBuilder`, click on **ProBuilder** when it appears in the list, and finally, click on **Install**.
3. Now that ProBuilder is installed, you can successfully import the provided `BasicSceneSetup.unitypackage` package into your Unity project.
4. Open the `BasicScene` scene.

5. In the **Hierarchy** panel, select the modernHouse GameObject. Then, in the **Inspector** panel, ensure its position is (7, 0, 20).

6. You should see a house that includes red floors and white walls. Note that there is also a second directional light to reduce shadows.

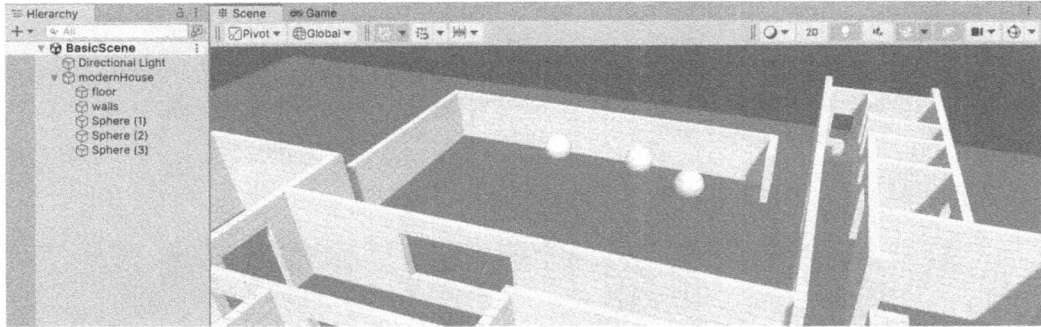

Figure 13.8: The modern house ProBuilder assets from the BasicSceneSetup package

7. Now let's first add the free Unity Technologies third-person character assets to this project. If the **Starter Assets Third Person Character Controller** is not already in your Asset Store assets, first open a web browser tab on the Unity Asset Store by choosing the **Asset Store | Asset Store Web** menu option at the top left of the Unity Editor application window. Search for the free **Starter Assets Third Person Character Controller**. Select the assets, and when viewing their details, click the **Add to My Assets** button.

8. In the Unity Editor, open the **Package Manager** to list your Asset Store assets by choosing menu: Window | Asset Store. In the Package Manager panel, choose My Assets from the Packages dropdown menu, then locate, download (if not done previously), and import the Start Assets Third Person Character Controller package into this project.

> **Note:** When importing the package, if asked, make sure to agree to additional Package Manager dependencies (click **Install/Upgrade**), and agree to enable the new input system backends and restart the Editor (click **Yes**). Note you may have to locate and click **Import** a second time after the Editor restarts.

9. You should now have a Starter Assets folder in your **Project** panel.

10. Let's now add a third-person controller to the scene from the third-person starter assets we imported earlier. In the **Project** panel, locate the Prefabs folder in StarterAssets | ThirdPersonController.

11. Drag the prefab named PlayerArmature from the **Project** panel into the scene. In the **Inspector** set the newly created PlayerArmature GameObject's position to (0, 1, 0).

12. Let's attach the scene's Main Camera to the character so that you'll be able to view this third-person controller character all the time as you move it around the house. Child the **Main Camera** to the PlayerArmature character GameObject, and in the **Inspector** panel, set its **Position** to (0, 2, -1) and **Rotation** to (20, 0, 0).

13. Save and run the scene.

14. As you use the arrow keys to move the character around the house, the Main Camera should move and rotate automatically with the character, and you should be able to see the back of the character at all times.

How it works...

The **BasicSceneSetup** Unity package contains a set of ProBuilder mesh objects for a red ground plane, and a white-brick style set of walls for a modern-style house. After first ensuring that the ProBuilder package was part of this project, you imported the package and loaded the pre-made scene containing these ProBuilder meshes.

Having added the **Starter Assets Third Person Character Controller** package to the project, you then extracted the PlayerArmature GameObject – so we have a keyboard-controlled character in the scene.

By childing the Main Camera to the character GameObject, the Main Camera maintains the same position and rotation relative to the character at all times. Therefore, as the character moves or rotates, so does the Main Camera, giving a simple, over-the-shoulder type of viewpoint for the game action.

Remember, because the Main Camera is now a child of the PlayerArmature character GameObject, so the **Transform** component's **Position** and **Rotation** values are relative to the **Position** and **Rotation** values of its parent. For example, the Main Camera position of (0, 2, -1) means that it is 2 Unity units above the PlayerArmature (Y = 3) and 1 Unity unit behind (Z = -1).

There's more...

There are some additional details that you don't want to miss.

Fixing pink textures

When importing assets into a project, sometimes there are issues when assets were set up for a different/older Render Pipeline. If you see any bright pink assets in your scenes, this indicates a rendering issue. However, Unity has made it very easy to automatically fix almost all the issues that arise. If you see lots of pink objects, such as when you import the Third Person prefab into the scene, then follow these steps to fix the problem:

1. Open the Render Pipeline Converter via the **Window** | **Rendering** | **Render Pipeline Converter** **menu**.

2. Check all four boxes and click the **Initialize And Convert** button.

3. Wait for about a minute – there's a progress bar at the bottom right of the footer of the application window.

4. That's it! Save and run the scene, and hopefully everything looks fine now!

Working with a fixed Main Camera

A quick way to begin creating a scene where the player controls a third-person character is to use a fixed **Main Camera** that faces the area the character will start moving around in.

We can position and orient the Camera to view most of the area that the player's character will be moving around in, and we can change the amount of what is in front of a camera, which the camera "sees," by changing its **Field of View**. In this recipe, we'll use the basic 3D scene from the previous recipe and work with the Main Camera so that we can see our character moving around. We'll delete the Main Camera from the existing scene and learn how to create a new Camera, which we will then turn into the Main Camera for a scene. Then, we'll work with the **Position, Rotation,** and **Field of View** properties so that the player can see much of the house scene from a single, fixed perspective.

In later recipes, we'll learn how to use more dynamic cameras with our third-person characters.

Getting ready

This recipe follows on from the previous one. So, either work from a copy of the previous recipe, or create a new 3D project and follow the steps in the previous recipe first.

How to do it...

To enable a fixed **Main Camera** to look at our third-person character, perform the following steps:

1. Open the **BasicScene** scene.

2. In the **Hierarchy** panel, delete the Main Camera **GameObject** – this is a child of the PlayerArmature GameObject.

3. Now, create a new **Camera** in the scene, naming it Main Camera. In the **Inspector** panel, give this GameObject the preset tag of Main Camera.

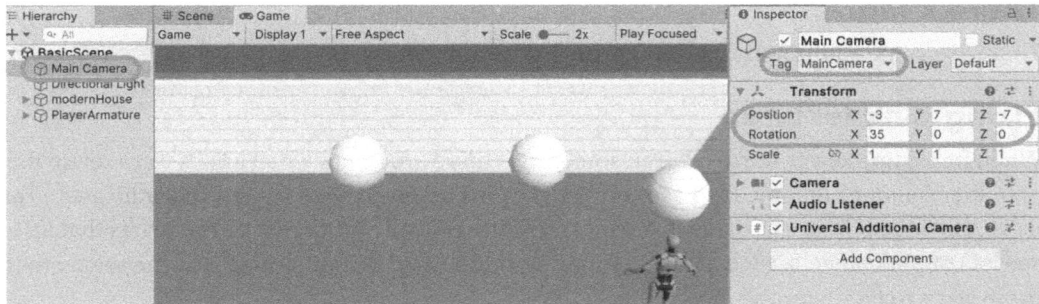

Figure 13.9: Tagging our new Camera as the Main Camera

4. Save and run the scene. You'll see the character move as you control the robot, but the Camera angle does not change.

5. In the **Inspector** panel, set the position of our new Main Camera to (-3, 7, -7). Additionally, set the rotation to (35, 0, 0). This positions the Camera above and behind the main room, in which the PlayerArmature character is located.

6. The default **Field Of View** for a new Camera is 60 degrees. However, we can change this to a larger value, for example, 120 degrees. This is so that our Camera can "see" much more of the scene. In the **Inspector** panel, change **Field of View** to 120, **Position** to (-3, 8, -6), and **Rotation** to (50, 0, 0).

When you run the scene, it should look something similar to the following diagram. Here, we can see our character move between rooms from this bird's-eye view Camera setup:

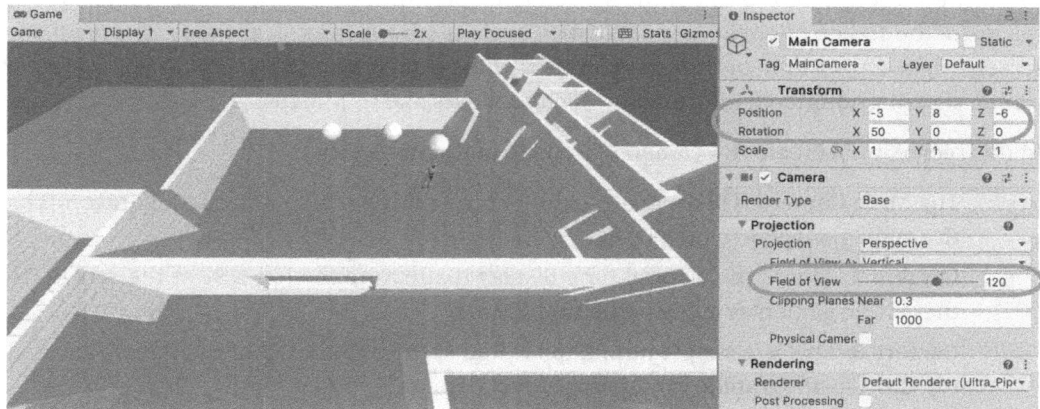

Figure 13.10: Using a larger Field of View and higher Camera for a bird's-eye view

7. Save and run the scene. You'll see the character move as you control the robot, but the Camera angle does not change. But this time you have a much wider viewing angle – although you'll see more perspective distortion due to this wide viewing angle.

How it works...

There is nothing special about a **Camera** that is the Main Camera – it's just a **Camera** that has been tagged MainCamera. It is important that the Main Camera of the scene is tagged with this special predefined tag.

> Some scripts explicitly search for the Main Camera of a scene using the Camera.main C# expression. So, we should always have the Main Camera in scenes using such scripts.
>
> You can learn more about this in the Unity manual at https://docs.unity3d.com/ScriptReference/Camera-main.html.

By changing the **Position** and **Rotation** values of this Camera, we positioned it so that the player can see most of the main room where the PlayerAvatar character starts off. This fixed Main Camera might be sufficient if all the action for a scene is inside a single room. However, by making the camera's **Field of View** larger, more of the scene can be seen in the camera's rendering to the **Game** panel. Additionally, by adjusting the **Position** and **Rotation** values of the Camera, many of the different rooms of the house in this scene can be seen from this single fixed Main Camera.

While it might not enough for a finished game, there are times when we might want the user to benefit from an unchanging camera view, perhaps even augmented with a minimap or a top-down view.

Changing how much of the screen a Camera renders

Often, we want different parts of the screen to display different things to the player. For example, the top of the screen might be game statistics, such as scores, lives, or pickups. The main area of the screen might show a first- or third-person view of the gameplay, and we might also have a part of the screen showing additional information, such as radars and minimaps.

In this recipe, we'll create two cameras to add to our player's view:

- The **Main Camera**: This is made a child of the third-person-controller character. This shows the main gameplay to the player.
- **Camera 2**: This is the elapsed time display that covers the top 15% of the screen. Note that there is no gameplay going on behind the text.
- **Camera 3**: This is a simple minimap located at the bottom left of the screen, created from a top-down **orthographic** (non-perspective) Camera:

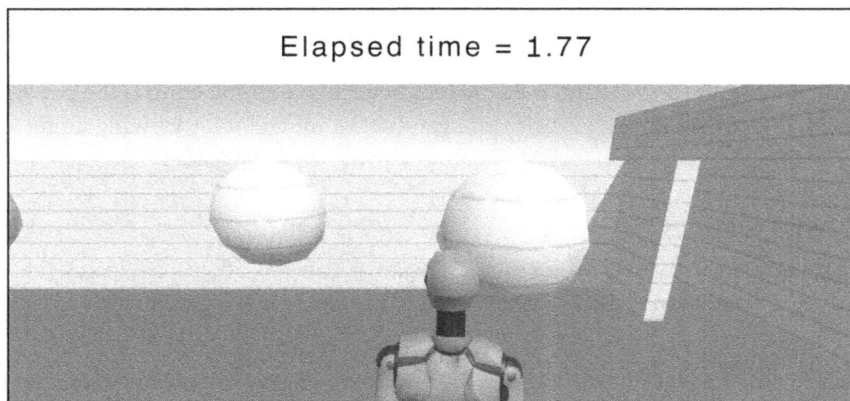

Figure 13.11: Cameras rendering to different parts of the screen

Getting ready

This recipe builds on the basic scene created in the first recipe of this chapter. So, make a copy of that to work on in this recipe.

How to do it...

To change how much of the screen a Camera renders, perform the following steps:

1. Open the **BasicScene** scene.

2. First, let's reduce the amount of the screen that the Main Camera renders. Select the Main Camera in the **Hierarchy** panel – this is a child of the PlayerArmature GameObject. Then, in the **Inspector** panel, go to the **Output** section of the **Camera** component, set the **H** (**for height**) value of the camera's **Viewport Rect** parameter to 0.85:

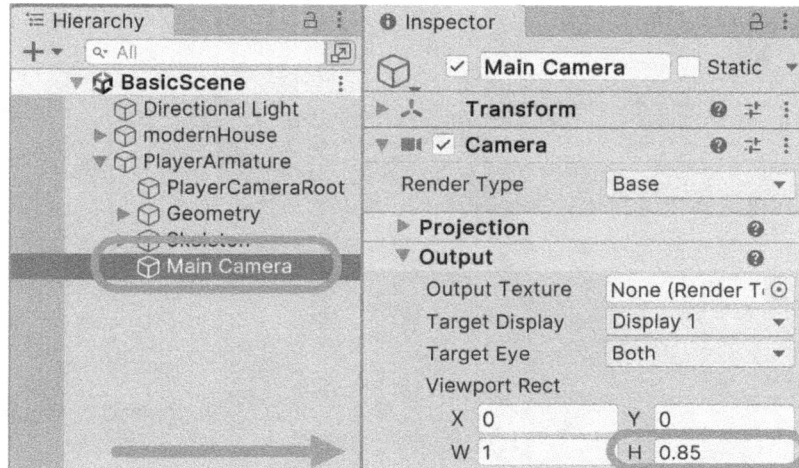

Figure 13.12: Setting the Viewport Rect of the Main Camera to 85% of the screen height

3. Let's now create a second **Camera** in the scene to display a rectangle in the top 15% of the screen. In the **Hierarchy** panel, create a new **Camera** called Camera-2 - timer. You can do this by navigating to **GameObject | Camera** and renaming it.

4. Delete the **AudioListener** component of Camera-2 - timer – since we should only have one in a scene, and there is already one in the Main Camera.

5. For **GameObject** Camera-2 - timer, in the **Inspector** panel, go to the **Output** section of the **Camera** component and set its **Y Viewport Rect** value to 0.85. It should have the values of **X =** 0, **W = 1, Y = 0.85, H = 1**. This means it will render to the full width of the screen, for the top 15% (from 0.85 to 1.0).

6. Additionally, for the **Environment** section of the **Camera** component, component, set **Background Type** to **Solid Color** and **Background** to a **White** color. This is so that the black timing text will now stand out within the bright white rectangle (rather than get lost in the **skybox** gradients).

7. Finally, we only want this new Camera to display UI content, so for the **Rendering** section of the **Camera** component, set the **Culling Mask** to only elements on the **UI** layer.

8. You should now see that the top 15% of the Game panel is just a white rectangle (where we'll soon add the timer UI text ...).

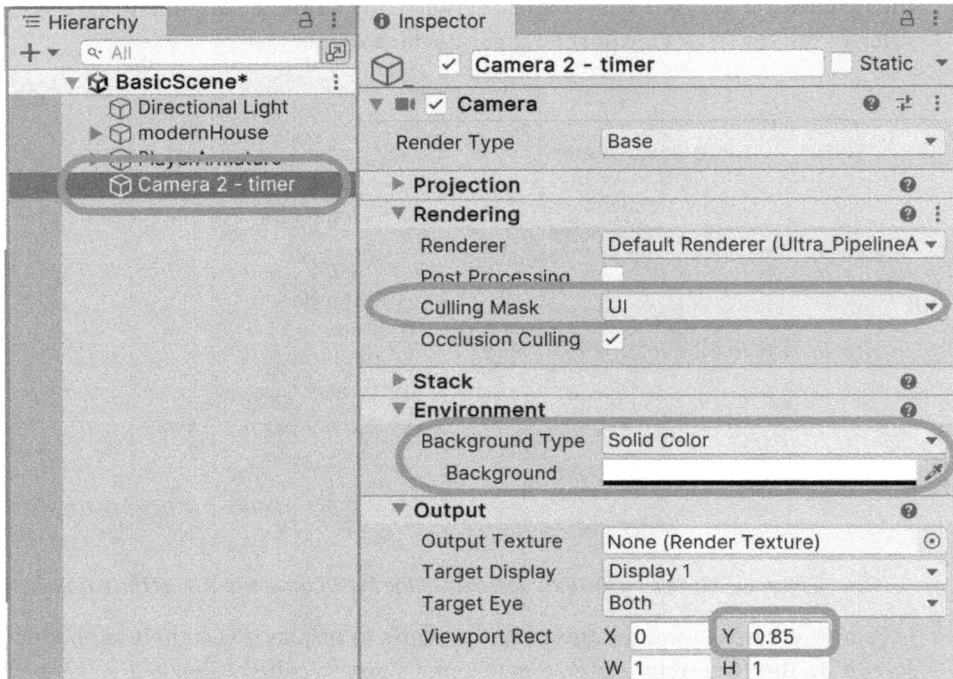

Figure 13.13: Setting the Viewport Rect, Solid Color, and UI Culling Mask for Camera–2 - timer

9. Now, let's create **Canvas** and UI **Text** GameObjects to display our game timings. In the **Hierarchy** panel, create a new UI **Text (TMP)** GameObject named Text-timer, by going to **Create | UI | Text – TextMeshPro**, and agree to **Import TMP Essentials.** You also see that a parent Canvas GameObject, as well as an EventSystem GameObject have automatically been created for when you created the Text-timer UI Text GameObject.

10. Ensure that both Canvas and Text-timer are on the **UI** layer – they should have been set to this layer automatically when created.

11. Select the Canvas GameObject selected in the **Inspector**. For the **Canvas** component, set its **Render Mode to Screen Space - Camera.** A **Render Camera** property slot should appear in the **Inspector** panel. Drag the Camera 2 - timer GameObject into this property slot.

Figure 13.14: Setting Camera–2 - timer for the Render Camera UI Canvas

12. Select the Text-timer **GameObject**. In the **Inspector** panel, set **Vertex Color** to **Black**. Then for the **Rect Transform** position this GameObject **middle-center** (while holding down the **SHIFT** and **ALT** keys). Set **Font Size** to **45**, make the text **Bold**, and set **Wrapping** to **Disabled**.

13. You should now see the words **New Text** in black, in the center of the white rectangle at the top of the **Game** panel.

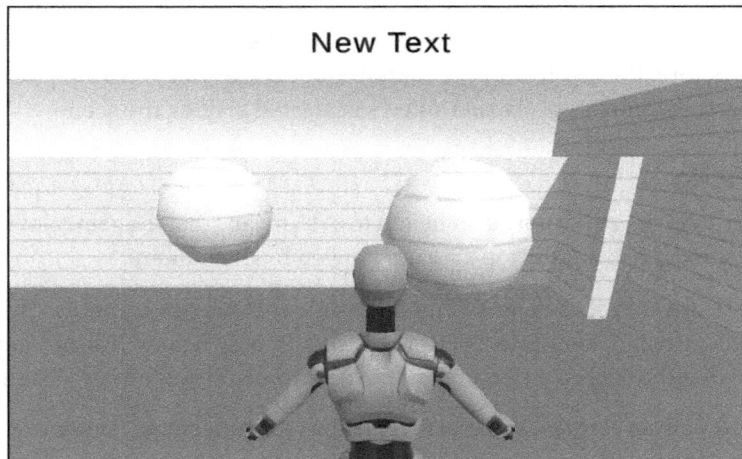

Figure 13.15: Our UI Text at the top of the Game panel

14. In the **Project** panel, create a new C# script class called GameTime.cs that contains the following code:

```
using UnityEngine;
using TMPro;

public class GameTime : MonoBehaviour {
    private TextMeshProUGUI uiText;
```

```
        private float startTime;

    void Awake() {
        uiText = GetComponent<TextMeshProUGUI>();
        startTime = Time.time;
    }

    void Update() {
        float elapsedSeconds = (Time.time - startTime);

        string timeMessage = "Elapsed time = " + elapsedSeconds.ToString
("F");
        uiText.text = timeMessage;
    }
}
```

15. Add an instance of our script class to our `Text-timer` UI Text GameObject by dragging the `GameTime.cs` C# script onto the `Text-timer` GameObject in the **Hierarchy** panel.

16. Save and run the scene. While being able to move around the scene using the WASD-arrow keys as before, the top 15% of the screen should be a white rectangle dedicated to displaying how many seconds have elapsed since you started playing the scene.

How it works...

Setting the `Main Camera`'s **Viewport Rect** height to `0.85` means that the gameplay will only be rendered for 85% of the screen, leaving the top 15% unchanged by the Camera.

We had to delete the `AudioListener` components in our two extra cameras. This is because, for the Unity 3D sound system to work effectively, there should only be one listener in a scene (at one location). This is so that the reproduction of the distance from and angle to the listener can be simulated.

By setting our UI Canvas' **Render Mode to Screen Spa-e - Camera** and its **Render Camera** to `Camera-2 - timer`, our timing text appears clearly displayed in the top 15% of the screen. The `GameTime.cs` C# script class gets a reference to the UI Text (TMP) component of the GameObject that the script has been added to (`_uiText`). It records the time the scene begins (in the `startTime` variable), and each frame subtracts this from the current time, giving an elapsed time. A string message is displayed by updating the value in the `_uitext` variable's `text` property. Note that the default string formatting for float (decimal) numbers is two decimal places.

Adding a top-down orthographic minimap Camera

We saw how we could improve the information shown to the player by using two Cameras displaying the screen. In this recipe, we'll create a third Camera to add to our player's view:

- **Camera 3:** A simple minimap located at the bottom left of the screen, created from a top-down **orthographic** (non-perspective) Camera

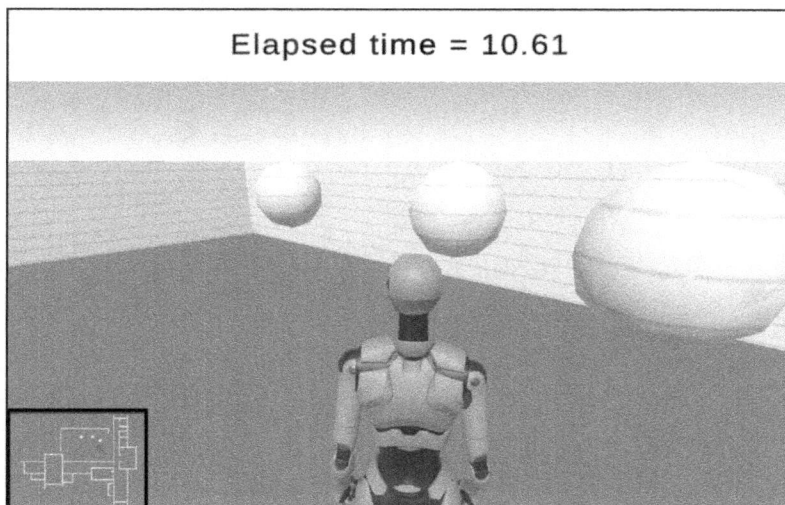

Figure 13.16: The game running showing the added minimap Camera in the bottom-left of the screen

Getting ready

This project follows on from the previous one, so make a copy of that and work on that copy.

How to do it...

To add a top-down orthographic minimap Camera, perform the following steps:

1. In the **Hierarchy** panel, create a third **Camera** named Camera-3 - minimap. Delete its **AudioListener** component (since we should only have one in a scene, and there is already one in the Main Camera).

2. Make Camera-3 - minimap render to the bottom-left corner of the screen. You can do this by setting its **Viewport Rect** values in the **Inspector** panel to X = 0, W = 0.2, Y = 0.0, and H = 0.2. This means it will render a small rectangle (that is, 20% of the screen size) at the bottom-left corner of the screen, from (0,0) to (0.2, 0.2).

3. Now, set this camera's **Projection** property to **Orthographic** (non-perspective), its **Size** to 14, its **Position** to (-5.5, 20, -4), and its **Rotation** to (90, 0, 0). Also, to ensure the output from this minimap Camera appears above that of the Main Camera, in the **Rendering** section set **Priority** to 1. So, this Camera is above the scene and is looking down at its contents (GameObjects).

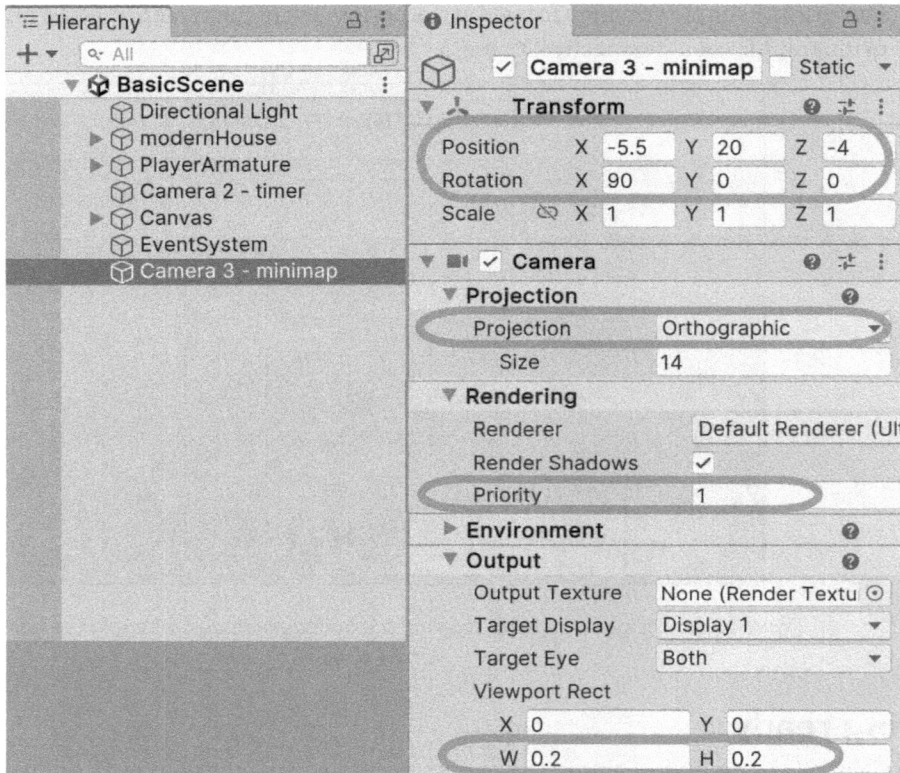

Figure 13.17: Setting up the properties for Camera 3 – minimap

4. The final thing we'll do is a little trick that will make it easier to see an indication of the player's character on the minimap, that is, a large blue floating **Sphere** above the head of PlayerArmature. Since this large Sphere will be seen first by our floating minimap Camera, we'll see a blue circle to mark wherever our Player's character is located in the minimap view. In the **Project** panel, create a new blue **Material** asset named m_blue.

5. Now, in the **Hierarchy** panel, create a new **Sphere** 3D object primitive (**Create | 3D Object | Sphere**). Drag the m_blue **Material** asset from the **Project** panel on the **Sphere** GameObject, and child this **Sphere** GameObject to the PlayerArmature GameObject.

6. Finally, set the **Position** of the **Sphere** to (0, 10, 0) and its **Scale** to (2, 2, 2).

7. When you save and run the scene, you should now be able to view a neat minimap in the bottom-left corner of the screen and a moving blue circle indicating where in the scene your player's character is located.

How it works...

An orthographic Camera is one where the size of objects does not change regardless of how near or far away they are – which is perfect for a minimap. By childing the Sphere to the `PlayerArmature` GameObject and setting the Sphere's (relative) Position to (0, 10, 0) we have made this Sphere "float" directly above the location of the player's character at a height of 10 Unity units (meters!). Since our minimap Camera is at a height of 20 meters, the Sphere will always be seen by it first, so we see a blue circle in the minimap at all times to indicate our player's character position in the scene.

When working with multiple cameras rendering (outputting) to the same parts of the screen, we need to ensure they appear in the correct order. For this project, we are using the **Universal Render Pipeline (URP)**, since that was enabled when the third-person controller package was installed. For the URP, we control the Camera order through the **Rendering Priority** property. Our **Main Camera** had the default priority of zero, so by setting the **Priority** of our minimap Camera to 1 we ensure that the minimap content will be rendered above that of the Main Camera.

> For the **Built-In Render Pipeline (BIRP)**, all the Camera properties are in a single section, and Camera stacking is controlled through a property called **Depth**. So, if we were using the **BIRP**, then we'd set the **Depth** property of the Main Camera to 0, and the **Depth** property of the minimap Camera to 1.

There's more...

There are some details that you don't want to miss.

Adding a floating arrow to indicate the direction the player is facing in the minimap

Another tweak we can make to our simple minimap is for an **arrow** rather than a **circle** to indicate the location of our player's character. This is so that we can view the direction that the player is facing in the minimap. In parts of some games, players do much of their gameplay by simply looking at the minimap for things such as solving mazes and avoiding enemies.

We can create an arrow effect by using three scaled cubes. Delete the blue sphere that is a child of the `PlayerArmature`. Create three blue cubes, and child them to the `PlayerArmature`. Set the properties of the three cubes as follows:

*   Cube 1: **Position** (0, 10, -1.2), **Rotation** (0, 0, 0), and **Scale** (0.3, 1, 3)
*   Cube 2: **Position** (-0.5, 10, -0.24), **Rotation** (0, 45, 0), and **Scale** (0.3, 1, 1.5)

- Cube 2: **Position** `(0.44, 10, -0.2)`, **Rotation** `(0, -45, 0)`, and **Scale** `(0.3, 1, 1.5)`

Figure 13.18: The giant floating "arrow" above the ThirdPersonController

Also, since we don't want a giant arrow-shaped shadow appearing in our game, ensure the **Cast Shadows** property of the **Lighting Section** of the **Mesh Renderer** component is set to **Off** for these three Cubes that form the arrow.

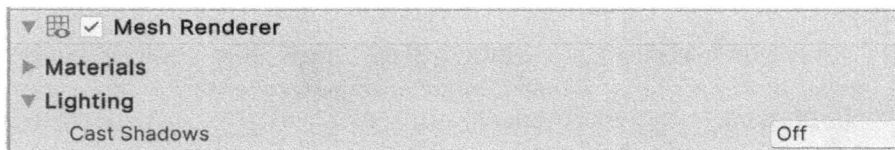

Figure 13.19: Ensuring our arrow Cubes do not cast shadows

Using a culling mask to avoid the sphere being rendered unintentionally

Although our settings for this recipe mean the user probably won't see the giant sphere or arrow in their Main Camera view, in scenes with more camera controls, unintended displays of these types of GameObjects are possible. A more sophisticated approach to ensure objects like these are never rendered in a first-person or third-person Camera would be to place the GameObjects to be hidden on a special layer, and then use a **culling mask** on the Camera to avoid rendering any content for the layer on which the sphere or arrow GameObjects have been placed.

Learn more about Unity culling masks here:

- `https://docs.unity3d.com/Manual/class-Camera.html`
- `https://www.youtube.com/watch?v=gRH9fWBjBhk`

Creating an in-game mirror using a RenderTexture to send Camera output to a Plane

Cameras do not have to output directly to the screen all the time. Different effects can be achieved by having the cameras send their output to a **RenderTexture** asset file. In the scene, 3D objects can be linked to a **RenderTexture** asset file, and so the output of a Camera can be directed to 3D objects such as **Planes** and **Cubes**.

In this recipe, first, we'll duplicate the over-the-shoulder **Main Camera** child of the PlayerArmature, and send the output of this duplicate Camera (via a **RenderTexture** asset file) to a **Plane** on one of the house walls. Then, we'll add a different Camera facing out from the wall. This is so that our Plane will act just like a mirror, rather than duplicating the over-the-shoulder Main Camera:

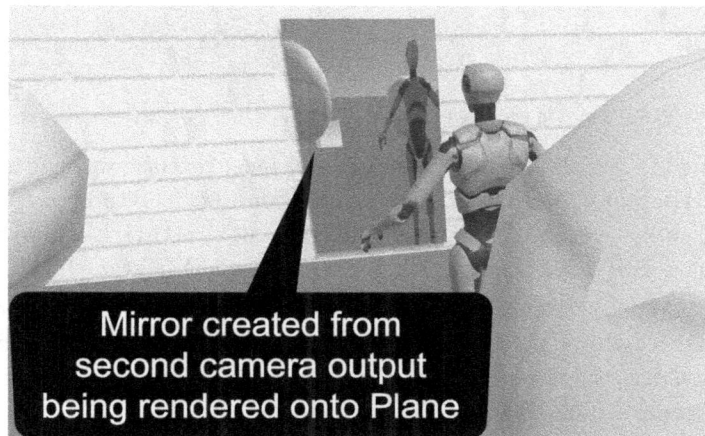

Figure 13.20: A copy of Main Camera rendering to RenderTexture, which is being displayed on a Plane in the scene

Getting ready

This recipe builds on the basic scene created in the first recipe of this chapter.

How to do it...

To create an in-game mirror using a **RenderTexture** to send Camera output to a Plane, perform the following steps:

1. Open the **BasicScene** scene.
2. All the work of this recipe requires a special **RenderTexture** asset file to be created, so let's do that next. In the **Project** panel, create a new **RenderTexture** file and rename it MyRenderTexture (**Create | RenderTexture**).
3. Duplicate the Main Camera child inside the PlayerArmature GameObject, naming the copy Main Camera CLONE. Delete its AudioListener component (since we should only have one in a scene, and there is already one in the Main Camera).

4. With `Main Camera CLONE` selected in the **Hierarchy** panel, drag the `MyRenderTexture` asset file from the **Project** panel to the **Output Texture** property slot of the **Camera** component inside the **Inspector** panel:

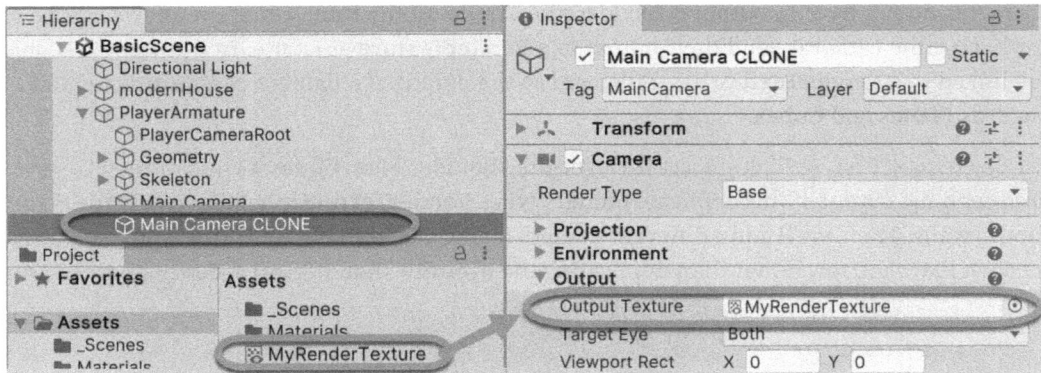

Figure 13.21: Sending the Main Camera CLONE output to the MyRenderTexture asset file

5. Now our cloned Camera is sending its output to `MyRenderTexture`, we need to create a 3D object to display this output. Add a new 3D **Plane** to the scene. Set its **Position** to (`-0.4`, `1`, `4.6`), its **Rotation** to (`-270`, `-90`, `90`), and its **Scale** to (`0.1`, `0.2`, `0.18`). It should now be positioned like a mirror on the wall of the house facing the player's character when the game runs.

6. Now we can assign the `MyRenderTexture` file to be the **Texture** to be displayed on the **Plane**. We do this through a **Material**. Create a new **Material**, `m_renderMaterial`. When this is selected in the **Hierarchy**, drag the `MyRenderTexture` file from the **Project** panel into the **Surface Inputs | Base Map** property of the **Material** in the **Inspector**.

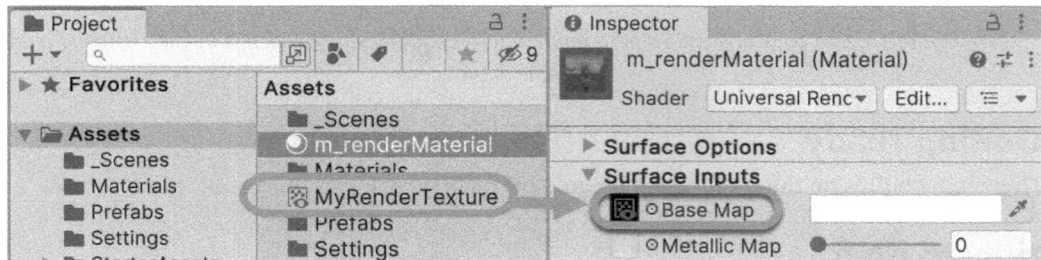

Figure 13.22: Setting the Texture for the Material to be MyRenderTexture

7. Select the **Plane** GameObject in the **Hierarchy**, and drag the m_renderMaterial Material asset file from the **Project** panel to **Element 0** of the **Materials** array of the **Mesh Renderer** component inside the **Inspector**.

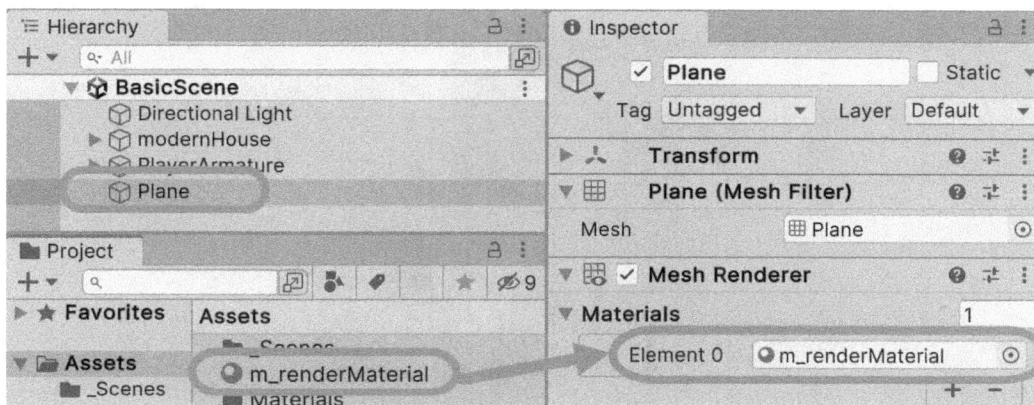

Figure 13.23: Setting the Material for the Plane GameObject in the scene

8. Save and run the scene. As your character approaches the **Plane** on the wall, you should also be able to view a rendering of the screen displayed in the **Plane**.

9. Now, let's create a better mirror effect. Delete the Main Camera CLONE inside the PlayerArmature GameObject.

10. Create a new Camera in the scene, naming it Camera-2 - mirror. Delete its AudioListener component (since we should only have one in a scene, and there is already one in the Main Camera).

11. Just as we did with the cloned Camera, we want this Camera to send its output to our MyRenderTexture asset file. With Camera-2 - mirror selected in the **Hierarchy** panel, drag the MyRenderTexture asset file from the **Project** panel into the **Target Texture** property slot of the **Camera** component inside the **Inspector** panel.

12. Now, all we have to do is locate the new Camera to face out from where our **Plane** is located. Set its **Position** to (-0.4, 1, 4.6), and its **Rotation** to (0, 180, 0). The Camera-2 - mirror **GameObject** should now be positioned so that it is facing anything standing in front of the Plane.

13. Save and run the scene. Now, when your character gets close to the **Plane**, it works like a mirror – except that things are not reversed left-to-right (please refer to the *There's more...* section to learn how to flip the image to be like a real-world mirror):

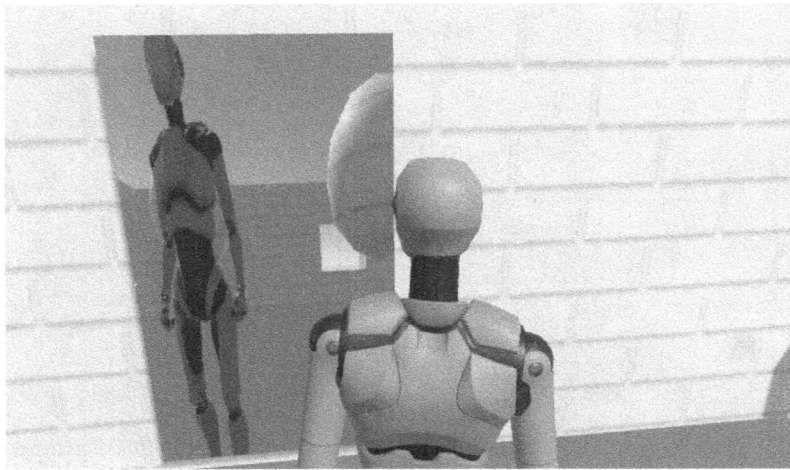

Figure 13.24: Our simulated mirror using a RenderTexture and a second Camera

How it works...

A **RenderTexture** asset file is a special asset file that can store images sent from cameras. Cameras usually render (that is, send their images) to all or part of the screen. However, as we've seen in this recipe, interesting effects are possible when we send a Camera output to a **RenderTexture** and then have a 3D object in our scene that displays the current image in the **RenderTexture** asset file.

In this recipe, we've simulated a kind of mirror effect. Another common use of cameras and **RenderTexture** asset files are simulated closed-circuit TVs, where a **Cube** (or perhaps a more detailed model of a TV monitor) can display what is being viewed in some other part of the scene by a Camera. So, you might have a game where a player is in a control room, changing between the cameras they are looking at to find out the location of enemies or teammates.

There's more...

There are some details that you don't want to miss.

Inverting our mirror Camera horizontally

Let's fix that mirror effect. We need to flip the image rendered by Camera 2 - mirror for the *X* axis (horizontally). We can do that by using a negative X-axis scaling value for the **plane** that is displaying the **RenderTexture** content. So, select **Plane** in the **Hierarchy**, and change its **Transform Scale** value to (-0.1, 0.2, 0.18).

Saving screenshots and RenderTextures to files

At each point in time, a **RenderTexture** contains an image – a screenshot from the point of view of the Camera rendering to that **RenderTexture**. Games can have features such as snapshots at exciting points in time, such as crossing a finish line, landing a difficult jump, or taking the lead in a race. Through scripting we can take the picture data from the **RenderTexture** and save it as an image file.

Also, if you simply want to be able to take a snapshot of the Game screen (Main Camera view) at any point during the game as a PNG file, this is something Unity has made really easy too. We'll explore both these features in this recipe.

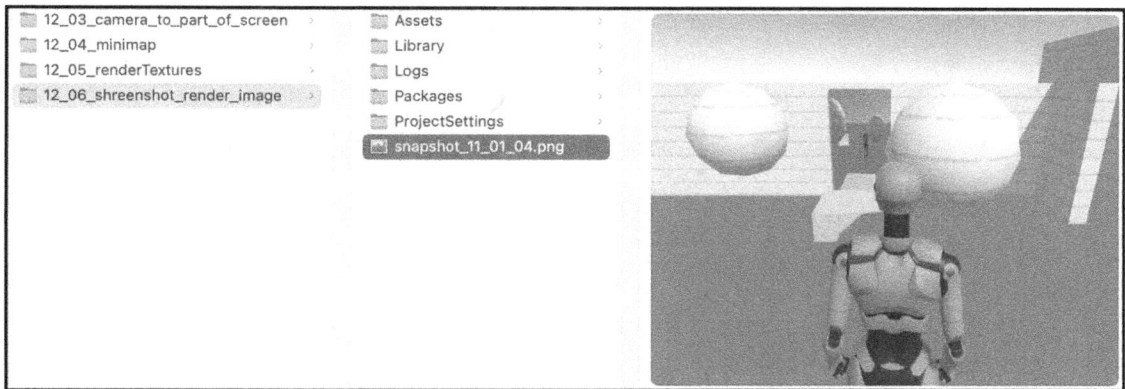

Figure 13.25: PNG screenshot image of Main Camera view taken while playing the game

Getting ready

This project follows on from the previous one, so make a copy of that one to work on in this recipe. Note – many thanks to **krzys-h** for posting the RenderTexture-to-image file code as a public GitHub gist: `https://gist.github.com/krzys-h/76c518be0516fb1e94c7efbdcd028830`.

How to do it...

To save screenshots and RenderTextures to files, perform the following steps:

1. In the **Project** panel, locate the `StarterAssets.inputsettings` asset file in the **StarterAssets / InputSystem** folder.
2. Open the asset file in the **Input Settings** window, either by clicking the button in the Inspector, or by double-clicking on the asset file in the **Project** panel.
3. Check the **Auto-Save** option in the title row of this panel.

4. Create a new **Player** action named **PrintScreen** by clicking the plus-sign (+) button in the **Actions** section. Select the **<No Binding>** child, then click the **Path** property in the **Binding Properties** area. In the **Listen** input box, press the Pkey, and choose **P [Keyboard]**.

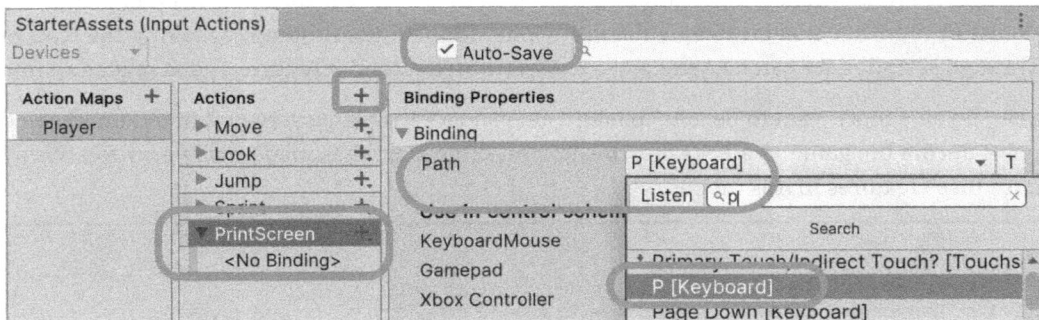

Figure 13.26: Setting up new Player Input Action PrintScreen for the P-key

5. Well done – you've now added a **PrintScreen** input action that you can test for through scripting.

6. Create a new C# script class named Screenshot, containing the following code:

```csharp
using UnityEngine;
using System;
using UnityEngine.InputSystem;

public class Snapshot : MonoBehaviour {
    public InputActionReference printScreen;

    private void Update() {
        if (printScreen.action.WasPressedThisFrame()) {
            SaveScreenSnapshot();
        }
    }

    private void SaveScreenSnapshot() {
        print("taking snapshot");
        string timeStamp = DateTime.Now.ToString("HH_mm_ss");
        ScreenCapture.CaptureScreenshot("snapshot_" + timeStamp +
".png");
    }
}
```

7. Create a new **Empty** GameObject named screenshot-object, and attach an instance of the Screenshot C# script class as a component of this GameObject.

8. Ensure the `screenshot-object` GameObject is selected in the **Hierarchy**. Then in the **Inspector**, click the selection circle button for the public **Print Screen** property, and select **Player | Print Screen**. You have now linked the P-key `PrintScreen` input action to the variable in the scripted `Screenshot` component of the `screenshot-object` GameObject.

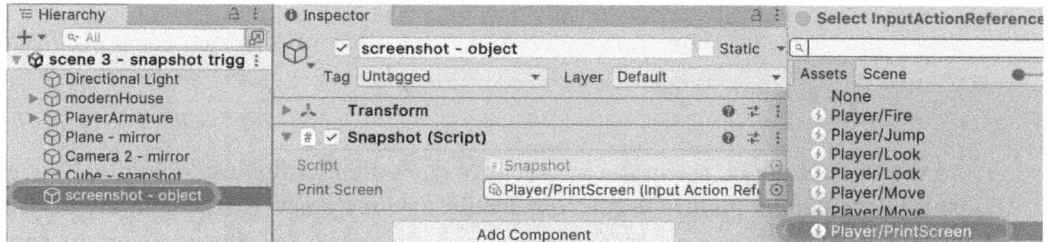

Figure 13.27: Selecting Player | PrintScreen input action for public scripted component variable

9. Save and run the scene. Each time you press the *P*key a new screenshot file will be created, whose filename is the time when the key was pressed.

10. Now let's take a different approach, and add an Editor menu item that, when a **RenderTexture** asset file is selected, allows us to make a copy of it as a PNG image file. In the **Project** panel, create a folder named `Editor`. In this `Editor` folder create a new C# script class named `Screenshot`, containing the following code:

```
using UnityEngine;
using UnityEditor;

public class RenderTextureToFile : MonoBehaviour {
    [MenuItem("Assets/Save RenderTexture to file")]
    public static void SaveRTToFile() {
        RenderTexture rt = Selection.activeObject as RenderTexture;

        RenderTexture.active = rt;
        Texture2D tex = new Texture2D(rt.width, rt.height, TextureFormat.
RGB24, false);
        tex.ReadPixels(new Rect(0, 0, rt.width, rt.height), 0, 0);
        RenderTexture.active = null;

        byte[] bytes;
        bytes = tex.EncodeToPNG();

        string path = AssetDatabase.GetAssetPath(rt) + ".png";
        System.IO.File.WriteAllBytes(path, bytes);
        AssetDatabase.ImportAsset(path);
```

```
        Debug.Log("Saved to " + path);
    }


    [MenuItem("Assets/Save RenderTexture to file", true)]
    public static bool SaveRTToFileValidation() {
        return Selection.activeObject is RenderTexture;
    }
}
```

11. In the **Project** panel, select the myRenderTexture **RenderTexture** asset file. Now go to **Assets | Save RenderTexture to file**. You should now see a new asset file in the **Project** panel named MyRenderTexture.renderTexture.png, which is a PNG image file of the contents of the myRenderTexture **RenderTexture** asset file.

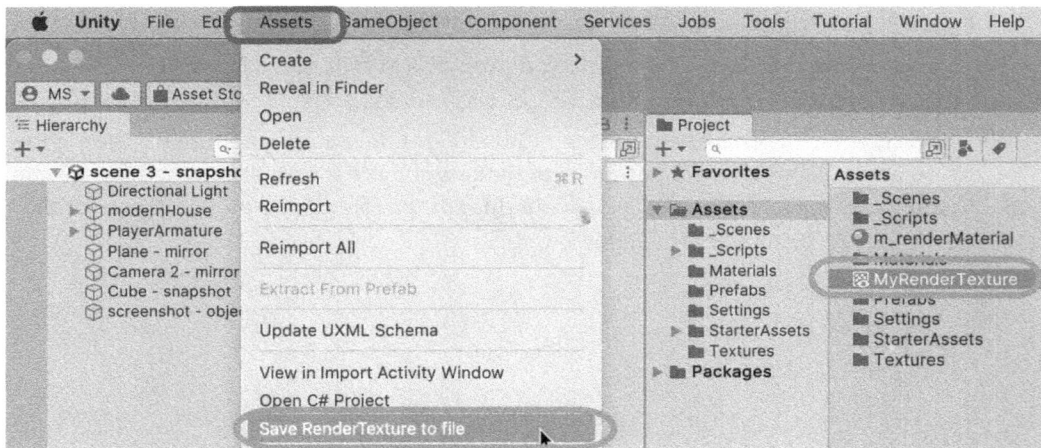

Figure 13.28: Adding a menu item to create a PNG from selected RenderTexture asset file

How it works...

The new input system (activated when we use the **Starter Assets Third Person Controller package**) separates input actions from scripts. So, in order to have a keypress during gameplay be detected by a script, we first have to register the keypress as a named input action. We did this by adding a new Player action named **PrintScreen**, bound to a *P* keypress in the **Input Actions** asset file, StarterAssets. Having created the Snapshot C# script class and added an instance as a component to an empty GameObject, we could then associate this Player | PrintScreen input action with a public variable from that script class.

The Update() method in the Snapshot C# script class checks every frame whether the **Player |
PrintScreen** input action (bound to a press of the P-key) has occurred. If so, the SaveScreenShot()
method is executed, which uses the time as a file name and the ScreenCapture.CaptureScreenshot()
method to create and save a PNG image file of the contents of the **Game** panel at that point in time.

The second part of this recipe used a script that adds a new menu item to the **Assets** menu when a
RenderTexture asset file has been selected in the **Project** panel. This script does some fancy logic, creating a Texture2D object from the pixel values of the **RenderTexture**, and then using an EncodeToPNG()
method to create and save a new PNG image file with a copy of the image inside the **RenderTexture**
at that point in time.

There's more...

There are some more details that you don't want to miss.

Automate screenshot capture with OnTriggerEnter()

If you add the following method to the Snapshot C# script class, you can create GameObjects with
Colliders that are triggers, and trigger a screenshot each time they are hit by a GameObject:

```
private void OnTriggerEnter(Collider other)
{
    SaveScreenSnapshot();
}
```

For example, you could have such a GameObject trigger at the finish line of a racing game, and so on...

Capture screenshots as a Texture or RenderTexture

You can also capture a screenshot as a **Texture** or **RenderTexture** for later use in your game. For example, you could take a screenshot as someone's character dies, or achieves a high score, or passes
the finish line as a **Texture,** and then make that **Texture** to appear in a photo frame or giant "poster"
later in the game...

Learn about the other methods for capturing screenshots in the Unity documentation on the
ScreenCapture built-in C# script class: https://docs.unity3d.com/ScriptReference/ScreenCapture.
html.

Using Cinemachine ClearShot to switch cameras to keep the player in shot

Cinemachine is a suite of tools that sets up a suite of cameras that improves player capture for dynamic view or improved cutscenes. This recipe implements Cinemachine **ClearShot** to switch between
different virtual cameras to keep the player within shot.

We'll do this by specifying that the center of the human character's chest should be kept in the center of the shot as much as possible, and the AI system will decide which Camera to select to best meet these criteria:

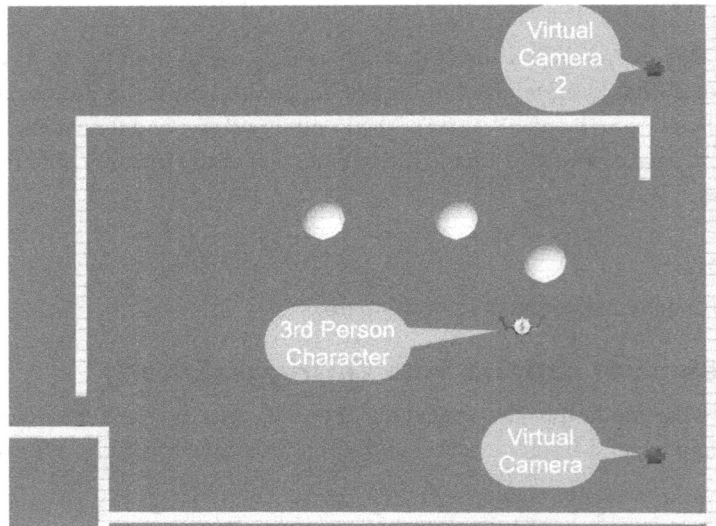

Figure 13.29: Two virtual cameras the Cinemachine Brain can switch between for the best shot of the character

Getting ready

This recipe builds on the basic scene created in the first recipe of this chapter, so make a copy of it to work on for this recipe.

How to do it...

To use **Cinemachine ClearShot** to switch cameras to keep the player in the shot, simply perform the following steps:

1. Open the **BasicScene** scene.
2. This scene uses multiple, fixed-position cameras, so unchild the Main Camera from the PlayerArmature GameObject. In the **Hierarchy** panel, you should see Main Camera as a separate GameObject, at the same level as the PlayerArmature GameObject.
3. Add a **Cinemachine ClearShot** Camera GameObject to the scene by selecting **Create | Cinemachine | ClearShot Camera**. You should now see a new GameObject in the **Hierarchy** panel named Clearshot Camera. Set the position of this new GameObject to (0,0,0).
4. You should also see that a Cinemachine Brain has been added to the Main Camera GameObject – this is indicated in the **Hierarchy** by a half-gray-cog and half-red-camera icon next to the GameObject.

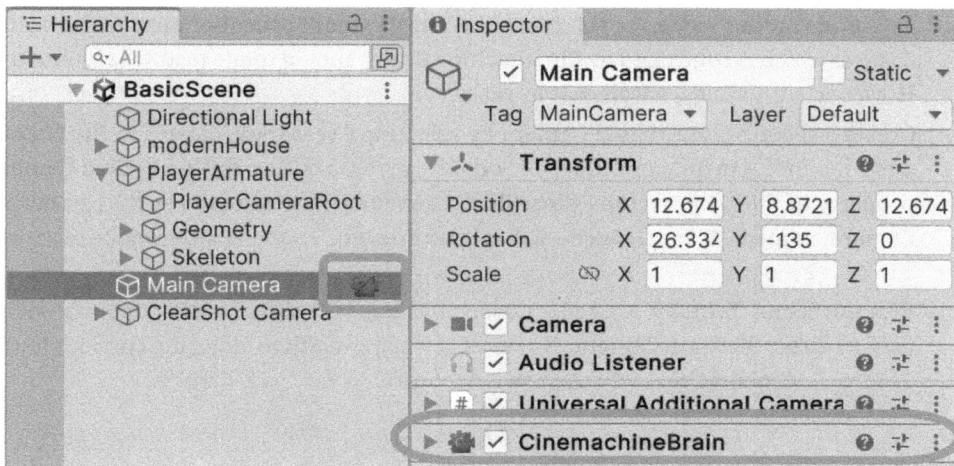

Figure 13.30: The Cinemachine Brain added automatically to our Main Camera GameObject

5. The Clearshot Camera GameObject should have a child GameObject: a Cinemachine virtual **Camera** called Virtual Camera. Set the **Position** of this virtual Camera to (3, 2, -3).

6. Locate the PlayerCameraRoot GameObject in the **Hierarchy** – this is a child of the PlayerArmature GameObject. If you double-click this GameObject you'll see that it is located in the center of the robot character's chest – which is a good location for a camera to focus on. We'll use this part of **the Third Person Character** as the target that our **Cinemachine** cameras will orient toward.

7. Select Clearshot Camera, and in the **Inspector** panel, populate the **Look At** property of the **Cinemachine ClearShot** component with a reference to the PlayerCameraRoot GameObject. Do this by dragging the GameObject from the **Hierarchy** into the **Look At** property inside the **Inspector** panel.

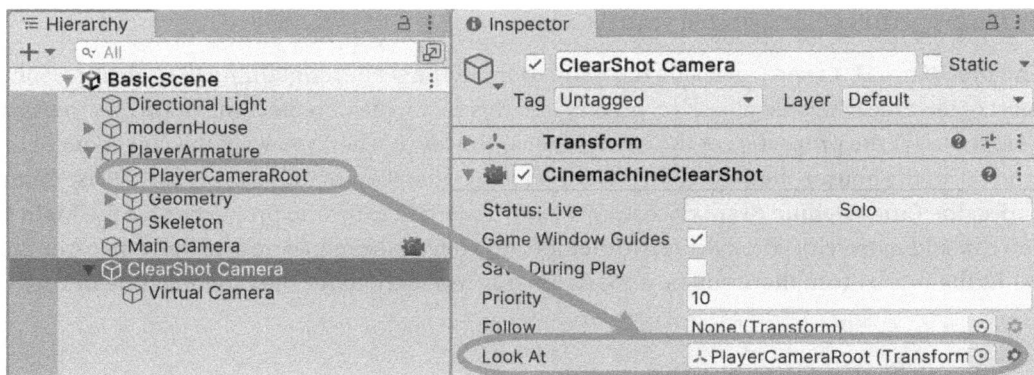

Figure 13.31: Setting the ClearShot Camera Look At property to PlayerCameraRoot child of PlayerArmature

8. Save and run the scene. As you move **the ThirdPersonController** around the scene, the Main Camera (which is controlled by **Cinemachine Brain**) should rotate to always look at the character. However, sometimes, a wall obscures the view.

9. Create a second child virtual Camera by selecting Clearshot Camera in the **Hierarchy** panel, and then in the **Inspector** panel, click on the plus (+) button for the **Virtual Camera Children** property of the Cinemachine **ClearShot** component. You should see that a new child virtual Camera, named Virtual Camera 2, has been created. Set the **Position** value of Virtual Camera 2 to (3, 2, 6).

10. Run the scene. Initially, Virtual Camera has the best shot, so this camera's position will be used to direct the **Main Camera**. However, if you move **Ethan** along the corridor toward Virtual Camera 2, **Cinemachine** will then switch control to Virtual Camera 2.

> If you find that the Virtual Camera 2 is displaying, initially, you can set the priority by selecting CM Clearshot 1 and increasing the priority of Virtual Camera 1.

How it works...

A **Cinemachine Brain** component was added to the scene when we added a **Cinemachine Clearshot Camera**. The **Cinemachine Brain** takes control of the Main Camera in the scene and uses properties of one or more **Virtual Cinemachine** cameras to decide what properties to apply to the Main Camera.

The **Look At** property of the **Clearshot Camera** was set to the center of the robot character's chest, which is the PlayerCameraRoot child of the PlayerArmature GameObject. The position of this PlayerCameraRoot GameObject is used by the ClearShot component to rank each virtual camera's quality of the shot. The **Cinemachine Brain** then chooses the best, and makes the Main Camera take on the properties of the selected virtual Camera.

The first Virtual Camera is inside the main room in the scene, containing the 3 floating spheres. So most of the time when the player's character is in this room, the **Cinemachine Brain** chooses this virtual Camera to set the properties of the Main Camera. However, when you walk the player out of the room (and out of the house), then Virtual Camera 2 has a better view of the PlayerCameraRoot GameObject, and so the **Cinemachine Brain** chooses Virtual Camera 2 to set the properties for the **Main Camera**. You can add more virtual cameras in other locations in the scene, to ensure that wherever the player walks their character, there will be a good camera view of them for the **Cinemachine Brain** to choose.

A Camera to always look at and follow the Third Person Controller

Our project uses the **Third Person Character Controller Starter Assets**, and this package includes mouse-control functionality for the character's direction, so in this recipe, we'll create a Camera to move with and look at the third-person character by creating a Cinemachine `Virtual Camera` that always follows our robot's `PlayerCameraRoot` GameObject.

It's always good to give players choices and control over their gaming experience. In this recipe, we'll set up a mouse-controllable **Cinemachine FreeLook** Camera and let the player switch to it.

Getting ready

This recipe builds on the basic scene created in the first recipe of this chapter. So, make a copy of that one to work on in this recipe.

How to do it...

To create a Camera that will always look at and follow the Third Person Controller, perform the following steps:

1. Open the **BasicScene** scene.
2. This scene uses multiple fixed-position cameras, so unchild the `Main Camera` from the `PlayerArmature` GameObject. In the **Hierarchy**, you should see `Main Camera` as a separate GameObject at the same level as the `PlayerArmature` GameObject.
3. Add a **Cinemachine Virtual Camera** to the scene. Do this by selecting **Create | Cinemachine | Virtual Camera**. You should see a new GameObject in the **Hierarchy** panel named `Virtual Camera`.
4. You should also see that a Cinemachine Brain has been added to the `Main Camera` GameObject – this is indicated in the **Hierarchy** by a half-gray-cog and half-red-camera icon next to the GameObject.
5. Locate the `PlayerCameraRoot` GameObject in the **Hierarchy** – this is a child of the `PlayerArmature` GameObject. If you double-click this GameObject you'll see that it is located in the center of the robot character's chest – which is a good location for a camera to follow. We'll use this part of the **Third Person Character** as the target for our **Cinemachine** `Virtual Camera` to focus on.
6. Select `Virtual Camera`, and in the **Inspector** panel, populate the **Follow** property of the **CinemachineVirtualCamera** component with a reference to the `PlayerCameraRoot` GameObject. Do this by dragging the GameObject from the **Hierarchy** into the **Follow** property inside the **Inspector**.

7. Also, for the **Body** property of the **CinemachineVirtualCamera** component choose **3rd Person Follow** from the drop-down menu.

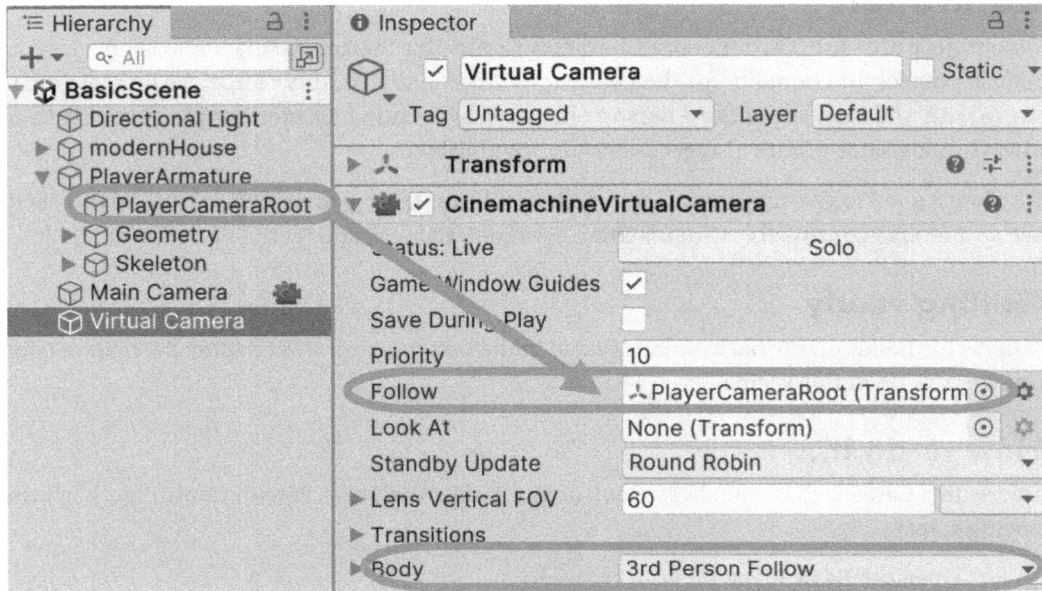

Figure 13.32: Setting the Virtual Camera to follow a third-person character

8. Save and run the scene. As you move **the ThirdPersonController** around the scene, the Main Camera (which is controlled by **Cinemachine Brain**) should rotate to always look at the character. However, sometimes, a wall obscures the view.

How it works...

In this recipe, you added a **FreeLook Cinemachine** GameObject, but with a priority of 0. So, initially, it will be ignored. When the 1 key is pressed, the script increases the **Priority** to 99 (much higher than the default of 10 for the **ClearShot** cameras), so then the **Cinemachine** Brain will make the **FreeLook** virtual Camera control the Main Camera. Pressing the 2 key reduces the **FreeLook** component's **Priority** back to 0. So, the **ClearShot** cameras will be used again.

There should be a smooth transition from **FreeLook** to **ClearShot** and back again since you set the **Default Blend** property of the **Cinemachine Brain** component in the Main Camera to **Ease In Out**.

> **Note:** If we were not already using a third-person character whose scripting uses the mouse to control where the player is facing, we could allow the mouse to control where the Virtual Camera is facing. To this end, we'd use a Cinemachine Free Look Camera rather than a Virtual Camera set to loop at a third-person body follow.

There's more...

There are some more details that you don't want to miss.

Adding the option to switch between a following camera and multiple ClearShot cameras

You may wish to give your players the choice of allowing the game to choose the best **ClearShot** Camera automatically, or allowing the player to control the single, always-following camera. The choice could be set through a button on a settings popup, or perhaps a special key to be pressed (such as the Ckey, for camera mode). All that is required to implement this feature is to have the **ClearShot** Camera with its multiple child virtual cameras, and a camera to follow the player, both present in the scene at the same time. You can then change the priority of the following Camera to -1 (to allow the **ClearShot** cameras to take over), or to 99, resulting in the following camera always having the highest priority.

You could use some variation of the following code to enable/disable the always-following camera via the two public methods, `SetHighestPriority()` and `SetLowestPriorioty()`:

```
using Cinemachine;
using UnityEngine;

public class AlwaysFollowCameraStatus : MonoBehaviour {
    private CinemachineVirtualCamera followCamera;

    private void Awake() {
        followCamera = GetComponent< CinemachineVirtualCamera >();
    }

    public void SetHighestPriority() {
        followCamera.Priority = 99;
    }
    public void SetLowestPriority() {
        followCamera.Priority = -1;
    }
}
```

Adding film grain and vignette effects with URP postprocessing

The **Universal Render Pipeline (URP)** is one of the scriptable Render Pipelines that Unity supports. While the **Built-In Render Pipeline (BIRP)** is fine for most projects, once you start to need very high-quality graphics or special visual effects, then you'll need to start using one of the Scriptable Render Pipelines.

> **Note:** Our project is already using the URP since it was added as a requirement of the **Starter Assets Third Person Controller** package. If you are working on a different project, you can choose a URP or HDPR project template rather than the default BIRP when creating a new Unity project.

URP provides workflows that let you quickly and easily create optimized graphics that improve the visual effect of your project. In this recipe, we'll use the basic scene created in the first recipe to add **Film Grain** as a postprocessing effect to our scene. The **Film Grain** effect simulates the random (grainy!) texture of a photographic film:

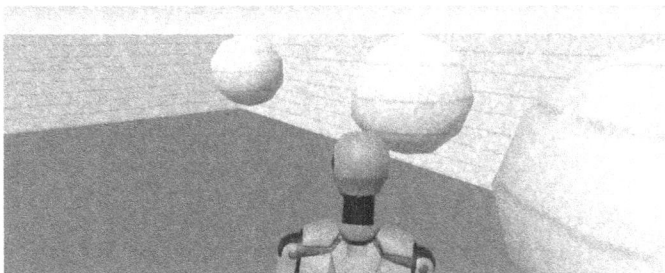

Figure 13.33: An example of the recipe with the Film Grain visual effect applied

Getting ready

This recipe builds on the basic scene created in the first recipe of this chapter. So, make a copy of that and work on the copy.

How to do it...

To add Film Grain and Vignette effects with URP postprocessing, perform the following steps:

1. First, we need to enable **Post Processing** on the Camera. With Main Camera selected in the **Hierarchy** panel, check the **Post Processing** option for the **Rendering** component.

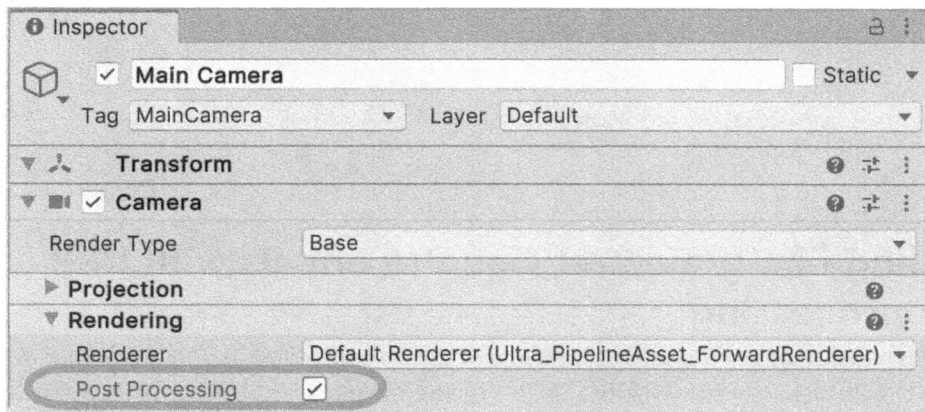

Figure 13.34: Enabling Post Processing for the Rendering component of the Main Camera

2. Create an empty **GameObject** and name it Post Processing.
3. We will now add a postprocessing **Volume** component to the Post Processing **GameObject**. Ensure Post Processing is selected in the **Hierarchy** panel. Then, in the **Inspector** panel, click on **Add Component** and search for **Volume**. Now create a postprocessing profile by clicking on the **New** button at the bottom-right of the **Volume** component in the **Inspector**. You should now

see that the value for the **Volume Profile** property is **PostProcessing Volume (Volume Profile)**.

> **Note:** When you create a new postprocessing profile, the overrides and their properties are stored in an asset file in the **Project** panel. You will see a new folder with the same name as the current scene, and inside that an asset file named **Post Processing Profile**, which gets its name from the name of the GameObject in which the Volume component was created.

Figure 13.35: Adding a new postprocessing profile

4. With Post Processing still selected inside the **Inspector**, add an override by clicking the **Add Override** button. From the drop-down menu, select **Post Processing | Film Grain.**

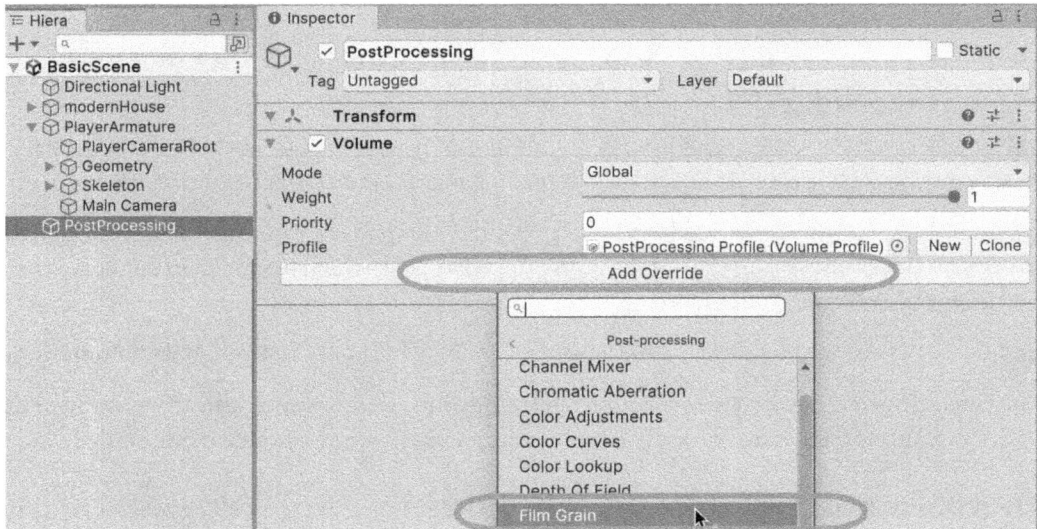

Figure 13.36: Adding a new postprocessing profile

5. For the **Film Grain** Volume Override, click the **ALL** text (shown in white) to allow property editing. Then choose **Large 01** from the drop-down menu for **Type**, and drag the **Intensity** slider all the way up to 1, and the **Response** slider all the way down to 0.

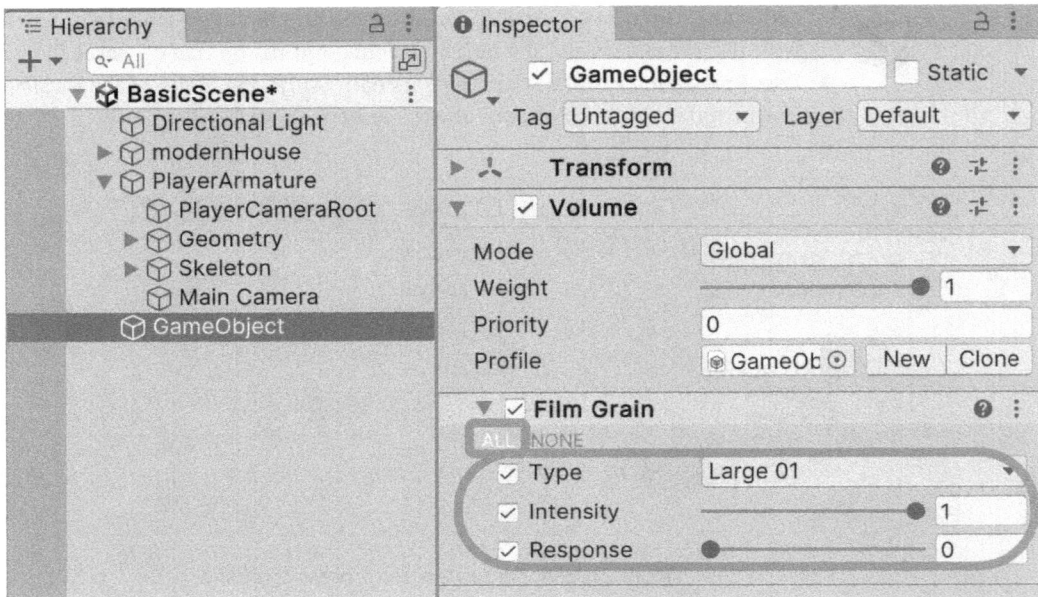

Figure 13.37: Setting the properties of the Volume Film Grain post-processing override

6. Save and run the game. You should be able to see an old-film-style grain effect across the gameplay screen.

How it works...

Each Camera whose rendering you wish to be affected by post-processing volumes must have its **Post Processing** option checked. We did this in the Main Camera GameObject **Camera** component properties.

A GameObject in the scene needs to have a **Volume** component with one or more post-processing overrides. One of these overrides is the **Film Effect** override, which we added to our Post Processing GameObject's **Volume** component.

Additionally, the Main Camera enables postprocessing visual effects through its **Volume** component.

The rendering of the Main Camera is affected by the Film Grain volume, and so we see the Film Grain effect globally applied to everything that is rendered by the Main Camera on the screen.

There's more...

There are some more details that you don't want to miss.

Adding a vignette effect

In the previous recipe, we set up a volumetric postprocessing effect for a scene on a **global** basis, that is, affecting the whole scene. This recipe will build on the previous one by creating a second postprocessing effect within a **local** area of a scene. In this recipe, we will apply a visual *vignette*. This adds a sort of dark fog around the edges of our view.

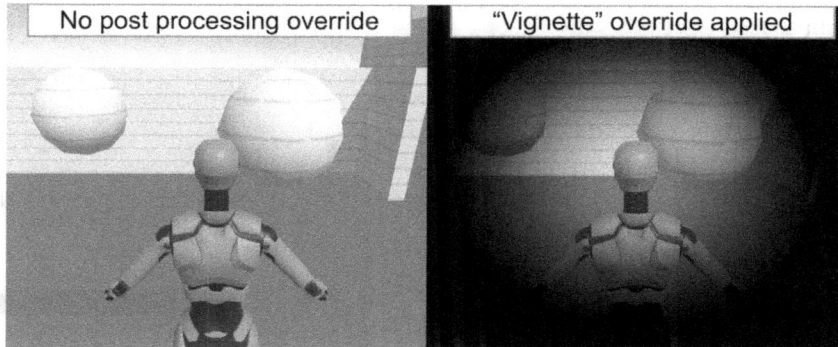

Figure 13.38: An example of the recipe demonstrating the effect of adding a vignette

> Vignetting refers to the darkening of the corners of an image compared to its center. In photography, vignetting can be caused by the camera optics. In Unity, you can use URP postprocessing to draw the viewer's eye away from the distractions in the corner and, instead, toward the center of the image. In games, vignetting can be used to add suspense or atmosphere to a scene, or perhaps to encourage the player to focus on a target in the very center of what is being viewed.

Just as we added a Film Grain override to the Volume property of our Post Processing GameObject, we can add an override for a vignette. Do the following:

1. Save the scene with a different name.
2. Delete the Post Processing GameObject.
3. Now create a new empty GameObject named Post Processing 2.
4. Now create a postprocessing profile for this object in this scene by clicking on the **New** button at the bottom-right of the **Volume** component in the **Inspector**. You should now see that the value for the **Volume Profile** property is **Post Processing 2 Volume (Volume Profile)**.
5. Ensure the Post Processing 2 GameObject is selected inside the **Inspector**.
6. Add a new override by clicking the **Add Override** button. From the drop-down menu, select **Post Processing | Vignette**.
7. For the **Vignette Volume Override**, click the **ALL** text (shown in white) to allow property editing. Then drag the **Intensity** slider to somewhere between **0.6** and **0.7**.

8. You may also wish to change the **Smoothness** slider to choose a nice mid-range value for the vignette, between a defined oval shape and one with very blurred edges between the black outer areas and unaffected inner areas of the Camera view.

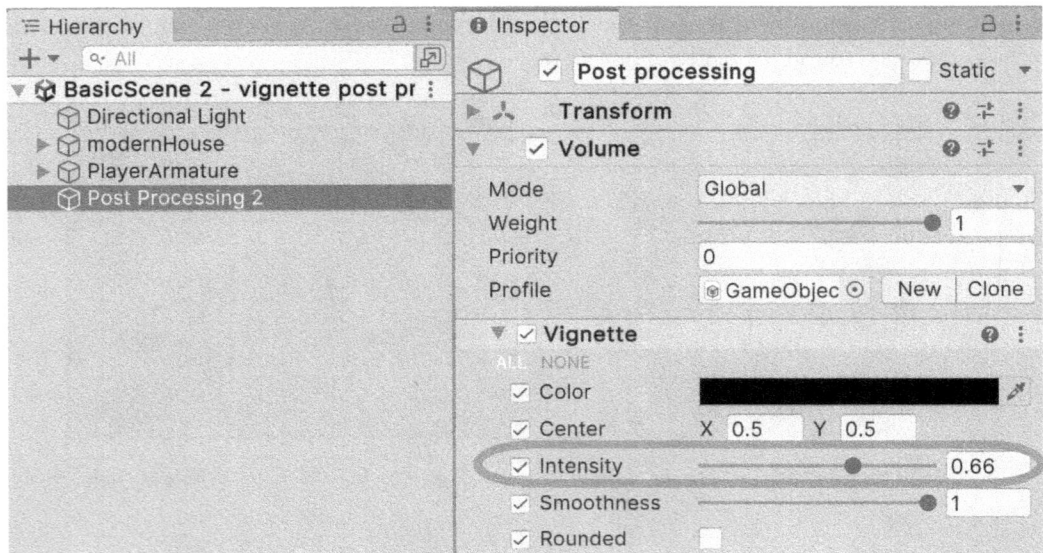

Figure 13.39: Setting the Intensity slider for the vignette effect

9. Save and run the game; you should see the vignette effect displayed across the gameplay screen, where the edges are dark but you can still see the details in the center.

Creating an HDRP project with an HDRI skybox

HDRI is a method of representing images with a high range of luminosity (human-perceived color brightness). In this recipe, we'll use a Creative Commons-licensed HDRI image to create a high-quality skybox in an HDRP project.

Figure 13.40: The detailed sun, clouds, and sky scene we see from the Koppenheim HRDI skybox

We'd like to thank Geg Zaal (http://gregzaal.com/) for publishing the Kloppenheim 06 HDRI assets under the CC0 license at HDRI Haven.

Getting ready

For this recipe, we have prepared the image that you need in a folder named HDRI_ Kloppenheim06, inside the 13_10 folder.

How to do it...

To create an HDRP project with an HDRI skybox, perform the following steps:

1. Create a new **3D (HDRP)** project.

2. Create a new **Basic Outdoors (HDRP)** scene named HDRI sky scene.

3. Import the provided HDRI_ Kloppenheim06 folder.

4. In the **Project** panel, select the kloppenheim_06_4k asset file from inside the HDRI_ Kloppenheim06 folder.

5. In the **Inspector** panel, change **Texture Shape** to **Cube**, **Max Size** to **4096 (you may need to check the Override checkbox to enable this)**, and then click on the **Apply** button. Depending on the speed of your computer, it might take a few seconds for these properties to be applied to the asset file.

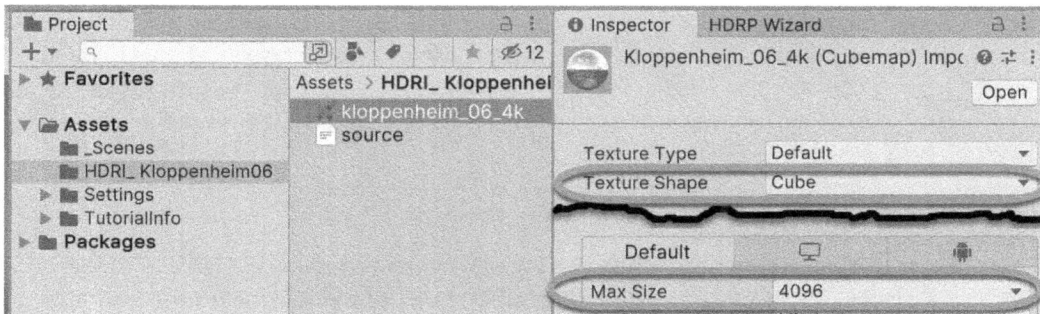

Figure 13.41: Setting Texture Shape to Cube and Max Size to 4096 for kloppenheim asset file

6. Select the Sky and Fog Volume **GameObject** in the **Hierarchy** panel.

7. In the **Inspector** panel, in the **Visual Environment** section, check **Sky type** and choose **Sky | HDRI Sky** from the drop-down menu.

8. Next, click the **Add Override** button and select **Sky | HDRI Sky**.

9. Ensure that the **HDRI Sky** property is checked.

10. Finally, drag the `kloppenheim_06_4k` asset file into the **HDRI Sky** property slot.

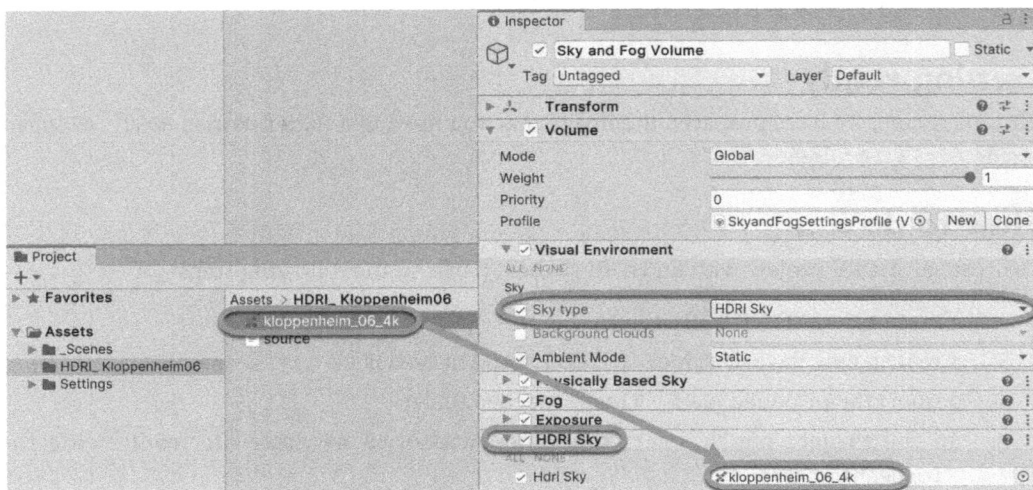

Figure 13.42: Updating Sky and Fog Volume for an HDRI skybox

11. Save and run the scene. Above the horizon, you should see a realistic outdoor sky featuring the Sun, a blue sky, and some white clouds.

> **Note:** Depending on your system, if the screen seems very white, then in the **Inspector** panel you may need to increase the **Limit Max** setting for the **Exposure** section of the **Visual Environment** component of the `Sky and Fog Volume` GameObject.

How it works...

For things to work properly here, Unity needs to know the way data is organized in each HDRI asset file. The `kloppenheim_06_4k` asset file was recorded as a Cube at 4K resolution, so we first needed to apply these settings to the asset file in the **Project** panel.

Having created an HDPR project, we changed the settings of the `Sky and Fog Volume` GameObject from the default **Physically Based Sky** to an **HDRI Skybox**. We were then able to assign the provided `kloppenheim_06_4k` asset file to achieve a realistic Skybox based on an HDRI image asset.

There's more...

There are some more details that you don't want to miss.

Using a skybox for default ground image

If there is sufficient detail in the HDRI asset, rather than just an above-the-horizon Skybox, we can also use an HDRI asset to provide a detailed ground image, too:

1. Delete the `Sky and Fog Volume` GameObject from the **Hierarchy**.

2. In the **Project** panel, create a new **Volume Profile** asset file named sky Volume Profile by selecting **Create | Volume Profile**. Having selected this new asset file, in the **Inspector** panel, add an **HDRI Sky override**. You can do this by clicking on the **Add Override button**, then selecting **Sky | HDRI Sky**.

3. With the sky Volume Profile asset file still selected, in the **Inspector** panel, click the **ALL** text, and ensure that the **Hdri Sky** property is checked. Finally, drag the kloppenheim_06_4k asset file into this property slot.

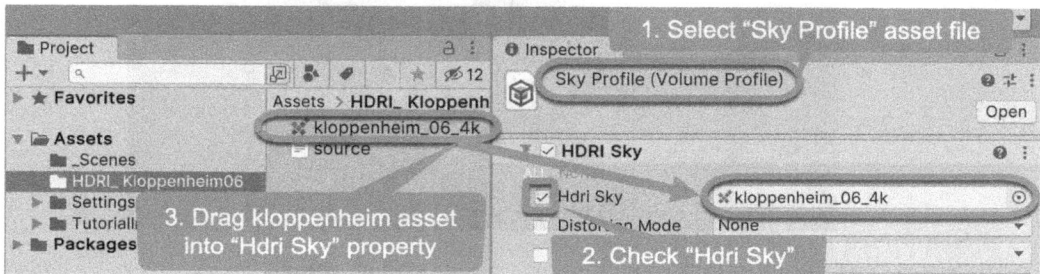

Figure 13.43: Setting up the Sky Profile asset properties

4. Select the Main Camera GameObject in the **Hierarchy**.

5. In the **Inspector** panel, add a new **Volume** component to the Main Camera by clicking on the **Add Component** button and selecting the **Volume** component. Drag the sky Profile asset file from the **Project** panel into the **Profile** property of this new **Volume** component.

Figure 13.44: Creating a new profile for the Volume component of the Main Camera

6. Add a 3D Cube and 3D Sphere to the scene.

7. Save and run the game. You should see a view of some rocks in a green, grassy landscape.

Everything rendered by the Main Camera is now affected by this **Volume** and is displayed in the HDRI image as a Skybox that extends above and below the horizon.

GameObjects in the scene will be rendered in front of the Skybox content, as we can see with a Cube and Sphere that were added in the following figure.

Figure 13.45: The rocky landscape (Koppenheim HRDI skybox) with a Cube and Sphere we added to the scene

Creating and applying a cookie texture to a spotlight

Lights can have a cookie **Texture** applied. "Cookies" are **Textures** used to cast shadows or silhouettes in a scene. They are produced by using the cookie **Texture** as a mask between the light source the surfaces being rendered. Their name, and usage, comes from physical devices called **cucoloris** (nicknamed cookies) used in theatre and movie productions to create shadow effects implying environments such as moving clouds, the bars of a prison window, or the sunlight broken up by a jungle leaf canopy, for example. **Cookie Textures** can work well with Unity **SpotLights** to simulate shadows for these effects as well.

In this recipe, we'll create and apply a cookie **Texture** suitable for use with a Unity **SpotLight**.

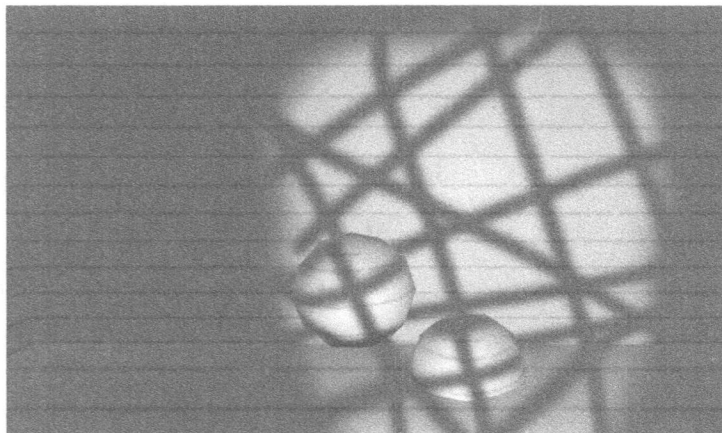

Figure 13.46: Cookie Texture as shadows thrown by a SpotLight

Getting ready

If you don't have access to an image editor, or prefer to skip the **Texture** map elaboration in order to focus on the implementation, we have provided the prepared cookie image file called spotCookie. tif inside the 13_11 folder.

However, if you wish to create your own cookie textures, try the following:

1. In your image editor, create a new 512 x 512 grayscale pixel image.
2. Ensure the border is completely black by setting the brush tool color to black and drawing around the four edges of the image. Then draw some crisscrossed lines (or a cloud shape, or prison bars, etc.). Save your image, naming it spotCookie.

Figure 13.47: Creating the cookie texture image in Photoshop

This recipe builds on the basic scene created in the first recipe of this chapter, so make a copy of that one to work on for this recipe.

How to do it...

To create and apply a cookie texture to a spotlight, perform the following steps:

1. Open the **BasicScene** scene.
2. This scene uses multiple, fixed-position cameras, so unchild the Main Camera from the PlayerArmature GameObject. In the **Hierarchy** panel, you should see Main Camera as a separate GameObject at the same level as the PlayerArmature GameObject. Set its **Position** value to (0, 2, 0) and its **Rotation** to (15, 180, 0).
3. Import the spotCookie.tif cookie image file into your project.

4. Select the spotCookie asset file in the **Project** panel. Then in the **Inspector** panel, change the **Texture Type** dropdown to **Cookie**, **Light Type** to **Spotlight**, and set **Alpha Source** to **From Gray Scale**. Then, click on **Apply**, as shown.

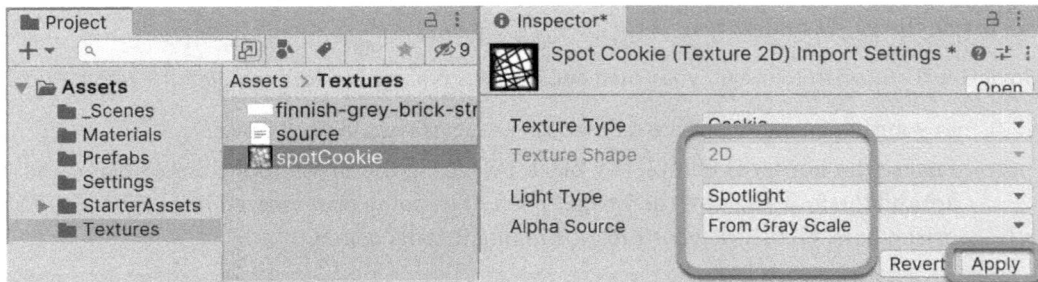

Figure 13.48: Setting the cookie texture import settings

5. Delete the **Directional Light**, since we'll be generating light in this scene from a **Spotlight**.

6. Add a **Spotlight** to the scene, select **Create | Light | Spotlight**. Set its **Position** value to (0, 1, 0) and its **Rotation** to (0, 180, 0).

7. Set **Shadow Type** to **No Shadows** and drag the spotCookie asset file from the **Project** panel into the **Cookie** slot. Set **Intensity** to 20.

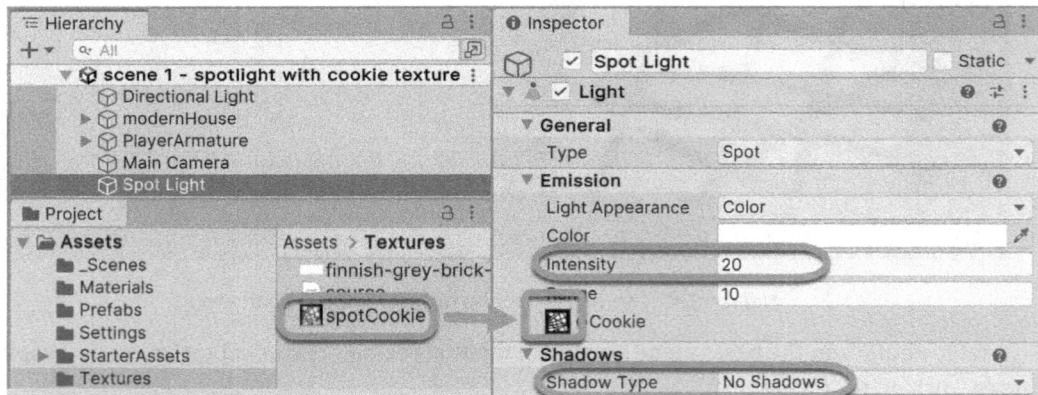

Figure 13.49: The Spotlight settings

8. Save and play the scene. You should now see the spotlight casting shadows on the wall as if shone through a grid of planks of wood or metal.

How it works...

We created a grayscale **Texture** for use with Unity **Spotlights** that is completely black around the edges so light does not bleed around the edge of our Spotlight emission. The black lines in the **Texture** are used by Unity to create shadows in the light emitted by the **Spotlight**, creating the effect that there are some straight beams of wood or metal through which the **Spotlight** is being shone.

There's more...

There are some more details that you don't want to miss.

Adding more content and creating ground shadows

We don't see any cookie shadows on the floor since that ProBuilder shape is not textured. Let's create a textured floor using a Cube, and we'll also move a couple of the Spheres in the scene to help demonstrate the atmospheric lighting effects possible with a cookie-textured spotlight. Do the following:

1. Set the **Position** value of the Sphere 1 GameObject (child of modernHouse) to (-6, 0, -23).

2. Set the **Position** value of the Sphere 2 GameObject (child of modernHouse) to (-7, -0.5, -23). Having a Y-value of -0.5 means half of this Sphere is embedded in the floor.

3. Add a 3D Cube to the scene. Set its **Position** value to (0, 0, -3), and its **Scale** to (10, 0.1, 3). Set the **Material** value for this **Cube** to m_white_brick.

Using Cookies with Directional Lights

Cookies can be applied to other kinds of light. In the following screenshot, the window-frame image from the Unity cookie manual page (https://docs.unity3d.com/Manual/Cookies.html) was applied to a **Directional Light** GameObject in our scene (and the **Spot Light** from earlier in this recipe was removed).

Figure 13.50: Cookie applied to a Directional Light

The window-frame cookie texture was applied to the **Cookie** property in the **Emission** section of the **Light** component of the **Directional Light**. You may have to play around with the **Rotation** value of the Directional Light and the **Cookie Size** settings to get the desired effect. Also, you may wish to set the **Shadows** property to **No Shadows**.

The window-frame image can be found on the following Unity cookie manual page: `https://docs.unity3d.com/2018.1/Documentation/Manual/Cookies.html`.

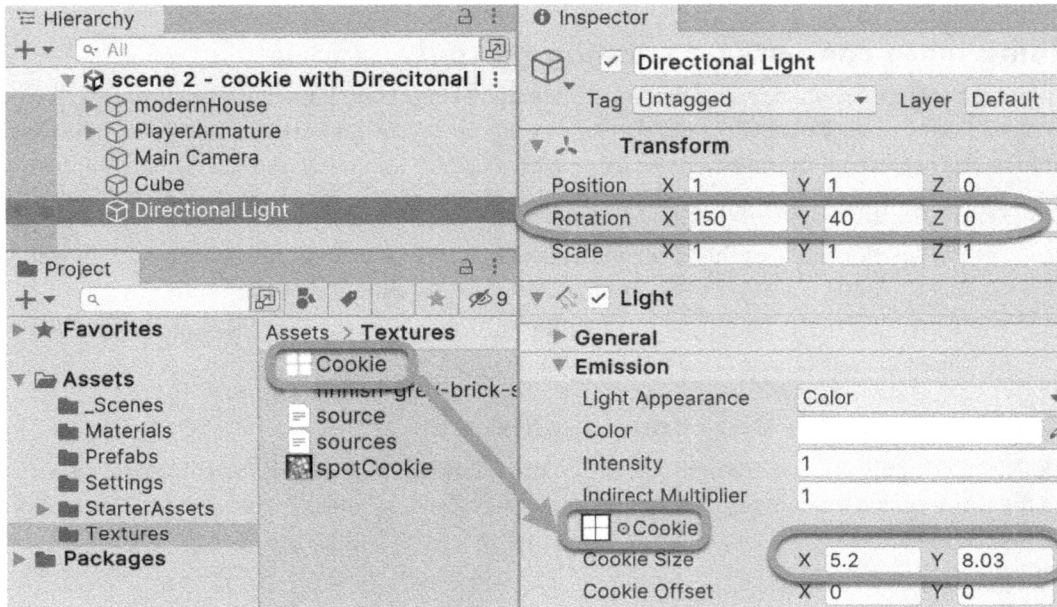

Figure 13.51: Cookie applied to a Directional Light

Baking light from an emissive material onto other scene GameObjects

As well as **Lights**, other objects can also emit light if their **Materials** have **Emission** properties (such as a **Texture**, and/or tint **Color**). In this recipe we'll create a lamp that glows green via its **Emission Texture**. The lamp and other 3D objects in the scene will be baked to create a pre-computed **Lightmap** for the scene.

Figure 13.52: Cube with green emissive Material

Getting ready

This recipe follows on from the previous one, so make a copy of that project to work on for this recipe.

How to do it...

To bake light from an emissive **Material** onto other scene GameObjects, follow these steps:

1. If there is a **Spot Light** in the scene, delete it.

2. Ensure your scene has a **Directional Light**, and that it has no **Cookie Texture**. Reduce its **Emission Intensity** value to 0.5 (so we can better see the glowing green Cube we are about to create!). Also in its **General** settings, set this light's **Mode** to **Baked**.

Figure 13.53: Setting the Directional Light's Mode to Baked

3. Next, ensure your scene has a **Directional Light** that has no **Cookie Texture**.

4. Rename the floor Cube to Cube-floor.

5. Add a 3D **Cube** to the scene named Cube-emissive. Set its **Position** value to (-1, 0, -3). Also check this GameObject's **Contribute Global Illumination** option.

6. Baked lighting only works for **Static** GameObjects, so select the **Cube**, the three **Spheres**, and the walls of the scene, and ensure their **Static** property is checked (at the top of the **Inspector**).

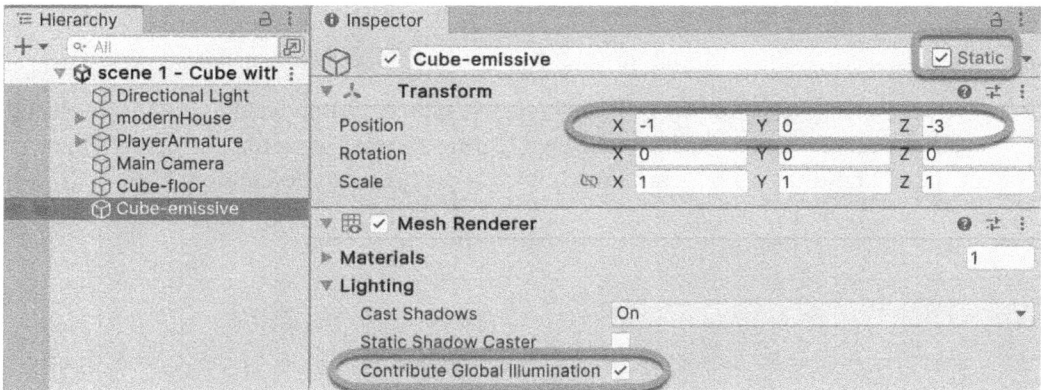

Figure 13.54: The properties for our emissive 3D Cube

7. Create a new **Material** named m_greenEmissive, and assign this **Material** to the Cube-emissive GameObject.

8. Select the **Material** asset file m_greenEmissive in the **Project** panel and check its **Emission** property. You can now click its **HDR** color picker and choose a bright green color. You can also set the **Intensity** value of this emission, either with the slider, by entering a number such as 1 or 2, or clicking the relevant buttons to increase/decrease the **Intensity** value by 1 or 2.

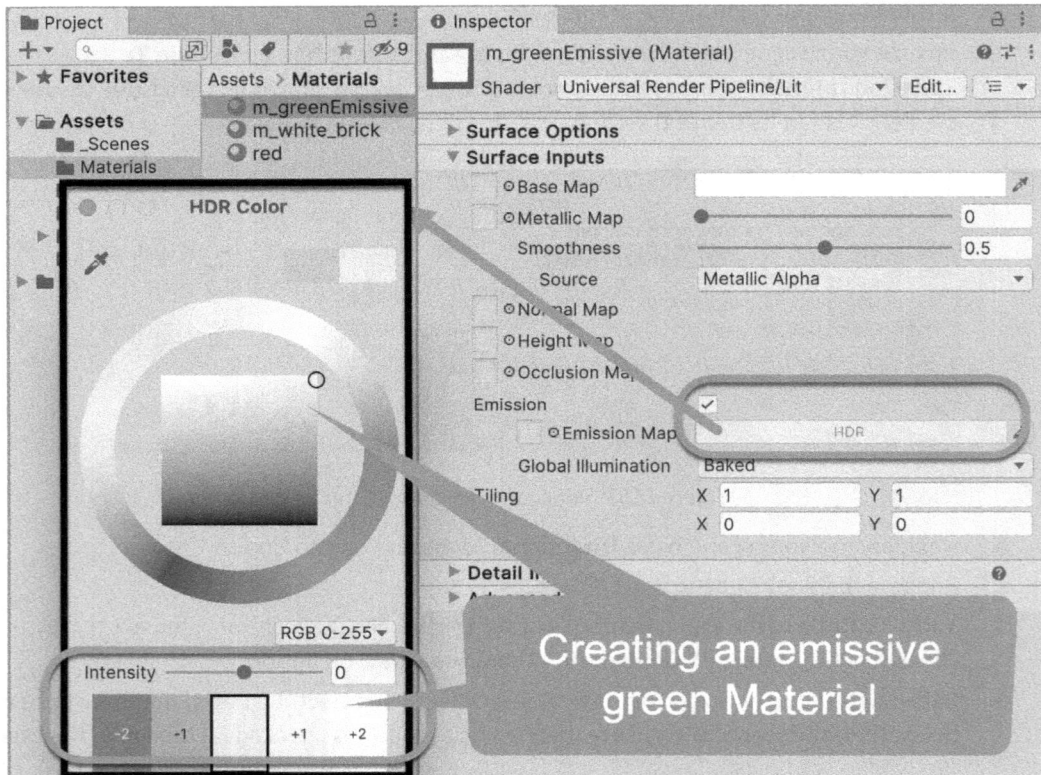

Figure 13.55: Creating an emissive green Material

9. Open the **Lighting Settings** window (via the **Window | Rendering | Lighting Settings** menu). Click the **New** button to create a **New Lighting Settings** asset for this scene. Then, under the **Mixed Lighting** section, check the **Baked Global Illumination** option:

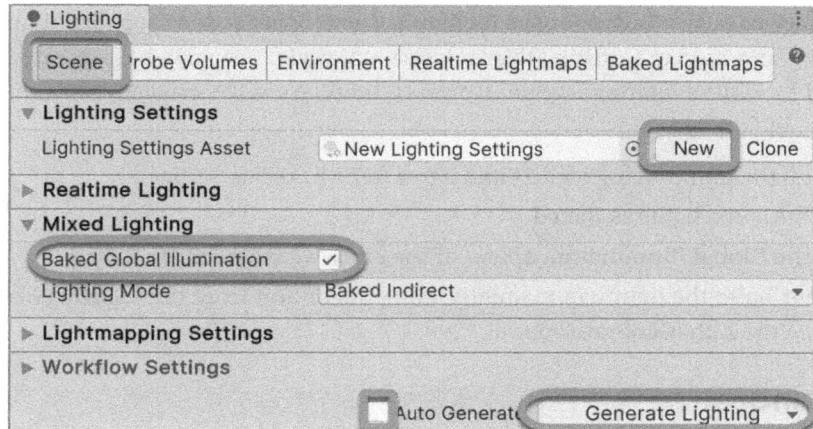

Figure 13.56: Setting up the Scene Lighting settings

10. Drag the **Directional Light** from the **Hierarchy** into the **Sun** slot under the **Scene Environment** properties. Also, set the **Environment Lighting Ambient Mode** drop-down menu to **Baked**. In **Debug Settings**, ensure that **Auto Generate** is unchecked. Finally, click the **Generate Lighting** button to bake the Ambient light and green lamp emission light into the scene.

11. For a few seconds (depending on the speed of your computer and the complexity of the scene), you'll see a progress bar of the lightmap baking process, at the bottom-right of the Unity Editor application window.

Figure 13.57: The lightmap baking progress bar in the Unity Editor window footer

12. Save and play your scene. You should see how the scene objects are lit both by the **Directional Light,** and by the green **Texture** emitted from the lamp.

13. Change the rotation of the **Directional Light,** and try setting its **Light Intensity** and **Indirect Multiplier** values to 0.5, and play with the HDR intensity of the m_greenEmissive **Material**, then re-bake the **Lightmap** to make the lamp emission more emphasized (and the **Directional Light** play a lesser role).

How it works...

You have added an emissive **Material** to a GameObject (the lamp) and baked a lightmap based on the static objects in the **scene** (which include the 3D Cube, 3D Spheres, walls, and a textured Cube for the floor).

The lighting was configured so that the scene was lit by Baked Global Illumination – so static objects were lit by the emissions from our Cube.

Lightmaps are basically **Texture** maps including **scene** lights/shadows, global illumination, indirect illumination, and objects featuring the **emissive** materials. They can be generated automatically or on-demand by Unity's lighting engine. However, there are some crucial points that you should pay attention to, including:

- Set all the non-moving objects and lights to be baked as **Static**
- Set the game lights as **Baked**
- Set the **Global Illumination** option of the emissive **materials** to Baked
- Either build the lightmap manually using the button from the **Lighting Settings** window, or check the **Auto Generate** option

There's more...

There are some more details that you don't want to miss.

Emissive materials that do not contribute to lighting other objects in the scene

Sometimes we want something to be emissive, such as a neon-style sign or object, but not actually contribute to lighting up other objects in the scene. In these cases, ensure that the GameObject with the emissive **Material** in question has its **Contribute Global Illumination** option unchecked.

Further reading

Projects made using the URP are not compatible with the HDRP or the BIRP. Before you start development, you must decide which Render Pipeline you want to use in your project. Some good sources about the different Render Pipelines can be found as follows:

- The Unity manual provides some information on how to choose a Render Pipeline: `https://docs.unity3d.com/Manual/render-pipelines.html`
- A great web article from Michael Besnard about Unity render pipelines: `https://myjl-besnard.medium.com/what-are-render-pipelines-in-unity-bb9bcd02cf98`
- You can learn more about the URP from the Unity manual, blog, and YouTube channel:
    - `https://docs.unity3d.com/Packages/com.unity.render-pipelines.universal@9.0/manual/index.html`
    - `https://blogs.unity3d.com/2020/08/26/learn-how-to-bring-your-game-graphics-to-life/`
    - `https://www.youtube.com/watch?v=m6YqTrwjpP0`

A complete list of postprocessing volume effects can be found here: `https://docs.unity3d.com/Packages/com.unity.render-pipelines.universal@7.1/manual/EffectList.html`.

Some great Cinemachine sources include:

- Lots of Unity Learn tutorials: `https://learn.unity.com/search/?k=%5B%221ang%3Aen%22%2C%22q%3Acinemachine%22%5D`
- The Unity blog about Cinemachine and the Timeline: `https://blog.unity.com/community/community-stories-cinemachine-and-timeline`
- The Unity Cinemachine documentation: `https://docs.unity3d.com/Packages/com.unity.cinemachine@2.1/manual/index.html`
- CG Auro's CutScene in Unity 3D video can be found at `https://www.youtube.com/watch?v=w6lc8svzBms`

Learn more about postprocessing and vignettes at:

- `https://docs.unity3d.com/Packages/com.unity.render-pipelines.high-definition@17.0/manual/index.html`
- `https://www.youtube.com/watch?v=cuvx9tWGlHc`

Some great sources of textures for your games:

- Poly Haven Creative Commons licensed high quality textures, HDRI cubes, and models (thanks guys!): `https://polyhaven.com/`
- Architextures is a great source for architectural textures: `https://architextures.org/`

Here are some more resources that provide more details about this chapter's topics:

- The Unity manual lighting page: `https://docs.unity3d.com/Manual/Lighting.html`
- Unity manual Skybox component: `https://docs.unity3d.com/Manual/class-Skybox.html`
- Unity lightmapping resources:
    - `https://docs.unity3d.com/Manual/Lightmappers.html`
    - `https://learn.unity.com/tutorial/configuring-lightmaps`
- Unity's **Global Illumination** (GI) pages:
    - `https://docs.unity3d.com/Manual/realtime-gi-using-enlighten.html`
    - `http://docs.unity3d.com/Manual/GlobalIllumination.html`
- Information about Unity's **Cookie Textures** can be found on the following manual page: `https://docs.unity3d.com/Manual/Cookies.html`
- Another source for Unity and **Cookie Textures** is the *CgProgramming WikiBook* for Unity: `https://en.wikibooks.org/wiki/Cg_Programming/Unity/Cookies`
- Unity manual page about choosing a color space: `https://unity3d.com/learn/tutorials/topics/graphics/choosing-color-space`
- Unity manual about the **Lighting Explorer** window: `https://docs.unity3d.com/Manual/LightingExplorer.html`

Unity offers a choice of two color spaces: **Gamma** (the default) and **Linear**. You can select your desired **color space** by going to **Edit | Project Settings | Player**. While **Linear** space has significant advantages, it isn't supported by all hardware (especially mobile systems), so the appropriate choice for you will depend on which platform you are deploying on. For more information about **Linear** and **Gamma** lighting workflows, go to: `https://docs.unity3d.com/Manual/LinearRendering-LinearOrGammaWorkflow.html`

Learn more on Discord

To join the Discord community for this book – where you can share feedback, ask questions to the author, and learn about new releases – follow the QR code below:

`https://packt.link/unitydev`

14

Shader Graphs and Video Players

Two powerful features of Unity are the Video Player API and the Shader Graph tool. Between them, they offer easy and configurable ways configurable ways to work with visual content in games. For example, they help with loading and playing videos on different visible objects, and they also provide a comprehensive way for non-shader programmers to construct sophisticated shader transformations using a visual graphing approach. This chapter explores the Unity Shader Graph and Video Player components.

The Shader Graph tool

Shader Graph is a tool that allows us to visually build shaders by creating and connecting the input and output of nodes.

Some great Shader Graph features include the following:

- An instant, visual preview of each node in the graph is provided so that you can see how different nodes are contributing to the final master output node.
- Properties can be publicly exposed in the graph (via the blackboard) so that they become customizable values in the **Inspector** panel for the material using Shader Graph.
- Publicly exposed properties can also be accessed and changed via scripts.

- The output of one node can become one of the inputs to another node, so a sophisticated shader can be created from many combined, simple component nodes:

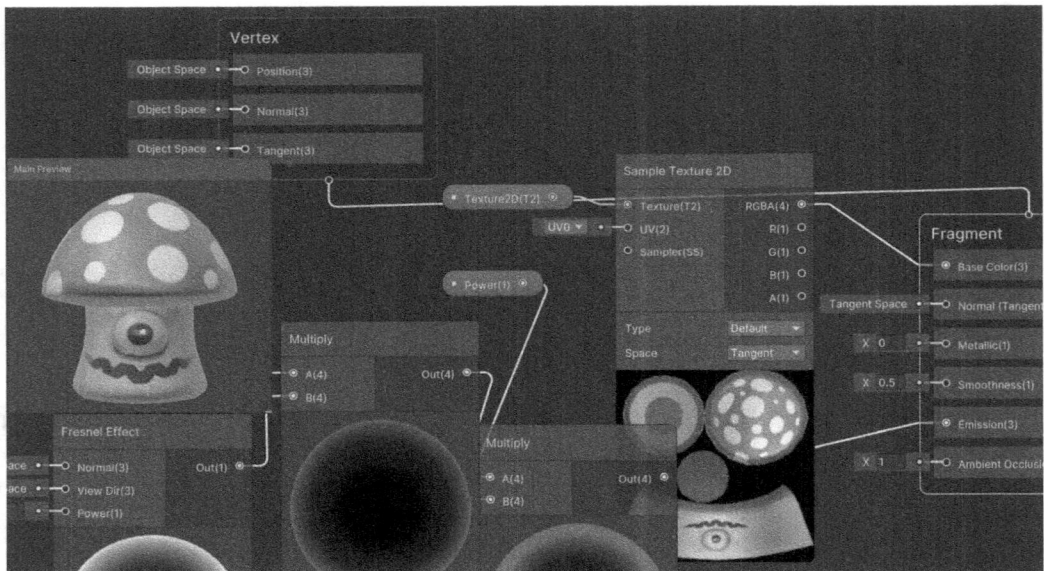

Figure 14.1: Example of how a Shader Graph is composed

As shown in the preceding screenshot, a Shader Graph is composed of a graph of connected nodes, where the output of one node becomes the input of another node. Node input/output can be numeric values, textures, noise, Boolean true/false values, and colors, and other Shader Graph files are created in the **Project** window and can be selected as a graph shader in the **Shader** property of a material component.

Unity uses a **Physically Based Rendering** (**PBR**) approach that attempts to render images that model the flow of light in the real world. This is achieved by using Shader Graph, and several recipes will be presented in this chapter to introduce some of the powerful features of Shader Graph and its workflow.

Playing videos with the Video Player API

Playing videos is as simple as manually adding a VideoPlayer component in the **Inspector** panel to a GameObject at design time and associating a VideoClip asset file from the **Project** panel, or providing the URL of an online resource.

Videos can be played on the camera's far plane (appearing behind the scene content) or near plane (appearing in front of content – often with semi-transparency). Video content can also be directed to a RenderTexture asset, which can then be displayed on a 2D or 3D object in the scene, via a **material** component. The internal texture that's used by a VideoPlayer component can also be mapped to a texture on screen – such as a UI RawImage.

Scripting can be used to manage video playback for single videos and arrays (sequences) of video clips. Several recipes will be presented in this chapter to introduce these different ways of working with VideoPlayer.

In this chapter, we will cover the following recipes:

- Playing videos by manually adding a VideoPlayer component to a GameObject
- Using scripting to control video playback on scene textures
- Ensuring a movie is prepared before playing it
- Outputting video playback to a render texture
- Using scripting to play a sequence of videos back to back
- Creating and using a simple shader graph
- Using Shader Graph to create a color glow effect
- Toggling a Shader Graph color glow effect through C# code

Playing videos by manually adding a VideoPlayer component to a GameObject

If you want complex animated materials in your games, you can play video files as texture maps – for example, you can simulate TV sets, projectors, and monitors by playing a video on their virtual screens. In this recipe, we will learn how to add and use VideoPlayer components to the Main Camera GameObject. By default, the Main Camera fills the entire visible game window and has a renderer component, making it an ideal location for adding a VideoPlayer component when we want the video images to fill the whole screen.

Getting ready

If you need a video file so that you can follow this recipe, please use the 54321TestVideo.mov file included in the 14_01 folder.

How to do it...

To place videos manually with a VideoPlayer component, follow these steps:

1. Create a new Unity 3D project.
2. Import the provided 54321TestVideo.mov file.
3. Add a 3D cube to the scene by going to **GameObject | 3D Object | Cube**.
4. Select the Main Camera GameObject and, in the **Inspector** panel, add a VideoPlayer component by clicking **Add Component** and choosing **Video | Video Player**. Unity will have noticed that we are adding the VideoPlayer component to a camera, so it will have set up the default properties correctly for us:

   - **Play On Awake** (Checked)
   - **Wait For First Frame** (Checked)

- **Render Mode: Camera Far Plane**
- **Camera: Main Camera (Camera)**

5. Drag the video clip asset file called `videoTexture` from the **Project** window into the **Video Clip** property slot in the **Inspector** panel:

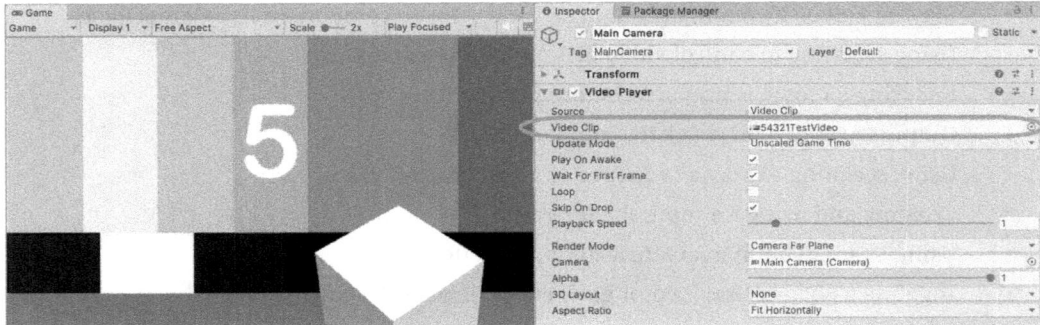

Figure 14.2: The Inspector panel of Main Camera showing the source Video Clip

6. Save and run your scene.

You should be able to see the movie being played behind the scene's content. Notice that there are a variety of options for how the video should be displayed on the screen. For example, you can choose whether to stretch the video content by changing the **Aspect Ratio** property in the **Inspector** panel.

How it works...

We gave the `VideoPlayer` component a reference to a `VideoClip` asset file. Since we added the `VideoPlayer` component to a camera (`Main Camera`, in this example), it automatically chose **Camera Far Play Render Mode**, which is linked to the `Main Camera` GameObject.

The default setting is **Play On Awake**, so as soon as the first frame has loaded (since **Wait For First Frame** is also checked by default), the video will start playing. The video is displayed behind all `Main Camera` content (the far plane). Because of this, we can see our 3D cube in the scene, with the video playing in the background.

There's more...

Sometimes, we may want to play a video so that it's the user's main focus, but allow them to see scene objects in the background.

To achieve this with the `VideoPlayer` component, we just need to make two changes:

- Change **Render Mode** to **Near Camera Plane** (so that the video content is played in front of the scene's content).
- To allow the user to partially see through the video, we need to make **Video Player** semi-transparent. To do so, change its **Alpha** property to `0.5`.

Now, when you run the scene, you'll see the video playing in front of the scene's content, but you will also be able to see the 3D cube in the background.

> At the time of writing this book, there seemed to be issues with the **direct** option for **Audio Output Mode** audio playback for some non-Apple systems. One solution is to add an AudioSource component to the same GameObject that has the VideoPlayer component and to set **Audio Output Mode** to AudioSource.

Using scripting to control video playback on scene textures

While the previous recipe demonstrated how we can place videos using the VideoPlayer component, which is set up at design time, much more is possible when controlling video playback through scripting – for example, allowing the user to interactively pause/play/stop video playback:

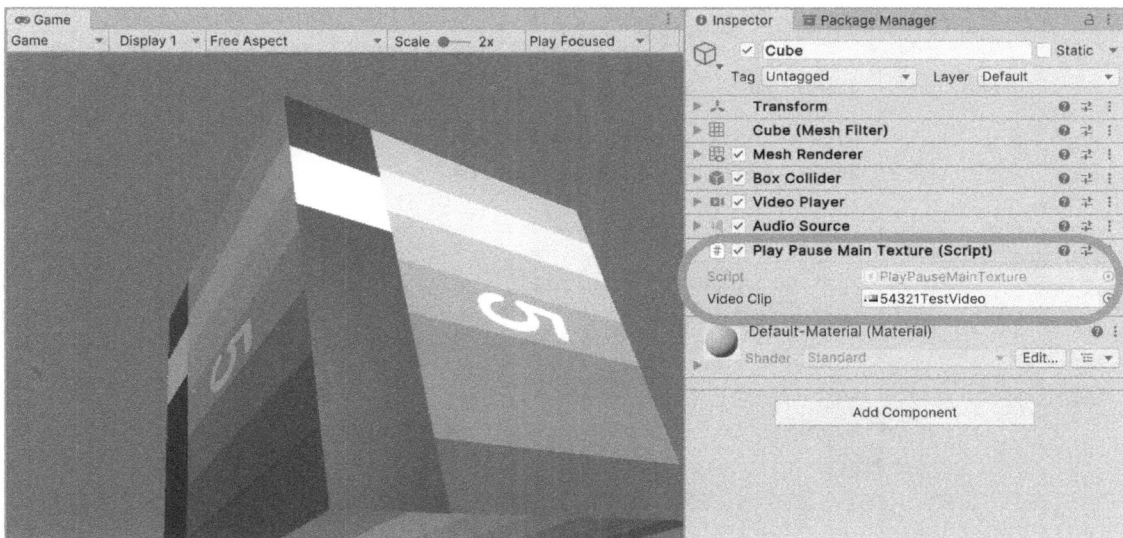

Figure 14.3: Example of using a script to control the video

In this recipe, we'll use scripting to play/pause the playback of a video rendered onto a 3D cube.

Getting ready

If you need a video file to follow this recipe, please use the 54321TestVideo.mov file included in the 14_02 folder.

How to do it...

To use scripting to control video playback, follow these steps:

1. Create a new Unity 3D project.

2. Import the provided 54321TestVideo.mov file.

3. Create a 3D cube by going to **GameObject | 3D Object | Cube**.

4. Create a C# script class named PlayPauseMainTexture and attach an instance object as a component to your 3D cube GameObject:

```csharp
using UnityEngine;
using UnityEngine.Video;

[RequireComponent(typeof(VideoPlayer))]
[RequireComponent(typeof(AudioSource))]

public class PlayPauseMainTexture : MonoBehaviour {
  public VideoClip videoClip;

  private VideoPlayer videoPlayer;
  private AudioSource audioSource;

  void Start() {
    videoPlayer = GetComponent<VideoPlayer>();
    audioSource = GetComponent<AudioSource>();

    // disable Play on Awake for both vide and audio
    videoPlayer.playOnAwake = false;
    audioSource.playOnAwake = false;

    // assign video clip
    videoPlayer.source = VideoSource.VideoClip;
    videoPlayer.clip = videoClip;

    // setup AudioSource
    videoPlayer.audioOutputMode = VideoAudioOutputMode.AudioSource;
    videoPlayer.SetTargetAudioSource(0, audioSource);

    // render video to main texture of parent GameObject
    videoPlayer.renderMode = VideoRenderMode.MaterialOverride;
    videoPlayer.targetMaterialRenderer = GetComponent<Renderer>();
    videoPlayer.targetMaterialProperty = "_MainTex";
  }

  void Update() {
    // space bar to start / pause
```

```
        if (Input.GetButtonDown("Jump"))
            PlayPause();
    }

    private void PlayPause() {
        if (videoPlayer.isPlaying)
            videoPlayer.Pause();
        else
            videoPlayer.Play();
    }
}
```

5. Ensure that your 3D cube is selected in the **Project** panel. Then, drag the Video Clip asset file called 54321TestVideo.mov from the **Project** panel into the **Video Clip** property slot of the PlayPauseMainTexture component (script) in the **Inspector** panel.

6. Run your scene. Pressing the *Spacebar* should play/pause playback of the video on the surfaces of the 3D cube. You should also hear the beeping audio for the video.

How it works...

We have explored the basics of using scripting and the VideoPlayer component. As well as defining and setting up where VideoPlayer will render, as we need to each time, in the steps above, we have completed the following:

1. Creating or getting references to the VideoPlayer and AudioSource components (we will automatically have both components for this recipe since we have the RequireComponent(...) script instructions immediately before our class declaration):

```
videoPlayer = GetComponent<VideoPlayer>();
audioSource = GetComponent<AudioSource>();
```

2. Setting their **Play On Awake** properties to true/false:

```
videoPlayer.playOnAwake = false;
audioSource.playOnAwake = false;
```

3. Defining where VideoPlayer will find a reference to the video clip to play:

```
videoPlayer.source = VideoSource.VideoClip;
videoPlayer.clip = videoClip
```

4. Defining the audio settings (so that you can output to the AudioSource component):

```
videoPlayer.audioOutputMode = VideoAudioOutputMode.AudioSource;
videoPlayer.setTargetAudioSource(0, audioSource);
```

In this recipe, we followed these four steps by adding the instance object of our `PlayPauseMainTexture` scripted class to the 3D cube and dragging a reference to a `Video Clip` asset file to the public slot. In the following code, we're telling the `VideoPlayer` component to override the material of the object it is a component of (in this case, the 3D cube) so that `VideoPlayer` will render (display) on the main texture of the 3D cube:

```
videoPlayer.renderMode = VideoRenderMode.MaterialOverride;
videoPlayer.targetMaterialRenderer = GetComponent<Renderer>();"
videoPlayer.targetMaterialProperty = "_MainTex";
```

There's more...

Here are some additional ways to work with video player scripting. Sometimes, the video we want is not available as a video clip file on the computer we are using – instead, it is available online via a URL. In this section, we will examine how to adapt our script so that the source of the video can be provided as a URL.

Downloading an online video (rather than a clip)

Rather than dragging an existing `Video Clip` asset file to specify which video to play, `VideoPlayer` can also download video clips from an online source. To do this, we need to assign a string URL to the video player's URL property.

To download a video, do the following:

1. Replace the property declaration line `public VideoClip videoClip;` with this public array of strings, in which one or more URLs can be defined:

    ```
    public string[] urls = {
        "http://mirrors.standaloneinstaller.com/video-sample/grb_2.mov",
        "http://mirrors.standaloneinstaller.com/video-sample/lion-
    sample.mov"
    };
    ```

2. Declare a new method that returns one URL string, randomly chosen from the array:

    ```
    public string RandomUrl(string[] urls)
    {
        int index = Random.Range(0, urls.Length);
        return urls[index];
    }
    ```

3. Finally, in the `SetupVideoAudioPlayers()` method, replace the line `videoPlayer.clip = videoClip;` with the following:

    ```
    // assign video clip
    string randomUrl = RandomUrl(urls);
    videoPlayer.url = randomUrl;
    ```

Ensuring a movie is prepared before playing

In the preceding recipe, the movie has time to be prepared since the game waits until we press the jump/*Spacebar* key. If we are using scripting to set up a video player for a video clip, we need to do some initial work before the video is ready to play. Unity provides the `prepareCompleted` event for this, which allows us to register a method to be invoked once `VideoPlayer` is ready to play.

Getting ready

This recipe follows on from the previous one, so make a copy of that and work on the copy.

How to do it...

To ensure a movie is prepared before it's played by subscribing to a `prepareCompleted` event, do the following:

1. Add a UI `RawImage` to the scene by going to **GameObject | UI | Raw Image**.
2. Create a new, empty **GameObject** named `video-object`.
3. Create a C# script class named `PrepareCompleted` and attach an instance of the script as a component to the `video-object` GameObject:

```csharp
using UnityEngine;
using UnityEngine.UI;
using UnityEngine.Video;

public class PrepareCompleted: MonoBehaviour {
  public RawImage image;
  public VideoClip videoClip;

  private VideoPlayer videoPlayer;
  private AudioSource audioSource;

  void Start() {
    SetupVideoAudioPlayers();
    videoPlayer.prepareCompleted += PlayVideoWhenPrepared;
    videoPlayer.Prepare();
    Debug.Log("A - PREPARING");
  }

  private void SetupVideoAudioPlayers() {
    videoPlayer = gameObject.AddComponent<VideoPlayer>();
    audioSource = gameObject.AddComponent<AudioSource>();

    videoPlayer.playOnAwake = false;
```

```
        audioSource.playOnAwake = false;

        videoPlayer.source = VideoSource.VideoClip;
        videoPlayer.clip = videoClip;

        videoPlayer.audioOutputMode = VideoAudioOutputMode.AudioSource;
        videoPlayer.SetTargetAudioSource(0, audioSource);
    }

    private void PlayVideoWhenPrepared(VideoPlayer theVideoPlayer) {
        Debug.Log("B - IS PREPARED");
        image.texture = theVideoPlayer.texture;
        Debug.Log("C - PLAYING");
        theVideoPlayer.Play();
    }
}
```

4. Ensure that the video-object GameObject is selected in the **Project** panel. Now, drag the Raw Image GameObject from the **Hierarchy** panel into the **Raw Image** slot. Then, drag the Video Clip asset file called 54321TestVideo.mov from the **Project** panel into the **Video Clip** property slot of the PrepareCompleted component (script) in the **Inspector** panel.

5. Save and run your scene. You should be able to see the movie being played behind the scene's content.

How it works...

As you can see, in the Start() method, we register a method named PlayVideoWhenPrepared with the videoPlayer.prepareCompleted event, before invoking the Prepare() method of the videoPlayer component:

```
videoPlayer.prepareCompleted += PlayVideoWhenPrepared;
videoPlayer.Prepare();
```

The PlayVideoWhenPrepared(...) method has to accept a parameter as a reference to a VideoPlayer object. A UI GameObject was added to the scene displaying a **RawImage** texture.

We directly assigned the VideoPlayer object's texture property to the UI GameObject's RawImage's texture. Then, we sent the Play() message.

You can track the progress of clip preparation and so on through the **log** messages in the **Console** panel.

Outputting video playback to a RenderTexture asset

Directly working with the VideoPlayer texture works for the previous recipe, but usually, setting up a separate RenderTexture asset is more reliable and flexible. This recipe will build on the previous one to demonstrate this approach.

A flexible way to work with video players is to output their playback to a `RenderTexture` asset file. A material can be created to get input from `RenderTexture` and, using that material, GameObjects will display the video. Also, some GameObjects can directly have `RenderTexture` assigned to their texture.

Getting ready

This recipe follows on from the previous one, so make a copy of that and work on the copy.

How to do it...

To output video playback to a `RenderTexture` asset, do the following:

1. In the **Project** panel, create a new `Render Texture` asset file named `myRenderTexture` by going to **Create | Render Texture**.
2. Select the UI RawImage in the **Hierarchy** panel, and assign its **Raw Image (Script)** texture property to the `myRenderTexture` asset file.
3. In the **Project** panel, create a new `Material` asset file named `m_video`. For this material, in the **Inspector** panel, set its **Albedo Texture** property to `myRenderTexture` (drag it from the **Project** panel into the **Inspector** panel).
4. Create a new 3D capsule in the scene and assign the `m_video` material to it.
5. Edit the C# script class called `PrepareCompleted` by replacing the public `rawImage` variable with a public `renderTexture` variable:

    ```
    public VideoClip videoClip;
    public RenderTexture renderTexture;
    ```

6. Edit the C# script class called `PrepareCompleted` by adding the following statements at the end of the `SetupVideoAudioPlayers()` method to output video to `RenderTexture`:

    ```
    videoPlayer.renderMode = VideoRenderMode.RenderTexture;
    videoPlayer.targetTexture = renderTexture;
    ```

7. Edit the `PlayVideoWhenPrepared()` method in the C# script class called `PrepareCompleted`. Remove the statement that directly assigns the `VideoPlayer` component's texture property to the RawImage's texture:

    ```
    private void PlayVideoWhenPrepared(VideoPlayer theVideoPlayer) {
        Debug.Log("B - IS PREPARED");

        // Play video
        Debug.Log("C - PLAYING");
        theVideoPlayer.Play();
    }
    ```

8. Ensure that the video-object GameObject is selected in the **Project** panel. Now, drag the myRenderTexture asset from the **Project** panel into the **Render Texture** public property of **Prepare Completed (Script)** in the **Inspector** panel:

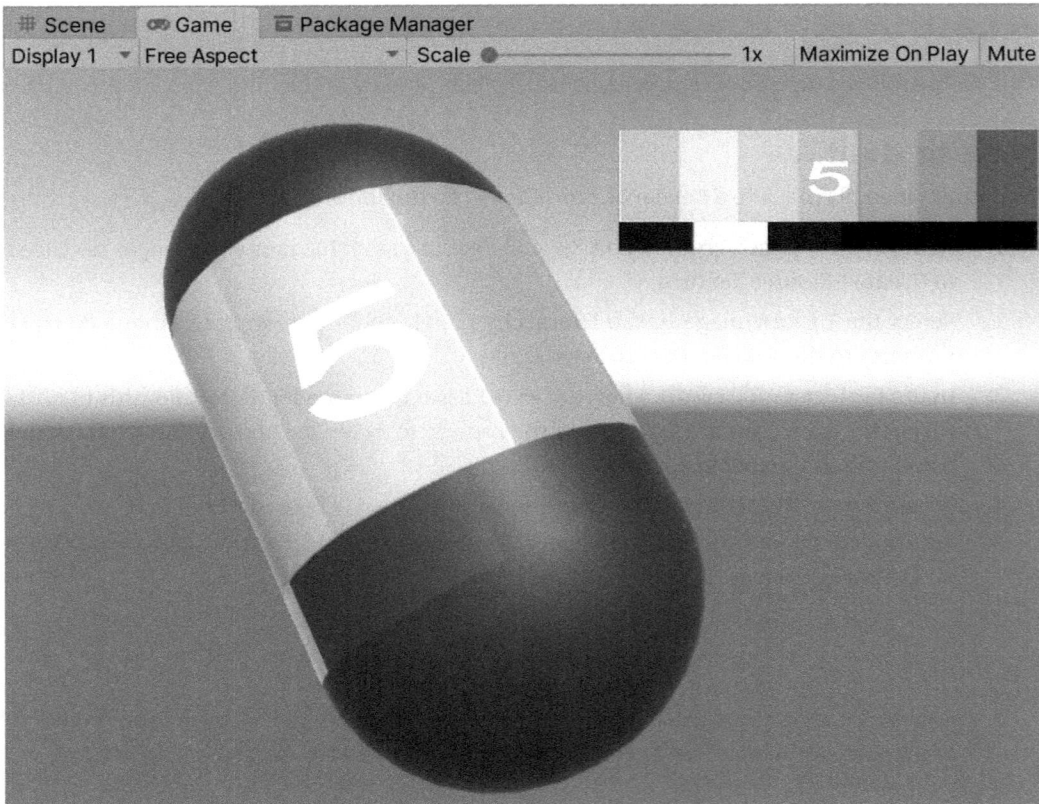

Figure 14.4: Example of a video being rendered onto a capsule

9. Save and run the scene. You should now see the video playing both in the UI RawImage and also rendered over the 3D capsule object.

How it works...

A render texture is an asset file (in the **Project** panel). This file can be written to with the current frame of a video being played. In this recipe, you created a render texture asset file named myRenderTexture. We had to set two properties of the VideoPlayer object in order to write the video to our myRenderTexture:

- renderMode was set to a public constant value called VideoRenderMode.RenderTexture
- targetTexture was set to a new public variable called renderTexture

We modified the UI RawImage in the canvas and set its **Raw Image (Script)** texture property to the myRenderTexture asset file. This means that whatever image is currently in myRenderTexture will be displayed on the screen as a flat, rectangular image in the UI RawImage.

Finally, we had to assign our asset file to the public variable. We did this by dragging myRenderTexture from the **Hierarchy** panel into the public variable named renderTexture in the **Inspector** panel. Having made these changes, the video plays both as the texture on the capsule GameObject and also via the Render Texture asset file called myRenderTexture to the UI RawImage in **the canvas**.

Using scripting to play a sequence of videos back to back

One of the advantages of scripting is that it allows us to easily work with multiple items through loops, arrays, and so on. In this recipe, we'll work with an array of Video Clip assets and use scripting to play them back to back (one starts as soon as the previous clip finishes), illustrating the use of the isPrepared and loopPointReached events to avoid complicated loops and coroutines.

Getting ready

If you need video files to follow this recipe, please use the 54321TestVideo.mov file included in the 14_01 folder and the jellyfish.wmv video clip in 14_04.

> The Standalone Installer website is a good online source of test videos: http://mirrors. standaloneinstaller.com/video-sample/

How to do it...

To play a sequence of videos using scripting, follow these steps:

1. Create a new Unity 3D project.
2. Import the provided 54321TestVideo.mov file, and perhaps a second video clip, so that we have a sequence of two different videos to test (although you can run the same one twice if you wish).
3. In the **Project** panel, create a new Render Texture asset file named myRenderTexture by going to **Create | Render Texture**.
4. Add a UI RawImage to the scene by going to **GameObject | UI | Raw Image**.
5. Select the UI RawImage in the **Hierarchy** panel and assign its **Raw Image (Script)** texture property to the myRenderTexture asset file.
6. In the **Project** panel, create a new Material asset file named m_video. For this material, in the **Inspector** panel, set its **Albedo Texture** property to myRenderTexture (drag it from the **Project** panel into the **Inspector** panel).
7. Create a 3D cube by going to **GameObject | 3D | Cube**. Assign the m_video material to your 3D cube.
8. Create a new, empty GameObject named video-object.
9. Create a C# script class named VideoSequenceRenderTexture and attach an instance of the script as a component of the video-object GameObject:

```
using UnityEngine;
```

```
using UnityEngine.Video;

public class VideoSequenceRenderTexture : MonoBehaviour {
    public RenderTexture renderTexture;
    public VideoClip[] videoClips;

    private VideoPlayer[] videoPlayers;
    private int currentVideoIndex;

    void Start() {
        SetupObjectArrays();
        currentVideoIndex = 0;
        videoPlayers[currentVideoIndex].prepareCompleted +=
PlayNextVideo;
        videoPlayers[currentVideoIndex].Prepare();
        Debug.Log("A - PREPARING video: " + currentVideoIndex);
    }

    private void SetupObjectArrays() {
        videoPlayers = new VideoPlayer[videoClips.Length];
        for (int i = 0; i < videoClips.Length; i++)
            SetupVideoAudioPlayers(i);
    }

    private void PlayNextVideo(VideoPlayer theVideoPlayer) {
        VideoPlayer currentVideoPlayer = videoPlayers[currentVideoIndex];

        Debug.Log("B - PLAYING Index: " + currentVideoIndex);
        currentVideoPlayer.Play();

        currentVideoIndex++;
        bool someVideosLeft = currentVideoIndex < videoPlayers.Length;

        if (someVideosLeft) {
            VideoPlayer nextVideoPlayer =
videoPlayers[currentVideoIndex];
            nextVideoPlayer.Prepare();
```

```
                       Debug.Log("A - PREPARING video: " + currentVideoIndex);
                       currentVideoPlayer.loopPointReached += PlayNextVideo;
                } else {
                       Debug.Log("(no videos left)");
                }
           }


     private void SetupVideoAudioPlayers(int i) {
           string newGameObjectName = "videoPlayer_" + i;
           GameObject containerGo = new GameObject(newGameObjectName);
           containerGo.transform.SetParent(transform);
           containerGo.transform.SetParent(transform);


           VideoPlayer videoPlayer = containerGo.
AddComponent<VideoPlayer>();
           AudioSource audioSource = containerGo.
AddComponent<AudioSource>();


           videoPlayers[i] = videoPlayer;


           videoPlayer.playOnAwake = false;
           audioSource.playOnAwake = false;


           videoPlayer.source = VideoSource.VideoClip;
           videoPlayer.clip = videoClips[i];


           videoPlayer.audioOutputMode = VideoAudioOutputMode.AudioSource;
           videoPlayer.SetTargetAudioSource(0, audioSource);


           videoPlayer.renderMode = VideoRenderMode.RenderTexture;
           videoPlayer.targetTexture = renderTexture;
           }
     }
```

10. Ensure that the video-object GameObject is selected in the **Project** panel. Now, drag the myRenderTexture asset from the **Project** panel into the **Render Texture** public property of VideoSequenceRenderTexture (Script) in the **Inspector** panel. For the **Video Clips** property, set its size to 2. You should now see two video clip elements (elements 0 and 1).

From the **Project** panel, drag a video clip into each slot:

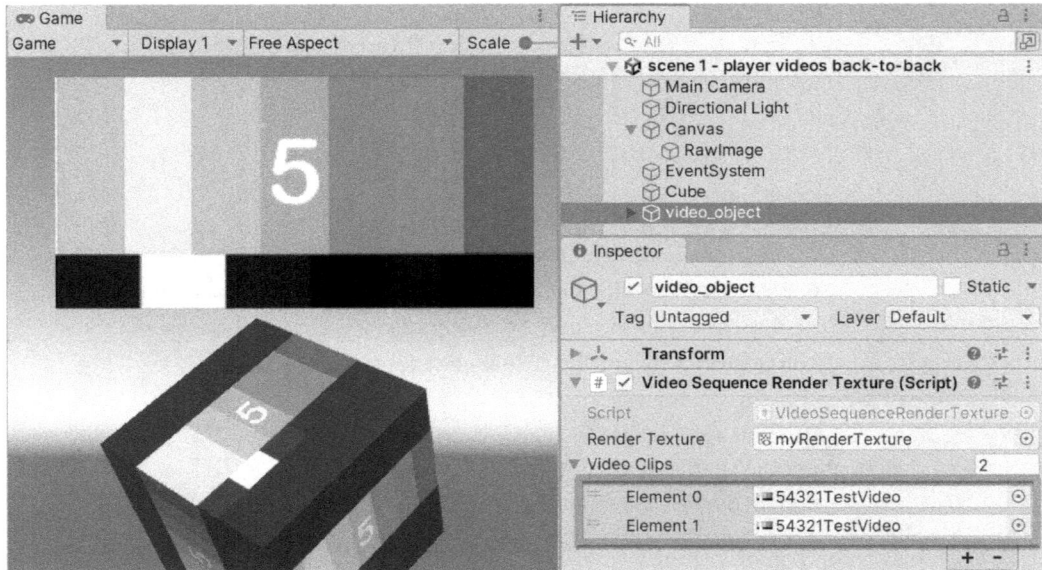

Figure 14.5: Settings for Video Sequence Render Texture of video-object

11. Save and run the scene. You should now see the first video clip playing both for the UI RawImage and the 3D cube surface. Once the first video clip has finished playing, the second video clip should immediately start playing.

You can track the progress of clip preparation and so on through the **Log messages** section of the **Console** panel.

How it works...

This script class makes the Video Player objects output their videos to the render texture's asset file, myRenderTexture. This is used by both the 3D cube and the UI RawImage for their surface displays.

The videoClips variable is a public array of video clip references.

The instance object of the C# script class called VideoSequenceRenderTexture was added as a component to the video-object GameObject. This script will create child GameObjects of the video-object GameObject, each containing a VideoPlayer and AudioSource component, ready to play each of the video clips that have been assigned in the public array's videoClips variables.

The SetupObjectArrays() method initializes videoPlayers to be an array that's the same length as videoClips. It then loops for each item, invoking SetupVideoAudioPlayers(...) by passing the current integer index.

The SetupVideoAudioPlayers(...) method creates a new child GameObject for the GameObject's video-object and adds the VideoPlayer and AudioSource components to that GameObject. It sets the Video Player clip property to the corresponding element in the public videoClips array variable. It also adds a reference to the new VideoPlayer component to the appropriate location in the videoPlayers array. It then sets the video player to output audio to the new AudioSource component, and to output its video to the renderTexture public variable.

The Start() method does the following:

- It invokes SetupObjectArrays().
- It sets the currentVideoIndex variable to 0 (for the first item in the arrays).
- It registers the PlayNextVideo method for the prepareCompleted event of the first videoPlayers object (currentVideoIndex = 0).
- It invokes the Prepare() method for the videoPlayers object (currentVideoIndex = 0).
- It logs a debug message stating that the item is being prepared.

The PlayNextVideo(...) method gets a reference to the Video Player element of the videoPlayers array that corresponds to the currentVideoIndex variable.

This method ignores the reference to the videoPlayer argument it receives – this parameter is required in the method declaration since it is the required signature to allow this method to register for the prepareCompleted and loopPointReached events. It performs the following steps:

1. First, it sends a Play() message to the current video player.
2. Then, it increments the value of currentVideoIndex and tests whether there are any remaining video clips in the array.
3. If there are remaining clips, then it gets a reference to the next clip and sends it a Prepare() message. Also, the video player that's currently being played has its loopPointReached event registered for the PlayNextVideo method (if there are no videos left, then a simple debug log message is printed and the method ends).

The clever bit is when the currently playing video player has its loopPointReached event registered for the PlayNextVideo method. The loopPointReached event occurs when a video clip has finished playing and will start to look again for the next video to play (if its loop property is true). What we are doing with this script is saying that when the video player's video clip has finished, the PlayNextVideo(...) method should be invoked again – once again using the value of currentVideoIndex to send a Play() message to the next video player, and then testing for any remaining video players, and so on until the end of the array has been reached.

This is a good example of conditions (if statements) being used with events, rather than coroutine while loops. So long as you're happy with how methods can be registered with C# events, then this approach allows our code to be less complex by avoiding loops and coroutine yield null statements.

In the following screenshot, we can see how our `video-object` GameObject, at runtime, ends up with `videoPlayer_0` and `videoPlayer_1` child GameObjects – one for each element in the array. This allows one `VideoPlayer` to be playing while the next is preparing, and so on:

Figure 14.6: Example highlighting the child elements at runtime, one for each element in the array

Creating and using a simple shader

The Shader Graph feature in Unity is a powerful and exciting feature that opens up shader creation and editing to everyone, without any need for complex mathematics or coding skills. In this recipe, we'll create a simple **shader** to generate a checkerboard pattern, create a material that uses that shader, and then apply it to a 3D cube. It is worth noting that the Shader Graphs known as PBR Shader Graphs in previous versions of Unity is now known as URP Lit shaders. The end result of this recipe will be as follows:

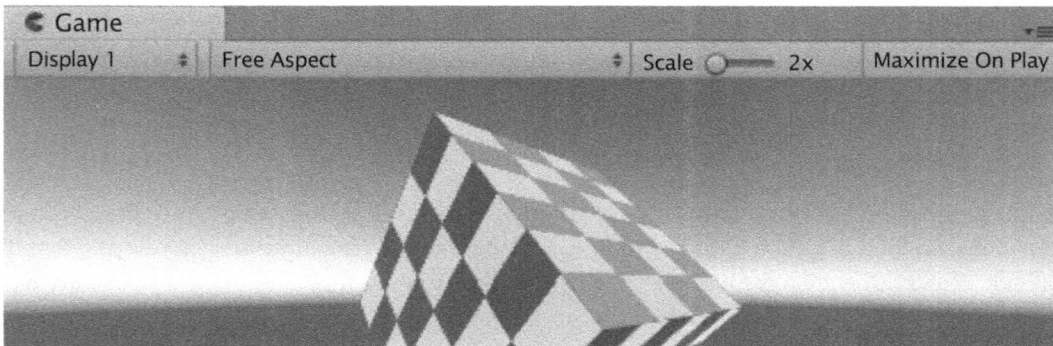

Figure 14.7: Example of a checkerboard pattern being applied to a 3D cube

How to do it...

To create and use a simple **shader**, follow these steps:

1. Create a new Unity 3D project – choose the **3D (URP)** template from the list of templates.

2. First, we need to install the **Universal Render Pipeline (URP) package**. Use Package **Manager** and make sure **Unity Registry** is checked. Search for **Universal RP** and install it.

3. Our project needs to be set up to use the Universal Render Pipeline. Right-click in the **Project** panel and choose **Create | Rendering | URP Asset (With Universal Render)**. Name the pipeline myRenderPipeline.

4. In the **Inspector** panel, display the project's graphics settings by going to **Edit | Project Settings | Graphics**. Then, drag myRenderPipeline from the **Project** window into the **Scriptable Render Pipeline Settings** property:

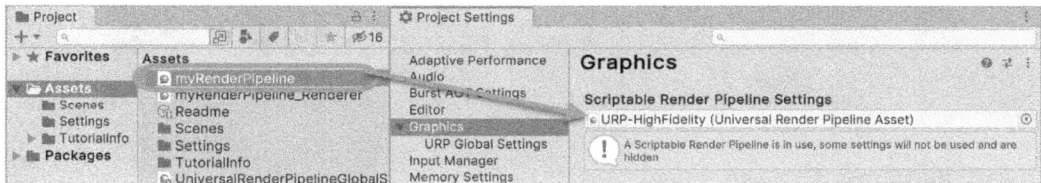

Figure 14.8: Dragging myRenderPipeline into the Scriptable Render Pipeline Settings

5. In the **Project** panel, create a new **Universal RP Lit** Shader Graph named myShaderGraph by going to **Create | Shader Graph | URP | Lit Shader Graph**.

6. In the **Project** panel, create a new material named m_cube by going to **Create | Material**.

7. With m_cube selected, in the **Inspector** panel, set its **Shader** property to myShaderGraph. For the material's **Shader** property, choose **Graphs | myShaderGraph**:

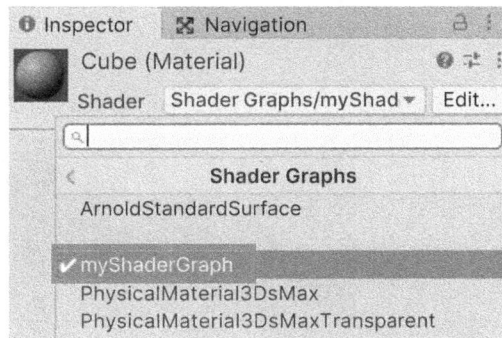

Figure 14.9: Setting m_cube's Shader property to myShaderGraph

8. Add a 3D cube to the scene by going to **GameObject | 3D Object | Cube**. Set the material for this 3D cube to m_Cube.

9. In the **Project** panel, double-click myShaderGraph to open the Shader Graph editing panel. A new Shader Graph will open with three components; that is, (1) the **Blackboard** (for publicly exposing parameters), (2) the **Graph Inspector**, and (3) **Preview,** the output previewer node:

Figure 14.10: The Shader Graph panel

> When editing a Shader Graph, you should maximize the **Shader Graph** panel so that it's easier to see.

10. Right-click the output previewer and select **Cube:**

> You can zoom in and rotate the preview mesh. You can also choose a custom mesh from within your project, allowing you to preview the Shader Graph on the intended destination's 3D object.

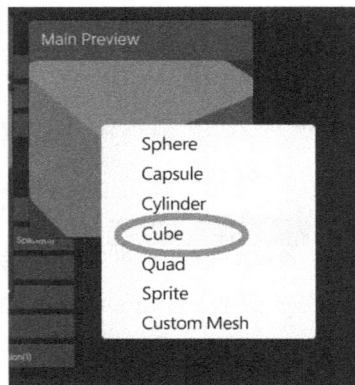

Figure 14.11: Selecting Cube in the output previewer

11. Let's flood the shader with a red color. Choose red from the color picker for the top property of the **Fragment** node; that is, **Base Color**

12. Create a new graph node by right-clicking and going to **Create Node | Procedural | Checkerboard**. You'll see the checkerboard pattern in the preview for this node. Set the X property to 2 and the Y property to 3.

13. Now, drag a link from the checkerboard's note output, **Out(3)**, to the **Emission (3)** input for the **PRB master** node. You should now see a red/pink checkerboard pattern in the **PBR master** node preview. You'll also see that the following output has been applied to the **cube** mesh in the output previewer node:

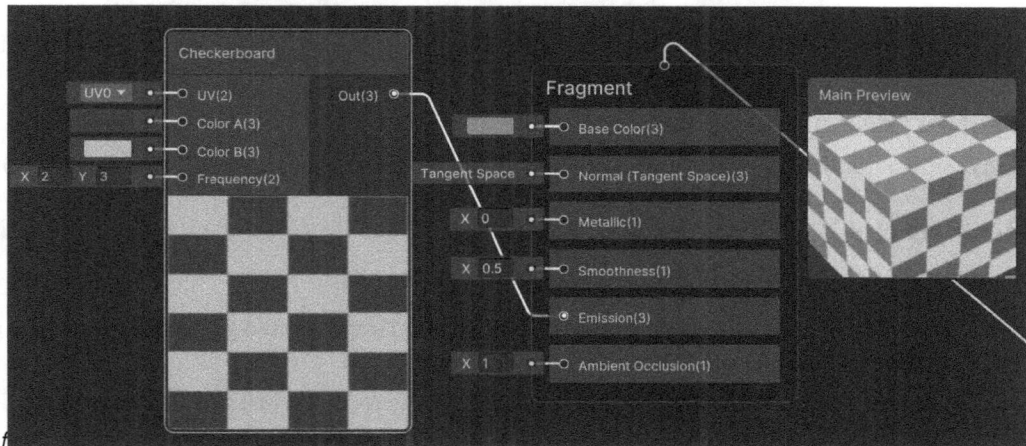

Figure 14.12: Linking the checkerboard's note output Out(3) to the Emission (3) input for the Fragment node

14. You must save the changes for your Shader Graph before they will be applied to the scene. Click the **Save Asset** button at the top left of the Shader Graph window to do so.

15. Save and run your scene. You should see a red/pink checkerboard 3D cube being displayed.

How it works...

In this recipe, you enabled the Universal Rendering Pipeline by installing the package, creating an asset called myRenderPipeline, and choosing that asset for the project's Scriptable Rendering Pipeline Graphics property.

You then created a new Shader Graph asset file and a new material that uses your shader.

Your Shader Graph feeds a procedurally generated checkerboard into the **Emission** property of the **Fragment** output node and also tints the output by choosing a red **Color** value for the **Base Color** property. You saved the changes to your Shader Graph asset so that they will be available when the scene runs.

Creating a glow effect with Shader Graph

In the previous recipe, we created a simple Shader Graph using a material for a primitive 3D cube mesh. In this recipe, we'll take things further by creating a Shader Graph that applies a parameterized glow effect to a 3D object. The end result will look as follows:

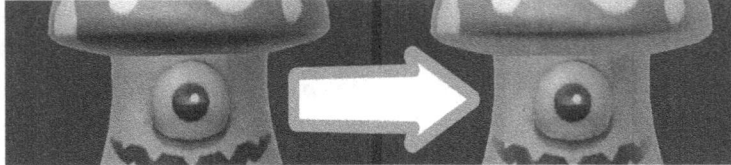

Figure 14.13: Example demonstrating a blue glow effect

Getting ready

This recipe builds on the previous one, so make a copy of that project and use the copy for this recipe.

How to do it...

To create a glow effect with a Shader Graph, follow these steps:

1. In the **Project** panel, create a new **URP Lit** Shader Graph named glowShaderGraph by going to **Create | Shader Graph | URP | Lit Shader Graph**.

2. We now need a 3D mesh object for the scene. While we could just use a 3D cube again, it's more fun to add a low-polygon textured character to the scene. For this recipe, we've used the free Unity Asset Store character pack called *Fantasy Mushroom Mon(ster)* from *AmusedArt*. Once the package has been added, drag the Mushroom Monster **Prefab** into the scene from the **Project** panel folder; that is, **amusedART | Mushroom Monster | Prefab**.

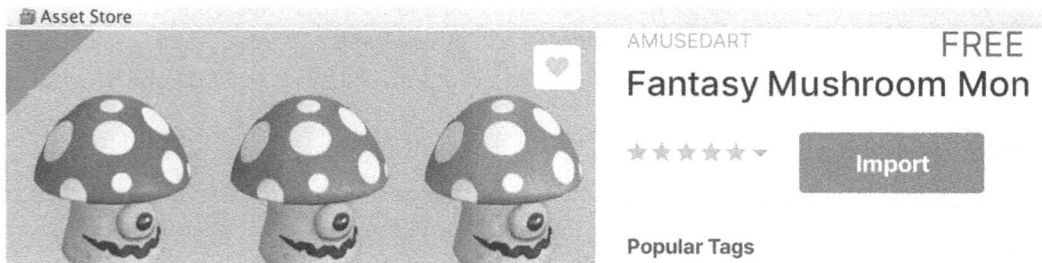

Figure 14.14: Free Fantasy Mushroom Mon(ster) asset from AmusedArt

3. Add `glowShaderGraph` to the `MushroomMon` material component. In the **Hierarchy** panel, expand the `MushroomMonster` object, and select `MushroomMon`. In the **Inspector** panel, under **Materials**, update the material from `MushroomMonGreen` to `Material/glowShaderGraph`.

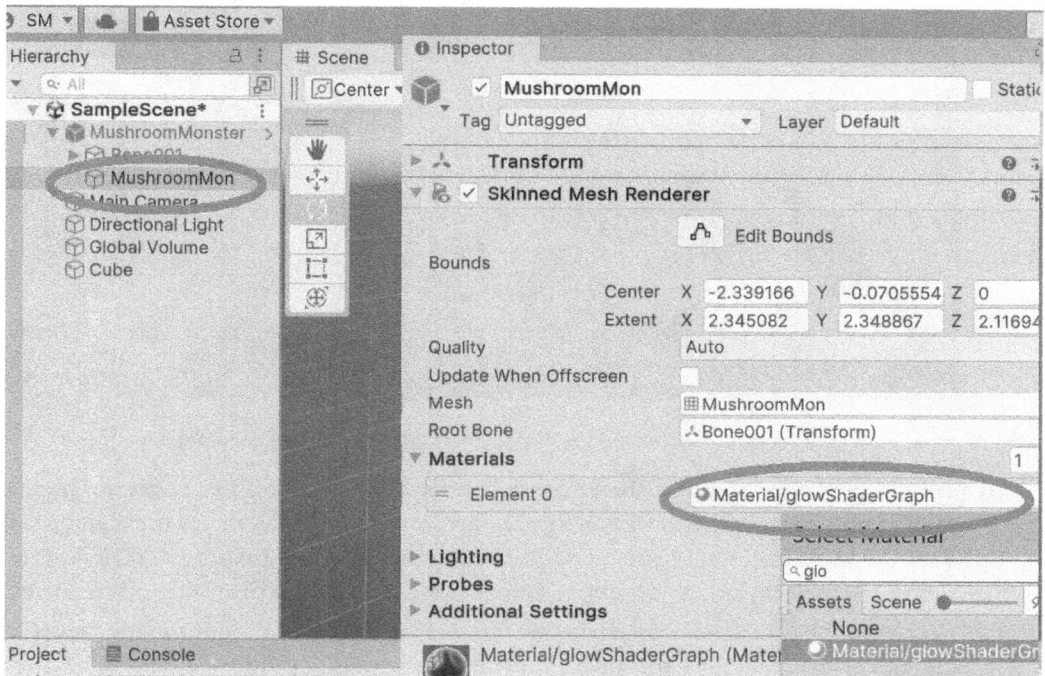

Figure 14.15: Adding glowShaderGraph as a material to the MushroomMon object

4. In the **Project** panel, double-click `glowShaderGraph` to open the **Shader Graph** editing panel.

5. Right-click the output previewer, select **Custom Mesh**, and choose `MushroomMon` from the selection dialog. As a workaround, select the **Cube** option from the list. This will give some indication of how the Shader Graph is working.

6. Add a new `Exposed` texture property to your Shader Graph by creating a new property texture in the **Shader Graph Blackboard**. Click the plus (+) button and choose the `Texture` 2D property type.

7. In **Blackboard**, change the **Default** value setting of the **Texture** 2D property from **None** to **Mushroom Green**.

8. To use a publicly exposed **Blackboard** property in our Shader Graph, we need to drag a reference to the property from the **backboard** into the **Graph** area.

9. Drag the **Blackboard** property's **Texture** into the **graph** area. You should see a new node with a title and value of **Texture2D**:

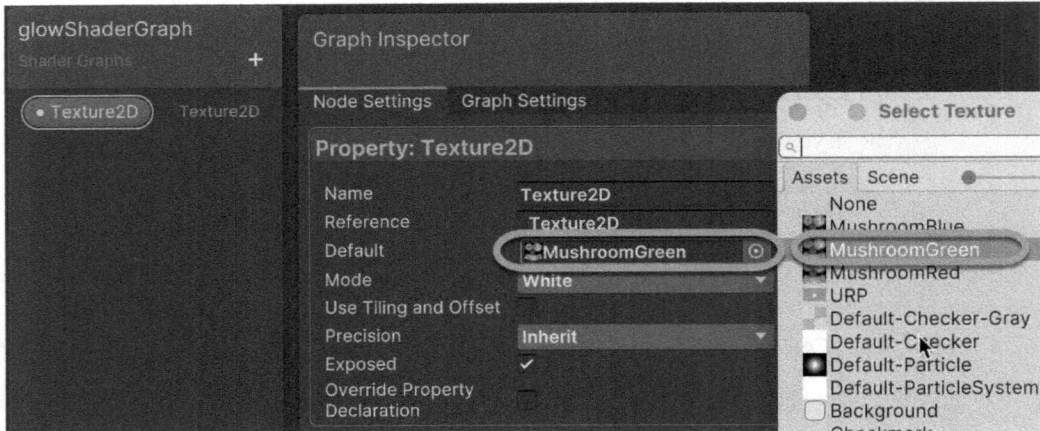

Figure 14.16: Dragging the blackboard property's texture to create a new node

10. There is no texture input to the **Fragment** node, so we need to add a converter node that can take sample data from a 2D texture image and turn it into RBG values that can be sent to the **Base Color** input of the **Fragment** node. Create a new **Sample Texture 2D** node in your Shader Graph by right-clicking and going to **Create Node | Input | Texture | Sample Texture 2D**.

11. Now, let's send the Mushroom Green texture into the **Fragment** node via the **Sample Texture 2D converter** node. Link the **Texture(T2)** output from the **Property** node to the **Texture(T2)** input of the **Sample Texture 2D** node. You should now see the Mushroom Green texture image appear in the 2D rectangular preview at the bottom of the **Sample Texture 2D** node.

12. Next, link the **RGBA(4)** output of the **Sample Texture 2D** node to the **Base Color(3)** input of the **Fragment** node (Unity will intelligently just ignore the fourth Alpha (A) values). You should now see the Mushroom Green texture image appear in the preview at the bottom of the **Fragment** node. You should also see the Mushroom Green texture being applied to 3D Mushroom Monster Mesh in the Shader Graph output previewer node:

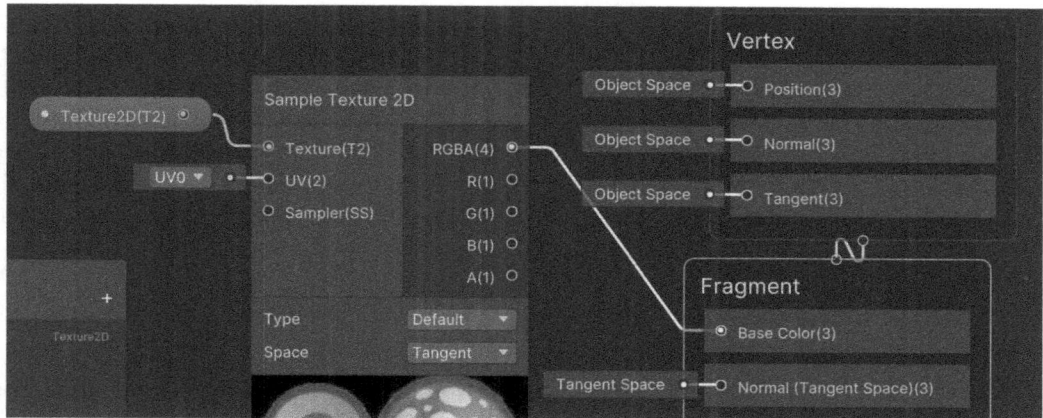

Figure 14.17: Shader Graph showing connections between Sample Texture 2D and Fragment node

13. One way to create a glow effect is by applying a **Fresnel Effect**. (By adjusting the viewing angle, it adjusts the reflection. For more, see the following information box.) Create a new **Fresnel Effect** node in our Shader Graph. Link the **Out(1)** output from the **Fresnel Effect** node to the **Emission(3)** input of the **Fragment** node. You should now see a brighter glow outline effect in the Shader Graph output previewer node. To better understand the Fresnel Effect, adjust the X variable to increase or decrease the **Power(1)** value and rerun the project.

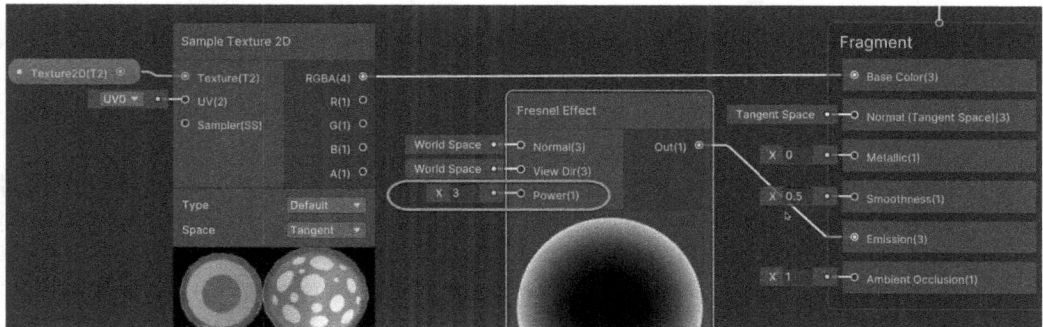

Figure 14.18: Adding the Fresnel Effect and adjusting the X variable

Augustin-Jean Fresnel (1788-1827) studied and documented how an object's reflection depends on the viewing angle – for example, looking straight down into still water, little sunlight is reflected and we can see into the water clearly. But if our eyes are closer to the level of the water (for example, if we are swimming), then much more light is reflected by the water. Simulating this effect in a digital shader is one way to make the edges of an object lighter, since light is glancing off the edges of the object and reflected in our game camera.

14. Let's tint our **Fresnel Effect** by combining it with a publicly exposed **Color** property, which can be set either in the **Inspector** window or through C# code.

15. First, delete the link of the **Out(1)** output from the **Fresnel Effect** node for the **Emission(3)** input of the **Fragment** node.

16. Add a new **Color** property to your Shader Graph by creating a new property called Color in the blackboard. Click the plus (+) button and choose the **Color** property type.

17. In the blackboard, set the default value of the **Color** property to red (using the color picker).

18. Drag the blackboard's **Color** property into the **Graph** area. You should see a new node called **Property** with a value of **Color(4)**.

19. Create a new **Multiply** node in your Shader Graph by right-clicking and going to **Create Node | Math | Basic | Multiply**.

The mathematical **Multiply** node is an easy way to combine values from two nodes, which are then passed to a single input of a third node.

20. Let's combine the **Color** property and **Fresnel Effect** by making both inputs of the **Multiply** node. Link the **Color(4)** output from the **Property (Color)** node to the **A(4)** input of the **Multiply** node. Next, link the **Out(1)** output from the **Fresnel Effect** node to the **B(4)** input of the **Multiply** node. Finally, link the **Out(4)** output from the **Multiply** node to the **Emission(3)** input of the **Fragment** node. You should now see a red-tinted glow outline effect in the Shader Graph output previewer node. The following screenshot shows these node connections for our Shader Graph:

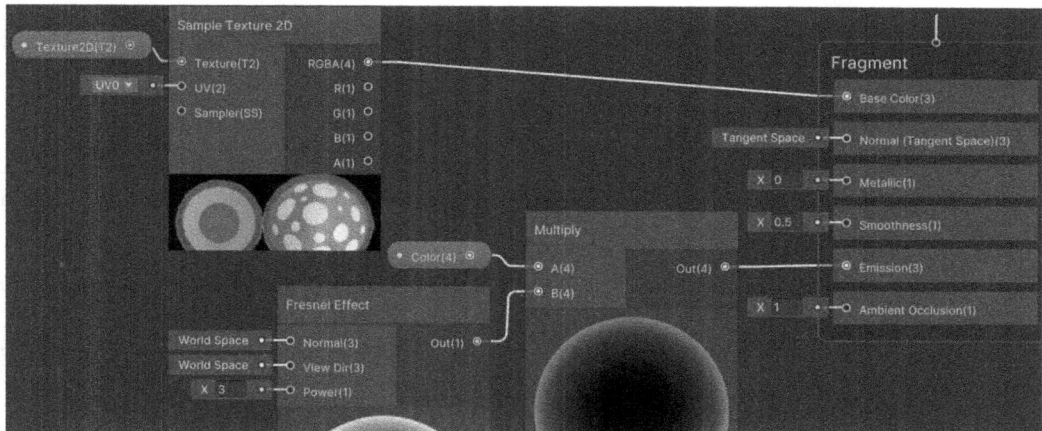

Figure 14.19: Final Shader Graph with connections for the glow effect

21. Save your updated Shader Graph by clicking the **Save Asset** button at the top right of the **Shader Graph** window.

22. Save and run your scene. You should see a red glow around the character:

How it works...

In this recipe, you created a new Shader Graph that had several connected nodes. The output(s) of one node becomes the input(s) of another node.

You created publicly exposed properties for **color** and **texture** using the Shader Graph Blackboard and introduced those properties as inputs in your Shader Graph.

You used a **Sample Texture 2D** node to convert the **2D texture** image into RBG values suitable for the **Base Color** input for the **Fragment** node and assigned the MushroomMonGreen materials to this texture. This step replaces the need to create a material object, which would usually be applied to a GameObject. The material is now controlled in the **Texture 2D** node in the Shader Graph.

You then created a **Fresnel Effect** node and combined this, via a **Multiply** node, with the publicly exposed **Color** property, sending the output into the **Emission(3)** input of the **Fragment** node. **Multiply** nodes can be used to combine different elements of a Shader Graph.

To apply a Shader Graph to a GameObject in a scene, you assign the Shader Graph to the GameObject or Prefab. In the GameObject, the rendering is then influenced by the Shader Graph.

Finally, you changed the publicly exposed property for **Color** in the **Inspector** window via the material's properties.

Toggling a Shader Graph color glow effect through C# code

Effects such as the glow effect from the previous recipe are often features we wish to toggle on and off under different circumstances. The effect can be turned on or off during a game to visually communicate the status of a GameObject – for example, an angry character might glow red, while a happy monster might glow green, and so on.

We'll add to the previous recipe to create a new publicly exposed Shader Graph blackboard property named Power. Then, we'll write some code that can be used to set this value to 0 or 5 in order to turn the glow effect on and off. We'll also access the **Color** property so that we can set what color the glow effect displays.

Getting ready

This recipe builds on the previous one, so make a copy of that project and use the copy for this recipe.

How to do it...

To toggle the glow effect from the Shader Graph, follow these steps:

1. First, delete the link from the **Out(4)** output from the **Multiply** node to the **Emission(3)** input of the **Fragment** node. We are doing this because the output of this **Multiply** node will become the input for a second **Multiply** node that we are about to create.

2. Create a new **Multiply** node in your Shader Graph by right-clicking and going to **Create Node | Math | Basic | Multiply**.

3. Link the **Out(4)** output from the original **Multiply** node to the **A(4)** input of your new **Multiply** node. Also, link the **Out(4)** output from the new **Multiply** node to the **Emission(3)** input of the **Fragment** node.

4. Add a new property of type float (decimal number), to your Shader Graph, click the plus (+) button, choose the **Float** property type, and rename this Power.

5. In the blackboard, set the default value of the **Power** property to 5. Also, set **Display mode** to **Slider** with the values of **Min** 0 and **Max** 5.

6. Drag the blackboard **Power** property into the **Graph** area. You should see a new node with the title of **Power**.

7. Finally, link the **Power** output from the **Power** node to the **B(4)** input of the new **Multiply** node:

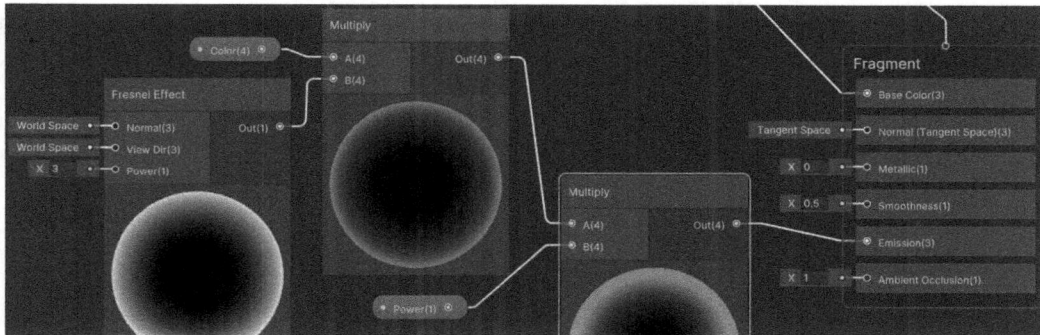

Figure 14.20: Settings of the Shader Graph with an additional Multiply node

8. Save your updated Shader Graph by clicking the **Save Asset** button at the top left of the **Shader Graph** window.

9. Create a new C# script class named `GlowManager` containing the following code:

```csharp
using UnityEngine;

public class GlowManager : MonoBehaviour {
    private string powerId = "_Power";
    private string colorId = "_Color";

    void Update () {
        if (Input.GetKeyDown("0"))
            GetComponent<Renderer>().material.SetFloat(powerId, 0);

        if (Input.GetKeyDown("1"))
            SetGlowColor(Color.red);

        if (Input.GetKeyDown("2"))
            SetGlowColor(Color.blue);
    }

    private void SetGlowColor(Color c) {
        GetComponent<Renderer>().material.SetFloat(powerId, 5);
        GetComponent<Renderer>().material.SetColor(colorId, c);
    }
}
```

10. Select `glowShaderGraph` in the **Project** panel and view its properties in the **Inspector** panel:

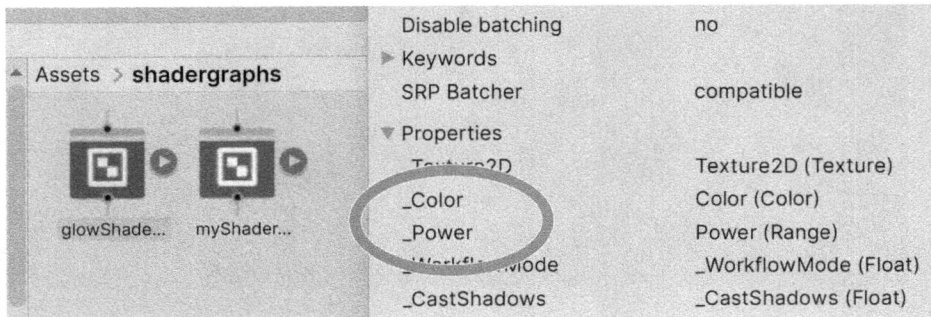

Figure 14.21: The properties of glowShaderGraph shown in the Inspector window

11. Find the internal IDs of the publicly exposed **Power** and **Color** properties – they will be something like _Power and _Color. Copy these IDs into C# script statements by setting the ID strings:

```
private string powerId = "_Power";
private string colorId = "_Color";
```

12. In the **Hierarchy** window, locate the component of your 3D GameObject that contains the Mesh Renderer component (for our Mushroom Monster example, this is the MushroomMon child of the Mushroom Monster GameObject). Add an instance object of the GlowManager script class as a component to this GameObject.

13. Save and run your scene. Pressing the *1* key should turn on the red glow effect, pressing the *2* key should turn on the blue glow effect, and pressing the *0* key should turn off the glow effect.

How it works...

In this recipe, you created a new **Float** property for your Shader Graph that combined with the Fresnel color effect so that a value of 0 will turn off the effect. You looked up the internal IDs of the **Power** and **Color** properties and updated the C# script so that it can update these properties.

The script class checks for the 0/1/2 keys and turns the effect off, to a red glow, or to a blue glow, respectively. The script class can influence the Shader Graph because we found the internal IDs that referred to the **Power** and **Glow** color variables.

By combining publicly exposed properties with code, we can change Shader Graph values at runtime through events detected by code.

There's more...

Here are some ways to take your Shader Graph features even further.

You could use **Sine Time** to create a pulsating glow effect, as follows:

- You could make the glow effect pulse by creating a **Time** node and then link the **Sine Time(1)** output to the **Fresnel Effect** node's input of **Power(1)**.

As the **Sine Time** value changes between - 1/0/+1, it will influence how strong the Fresnel effect is, changing the brightness of the glow effect.

Another way to find exposed property IDs (and to get more of an idea of how Unity sees Shader Graphs internally) is to use the **Compile and Show Code** button.

When you are viewing the properties of a Shader Graph asset in the **Inspector** panel, you'll see a button entitled **Compile and Show Code**. If you click this, you'll see a generated ShaderLab code file in your script editor.

This isn't the actual code used by Unity, but it provides a good idea of the code that is generated from your Shader Graph. The internal IDs for your publicly exposed blackboard properties are listed in the Properties section, which is at the beginning of the generated code:

```
Shader "graphs/glowShaderGraph" {
    Properties {

    [NoScaleOffset]  Texture_C5AA766B ("Texture", 2D) = "white" { }
    Vector1_AA07C639 ("Power", Range(0.000,5.000)) = 5.000

    Color_466BE55E ("Color", Color) = (1.000,0.000,0.038368,0.000)
    }

    etc.
```

Shader Graphs are very powerful ways to influence the visual output of your games. We've explored a few of their features in this chapter. Spending some time getting to know them better will result in increased ways to add visual effects to your games.

Further reading

The following are some useful sources for the topics that were covered in this chapter.

Shader Graph online resources

The Unity documentation and third-party articles about Shader Graphs can be found online at the following links:

- Introduction to Shader Graph: https://docs.unity3d.com/Packages/com.unity.shadergraph@15.0/manual/Getting-Started.html
- The Shader Graph Example Library on GitHub: https://github.com/UnityTechnologies/ShaderGraph_ExampleLibrary
- Video Tutorial of Shader Graph from Unity Technology's Andy Tough at GDC 2018: https://www.youtube.com/watch?v=NsWNRLD-FEI
- Unity's Manual Page for the Universal Render Pipeline: https://docs.unity3d.com/Manual/com.unity.render-pipelines.universal.html

- Universal Render Pipeline tutorial from SyntxStudios: `https://www.youtube.com/watch?v=KpTK-OraZ-g`
- A tutorial on adding a glitter effect in Shader Graphs: `https://danielilett.com/2021-11-06-tut5-19-glitter/`

Video player online resources

The Unity documentation and third-party articles about the `VideoPlayer` component can be found online at the following links:

- The Unity Video Player Manual Page: `https://docs.unity3d.com/Manual/class-VideoPlayer.html`
- The Unity Scripting Reference for the VideoPlayer Class: `https://docs.unity3d.com/ScriptReference/Video.VideoPlayer.html`

15

Particle Systems and Other Visual Effects

While a lot of content in scenes is from terrain and 3D meshes, and other effects from light sources, some fantastic effects can be created from Particle Systems. Particle systems are basically a collection of points whose creation changes over time, and whose lifetime can be controlled through parameters and rules. Also, how each particle is rendered can be controlled over time, sometimes using colors and other times using sprite images. Effects such as gravity and collisions with surfaces can be applied to make it more convincing to the player that the particles are part of the scene and interacting with other scene objects.

Some possible effects of Particle systems include:

- Fire, explosions, smoke, and sparks
- Water drops, moving water, and waterfalls
- Other effects such as dust, fish, glowing effects from magic potions, and so on

Figure 15.1: Particle systems used to great effect in the Witcher computer game modification

In this chapter, we will cover the following recipes:

- Exploring Unity's Particle Pack and reusing samples for your own games
- Creating a simple Particle System from scratch
- Using Texture Sheets to simulate fire with a Particle System
- Making particles collide with scene objects
- Simulating an explosion
- Using Invoke to delay the execution of an explosion
- Using a Line Renderer to create a spinning laser trap

Exploring Unity's Particle Pack and reusing samples for your own games

Unity has published a great demonstration of Particle System effects as a package in the Asset Store. While it's fine to explore the samples, by looking at the GameObjects in the scene and the asset files in the **Project** window, we can exploit these free resources and adapt them for our own projects.

In this recipe, we'll explore the sample scene and create two versions of a new scene. One will give the effect of fire on the ground, while the second will look like a flaming torch attached to a concrete wall. In both cases, we'll make use of fire Prefabs from the Unity examples.

Figure 15.2: Part of the Unity Particle System demo scene

How to do it...

To reuse Unity particle examples for your own games, follow these steps:

1. Start a new **Unity 3D project** and ensure you are logged in to your Unity account in the Unity Editor.

2. Open the **Unity Asset Store** (click the **Asset Store Web** button at the top left of the Unity Editor).

3. In the Asset Store, check **Free Assets** and search for and select the free **Unity Particle Pack** (not the Legacy version, but the one that was updated in 2023).

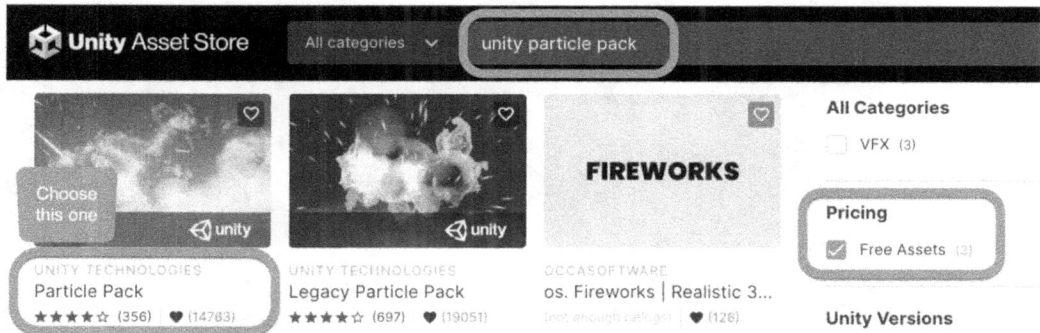

Figure 15.3: The Unity Particle Pack on the Asset Store website

4. Click **Add to My Assets**. Then, after the button changes, click **Open in Unity**.

5. In the Unity Editor, the **Package Manager** window should open, and **Particle Pack** should be selected in the **My Assets** list. Click **Download** and, once downloaded, click **Import**. You should now see several folders appear in the **Project** window. The import folder is `EffectExamples`.

6. There should also be a new scene named `Main` in the **Scenes** folder in the **Project** window. Open this scene.

7. Run the game and, using **First Person Controller** (*arrow keys | WASD*), explore all the different Particle System examples from Unity.

8. Stop the game and create a new **Basic (Built In)** scene. In this scene, add a 3D **Plane** at **Position** (`0, 0, 0`).

9. In the **Inspector** window for the **Mesh Renderer** component, select the `Concrete Floor` material (click the circle asset file selector for **Element 0** of the **Materials** property). This will act as a floor that we can create some fire on.

10. Select `Main Camera` in the **Hierarchy** window and, in the **Inspector** window, set its **Position** to (`0, 3, -3`) and its **Rotation** to (`45, 0, 0`).

11. In the **Project** window, locate the `LargeFlames` prefab (in **EffectExamples | Fire & Explosion Effects | Prefabs**). Drag this prefab into the **Hierarchy** window to add a new GameObject clone of this prefab in the scene. In the **Inspector** window, ensure the **Position** of this `LargeFlames` GameObject is (`0, 0, 0`).

12. Save and run the scene. You should see some realistic animated flames in the center of the concrete floor:

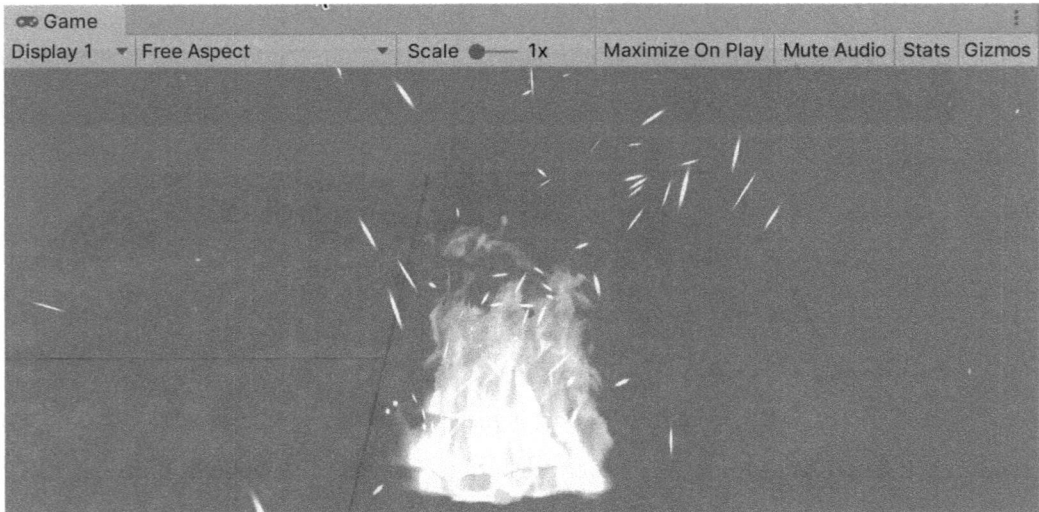

Figure 15.4: Reusing the LargeFlames prefab on a concrete-styled 3D plane

13. Stop the scene.

14. Make a copy of the scene and, in this copy, delete the LargeFlames GameObject.

15. Now, let's create a wall with a flaming torch attached to it.

16. To create the wall, add a 3D **Cube** to the scene named Cube-wall with a **Position** of (0.5, -1, 4), a **Rotation** of (0, 25, 0), and a **Scale** of (7, 7, 2.5). In the **Inspector** window for the **Mesh Renderer** component, set **Element 0** to Concrete Wall (in the **Project** window, go to **Assets | Shared | Environment | Materials**).

17. To create the head of the torch, add a 3D **Cube** to the scene named Cube-torch with a **Position** of (1.15, 1.5, 1.35), a **Rotation** of (5, 120, -40), and a **Scale** of (0.15, 0.05, 0.15).

18. To create the pole of the torch, add a 3D **Cylinder** to the scene named Cylinder-torch with a **Position** of (1.4, 0.5, 1.95), a **Rotation** of (154, -17, -20), and a **Scale** of (0.1, 1.15, 0.1).

19. To see the flaming torch close up, move the **Main Camera** closer by setting its **Position** to (1, 2, 0) and its **Rotation** to (20, 0, 0) in the **Inspector** window.

20. Now, let's add an animated flame to our torch. From the **Project** window, locate the Candles prefab (in **EffectExamples | Misc Effects | Prefabs**) and drag it into the **Hierarchy** window to create a clone of it as a GameObject.

21. We only want a part of this, so unlink this from its prefab parent by choosing **Prefab | Unpack Completely** from the right-mouse-button menu for this GameObject in the **Hierarchy** panel.

22. Drag one of the CandleFlame child objects away from this parent GameObject, and then delete the Candles GameObject. You should now have a GameObject named CandleFlame, which is a small, flickering flame.

23. Then, set the **Position** of this `CandleFlame` GameObject to `(1.15, 1.52, 1.3)`, and its **Scale** to `(0.7, 0.7, 0.7)`.

24. Save and run the scene. You should see your torch object with a nice animated flame:

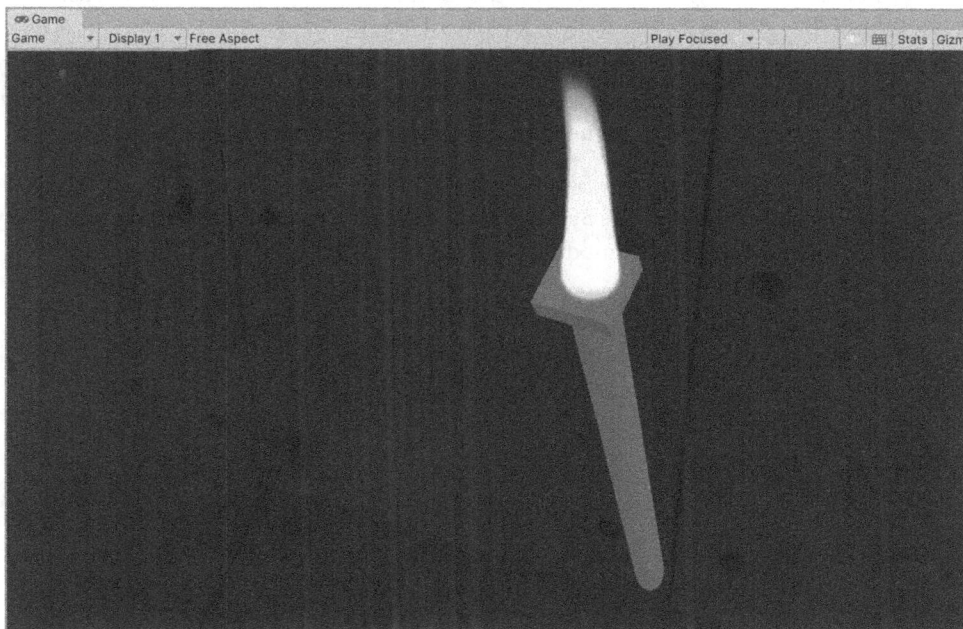

Figure 15.5: Reusing the CandleFlame prefab for our simple wall torch object

How it works...

In this recipe, we learned how to add the Unity **Particle Pack** asset package to a new project and explored the included sample scene.

The **Particle Pack** package includes many Prefabs for each of the sample Particle Systems. We were able to select one prefab, `LargeFlame`, and create our own scene with a concrete floor **Material** for a 3D **Plane**, and then make this floor look like it was on fire by adding a clone of the `LargeFlame` prefab.

Prefabs can be used for different purposes, as shown in this demo scene. So, we were able to make use of a `CandleFlame` child GameObject inside an instance of the `Candles` prefab from the Unity examples and use it as the flame for a simple model of a flaming torch. We did this by using a 3D **Cube** and a 3D **Cylinder**.

Unity provides many Prefabs of ready-to-use particle effects. In many cases, as in this recipe, we just need to import the pack from Unity and then create GameObjects in our own scenes using the assets provided by Unity.

Creating a simple Particle Systems from scratch

Making use of Prefabs from Unity, as we did in the previous recipe, is fine if we're happy to use such Prefabs without any *tweaking*. However, we often want to adjust the look and feel of assets and Particle Systems so that they fit in with the style of a particular game or scene. Therefore, it is useful to learn how to create and customize Particle Systems from scratch in order to learn about the different parameters and modules that make up Unity Particle Systems. Knowing how to adjust these values means we are able to either create what we need or customize a Particle Systems prefab from a third-party source, such as the Unity examples.

In this recipe, we'll create a Particle Systems from scratch and customize it in several ways to learn about some of the most important modules and parameters:

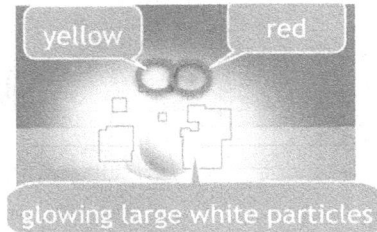

Figure 15.6: The scene we'll create in this recipe with white, red, and yellow particles shimmering around a sphere

How to do it...

To create a Particle Systems from scratch, follow these steps:

1. Start a new **Unity 3D project**.
2. Let's create a floor and wall from scaled 3D Cubes to create a location where we can view our Particle System.
3. Create a 3D **Cube** GameObject named Cube-wall. In the **Inspector** window, set its **Position** to (1, 0, 3) and its **Scale** to (20, 20, 1).
4. Create a second 3D **Cube** GameObject named Cube-floor. In the **Inspector** window, set its **Position** to (0, -1, 0) and its **Scale** to (20, 1, 20).
5. Create a new **Material** asset file in the **Project** window named m_grey. Set the **Albedo** property of this **Material** asset to a mid-gray color (halfway between white and black). Now, drag the m_grey asset file from the **Project** window onto Cube-wall in the **Hierarchy** window. The wall should now be gray.
6. Often, the visual effect of a Particle System comes from a GameObject, so let's create a 3D **Sphere** at a **Position** of (0, 0, 0) so that it appears as the source of our particles.
7. Now, let's add a GameObject containing a **Particle System** component by choosing **Create | Effects | Particle System**. You should now see a new GameObject in the **Hierarchy** window named **Particle System**. In the **Inspector** window, you'll see it has just two components: a **Transform** and a **Particle System**.

Ensure the **Position** of the **Transform** is (0, 0, 0) so that the particles appear to be coming out of the sphere at that same location.

> When the **Particle System** GameObject is selected in the **Hierarchy**, you will see an animated preview of the Particle System in the **Scene** panel. You'll also have a **Particle Effect** control panel that allows you to **Pause/Restart/Stop** the Particle System. You can minimize the **Particle System** control panel by clicking the small triangle icon in its top-left corner.

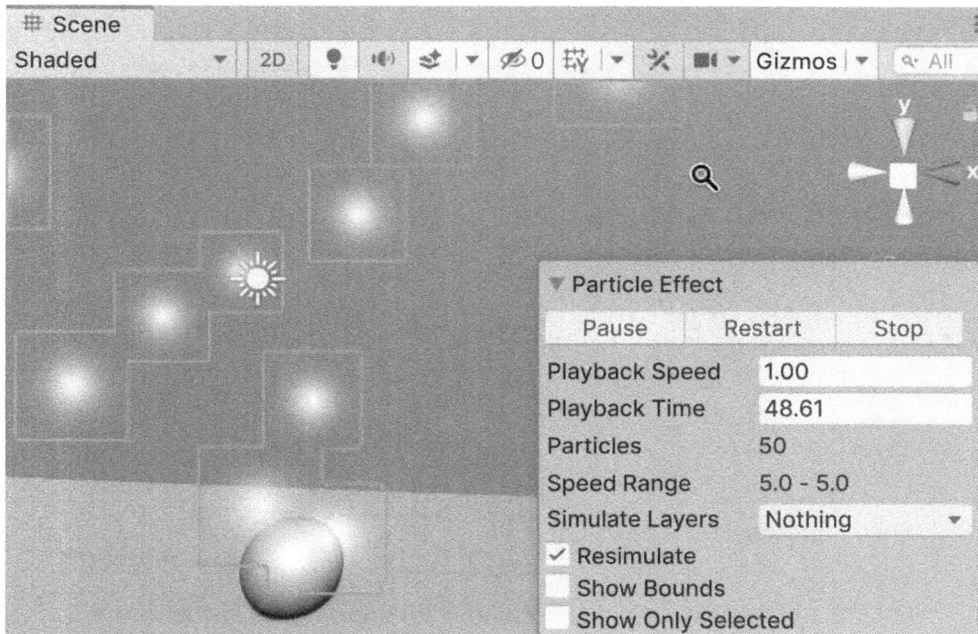

Figure 15.7: The Scene window's Particle System previewer

> Sometimes, we'll want to add a **Particle System** component to an existing GameObject. When we do so, there is no default setting for the renderer's material, so we'll see pink squares indicating there's no **material**. Let's learn how to do this and assign a **Material** to get rid of the pink squares.

8. Delete the **Particle System** GameObject from the scene. Now, create a new, empty GameObject in the scene (**GameObject | Create Empty**) named MyParticles. In the **Inspector** window, add a **Particle System** component by clicking the **Add Component** button and searching for **Particle System**.

9. Ensure that the `MyParticles` GameObject is selected in the **Hierarchy** window. You should see a preview of the animated **Particle System** in the **Scene** window. If you are seeing pink squares rather than white circles, then the **Particle System** component's **Renderer** module is missing the link to the **Default-Particle** material. Let's fix that.

10. In the **Inspector** window, click the **Renderer** module title bar to open the properties of this part of the **Particle System** component. Then, for the **Material** property, click the **selection** button (the small circle with the dot in the middle), and from the list of materials, select **Default-Particle**. You should now see "glowing" white circles being generated from your particle system (and no more pink squares):

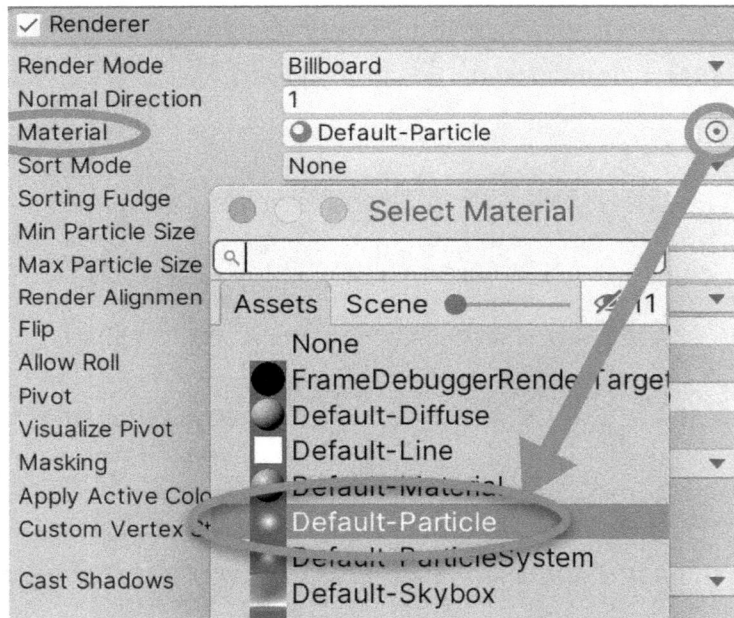

Figure 15.8: Setting Material in the Renderer module to Default-Particle

11. Now, since we usually want particles to be initially floating upward, in the **Inspector** window, set **Rotation** to (`-90, 0, 0`). With that, we have recreated the same GameObject containing a **Particle System** component that Unity creates for us when we go to **GameObject | Effects | Particle System**.

> As can be seen in the **Inspector,** there are quite a few modules in the **Particle System** component. Particle systems are powerful components that can generate a wide range of visual effects. Click the headings of different modules to see the different properties for each module to get to know some of the customizable parameters.

12. Some of the most influential properties are in the top section of the component, under the name of the GameObject, so in this case, it will be MyParticles.

13. With MyParticles selected in the **Hierarchy** window, stop the **Particle System** preview. Then, play it again.

    You should see that when you press **Play**, there are initially no particles in the scene, and that particles begin to appear from the location of the sphere. After about 3 or 4 seconds, as the first particles start to disappear above our **wall** GameObject, the scene settles into a steady state, with new particles appearing at the same rate as old particles disappear.

14. Now, check the **Prewarm** option in the **Inspector** window. Then, in the **Scene** window, stop and play the Particle System again.

    You'll see that the scene is in a steady state, with new particles appearing at the same rate as old particles disappear. This **Prewarm** option makes Unity store the **Particle System** simulation status after running it for a little bit, so as soon as we enable **Particle System**, it will be prewarmed into its steady state. This is useful for visual effects such as flames or fog, where we don't want the user to see the beginning of the Particle System from its source. However, for visual effects such as explosions, we will want the user to see the beginning and growth of the Particle System, so in those cases, we would **not** check the **Prewarm** option:

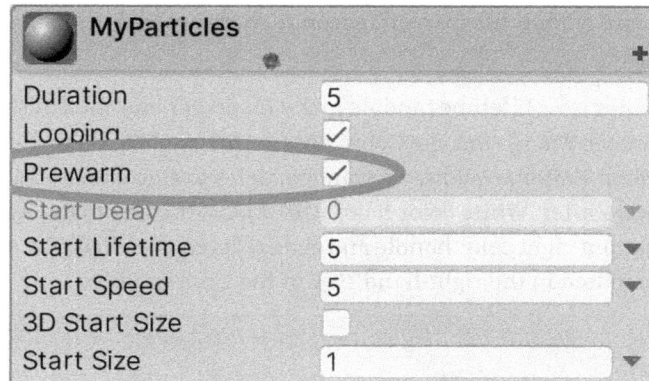

Figure 15.9: Checking the Particle System Prewarm option

15. The initial **Start Lifetime** of a Particle System is 5 seconds. Let's reduce that to 1 second. Also, let's reduce **Start Speed** from 5 to 1. This means we now have particles that rise up slowly and die after 1 second. Finally, let's increase **Start Size** from 1 to 5.

We should now have large white glowing particles hovering around the top of our sphere – a bit like our sphere is being **haunted**:

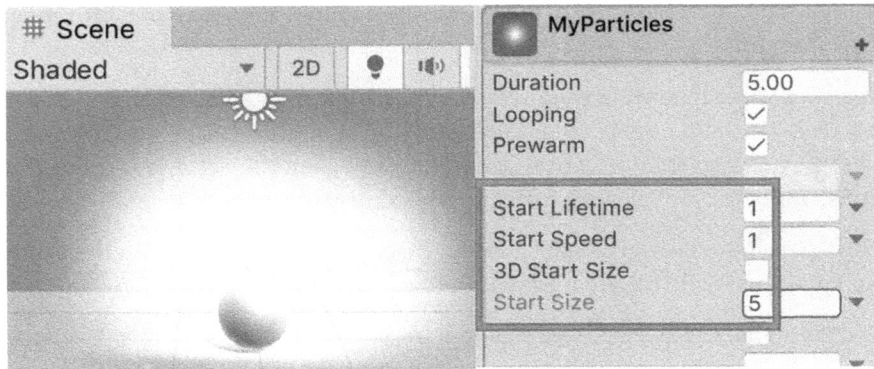

Figure 15.10: Big, short-lived, slow-rising particles around our sphere produce a ghostly glowing effect

Now, let's add a nice red spot effect. We'll make the particles turn red just before they disappear, and also get smaller at the same time – so, we'll see little red dots around the top of the white glowing Particle System. To achieve what we want, our particles should change color and size toward the end of their lifespan of 1 second. There are dedicated modules to customize these parameters.

16. Check the **Color over Lifetime** module in the **Inspector** and click this module's title to reveal its **Color** parameter. Then, click the **Color** selector rectangle to open the **Gradient Editor** window. In the **Gradient Editor** window, we want particles to stay **white** until their last 10% of life. So, move the bottom-left **White** color handle 9/10ths of the way to the right (**Location: 90%**). Then, select the bottom-right color handle and change its color to **Red**. You should now see a gradient from **White** to **Red** in the right-hand 10% of the color gradient:

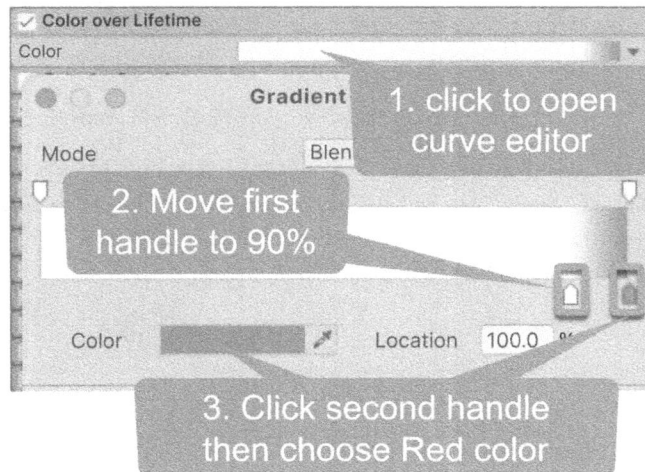

Figure 15.11: Making particles turn red toward the end of their life

If you look at the scene, you should now see occasional large flickers of **red** light in the Particle System around our sphere.

Now, let's change the color of the smaller balls to **Red** by changing the **Size over Lifetime** parameter curve.

> Many parameters of Particle Systems are defined by curves. When a GameObject containing a **Particle System** component is selected at the bottom of the **Inspector** window, the **Particle System Curves** window will appear. This panel displays each of the active (checked) modules that are controlled by curves. Each active curve is assigned a unique color, and when it is being edited, only that curve will be displayed as a bright line (any other curves will be faintly drawn in their colors behind the current curve being edited).

17. Check the **Size over Lifetime** module in the **Inspector** window and check this module's title to enable and reveal its **Size** curve parameter. Leave the **Separate Axes** option unchecked. Click inside the **Size** curve parameter rectangle to make the **Particle System Curves** window display the curve controller of this **Size over Lifetime** module.

18. From the selection of preset curves at the bottom of the **Particle System Curves** window, choose the third one – from high to low. This should leave the beginning size at **1.0** (the full size at the beginning of the particle's life) and make the particle disappear (size **0.0**) at the end of its life. Those red dots are a bit small, so let's have the particles go no smaller than 20% before they disappear by dragging the right-hand end of the curve up from **0.0** to **0.2**. You should now see small (but not too small) red light particles flashing just before each particle dies:

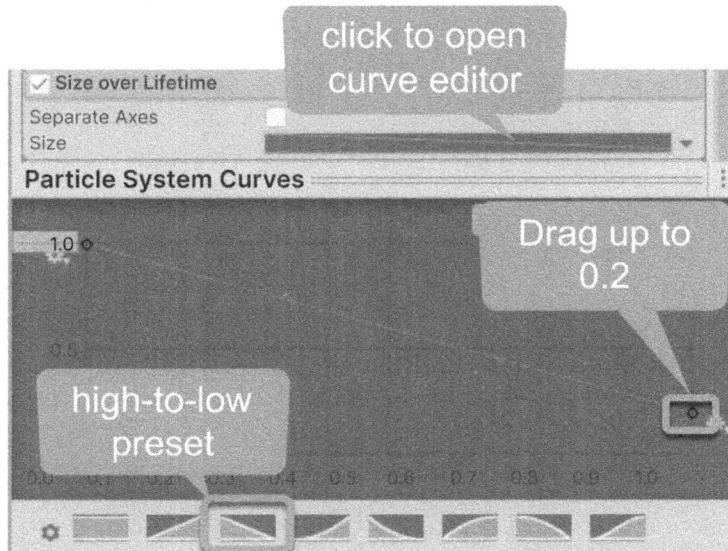

Figure 15.12: Editing the Size over Lifetime curve to make particles smaller toward the end of their life

19. Finally, let's add some yellow dots of light as well. First, rename our particle GameObject `MyParticles-red` and then duplicate it, naming the copy `MyParticles-yellow`. With `MyParticles-yellow` selected in the **Hierarchy** window, in the **Inspector** window, click the **Color over Lifetime** module's title to reveal its **Color** parameter. Then, click the **Color** selector rectangle to open the **Gradient Editor** window. Click the (red) bottom-right color handle and change its color to yellow. Select both `MyParticles-red` and `MyParticles-yellow` in the **Hierarchy** window to see both of their previews at the same time.

    While we can see the occasional red and yellow flash of light, we now have twice as many particles, emitting white for most of their life, so it's harder to see the colored flashes now. Let's solve this by making the particles from `MyParticles-yellow` mostly transparent during their lifetime, except becoming yellow just before they die. We can do this easily in the **Gradient Editor** window by making use of the upper two color handles.

20. Click the top-right color handle and then use the slider (or text box) to set the **Alpha** value to 0. Now, our yellow particles will start as invisible, and then become visible and yellow at the end of their lifetime:

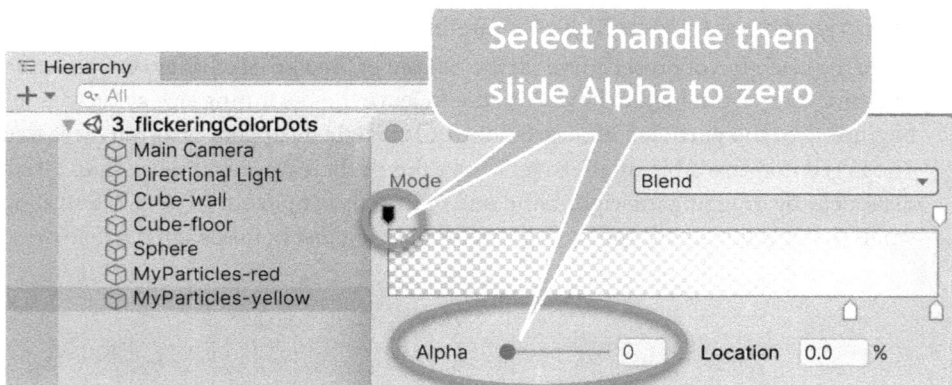

Figure 15.13: Setting the yellow particles to start life as invisible (Alpha = 0)

21. Save and run the scene. You should see large white shimmering particles around the sphere, and occasional flashes of small red and yellow lights as those particles die.

How it works...

In this recipe, we explored a range of the different modules and parameters of Unity Particle System. The core of a Particle System is based on the **Material** particles they are displayed (rendered) with. By changing the material of the **Renderer** module, we can influence how the system looks – in this case, we ensured the **Default-Particle** material was being used, solving the issue of pink squares for no **Material**.

Changing the **Start Lifetime** and **Start Speed** properties of particles determines how far the particles will go from their starting point – the shorter the lifetime and the slower they move, the less distance they will travel. Changing the color and size of particles during their life allows effects such as the red glows/spots we created in the recipe to be implemented. Having multiple Particle Systems and varying the **Alpha** transparency of particles allows us to customize how they may interact; so, our yellow particles were only visible toward the end of the life of each particle.

Many aspects of **Particle System** module parameters are controlled by curves, as we explored with the **Size over Lifetime** parameter. So, it is worthwhile getting familiar with the **Particle System Curves** window, both for selecting preset curves and then customizing them as required.

Using Texture Sheets to simulate fire with a Particle System

Much of the work of creating visual effects can be achieved if we have a multiple-image **sprite sheet**. Then, the visual form of each particle, as it changes over its lifetime, can be driven as a 2D animation, looping through the sequence of images from the sprite sheet.

In this recipe, we'll create a flickering blue flame effect using a Particle System that animates its particles from a multiple-image sprite sheet from the Open Game Art website:

Figure 15.14: The flickering blue flames Particle System we'll create in this recipe

Getting ready

For this recipe, we need a sprite sheet with multiple images on a black background. Either use one of your own or use the fire2_64.png blue flame sprite sheet from Open Game Art. Thanks to *Ben Hickling* for sharing this asset for game developers to use.

We've prepared the file you need in the 14_03 folder. Alternatively, you can download the sprite sheet image file directly from the Open Game Art website at https://opengameart.org/content/animated-fire:

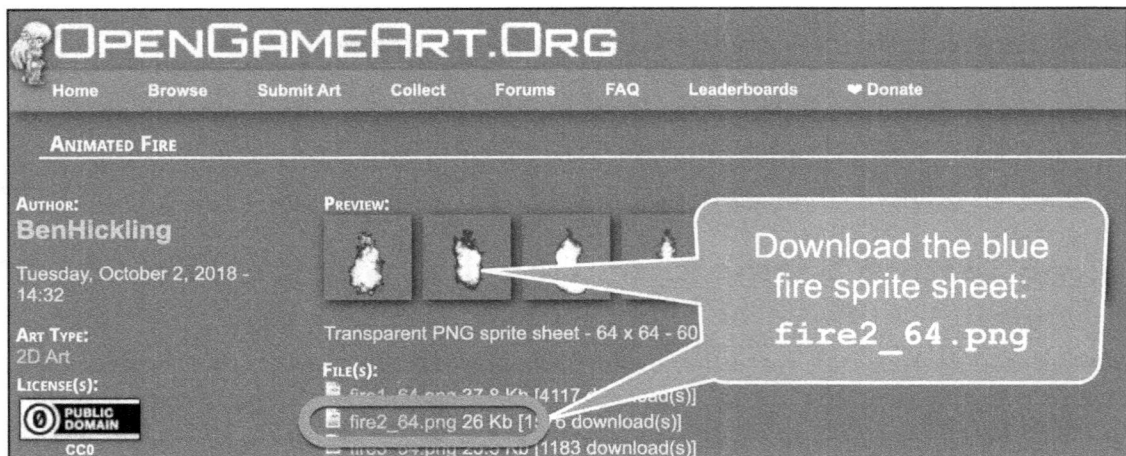

Figure 15.15: Downloading the blue fire sprite sheet from the Open Game Art website

How to do it...

To simulate fire using Texture Sheets, follow these steps:

1. Create a new **Unity 3D project.**

2. Import the blue2_64.png flame sprite sheet texture image asset file. Ensure this image asset has its **Texture Type** set to **Sprite 2D and UI** in the **Inspector** window.

3. Examine the sprite sheet, making a note of the number of columns (X-tiles) and rows (Y-tiles). We need these numbers to customize our Particle System later in this recipe.

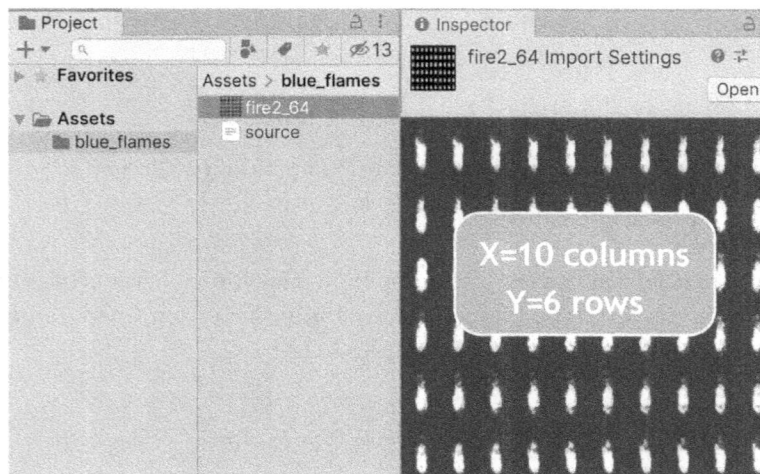

Figure 15.16: The blue flames sprite sheet (X = 10 columns, Y = 6 rows)

4. To use sprite sheet texture images, we must create a **Material** to assign to the Particle System's **Renderer** module. In the **Project** window, create a new **Material** named m_blueFire (**Create | Material**).

5. With m_blueFire selected in the **Project** window, in the **Inspector** window, select the **Particles/ Standard Unlit** shader, set **Rendering Mode** to **Additive**, and check the **Soft Particles** option.

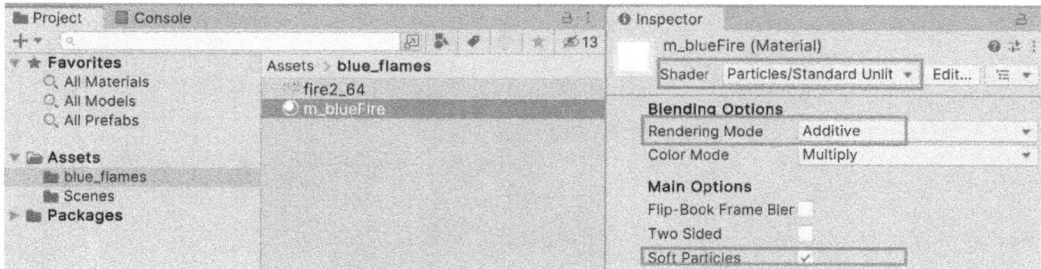

Figure 15.17: Customizing m_blueFire for the Particle System's Renderer component

6. Then, in the **Inspector** window, set the **Albedo** property of the material to the blue2_64 sprite asset.

   First, let's create a floor and wall from scaled 3D Cubes as a location to view our Particle System.

7. Create a 3D **Cube** GameObject named Cube-wall. In the **Inspector** window, set its **Position** to (1, 0, 3) and its **Scale** to (20, 20, 1).

8. Create a second 3D **Cube** GameObject named Cube-floor. In the **Inspector** window, set its **Position** to (0, -1, 0), and its **Scale** to (20, 1, 20).

9. Now, let's create a new **Particle System** in the scene by choosing **Create | Effects | Particle System**. Set **Start Lifetime** to 2 and **Start Speed** to 0 so that particles live for 2 seconds and do not move from the place they are created. Set **Start Size** to 5 and **Max Particles** to 1 so that we have just one large (size 5) flickering flame based on our animated sequence of images.

10. For the Particle System's **Renderer** module, set **Material** to m_blueFire.

11. At the moment, we can see the entire sprite sheet as a particle (all 60 images!). For our final customization, we need to check the **Texture Sheet Animation** module of the **Particle System** component in the **Inspector** window and set the grid side to X = 10 and Y = 6 (since our sprite sheet has 10 columns and 6 rows).

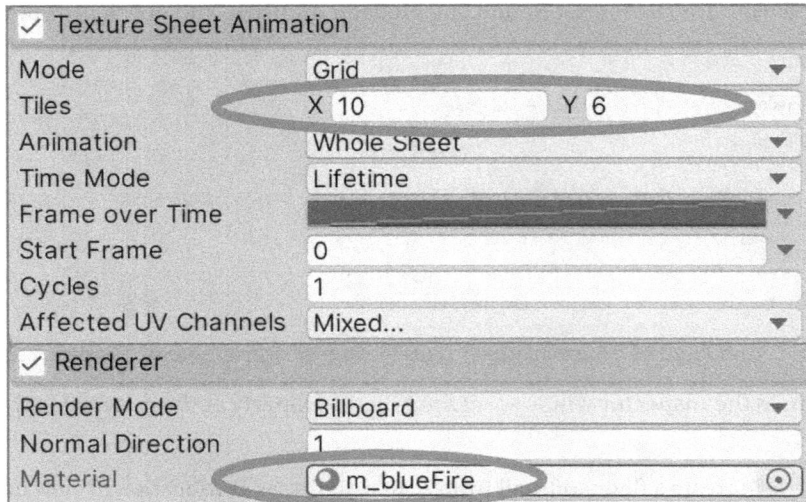

Figure 15.18: Setting the 10 x 6 Texture Sheet grid and the Renderer material

12. Save and run the scene. You should see a nicely animated fire with blue flames. You may wish to reduce the **Intensity** of **Directional Light** in the scene to 0.5 to see the blue flames more easily.

How it works...

By creating a **material** using one particle that's been rendered and linked to a multiple-image sprite sheet, we are able to take advantage of the **Texture Sheet Animation** module feature of Unity's **Particle System**.

> The default **Render Mode** for Particle Systems is **Billboard**. This means that the image for each particle is always facing toward the camera. In this way, a 2D animation of a flame works very well since whatever direction the user is looking at the Particle Systems from, they will always see the intended 2D flame animation, giving the visual effect of a 3D fire.

The Particle System's **Rendered** component needs to be linked to the material for the sprite sheet, and the X- and Y-tile grid needs to match the number of columns (X) and rows (Y) in the sprite sheet file.

By having 0 for **Start Speed** and only one particle (**Max Particles** = 1), we ensure we have a single animated particle staying in one spot. The size of the particle may be limited by the pixel density of your sprite sheet images – less detailed images will only work well with smaller particles.

We selected the **Particles/Standard Unlit** shader for our **Material**. This is one of the shaders that's been designed for particles, and it is faster than the **Particles/Standard Surface** Shader since it does not perform any lighting calculations.

The flames of our system do not need to show any interaction with any light sources in our scene.

For our material's **Rendering Mode**, we chose the **Additive** option. This adds the particle's color and background pixel colors together, which is great for glowing effects if we wish to simulate fire, for example. We could have tweaked the way the final colors for pixels are chosen even further by trying out different **color modes**. Color modes determine how particle colors are combined with the material's text colors. The default color mode is **Multiply**, which, again, works fine for fire effects. Other texture/particle color combinations include additive/subtractive and so on. The Unity documentation on particle Shaders (`https://docs.unity3d.com/Manual/shader-StandardParticleShaders.html`) explains the different **color mode** options well:

We enabled **Soft Particles** for our **Material**, which reduces hard edges when particles are close to surfaces in the scene – again, this is good for flame effects.

While the visual effect from this recipe is rather modest, when combined with other sprite systems and 3D modules, animated Particle Systems with sprite sheets can speed up your scene development process. This is because a lot of the work has already been completed through the animations that were achieved in the multiple-image sprite sheets.

Making particles collide with scene objects

A great way to enhance the visual effects of Particle Systems is for the particles to behave as if they are a real part of the 3D world. We can easily give this impression by enabling the **Collision** module, which makes particles change their direction of movement when they "hit" the colliders of 3D objects in the scene.

In this recipe, we'll create a scene containing some 3D objects and create two Particle Systems – one like a fountain of ping-pong balls and another like a gun firing many small balls in a line. Both sets of particles will bounce and change direction when they collide with the objects in the scene:

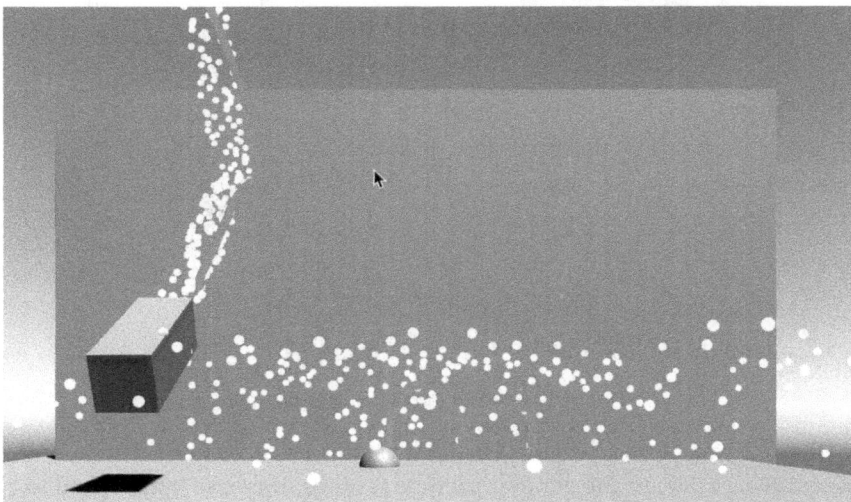

Figure 15.19: The two bouncing Particle Systems in the collision scene we'll create in this recipe

How to do it...

To create a Particle System from scratch, follow these steps:

1. Create a new Unity 3D project.

2. Move MainCamera to **Position** (0, 1, -20) so that we can see more of the scene GameObjects.

3. Now, let's create a floor and wall from scaled 3D Cubes, as a location to view our Particle System. First, create a 3D **Cube** GameObject named Cube-wall. In the **Inspector** window, set its **Position** to (1, 0, 3) and its **Scale** to (20, 20, 1).

4. Create a second 3D **Cube** GameObject named Cube-floor. In the **Inspector** window, set its **Position** to (0, -1, 0), and its **Scale** to (20, 1, 20).

5. Often, the visual effect from a **Particle System** comes from a GameObject, so let's create a 3D **Sphere** at a **Position** of (0, -0.5, 0) so that it appears to be embedded in the ground and the source of our particles. Check the **Is Trigger** setting for the sphere's **Collider** so that it will not be something our particles will bounce into.

6. Now, let's add a GameObject containing a **Particle System** component by choosing **Create | Effects | Particle System**. Rename this GameObject Particle System-fountain. In the **Inspector** window, set this new GameObject's **Position** to (0, 0, 0) so that the particles appear to be coming out of the sphere at that same location.

7. Let's create a circle-like **Material** for our fountain particles. In the **Project** window, use the right-mouse-button menu to create a new **Material** named m_circleParticle.

8. Select m_circleParticle. Then, in the **Inspector** window, change its **Renderer** to **Particles/ Standard Unit** and its **Rendering Mode** to **Additive**. Click the texture section circle for **Albedo** and choose the built-in **Knob** texture from the **Select Texture** window.

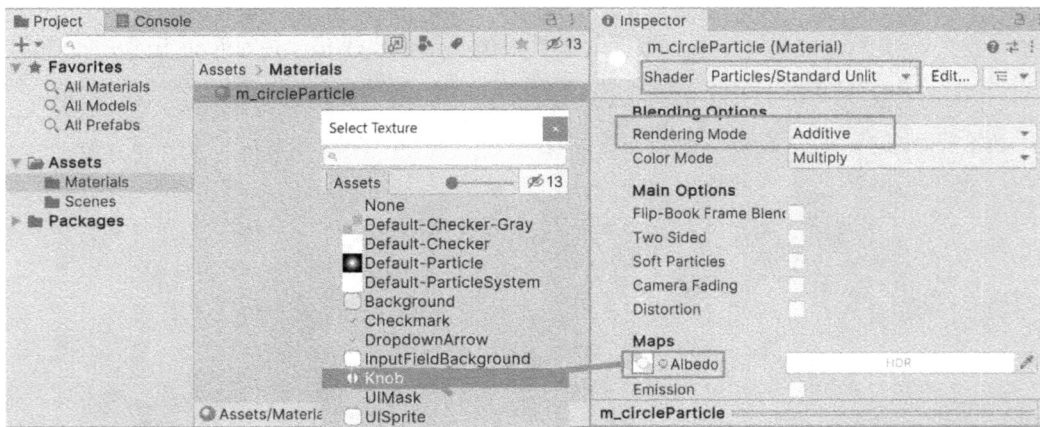

Figure 15.20: Creating our circle particle material

9. In the **Hierarchy** window, select this **Particle System**, and in its **Renderer** module, set **Material** to m_circleParticle. In the general particle settings, increase **Start Speed** to 7 and reduce **Size** to 0.3. Check the **Collision** module to enable it, and change **Type** to **World** (so particles bounce off the colliders in the scene).

10. Save and run the scene. It should look as if lots of small ping-pong balls are being emitted from the sphere.

11. Now, let's try to add some gravity and increase the number of particles being emitted to make it more fountain-like. In the **Hierarchy** window, select this **Particle System**, and in the **Inspector** Particle System general settings (labeled **Particle System-foundation**), set **Gravity Modifier** to 1 – these particles should now go up and then fall toward the ground. In the **Emission** module, increase **Rate over Time** to 50 – we should now see lots of particles being "sprayed" upward from our sphere, then falling and bounding off the "ground."

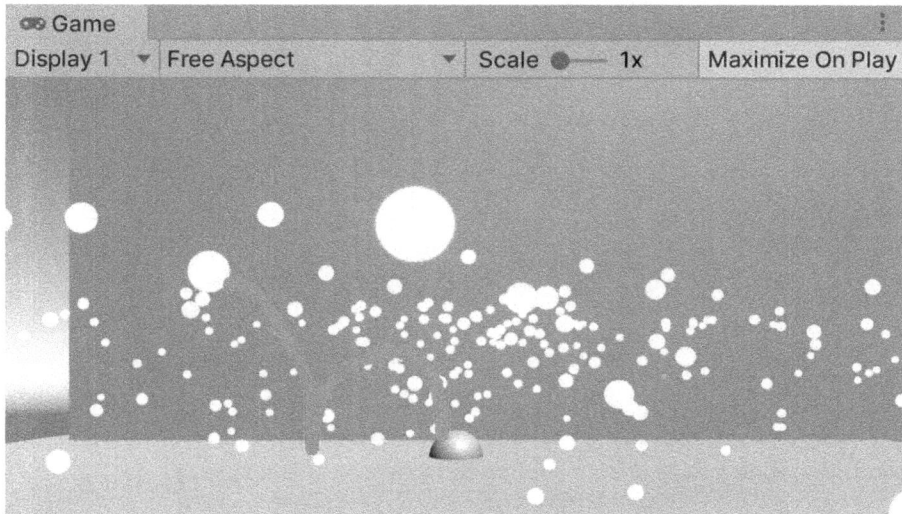

Figure 15.21: The ping-pong ball foundation from our half-sphere

12. Save and run the scene. You should see lots of particles being emitted and bouncing off the ground like ping-pong balls.

13. Let's add a ping-pong hose/gun effect. Add a 3D **Cube** to the scene named Cube-hose. In the **Inspector** window, set its **Position** to (-4, 2, -6), **Rotation** to (-25, 0, 0), and **Scale** to (1, 1, 3) to give the effect it is pointing toward the wall. Check the **Is Trigger** setting for the Cube's **Collider** so that this GameObject will not be something our particles will bounce into.

14. Create a new **Particle System** and name it Particle System-hose. Set this Particle System's **Position** to (-4, 2.5, -5) and **Rotation** to (-125, 0, -180).

15. Ensure that the Particle System-hose GameObject is selected in the **Hierarchy** window. In the **Inspector** window, for the **Renderer** module, set **Material** to m_circleParticle. Increase **Start Speed** to 10 and reduce **Size** to 0.3. Check the **Collision** module to enable it and change **Type** to **World** (so that particles bounce off the colliders in the scene).

16. To get all our ping-pong ball's particles to be fired out in a straight line, we need to set the **Shape** module of our **Particle System** to **Shape** = **Edge**, with a **Radius** of 0.5.

> Another way to achieve a less flattened hose effect would be to use a **cone** shape with an angle of **zero**.

17. Set the **Emission** module's **Rate over Time** to 100 so that there is a dense steady stream of ping-pong ball-type particles being emitted:

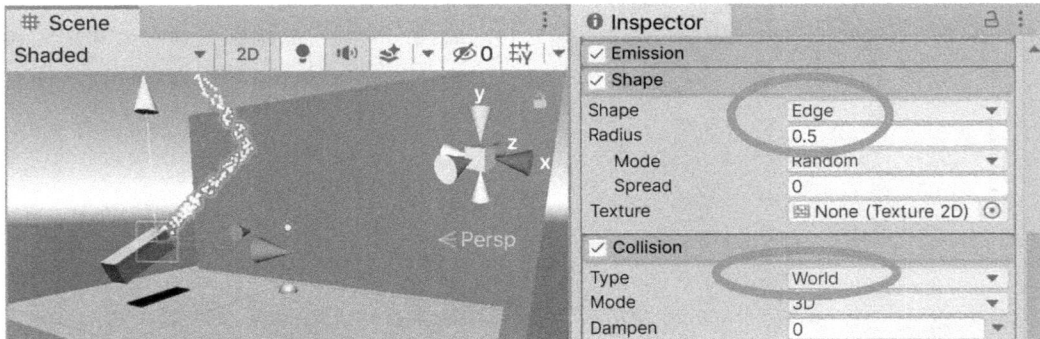

Figure 15.22: Collision with the World GameObject and the edge-shaped Particle System emitter

When you save and run the scene, you should see many ping-pong balls being emitted as if from a hose toward the wall, and as if from a fountain from the sunken sphere.

How it works...

When a 3D scene contains objects with **Collider** components, they define 3D spaces where there are surfaces that can be collided with. By enabling the **Collision** module of our **Particle System**, we are able to create more convincing visual effects since the particles "bounce" off the surfaces of objects that have colliders.

We used an edge-shaped **Particle System** that emitted hundreds of particles to create a steady stream of particles as if they had been fired from a gun or pumped out from a hose.

There's more...

Interesting effects can be achieved when a "trail" is added to some or all particles, visually marking the path the particle has followed during its lifetime. Explore this by making the following changes to the fountain scene:

1. Create a new yellow material named m_yellow.
2. In the **Hierarchy** window, select the fountain's **Particle System**. Then, in the **Inspector** window, enable the **Trails** module.
3. You'll now see a new property for the **Renderer** module named **Trail renderer**. For this property, select the new m_yellow **material**.

4. For the **Trails** module, set the following options to make every hundredth particle leave a thin trail behind it:

- **Ratio:** 0.01 (1 out of 100 get a trail)
- **Width over Trail:** 0.1 (narrow):

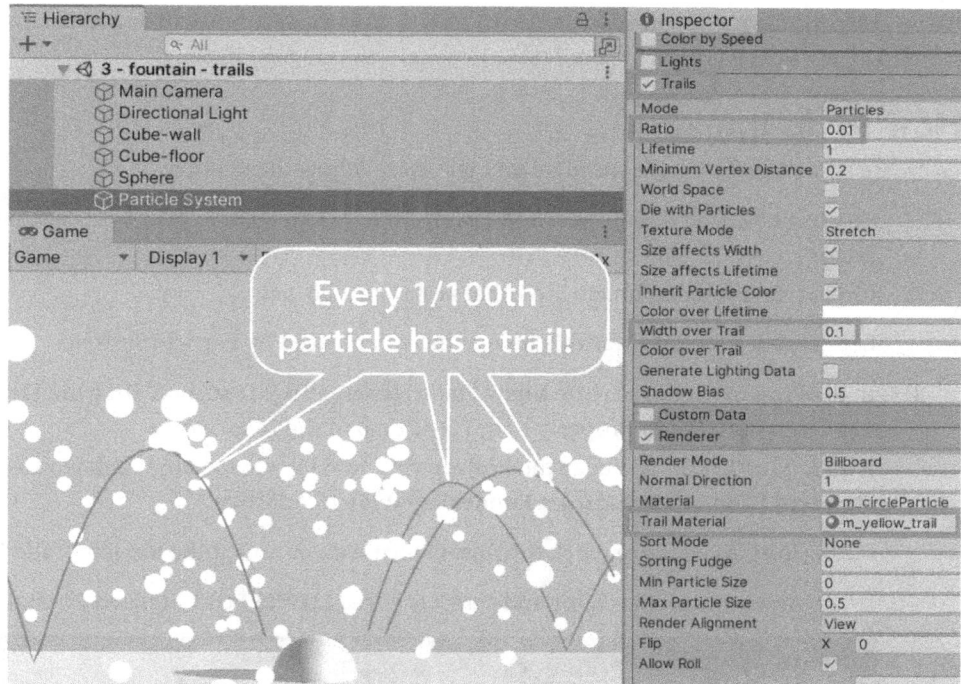

Figure 15.23: Adding occasional trails to particles from our fountain

Simulating an explosion

For many games, there will be events where we want to display an explosion in the scene, such as when the player's character dies or when they collide with an object that deals damage. We can quite easily create a visual explosion effect using an appropriate 2D image and a Particle System that fires one with a burst of many fast-moving, short-lived particles, in all directions. That's what we'll create in this recipe.

Figure 15.24: The explosion effect we'll create

Getting ready

For this recipe, we need an image of an explosion, so we'll use the `firecloud.png` image that's been published on the PNG-EGG website. Thanks to *GAMEKILLER48* for sharing this asset for game developers to use.

We've prepared the file you need in the 15_05 folder. Alternatively, you can download the image directly from the PNG-EGG website: `https://www.pngegg.com/en/png-ffmwo`.

How to do it...

To create a Particle System to simulate an explosion, follow these steps:

1. Create a new Unity 3D project and ensure you are logged in to your Unity account in the Unity Editor.
2. Import the provided image file – that is, `firecloud.png`.

   Let's create a **Material** for this image that can be used by Particle Systems.

3. In the **Project** window, create a new **Material** named m_fireCloud. Drag the `firecloud` texture asset into the **Albedo** property of m_fireCloud in the **Inspector.**
4. With m_fireCloud selected in the **Project** window, in the **Inspector** window, select the **Particles/ Standard Unlit** Shader and set **Rendering Mode** to **Additive.**

   Open **Unity Asset Store** (click the **Asset Store Web** button at the top left of the Unity Editor).

   In the **Asset Store**, search for and select the free **LOWLYPOLY Stylized Crystal** asset.

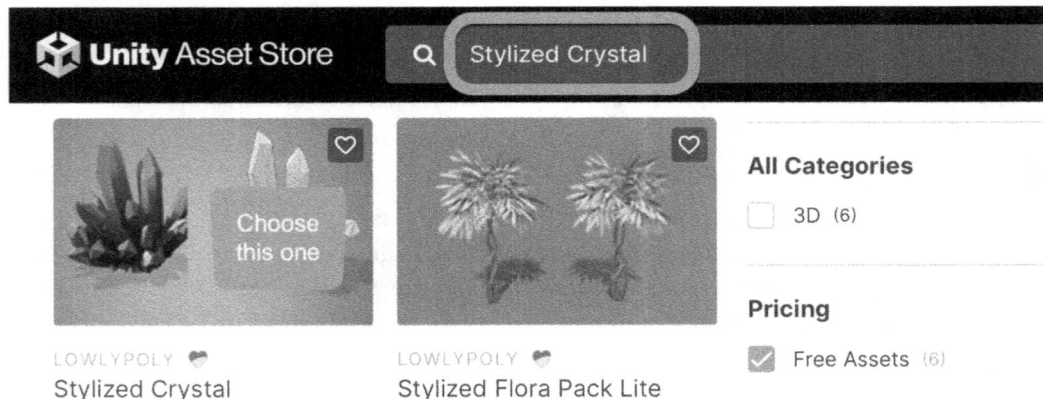

Figure 15.25: The free Stylized Crystal asset in the Asset Store

5. Click **Add to My Assets**. Then, after the button changes, click **Open in Unity**.
6. In the Unity Editor, the **Package Manager** window should open, and the **Stylized Crystal** assets should be selected in the **My Assets** list. Click **Download** and, when downloaded, click **Import**. In the **Project** window, you should now see a new folder called Stylized Crystal.

We are also going to need a 3D character and scene for this project, so we'll use the third-person character and demo scene from one of the free Unity **Starter Assets** packages.

7. Visit the Asset Store again and search for the free **Starter Assets – Third Person Character Controller** package from Unity Technologies.

8. Click **Add to My Assets**. Then, after the button changes, click **Open in Unity**.

9. In the Unity Editor, the **Package Manager** window should open, and the **Starter Assets - Third Person Character Controller** package should be selected in the **My Assets** list. Click **Download** and, when downloaded, click **Import**. Say **Yes** to any popup about resetting the project to use the new input system.

10. In the **Project** window, you should now see a new folder called StarterAssets.

11. Open the provided **Playground** scene, which you can find in the **Project** window (Assets | StarterAssets | ThirdPersonController | Scenes).

> **Note:** If you see lots of pink textures, it means some of the assets need to be converted to the URP pipeline for this project. Do this by opening the **Render Pipeline Converter** panel by choosing **Window | Rendering | Render Pipeline Converter**. Then check all 4 sections, and click the **Initialize And Convert** button at the bottom right. Then wait a few seconds – you can watch a progress bar of the conversion at the bottom right of the Unity Editor window. Once completed, the textures in the scene should render correctly.

12. Locate PlayerArmature in the **Hierarchy** window and move this character near the bottom of the stairs in the scene. Set its **Position** to (13, 0.3, 16) and its **Rotation** to (0, 90, 0).

13. Now, let's add a spiky crystal to this scene. From the **Project** window, drag the crystal_17_2 prefab (**Stylized Crystal | Prefab**) to a position just in front of the PlayerArmature character. Set its **Position** to (14, 0.3, 16).

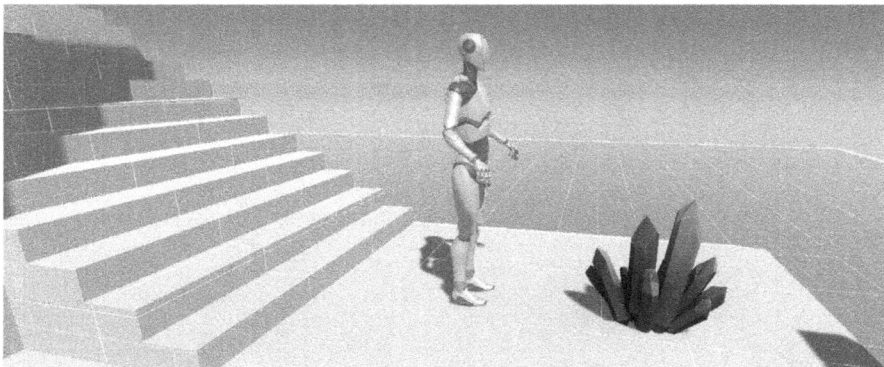

Figure 15.26: Character standing in front of the spiky Stylized Crystal GameObject

14. In the **Inspector** window, add a **Box Collider** to `crystal_17_2` and check its **Is Trigger** option. This means that `OnTriggerEnter(...)` event messages will be received by both the `crystal_17_2` and `PlayerArmature` GameObjects when the robot walks into the crystal.

15. Now, let's create an explosion-style **Particle System** that we'll then make into a prefab and instantiate from code when that collision occurs. Add a new **Particle System** to the scene named `Particle System-explosion` (**Create | Effects | Particle System**).

16. Now, let's ensure that this **Particle System** is located at the same place as `crystal_17_2`. Focus the **Scene** window on the location of `crystal_17_2` by double-clicking this GameObject in the **Hierarchy** window. Then, single-click to select the `PlayerArmature` GameObject in the **Hierarchy** window and choose **GameObject | Move To View**. This moves the selected GameObject (`Particle System-explosion`) to the location the **Scene** window is focused on.

17. We want particles to be single (not looped), short-lived, and moving quickly due to gravity. So, in the **Inspector** window for the general properties of `Particle System-explosion`, set **Duration** to `0.50` and uncheck **Looping**. Then, set **Start Lifetime** to `0.5`, **Start Speed** to `10`, **Start Size** to 2, and **Gravity Modifier** to 1.

18. We want the explosion to go in **all** directions, so enable the **Shape** module and set **Shape** to **Sphere**.

19. We want our **Particle System** to use the `m_fireCloud` material we created at the beginning of this recipe, so in the **Renderer** module, set **Material** to `m_fireCloud`:

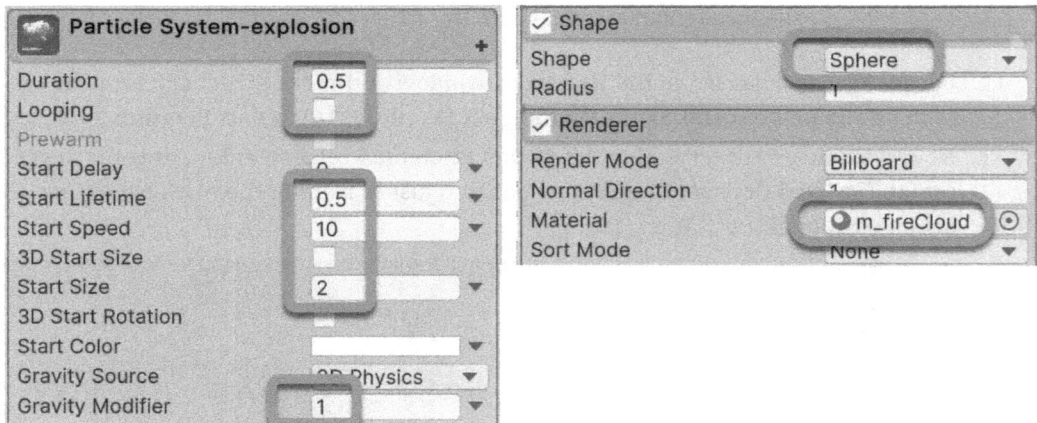

Figure 15.27: The General, Emission, and Shape settings for our explosion Particle System

20. We want a burst of many particles as the Particle System begins, which we can achieve in the **Emission** module. Click the plus (+) button to create an emission burst and set the count to 200:

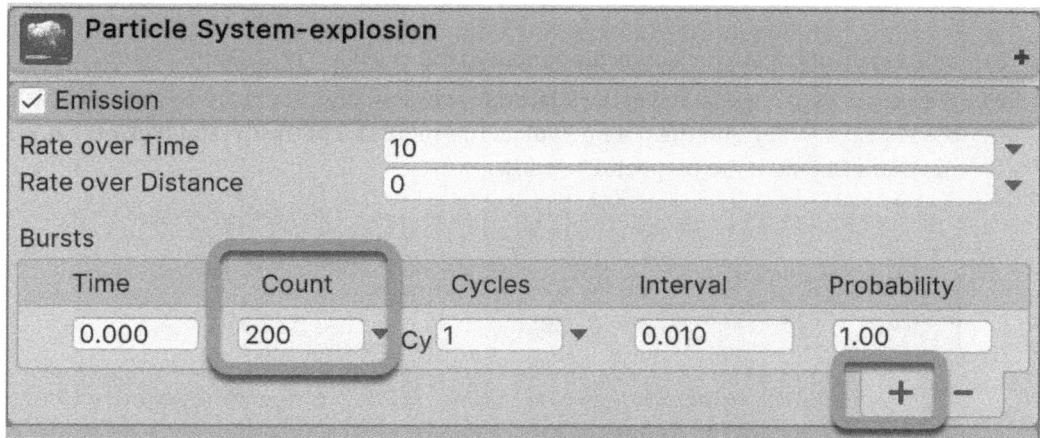

Figure 15.28: Setting an emission burst of 200

Now that we've customized our **Particle System** in the scene to behave as the explosion we need, let's turn this into a Prefab that we can instantiate at runtime when the player's character dies by hitting a spiky crystal.

21. Create a folder in the **Project** window named gre. Drag Particle System-explosion into this folder from the **Hierarchy** window. You should now see a new prefab named Particle System-explosion in the Prefabs folder.

22. Delete the Particle System-explosion GameObject from the scene.

23. Create a new C# script class called Crystal.cs containing the following code:

```csharp
using UnityEngine;

public class Crystal : MonoBehaviour {
    public GameObject explosionPrefab;

    private void OnTriggerEnter(Collider other) {
        // create explosion at same location as this Crystal
        GameObject explosion = Instantiate(explosionPrefab, transform.position,
            transform.rotation);

        // destroy particle system after 1 second
        Destroy(explosion, 1);

        // remove this Crystal
```

```
        Destroy(this.gameObject);
    }
}
```

24. Add a copy of `Crystal.cs` as a component of the `crystal_17_2` GameObject.

25. With `crystal_17_2` selected in the **Hierarchy** window, drag `Particle System-explosion` from the `Prefabs` folder into the public **Explosion Prefab** slot of the **Crystal (Script)** component of the GameObject in the **Inspector** window:

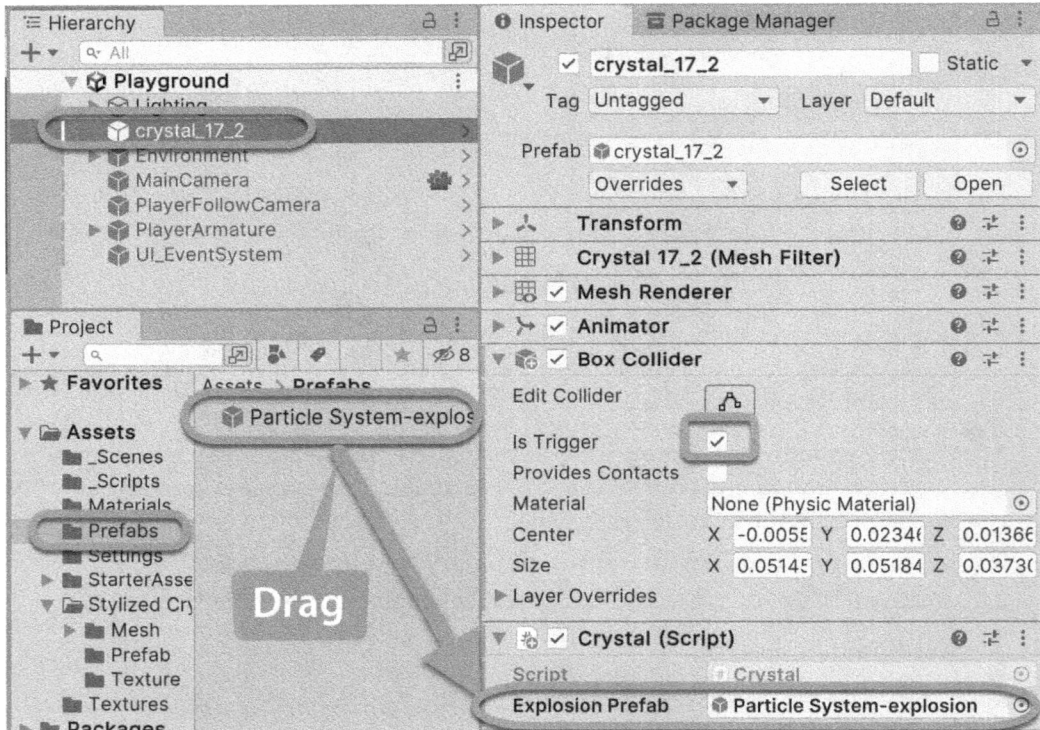

Figure 15.29: Assigning the explosion prefab to the Crystal (Script) component

26. Run the scene. When the player's character collides with the crystal, the crystal is destroyed and an explosion takes place (before being destroyed itself after 1 second).

How it works...

Using a Material from a related still image is a good way to improve the visual effect of a Particle System and saves us from having to do extra work, such as changing the colors of particles over time. The **Particles/Standard Unit** Shader in the **Additive** mode is all we needed for this example to make the material ready to be selected by the Particle System's **Renderer** module.

We were able to create the quick and striking visual effect of an explosion by changing the Particle System's **Shape** to **Sphere** so that the particles go everywhere, and also by creating a burst **Emission** set to 200. We made our particles move fast, live a short time, and be affected by a **Gravity Modifier**.

We only wanted the explosion to happen once, so we disabled the **Loop** option in the general **Particle System** parameters.

We were able to create a reusable Prefab from the Particle System in the scene simply by dragging the GameObject into our **Project** window and then removing the original **Particle System** GameObject from the scene.

The Crystal.cs script class we created was quite straightforward as it was designed to have instance objects attached as components to the crystal GameObjects in the scene. When a crystal detects a collision receiving an OnEnterFrame(...) message, it creates (instantiates) a new GameObject instance of the explosion Particle System from the Prefab asset file we created. The explosion GameObject is created at the same location and rotation as the crystal, invoking Destroy(...) on the Particle System but with a delay of 1 second (so that the explosion will be completed), and then also destroying the GameObject that was collided with.

You may wish to add an audio effect to increase the impact of the explosion on the player – you learned how to add sound effects in *Chapter 4, Playing with and Manipulating Sounds*.

Using Invoke to delay the execution of an explosion

In the previous recipe, we wanted to immediately instantiate and play a Particle System as soon as the player's character collided with a crystal GameObject. However, there are times when we want to delay for a few moments when we want a prefab to be instantiated. This recipe customizes the previous one, in that we'll allow our player to move around the scene while dropping bombs. Three seconds after being dropped, that bomb will be replaced by an explosion – an instantiation of our explosion **Particle System** from the previous recipe. This recipe is inspired, although in a much-simplified form, by games such as **BomberMan**.

Figure 15.30: Our Unity BomberMan

Getting ready

This recipe follows on from the previous one, so make a copy of that and work on the copy.

How to do it...

To delay an explosion using the Invoke(...) method, follow these steps:

1. Open a copy of the previous recipe.
2. Ensure you are logged in to your Unity account in the Unity Editor.
3. Open **Unity Asset Store** (click the **Asset Store Web** button at the top left of the Unity Editor).
4. In the Asset Store, check **Free Bombs** and search for and select the **Yughues Free Bombs** asset.

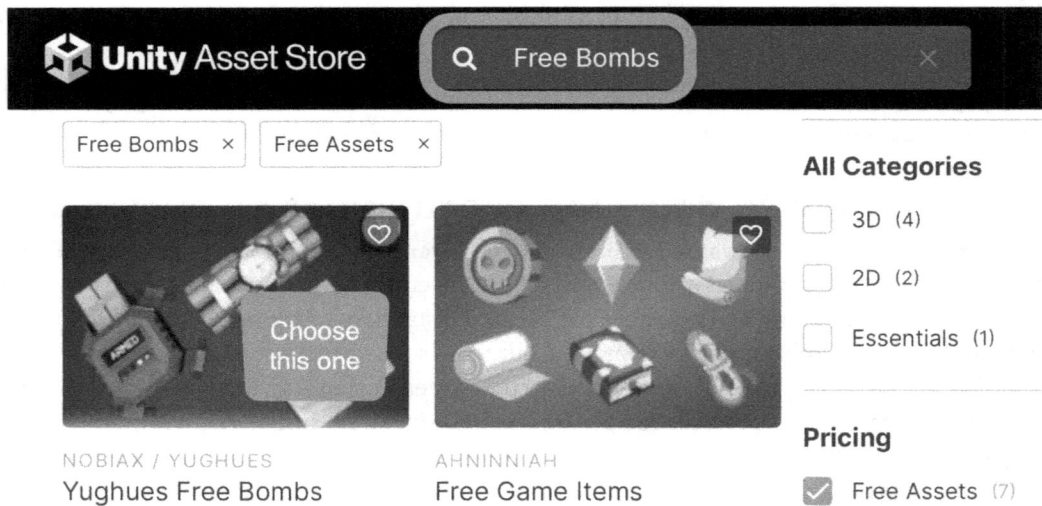

Figure 15.31: The free Yughues Free Bombs asset in the Asset Store

5. Click **Add to My Assets**. Then, after the button changes, click **Open in Unity**.
6. In the Unity Editor, the **Package Manager** window should open, and the **Free Bombs** asset should be selected in the **My Assets** list. Click **Download** and, when downloaded, click **Import**. In the **Project** window, you should now see a new folder called Meshes containing another folder called Bomb Packs.
7. Delete the **crystal_17_2** GameObject from the scene.
8. In the **Project** window, locate the Old-time bomb prefab asset file (folder: **Meshes | Bomb Packs | Old-time bomb**) and create a GameObject in the scene by dragging the prefab from the **Project** window into the **Scene** window. Rename the new GameObject bomb-small.

Note: If you see lots of pink textures, it means some of the assets need to be converted to the URP pipeline for this project. Do this by opening the Render Pipeline Converter panel by choosing Window | Rendering | Render Pipeline Converter. Then check all 4 sections, and click the Initialize And Convert button at the bottom right. Then wait a few seconds – you can watch a progress bar of the conversion at the bottom right of the Unity Editor window. Once completed, the textures in the scene should render correctly.

9. With this new bomb-small GameObject selected in the **Hierarchy** window, make it 100 times smaller by changing its **Scale** to (0.01, 0.01, 0.01) in the **Inspector** window – the bomb should now be about the same size as the PlayerArmature character.

Figure 15.32: The bomb-small GameObject in the Scene window

10. Create a new C# script class called BombBehaviour.cs containing the following code:

```csharp
using UnityEngine;

public class BombBehaviour : MonoBehaviour {
    public GameObject explosionPrefab;

    void Start() {
        // after 3 seconds instantiate the explosion
        float delay = 3;
        Invoke(nameof(Explode), delay);

    }

    private void Explode() {
        // create explosion at same location as this player
        GameObject explosion = Instantiate(explosionPrefab,
            transform.position, transform.rotation);
```

```
        // destroy particle system after 1 second
        Destroy(explosion, 1);

        // destroy this bomb GameObject
        Destroy(gameObject);
    }
}
```

11. Add an instance object script class, BombBehaviour, as a component to bomb-small. With bomb-small selected in the **Hierarchy** window, drag **Particle System-explosion** from the Prefabs folder into the public **Explosion Prefab** slot of the **Bomb Feature (Script)** component of the GameObject in the **Inspector** window:

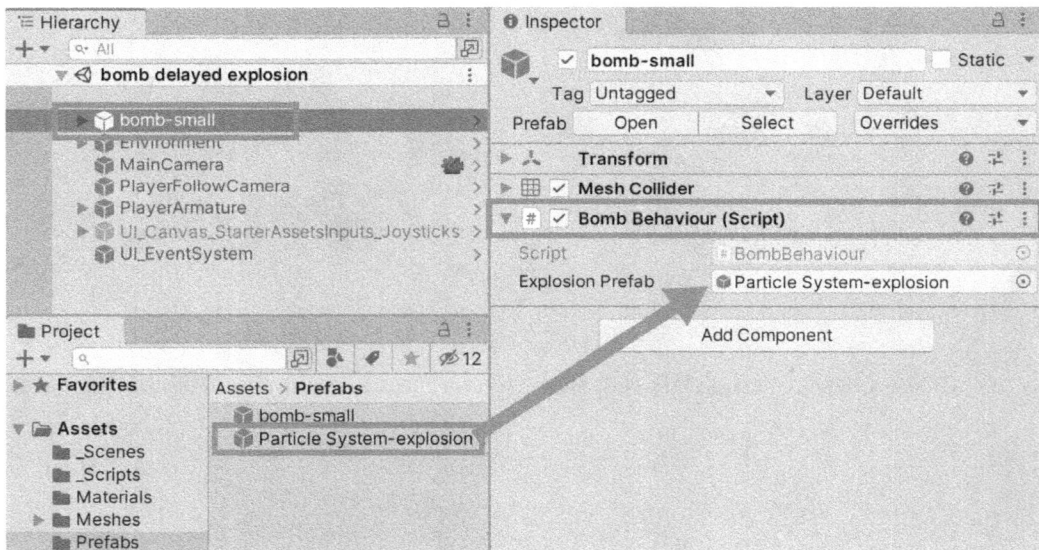

Figure 15.33: Creating our bomb-small prefab

12. Now, let's turn this into a prefab that we can instantiate at runtime so that when the player presses the *B* key, a bomb-small GameObject will be instantiated. Drag the bomb-small Game-Object into the Prefabs folder in the **Project** window (choose **Original Prefab** when asked what type to create). You should now see a new prefab named bomb-small in the Prefabs folder.

13. Now, delete the bomb-small GameObject from the **Hierarchy** window.

14. Now, we need to write more C# code to detect when the player presses the *B* key to drop a new bomb. Create a new C# script class called BombFeature.cs containing the following code:

```
using UnityEngine;
using UnityEngine.InputSystem;

public class BombFeature : MonoBehaviour
```

```
{
    public GameObject bombPrefab;

    void Update()
    {
        if (Keyboard.current[Key.B].wasReleasedThisFrame)
        {
            // create bomb at same location as this player
            Instantiate(bombPrefab, transform.position, transform.
rotation);
        }
    }
}
```

15. Add an instance object script class, BombFeature, as a component of the PlayerArmature GameObject.

16. With PlayerArmature selected in the **Hierarchy** window, drag bomb-small from the Prefabs folder into the public **Bomb Prefab** slot of the **Bomb Feature (Script)** component of the GameObject in the **Inspector** window.

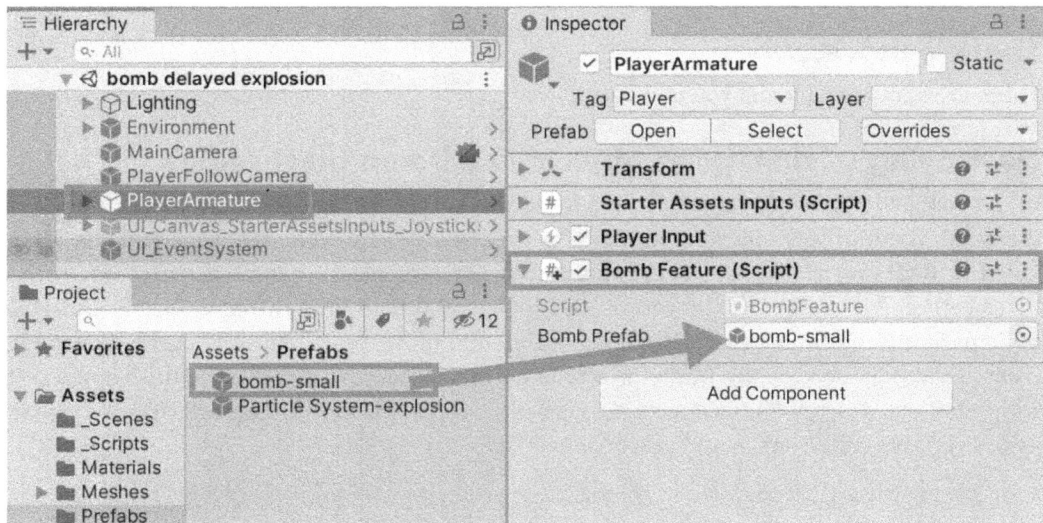

Figure 15.34: Populating Bomb Feature (Script) on ThirdPersonController

17. Save and run the scene. As you walk around the scene with PlayerArmature, you can press the *B* key to drop a bomb. After 3 seconds, the bomb will disappear and be replaced by an explosion – this is due to our explosion prefab!

How it works...

Having imported a bomb model into our project, we used a 100-times smaller (scaled **0.01**) version to create a prefab for our game. Before saving the small bomb as a prefab, an instance of the BombBehaviour.cs C# script class was added as a component of the GameObject, and its public **Explosion Prefab** property was linked to the explosion prefab we created in the previous recipe.

A second script class called BombFeature.cs was added as a component to PlayerArmature, and its public **Bomb Prefab** property was linked to our new bomb-small prefab. Since the starter assets our project is based on use the new Unity input system, we needed to write if (Keyboard.current[Key.B]. wasReleasedThisFrame) to detect the *B* key being pressed and released. We also needed to add a using statement to import the UnityEngine.InputSystem package.

When a new bomb-small GameObject was created, its Start() method was executed. In that method, we used an Invoke() statement to create a 3-second delay before executing the Explode() method. The Invoke() statement needs just the name of the method and the length of the delay in seconds (3, in our case). Most of this work was achieved by the Explode() method of the BombBehaviour.cs script class. Just as in the previous recipe, an explosion GameObject is created from its prefab, and a delayed Destroy() statement is used to delete the explosion GameObject after 1 second, which is more than enough time for us to see the explosion effect from the Particle System. Having created the explosion and its delayed Destroy() statement, we destroy the bomb GameObject itself.

There's more...

There are some details that you don't want to miss.

Destroying or damaging objects in a bomb blast radius

In the BombBehaviour.cs C# script class, we could add some logic to the Explode() method before the bomb GameObject is destroyed. Since we've displayed an explosion, we could use, for example, Physics.OverlapSphere(...) to get an array of all the colliders that are in the radius of the sphere that's centered on our bomb's location. Our logic could then loop through those colliders, sending their parent GameObjects (or scripted components) messages about how much damage has been taken, or simply destroy the GameObjects of all the colliders in the blast radius's sphere. Here is a script that will destroy GameObjects tagged Destroyable whose colliders are within the distance set by the variable called destructionSphereRadius:

```
private void Explode() {
    GameObject explosion = Instantiate(explosionPrefab, transform.position,
transform.rotation);
    Destroy(explosion, 1);

    // --- do any collision logic for bomb here -----
    Vector3 center = transform.position;
    Collider[] hitColliders = Physics.OverlapSphere(center,
destructionSphereRadius);
```

```
    // loop for all colliders within "blast radius"
    foreach (var collider in hitColliders) {
        // if collider in a GameObject tagged Destroyable, then destroy it !
        if (collider.CompareTag("Destroyable")) {
            print("destroying GameObject name = " + collider.name);
            Destroy(collider.gameObject);
        }
    }

    Destroy(gameObject);
}
```

Adding camera shake when the bomb explodes

Add some realism to the player's experience by making the camera shake when the bomb explodes! Some scripts to do this can be found at this GitHub Gist location: https://gist.github.com/ftvs/5822103.

Using a Line Renderer to create a spinning laser trap

Interesting visual effects can be achieved through a scripted approach by creating Line Renderers. Unity Line Renderers can draw lines in the scene using assigned materials and their colors and **Alpha Transparency**. Lines can be drawn for many different reasons in games, such as to simulate laser beams, provide floating arrows to give the player hints on the direction to follow, and so on.

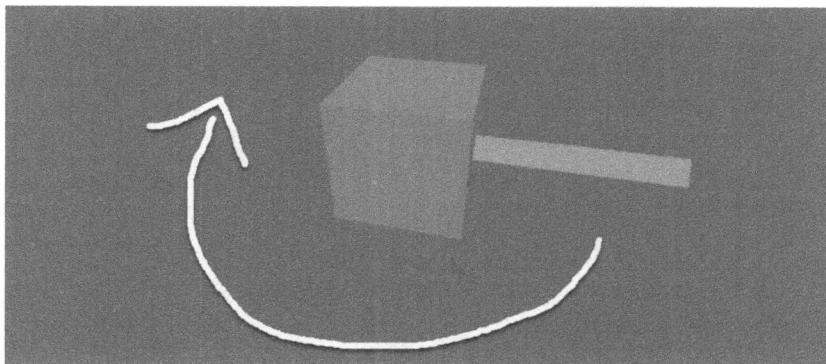

Figure 15.35: The rotating cube with the laser being emitted

In this recipe, we'll create the effect of a rotating 3D Cube emitting a laser beam for a short distance. We'll also learn how to create a scaled invisible cube in the same place as the laser, whose **Box Collider** can be used for collisions (as if they were with the laser).

Finally, in the *There's more...* section, we'll learn how to add these effects to the scene from the previous recipe to then create explosions when the player's character collides with the laser (the invisible cube's) collider.

How to do it...

To use a Line Renderer to create a spinning laser trap, perform the following steps:

1. Create a new 3D project.

2. Our **Line Renderer** will need a material to work with. Create a new material named m_laser (**Create | Material**).

3. With m_laser selected in the **Project** window, in the **Inspector** window, select the **Particles/ Standard Unit** Shader and set **Rendering Mode** to **Additive**.

4. Add a 3D **Cube** to the scene (**Create | 3D Object | Cube**) called Cube-laser.

5. Create a new C# script class called LaserDisplay containing the following code and add an instance object as a component to the Cube-laser GameObject:

```csharp
using UnityEngine;
public class LaserDisplay : MonoBehaviour {
    public float lineWidth = 0.2f;
    public float lineLength = 2;
    public Color color = Color.white;
    public Material material;
    public float rotationSpeed = 0.1f;
    private LineRenderer _lineRenderer;

    void Awake() {
        _lineRenderer =
      gameObject.AddComponent<LineRenderer>();
        _lineRenderer.material = material;
        _lineRenderer.positionCount = 2;
        _lineRenderer.startWidth = lineWidth;
    }
    void Update () {
        _lineRenderer.material.SetColor("_Color", color);
        Vector3 forward =
                transform.TransformDirection(Vector3.forward);
        Vector3 lineStart = transform.position;
        Vector3 lineEnd = transform.position + forward *
      lineLength;
        _lineRenderer.SetPosition(0, lineStart);
        _lineRenderer.SetPosition(1, lineEnd);
        transform.Rotate(0, rotationSpeed, 0);
    } }
```

6. With the Cube-laser GameObject selected in the **Hierarchy** window, from the **Project** window, drag the **m_laser** material into the **Material** variable slot in the **Inspector** window for the **Laser Display (Script)** component.

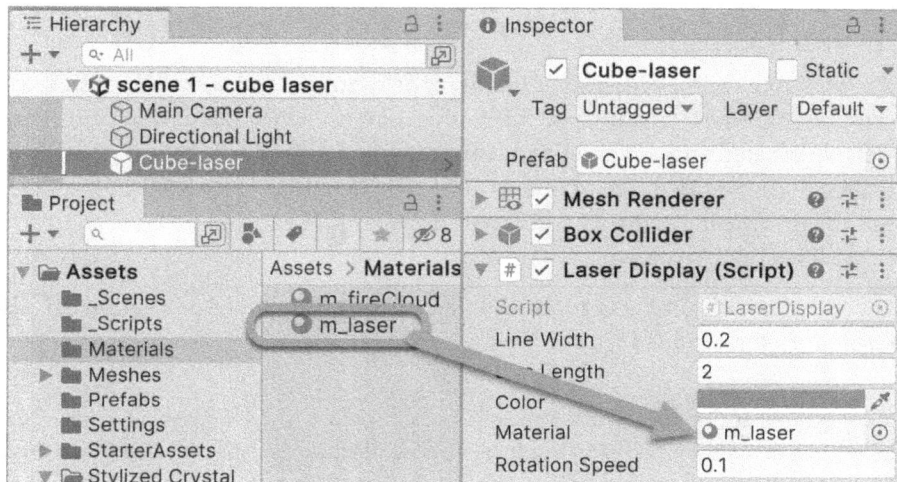

Figure 15.36: Assigning the m_laser material for the LaserDisplay scripted component

7. In the **Inspector** window, use the **Color** selector popup to set a red color for the **Laser Display (Script)** component. Of course, you can play around with the color and its **Alpha Transparency** to get the effect you want for your game.

8. Save and run the scene; you should see a slowly spinning cube with a laser beam coming out of one of its sides.

9. Let's also prepare this GameObject for collision usage by creating a collider that matches the space in which the beam is drawn.

10. Create a 3D **Cube** named Cube-collider. Make this a child of the Cube-laser GameObject and set its **Position** to (0, 0, 1) and its **Scale** to (0.1, 0.1, 2). Disable its **Mesh Renderer** (so that the player won't see it) and check the **Box Collider Is Trigger** option so that collision events will be generated.

Figure 15.37: Creating the Cube-collider so that it matches the laser space

11. Run the scene again – you won't see any difference, but we've now created an object with a collider that matches the laser beam so that OnTriggerEnter(...) collision event messages will be generated when the GameObjects, such as a player character, collide with the beam.

How it works...

The script we wrote draws with a Line Renderer at runtime. This draws lines in the scene that the player sees, but the lines don't interact with other GameObjects. Our Line Renderer needed a material, which we created and assigned and which can be customized through the public variables of the scripted component in terms of **Line Width**, **Line Length**, and **Color**. Different colors can be set at design time or by using code, with different transparency settings being used to achieve different looks for the lines that are drawn.

The script contains a statement to rotate the GameObject. The speed of rotation can also be adjusted or set to zero for no rotation.

> **Note:** If the degree of rotation is quite high (above 10 degrees, for example), there is a chance that a collision might not occur because the player's character might be in the "gap" as the laser jumps from one rotated position to another. This could be solved with a slightly larger collider, beyond the visible laser area.

Finally, we created an invisible Box Collider that took up the same space as the drawn line so that OnTriggerEnter(...) collision event messages would be generated when the GameObjects, such as a player character, collide with the beam. There are two values of this Cube-collider GameObject that are important to ensure the collider matches the laser being drawn. The position of the Cube-collider child GameObject should be (0, 0, laserLength/2). This ensures that the collider starts in the center of Cube-laser, just like the Line Renderer. So, with a laser length of 2, the **Position** value of Cube-collider was (0, 0, 1). The second important value is the **Scale** value of Cube-collider – the Z-value should match the laser beam's length – that is, (0.1, 0, 1, laserLength). So, if we wish to change the laser beam's length, then these values need to change too.

There's more...

There are some details that you don't want to miss.

Triggering an explosion when the player's character collides with the beam

With some scripting and an explosion Particle System, we can trigger an explosion when the player's character collides with our laser-style Line Renderer.

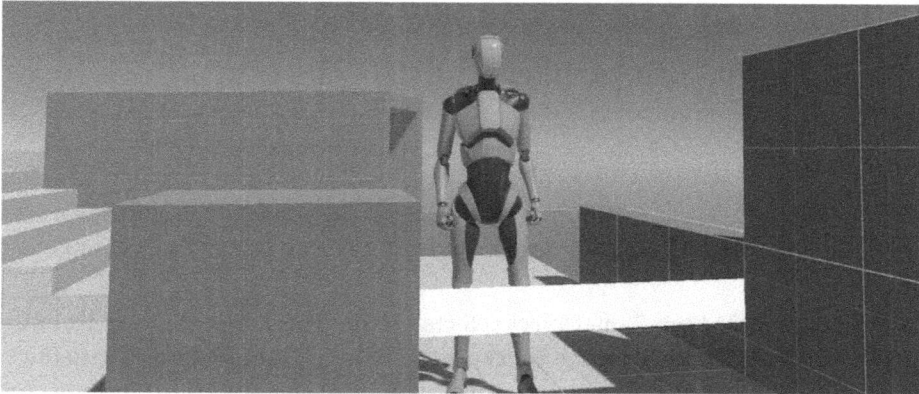

Figure 15.38: Triggering an explosion when a character collides with the laser

Do the following:

1. Create a copy of the project from the *Simulating an explosion* recipe.
2. Delete the crystal GameObject from the scene.
3. Create a rotating 3D Cube with a laser-style Line Renderer, as described in this recipe.

> **Note:** Alternatively, you could export the whole project as a Unity package, and import it into the copy of the *Simulating an explosion* recipe – this would save recreating the steps to create the rotating laser cube.

4. With the `Cube-laser` GameObject selected in the **Hierarchy** window, set the **Rotation Speed** value to 0 for the **Laser Display (Script)** component in the **Inspector** window.
5. Create a new C# script class called `LaserCollision` containing the following code:

```csharp
using UnityEngine;
public class LaserCollision : MonoBehaviour {
    public GameObject explosionPrefab;
    private void OnTriggerEnter(Collider other)
    {
        // create explosion at same location as this Crystal
        GameObject explosion = Instantiate(explosionPrefab,
            other.transform.position,
other.transform.rotation);
```

```
            // destroy particle system after 1 second
            Destroy(explosion, 1);
            // do other logic here - e.g. reduce player health
      etc. ...
      } }
```

6. Select the Cube-collider GameObject in the **Hierarchy** window (the child object inside Cube-laser). Add an instance object of the script class LaserCollision as a component to Cube-collider.

7. With the Cube-collider GameObject selected in the **Hierarchy** window, from the **Project** window's Prefabs folder, drag the Particle System- explosion prefab into the **Explosion Prefab** variable slot into the **Inspector** window for the **Laser Collision (Script)** component.

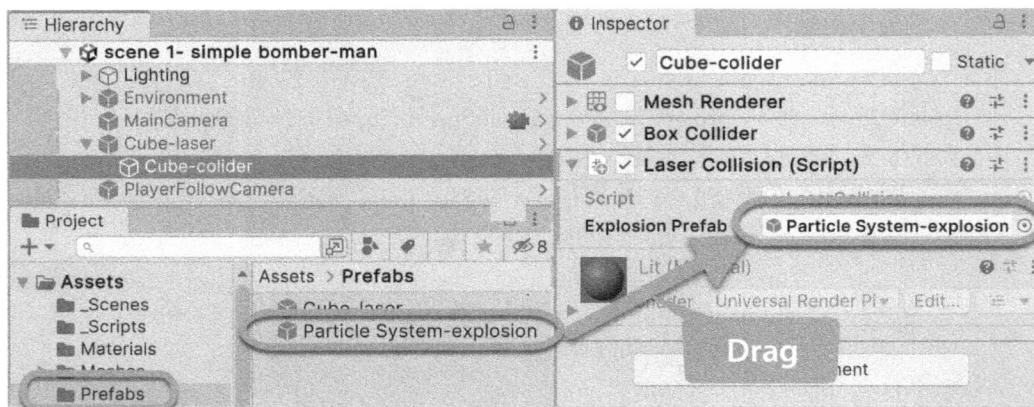

Figure 15.39: Assigning the explosion Prefab to the LaserCollision scripted component

8. Run the scene. Each time the player's character hits the laser, an explosion will be created.

In the preceding script, an explosion is instantiated at the location of the object that collided with the laser GameObject (other.transform.position).

You can add logic either to the LaserCollision script or a script on the third-person controller to reduce the score, reduce the player's health, or whatever else is appropriate in your game for when the player hits a laser.

Further reading

You can learn more about Particle Systems at the following links:

*   Kodeco's (formerly Ray Wenderlich's) introduction to Particle Systems: https://www.kodeco.com/138-introduction-to-unity-particle-systems

*   PolyToot's *Advanced Fire Effect Tutorial With Unity Particles*: https://www.youtube.com/watch?v=G0R7MIbX3MU

Learn more about Line Renderers in the Unity manual pages:

- https://docs.unity3d.com/Manual/class-LineRenderer.html

An exciting package now out of preview is **Visual Effects Graph**. This offers powerful ways to create stunning visual effects in Unity projects. Learn more at these links:

- Kodeco's (formerly Ray Wenderlich's) Visual Effects Graph tutorial: https://www.kodeco.com/9261156-introduction-to-the-visual-effect-graph
- The Unity Visual Effects Graph samples project to download (it's over 1 GB): https://github.com/Unity-Technologies/VisualEffectGraph-Samples
- A Unity blog post about Visual Effects Graph samples: https://blog.unity.com/technology/visual-effect-graph-samples

Learn more on Discord

To join the Discord community for this book – where you can share feedback, ask questions to the author, and learn about new releases – follow the QR code below:

https://packt.link/unitydev

16

Mobile Games and Applications

In this chapter, we will present a set of recipes introducing the techniques you will need to build and deploy Unity games and applications to mobile devices.

Mobile projects might be games for an Apple or Android cell phone, an **Augmented Reality (AR)** app for an Apple iOS iPad tablet, or perhaps a **Virtual Reality (VR)** game for a dedicated VR headset. The concepts and skills for deploying to mobile devices are the same for all these uses, so in this chapter, we'll explore how to set up, build, and deploy games for Android and Apple mobile devices, and you can then use these skills in the following two chapters when you learn how to develop and deploy AR and VR applications.

There are several steps to get from a project in the Unity Editor to a mobile app running on a device. For Android, we need to use the Unity Editor to build an APK Android app executable, and then get that APK file onto the device to install and run. For Apple iOS, there is an extra step since what the Unity Editor creates is a project for the Apple Xcode editor. The Xcode editor then builds an IPA iOS app executable, which we then need to get onto an iOS device.

The simplest way to get app executables onto a mobile device is through a USB cable connecting the device to our development computer.

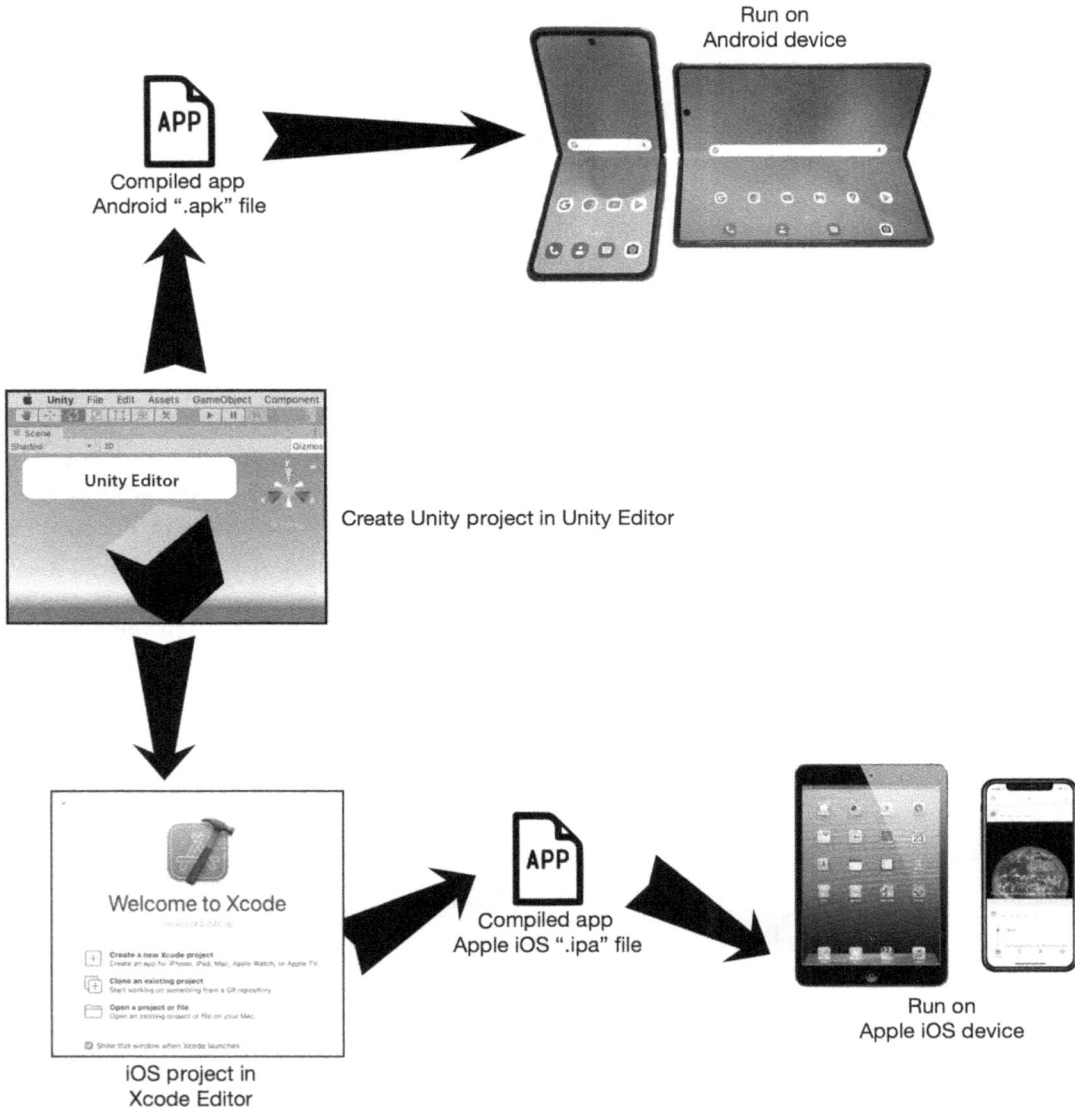

Figure 16.1: App executable build process for Android and Apple iOS devices

In this chapter, we will cover the following recipes:

- Setting up your system for Android mobile app development
- Setting up your system for Apple iOS mobile app development
- Deploying the Third Person Character Controller Starter project for an Android device
- Deploying the Third Person Character Controller Starter project for an Apple device
- Creating and deploying a mobile game with Unity's Runner Game project template

Setting up your system for Android mobile app development

There are some basic steps you need to perform to add the modules you need to your editor application so that your Unity Editor application is set up for Android mobile development.

How to do it...

To set up your system for Android mobile app development, do the following:

1. Open the **Unity Hub** application and choose the **Installs** tab on the left.
2. Now, for the version of the Unity Editor you wish to develop a mobile project for, click the settings "cog" button and click the **Add modules** button.

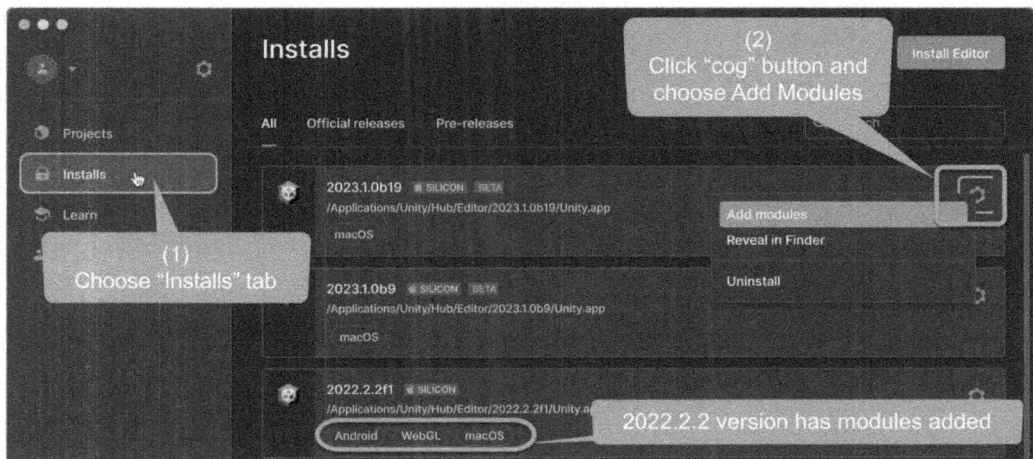

Figure 16.2: The Add modules button for an installed Unity Editor version in the Unity Hub application

3. For Android development, check the **Android Build Support** modules for your application. Then ensure that both the **OpenJDK** and **Android SDK & NDK Tools** sub-modules are also checked, and click the **Continue** button.

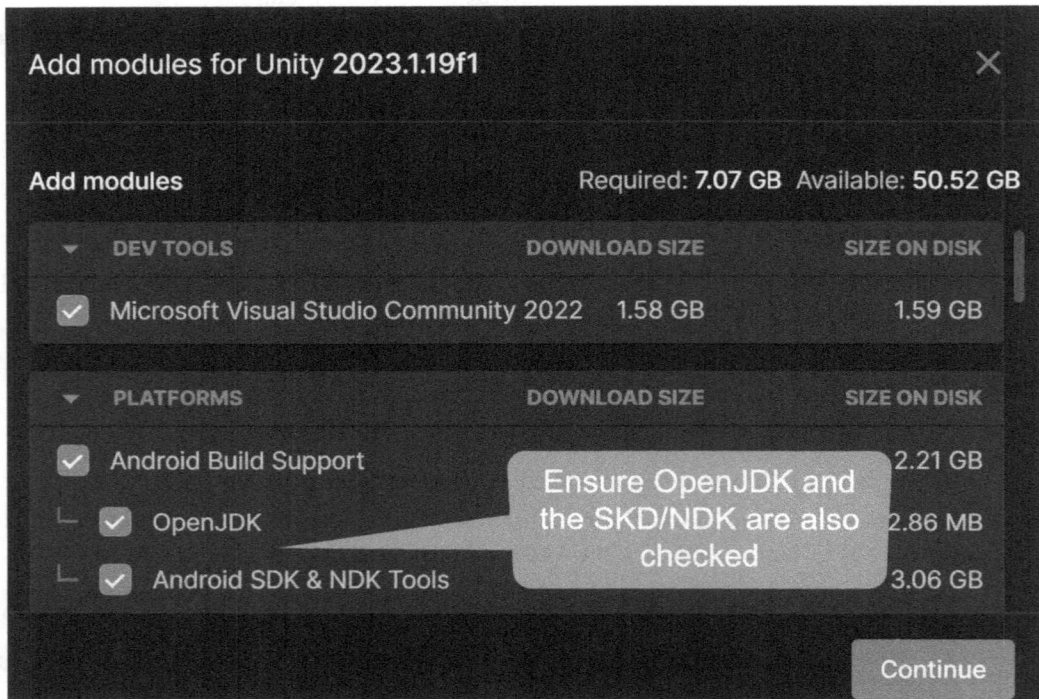

Figure 16.3: Adding the Android Build Support with OpenJDK and SDK/NDK modules to an Editor version

4. You may have to wait a while for the selected module(s) to download and be added to your Unity Editor installation.

    Congratulations! You have added Android mobile development build support for your Unity application.

How it works...

Unity Hub allows you to add additional modules for each version of the Unity Editor installed on your computer system. If you're working with Android mobile app builds, you need to add **OpenJDK** and **Android SDK & NDK Tools** to your system.

Setting up your system for Apple iOS mobile app development

There are some basic steps you need to perform so that your Unity Editor and Mac computer system are set up for Apple mobile development. In this recipe, you'll learn how to add the modules you need to your editor application for mobile development, as well as how to link an Apple ID to the Mac **Xcode** program editor, which is used to build Apple iOS applications.

Getting ready

If you don't have one already, create a free Apple ID. You can do this when you set up an Apple device, or you can visit the Apple ID section of Apple's website: `appleid.apple.com`.

How to do it...

To set up up your system for Apple iOS mobile app development, do the following:

1. Open the **Unity Hub** application and choose the **Installs** tab on the left.
2. Now, for the version of the Unity Editor you wish to develop a mobile project for, click the settings "cog" button and click the **Add modules** button.

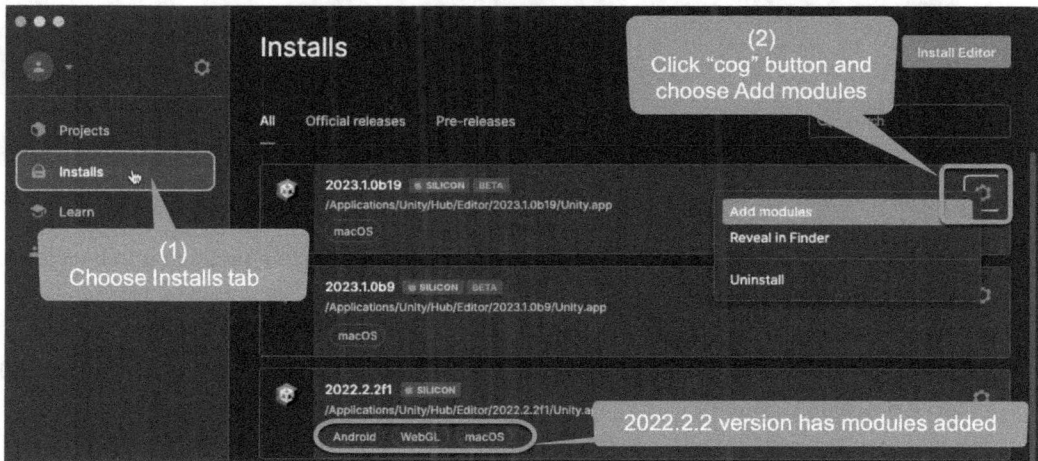

Figure 16.4: The Add modules button for an installed Unity Editor version in the Unity Hub application

3. For Apple development, check the **iOS Build Support** module for your application, and click the **Continue** button.

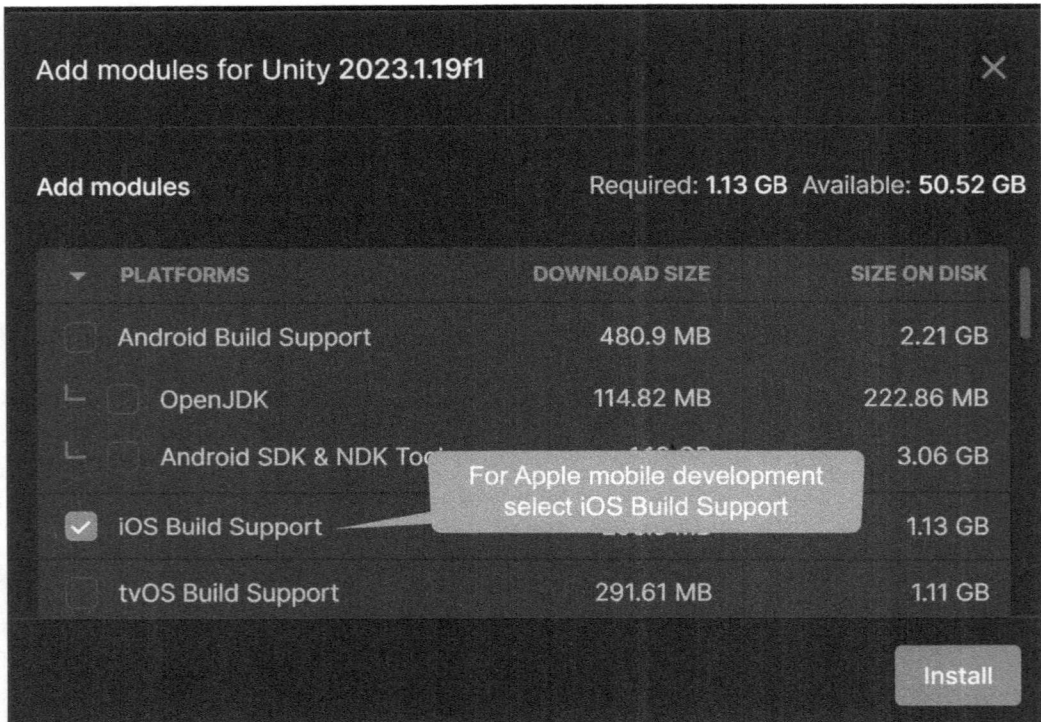

Figure 16.5: Adding the iOS Build Support module to an Editor version

4. You may have to wait for a while for the selected module(s) to download and be added to your Unity Editor installation.

5. Next, open the Xcode editor on your Mac.

6. From the **Xcode** menu, choose **Preferences...**, and then click the **Accounts** tab when the **Preferences** dialog appears.

7. Click the plus sign (+) button at the bottom left to add a new Apple ID account.

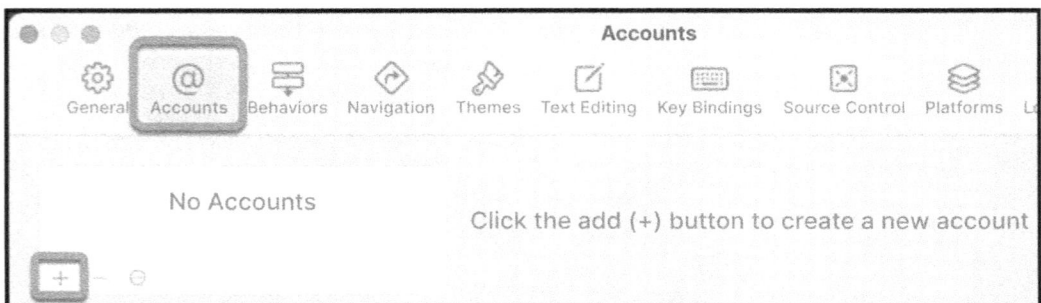

Figure 16.6: The Xcode Accounts section of Xcode preferences

8. When asked which type of account to add, select the first option, which is **Apple ID**, then click the **Continue** button.

9. Next, enter your **Apple ID** and password.

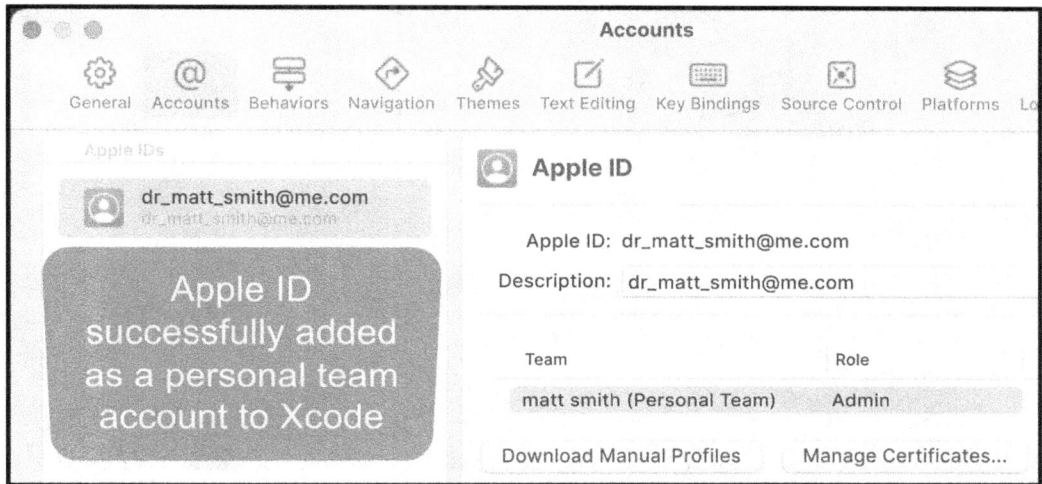

Figure 16.7: Apple ID successfully added as a personal team account to Xcode

Congratulations! You have added Apple iOS mobile development build support for your Unity application.

How it works...

Unity Hub allows you to add additional modules for each version of the Unity Editor installed on your computer system. If you're working with iOS mobile app builds, you need to add the **iOS Build Support module** to your system. Finally, you need to have an Apple ID and to add it to a team in the Apple Xcode application editor.

Deploying the Third Person Character Controller Starter project for an Android device

In this recipe, you'll add the Unity **Third Person Character Controller** Starter project, configure it for mobile, and then build and run the game on an Android device attached to your computer with a USB cable.

Other ways to get Android APK application files onto a device are explored in the *There's more...* section of this recipe.

Figure 16.8: Third-Person Playground scene running on an Android mobile device

Getting ready

Before starting this recipe, ensure you have completed the first recipe in this chapter: *Setting up your system for Android mobile app development.*

How to do it...

To deploy the Third Person Controller Starter project for an Android device, perform the following steps:

1. Create a new **Unity 3D** project.
2. Open a web browser tab to the Unity Asset Store, by choosing **Asset menu: Window | Asset Store.**
3. Search for the free **Starter Assets Third Person Character Controller.**
4. Select the assets, and when viewing their details, click the **Add to My Assets** button.

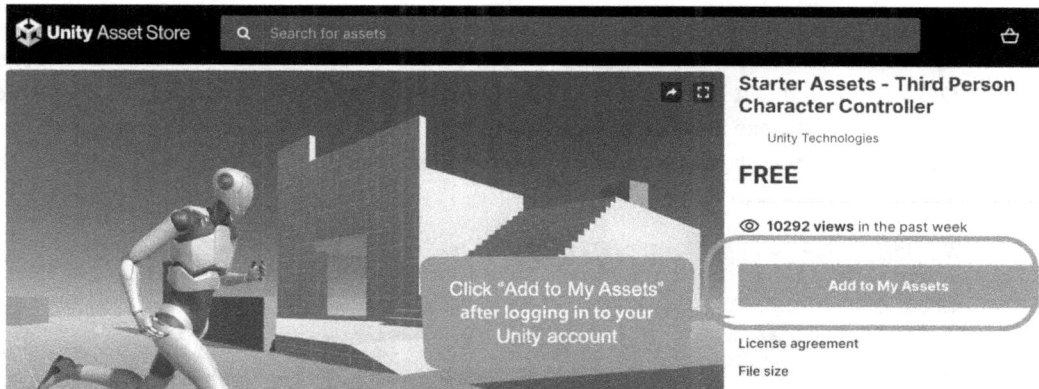

Figure 16.9: Adding Third Person Character Controller to your assets on the Unity Asset Store website

5. In the Unity Editor, open the **Package Manager** to list your Asset Store assets by choosing **Window | Package Manager** From the Packages dropdown choose My Assets (see the following screenshot). Dock the **Package Manager** panel alongside the **Inspector** panel.

Figure 16.10: Option to view My Assets in Package Manager in Unity Editor

6. In the **Package Manager** panel, locate and import the **Starter Assets Third Person Character Controller** package.

> **Note:** When importing the package, if asked, agree to additional Package Manager dependencies (click **Install/Upgrade**), and agree to enable the new input system's backend and restart the editor (click **Yes**).

7. You should now have a **Starter Assets** folder in your **Project** panel.

8. Open the Playground scene, in the **Project** panel folder: **Starter Assets | Third-Person Starter | Scenes**.

9. Since we are going to be deploying to a mobile device, we need to enable the mobile touch controls. So, in the **Hierarchy**, select the UI_Canvas_StarterAssetsInputs_Joysticks Game-Object and enable it by checking its checkbox at the top of the **Inspector**. In the **Game** panel, you should now see several circular touch controls, which will allow camera and character control when the project is running on a mobile device.

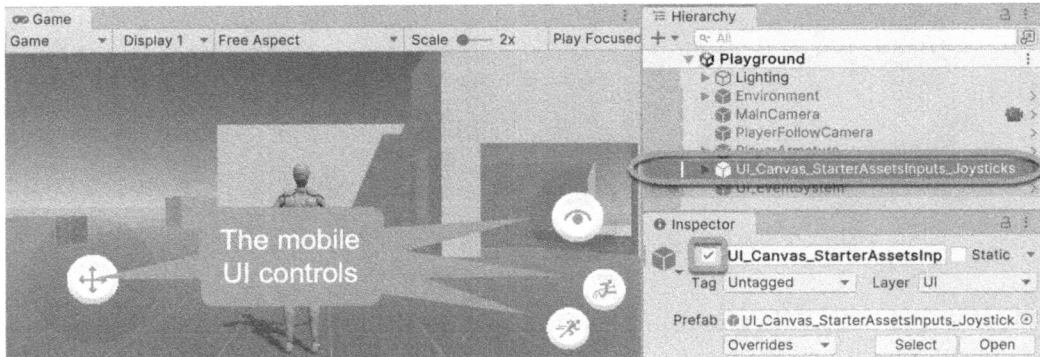

Figure 16.11: The enabled UI mobile touch controls displayed in the Game panel

10. Save and run the scene on your computer. You can move your robot character around the 3D environment using the arrow keys/*WASD*, and use the mouse to control the camera as usual. However, you can also see the UI touch controls.

11. Open the **Build Settings** dialog panel, by choosing **Edit | Build Settings....** Then add the current (Playground) scene to the build, by clicking the **Add Open Scenes** button.

12. Next switch the build target platform to Android, by clicking **Android** in the list of platforms, and clicking the **Switch Platform** button. You'll then see a pop-up dialog with a progress bar while Unity configures/compresses assets ready for the Android application build – wait for this to finish and disappear before moving on to the next step in this recipe.

13. Connect your phone/device to your computer with an appropriate USB cable.

> **Note:** You'll need to enable USB debugging and give permission for the computer to communicate with your Android device, including setting it to Developer Mode.

14. In Unity, click the **Refresh** button in the **Build Settings** panel.

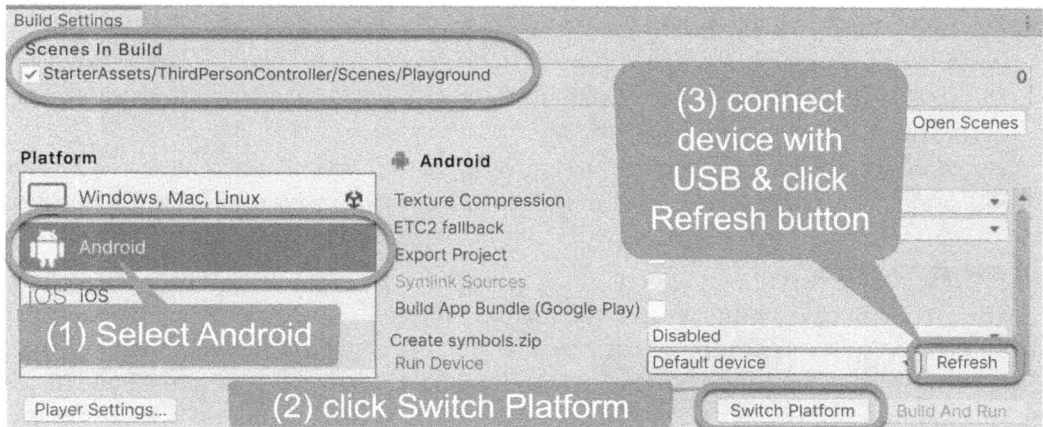

Figure 16.12: Adding Third-Person Starter to your assets on the Unity Asset Store website

15. Click to see the options of the **Run Device** drop-down menu and you should see a list of devices, from which you can select your connected device.

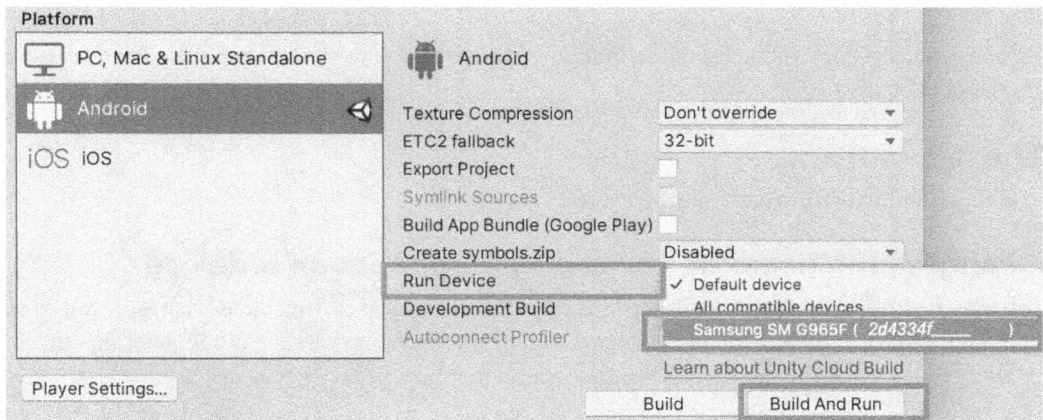

Figure 16.13: Selecting a Samsung Android phone from the list of devices

16. Build the project and send it to your device by clicking the **Build And Run** button. You'll be asked to enter an appropriate folder name for the project build – Build is a good name since this will become the name of a new folder in this project.

17. Unity will then build the project as an Android APK application – this may take a few minutes, and you'll see a pop-up dialog showing the progress.

18. Once the build is complete, the APK file should be sent to the Android device through the USB cable, and the application should start running.

19. Since we only have one scene (the Playground scene), you should immediately be able to play the game on your device, using the UI touch controls on your Android mobile device!

How it works...

The first step is to have a Unity project with mobile input controls enabled (such as the **Starter Assets Third Person Character Controller** project). The app APK installer will only contain the scenes that have been added to the build settings – so we added the Playground scene to the build in this recipe. As long as you have added the Android modules for the Unity Editor version you are using, you should be able to switch the build platform to **Android** for the project. Once your Android device is connected via USB, it should appear as a selectable device in the **Build Settings** panel. When you click **Build And Run**, you instruct Unity to create an APK app installation file, transfer it to the Android device, and then install and run the app. When this is successfully completed, you will be able to play the game on your mobile device.

Finally, you can customize the name that will appear on the device through the **Product Name** value in **Project Settings**.

There's more...

There are some details that you won't want to miss.

Changing the name of the app as it appears on a device

You may have noticed an odd name for the app installed on your Android device. You can choose your own name for the app in **Project Settings** (**Edit | Project Settings**). Select the **Player** tab, and enter your preferred name for the app in the **Product Name** text box.

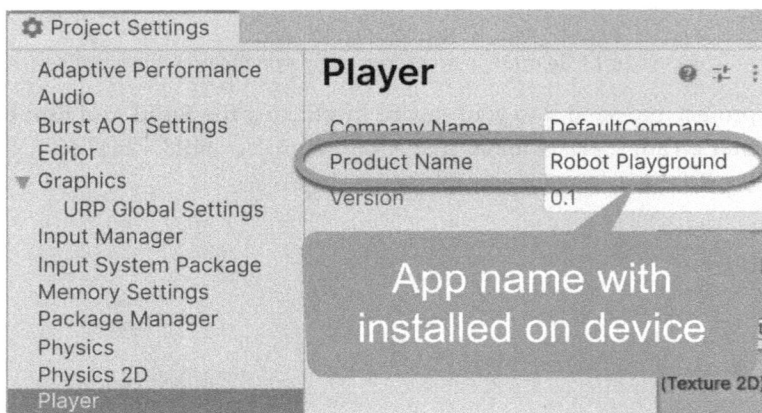

Figure 16.14: Setting the app name in Project Settings

Copying APK files from a Windows PC to an Android device

An alternative to **Build And Run** is to just build the APK Android app file and then "sideload" it to the Android device. Do the following:

1. Connect a mobile device with a USB cable to a Windows PC, then locate the device in **File Explorer** in the **This PC (My Computer)** location. Do this by opening **This PC** / **My Computer** from the **Start** menu.

2. Select (double-click) the icon representing your mobile device. You are now viewing the contents of your device's storage.

3. Then navigate to the device folder into which you wish to copy the APK app installation file.

4. Drag or copy and paste the APK app file to your device.

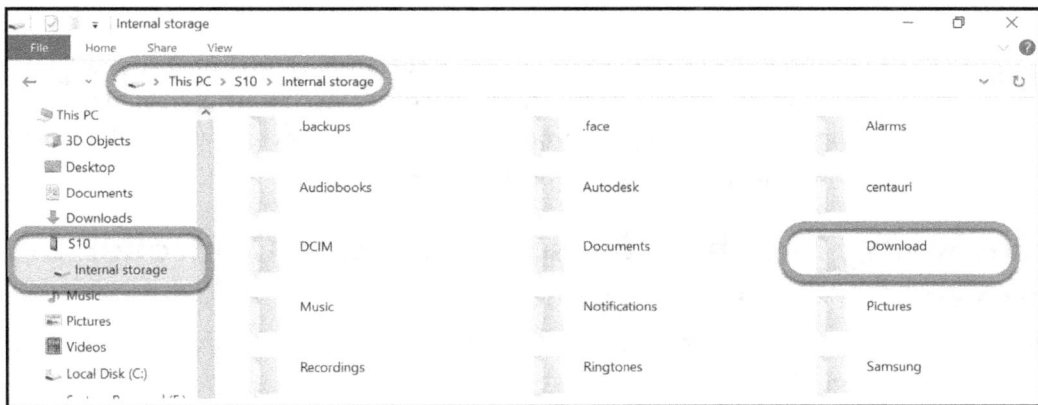

Figure 16.15: Locating the mobile device Download folder in Windows Explorer

I recommend you copy the APK file to your device's Download phone storage folder. Then on the mobile device, locate the file by navigating to **My Files | APK Installation Files | Download**. You can then install the app.

Sideloading APK files from a Mac to an Android device

An alternative to **Build And Run** is to just build the APK Android app file and then "sideload" it to the Android device. To do this on an Apple Mac computer system, you can use one of the following applications:

- The Android File Transfer app: https://www.android.com/filetransfer/.
- The command-line **Android Debug Bridge** (ADB). You can find a good ADB tutorial here: https://androidcommunity.com/how-to-getting-adb-on-your-pc-without-installing-full-android-sdk-20180307/.

I recommend you copy the APK file to your device's Download phone storage folder. Then, on the mobile device, locate the file by navigating to **My Files | APK Installation Files | Download**. You can then install the app.

Downloading Android APK files on your phone

An alternative to using a USB cable to transfer a build to your Android device is to publish the build APK file on a website, and then install it on your device via a mobile web browser. We have provided some steps for downloading Android APK builds using a web browser app.

> **IMPORTANT:** Only download and install APK files you have created yourself or from sources you trust!

To allow the download of APK files from a mobile browser app, do the following:

1. Go to your phone's **settings**.
2. Go to **Security & Privacy | More Settings**, then choose **Install Apps from External Sources**. Note that for Samsung devices you need to go to **Biometrics** then **Security | Install Unknown Apps**.
3. Select the browser (for example, **Chrome** or **Firefox**) you want to download the APK files from.
4. Enable **Allow App Installs / Allow from this source**.

Deploying the Third Person Character Controller Starter project for an Apple device

In this recipe, you'll adapt the previous recipe to build and run the game on an Apple device attached to your computer with a USB cable.

Getting ready

Before starting this recipe, ensure you have completed the second recipe in this chapter: *Setting up your system for Apple iOS mobile app development*.

This project follows on from the previous one – we'll just be changing the build target to Apple iOS, so you can simply open your existing project.

How to do it...

To deploy the Third Person Controller Starter project for an Apple device, perform the following steps:

1. Open the **Third Person Controller** Starter project created in the previous recipe.
2. Open the **Build Settings...** panel by choosing **Edit | Build Settings....** Switch the build platform to **iOS** (select **iOS** and click the **Switch Platform** button). You may need to wait a few minutes for Unity to convert the project assets ready for an Apple iOS build.
3. Open the **Project Settings** panel (**Edit | Project Settings...**). In the **Player** tab, under the **Identification** section, check the **Override Default Bundle Identifier** option, and enter a bundle identifier. Apple bundles are usually named as the reverse of a unique website address followed by a name for the app, for example, `com.mattsmithdev.Robot-Playground`.

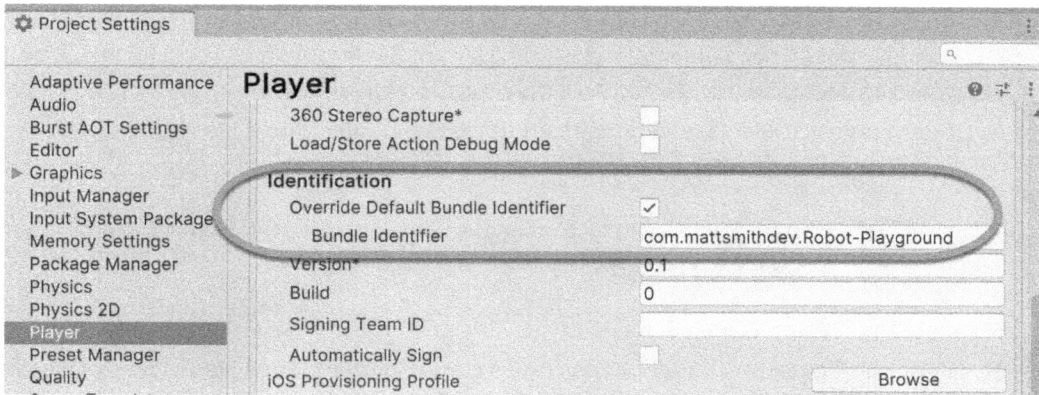

Figure 16.16: Setting the app bundle identifier in the Player settings

4. In the **Build Settings** panel, click the **Build** button. You'll be asked to select a folder. I suggest `BuildApple` if creating Android and Apple builds from the same project. Unity will then spend a while building an Xcode project.

5. In your build folder (such as `BuildApple`), you'll then see a file ending with the `.xcodeproj` extension (`Unity-iPhone.xcodeproj`). Unity does not build iOS apps – it builds Xcode projects for your iOS app.

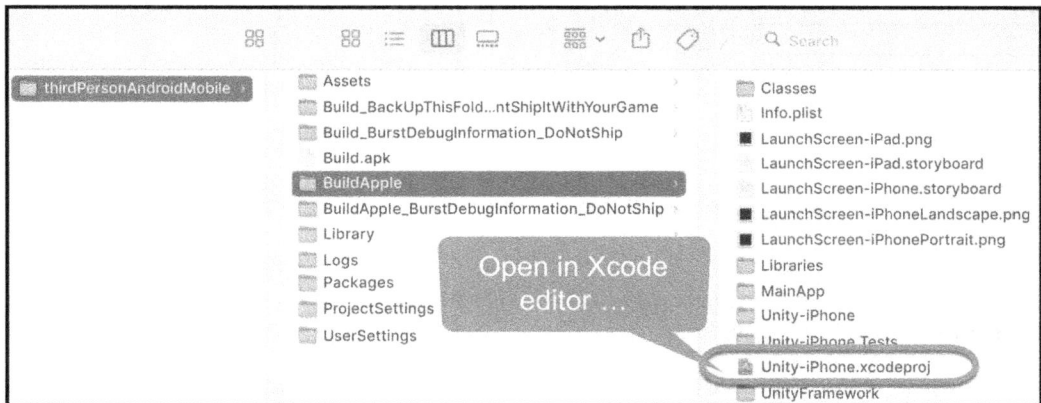

Figure 16.17: The generated Xcode project files

6. Open the `.xcodeproj` file to open this project in the Xcode editor by double-clicking it.

7. Select `Unity-iPhone` in the Xcode project content hierarchy at the top left of the editor and fix any issues.

> **Note:** You may have to download certificates and a provisioning profile the first time you use Xcode with your Apple ID.

8. Enable **Developer Mode** on your iOS device, by navigating to **Settings | Privacy & Security** and toggling **Developer Mode** on. Acknowledge that this is a reduced security mode, and restart your iOS device.

9. Connect your iOS device (iPhone/iPad) to your computer with a USB cable.

> It's a good idea to turn off auto-locking on your device since you don't want your phone to lock while the Xcode build is taking place.

10. Build the Xcode project (click the "run" triangle button). Then select your device from the dropdown menu at the top of the Xcode application window.

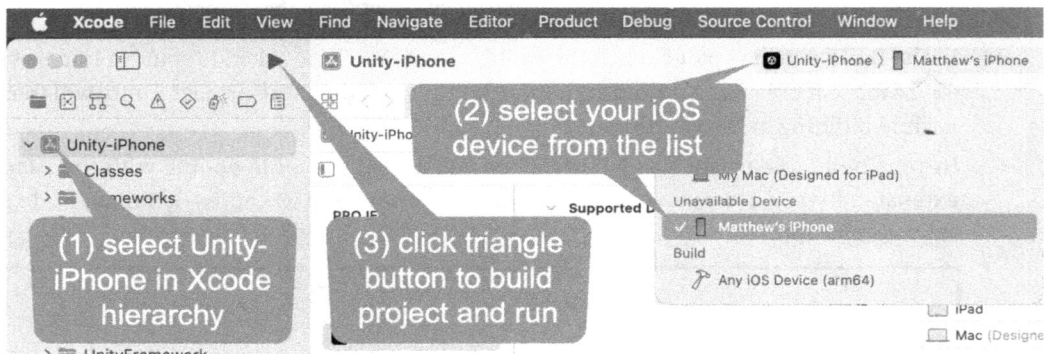

Figure 16.18: Building and running the Xcode project

11. When complete, the app will be on your iOS device, ready to run.

How it works...

To build and deploy an app for an Apple iOS device, several things are required, which were performed during this recipe:

1. You have to set an appropriate bundle identifier in the Unity project settings.

2. You switch the build platform to iOS and build an Xcode project.

3. You set up your iOS device in **Developer Mode** and then restart it. (Disable auto-locking to avoid problems if the build/deployment takes a few minutes.)

4. You connect the iOS device to your computer via a USB cable.

5. You open the Xcode project in the Xcode editor and select your iOS device as the target.

6. You build and run the Xcode project.

Once you've gone through these steps a few times, it should become a straightforward process, but the first time, it's worth taking your time and ensuring you complete each step that is required.

There's more...

There are some details that you won't want to miss.

The device will not prepare for development

You have to have a matching version of Xcode for the version of iOS on your device. Visit this Apple iOS release notes website to check the Xcode version for your device's iOS version: `https://developer.apple.com/documentation/ios-ipados-release-notes/`.

For example, when preparing this recipe initially, my version of Xcode was 14.2, and I was trying to deploy to an iPhone running iOS 16.5.1. In Xcode, each time I tried to select my device, it was labeled `Failed to prepare the device for development`. Looking up the notes for iOS 16.5, I found that it needed Xcode 14.3 or later. The solution was to install a newer version of the Xcode editor.

Figure 16.19: iOS and Xcode version compatibility in iOS release notes

Creating and deploying a mobile game with Unity's Runner Game project template

Unity has published a great Runner Game mobile project template. In this recipe, we'll create a project with that template, customize the scene, and deploy it to a mobile device.

Figure 16.20: Photos of the mobile Runner Game deployed and running on an Android phone

How to do it...

To create and deploy a mobile runner game with the Runner Game project template, perform the following steps:

1. Open the Unity Hub application. Click **New Project**, and select the version of the Unity Editor you wish to develop with.

2. Locate the **Runner Game** template in the list of templates (you'll need to download it the first time you use the template).

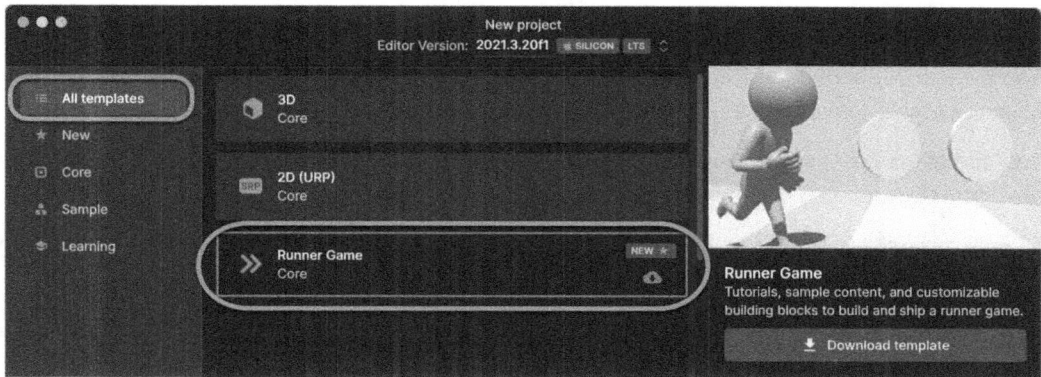

Figure 16.21: Selecting the Runner Game project template in Unity Hub

3. Create and open the new project.

4. When you open the project, you may wish to explore this tutorial template by clicking through the tutorial panel that should appear when the project opens.

5. Open the **Level Editor** panel by choosing **Window | Runner Level Editor**.

6. Locate the Level_000 asset file from the **Project** panel (**Assets | Runner | Environment | Levels**). Drag the Level_000 asset file into the **Level Definition** variable slot in the **Level Editor** panel.

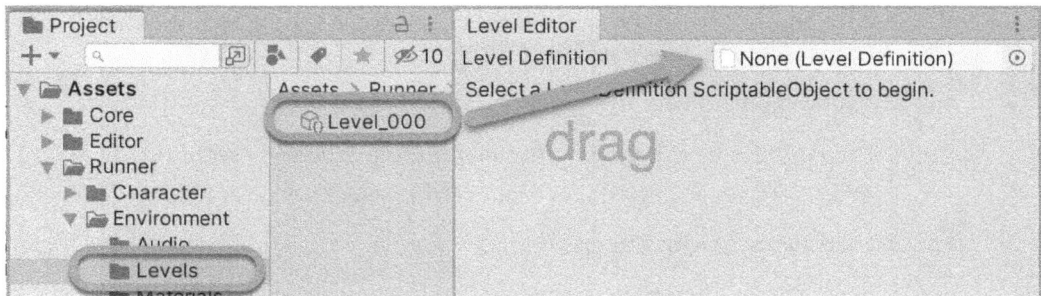

Figure 16.22: Assigning the level data file in the Level Editor panel

7. You'll want to drag some obstacles into the scene. You can drag the Obstacle Prefab asset file from the **Project** panel (from the **Assets | Runner | Environment | Prefabs | Spawnables folder**) into the path of the player's character.

8. Now add the current scene to the build settings, switch the target platform to Android, connect your Android device to the computer, and locate it on the list of devices, then click **Build And Run**.

9. Once the project has been built, it should be transferred via USB and you should be able to now run the game on your device. You should be able to move your character's player with left and right swipes on your device's touchscreen.

How it works...

You have downloaded the Runner Game mobile project template. Different levels can be customized as game assets – we made use of the default `Level_000` asset file in this recipe.

You can learn more about this game template by following the tutorial in the Unity Editor, and also from these Unity web pages:

*   `https://blog.unity.com/games/rapid-prototyping-with-new-mobile-runner-game-template`
*   `https://unity.com/features/build-a-runner-game`

> **Note:** At the time of writing, the Runner Game template was not available for Unity 2023; however, it was working fine for Unity 2021 LTS. Since this is a relatively new and fully featured project template, it is hoped the template will be available for use in Unity 2023 by the time you read this book!

Further reading

The following are some useful sources for the topics that were covered in this chapter:

*   Unity mobile development:

    *   `https://unity.com/mobile-solution-guide`

*   Apple iOS development with Unity:

    *   `https://docs.unity3d.com/Manual/ios-environment-setup.html`
    *   `https://docs.unity3d.com/Manual/iphone-GettingStarted.html`

*   Android development with Unity:

    *   `https://docs.unity3d.com/Manual/android-sdksetup.html`
    *   `https://docs.unity3d.com/Manual/android-getting-started.html`

If you wish to publish your games on the Apple **App Store** or Google **Play Store**, you'll need to pay and register as an Android/Apple developer. Learn more here:

*   `https://developer.apple.com/`
*   `https://developer.android.com/`

Unity Technologies published a free runner game on the Asset Store for Unity 2020. It may still be useful for assets and ideas to inspire a fully-working endless runner game:

*   `https://assetstore.unity.com/packages/templates/tutorials/endless-runner-sample-game-87901`

17

Augmented Reality (AR)

Augmented Reality (AR) involves adding computer-generated content to the user's real-world experience in real time. As discussed in the previous chapter, the device may be a dedicated AR system, such as Microsoft's HoloLens, or the user may be using an Apple or Android mobile device (phone or tablet) as the interface between them and the real-world environment they are viewing with augmentations. We'll look at mobile phone-based AR devices in this chapter, so before beginning the recipes you should have explored the Android and/or Apple iOS recipes in the previous chapter (*Chapter 16, Mobile Games and Applications*) to be familiar with the procedure for switching build targets to mobile devices, and running games on USB-connected devices. Also, since you'll be running your AR projects on either an Android or Apple iOS device, ensure you have completed the first and/or second recipe of the previous chapter so that your system is set up for mobile device development by installing the appropriate Build Support plugin. These recipes were:

- *Chapter 16, Mobile games and applications*

  - *Setting up your system for Android mobile app development*
  - *Setting up your system for Apple iOS mobile app development*

In this chapter, we will cover the following recipes:

- Exploring the Unity AR samples
- Creating an AR project with the AR (Core) template
- Adding the AR Foundation package and GameObjects to a 3D scene
- Detecting and highlighting planes with AR Foundation
- Creating an AR furniture previewer by detecting horizontal planes
- Creating a floating 3D model over an image target

Special thanks to Nina Lyons from TU Dublin for helping out with the AR recipes.

Exploring the Unity AR samples

Unity provides an up-to-date set of sample AR Scenes in a project you can download from the https://github.com/Unity-Technologies/arfoundation-samples website. In this recipe, we'll download the project, build an AR phone app, and try some of the samples from Unity.

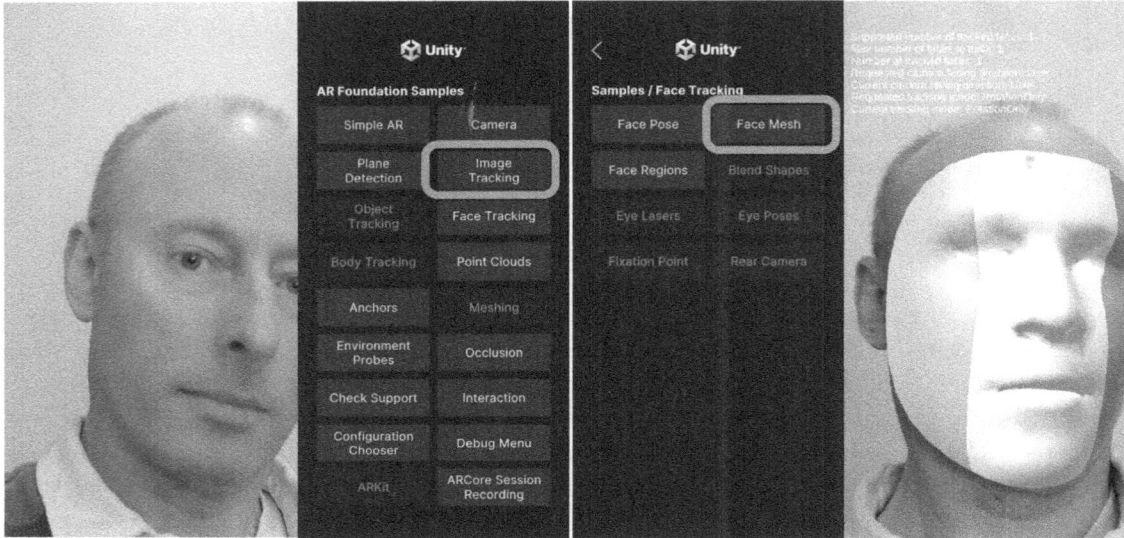

Figure 17.1: Running the face mesh scene, which maps a flag image over your face!

How to do it...

To download and explore the Unity AR samples, do the following:

1. Go to https://github.com/Unity-Technologies/arfoundation-samples and download the files as a ZIP file.

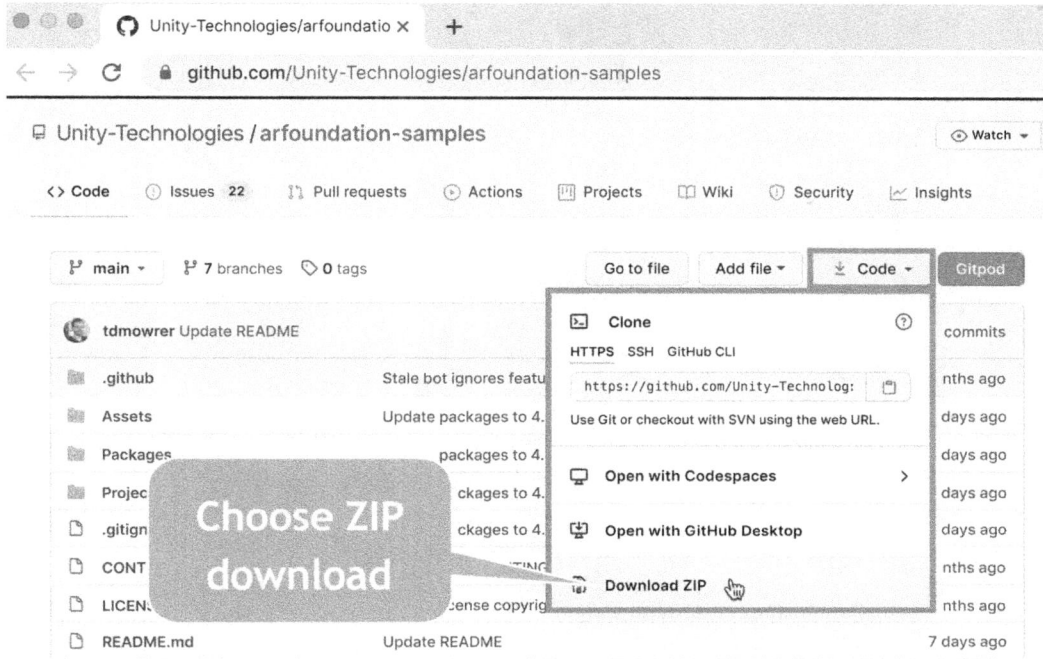

Figure 17.2: Downloading the AR samples from GitHub

2. Unzip the project files and open them in Unity.

3. Open the **Build Settings** dialog window and switch **Build Target** to **Android** (or **iOS** if appropriate).

4. You'll see that all the scenes have already been added to the build for this project, with the **AR Foundation Menu** scene listed first.

5. Connect your computer to your phone/device with an appropriate USB cable and on your device, agree to permit the computer to communicate with it. Then, **Refresh** the list of devices and select your device.

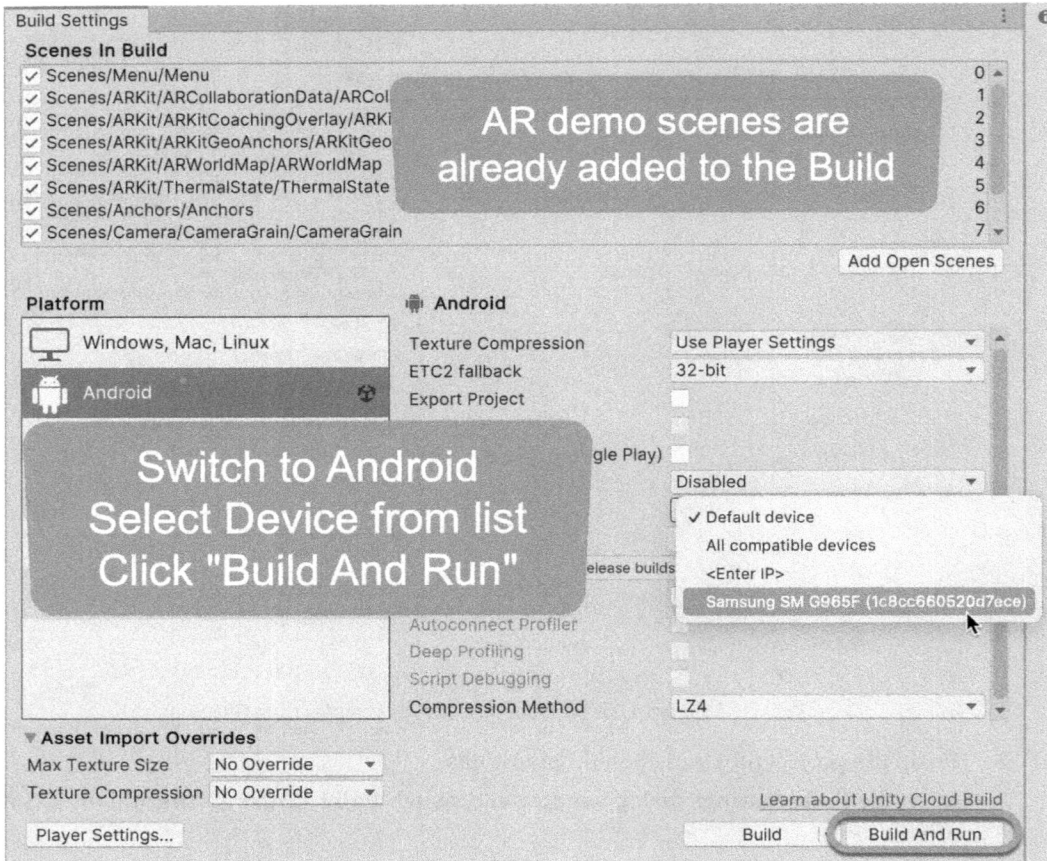

Figure 17.3: Building and running the AR Foundation samples project

6. Click the **Build And Run** button. You'll be asked to enter an appropriate name for the project build, such as ARSamples.

7. Once Unity has finished building the app, on your mobile device you should now see a menu listing all the runnable scenes, including **Face Tracking**, **Plane Detection**, and so on.

8. Run some of the scenes to get a feel for how AR apps are able to identify faces, flat surfaces, and so on.

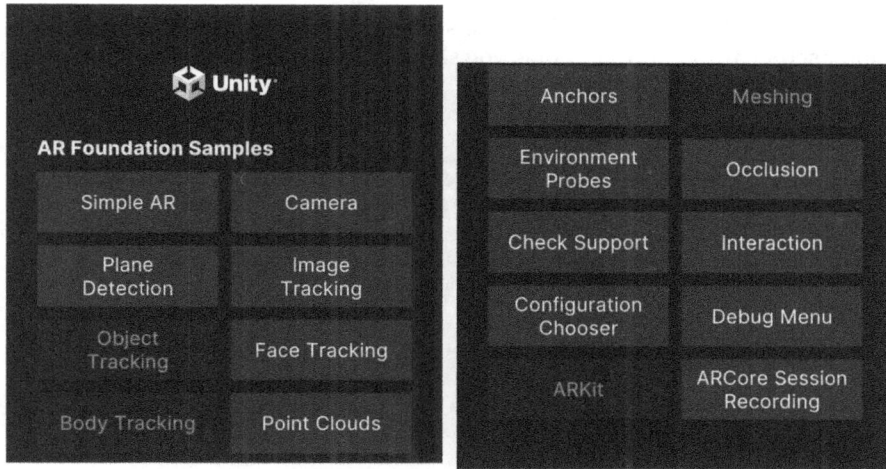

Figure 17.4: The main menu list of scenes running on an Android phone

9. On your computer, select the Rafflesia image in the **Project** panel (**Scenes | ImageTracking | Images**) and display its details along with a preview in the **Inspector** window.

Figure 17.5: Displaying the Rafflesia image on your computer screen

10. On your phone, choose **Image Tracking | Basic Image Tracking**. Now, when you view the computer screen through your phone, you should see annotations overlaid on the **Rafflesia** image:

Figure 17.6: The AR information overlay when viewing the Rafflesia flower image with the phone app

11. Load an image asset file of your country's flag into the Textures folder in the **Project** panel. I live in Ireland, so I loaded the Irish tricolor flag – I got the Flag_of_Ireland image from Wikipedia and copied it into the Unity project as a PNG image file asset.

12. In the **Project** panel, select the FaceMeshPrefab asset file (this can be found in the Prefabs folder).

13. In the **Inspector** panel, change the **Albedo Main Map** property to your map image. Do this by clicking the circle asset selected next to the albedo image, then searching for your flag's image name in the popup.

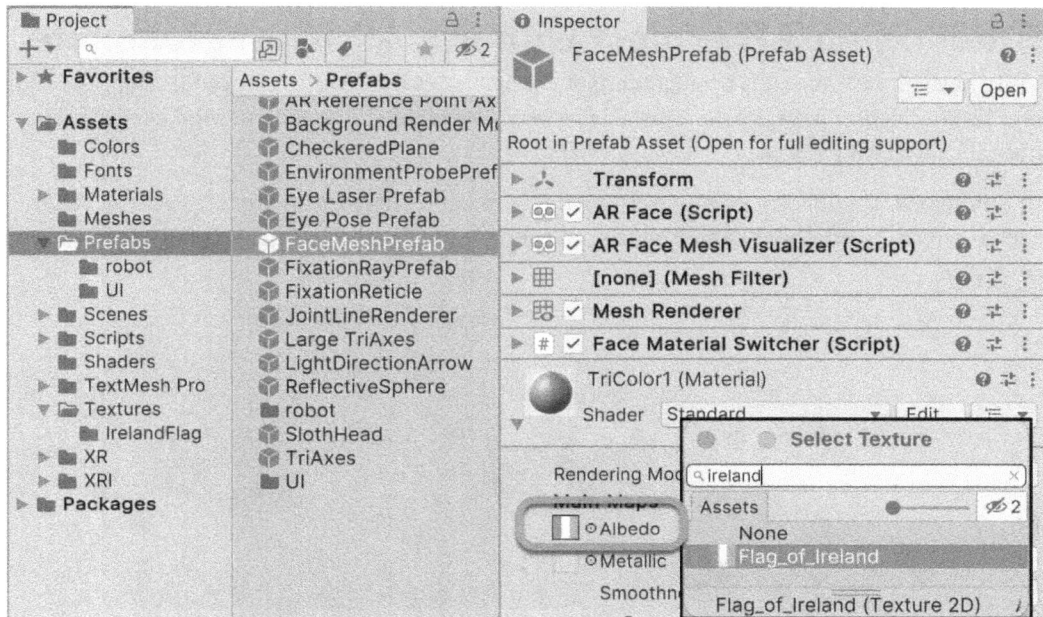

Figure 17.7: Setting the image to texture the face mesh mask

14. Now, build the app again and load the updated app onto your phone.

15. Run the phone app and choose **Face Tracking | Face Mesh**. The app should use the front phone camera to view your face. After a second or two, it will overlay a mask textured with your map image over your face. As you move your head around and open and close your mouth, the mask should move accordingly, as if you were really wearing a flag mask!

How it works...

We downloaded the Unity AR demo project from the GitHub repo. We had to ensure the correct modules were installed by **Unity Hub** for the target device to be used for running the AR project (that is, **Android** or **iOS**). Once a project is open in Unity, the required **Build Target** must be switched to, and all the scenes that will be part of the app need to be added to the Build Settings.

The Unity project contains many (over 20) scenes, each illustrating different AR features. Once the app was built and transferred to our device, the buttons on the menu scenes allowed us to navigate to each of the sample scenes.

By changing the settings of a prefab – the FaceMeshPrefab asset file (in the Prefabs folder) – and rebuilding the app, we were able to see how the **Material** setting of the prefab influenced which image was dynamically drawn as a face mask overlay using Unity's **AR Foundation Face Mesh** detection and tracking functionality.

Creating an AR project with the AR (Core) template

Unity have kindly created a project template, which does the required work for AR projects of adding the AR Foundation package, and provides a sample scene that already has the core AR GameObjects needed for an AR project.

In this recipe, you'll learn how to create a new project from the provided template, open the sample scene and add a 3D object, and then set up the Build Settings config for deployment on an Android device.

Figure 17.8: The AR project we'll create, with just a Cube overlaid on the device camera view

How to do it...

To create an AR project with the AR (Core) template, do the following:

1. In Unity Hub, click the **New Project** button.
2. Select your version of Unity, then scroll down to find the AR (Core) template. If you've not used this before you may need to first download this template.
3. Provide a project name and destination and click the **Create project** button.

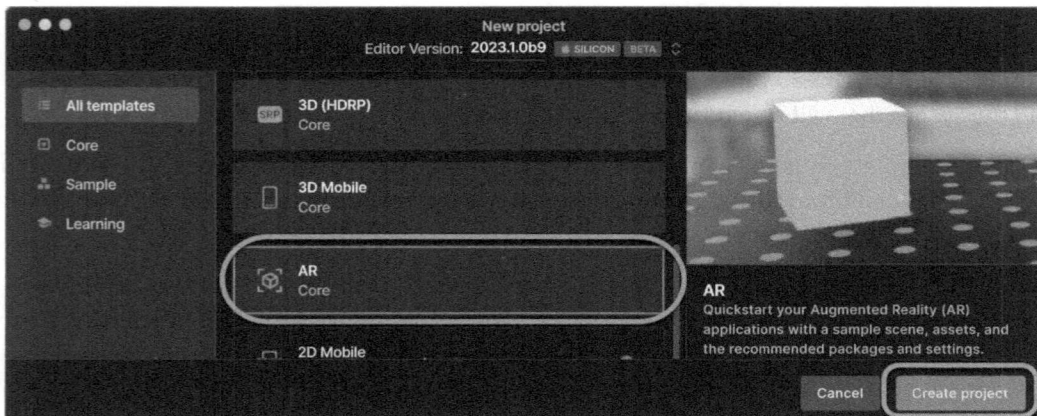

Figure 17.9: Selecting the AR (Core) project template from Unity Hub

4. After a short time, Unity should start, with a pop-up window welcoming you to the **AR Project Template,** and reminding you to select your chosen platform(s) from the **XR Plug-in Management** Project Settings panel.

5. In the **Project** panel, you'll see a Scenes folder, as well as two folders named XR and ExampleAssets – these are part of the AR template resources.

6. Open the **Project Settings** panel (via the **Edit | Project Settings** menu).

7. For the **XR Plug-in Management** settings, choose the **Android** (robot icon) tab and then select the tab for your target platform, **Android.** Then, check the plugin provider for the platform: **Google ARCore.** Then wait a few seconds while Unity downloads and installs the plug-in package for this project.

Figure 17.10: Adding ARCore via the XR Plugin Management plugin settings

8. If it's not already open, open the provided SampleScene (in the **Project** panel, go to the **Assets | ExampleAssets | Scenes** folder). In the **Hierarchy** panel, you'll see that it has an **AR Session Origin** GameObject, which has an AR Camera child. There is also an **AR Session** GameObject:

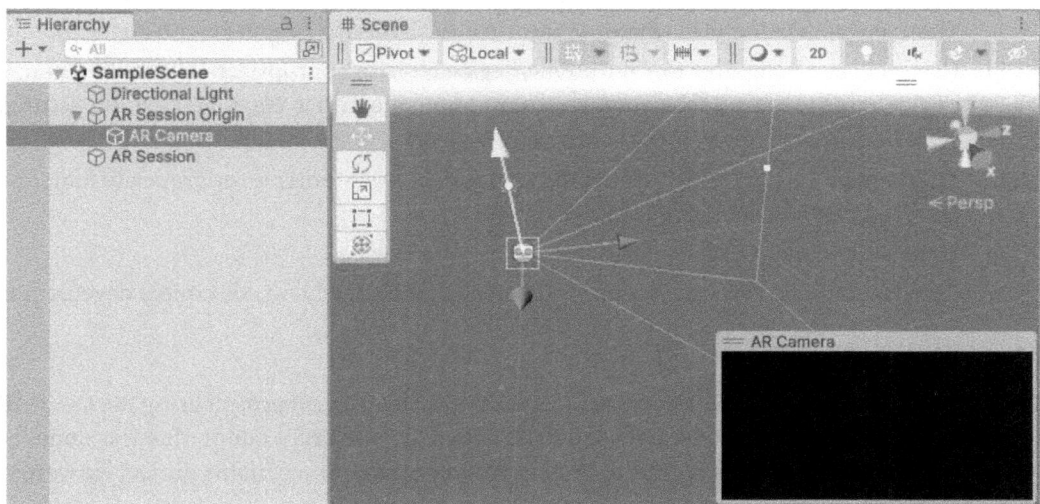

Figure 17.11: The AR GameObjects in the provided SampleScene

9. Let's add a **Cube** to this scene so that we have something to see when we build and run this app. Create a 3D Cube with a position of (0, 0, 6) and a rotation of (45, 90, 45). If you select the **AR Camera** GameObject child of **AR Session Origin**, you should be able to see the cube in the camera preview rectangle in the **Scene** panel.

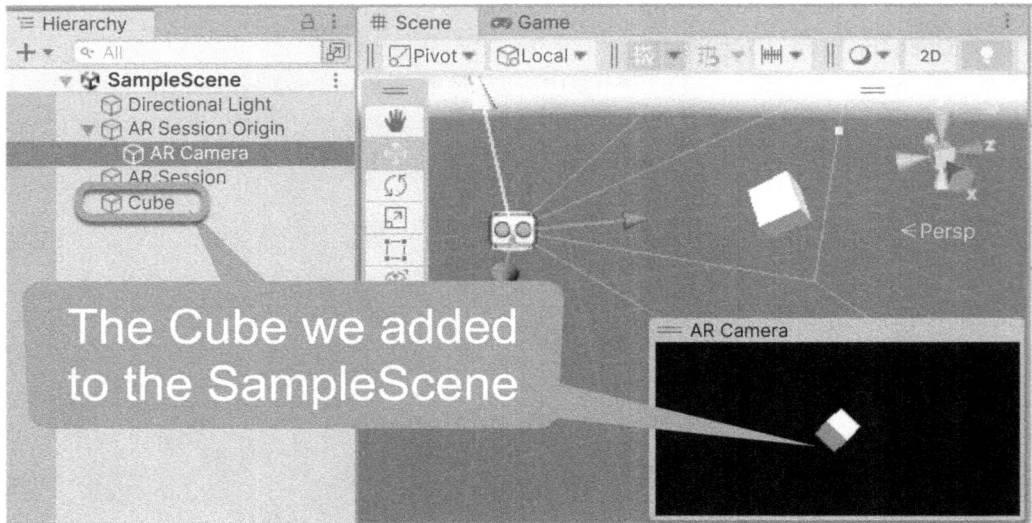

Figure 17.12: The Cube added to our scene

10. Open the **Build Settings** panel (via the **Edit | Build Settings...** menu).

11. Add this **SampleScene** to the build by clicking the **Add Open Scenes** button.

12. Switch the build target **Platform** to **Android**.

13. Open the **Project Settings** panel (via the **Edit | Project Settings...** menu) and select the **Player** tab. Then set **Product Name** to whatever name you wish the app to appear with on the mobile device.

14. Connect your computer to your phone/device with an appropriate USB cable. Then, on your device agree to permit the computer to communicate with it. Next, in the **Build Settings** panel, refresh the list of devices and select your device.

15. Build the project by clicking the **Build And Run** button – enter an appropriate folder name for the project build, such as Build1.

16. Unity should build and then run the app on your device.

17. Congratulations! You have set up a project so that it's ready for AR project development.

How it works...

This recipe illustrated the basic steps you must follow to set up an AR project using the Unity AR (Core) template. The appropriate module needed to be added to your Unity Editor, this was done in the **XR Plug-in Management settings** tab in the **Project Settings** panel. Since in this project we were working with an Android device we added the **Google ARCore XR Plug-in**.

The provided SampleScene gave us an AR Session Origin GameObject, with a child AR Camera. We added a 3D Cube to the Scene, and positioned it so it could be seen from the starting position of the AR Camera. This means when we run the AR app on our phone, wherever we are pointing the camera as it starts, we'll have fixed a floating 3D Cube at that location.

There's more...

The first time we create and build a new type of project, there may be some issues. Here are some troubleshooting tips. If Unity fails to build, then check the error message in the **Console** Window.

Deprecation warnings with the AR (Core) template

At the time of writing there were some deprecation warnings when using Unity's AR (Core) template. If you get such warnings, the just skip to the next recipe, where you'll learn how to manually install the AR Foundation package, and add the required AR components to an empty scene.

Build failure due to Vulkan graphics API settings

AR projects don't work with the Vulkan graphics API, so if this is listed in your **Player** settings, you'll get an error.

To resolve this error, open the **Project Settings** window (via the **Edit | Project Settings menu**) and select the **Player** tab. Then, under **Graphics API**, ensure the **Auto Graphics API** option is unchecked, then select the **Vulkan** row and delete it by clicking the minus (-) button.

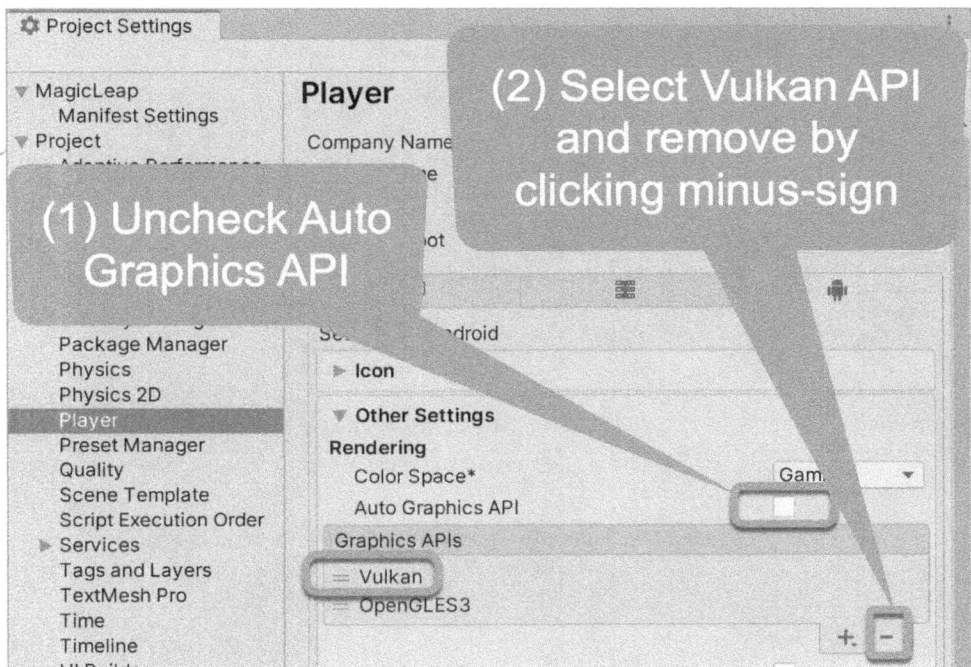

Figure 17.13: Removing the Vulkan graphics system from the Player settings

Build failure due to Android version

A common error is that the version of Android in the **Player** settings is too low for the version of Android on the device.

For example, the first time I tried to build for an Android AR project, I got the following error:

```
Error building Player: Android Minimum API Level must be set to 23 or higher.
```

> By the time you run this project, the required Android version may be 24 or higher – read the message and follow its requirements. There may be some setting up the first time you build for Android, but once you have it working, you'll be able to develop and publish VR and AR projects on your Android devices.

This can be resolved by opening the **Project Settings** window (via the **Edit | Project Settings** menu) and increasing **Minimum API Level** to the number stated in the error message, or a higher version. You can do this in the **Player** section and by going to the **Other Settings** values:

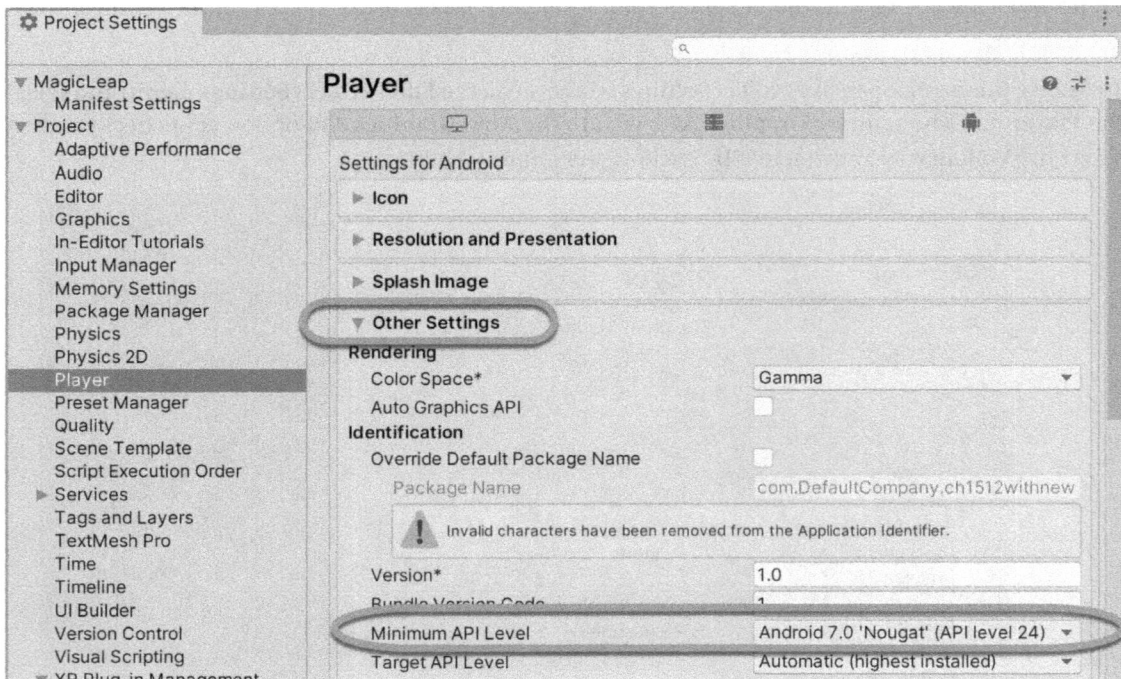

Figure 17.14: Setting Minimum API Level in the Player settings

Building for Apple iOS

This recipe gave examples of building for an Android device. If you're working with an Apple iOS device, here are the main differences:

- When selecting the **XR Plugin Management** settings, choose the **Apple** (Apple icon) tab and then select the tab for your target platform: **iOS**. Then, check the plugin provider for the **ARKit** platform for **iOS**.

- In **Project Settings** (via the **Edit | Project Settings...** menu), ensure that you have NOT checked the **Auto Graphics API** option.

Adding the AR Foundation package and GameObjects to a 3D scene

If you already have a 3D project, then there are some basic steps you need to perform so that your Unity Editor application is set up for AR development. AR projects created with Unity have, at their core, the **AR Foundation** package and **XR Plugin Manager**. In the previous recipe, things were set up for us already as part of the Unity AR (Core) template project. In this recipe, we'll create a new Unity 3D project and then manually add the **AR Foundation** package to the project. In this recipe, we'll create the same scene setup, but will be building a scene from scratch to get to know the **AR Foundation** GameObjects and how to add them to any scene.

This recipe can be used as a guide on how to add AR capabilities to any existing 3D project you might already have. As part of this recipe, we'll also be adding the appropriate plugin for the **Android** target device on which the AR project is going to be developed.

Figure 17.15: Running this recipe's AR project on my Android phone – the two Cubes are floating in front of my window!

How to do it...

To add the **AR Foundation** package and GameObjects to a 3D scene, do the following:

1. Create a new 3D Unity project – just a simple 3D project, do NOT use the AR (Core) template.
2. Open the **Package Manager** panel (via the **Window | Package Manager** menu).
3. Choose to search packages from the Unity Registry.
4. Search for the AR Foundation package from Unity Technologies, then click the **Install** button. Once it's installed, you should see an XR folder added to the Project panel.

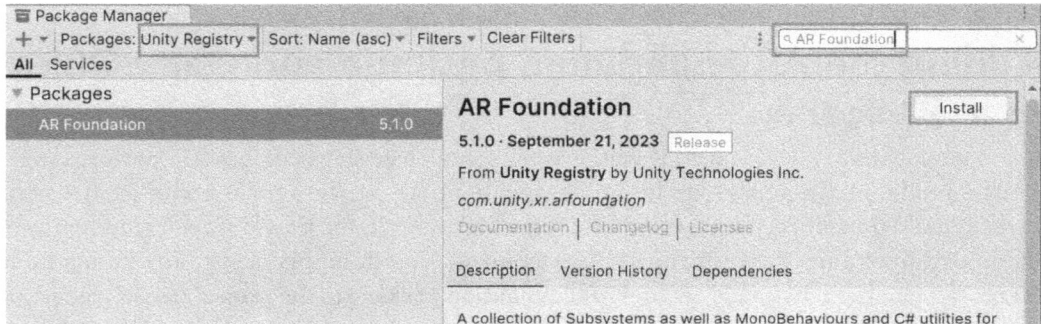

Figure 17.16: Installing the AR Foundation package from the Package Manager

5. Open the **Project Settings panel** by chosing **Edit | Project Settings** menu.
6. For the XR Plugin Management settings, choose the Android (robot icon) tab and then select the tab for your target platform, Android. Then, check the plugin provider for the platform: **Google ARCore**. Then wait a few seconds while Unity downloads and installs the plug-in package for this project.
7. Create a new empty scene.
8. Add to the scene an XR Origin GameObject by chosing **GameObject | XR | XR Origin** menu.
9. Add the AR content to your scene. In this simple recipe, we'll add two floating cubes in front of the camera to the lower right of the view. Create two 3D **Cubes** (via the **Create | 3D Object | Cube** menu) and position them at (1, -0.25, 3) and (1.5, -1.5, 4):

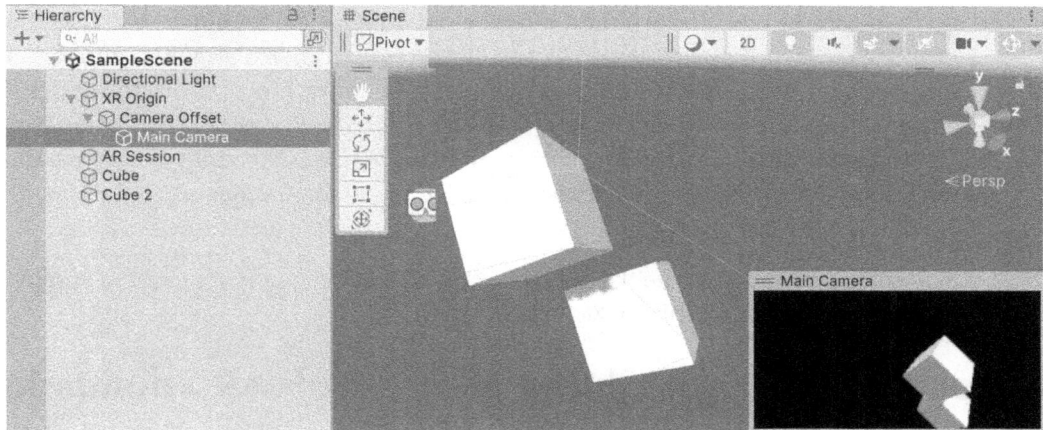

Figure 17.17: The two Cubes that will appear in our AR application

10. Save your scene, naming it Two Cubes.

11. Open the **Build Settings** panel by chosing **File | Build Settings...** and add your scene to the build.

12. Set the **Build** target to **Android** as appropriate and click **Switch Target**.

13. At this point, you may wish to set the **Product Name** value in the **Player** settings to give the app an appropriate name when it is installed on your device.

14. In the **Build Settings** panel, refresh the list of devices and select your device.

15. Now, either **Build And Run** your app on your device, or just **Build** the app and copy/install it onto your device.

16. When you run it, you should see a cube floating in front of you wherever you look. Congratulations – you've just created an AR scene from scratch!

How it works...

The main difference from the normal working environment is that we have added two GameObjects called XR Origin and AR Session to the **Hierarchy** window. XR Origin and AR Session are important, so it's worth understanding a little bit about them:

- XR Origin: This is the parent for an AR setup. It contains a camera and any GameObjects that have been created from detected features such as planes and point clouds. XR Origin defines the origin of the project. If you select this object in the **Hierarchy** panel, in the **Inspector**, you will see that it has the **XR Origin (Script)** component. It also has a child GameObject named Camera Offset, containing Main Camera and its special AR camera scripts. This is the AR camera that we see. It will render what the camera sees behind it, and any AR features inside of Unity will be on top of that.

- AR Session: As described in the Unity documentation, AR Session coordinates the major processes that ARKit performs on its behalf to create an AR experience. These processes include reading data from the device's motion sensor hardware, controlling the device's built-in camera, and performing image analysis on captured camera images. The session synthesizes all of these results to establish a correspondence between the real-world space the device inhabits and a virtual space where you model AR content. Another way to look at this is that AR Session is what *maps* our AR world.

Note that we didn't need a separate Main Camera GameObject in our scene since AR projects use a Main Camera instance that is a child of XR Origin.

Detecting and highlighting planes with AR Foundation

One of the features of AR is its ability to detect flat surfaces through device cameras and other sensors (such as the LIDAR sensors on some high-end iPads and iPhones).

This recipe illustrates how a Unity **AR Raycast**, combined with some **AR Foundation** objects and scripts, can work to highlight detected planes (both horizontal and vertical) with a circular highlight:

Figure 17.18: A plane being detected in an indoor scene

Getting ready

This recipe follows the previous one, so make a copy of it and work on that copy.

We also need an image to display on the detected planes. We have provided a suitable circular image called plane_highligher.png in the 17_04 folder.

How to do it...

To create an AR plane highlighter app, do the following:

1. Delete the two cubes in the **Hierarchy** panel.

2. In the **Hierarchy** panel, select the XR Origin GameObject. In the **Inspector** panel, add an **AR Plane Manager (Script)** component by clicking the **Add Component** button and searching for **AR Plane** in the **Inspector**. Ensure that **Detection Mode** is set to **Everything**.

3. Now, add an **AR Raycast Manager (Script)** component to this GameObject by clicking the **Add Component** button and searching for **AR Raycast** in the **Inspector** panel.

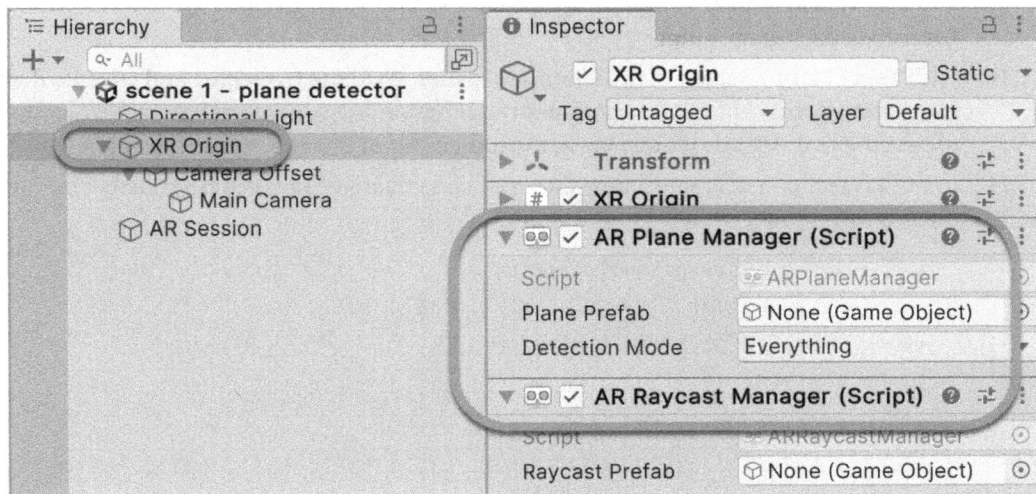

Figure 17.19: The AR Plane Manager and AR Raycast Manager components of XR Origin

4. Import the provided image; that is, plane_highligher.png.

5. Add a 3D **Plane** named PlaneHighlighter to the scene. To do so, in the **Hierarchy** panel, select **Create | 3D Object | Plane**.

6. Set the position of the PlaneHighlighter GameObject to (0, 0, 0) and make it suitably small for your AR device camera by setting its scale to (0.01, 0.01, 0.01).

7. Drag the plane_highligher image asset file from the **Project** panel onto the PlaneHighlighter GameObject in the scene.

8. In the **Inspector** panel, set the **Shader** property of Plane_highlighter (Material) to **Unlit/Transparent** (we do not need to provide a value for the **Text Image Data** field – this will be set by the script class we'll create next):

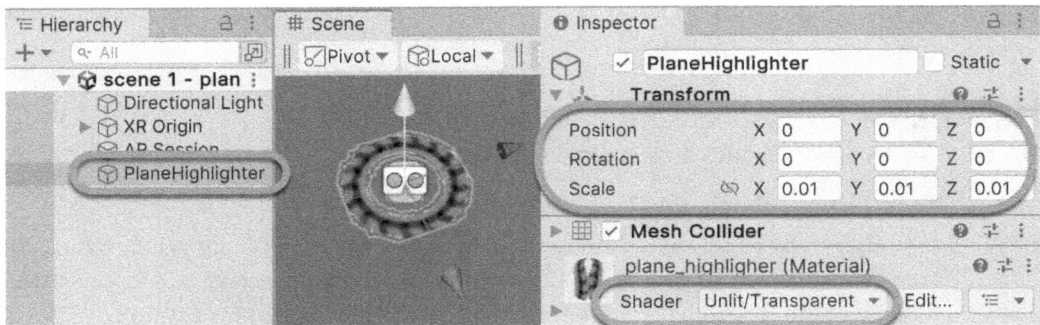

Figure 17.20: Setting the PlaneHighlighter shader to Unit/Transparent

9. Add a **Text (TMP)** UI text GameObject to the scene. You'll need to agree to the **Import TMP Essentials** package prompt.

10. With **Text (TMP)** selected in the **Hierarchy** panel, go to the **Inspector** and click its **Rect Transform,** then press *shift+alt* for the **top-stretch** option. Also, center-align its text content. This means the text will always display at the top-center of the app when running.

11. In the **Project** panel, create a new C# script class named `PlacementIndicator.cs` containing the following code:

```csharp
using System.Collections.Generic;
using UnityEngine;
using UnityEngine.XR.ARFoundation;
using UnityEngine.UI;

public class PlacementIndicator : MonoBehaviour {
    public GameObject planeVisual;
    public Text textImageData;
    private ARRaycastManager _raycastManager;
    private List<ARRaycastHit> hits = new List<ARRaycastHit>();

    void Awake() {
        _raycastManager = GetComponent<ARRaycastManager>();
    }

    private void Update() {
        Vector2 screenCenter = new Vector2(Screen.width / 2, Screen.height / 2);
        _raycastManager.Raycast(screenCenter, hits);
```

```
                  if(hits.Count >0) {
                      Pose hitLocation = hits[0].pose;

                      planeVisual.transform.position = hitLocation.position;
                      planeVisual.transform.rotation = hitLocation.rotation;
                      if(!planeVisual.activeInHierarchy)
                          planeVisual.SetActive(true);

                      textImageData.text = "plane WAS detected - should show
             highlighter!";
                  } else {
                      textImageData.text = "(no plane detected)";
                      planeVisual.SetActive(false);
                  }
              }
          }
```

12. Add an instance object of this script class to the XR Origin GameObject in the **Hierarchy** window by dragging PlacementHighlighter from the **Project** panel onto the AR Session Origin GameObject in the **Hierarchy** panel.

13. Now, in the **Inspector** panel, drag the PlaneHighlighter GameObject into the **Plane Visual** public script variable for the **Placement Indicator (Script)** component. Also, drag the Text (TMP) UI text GameObject into the public **Text Image Data** public script variable.

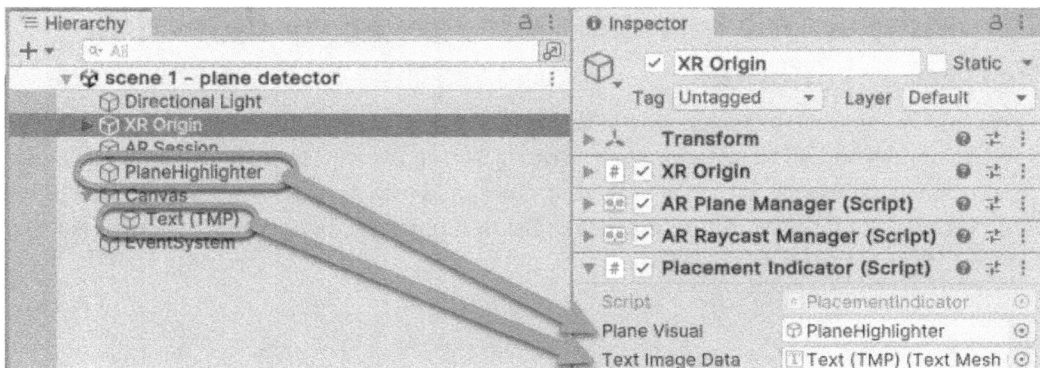

Figure 17.21: Assigning our PlaneHighlighter GameObject to the public Plane Visual public script variable

14. Now, save the scene and add it to the build. Remove any other scenes from the build.

15. At this point, you may wish to set the **Product Name** value in the **Player Settings** to give the app an appropriate name when it is installed on your device.

16. Now, either **Build And Run** your app on your device, or just **Build** the app and copy/install it onto your device.

17. When you run the app on the mobile device, each time **AR Foundation** identifies a **horizontal** or **vertical** plane, the circle highlighter image should be displayed, showing where the plane is being detected. Also, a text message should display on the screen stating that either no plane was detected, or that one was and should be highlighted.

How it works...

The AR Plane Manager allows the app to detect horizontal and vertical planes in the device's camera's view. The **AR Raycast Manager** allows the app to identify which horizontal or vertical plane a ray cast from the direction of the camera would strike. Our script fires a ray every frame via the Update() method.

Since the Main Camera GameObject is tagged Main Camera, when scripting we can easily get a reference to the camera using Camera.main.

By placing the instance object of our script class as a component of the XR Origin GameObject, we can get a reference to the AR Ray Cast Manager sibling script component by using GetComponent<class>() in the Awake() method.

We set two public variables for our scripted component. First, we link our Text (TMP) UI text GameObject to the **Text Image Data** public variable. Second, we link GameObject PlacementIndicator to the **Plane Visual** public variable. This enables our scripted component to display a message about the plane detection status, and if a plane has been detected, it can display the plane to the user using the PlacementIndicator plane circle texture.

Creating an AR furniture previewer by detecting horizontal planes

As we saw in the previous recipe, one of the features of AR is its ability to detect flat surfaces through device cameras and other sensors (such as the LIDAR sensors on some high-end iPads and iPhones). This recipe will illustrate how a Unity raycast, combined with some AR Foundation objects and scripts, lets us create a basic AR furniture previewer just by tapping on the screen where we want a virtual 3D piece of furniture to be displayed.

Figure 17.22: Adding 3D sofas to a back yard

Getting ready

This recipe builds upon the previous recipe, so make a copy of that project to work on in this recipe.

We'll be making use of a free model from the Unity Asset Store, which can be found here:

- https://assetstore.unity.com/packages/3d/props/furniture/patio-sofa-set-227952

How to do it...

To create an AR furniture previewer, do the following:

1. In the **Hierarchy** panel, select the XR Origin GameObject, and remove its **Placement Indicator (Script)** component.
2. Next, delete the Canvas and PlaneHighlighter GameObjects from the **Hierarchy**.
3. Ensure you are logged into your Unity account in the Unity Editor.
4. Open the **Unity Asset Store** web page (by choosing menu: **Window | Asset Store**).

5. In the Asset Store, check **Free Assets** and search for and select the free **Patio sofa set from Ivan Loginov.**

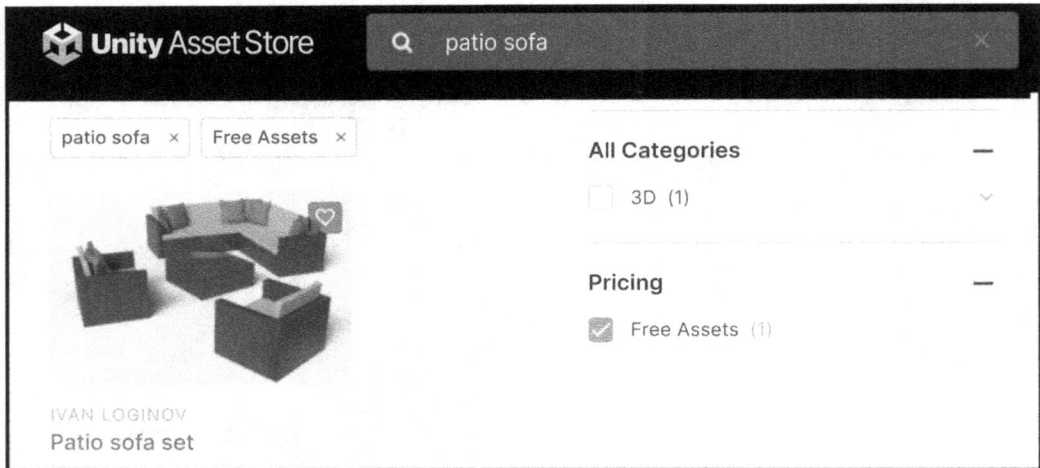

Figure 17.23: The free Patio sofa set assets on the Unity Asset Store

6. Click **Add to My Assets.** Then, after the button changes, click **Open in Unity.**

7. In your Unity Editor, the **Package Manager** panel should open, and **Patio sofa set** should be selected in the list of **My Assets.** Click **Download** and, once it's downloaded, click **Import.** You should now see the `Patio Furniture` folder appear in the **Project** panel.

8. In the **Project** panel, create a new C# script class named `FurnitureManager.cs` containing the following code:

```
using System.Collections.Generic;
using UnityEngine;
using UnityEngine.XR.ARFoundation;

public class FurnitureManager : MonoBehaviour {
    public GameObject furniturePrefab;
    private ARRaycastManager _raycastManager;
    public List<ARRaycastHit> hits = new List<ARRaycastHit>();

    void Awake() {
        _raycastManager = GetComponent<ARRaycastManager>();
    }

    void Update() {
        if(Input.GetMouseButtonDown(0))
        {
            Ray ray = Camera.main.ScreenPointToRay(Input.mousePosition);
```

```
                    if(_raycastManager.Raycast(ray, hits))
                    {
                        Pose pose = hits[0].pose;
                        Instantiate(furniturePrefab, pose.position, pose.
            rotation);
                    }
                }


            }
        }
```

9. Add an instance object of this script class to the XR Origin GameObject in the **Hierarchy** by dragging FurnitureManager from the **Project** panel onto the XR Origin GameObject in the **Hierarchy** panel.

10. Now, in the **Inspector**, drag the expensive_sofa_full prefab asset file from the **Project** panel into the **Furniture Prefab** public variable in the **Inspector** for the **Furniture Manager (Script)** component.

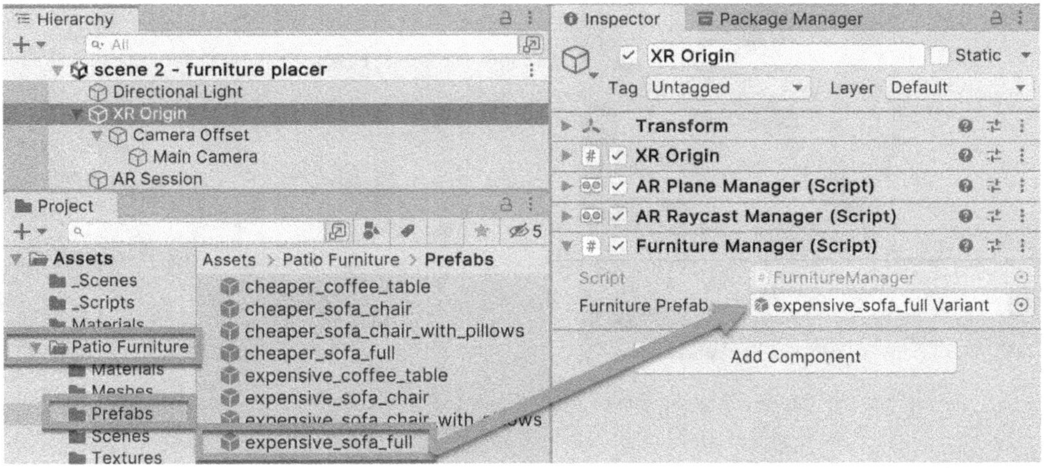

Figure 17.24 : Assigning our prefab to the Furniture Prefab public script variable

11. Now, save the scene and add it to the build. Remove any other scenes from the build.

12. At this point, you may wish to set the **Product Name** in the **Player** settings to give the app an appropriate name when it is installed on your device.

13. Now, either **Build And Run** your app on your device, or just **Build** the app and copy/install it onto your device.

14. When you run it, each time you tap the screen on a floor (horizontal) surface, a clone of the yellow cube-sphere chair model should be added to that location in the room.

How it works...

In this recipe, we adapted the previous project so that we can now detect screen taps, and then ray-cast to that point in the environment to see if it's a plane. If it is a plane, then we instantiate our sofa prefab at that location.

This is a simple app, but shows the basis of how AR can be used to create apps where we add 3D models into the view of our real-world environments.

Creating a floating 3D model over an image target

A common way to use AR to enhance the world that's seen through the device's camera is to recognize a *target* image and either display a different image or animation in its place, or add a 3D model to the view that tracks the location of the target image.

We'll do the latter in this recipe, making a 3D model appear over a book's cover image that we will add to a library.

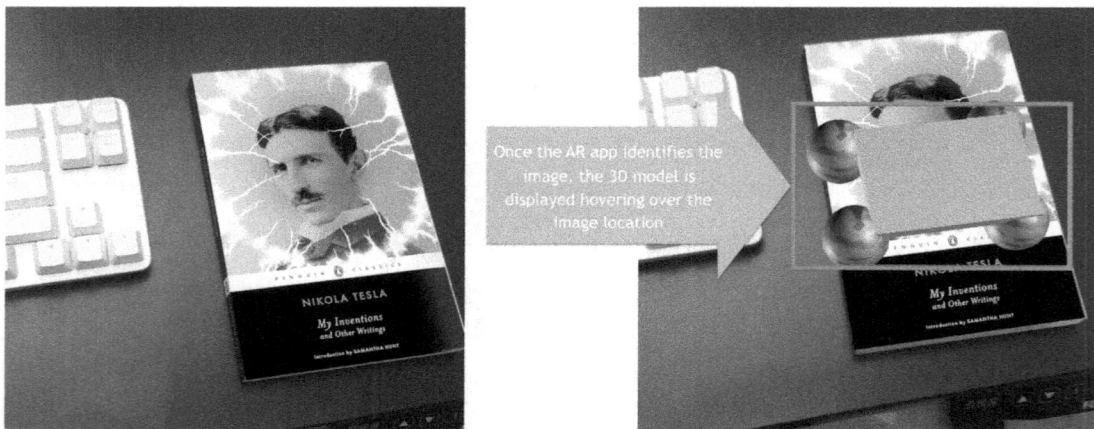

Figure 17.25: A 3D "car" model displayed over the book cover image in our AR phone app

Getting ready

This recipe builds upon the previous recipe, so make a copy of that project and use the copy.

You'll also need a target image. Either take a photo of an object in your home/office or use the provided book cover image called `tesla_book.jpg` in the `17_06` folder. You'll also need to print the image out or display the image onscreen while running the AR app on your phone.

How to do it...

To create a floating 3D model over an image target, do the following:

1. In the **Hierarchy** panel, select the `XR Origin` GameObject, and remove its **Furniture Manager (Script)** component.

2. Add a new, empty GameObject to the scene named car by chosing **GameObject | Empty**.

3. Now, add a 3D **Cube** GameObject to the scene by chosing **GameObject | 3D | Cube** and make it a child of car. In the **Inspector** panel, set the position of the Cube to (0, 0, 0) and its scale to (0.07, 0.01, 0.1).

4. Now, add a 3D **Sphere** GameObject to the scene by chosing **GameObject | 3D | Sphere** and make it a child of car. In the **Inspector** panel, set the position of the Sphere to (0.03, 0, 0.05) and its scale to (0.03, 0.03, 0.03).

5. Now, texture this **Sphere** with our Tesla book image (or some other image) by dragging the image asset from the **Project** panel onto the Sphere GameObject in the **Hierarchy**.

6. Duplicate the sphere three times, setting the positions of these duplicates like so:

    *   (-0.03, 0, 0.05)
    *   (0.03, 0, -0.05)
    *   (-0.03, 0, -0.05)

7. Finally, add an **AR Anchor** component to this car GameObject, by clicking the **Add Component** button in the **Inspector** panel and choosing **AR Anchor**.

8. Create a **Prefab** of this car GameObject by dragging the car GameObject into the **Project** window.

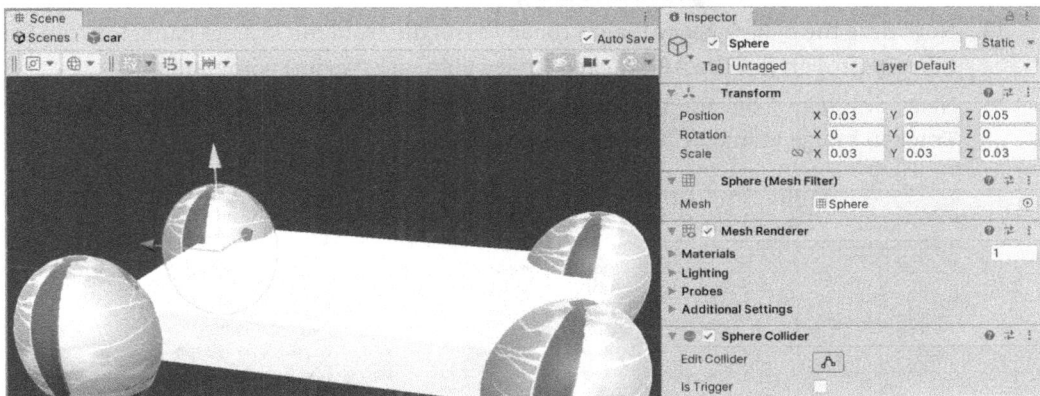

Figure 17.26: Our Cube-Sphere car Prefab 3D object

9. Now, remove the object from the scene by deleting the car GameObject from the **Hierarchy** panel.

10. Import the image you are using as a target for this project – that is, tesla_book.jpg, or some other image you have chosen.

11. Create a **Reference Image Library** asset file named myReferenceImageLibrary in the **Project** panel by chosing **XR | Reference Image Library**.

12. With the `myReferenceImageLibrary` asset file selected in the **Project** panel, in the **Inspector** panel click the **Add Image** button. Now, drag your image file into the image slot in the **Inspector** panel:

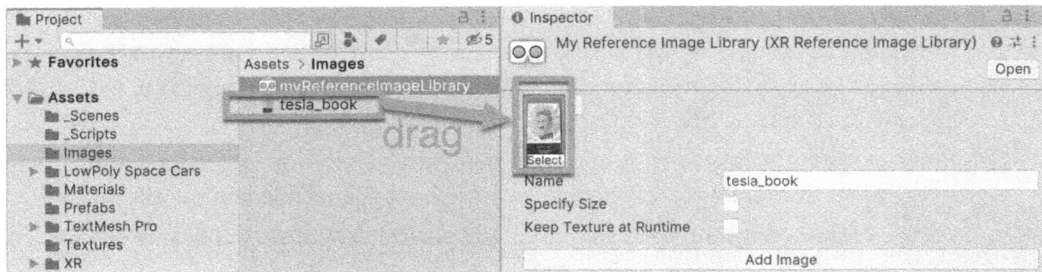

Figure 17.27: Adding our image to the Reference Image Library

13. In the **Hierarchy** panel, select the `XR Origin` GameObject. In the **Inspector** panel, click the **Add Component** button and locate the **AR Anchor Manager** component and add it to this GameObject.

14. With the **XR Origin GameObject** still selected in the **Inspector** panel, click the **Add Component** button and locate the **AR Tracked Image Manager** component and add it to this GameObject.

15. Now, drag the `myReferenceImageLibrary` asset file from the **Project** panel into the **Serialized Library** property of the **AR Tracked Image Manager (Script)** component.

16. Next, drag `car` from the **Project** panel into the **Tracked Image Prefab** property of the **AR Tracked Image Manager (Script)** component.

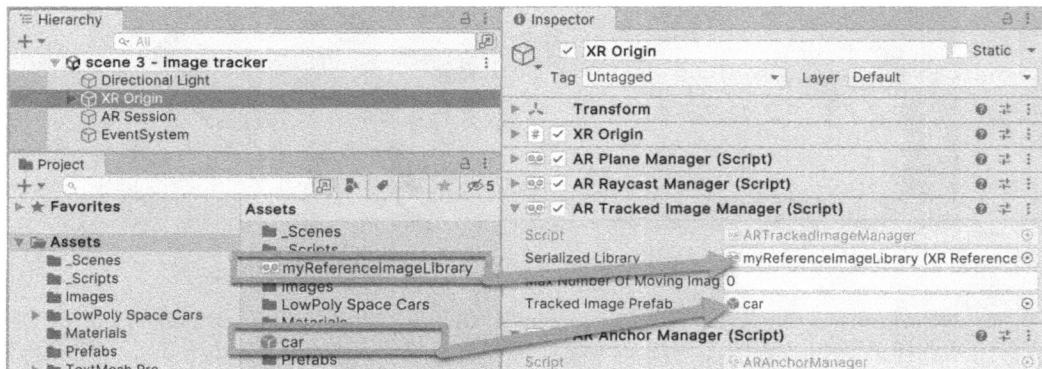

Figure 17.28: Adding our Reference Image Library to the Tracked Image Manager component of the AR Session Origin GameObject

17. Now, save the scene and add it to the build. Remove any other scenes from the build.

18. At this point, you may wish to set the **Product Name** in the **Player Settings** to give the app an appropriate name when it is installed on your device.

19. Now, either **Build And Run** your app on your device, or just **Build** the app and copy/install it onto your device.

20. Once you've copied/installed the app onto your device, then when you run it and when the image (for example, the Tesla book cover) from your reference library is visible to the device camera, you'll see the 3D car model floating above the book on your phone screen. If you change your position or move the book, the 3D model should move (track) with the target image.

How it works...

In this recipe, we created a **Reference Image Library** asset file containing one image – the book cover. The Unity **AR Foundation** package offers the scripted **AR Tracked Image Manager (Script)** component. This can use a library of images (in our case, **Reference Image Library**) to look for, identify, and track the position and orientation of the image of interest. It can then display a prefab instance at the location of a recognized image from the library.

For the prefab to stay fixed to the location and orientation of the recognized image, the GameObject needs an **AR Anchor** component. The **AR Anchor Manager** component of the **XR Origin** GameObject keeps track of anchors relative to the AR device viewing the real-world scene.

Further reading

Learn about basic AR Scenes and the XR Interaction Toolkit requirements from the Unity documentation:

- Basic AR scene requirements:

    https://docs.unity3d.com/Manual/AROverview.html

- Example XR Interaction Toolkit project:

    https://github.com/Unity-Technologies/XR-Interaction-Toolkit-Examples

- XR Interaction Toolkit Blog:

    https://forum.unity.com/threads/xr-interaction-toolkit-1-0-0-pre-2-pre-release-is-available.1046092/

Learn about the AR Foundation package:

- https://docs.unity3d.com/Packages/com.unity.xr.arfoundation@5.0/manual/index.html
- https://unity.com/mobile-solution-guide
- http://virtualxdesign.mit.edu/blog/2019/6/22/viewing-your-models-in-ar

Learn about Unity and developing for the Microsoft HoloLens AR headset here:

- Video tutorial on setting up development for HoloLens

    https://www.youtube.com/watch?v=dOsYerpKloY

- Microsoft recommend Unity version for HoloLens development

    https://learn.microsoft.com/en-us/windows/mixed-reality/develop/unity/choosing-unity-version

- The Microsoft HoloLens emulator

  `https://learn.microsoft.com/en-us/windows/mixed-reality/develop/advanced-concepts/hololens-emulator-archive`

- Learn how to use an Android emulator to test AR apps:

  `https://developers.google.com/ar/develop/c/emulator.`

Learn more on Discord

To join the Discord community for this book – where you can share feedback, ask questions to the author, and learn about new releases – follow the QR code below:

`https://packt.link/unitydev`

18

Virtual and Extended Reality (VR/XR)

In this chapter, we will present a set of recipes introducing **Virtual Reality (VR)** game development in Unity. **XR** stands for **eXtended Reality**. This is an umbrella term encompassing both fully immersive VR systems and **Augmented Reality (AR)** systems that allow users to sense the world around them. Some of the recipes in this chapter cover how to publish projects using the WebXR standards.

VR is about presenting an immersive audio-visual experience to the player, engaging enough for them to lose themselves in exploring and interacting with the game world that has been created. As well as gaming, VR applications offer amazing experiences to help us explore graphics and videos in immersive 3D, such as **Google Earth VR** and **Oculus Quest Wander**:

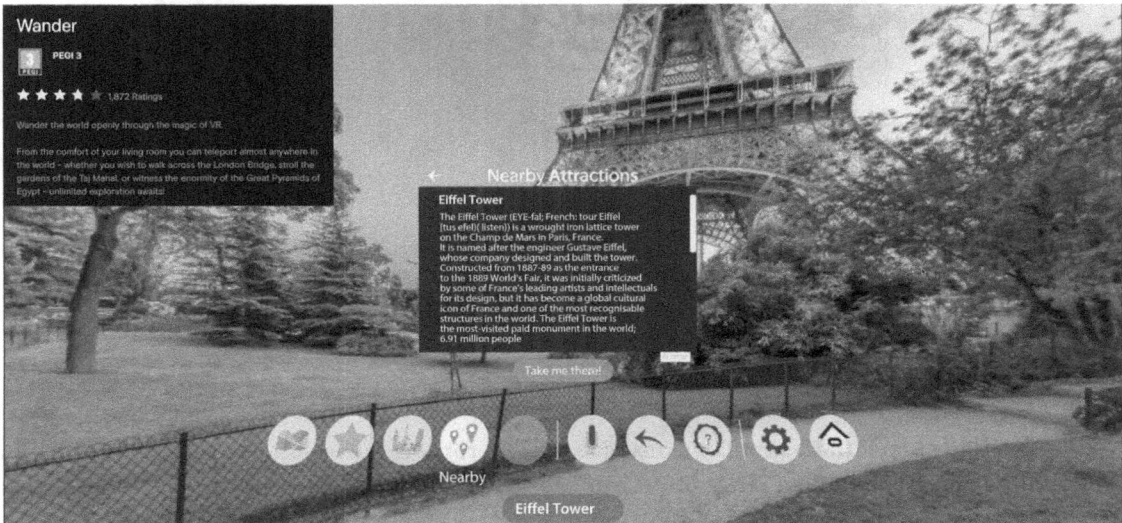

Figure 18.1: The Oculus Wander application showing the Eiffel Tower in Paris, France

From one point of view, VR simply requires two cameras in order to generate the images for each eye to give a 3D effect. But effective VR needs content, UI controls, and tools to help create it. In this chapter, we will explore recipes that work with 360-degree videos and pre-made assets for VR projects.

VR systems can be categorized as either desktop (tethered) or mobile. There are several types of VR devices, all of which we can develop for using the Unity Editor:

- **Desktop (tethered) VR:** A VR headset that's "tethered" via a high-speed cable to a computer with powerful graphics. Examples include the Oculus Rift, HTC Vive, and Windows Mixed Reality. Many early research VR systems were tethered but these days, many VR systems are mobile. Some standalone devices can also work tethered to make use of a more powerful desktop computer to run more processor-intensive VR applications, such as an Oculus Quest with the Oculus Link cable (there is also the wireless Air Link).

- **Mounted smartphone VR:** A form of VR housing for a mobile device. These are headsets that house an Android/iOS smartphone, of which the two best known are Google Cardboard and Samsung Gear. Interestingly, some proposed AR systems link digital glasses via a cable to a smartphone, so while they are mobile and smartphone-based, they won't be a single head-mounted display unit.

- **Standalone mobile VR headsets:** Many recent mobile VR headsets have Android mobile devices built into them. Examples include Google Daydream, Oculus Quest 1/2/3 and Go, HTC Vive Focus, and Pico Neo.

Figure 18.2: The low-cost Google Cardboard headset for mobile cell phones

There are basically five ways to get a Unity VR application running on a device to experience it, as shown in the following diagram. They are as follows:

- **The Unity Mock HMD simulated headset:** This is probably the fastest way to test a VR/AR app, although only a limited range of VR game features can be tested when running in 2D in the Unity Editor. We won't cover this in this book, but you can learn more about it here: https://docs.unity3d.com/Packages/com.unity.xr.mock-hmd@1.0/manual/index.html.

- **Tethered headset:** Desktop VR – a VR headset tethered via a high-speed cable (or Wi-Fi) to a computer with powerful graphics resources. If you use the Oculus Link/Air Link for your Unity VR Quest development then you'll be using your Quest as a tethered headset.

- **Compiled application:** Compile the Unity project into an executable app (usually Android or iOS) and copy it via USB cable, Wi-Fi, or web download onto the device. Then, run the app and experience the VR/AR application running.

- **Internet-published WebXR project:** Build a Unity project as a WebXR/WebGL project, and then publish this build on an internet server. The AR application can then be experienced when using an XR browser (such as **Wolvic**) on a device.

- **VR stores:** There are now VR stores, such as the Oculus Rift/Quest stores, so publishing a built project to a store is another route from Unity development to downloading and running a VR app on a device:

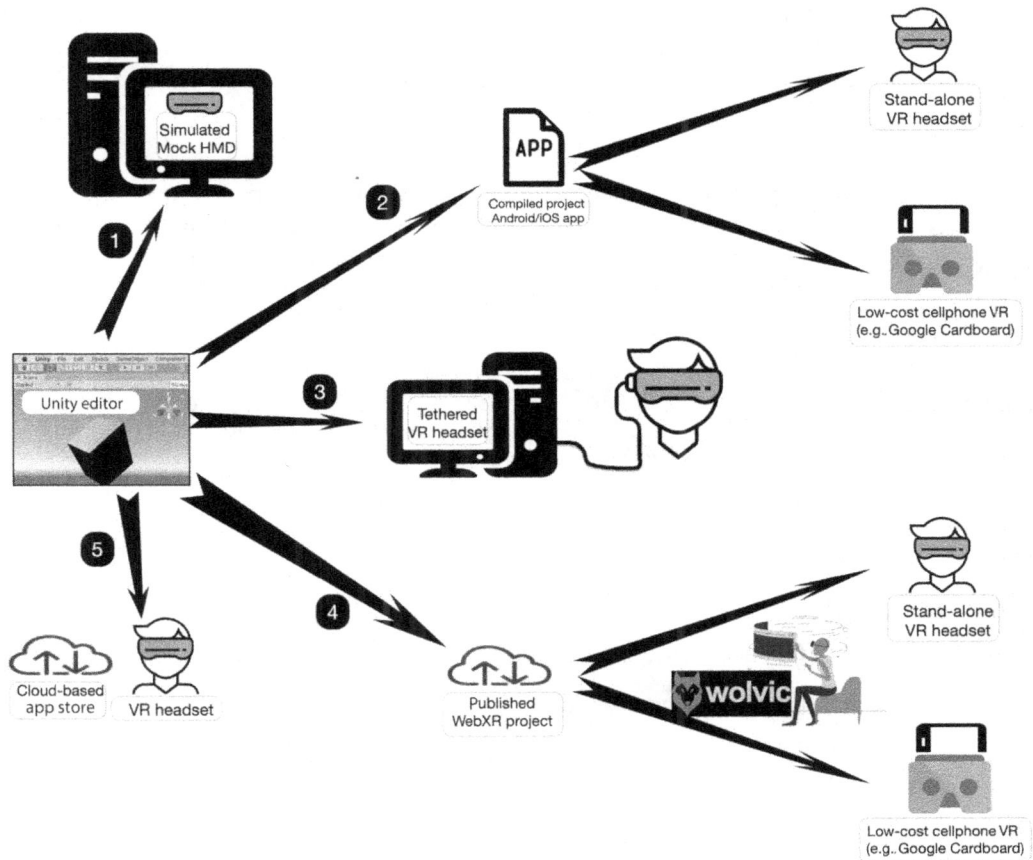

Figure 18.3: Methods of publishing and experiencing VR/AR

This chapter will help you become familiar with working with VR projects and some of the interactions you can present to your players. It will cover how to create VR projects with Unity and deploy those projects onto devices or, if you choose to publish them as XR projects, using a VR-enabled browser.

In this chapter, we will cover the following recipes:

- Setting up the Oculus Quest 2 for Unity development
- Creating and running a Unity project on a VR headset
- Beginning with the XR Interaction Toolkit
- Creating a 360-degree video VR project
- Exploring and building the **Desert** WebXR demo project
- Using GitHub Pages to publish your WebXR project for free
- Creating a simple WebXR scene from scratch

Setting up the Oculus Quest 2 for Unity development

The current most popular VR headset is probably Meta's **Oculus Quest 2**. These are low-cost, walk-around headsets that provide a great introduction to VR for many people. It can be a bit fiddly to set up a Quest for Unity project development, so we'll go through the steps required in this short recipe. Between this and the next recipe, you'll have the satisfaction of creating VR projects and loading them onto your headset.

> These steps were performed at the time the book was being written. But things change quickly, and it may be that not all the steps/screenshots are the same when you attempt this developer mode setup. However, we have included this recipe since we believe it will still be helpful to those beginning development for the Quest 2 to see the kinds of actions they'll need to take.

Getting ready

For this recipe, you'll need the following:

- A mobile phone on which to install the Meta Quest app
- An Oculus Quest 2 VR headset

How to do it...

To pair your Quest VR headset with the Oculus mobile phone app, do the following:

1. Fully charge the Quest 2 headset (until the LED is green!)
2. On a mobile phone, install the Meta Quest app and log in with your Facebook/Oculus account (you can create an account if you don't have one).
3. Enable Bluetooth on your mobile cell phone.
4. Pair your phone with your VR headset – choose the device (Quest 2). Pairing will either be automatic or you'll be asked to enter the device code (displayed in the Quest 2 view) into the cell phone app to pair the VR headset to your Meta cell phone app account.

5. Connect your Quest 2 headset to a Wi-Fi network. Then, after viewing the safety video, your headset will auto-update to download the latest Quest 2 operation software (for the headset and the left/right controller devices). While it is updating software, you can remove the headset and wait for it to play a sound to confirm the software has been downloaded and installed.

6. On the Quest 2 headset, enable the USB connection. Do this by choosing **Settings | System | Developer**, then enable the **USB Connection Dialog** option.

7. On your phone, now enable **Developer mode** for the Meta cell phone app. Choose the **Devices** tab – you should see the headset paired to the phone. Choose **Headset Settings** and enable the **Developer Mode** option. The app will display a prompt to create a developer account. Follow the instructions from Oculus to fully enable developer mode:

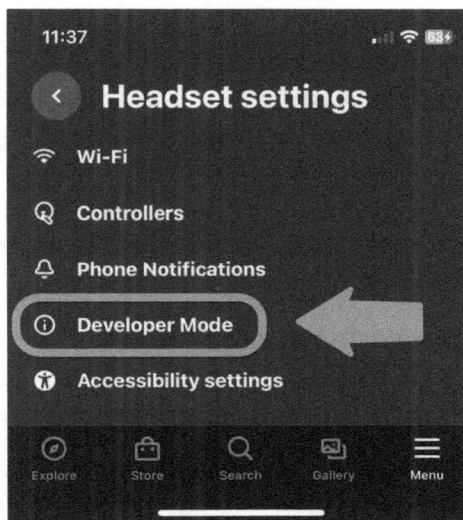

Figure 18.4: Choosing Developer Mode on the Meta phone app

How it works...

The mobile phone app connects to your Quest headset via Bluetooth. Each Quest headset needs to be linked to an Oculus/Facebook account. Android devices, including Quest headsets, need to go through USB debugging/Developer Mode to allow us to load our own compiled Android applications onto them, rather than only being able to download apps from the web or Oculus Store.

Having followed the steps in this recipe, your Quest headset should now be ready to run Unity VR projects that we will develop and build ourselves.

There's more...

Here are some suggestions for troubleshooting this recipe.

Speed up Windows development with Oculus Link/Air Link

If you have a Windows PC that meets the minimum VR requirements, you may be able to speed up your development using the Oculus Link connection. This uses either a high-speed USB cable or Wi-Fi to run the VR project on your PC, and then communicate with the Quest headset. In essence, you are running the VR app on the PC, and the headset is tethered to the PC. In terms of development with Unity, this means you can just play a scene and experience it on your VR headset immediately, rather than having to wait for an application APK to be built and then transferred to the headset.

Basically, you need to install the Oculus Quest application on your Windows PC and pair it with your headset. Here are some Quest Link guides (USB and Air Link):

* `https://www.uploadvr.com/how-to-air-link-pc-vr-quest-2/`

* `https://www.youtube.com/watch?v=25uae1jxk-E`

* `https://www.youtube.com/playlist?list=PLX8u1QKl_yPD4IQhcPlkqxMt35X2COvm0`

* `https://www.oculus.com/blog/introducing-oculus-air-link-a-wireless-way-to-play-pc-vr-games-on-oculus-quest-2-plus-infinite-office-updates-support-for-120-hz-on-quest-2-and-more/`

* Some tips for reducing frame rate lag with Link: `https://smartglasseshub.com/oculus-link-lag-stuttering-issues-quest-2/`

Oculus Link is not possible on most Macs

Note that this is pretty much for Windows users only. Sorry – unless you have an old, but powerful, Intel-based Mac with a suitable graphics card, then Link development is not available for Mac users.

One way to use Link with Intel-based Macs is to use Bootcamp to install a version of Windows. Here's a guide for this: `https://stealthoptional.com/gaming/how-to-connect-oculus-quest-to-mac-can-you-connect-oculus-quest-2-to-apple-mac-or-macbook/`.

Oculus Link Linux

There is a way to use your Oculus/Quest as a tethered VR headset if using Linux. See this GitHub project for an up-to-date version of the **OculusLinkLinux** software: `https://github.com/MilkJug1/OculusLinkLinux`.

Online guides to help you set up your Quest for VR development with Unity

Here are some online tutorials for setting up Quest 2 and Unity for VR development:

* ZyberVR tutorial: `https://zybervr.com/blogs/news/how-to-set-up-oculus-quest-2-step-by-step-beginner-guide-2022`.

* Circuit Stream 10-step setup guide: `https://circuitstream.com/blog/oculus-quest-unity-setup`

- Allen Janyska's setup guide on the Atomic Object website: `https://spin.atomicobject.com/2022/04/15/oculus-quest-2/`
- XR Terra's Unity and Quest guide: `https://www.xrterra.com/developing-for-vr-with-quest-2-unity-for-the-first-time-a-step-by-step-guide/`

Creating and running a Unity project on a VR headset

VR devices are mobile devices, so if you've created some mobile apps and loaded them onto a cell phone you'll be familiar with the procedure for working with VR headsets. Most VR devices run using the Android mobile operating system (including the Oculus Quest/2/3 and Pico VR headsets). This recipe is based on working with a Quest 2 headset, but the steps are similar for other headsets.

In this recipe, you'll learn how to create a project containing an empty, all-green scene, and build and run this project on a Quest 2 headset.

This recipe will be the foundation for the other recipes in this chapter.

Getting ready

Before beginning the recipes in this chapter, you should have explored the Android recipes in *Chapter 15, Mobile Games and Applications*, to be familiar with the procedure for switching build targets to mobile devices and running games on USB-connected devices. Also, since you'll be running your AR projects on an Android device, before starting any of the recipes in this chapter ensure you have completed the first and/or second recipe of the previous chapter so that your system is set up for mobile device development by installing the appropriate Build Support plugin.

How to do it...

To create and run a Unity project on a VR headset, do the following:

1. Create a new 3D (Core) Unity project.
2. When the new project has loaded, open the sample scene (from the `Scenes` folder in the **Project** panel). Save this scene with the name `scene1 - all green`.
3. Select the `Main Camera` in the **Hierarchy**, and in the **Inspector** for its **Camera** component, change its **Clear Flags** property from **Skybox** to **Solid Color**. Choose a **light green** color (this is to confirm that our scene is running on our Quest 2 headset later in this recipe).
4. Open the **Build Settings** panel (**File | Build Settings...**), select the **Android Platform**, and click the **Switch Platform** button.
5. Open the **Project Settings** panel, by choosing **Edit | Project Settings...**.
6. On the **XR Plugin Management** tab, click the button to install **XR Plugin Management**.
7. Once installed, choose the **Android** (robot icon) tab and then check the box for **Oculus**.

8. You'll now see a sub-tab of **XR Plug-in Management** named **Oculus**. Select that, and ensure your **Quest 2** device is checked.

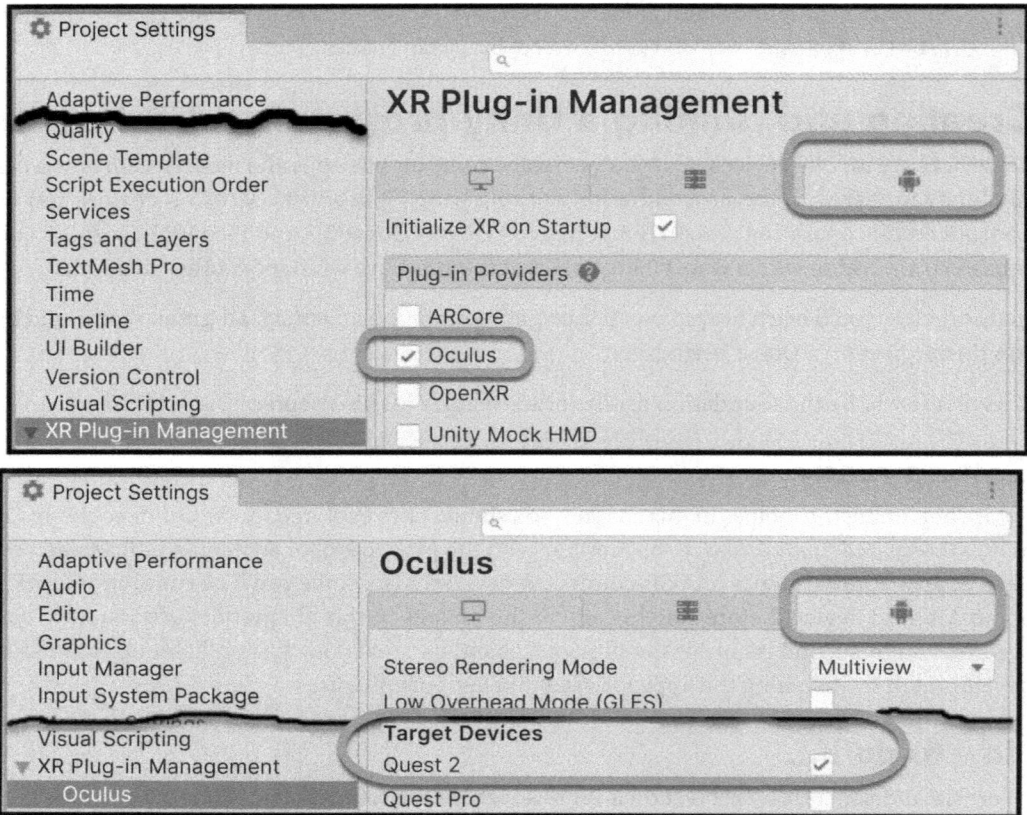

Figure 18.5: Selecting Oculus and Quest 2 as active Android XR plugins

9. Add the scene1 - all green scene to the build, and remove any other scenes from this build.

10. Ensure the **Texture Compression** for this build is **ASTC**. You can set this in the **Build Settings** panel, or through the **Other Settings** section of the **Player** tab in the **Project Settings** panel.

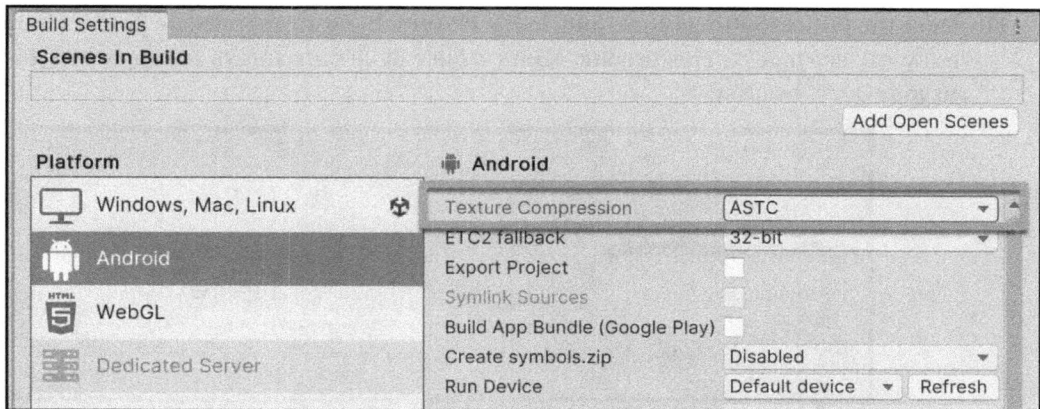

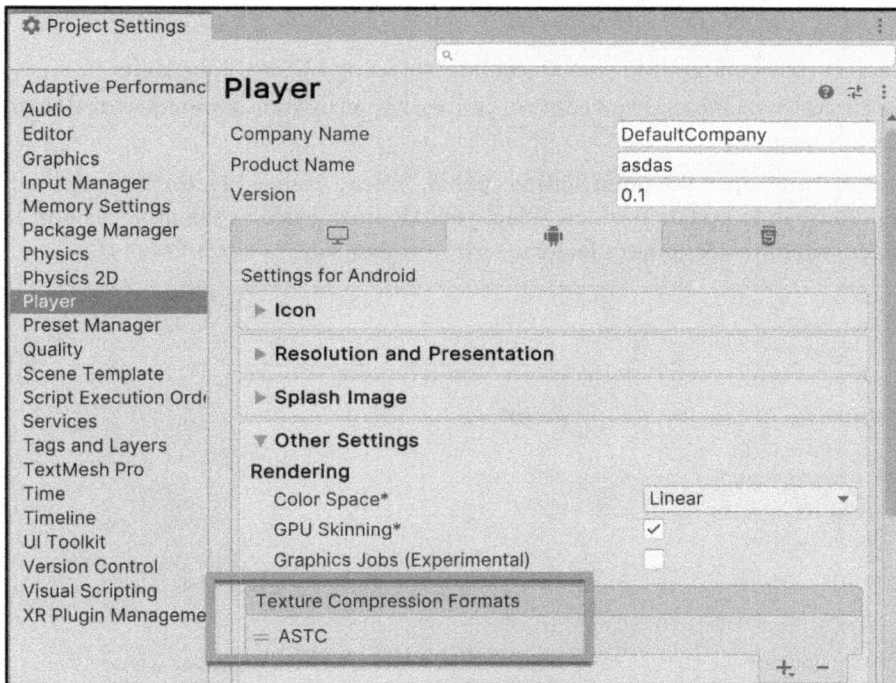

Figure 18.6: Ensuring the Texture Compression is set to ASTC

11. Open the **Project Settings** panel and, in the **Player** tab, set an appropriate **Product Name**, such as My VR Project 1. This **Product Name** value will be the name of the app when it is copied to your Quest headset.

Figure 18.7: Setting the project's name and other details in the Player settings

12. Connect your computer to your Quest headset using a USB-C data cable.

13. Put on the Quest headset and confirm you are happy to **Allow connected device to access files** for your headset.

14. Back in Unity, open the **Build Settings** panel. For the **Android Platform,** you should be able to click **Refresh** to update devices. Select your **Oculus** device as the project's **Run Device.** Your device will have a long hexadecimal serial number such as 1PASH23423423:

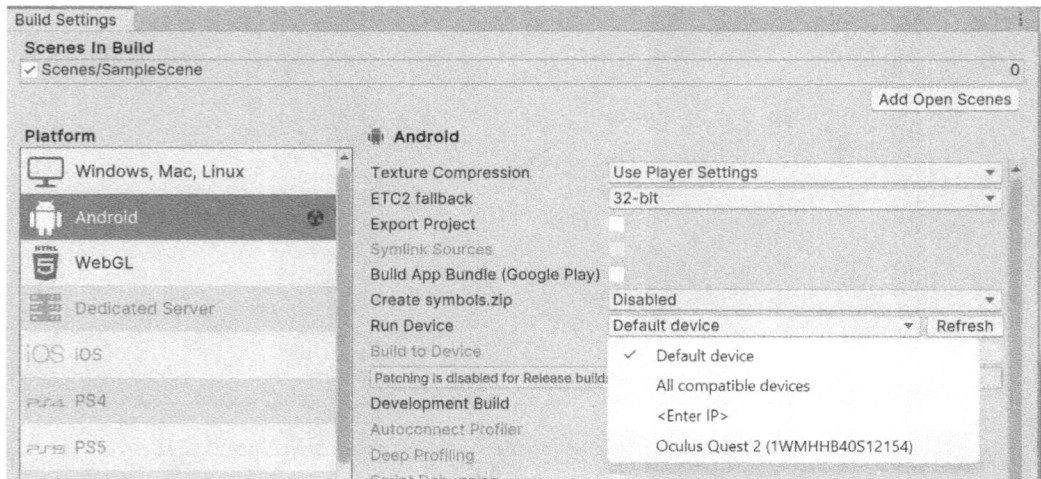

Figure 18.8: Selecting Oculus Quest as the run device from the Android Build Settings

15. In the **Build Settings** panel, click the **Build And Run** button (choosing a suitable name for the APK app installer, such as VR-from-scratch). After a few seconds/minutes of compiling, your Unity project will run on the connected Oculus Quest headset. Put on the headset and explore!

16. While the scene is running correctly on the Quest, there is **nothing to interact with – it's just all green!** Quit the Quest app by pressing the Oculus button on your controller and clicking the **Quit** application button.

How it works...

The build target needed to be set to Android. This is because our Quest 2 device runs a version of the Android mobile operating system. By connecting our Quest headset to the computer with a USB cable and enabling computer device access to the headset, we were able to set the Unity project's Android build target to be our specific connected Quest 2 headset.

The scene we created just contained a camera showing lots of green. But that was enough to confirm that we were able to build an APK Android application and deploy it to our Quest 2 headset.

We set the name of the app that will appear on the VR device by setting its **Product Name** in the **Player Settings**. On the Quest, once an application has been installed onto your Quest headset, it can be found in the **App Library** – we just need to filter for apps from **Unknown sources**.

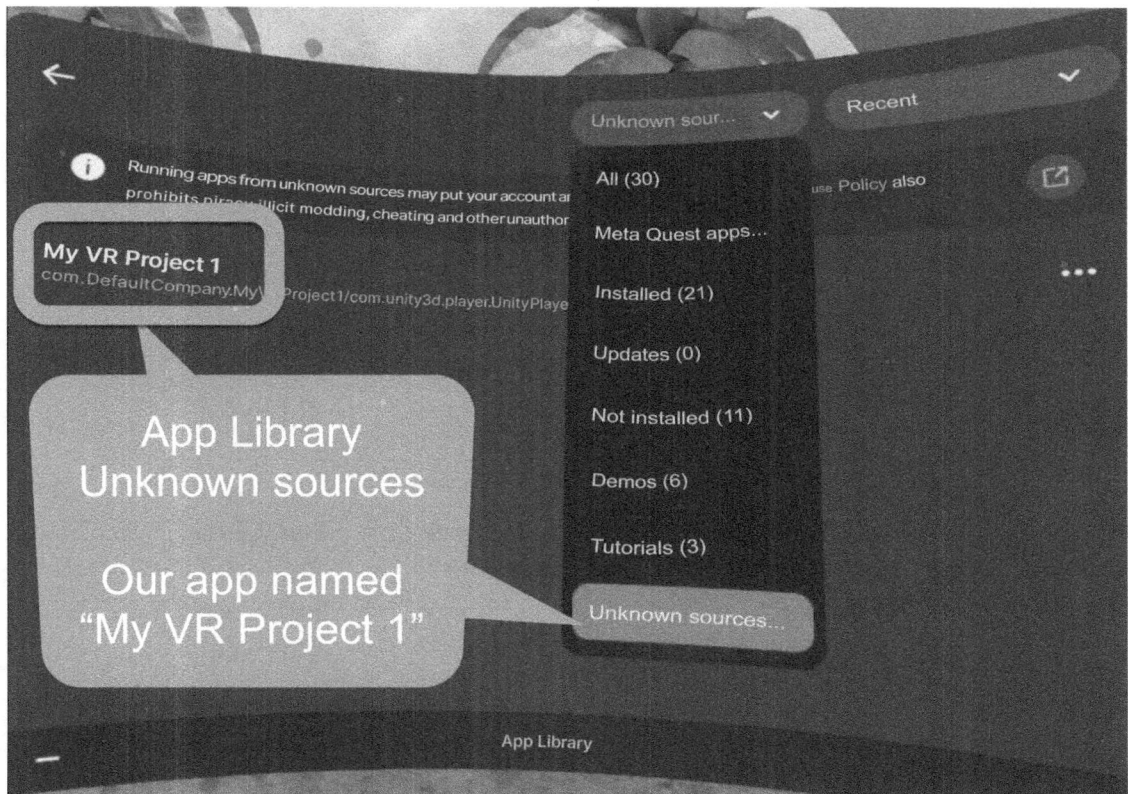

Figure 18.9: Locating your installed Quest app under Unknown sources in the App Library from the Product Name set in the Player Settings

There's more...

Here are some suggestions for troubleshooting this recipe.

Cannot choose the Quest device due to Android tools unknown location

Sometimes, Unity loses track of the Android JDK plugins compatible with the current Unity Editor version. For example, this might happen if you install a new version of the Unity Editor (and its plugins) but the preferences still link to the Android plugins for an older version of the editor. To solve this problem, first, ensure you have added the Android plugins to the Editor via the **Unity Hub**. Then, through the **Preferences | External Tools** panel you can set the correct paths (**File | Preferences...**).

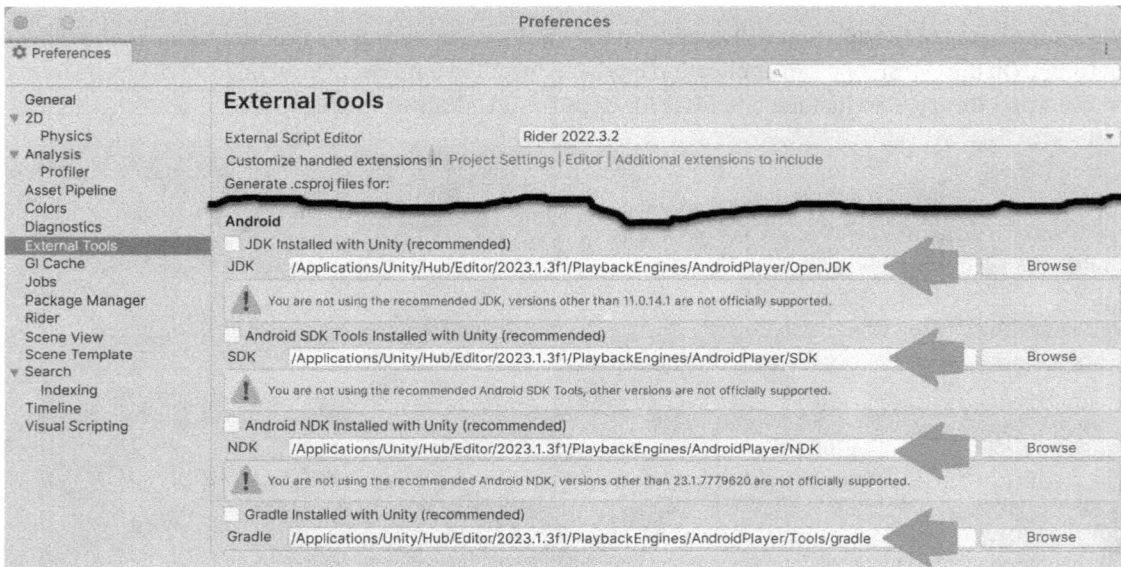

Figure 18.10: Fixing the path to the Android tools editor plugins

If there is already a path entered (the **Unity recommended** path), and the only part of the path that is wrong is the Editor version number, then click the **Copy Path** button, then uncheck the **Unity recommended** option, then paste in the path and update the Editor version number.

Build failure due to Android version

If Unity fails to build when your **Quest** is set as **Run Device** in **Build Settings**, check the error message in the **Console** panel. A common error is that the version of Android in the **Player** settings is too low for the version of Android on the device.

For example, the first time I tried to build for a Quest, I got this error:

```
Error building Player: Android Minimum API Level must be set to 23 or higher
for the Oculus XR Plugin.
```

It was resolved by opening the **Project Settings** panel by choosing **Edit | Project Settings...** and increasing **Minimum API Level** to at least the number stated in the error message. Note that at the time of writing (mid-2023), setting the minimum API to 29 worked for a Quest 2 headset with up-to-date software.

You can do this in the **Player** section and by modifying the **Other Settings** values:

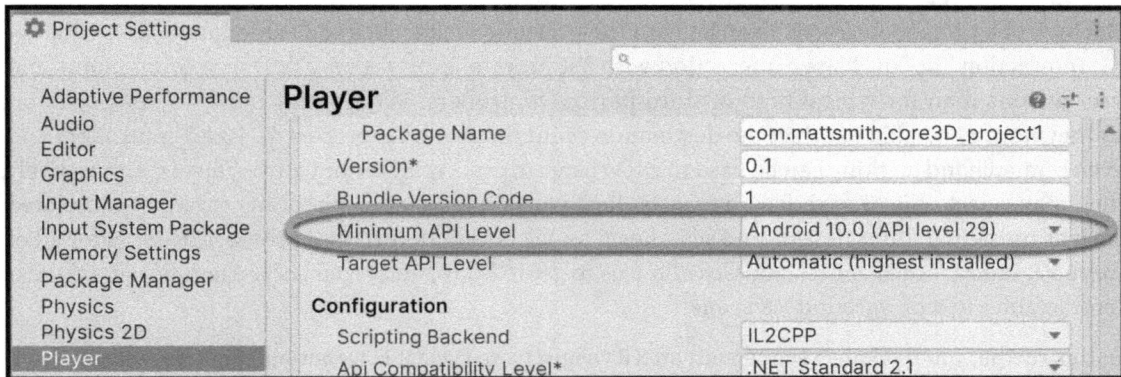

Figure 18.11: Setting Minimum API Level of the Android version in the Project Settings

Build failure due to "namespace cannot be found" error

When you are writing or using code with namespaces and assembly definitions, Unity may fail to build due to a compiler error in the following form:

```
The type or namespace name '<<someNamespace>>' could not be found
```

If this occurs, then you will need to create/edit the **Assembly Definition** asset file in the **Project** panel folder where the scripts are located. If an **Assembly Definition** already exists, select it (otherwise, create a new one and select that). Then, in the **Inspector**, check the **Android** option for **Include Platforms** and click **Apply**.

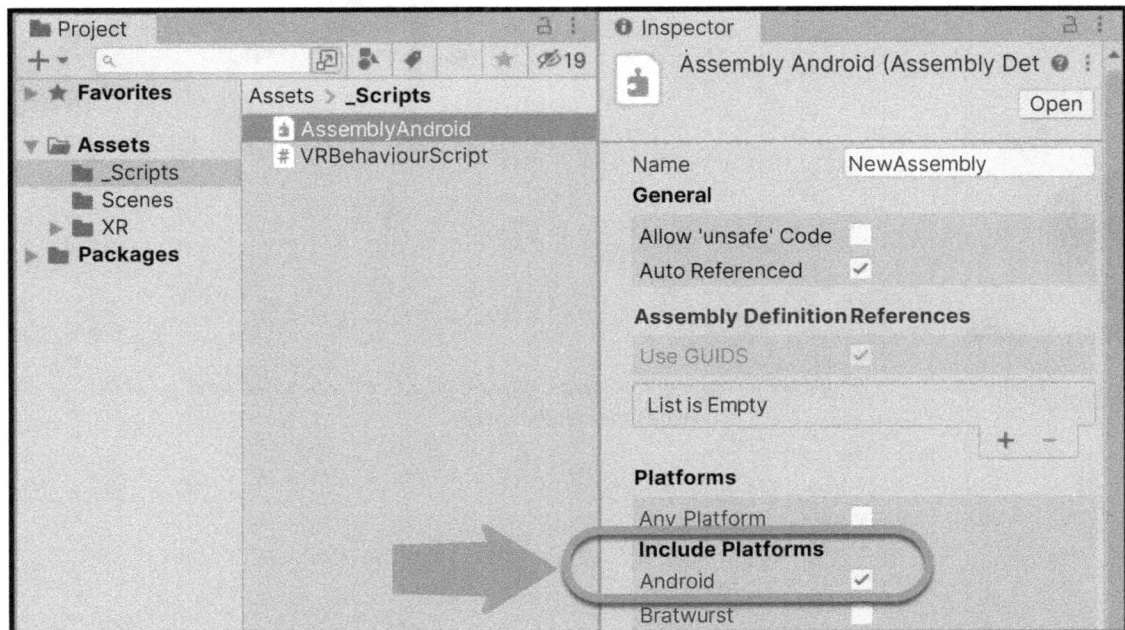

Figure 18.12: Adding the Android platform to the WebXR settings

Beginning with the XR Interaction Toolkit

Having got a project onto a VR headset with the previous recipe, let's add some interaction with the XR Interaction Toolkit. Player interaction with VR projects involves very different components and user actions than for typical first- or third-person controllers. Players may move via teleportation, so they have to be able to select the destination point they wish to teleport to. Hand controllers and even human hand tracking can be used to allow users to grab, release, and move objects. Alternatively, they may select objects with laser raycasts. Real-world interactable components can be simulated, such as press-in buttons, levers, and switches. The **XR Interaction Toolkit** provides features for all of these VR interactions, and its samples allow us to get to know, and repurpose, examples of different interactables in a playground VR scene.

In this recipe, you'll create a scene with an **XR Origin** from the **XR Interaction Toolkit** package. You'll also add a black sphere with which your VR controller rays can interact!

Figure 18.13: The VR app we'll create in this recipe, running on a Quest 2 headset. Rays go white when they hit the sphere

Getting ready

This recipe builds on the previous one, so make a copy of that and work on the copy.

How to do it...

To begin working with the XR Interaction Toolkit, do the following:

1. Open your copy of the project from the previous recipe.
2. Open the **Package Manager** panel (**Window | Package Manager**). Select **Unity Registry** from the dropdown and type XR I into the search bar. Select **XR Interaction Toolkit** and download and import this package. Agree to any warning about the new **Input System** and when you're asked if you want to enable backends. The Unity Editor may need to restart.
3. Then click the **Samples** tab for **XR Interaction Toolkit**, and click the **Import** button to import the **Starter Assets** for this package.

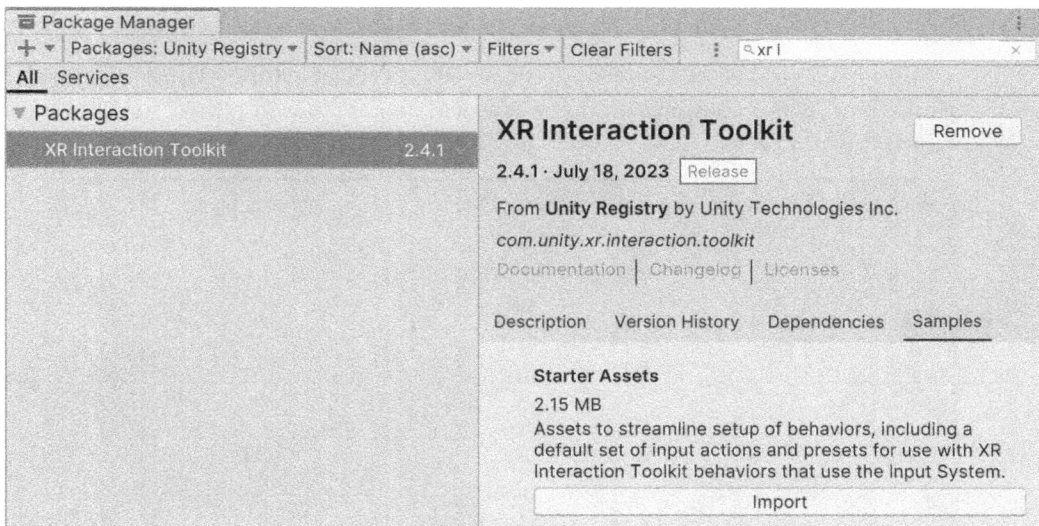

Figure 18.14: Importing the XR Interaction Toolkit with the Package Manager

4. You should now see the `Samples` and `XRI` folders in the **Project** panel.
5. We need to set up the default input actions for our **left** and **right** simulated controllers. In the **Project** panel, select the **XRI Default Left Controller** asset file in the **Assets/Samples/XR Interaction Toolkit/2.x.(version number)/Starter Assets** folder. Then, in the **Inspector**, click the **Add to ActionBasedController default** button.

6. Repeat this for the **XRI Default Right Controller** asset file.

Figure 18.15: Adding the XRI input actions to the project defaults

7. We now need to label the filters for these presets in **Project Settings**. Open the **Project Settings** panel (**Edit | Project Settings...**). In the **Preset Manager** tab, enter Right for **XRI Default Right Controller** and Left for **XRI Default Left Controller**.

Figure 18.16: Naming the Right/Left filters for the controller presets

8. Delete the Main Camera GameObjects from the **Hierarchy**.

9. We'll now add an **XR Origin** GameObject to the scene. Add an XR Origin (XR Rig) to the scene, by choosing **GameObject | XR | XR Origin (XR Rig)**. An XR Interaction Manager GameObject has been added to the **Hierarchy**.

10. Add to the scene **Hierarchy** a 3D cube, named Cube-table, with a position of (0, 0, 3) and a scale of (3, 0.1, 3).

11. Add to the scene **Hierarchy** a 3D sphere with a position of (0, 1, 4).

12. Create a black **Material** named m_black and apply it to the Sphere GameObject.

13. With the the Sphere GameObject still selected, in the **Inspector** add an **XR Simple Interactable** component to the Sphere GameObject.

14. Save this updated scene naming it: scene2 - XR Interaction.

15. In the **Build Settings** panel ensure your Quest device is still selected as the **Run Device**, and click the **Build And Run** button (you can simply choose the same APK app installer filename; overwrite the existing file).

16. Put on your Quest headset and explore! You should now see the hand controller models firing red rays, but the rays will turn white when they hit the interactable Sphere GameObject.

17. Quit the app on the Quest and return to Unity.

18. Open the DemoScene, which you'll find in the **Project** panel, in the Assets/Samples/XR Interaction Toolkit/2.4.1 (version number)/Starter Assets folder.

19. Add this scene to the **Build Settings** (and remove any other scenes).

20. At this point, it's useful to change the product name in the **Player Settings Player** tab, so when this project is loaded it has an appropriate app name on the Quest, such as XRI-Demo.

21. Build and run the scene. On the Quest, you can now explore the **XR Interaction Toolkit** demo. Some useful controls to know are:

    • Left thumbstick – move forward, backward, and left and right
    • Right thumbstick – rotate view (left/right), or push forward to get a teleport control

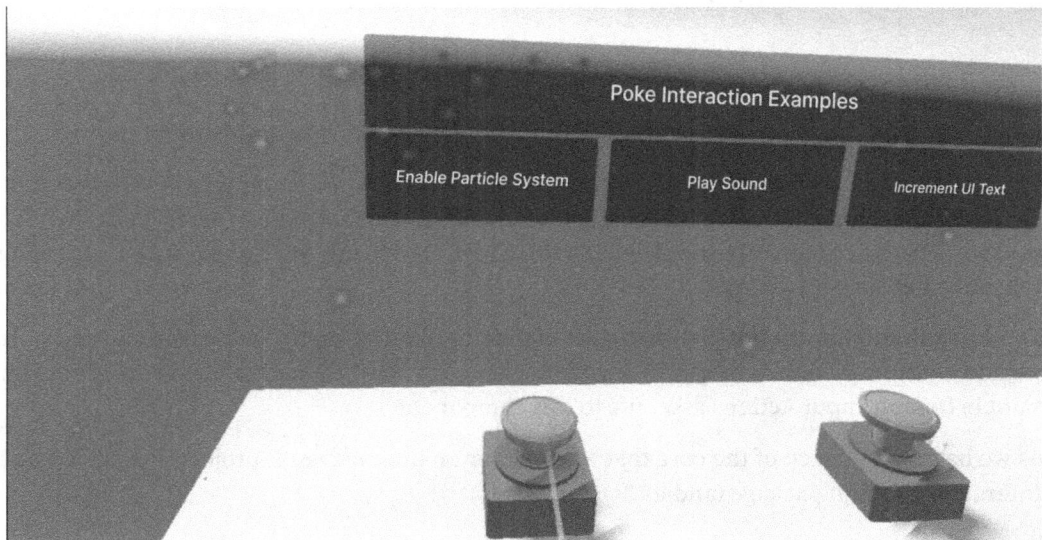

Figure 18.17: Exploring the XR Interaction Toolkit demo scene

22. Finally, there are many more examples, with documentation available from the **XR Interaction Example** project. This project can be found at: https://github.com/Unity-Technologies/XR-Interaction-Toolkit-Examples.

23. Download this project from the GitHub website (click **Code** and then **Download ZIP**), open it in Unity, Build and run it on your device, and spend many minutes/hours exploring a fantastic array of VR interactions, from smashing piggy banks with a hammer to shooting red balls into rotting targets, and many more.

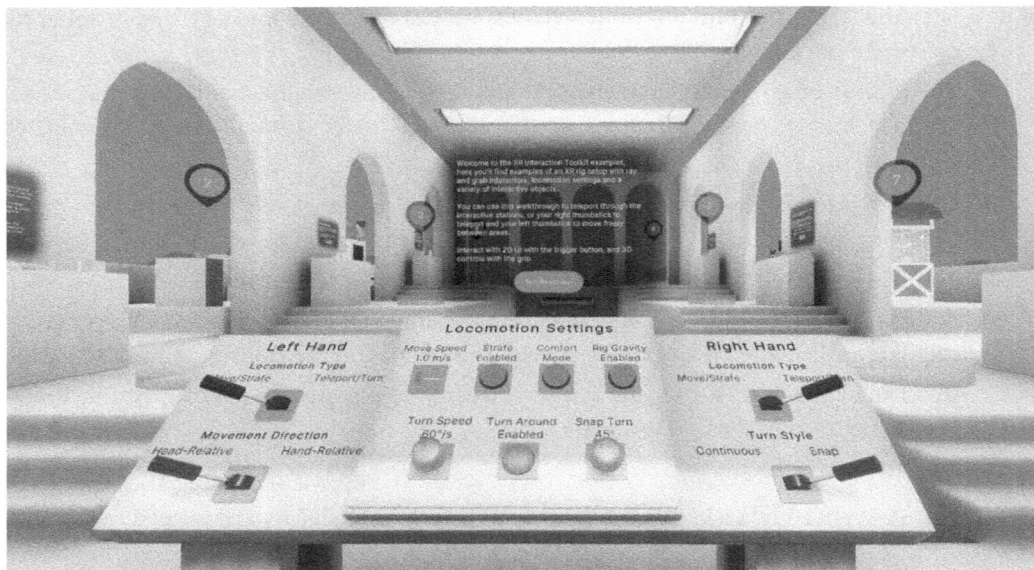

Figure 18.18: Exploring the fantastic XR Interaction Example project

How it works...

Our scene had an **XR Origin** and interaction system, which, when running on a VR headset, controls cameras for each eye based on the location, orientation, and movements detected by the VR headset sensors. The rig comes with some default controller 3D objects, which are moved according to the VR controller sensor readings.

The **XRI Interaction Toolkit** provides a set of default input actions that our app can recognize once we add an **Input Action Manager** to our XR Origin (XR Rig) GameObject and assign the imported sample **Default Input Actions** asset file to this component.

As we have seen, much of the core that we need for an interactive VR project is provided by the **XRI Interaction Toolkit** package (and its **Starter Assets**).

You can begin to experience the power of this toolkit when exploring the demo scene, and the scenes in the GitHub example project.

Creating a 360-degree video VR project

The prevalence of affordable 360-degree cameras means that it's easy to create your own, or find free online, 360-degree images and video clips. In this recipe, we'll learn how to add a 360-degree video clip as a skybox to a VR project.

You will also learn how 360-degree video clips can be played on the surface of 3D objects, including the inside of a sphere – a bit like **Google Earth** VR mode when you raise the sphere to your head to view its 360-degree image contents:

Figure 18.19: A 360-degree video sphere object in Google Earth VR

Getting ready

This recipe builds on the previous recipe. That project was based on a **3D (Core)** Unity project, with the **XR Interaction Toolkit** package and **Starter Assets** installed. So, make a copy of that project and work on the copy.

For this recipe, we have provided a short (royalty-free!) video called `Snowboarding_Polar.mp4` in the 18_04 folder.

> Special thanks to **Kris Szczurowski** from TU Dublin for permission to use his snowboarding 360-degree video clip, and for his help with these VR project recipes.

How to do it...

To create a 360-degree video VR project, follow these steps:

1. In Unity create a new **Basic (Built-in)** scene – this should contain a Main Camera and Directional Light.
2. Delete the `Main Camera` GameObject from the **Hierarchy**.
3. We'll now add an **XR Origin** GameObject to the scene. Add an `XR Origin (XR Rig)` to the scene by going to **GameObject | XR | XR Origin (XR Rig)**. An `XR Interaction Manager` GameObject has been added to the **Hierarchy**.
4. Import your 360-degree polar format video clip into your Unity project (in our example, this is **Snowboarding_Polar.mp4**).

5. Select the video asset in the **Project** panel. At the lower part of the **Inspector** panel is the asset preview. Select **Source Info** from the asset name dropdown and instead of an image preview, you can now see data about this asset, including its pixel dimensions. The provided **Snowboarding_Polar.mp4** 360 video has dimensions of 2560 x 1280.

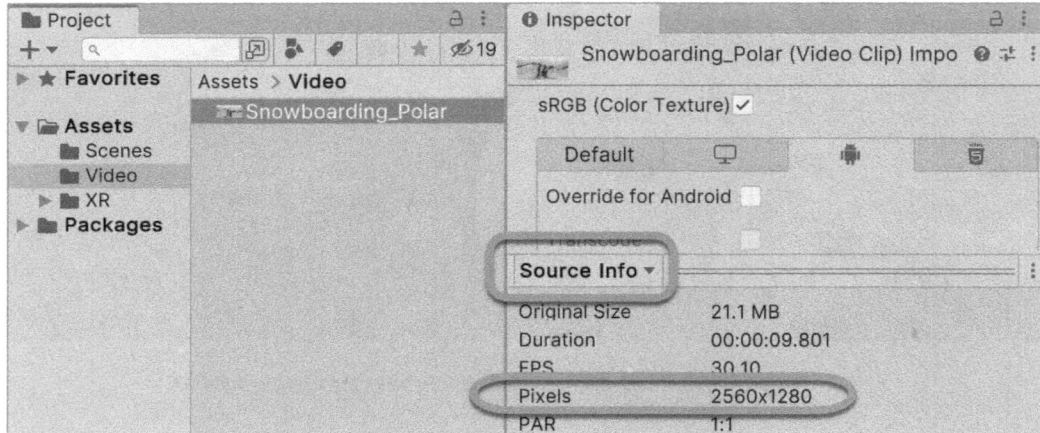

Figure 18.20: Noting the dimensions of your 360-degree video clip from the Inspector

6. Create a new empty GameObject named `video-player` by choosing **GameObject | Empty**.

7. With the `video-player` **GameObject** selected in the **Hierarchy**, in the **Inspector** add a **Video Player** component by clicking the **Add Component** button and choosing **Video Player**.

8. From the **Project** panel, drag your video asset file (for example, **Snowboarding_Polar**) into the **Video Clip** property of the **Video Player** component in the **Inspector**.

9. Also, check the **Loop** property of the **Video Player** component.

10. In the **Project** panel, create a new **Render Texture** asset file named `VideoRenderTexture` by choosing **Create | Render Texture**.

11. Set the **Size** property of the **VideoRenderTexture** asset file so that it matches the video asset resolution; for example, 2560 x 1280.

12. Set the **Depth Stencil** Format to None.

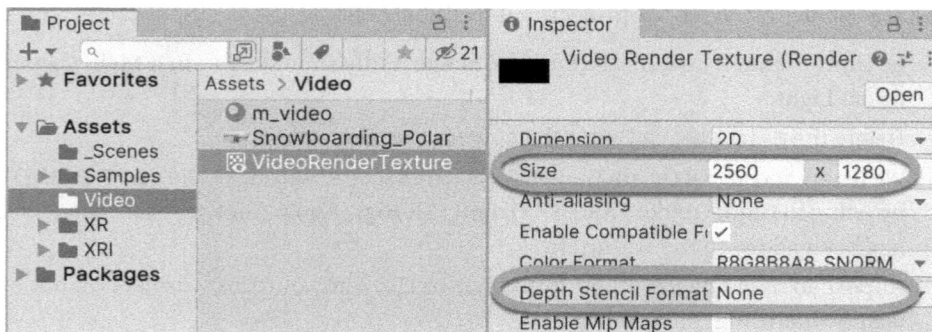

Figure 18.21: Matching the size of your Render Texture asset file to the 360 video dimensions

13. In the **Hierarchy** panel, select the `video-player` GameObject. For the **Video Player** component, set the **Target Texture** property to the **VideoRenderTexture** asset file from the **Project** panel.

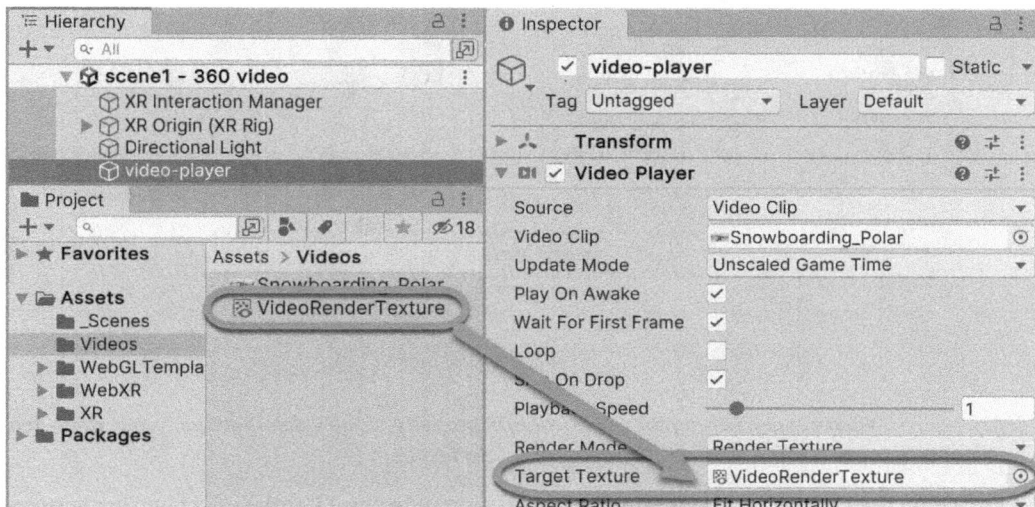

Figure 18.22: Configuring the Video Player component layer to send the video to the Video-RenderTexture asset file

14. In the **Project** panel, create a new **Material** asset file named `m_video` by choosing **Create | Material**.

15. With the **m_video Material** asset file selected in the **Project** panel, in the **Inspector** change its **Shader** property to **Skybox | Panoramic**.

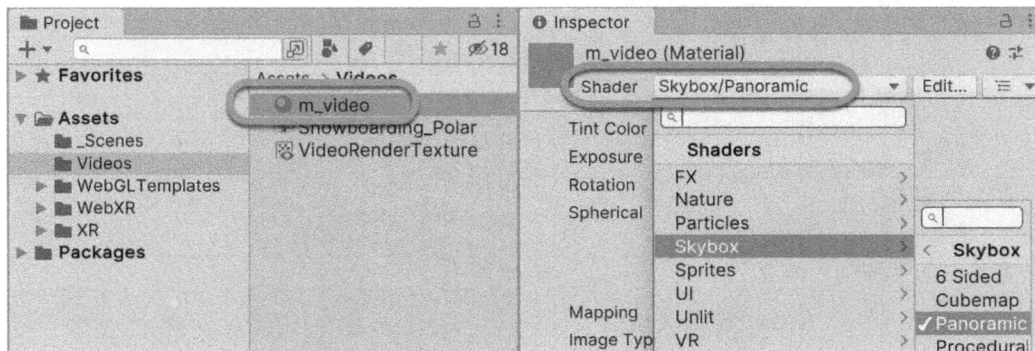

Figure 18.23: Setting the video Material shader material to a panoramic Skybox

16. In the **Inspector**, for the **Spherical HDR** property, click the **Select** button, enter **Video** in the search filter box, and select **VideoRenderTexture**.

Figure 18.24: Setting the Spherical HDR property of the video material to our video RenderTexture

17. Open the **Lighting Settings** window by choosing **Window | Rendering | Lighting Settings**. Select the **Environment** tab, and set the **Skybox Material** to m_video. You can do this by clicking the circle target asset selection button and then filtering for asset files beginning with m_.

18. Also ensure there is no **Sun Source**.

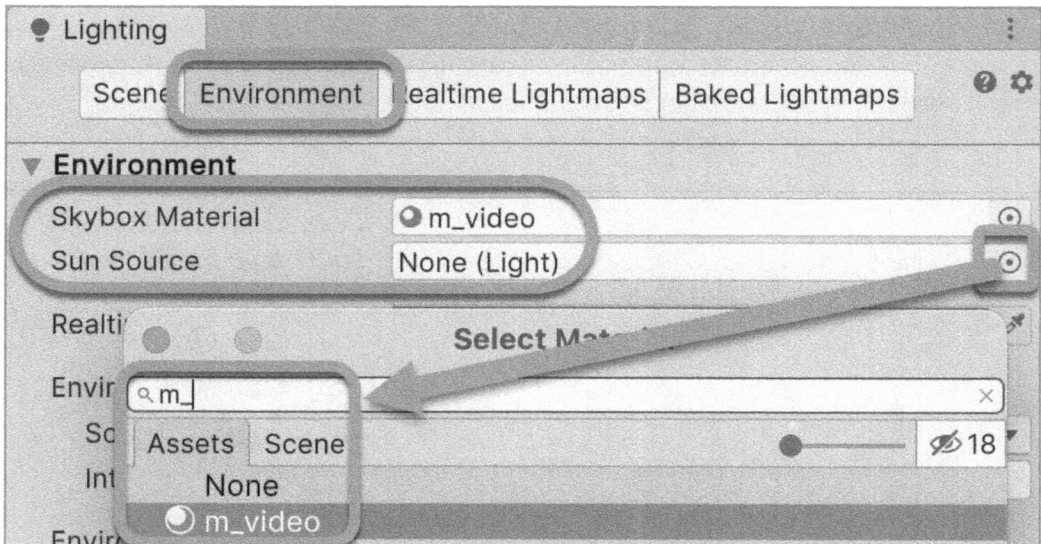

Figure 18.25: Setting the Environment Lighting settings for the Skybox material to our video material

19. Save the current scene as `scene1 - 360 video`.

20. In the **Build Settings** panel ensure your Quest device is still selected as the **Run Device**, and click the **Build And Run** button (you can simply choose the same APK app installer filename and overwrite the existing file).

21. Put on your Quest headset and explore! You should see and hear the 360-degree video playing all around you.

How it works...

This recipe uses the strategy of playing a 360-degree video as the panoramic Skybox material in a scene.

We use the **Lighting Environment** settings to define which material is used for our **Skybox**. By creating a new material called m_video, we were able to set this as the material to be used as the Skybox when the scene runs. For it to be used as a **Skybox** material, we had to set the **Shader** property of video_m to **Skybox | Panoramic**.

To play a video, we need a GameObject in the scene with a **Video Player** component, and we can make this play the video at runtime in a **Render Texture** asset file. The **Render Texture** needs to have the same dimensions as the resolution at which we wish to play the 360-degree video – usually, we'll set this to be the same dimensions as the video, for maximum quality. So, we set the dimensions of our **VideoRenderTexture** to 2560 x 1280.

To connect the playing video to our **Skybox** material, we made the m_video material link its **Spherical HDR** property to the **VideoRenderTexture** asset file, which is where the video plays at runtime.

Exploring and building the Desert WebXR demo project

One likely future of VR is through the web, using the **WebXR** standard. WebXR uses OpenGL as a way to publish AR and VR applications through websites. At present, WebXR projects can be published as websites, and then VR-enabled browsers running on headsets can be used to enter those interactive 3D scenes. The current free up-to-date WebXR browser is **Wolvic** from Igalia (see https://wolvic.com/).

> **Note:** The Wolvic project has superseded the previous Firefox Reality browser from Mozilla Corporation.

For several years, the Desert open-source WebXR demo project around which this recipe is based has been maintained for up-to-date versions of Unity by De-Panther – GitHub URL: https://github.com/De-Panther.

In this recipe, you'll first explore De-Panther's published version of the WebXR project, then you'll download the project source and build the website folder on your own computer using Unity and its WebGL build module.

Figure 18.26: The Desert WebXR sample project

How to do it...

To explore and build the Desert WebXR demo project, follow these steps:

1. On your VR headset, ensure you have the free Wolvic XR browser app installed.

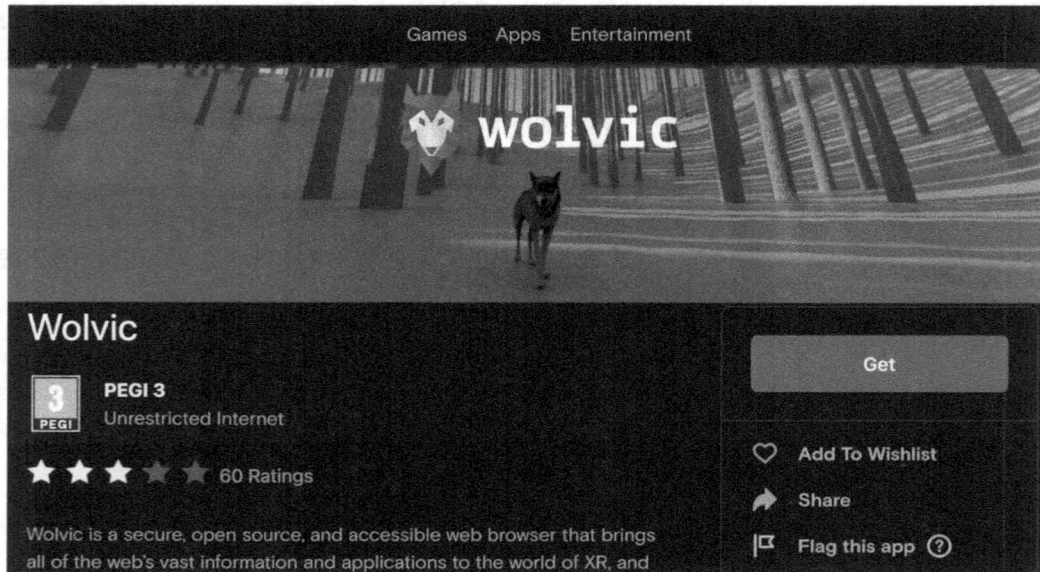

Figure 18.27: The Wolvic free WebXR browser app for Quest

2. In the Wolvic browser, search for `Github depanther`. When you find the WebXR exporter project, click the link for the live demo. Or just enter this URL directly: `https://de-panther.github.io/unity-webxr-export/Build/`. When you view the desert website, click the blue VR button at the bottom left to make your headset enter the 3D scene.

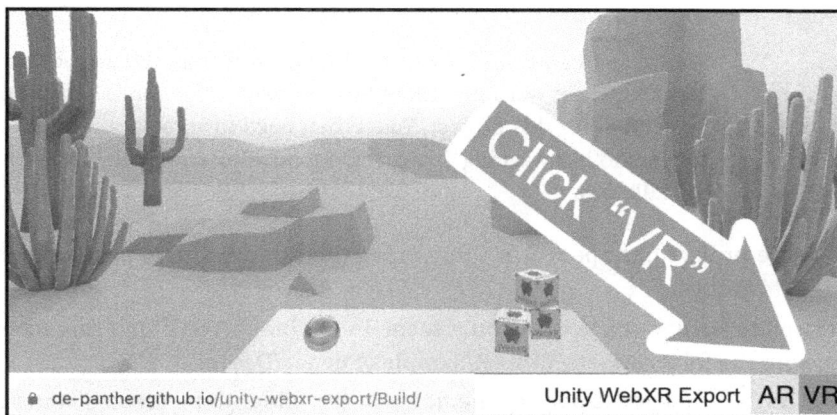

Figure 18.28: Entering the Desert WebXR project through the Wolvic browser

3. With your Quest, you can now walk around the scene and use your hand controllers to pick up the metal sphere and cubes.

4. Quit the application on your VR headset.

5. **WebXR** is built on top of **WebGL**, so you need to ensure the version of the Unity Editor you are using can build to a WebGL target platform. The WebGL module can be added to your Unity Editor version with **Unity Hub**. Open the **Unity Hub** application and click **Installs**. Then click the settings cog button for the version of the Unity Editor you are changing. Click **Add Modules**, then check the **WebGL Build Support** option and click the **Install** button.

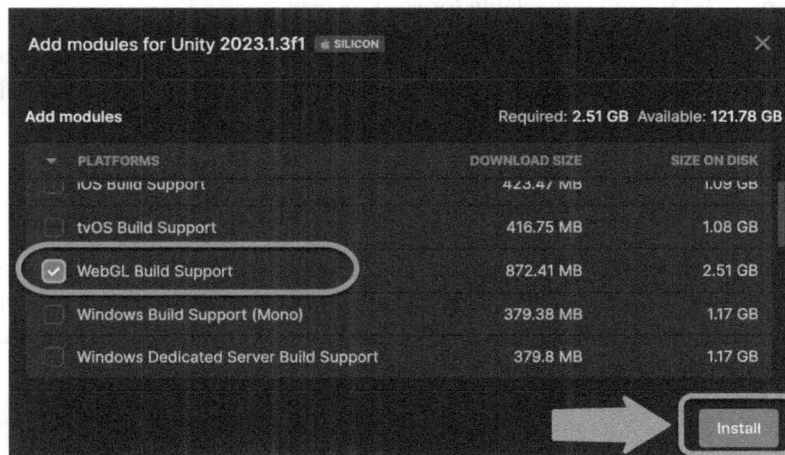

Figure 18.29: Adding WebGL Build Support for the selected Unity Editor with Unity Hub

6. On your computer, open a web browser at https://github.com/De-Panther/unity-webxr-export. Download this GitHub project by clicking **Code** and then **Download ZIP.**

7. Unzip the project on your computer and open the MainProject folder via the Unity Hub application.

> **Note:** If you a using a newer version of Unity than the project was created with, then you'll get warnings about **Missing Editor Version.** That's all fine, and agree to version for this project. You'll then need to agree again by clicking **Continue** when another warning appears about opening a project in **Non-Matching Editor Installation.**

8. Once the project has opened in Unity, open the **Build Settings** panel and switch the **Build Target** to **WebGL.**

9. Ensure the Desert scene is the only scene in the build. You'll find this scene in the **Project** panel, in the folder **Assets/WebXR/Samples/Scenes/Desert.**

10. Click **Build**, and choose docs as the name of the folder to be created containing the WebXR project. A folder will now be created for your build containing an index.html web page and the associated WebXR/WebGL and Unity assets.

11. If you have the facility to publish/host websites, you can now publish the contents of the created docs folder, and then visit the site on your Quest headset using the Wolvic browser. If not, then see the next recipe for a way to get GitHub to publish a website folder for you for free.

How it works...

When a WebXR project is viewed on a VR headset with a compatible browser, the code within the WebXR assets detects the type of VR device and communicates with the Unity web asset. It runs like any other VR project, allow full 3D immersion and interaction with controllers (and hands if hand tracking has been set up and is available for your headset).

The Desert project published by De-Panther is a full WebXR project, containing the interactable Desert scene. It also includes the WebXR Export and WebXR Interactions packages. These packages are available separately to this project as Unity packages from the OpenUPM website:

*   https://openupm.com/packages/com.de-panther.webxr/
*   https://openupm.com/packages/com.de-panther.webxr-interactions/

There are more and more WebXR games and apps now available. Some can be found on the start page when you open a new browser tab in Wolvic (https://wolvic.com/en/start). Another great selection can be found on the WebXR Showcase page at the ExposéXR website: https://www.exposexr.com/webxr-showcase.html. The VR climbing game **Boulderworld** is a great one!

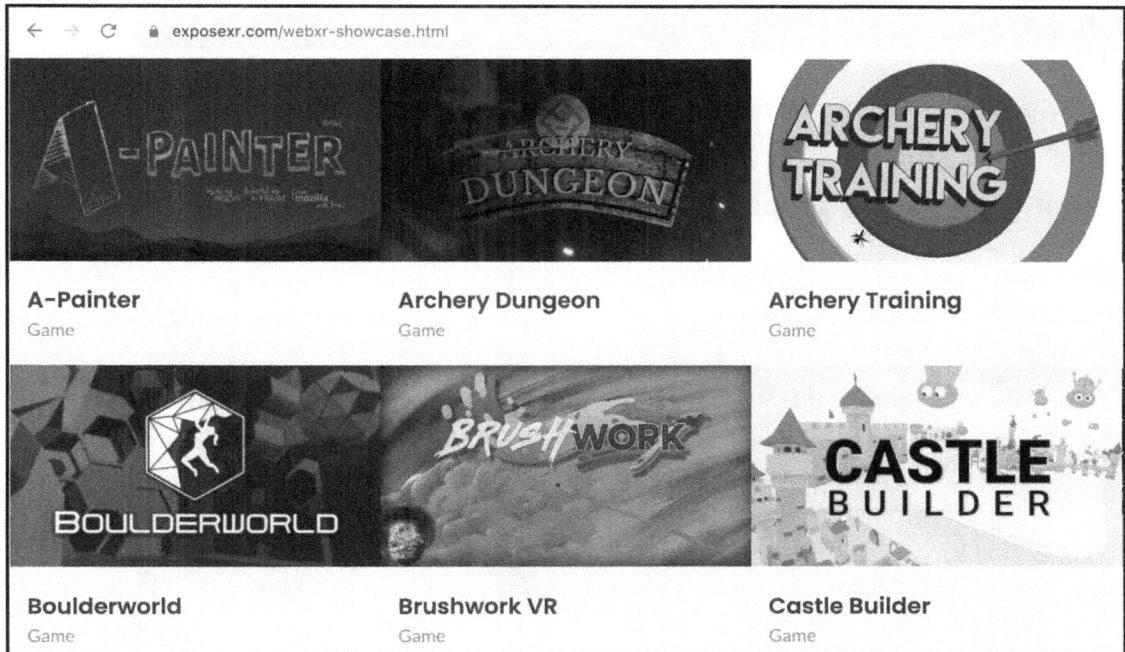

Figure 18.30: ExposéXR WebXR Showcase

There's more...

Here are some suggestions for troubleshooting this recipe.

Fixing pink (shader/material) problems

If there is a mismatch between the Unity version that the WebXR package was created for and the version of Unity you are using, there may be an issue with some of the materials in the project. The result is a shader error and objects that look **pink** in Unity. If an object is missing a material, then just choosing **Default-Diffuse** can be a quick fix for that object. You may be able to fix everything using the **Rendering Pipeline Converter**. First, choose **Window | Rendering | Rendering Pipeline Converter**. Then check all four boxes and click the **Initialize And Convert** button.

The OpenUPM community maintains more up-to-date XR resources

De-Panther's Desert project includes the **WebXR Export** and **WebXR Interactions** packages. These packages are available separately to the Desert project for use in your own projects. They can be added as Unity packages from the OpenUPM website:

- `https://openupm.com/packages/com.de-panther.webxr/`
- `https://openupm.com/packages/com.de-panther.webxr-interactions/`

Latest releases from De-Panther

De-Panther also creates a test release of the Desert project for different versions of the Unity Editor. For example, the 2023.1 releases can be found at `https://github.com/De-Panther/unity-webxr-export/tree/master/DebugProjects/Unity2023.1`.

Missing packages error

To save space, De-Panther has not copied the `webxr` and `webxr-interactions` package folders into the `MainProject` folder of their GitHub repository. So if you want to separate the `MainProject` folder from the other files downloaded, you need to do the following to prevent the missing package error when opening in Unity:

1. Delete the shortcut files `webxr` and `webxr-interactions` in the `MainProject/Packages` folder.
2. Move the folders `webxr` and `webxr-interactions` from the `Packages` folder of the parent of `MainProject` into the `MainProject/Packages` folder.

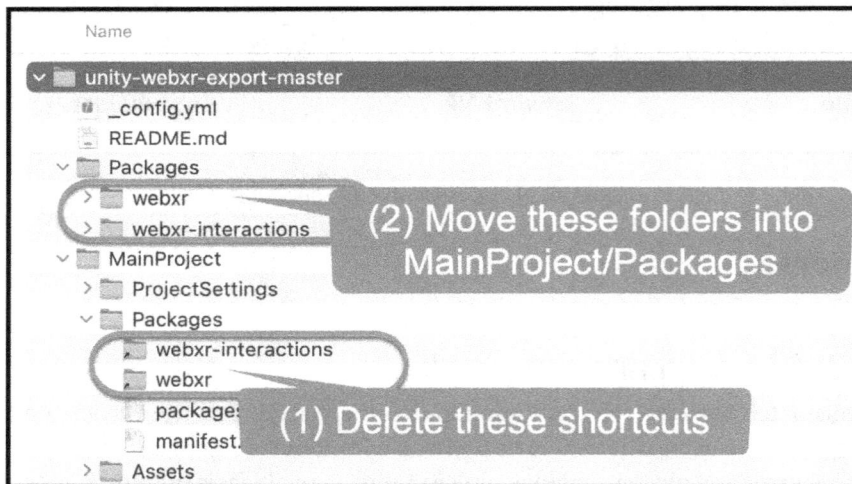

Figure 18.31: Removing shortcuts and copying in the actual package folders

There are also two further shortcuts that you need to replace with the actual folders. Both of these are in the **Main Project/Assets** folder.

The shortcuts to be deleted are:

* **Main Project | Assets | WebGLTemplates**
* **Main Project | Assets | WebXR | Samples | Desert**

The full folders to replace them with are:

* **Packages | webxr | Hidden~ | WebGLTemplates**
* **Packages | webxr-interactions | Samples~ | Desert**

Using GitHub Pages to publish your WebXR project for free

GitHub offers us a way to publish a free website for each project with a feature called **GitHub Pages**. The simplest way to do this is to have a docs folder in your GitHub project such as the one we created in the previous recipe.

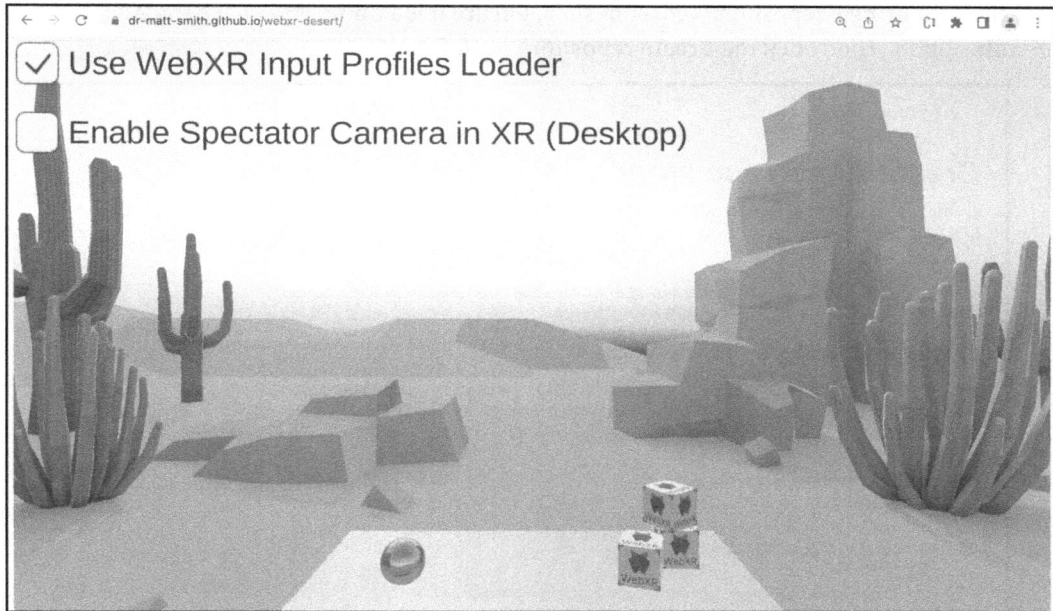

Figure 18.32: The published build via the free GitHub Pages service

Getting ready

This recipe makes use of the docs folder you built in the previous recipe. This is the content you'll be hosting for free with GitHub Pages in this recipe.

Since you'll be creating and making changes to a GitHub project, you will need to have a working GitHub account. If you don't already have such an account, you can create one for free here: https:// github.com/join.

> **Note:** GitHub projects are called "repositories," often abbreviated to "repos."

How to do it...

To use GitHub Pages to publish your WebXR project for free, perform the following steps:

1. Move your Unity WebXR to a new folder named `docs`.
2. In a web browser, log in to your GitHub account.
3. Create a new GitHub repository by clicking the green **New** button on the **Repositories** page. Name your new repository `webxr desert`, ensure it is **Public**, and check the **Add a README file** option. Then click the **Create repository** button.

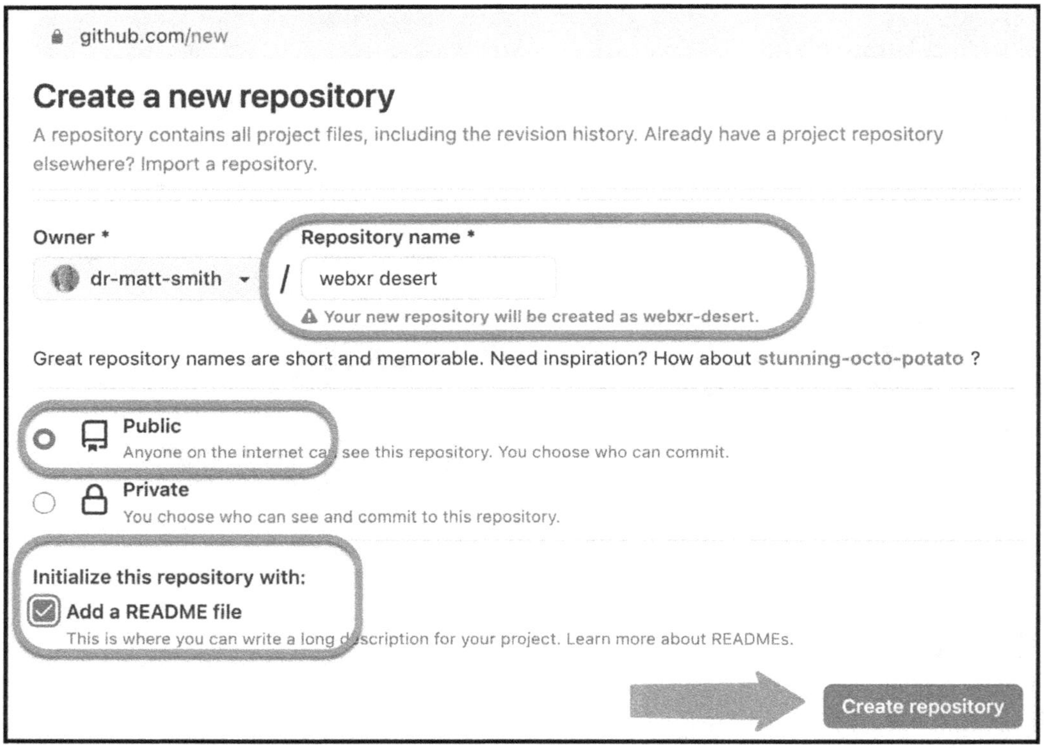

Figure 18.33: Creating a public GitHub repository to host our WebXR website

> **Note:** GitHub repository names cannot contain spaces, so any spaces will automatically be changed to minus signs. So our new repo is named `webxr-desert`.

4. Click on the **Add File** button and choose the **Upload Files** drop-down option.
5. Now drag the `docs` folder from your computer's file system into the repo drag target box.

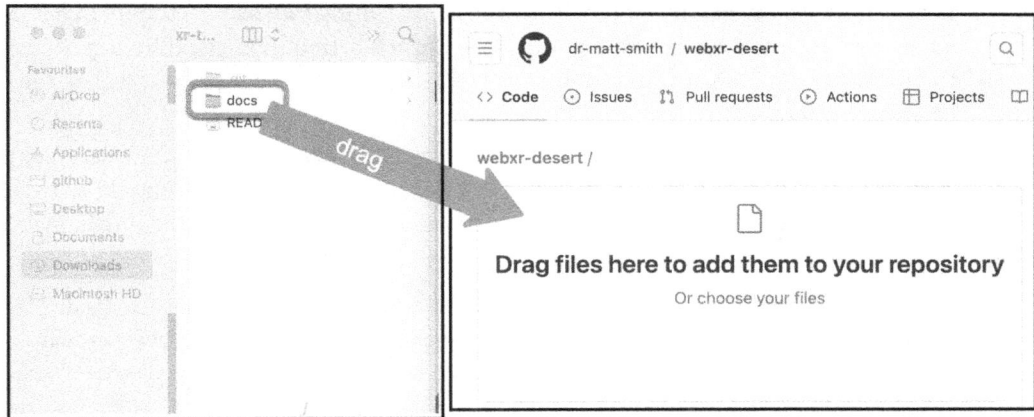

Figure 18.34: Uploading the docs folder into your GitHub repo

6. When you see that your folder contents have finished uploading, scroll to the bottom of the page and click the green **Commit** button to add this folder and its contents to your GitHub repo.

7. On the **Settings** tab, click **Pages**. This is where the website publishing settings are managed.

8. For **Branch**, choose **main**. Then, in the folder dropdown, choose /docs. Then click the **Save** button. This instructs GitHub to publish a website from the contents of the docs folder you just uploaded.

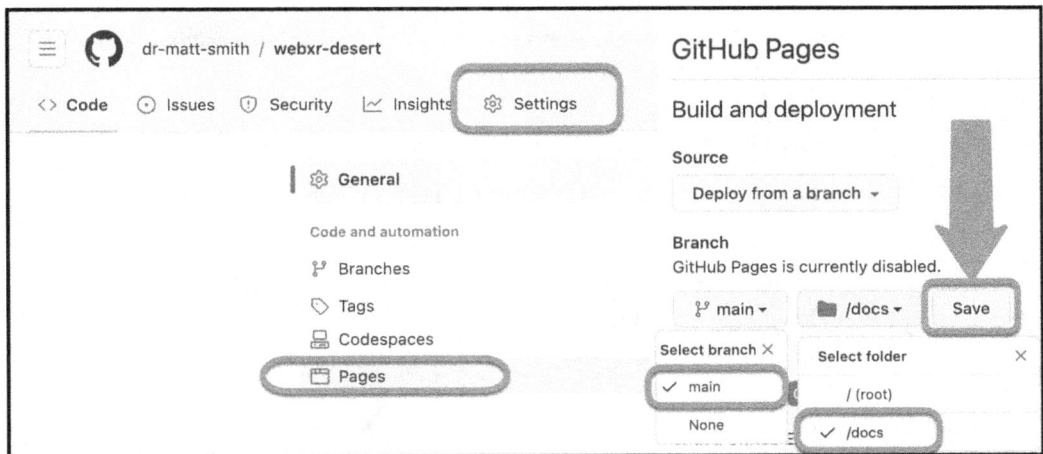

Figure 18.35: Uploading the docs folder into your GitHub repo

9. Open a browser to the GitHub Pages published docs folder for your project. The URL will be in the form https://<<github-user-name>>.github.io/<<github-project-name>>.

10. For example, since my GitHub username is **dr-matt-smith** and the GitHub repository name is **webxr-desert,** then the URL for my website is https://dr-matt-smith.github.io/webxr-desert.

11. Let's make it easy to get to this URL by adding it to the **README** text document that is displayed when viewing a GitHub repository. Click the **Code** tab, and then click the pencil icon to edit the **README** file being displayed. Add the URL to the README contents and click the **Commit Changes** button.

12. On your Quest, you can now visit this WebXR site in the Wolvic browser. An easy way to locate the site is to search for GitHub and your GitHub username. Then you should see your newly updated repo listed first in your list of repositories. You can then click the link in the README.

> **Note:** Once published the link for the GitHub Pages site will also be displayed in the Pages information on the **Settings** tab (although that won't be viewable to other GitHub viewers of the repo – only you as the owner of the project).

13. Once you have located the page, having clicked the **VR** button at the bottom left of the page, enter your published WebXR scene.

How it works...

Often, we want to provide information and documentation about our programming projects. For this purpose, GitHub created the Pages website publishing feature for repos. In this recipe, we've made use of this feature by building our WebXR project into a folder named docs, which is one of the folders that GitHub allows for the website to accompany a GitHub repo.

Since we made the repo public, it means that anyone using a WebXR browser such as Wolvic on a VR headset can now visit and enter your published Desert project!

Creating a simple WebXR scene from scratch

Rather than just exploring the Desert scene from De-Panther's WebXR project, let's create and publish a new scene from scratch, making use of the WebXR plugins already added as part of that project.

In this recipe we'll create a new scene containing a Cube, acting as a table, and two orange Spheres that can be picked up and interacted with.

Getting ready

This recipe builds on the *Exploring and building the Desert WebXR demo project* recipe. So, make a copy of that project and work on the copy.

How to do it...

To create a simple WebXR scene from scratch, perform the following steps:

1. In Unity create a new Basic (Built-in) scene – this should contain a **Main Camera** and **Directional Light**.

2. Delete the **Main Camera** GameObject from the scene (since we'll be adding a WebXR camera set next).

3. From the **Project** panel, drag the **WebXRCameraSet** asset file into the scene from the folder **Assets/WebXR/Samples/Desert/Prefabs**. This creates the core GameObject for our WebXR experience. You'll see this GameObject has three children – two hands (**handL** and **handR**) and a **Cameras** GameObject. Ensure the **Position** of this GameObject is (0, 0, 0).

4. Add a 3D **Cube** to the scene with **Position** (0, 0, 0) and **Scale** (1, 1, 1).

5. Add a 3D **Sphere** to the scene that's positioned above the Cube, with **Position** (0, 0.7, 0) and **Scale** (0.3, 0.3, 0.3). Tag this GameObject as **Interactable** (this tag should already be available in the list of tags).

6. With the Sphere selected in the **Hierarchy**, in the **Inspector** use the **Add Component** button to add a **Rigidbody** component to this GameObject.

7. Now use the **Add Component** button to add a **Mouse Drag Object (Script)** component to this **Sphere** GameObject.

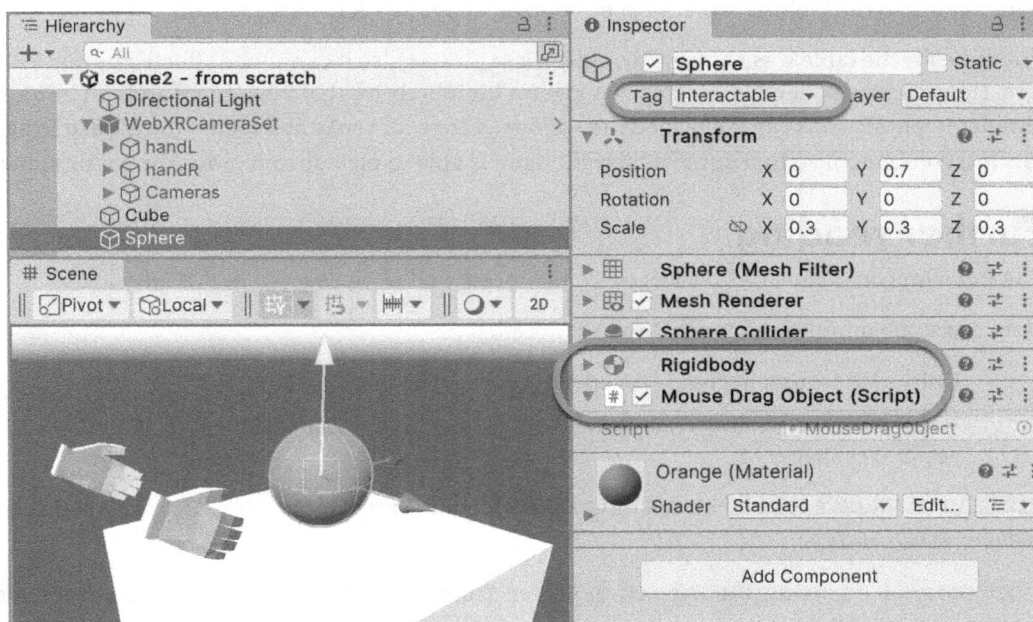

Figure 18.36: Making the Sphere GameObject an interactable Rigidbody GameObject with a Mouse Drag script component

8. Make the **Sphere** orange by choosing the **Orange** material for its **Mesh Renderer Materials** property in the **Inspector**.

9. Duplicate the Sphere, setting the second Sphere with **Position** (0.5, 0.7, 0).

10. Save the current scene as scene2 - from scratch.

11. Now, open the **Build Settings** dialog (**File | Build Settings**) and add the current scene to the build. Also, remove any other scenes for this build.

12. Click the **Build** button and choose docs as the name of the folder to be created containing the WebXR project. A folder will now be created for your build containing an index.html web page and the associated WebXR/WebGL and Unity assets.

13. Finally, publish it on the web (such as using the GitHub Pages docs folder technique from the previous recipe).

14. When you visit the page with a VR headset, you should have two virtual hands, two orange Spheres, and a Cube acting as a table. You should be able to pick up and drop/throw the spheres.

How it works...

As we saw when creating a scene from scratch, the core of the WebXR project is the GameObject combining cameras and hand models – this GameObject was added to the scene from the **WebXRCameraSet** Prefab in the Assets/WebXR/Samples/Desert/Prefabs folder. This provided us with a left and right hand, and also the WebXR cameras to be controlled by the VR headset sensors.

The cube for the table was not tagged or given a RigidBody, so it cannot be affected by our VR controllers. However, the sphere we created was given a **Rigidbody** (so that it has mass and can be affected by simulated physics forces). We tagged the sphere as Interactable and added the **Mouse Drag Object (Script)** component so that our virtual right hand is able to pick up and move/throw the sphere.

Further reading

The following are some useful sources for the topics that were covered in this chapter.

Here are some online forums for Unity VR developers:

- Unity's VR forum: https://forum.unity.com/forums/vr.160/
- Meta Unity forum: https://communityforums.atmeta.com/t5/Unity-VR-Development/bd-p/dev-unity/page/41

Rajat Kumar Gupta's tutorial on creating a 360 video in Unity: https://blog.logrocket.com/make-360-vr-video-unity/

Video tutorial on using the WebXR exporter from De-Panether: https://www.youtube.com/watch?v=e7hVROj9qm4

Video tutorial on XR Interaction Toolkit hand tracking: https://www.youtube.com/watch?v=mJ3fygb9Aw0

Unity has a guide to playing 360 videos as skyboxes: https://learn.unity.com/tutorial/play-360-video-with-a-skybox-in-unity

Although we don't have the space to explore it in this chapter, Oculus has published an integration package for Unity, which provides several Prefabs and demo scenes especially for Quest VR development.

- Get the package: https://developer.oculus.com/documentation/unity/unity-package-manager/

On the Asset Store, there is a free VR Escape Room! It's a couple of years old, but looks like great fun if it works on your version of Unity: `https://assetstore.unity.com/packages/templates/tutorials/vr-beginner-the-escape-room-163264`

The `VR Toolkit` (`VRTK`) looks like it could be a great resource for Unity VR developers:

- The VRTK Tiklia package importer on the Asset Store: `https://assetstore.unity.com/packages/tools/utilities/vrtk-v4-tilia-package-importer-214936`
- The VRTK website: `https://www.vrtk.io/`

The YouTube channel for Fist Full of Shrimp offers many great video tutorials on Unity VR development: `https://www.youtube.com/@FistFullofShrimp/videos`

19

Advanced Topics — Gizmos, Automated Testing, and More

This chapter will cover three sets of advanced recipes:

- Gizmos
- Automated testing
- An introduction to Unity Python

Gizmos facilitate Unity Editor customization. Gizmos are visual aids for game designers that are provided in the **Scene** panel. They can be useful as setup aids (to help us know what we are doing) or for debugging (understanding why objects aren't behaving as expected).

Gizmos are not drawn through **Editor Scripts**, but as part of the MonoBehaviour class, so they only work for GameObjects in the current scene. Gizmo drawing is usually performed with two methods:

- OnDrawGizmos(): This is executed at every frame or Editor window repaint, for every Game-Object in the **Hierarchy** panel.
- OnDrawGizmosSelect(): This is executed at every frame, for just the GameObject(s) that is currently selected in the **Hierarchy** panel.

Gizmo graphical drawing makes it simple to draw lines, cubes, and spheres. More complex shapes can also be drawn with meshes, and you can also display 2D image icons (located in the **Project** folder: Assets | Gizmos).

Several recipes in this chapter will illustrate how gizmos can be useful. Often, new GameObjects created from Editor extensions will have helpful gizmos associated with them.

Automated testing offers a way to formalize and structure how we test and develop our code. For a very simple computer program, we can write code, run it, enter a variety of valid and invalid data, and see whether the program behaves as we expect it to.

This is known as a code-then-test approach. However, this approach has several significant weaknesses:

- Each time we change the code, as well as run new tests relating to the code we are improving, we have to run all the old tests to ensure that no unexpected modified behavior has been introduced (in other words, our new code has not **broken** another part of our program).
- Running tests manually is time-consuming.
- We are relying on a human to rerun the test each time. However, this test may be run using different data, some data may be omitted, or different team members may take a different approach to run tests.

Therefore, even for simple programs (and most are not simple), some kind of fast, automated testing system makes a lot of sense.

There is an approach to software development called **test-driven development** (TDD), whereby code is only written until all tests pass. So, if we want to add or improve the behavior of our game program, we must specify what we want in terms of tests, and then the programmers write code to pass the tests. This avoids a situation whereby programmers write code and features that are not needed, spend time over-optimizing things that would have been fine, and so on. This means that the game development team directs its work toward agreed goals understood by all since they have been specified as tests.

The following diagram illustrates a basic TDD process in that we only write code until all tests pass. Then, it's time to write more tests:

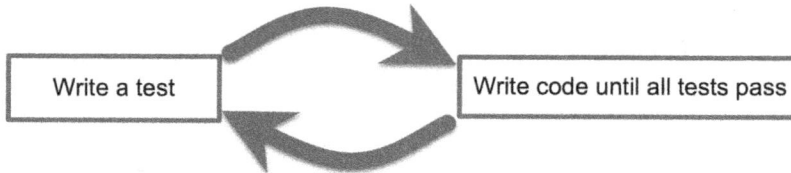

Figure 19.1: Create a test then write code to pass the test

Another way that TDD is often summarized is as red-green-refactor:

- **Red:** We write a new test for which code is needed, so initially, our test fails (in other words, we write a test for the new feature/improved behavior we wish to add to our system).
- **Green:** We write code that passes the new test (and all the existing ones).
- **Refactor:** We (may) choose to improve the code (and ensure that the improved code passes all the tests).

The two main kinds of automated software tests are as follows:

- Unit tests
- Integration tests

A **unit test** tests a "unit" of code, which can be a single method, but may include some other computer work being executed between the method being tested and the end result(s) being checked.

> *"A unit test is a piece of code that invokes a unit of work and checks one specific end result of that unit of work. If the assumptions on the end result turn out to be wrong, the unit test has failed."*
>
> —*Roy Oshergrove (p. 5, The Art of Unit Testing (Second Edition)*

Unit tests should be as follows:

- Automated (runnable at the "push of a button")
- Fast
- Easy to implement
- Easy to read
- Executed in isolation (tests should be independent of one another)
- Assessed as either having passed or failed
- Relevant tomorrow
- Consistent (the same results each time!)
- Able to easily pinpoint what was at fault for each test that fails

Most computer languages have an xUnit unit testing system available; for example:

- **C#**: NUnit
- **Java**: JUnit
- **PHP**: PHPUnit

Unity offers an easy way to write and execute NUnit tests in its Editor (and at the command line).

Typically, each unit test will be written in three sections, like so:

- **Arrange**: Set any initial values needed (sometimes, we are just giving a value to a variable to improve code readability)
- **Act**: Invoke some code (and, if appropriate, store the results)
- **Assert**: Make assertions for what should be true about the code that's been invoked (and any stored results)

Observe that the naming of a unit test method (by convention) is quite verbose – it is made up of lots of words that describe what it does. For example, you might have a unit test method named `TestHealthNotGoAboveOne()`. The idea is that if a test fails, the name of the test should give a programmer a very good idea of what behavior is being tested and, therefore, how to quickly establish whether the test is correct. If so, it tells you where to look in your program code for what was being tested. Another part of the convention of naming unit tests is that numerals are not used, just words; so, we write "one," "two," and so on in the name of the test method.

An **integration test** (PlayMode tests in Unity) involves checking the behavior of interacting software components, for example, ones that are real time, or a real filesystem, or that communicate with the web or other applications running on the computer. Integration tests are usually not as fast as unit tests, and may not produce consistent results (since the components may interact in different ways at different times).

Both unit and integration tests are important, but they are different and should be treated differently.

Unity offers **PlayMode** testing, allowing integration testing as Unity scenes execute with testing code in them.

Unity Python is a package published by Unity Technologies. It allows Python code to be executed as part of a Unity project. The final recipe in this chapter will help you install and begin using this scripting tool in a Unity project.

In this chapter, we will cover the following recipes:

- Using a gizmo to show the currently selected object in the Scene panel
- Creating an Editor snap-to-grid drawn by a gizmo
- Generating and running a default test script class
- Making a simple unit test
- Parameterizing tests with a DataProvider
- Unit testing a simple health script class
- Creating and executing a unit test in PlayMode
- PlayMode testing a door animation
- PlayMode and unit testing a player health bar with events, logging, and exceptions
- Reporting Code Coverage testing
- Running simple Python scripts inside Unity

Using a gizmo to show the currently selected object in the Scene panel

Gizmos are visual aids that are provided to game designers in the **Scene** panel. In this recipe, we'll highlight the GameObject that is currently selected in the **Hierarchy** panel in the **Scene** panel.

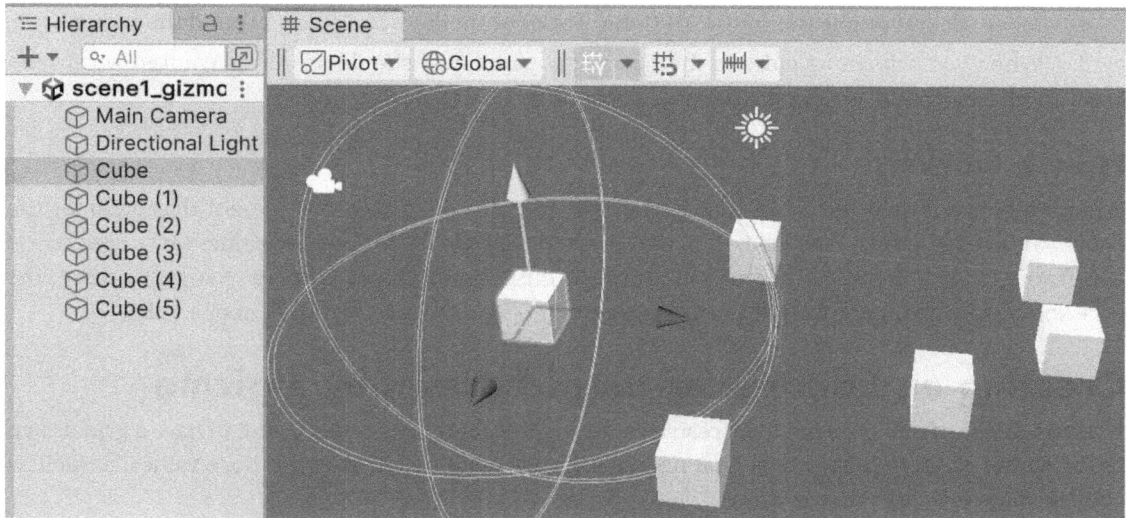

Figure 19.2: Wireframe spheres around the selected GameObject

How to do it...

To create a gizmo to show the selected object in the **Scene** panel, follow these steps:

1. Create a new **Unity 3D** project.
2. Create a **3D Cube** by going to **Create | 3D Object | Cube**.
3. Create a C# script class called GizmoHighlightSelected and add an instance object as a component to the 3D **Cube**:

```
using UnityEngine;

public class GizmoHighlightSelected : MonoBehaviour {
    public float radius = 5.0f;

    void OnDrawGizmosSelected() {
        Gizmos.color = Color.red;
        Gizmos.DrawWireSphere(transform.position, radius);

        Gizmos.color = Color.yellow;
        Gizmos.DrawWireSphere(transform.position, radius - 0.1f);

        Gizmos.color = Color.green;
        Gizmos.DrawWireSphere(transform.position, radius - 0.2f);
    }
}
```

4. Make several duplicates of the **3D Cube**, distributing them randomly around the **scene**.
5. When you select one **cube** in the **Hierarchy** panel, you should see three colored wireframe spheres drawn around the selected GameObject in the **Scene** panel.

How it works...

When an object is selected in a **Scene** panel, if it contains a scripted component that includes the `OnDrawGizmosSelected()` method, then that method is invoked. Our method draws three concentric wireframe spheres in three different colors around the selected object. You can change the size of the wire spheres by changing the public radius property of the scripted component of a cube.

Creating an Editor snap-to-grid drawn by a gizmo

If the positions of objects need to be restricted to specific increments, it is useful to have a grid drawn in the **Scene** panel to help ensure that new objects are positioned based on those values, as well as code to snap objects to that grid.

In this recipe, we'll use gizmos to draw a grid with a customizable grid size, color, number of lines, and line length. The result of following this recipe will look as follows:

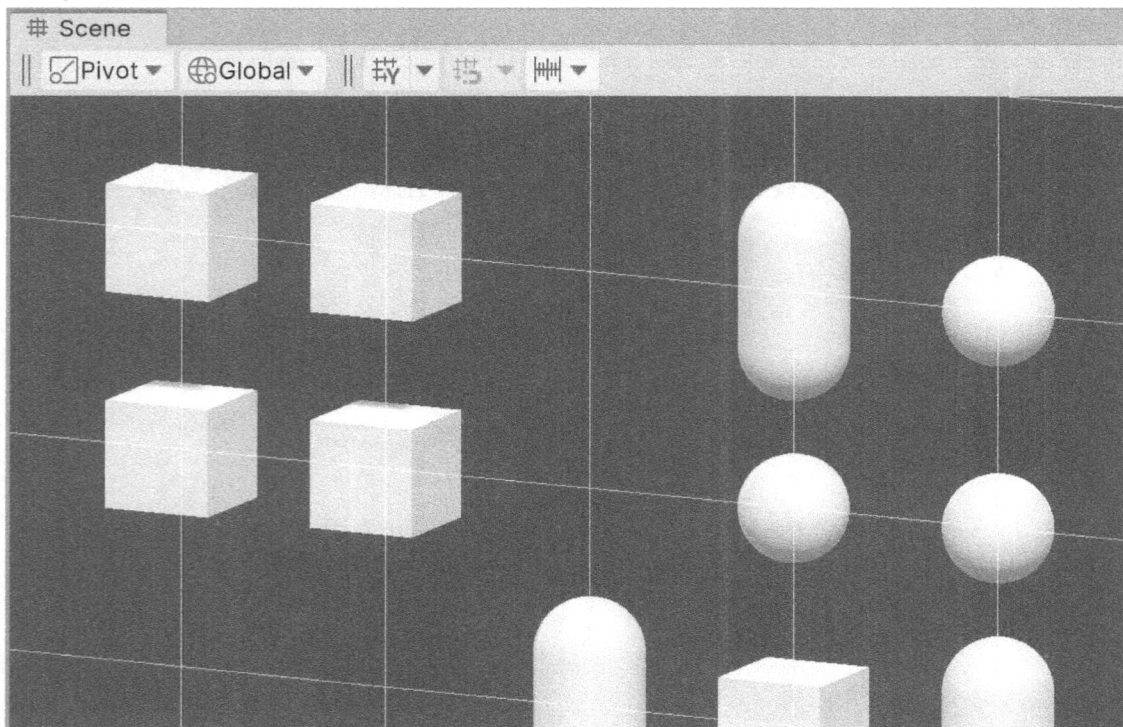

Figure 19.3: Example of a visible grid to which objects have been snapped

How to do it...

To create an Editor snap-to-grid drawn by a gizmo, follow these steps:

1. Create a new **Unity 3D** project.

2. In the **Scene** panel, turn off the **Skybox** view (or simply toggle off all the visual settings) so that you have a plain background for your grid:

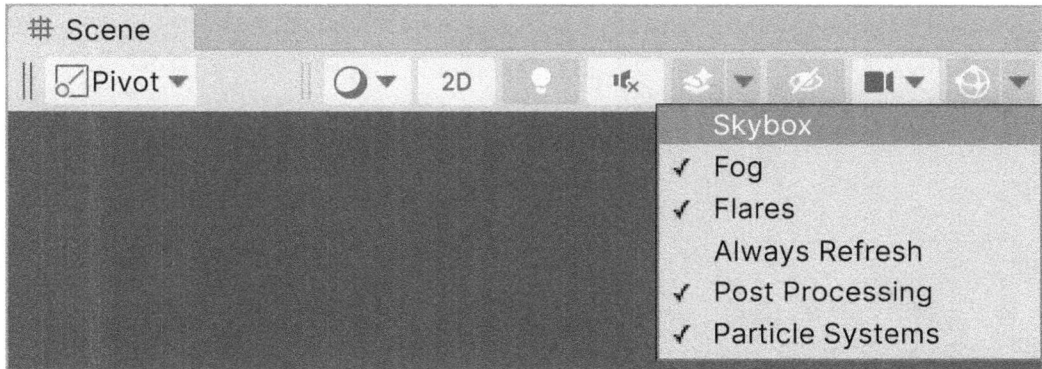

Figure 19.4: Turning off the Skybox view in the Scene panel

3. Updating the display and the child objects will be performed by a script class called `GridGizmo`. Create a new C# script class called `GridGizmo` that contains the following code:

```csharp
using UnityEngine;

public class GridGizmo : MonoBehaviour {
    public int grid = 2;

    public void SetGrid(int grid) {
        this.grid = grid;
        SnapAllChildren();
    }

    public Color gridColor = Color.red;
    public int numLines = 6;
    public int lineLength = 50;

    private void SnapAllChildren() {
        foreach (Transform child in transform){
        SnapPositionToGrid(child);
    }
    }
```

```
void OnDrawGizmos() {
    Gizmos.color = gridColor;

    int min = -lineLength;
    int max = lineLength;

    int n = -1 * RoundForGrid(numLines / 2);
    for (int i = 0; i < numLines; i++) {
        Vector3 start = new Vector3(min, n, 0);
        Vector3 end = new Vector3(max, n, 0);
        Gizmos.DrawLine(start, end);

        start = new Vector3(n, min, 0);
        end = new Vector3(n, max, 0);
        Gizmos.DrawLine(start, end);

        n += grid;
    }
}

public int RoundForGrid(int n) {
    return (n/ grid) * grid;
}

public int RoundForGrid(float n) {
    int posInt = (int) (n / grid);
    return posInt * grid;
}

public void SnapPositionToGrid(Transform transform) {
    transform.position = new Vector3 (
        RoundForGrid(transform.position.x),
        RoundForGrid(transform.position.y),
        RoundForGrid(transform.position.z)
    );
}
}
```

4. We can create an **Editor Script** to add a new menu item to the **GameObject** menu. Create a folder named `Editor` and, in that folder, create a new C# script class called `EditorGridGizmoMenuItem` that contains the following code:

```
using UnityEngine;
using UnityEditor;

public class EditorGridGizmoMenuItem : Editor {
    const string GRID_GAME_OBJECT_NAME = "___snap-to-grid___";

    [MenuItem("GameObject/Create New Snapgrid", false, 10000)]
    static void CreateCustomEmptyGameObject(MenuCommand menuCommand) {
        GameObject gameObject = new GameObject(GRID_GAME_OBJECT_NAME);

        gameObject.transform.parent = null;
        gameObject.transform.position = Vector3.zero;
        gameObject.AddComponent<GridGizmo>();
    }
}
```

5. Now, let's add another **Editor Script** for a custom **Inspector** display (and updater) for the `GridGizmo` components. Also, in your `Editor` folder, create a new C# script class called `EditorGridGizmo` that contains the following code:

```
using UnityEngine;
using UnityEditor;

[CustomEditor(typeof(GridGizmo))]
public class EditorGridGizmo : Editor {
    private GridGizmo gridGizmoObject;
    private int grid;
    private Color gridColor;
    private int numLines;
    private int lineLength;

    private string[] gridSizes = {
        "1", "2", "3", "4", "5"
    };

    void OnEnable() {
        gridGizmoObject = (GridGizmo)target;
```

```
            grid = serializedObject.FindProperty("grid").intValue;
            gridColor = serializedObject.FindProperty("gridColor").
colorValue;
            numLines = serializedObject.FindProperty("numLines").intValue;
            lineLength = serializedObject.FindProperty("lineLength").
intValue;
        }

    public override void OnInspectorGUI() {
            serializedObject.Update ();

            int gridIndex = grid - 1;
            gridIndex = EditorGUILayout.Popup("Grid size:", gridIndex,
gridSizes);
            gridColor = EditorGUILayout.ColorField("Color:", gridColor);
            numLines = EditorGUILayout.IntField("Number of grid lines",
numLines);
            lineLength = EditorGUILayout.IntField("Length of grid lines",
lineLength);

            grid = gridIndex + 1;
            gridGizmoObject.SetGrid(grid);
            gridGizmoObject.gridColor = gridColor;
            gridGizmoObject.numLines = numLines;
            gridGizmoObject.lineLength = lineLength;
            serializedObject.ApplyModifiedProperties ();
            SceneView.RepaintAll();
        }
    }
```

6. Add a new `GizmoGrid` GameObject to the scene by navigating to **GameObject | Create New Snap-grid**. You should see a new **GameObject** named ___snap-to-grid___ added to the **Hierarchy**.

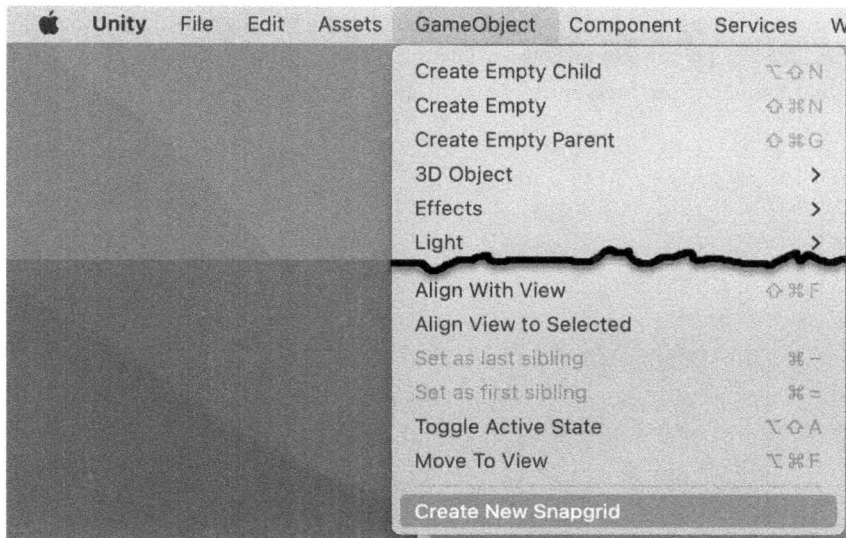

Figure 19.5: Our new menu item at the bottom of the GameObject menu

7. Select the ___snap-to-grid___ GameObject and set its position to (0, 0, 0). Now, modify some of its properties in the **Inspector** panel. You can change the grid's size, the color of the grid's lines, the number of lines, and their length.

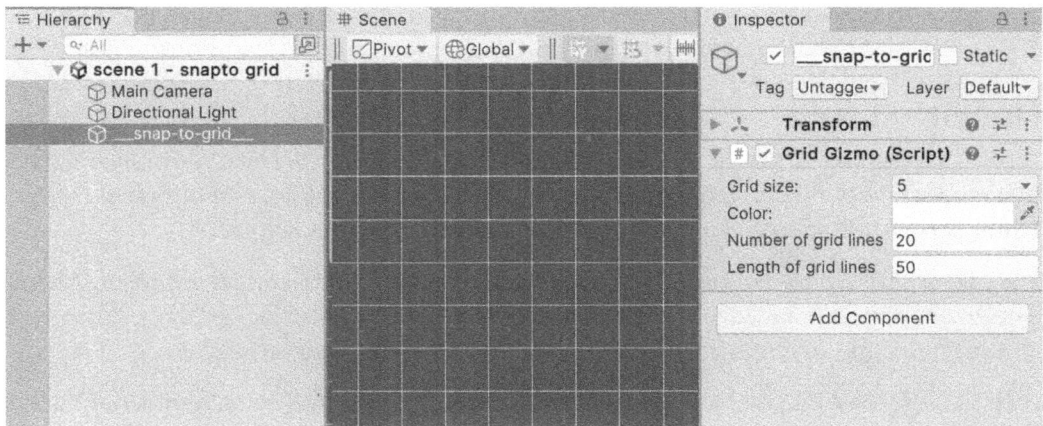

Figure 19.6: Changing the properties of the snap grid

8. Create a 3D **Cube** by choosing **GameObject | 3D Object | Cube**. Set its position to (0, 0, 0).

9. Snapping will work for all children of ___snap-to-grid___ in the **Hierarchy** panel. So, make this new Cube a child of ___snap-to-grid___ by dragging it onto ___snap-to-grid___.

10. We now need a small script class so that each time the GameObject is moved (in **Editor** mode), it asks for its position to be snapped by the parent scripted component. Create a C# script class called `SnapToGridGizmo` containing the following:

```
using UnityEngine;


[ExecuteInEditMode]
public class SnapToGridGizmo : MonoBehaviour {
    public void Update()
    {
#if UNITY_EDITOR
        transform.parent.GetComponent<GridGizmo>().
SnapPositionToGrid(transform);
    #endif
    }
}
```

11. Add an instance of the script class `SnapToGizmoGrid` object as a component to the 3D **Cube**.

12. Make some duplicates of the **3D Cube**, using the move tool in the **Scene** to distribute them randomly around the Scene – you'll find that they snap to the grid.

13. Select ___snap-to-grid___ again and modify some of its properties in the **Inspector** panel. You'll see that the changes are instantly visible in the **scene** and that all the child objects that have a scripted component of `SnapToGizmoGrid` are snapped to any new grid size changes.

How it works...

Scripts that we place in a folder named `Editor` are known as **Editor Scripts**, and in such scripts, we can customize and extend the features and look and feel of the Unity Editor. In this recipe, we've created **Editor Scripts** to display and limit the movement of scene objects in a grid.

The `EditorGridGizmoMenuItem` script class adds a new item to the **GameObject** menu. When selected, a new GameObject is added to the **Hierarchy** panel named ___snap-to-grid___, positioned at (0, 0, 0), and containing an instance object component of the `GridGizmo` **script class**.

`GridGizmo` draws a 2D grid based on public properties for grid size, color, number of lines, and line length. Regarding the `SetGrid(...)` method, as well as updating the integer grid size variable grid, it also invokes the `SnapAllChildren()` method so that each time the grid size is changed, all child GameObjects are snapped into the new grid positions.

The `SnapToGridGizmo` script class includes an **Editor** attribute called `[ExecuteInEditMode]` so that it will receive `Update()` messages when its properties are changed at **design time** in the **Editor**. Each time `Update()` is invoked, it calls the `SnapPositionToGrid(...)` method in its parent `GridGizmo` instance object so that its position is snapped based on the current settings of the grid. To ensure this logic and code are not compiled into any final **build** of the game, the contents of `Update()` are wrapped in an `#if UNITY_EDITOR` compiler test. Such content is removed before a build is compiled for the final game.

The `EditorGridGizmo` **script class** is a custom **Editor Inspector** component. This allows us to control which properties are displayed in the **Inspector** window and how they are displayed, and it allows actions to be performed when any values are changed. So, for example, after changes have been saved, the `SceneView.RepaintAll()` statement ensures that the grid is redisplayed since it results in an `OnDrawGizmos()` message being sent.

There's more...

The preceding implementation works fine when the only objects we want to tie to the grid are children of our ___snap-to-grid___ GameObject. However, if we don't want to require affected objects to be these children, then we can use the singleton pattern to allow a GameObject anywhere in the **Hierarchy** panel to get a reference to the **GridGizmo** component. Do the following to adapt this recipe to use this approach:

1. Update the contents of the `GridGizmo.cs` #C script class so that it matches the following:

```csharp
using System.Collections;
using System.Collections.Generic;
using UnityEngine;

public class GridGizmo : MonoBehaviour
{
  private static GridGizmo _instance = null;

  public static void SetInstance(GridGizmo instance) {
    _instance = instance;
  }

  public static GridGizmo GetInstance() {
    if(_instance == null){
      throw new System.Exception("error - no GameObject has GridGizmo component to snap to ...");
    }

    return _instance;
  }

  public int grid = 2;

  public void SetGrid(int grid) {
    this.grid = grid;
    SnapAllChildren();
```

```
    }
    ... the rest of the script is unchanged ...
```

2. Update the contents of the `SnapToGizmoGrid.cs` C# script class so that it matches the following:

```csharp
using System.Collections;
using System.Collections.Generic;
using UnityEngine;

[ExecuteInEditMode]
public class SnapToGridGizmo : MonoBehaviour
{
    /**
     * we've moved!
     * snap position to its grid
     */
    public void Update()
    {
#if UNITY_EDITOR
        GridGizmo.GetInstance().SnapPositionToGrid(transform);
#endif
    }

}
```

3. Update the contents of the `EditorGridGizmoMenuItem.cs` C# script class to match the following:

```csharp
using UnityEngine;
using UnityEditor;
using System.Collections;

public class EditorGridGizmoMenuItem : Editor
{
    const string GRID_GAME_OBJECT_NAME = "___snap-to-grid___";

    /**
     * menu item to create a GridSnapper
     */
    [MenuItem("GameObject/Create New Snapgrid", false, 10000)]
    static void CreateCustomEmptyGameObject(MenuCommand menuCommand)
    {
        GameObject gameObject = new GameObject(GRID_GAME_OBJECT_NAME);
```

```
        // ensure not a child of any other object
        gameObject.transform.parent = null;

        // zero position
        gameObject.transform.position = Vector3.zero;

        // add Scripted component
        GridGizmo newInstance = gameObject.AddComponent<GridGizmo>();
        GridGizmo.SetInstance(newInstance);
    }
}
```

When the menu item is chosen, the method in the `EditorGridGizmoMenuItem` C# script class creates the game grid GameObject, adds a `GridGizmo` scripted component, and also uses the new `SetInstance(...)` method of the `GridGizmo` script class to store the reference to the `GridGizmo` scripted component in its static variable, called `_instance`. This means that when we add GameObjects such as cubes or cylinders anywhere in the **Hierarchy** panel, we can add to them a `SnapToGizmoGrid` scripted component that can call the `GridGizmo.GetInstance()` public state method. To summarize what we have done, by using the singleton pattern, we have allowed any GameObject, anywhere in the **Hierarchy** panel, to access the `GridGizmo` component.

Generating and running a default test script class

Unity can create a default C# test script for you, thereby enabling you to quickly start creating and executing tests on your project. In this recipe, we will add the Unity Test Framework to a project and use it to automatically generate a default test script for us. This will be the basis for several of the following recipes:

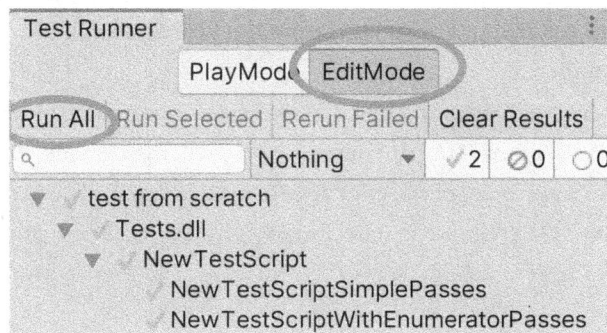

Figure 19.7: Tests passing (indicated by green ticks)

How to do it...

To generate a default test script class, follow these steps:

1. Create a new **3D Unity** project.
2. Display the **Test Runner** panel by going to **Window | General | Test Runner.**

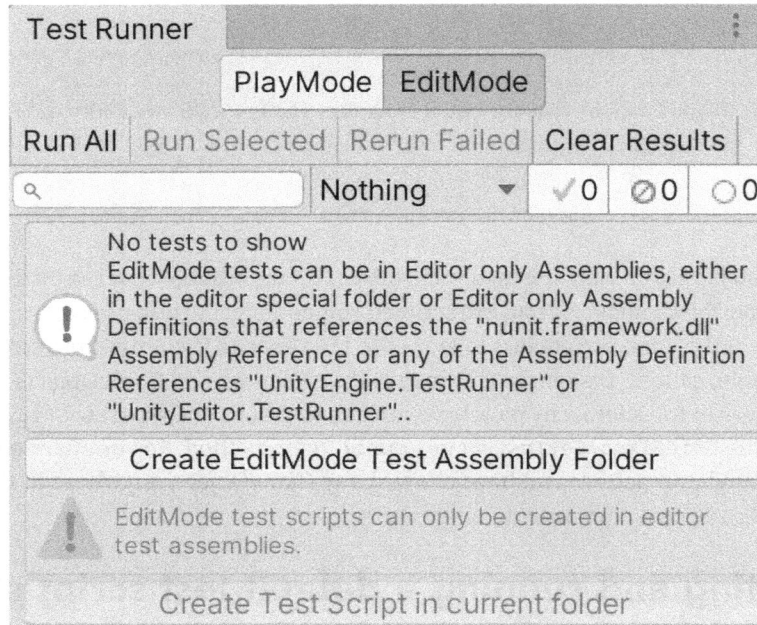

Figure 19.8: The Test Runner panel

3. Ensure that the **EditMode** button is selected in the **Test Runner** panel.
4. In the **Test Runner** panel, click the **Create EditMode Test Assembly Folder** button. You'll now see a folder called Tests that's been created in the **Project** panel.
5. Select the Tests folder.
6. In the **Test Runner** panel, click the **Create Test Script in current folder** button.
7. You should now have a new C# script named NewTestScript inside your **Tests folder.**
8. To run the tests in your script class, click the **Run All** button in the **Test Runner** panel.
9. You should now see all green ticks (checkmarks) in the panel, as shown in the figure at the beginning of this recipe.

How it works...

Unity checks that you have a folder named Editor selected in the **Project** panel, and then creates a NewTestScript C# script class for you containing the following:

```
using System.Collections;
using System.Collections.Generic;
```

```
using NUnit.Framework;
using UnityEngine;
using UnityEngine.TestTools;

public class NewTestScript {
    // A Test behaves as an ordinary method
    [Test]
    public void NewTestScript1SimplePasses() {
        // Use the Assert class to test conditions
    }

    // A UnityTest behaves like a coroutine in Play Mode. In Edit Mode you can use
    // `yield return null;` to skip a frame.
    [UnityTest]
    public IEnumerator NewTestScriptWithEnumeratorPasses() {
        // Use the Assert class to test conditions.
        // Use yield to skip a frame.
        yield return null;
    }
}
```

In the **Test Runner** panel, you should see the script class and its two methods listed – for example, see *Figure 19.7* above. Note that the first line in the **Test Runner** panel is the Unity project name; the second line will say <projectName>.dll, followed by the script class called NewTestScript, and, finally, each of the test methods. The generated default script creates two methods:

- NewTestScriptSimplePasses(): This is a simple, empty test method (since it's empty, it will always pass).
- NewTestScriptWithEnumeratorPasses(): This is a coroutine test method that always yields a null return; so, again, it will not create a failed event and will always pass.

The generated script and two methods are a basic skeleton that we can populate with simple methods and coroutine methods as appropriate for each project's testing requirements.

There are three symbols to indicate the status of each test/class:

- **Empty circle:** The test hasn't been executed since the script class was last changed.
- **Green tick** (checkmark): The test passed successfully.
- **Red cross:** The test failed.

There's more...

Here are some details that you won't want to miss.

Creating a default test script from the Project window's Create menu

Another way of creating a default **unit test** script is by going to the **Project** window and then to **Create | Testing | C# Test Script**.

EditMode minimum skeleton unit test script

If you are only going to use this script class for testing in **EditMode**, you can delete the second method using statements such as the following to give you a minimal skeleton to work from:

```
using NUnit.Framework;

public class UnitTestSkeleton
{
    [Test]
    public void NewTestScriptSimplePasses()
    {
        // write your assertion(s) here
    }
}
```

This simpler skeleton testing class is for when we are writing code-only unit tests.

Making a simple unit test

In the same way as printing "hello world" is most programmers' first program statement, asserting that 1 + 1 = 2 is perhaps the most common first test that's executed for those learning unit testing. That's what we'll create in this recipe:

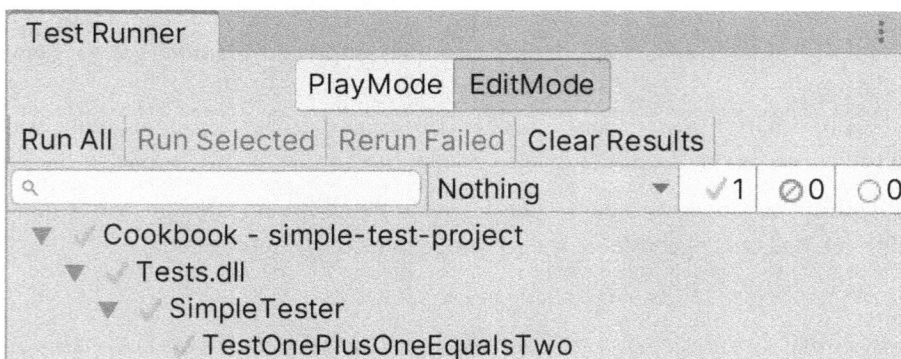

Figure 19.9: Our simple numeric test method passing with a green tick

How to do it...

To create and execute a simple unit test, follow these steps:

1. Create a new **3D Unity project**.

2. Display the **Test Runner** panel by going to **Window** | **General** | **Test Runner**.

3. Ensure that the **EditMode** button is selected in the **Test Runner** panel.

4. In the **Test Runner** panel, click the **Create EditMode Test Assembly Folder** button. You'll now see a folder called `Tests` that's been created in the **Project** panel.

5. Select the `Tests` folder.

6. In the **Test Runner** panel, click the **Create Test Script in current folder** button.

7. You should now have a new C# script named `NewTestScript` inside your `Tests` folder.

8. Rename the script class from `NewTestScript` to `SimpleTester` and replace its contents with the following:

```
using NUnit.Framework;

class SimpleTester
{
    [Test]
    public void TestOnePlusOneEqualsTwo()
    {
        // Arrange
        int n1 = 1;
        int n2 = 1;
        int expectedResult = 2;

        // Act
        int result = n1 + n2;

        // Assert
        Assert.AreEqual(expectedResult, result);
    }
}
```

9. Click the **Run All** button in the **Test Runner** panel.

10. You should see the results of your unit test being executed – if the test concluded successfully, it should have a green "tick" next to it.

How it works...

In this recipe, you declared that the `TestOnePlusOneEqualsTwo()` method in the `SimpleTester.cs` C# script class is a test method. When executing this test method, the **Unity Test Runner** executes each statement in sequence, so variables n1, n2, and expectedResult are set, then the calculation of 1 + 1 is stored in the variable result, and, finally (the most important bit), we make an assertion of what should be true after executing that code. Our assertion states that the value of the expectedResult variable should be equal to the value of the variable result.

If the assertion is `true`, the test passed; otherwise, it failed. Generally, as programmers, we expect our code to pass, so we inspect each fail very carefully, first to see whether we have an obvious error, then perhaps to check whether the test itself is correct (especially if it's a new test), and then to begin to debug and understand why our code behaved in such a way that it did not yield the anticipated result.

There's more…

Here are some details that you won't want to miss.

Shorter tests with values in the assertion

For simple calculations, some programmers prefer to write less test code by putting the values directly into the assertion. So, as shown here, our 1 + 1 = 2 test could be expressed in a single assertion, where the expected value of 2 and the expression 1 + 1 are entered directly into an `AreEqual(...)` method's invocation:

```
using NUnit.Framework;

class SimpleTester
{
    [Test]
    public void TestOnePlusOneEqualsTwo()
    {
        // Assert
Assert.AreEqual(2, 1 + 1);
    }
}
```

However, if you are new to testing, you may prefer the previous approach, whereby the way you prepare, execute, and store the results, as well as the property assertions about those results, is structured clearly in a sequence of **Arrange/Act/Assert**. By storing values in meaningfully named variables, what we are asserting is very clear.

Expected value followed by the actual value

When comparing values with assertions, it is customary for the expected (correct) value to be given first, followed by the actual value:

```
Assert.AreEqual( <expectedValue>, <actualValue> );
```

While it makes no difference to the true or false nature of equality, it can make a difference to messages when tests fail with some testing frameworks (for example, "got 2 but expected 3" has a very different meaning to "got 3 but expected 2"). Hence, the following assertion would output a message that would be confusing, since 2 was our expected result:

```
public void TestTwoEqualsThreeShouldFail() {
    // Arrange
```

```
        int expectedResult = 2;

        // Act
        int result = 1 + 2; // 3 !!!

        // Assert
        Assert.AreEqual(result, expectedResult);
    }
```

The following screenshot illustrates how we will get a misleading message when the arguments are the wrong way around in our assertion method:

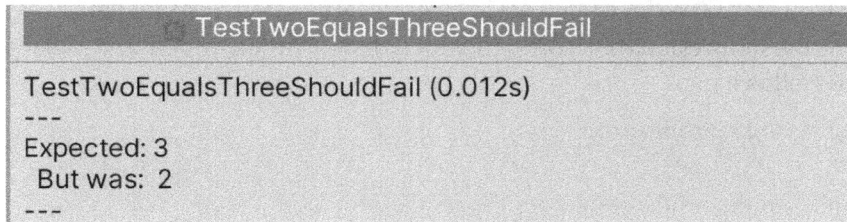

Figure 19.10: Confusing message due to an incorrect argument sequence in the assertion

Parameterizing tests with a DataProvider

If we are testing our code using a range of test data, then sometimes, there is little difference between each test apart from the values. Rather than duplicating our Arrange/Act/Assert statements, we can reuse a single method, and the Unity Test Runner will loop through a collection of test data, running the test method for each set of test data. The special method that provides multiple sets of test data to a test method is known as a **DataProvider**, and we'll create one in this recipe:

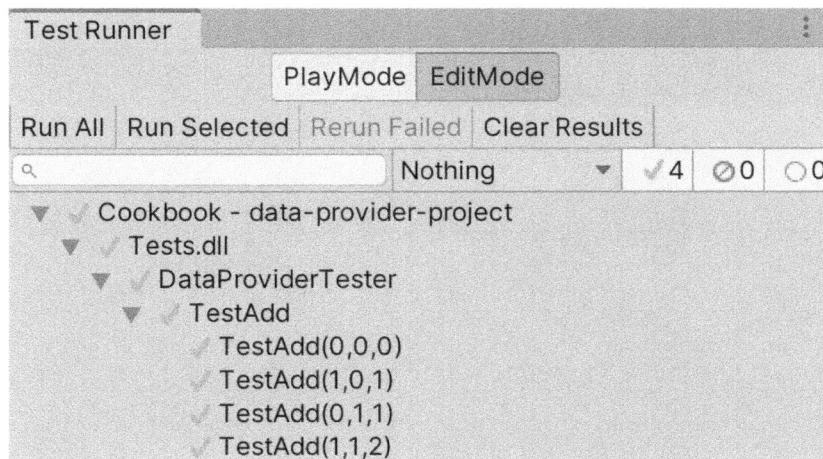

Figure 19.11: Running the test method with many sets of values with a DataProvider method

How to do it...

To parameterize tests with a **DataProvider** method, follow these steps:

1. Create a new **3D Unity** project.

2. Display the **Test Runner** panel by going to **Window | General | Test Runner.**

3. Ensure that the **EditMode** button is selected in the **Test Runner** panel.

4. In the **Test Runner** panel, click the **Create EditMode Test Assembly Folder** button. You'll now see a folder called Tests that's been created in the **Project** panel.

5. Select the Tests folder.

6. In the **Test Runner** panel, click the **Create Test Script in current folder** button. You should now have a new C# script named NewTestScript inside your **Tests folder.**

7. Rename the script class from NewTestScript to DataProviderTester and replace its contents with the following:

```csharp
using NUnit.Framework;

class DataProviderTester
{
    [Test, TestCaseSource("AdditionProvider")]
    public void TestAdd(int num1, int num2, int expectedResult)
    {
        // Arrange
        // (not needed - since values coming as arguments)

        // Act
        int result = num1 + num2;

        // Assert
        Assert.AreEqual(expectedResult, result);
    }

    // the data provider
    static object[] AdditionProvider =
```

```
        {
            new object[] { 0, 0, 0 },
            new object[] { 1, 0, 1 },
            new object[] { 0, 1, 1 },
            new object[] { 1, 1, 2 }
        };
    }
```

8. Display the **Test Runner** window by going to **Window | General | Test Runner**.

9. Ensure that the **EditMode** button is selected in the **Test Runner** panel.

10. Click **Run All**.

11. You should see the results of your unit test being executed. You should see four sets of results for the TestAdd(...) test method, one for each of the datasets provided by the AdditionProvider method.

How it works...

We have indicated that the TestAdd(...) method is a test method with a compiler attribute called [Test]. However, in this case, we have added additional information to state that the [TestCaseSource(...)] data source for this method is the AdditionProvider method.

This means that the **Unity Test Runner** (see the preceding screenshot) will retrieve the data objects from the additional provider and create multiple tests for the TestAdd(...) method, one for each set of data from the AdditionProvider() method.

In the **Test Runner** panel, we can see a line for each of these tests:

```
TestAdd(0, 0, 0)
TestAdd(1, 0, 1)
TestAdd(0, 1, 1)
TestAdd(1, 1, 2)
```

Unit testing a simple health script class

Let's create something that might be used in a game and that can easily be unit tested. Classes that do **not** subclass from MonoBehaviour are much easier to unit test since instance objects can be created using the new keyword. If the class is carefully designed with private data and public methods with clearly declared dependencies as parameters, it becomes easy to write a set of tests to make us confident that objects of this class will behave as expected in terms of default values, as well as valid and invalid data.

In this recipe, we will create a health script class and a set of tests for this class. This kind of class can be reused for both the health of human players and **Artificial Intelligence (AI)**-controlled enemies in a game:

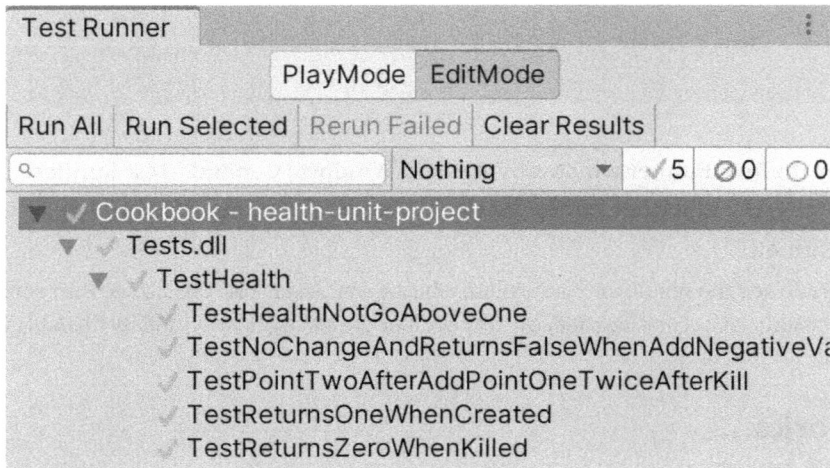

Figure 19.12: Passing tests for our health script class

How to do it...

To unit test a health script class, follow these steps:

1. Create a new **3D Unity** project.

2. Create a new folder named _Scripts.

3. Inside your _Scripts folder, create a new Health.cs C# script class containing the following:

```
using UnityEngine;
using System.Collections;

public class Health
{
    private float _health = 1;

    public float GetHealth()
    {
        return _health;
    }


    public bool AddHealth(float heathPlus)
```

```
    {
        if(heathPlus > 0){
            _health += heathPlus;

            // ensure never more than 1
            if(_health > 1) _health = 1;
            return true;
        } else {
            return false;
        }
    }

    public bool KillCharacter()
    {
        _health = 0;
        return true;
    }
}
```

4. Since we want to test scripts in this folder, we need to add an Assembly Definition here. Ensure the _Scripts folder is selected in the Project panel, then go to **Create | Assembly Definition** and rename the new Assembly Definition HealthScriptAssembly.

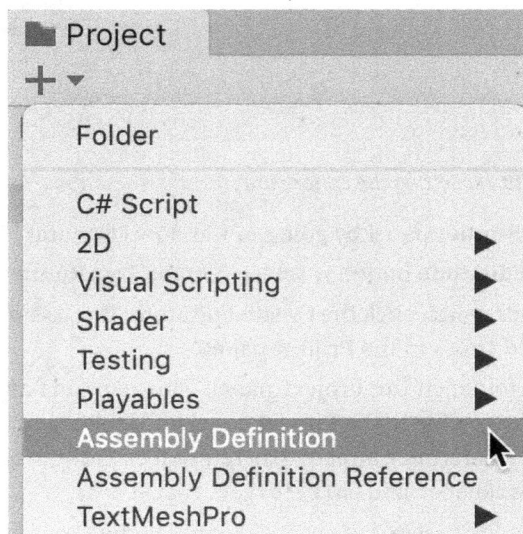

Figure 19.13: Creating an Assembly Definition in our _Scripts folder

5. Select HealthScriptAssembly in the **Project** panel. In the **Inspector,** ensure the **Name** property is HealthScriptAssembly, and in the list of **Platforms,** ensure that only the **Editor** is checked; then click the **Apply** button at the bottom of the **Inspector** panel.

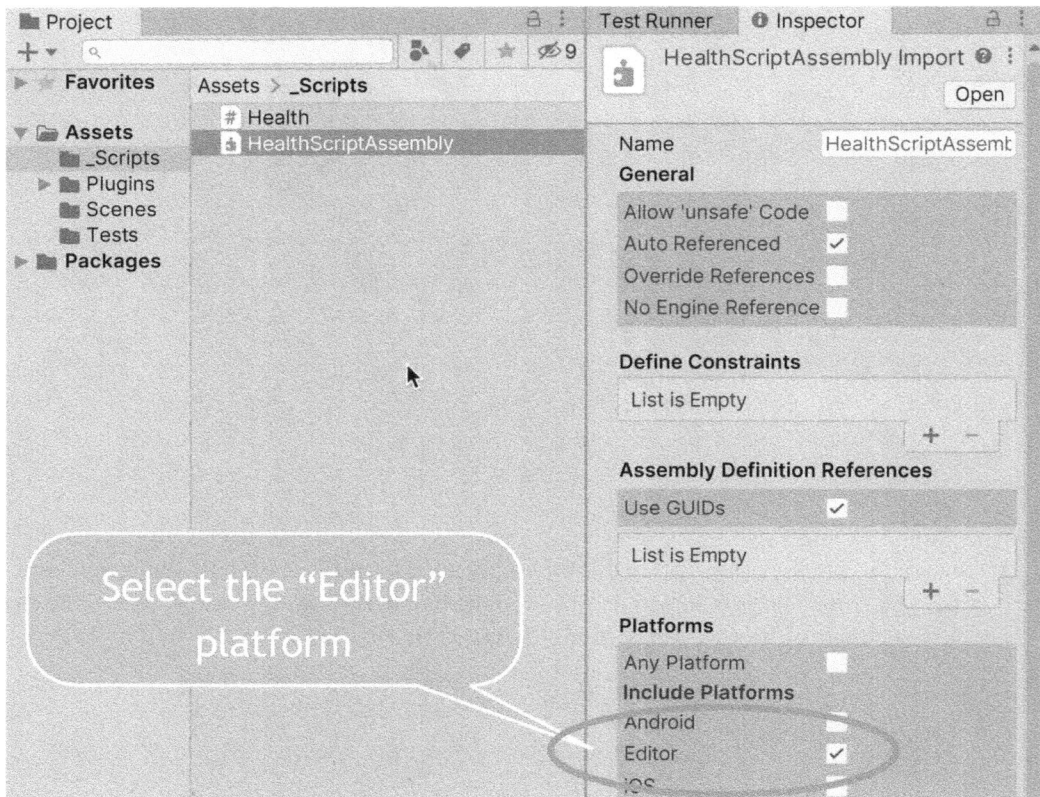

Figure 19.14: Setting the Editor platform and a new name for HealthScriptAssembly

6. Display the **Test Runner** panel by going to **Window | General | Test Runner.**

7. Ensure that the **EditMode** button is selected in the **Test Runner** panel.

8. In the **Test Runner** panel, click the **Create EditMode Test Assembly Folder** button. You'll now see a folder called Tests in the **Project** panel.

9. Open the Tests folder in the **Project** panel. This contains an **Assembly Reference** named Tests. Select this asset file in the **Project** panel and look at its properties in the **Inspector** panel. This **Assembly Reference** already has references to two **Assembly Definition References:** UnityEngine.TestRunner and UnityEditor.TestRunner.

10. Now, we need to add our HealthScriptAssembly to the Tests assembly so that we can test our Health script class. In the **Inspector** panel, click the plus (+) button to add a new **Assembly Definition Reference** to the slot that was created to locate and drag the **HealthScript Assembly Definition,** and then click the **Apply** button at the bottom of the **Inspector** panel.

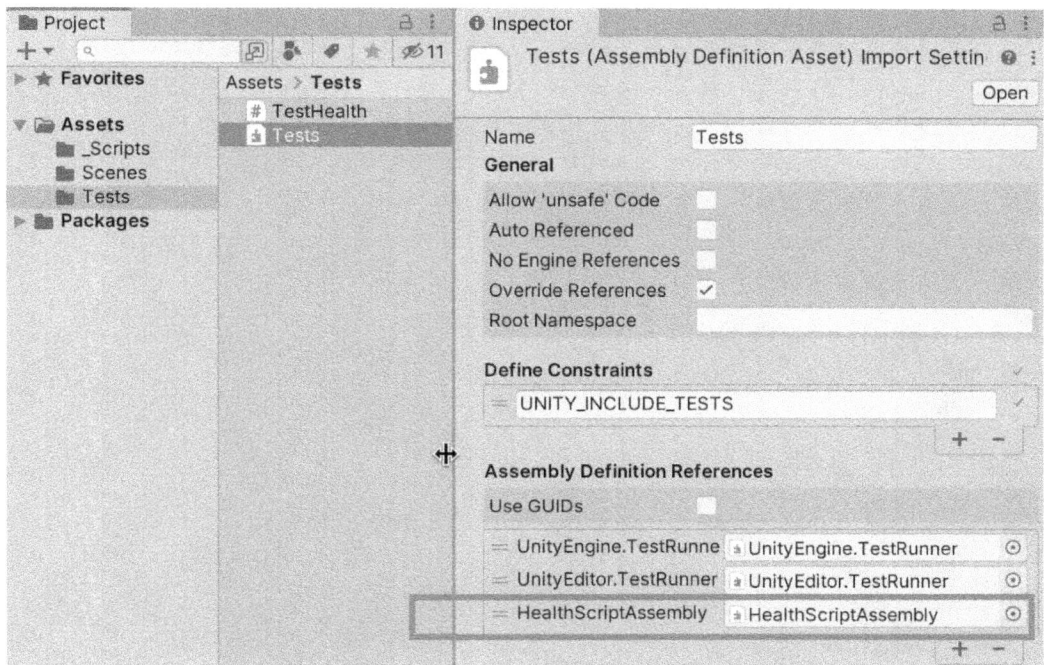

Figure 19.15: Adding HealthScriptAssembly to the Tests Assembly Definition

11. In the **Test Runner** window, click the **Create Test Script in current folder** button. Rename the script class `TestHealth.cs` and ensure it contains the following code:

```
using NUnit.Framework;

class TestHealth {
    [Test]
    public void TestReturnsOneWhenCreated()    {
        // Arrange
        Health h = new Health ();
        float expectedResult = 1;

        // Act
        float result = h.GetHealth ();

        // Assert
        Assert.AreEqual (expectedResult, result);
    }

    [Test]
```

```
public void TestPointTwoAfterAddPointOneTwiceAfterKill()    {
    // Arrange
    Health h = new Health();
    float healthToAdd = 0.1f;
    float expectedResult = 0.2f;

    // Act
    h.KillCharacter();
    h.AddHealth(healthToAdd);
    h.AddHealth(healthToAdd);
    float result = h.GetHealth();


    // Assert
    Assert.AreEqual(expectedResult, result);
}

[Test]
public void TestNoChangeAndReturnsFalseWhenAddNegativeValue()      {
    // Arrange
    Health h = new Health();
    float healthToAdd = -1;
    bool expectedResultBool = false;
    float expectedResultFloat = 1;

    // Act
    bool resultBool = h.AddHealth(healthToAdd);
    float resultFloat = h.GetHealth();

    // Assert
    Assert.AreEqual(expectedResultBool, resultBool);
    Assert.AreEqual(expectedResultFloat, resultFloat);
}

[Test]
public void TestReturnsZeroWhenKilled()    {
    // Arrange
    Health h = new Health();
    float expectedResult = 0;

    // Act
```

```
        h.KillCharacter();
        float result = h.GetHealth();

        // Assert
        Assert.AreEqual(expectedResult, result);
    }


    [Test]
    public void TestHealthNotGoAboveOne()     {
        // Arrange
        Health h = new Health();
        float expectedResult = 1;

        // Act
        h.AddHealth(0.1f);
        h.AddHealth(0.5f);
        h.AddHealth(1);
        h.AddHealth(5);
        float result = h.GetHealth();

        // Assert
        Assert.AreEqual(expectedResult, result);
    }
}
```

12. Display the **Test Runner** window by going to **Window | General | Test Runner.**

13. Ensure that the **EditMode** button is selected in the **Test Runner** window.

14. Click **Run All.**

> **Note:** Sometimes Unity seems to lose track of scripts. If you don't see a **Run All** button in the **Test Runner** panel, and/or you get a Console error about not being able to find a reference to the Health script class, try moving the `Health.cs` script file into the Tests folder, this lets Unity re-load all the scripts. Once this is done move Health.cs back into the `_Script` folder. This seems to solve the issue by allowing Unity to reference the scripts and Assembly definitions together.
>
> You can read more about this issue at:
>
> ```
> https://forum.unity.com/threads/cant-properly-reference-
> public-classes-from-the-assembly-csharp-into-assembly-csharp-
> editor.884593/
> ```

15. You should see the results of your unit tests being executed.

How it works...

Each of the C# script classes is described here.

Health.cs

This script class has one private property. Since it is private, it can only be changed by methods. Its initial value is 1.0 – in other words, 100% health:

- health (float): The valid range is from 0 (dead!) to 1.0 (100% health).

There are three public methods:

- GetHealth(): This returns the current value of the health float number (which should be between 0 and 1.0).
- AddHealth(float): This takes a float as input (the amount to add to the health) and returns a Boolean true/false regarding whether the value was valid. Note that the logic of this method is that it accepts values of 0 or more (and will return true), but it will ensure that the value of health is never more than 1.
- KillCharacter(): This method sets health to 0 and returns true since it is always successful in this action.

TestHealth.cs

This script class has five methods:

- TestReturnsOneWhenCreated(): This tests that the initial value of health is 1 when a new Health object is created.
- TestPointTwoAfterAddPointOneTwiceAfterKill(): This tests that after a kill (health set to 0), and then adding 0.1 on two occasions, the health is 0.2.
- TestReturnsZeroWhenKilled(): This tests that the health value is set to 0 immediately after the KillCharacter() method has been called.
- TestNoChangeAndReturnsFalseWhenAddNegativeValue(): This tests that attempting to add a negative value to health should return false and that the value of health should not have changed. This method is an example of a test with more than one assertion (but both are related to the actions).
- TestHealthNotGoAboveOne(): This test verifies that even when lots of values are added to health that total more than 1.0, the value that's returned from GetHealth() is 1.

Hopefully, all the tests pass when you run them, giving you some confidence that the logic that's been implemented in the Health.cs script class behaves as intended.

Creating and executing a unit test in PlayMode

It's a good idea to write as much of the logic for a game as isolated, non-MonoBehaviour classes that are easy to unit test in EditMode as possible.

However, some of the logic in a game relates to things that happen when the game is running. Examples include physics, collisions, and timing-based events. We test these parts of our games in **PlayMode**.

In this recipe, we'll create one very simple **PlayMode** test to check that the physics affect a **Rigidbody** (based on an example from the Unity documentation):

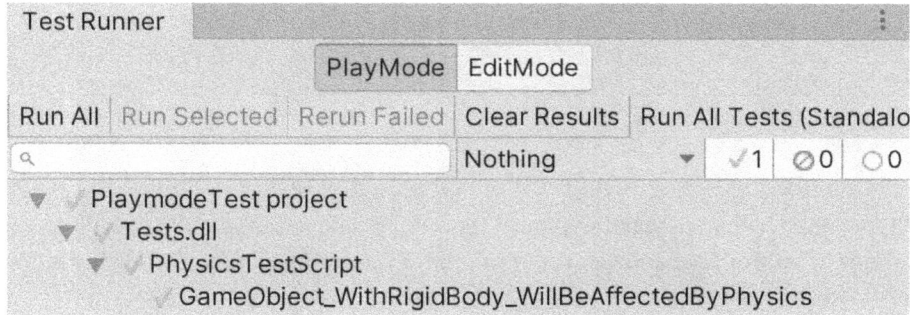

Figure 19.16: Running a physics PlayMode test

How to do it...

To create and execute a unit test in **PlayMode**, follow these steps:

1. Create a new **3D Unity** project.
2. Display the **Test Runner** panel by going to **Window | General | Test Runner.**
3. Ensure that the **PlayMode** button is selected in the **Test Runner** panel.
4. In the **Test Runner** panel, click the **Create PlayMode Test Assembly Folder** button. You'll now see a folder called **Tests** in the **Project** panel.
5. In the **Test Runner** panel, click the **Create Test Script in current folder** button. Rename the script class PhysicsTestScript.cs and ensure it contains the following code:

```
using UnityEngine;
using UnityEngine.TestTools;
using NUnit.Framework;
using System.Collections;

public class PhysicsTestScript
{
    [UnityTest]
    public IEnumerator GameObject_WithRigidBody_WillBeAffectedByPhysics()
    {
        // Arrange
        var go = new GameObject();
        go.AddComponent<Rigidbody>();
```

```
            var originalPosition = go.transform.position.y;

            // Act
            yield return new WaitForFixedUpdate();

            // Assert
            Assert.AreNotEqual(originalPosition, go.transform.position.y);
        }
    }
```

6. Click **Run All**.

7. In the **Hierarchy** panel, you'll see that a temporary scene is created (named something along the lines of **InitTestScene6623462364**) and that a GameObject named **Code Based Test Runner is created**.

8. In the **Game** panel, you will briefly see the message **Display 1 No Cameras Rendering:**

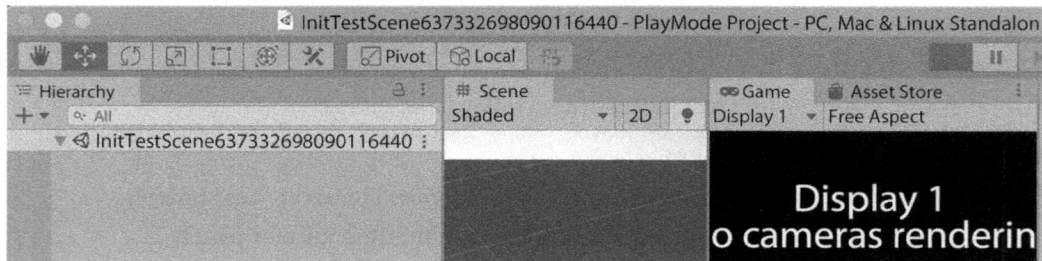

Figure 19.17: Temporary scene and GameObject for the runtime test

9. You should see the results of your unit test being executed – if the test is concluded successfully, it should have a green tick next to it.

How it works...

Methods marked with the [UnityTest] attribute are run as coroutines. A coroutine has the ability to pause execution (when it meets a yield statement) and return control to Unity, but then continue where it left off when called again (for example, the next frame, second frame, and so on). The yield statement indicates the statement after which, and for how long, execution of the method is to be paused. Examples of different types of yield include the following:

- Waiting until the next frame: null
- Waiting for a given length of time: WaitForSeconds(<seconds>)
- Waiting until the next fixed-update time period (physics is not applied to each frame (since the framerate varies) but after a fixed period of time): WaitForFixedUpdate()

The GameObject_WithRigidBody_WillBeAffectedByPhysics() method creates a new GameObject and attaches to it a **Rigidbody**.

It also stores the original Y position. The yield statement makes **PlayMode Test Runner** wait until physics has begun at the next fixed update period. Finally, an assertion is made that the original Y position is not equal to the new Y position (after the physics fixed update). Since the default for a **Rigidbody** is that gravity will be applied, this is a good test that physics is being applied to the new object (in other words, it should have started falling once physics had been applied).

PlayMode testing a door animation

Having learned the basics of **PlayMode** testing in the previous recipe, let's test something non-trivial that we might find in a game. In this recipe, we'll create a **PlayMode** test to ensure that a door opening animation plays when the player's sphere object enters a collider.

A scene has been provided with the player's sphere initialized to roll toward a red door. When the sphere hits the collider (the OnTriggerEnter event), some code sets the door's **Animator Controller Opening** variable to true, which transitions the door from its closed state to its open state, as shown in the following screenshot:

Figure 19.18: The door will open (upward) when hit by the sphere

Thanks to the creator of the ground texture; it was designed by *Starline* and published at freepik.com.

Getting ready

For this recipe, a Unity package has been provided (doorScene.unitypackage) in the 19_07 folder.

How to do it...

To **PlayMode** test a door animation, follow these steps:

1. Create a new **3D** project and delete the default folder, called Scenes.
2. Import the Unity package provided (doorScene.unitypackage).
3. Add the doorScene and menuScene scenes to the project build (the sequence doesn't matter), by going to **File | Build Settings...**, opening each scene and clicking the **Add Open Scene** button in the **Build Settings** panel.
4. Ensure that the scene that's currently open is menuScene.

5. Display the **Test Runner** panel by going to **Window | General | Test Runner.**

6. Ensure that the **PlayMode** button is selected in the **Test Runner** panel.

7. In the **Project** panel, select the top-level folder called Assets.

8. In the **Test Runner** panel, click the **Create PlayMode Test Assembly Folder** button. A new folder, named Tests, should have been created.

9. In the **Project** panel, open the Tests **folder.**

10. In the **Test Runner** panel, click the **Create Test Script in the current folder** button. Rename this script class DoorTest.

11. Edit the DoorTest.cs script class by replacing its content with the following:

```
using System.Collections;
using NUnit.Framework;
using UnityEngine;
using UnityEngine.SceneManagement;
using UnityEngine.TestTools;

public class DoorTest
{
    const int BASE_LAYER = 0;
    private string initialScenePath;
    private Animator doorAnimator;
    private Scene tempTestScene;

    // name of scene being tested by this class
    private string sceneToTest = "doorScene";

    [SetUp]
    public void Setup()
    {
        // setup - load the scene
        tempTestScene = SceneManager.GetActiveScene();
    }
}
```

12. Add the following test method to DoorTest.cs:

```
[UnityTest]
public IEnumerator TestDoorAnimationStateStartsClosed()
{
    // load scene to be tested
```

```
        yield return SceneManager.LoadSceneAsync(sceneToTest, LoadSceneMode.
    Additive);
        SceneManager.SetActiveScene(SceneManager.
    GetSceneByName(sceneToTest));

        // Arrange
        doorAnimator = GameObject.FindWithTag("Door").
    GetComponent<Animator>();
        string expectedDoorAnimationState = "DoorClosed";

        // immediate next frame
        yield return null;

        // Act
        AnimatorClipInfo[] currentClipInfo = doorAnimator.
    GetCurrentAnimatorClipInfo(BASE_LAYER);
        string doorAnimationState = currentClipInfo[0].clip.name;

        // Assert
        Assert.AreEqual(expectedDoorAnimationState, doorAnimationState);

        // teardown - reload original temp test scene
        SceneManager.SetActiveScene(tempTestScene);
        yield return SceneManager.UnloadSceneAsync(sceneToTest);
    }
```

13. Add the following test method to DoorTest.cs:

```
[UnityTest]
public IEnumerator TestIsOpeningStartsFalse()
{
    // load scene to be tested
    yield return SceneManager.LoadSceneAsync(sceneToTest, LoadSceneMode.
Additive);
    SceneManager.SetActiveScene(SceneManager.
GetSceneByName(sceneToTest));

    // Arrange
    doorAnimator = GameObject.FindWithTag("Door").
GetComponent<Animator>();

    // immediate next frame
```

```
        yield return null;

        // Act
        bool isOpening = doorAnimator.GetBool("Opening");

        // Assert
        Assert.IsFalse(isOpening);

        // teardown - reload original temp test scene
        SceneManager.SetActiveScene(tempTestScene);
        yield return SceneManager.UnloadSceneAsync(sceneToTest);
    }
```

14. Add the following test method to `DoorTest.cs`:

```
[UnityTest]
public IEnumerator TestDoorAnimationStateOpenAfterAFewSeconds()
{
    // Load scene to be tested
    yield return SceneManager.LoadSceneAsync(sceneToTest, LoadSceneMode.
Additive);
    SceneManager.SetActiveScene(SceneManager.
GetSceneByName(sceneToTest));

    // wait a few seconds
    int secondsToWait = 3;
    yield return new WaitForSeconds(secondsToWait);

    // Arrange
    doorAnimator = GameObject.FindWithTag("Door").
GetComponent<Animator>();
    string expectedDoorAnimationState = "DoorOpen";


    // Act
    AnimatorClipInfo[] currentClipInfo = doorAnimator.
GetCurrentAnimatorClipInfo(BASE_LAYER);
    string doorAnimationState = currentClipInfo[0].clip.name;
    bool isOpening = doorAnimator.GetBool("Opening");

    // Assert
    Assert.AreEqual(expectedDoorAnimationState, doorAnimationState);
```

```
        Assert.IsTrue(isOpening);

        // teardown - reload original temp test scene
        SceneManager.SetActiveScene(tempTestScene);
        yield return SceneManager.UnloadSceneAsync(sceneToTest);
    }
```

15. Click **Run All**.

16. As the tests run, you will see that, in the **Hierarchy**, **Game**, and **Scene** panels a temporary scene is created, and **doorScene** is running, with the sphere rolling toward the red door.

17. You should see the results of your unit test being executed – if all the tests conclude successfully, there should be green ticks (checkmarks) next to each test.

How it works...

In this recipe, you added two scenes to the build so that they can be selected in our scripts using **SceneManager** during **PlayMode** testing.

We opened menuScene so that we can clearly see when Unity runs different scenes during our **PlayMode** testing – and we'll see the menu scene reopened after testing takes place.

There is a SetUp() method that is executed before each test. The SetUp() and TearDown() methods are very useful for preparing things before each test and resetting things back to how they were before the test took place. Unfortunately, aspects such as loading our door scene before running each test, and then reloading the menu after each test, involve waiting until the scene load process has completed. We can't place yield statements in our SetUp() and TearDown() methods, so you'll see that each test has repeated scene loading at the beginning and end of each test:

```
    // load scene to be tested
    yield return SceneManager.LoadSceneAsync(sceneToTest, LoadSceneMode.Additive);
        SceneManager.SetActiveScene(SceneManager.GetSceneByName(sceneToTest));

    // Arrange-Act-Assert goes here

    // teardown - reload original temp test scene
    SceneManager.SetActiveScene(tempTestScene);
    yield return SceneManager.UnloadSceneAsync(sceneToTest);
```

For each test, we wait, either for a single frame (yield null) or a few seconds (yield return new WaitForSeconds(...)). This ensures that all objects have been created and physics has started before our test starts running. The first two tests check the initial conditions – in other words, that the door begins in the DoorClosed animation state and that the animation controller's isOpening variable is false.

The final test waits a few seconds (which is enough time for the sphere to roll up to the door and trigger the opening animation) and tests that the door is entering/has entered the `DoorOpen` **animation state** and that the animation controller's `isOpening` variable is `true`.

As can be seen, there is quite a bit more to **PlayMode** testing than unit testing, but this means that we have a way to test actual GameObject interactions when features such as timers and physics are running. As this recipe demonstrates, we can also load our own scenes for **PlayMode** testing, be they special scenes that have been created just to test interactions or actual scenes that are to be included in our final game build.

There's more...

There seem to be some changes in how to enable PlayMode tests for all assemblies. The issue is that enabling PlayMode tests can increase the size of build projects – the default setting is to disable Play-Mode tests for all assemblies. If your version of the **Unity Editor Test Runner** window does not offer a menu option to **enable playmode tests for all assemblies**, then you can enable these by setting the `playModeTestRunnerEnabled` setting to 1 in the `ProjectSettings/ProjectSetting.asset` file. You can learn more about this in the Unity Test Framework documentation: `https://docs.unity3d.com/Packages/com.unity.test-framework@1.1/manual/workflow-create-playmode-test.html`.

PlayMode and unit testing a player health bar with events, logging, and exceptions

In this recipe, we will combine many different kinds of tests for a feature that's included in many games – a visual health bar representing the player's numeric health value (in this case, a float number from 0.0 to 1.0). Although it doesn't comprehensively test all aspects of the health bar, this recipe will provide a good example of how we can go about testing many different parts of a game using the Unity testing tools.

A Unity package has been provided that contains the following:

- `Player.cs`: A player script class for managing values for player health that uses delegates and events to publish health changes to any listening `View` classes.
- Two `View` classes that register to listen for player health change events:
    - `HealthBarDisplay.cs`: This updates `fillAmount` for a UI image for each new player health value that's received.
    - `HealthChangeLogger.cs`: This prints messages about the new player health value that's received by the `Debug.Log` file.

- `PlayerManager.cs`: A manager script that initializes player and `HealthChangeLogger` objects, and also allows the user to change the health of the player by pressing the *Up* and *Down* arrow keys (simulating healing/damage during a game).

- A scene that has two UI images – one is a health bar outline (red heart and a black outline), while the second is the filler image, showing dark blue to light blue to green, for weak to strong health values.

This recipe allows several different kinds of testing to be demonstrated:

- **PlayMode** testing, to check that the actual `fillAmount` of the UI image displayed matches the 0.0 to 1.0 range of the player's health.

- **Unit testing, to check** that the player's health starts with the correct default value and correctly increases and decreases after calls to the `AddHealth(...)` and `ReduceHealth(...)` methods are made.

- Unit testing, to check that health change events are published by the player object.

- Unit testing, to check that expected messages are logged in `Debug.Log`.

- Unit testing, to check that argument out-of-range exceptions are thrown if negative values are passed to the player's `AddHealth(...)` or `ReduceHealth(...)` methods. This is demonstrated in the following screenshot:

Figure 19.19: The graphical heath bar we'll be testing in PlayMode

Thanks to *Pixel Art Maker* for the health bar image: `http://pixelartmaker.com/art/49e2498a414f221`.

Getting ready

For this recipe, a Unity package has been provided (`healthBarScene.unitypackage`) in the 19_09 folder.

How to do it...

To **PlayMode and unit test** a player health bar, follow these steps:

1. Create a new **3D Unity** project and delete the default **Project** panel folder `Scenes`.

2. Import the Unity package provided (`healthBarScene.unitypackage`).

3. Open the `HealthBar` **scene and add this scene to** the project's **Build**, by going to **File | Build Settings...**.

4. Display the **Test Runner** panel by going to **Window | General | Test Runner**.

5. Since our **PlayMode** tests make use of the `Player` script class in the **Project** folder (**Assets |
HealthBarScene | _Scripts**), we need to add an **Assembly Definition** there. Select this folder
in the **Project** panel, then from the **Create** menu, create a new **Assembly Definition**, naming
it `PlayerAssembly`.

Figure 19.20: PlayerAssembly created in the _Scripts folder

6. Now, select **PlayMode** in the **Test Runner** panel.

7. In the **Project** panel, select the top-level folder, called `Assets`.

8. In the **Test Runner** panel, click the **Create PlayMode Test Assembly Folder** button. A new folder,
named `Tests`, should have been created.

9. Select the `Tests` folder. This contains an **Assembly Reference** named **Tests**. Select this asset file
in the **Project** panel and look at its properties in the **Inspector** panel. This Assembly Reference
already has references to two Assembly Definition References: `UnityEngine.TestRunner` and
`UnityEditor.TestRunner`.

10. We now need to add our `PlayerAssembly` asset to the `Tests` assembly so that we can perform
our **PlayMode** tests that make use of the `Player` script class. In the **Inspector** panel, click the
plus (+) button to add a new **Assembly Definition Reference**. In the slot that's been created,
locate and drag the `PlayerAssembly` Assembly Definition asset file, then click the **Apply** button
at the bottom of the **Inspector** panel.

11. In the **Test Runner** panel, click the **Create Test Script in current folder** button. Rename this
script class `HealthBarPlayModeTests` and replace its content with the following code:

```
using UnityEngine;
using UnityEngine.UI;
using UnityEngine.TestTools;
using NUnit.Framework;
using System.Collections;
using UnityEngine.SceneManagement;

[TestFixture]
public class HealthBarPlayModeTests
```

```
{
    private Scene tempTestScene;

    // name of scene being tested by this class
    private string sceneToTest = "HealthBar";

    [SetUp]
    public void Setup()
    {
        // setup - load the scene
        tempTestScene = SceneManager.GetActiveScene();
    }
}
```

12. Add the following test to `HealthBarPlayModeTests.cs`:

```
[UnityTest]
public IEnumerator TestHealthBarImageMatchesPlayerHealth()
{
    // load scene to be tested
    yield return SceneManager.LoadSceneAsync(sceneToTest, LoadSceneMode.
Additive);
    SceneManager.SetActiveScene(SceneManager.
GetSceneByName(sceneToTest));

    // wait for one frame
    yield return null;

    // Arrange
    Image healthBarFiller = GameObject.Find("image-health-bar-filler").
GetComponent<Image>();
    PlayerManager playerManager = GameObject.
FindWithTag("PlayerManager").GetComponent<PlayerManager>();
    float expectedResult = 0.9f;

    // Act
    playerManager.ReduceHealth();

    // Assert
    Assert.AreEqual(expectedResult, healthBarFiller.fillAmount);

    // teardown - reload original temp test scene
```

```
        SceneManager.SetActiveScene(tempTestScene);
        yield return SceneManager.UnloadSceneAsync(sceneToTest);
    }
```

13. Click **Run All**.

14. As the tests run, you will see that, in the **Hierarchy, Game,** and **Scene** panels, a temporary scene is created, and **HealthBarScene** is running, with the visual health bar. You should see the results of your **PlayMode** test being executed – if the test concludes successfully, there should be a green tick (checkmark).

15. Now, let's add some unit tests to our player health feature.

16. Select **EditMode** in the **Test Runner** window.

17. In the **Project** window, select the top-level folder, called `Assets`.

18. In the **Test Runner** window, click the **Create EditMode Test Assembly Folder** button. A new folder, named `Tests 1`, should have been created; rename this `TestsEditMode`.

19. Select the `TestsEditMode` folder. This contains an **Assembly Reference** named `Tests`. Select this asset file in the **Project** window and look at its properties in the **Inspector** window. This Assembly Reference already has references to two Assembly Definition References: `UnityEngine.TestRunner` and `UnityEditor.TestRunner`.

20. As before, we need to add our `PlayerAssembly` asset to the `Tests` assembly so that we can perform our **PlayMode** tests that make use of the `Player` script class. In the **Inspector** window, click the plus (+) button to add a new **Assembly Definition Reference**. In the slot that's been created, locate and drag the `PlayerAssembly` Assembly Definition asset file, then click the **Apply** button at the bottom of the **Inspector** panel.

21. In the **Test Runner** window, click the **Create Test Script in the current folder** button. Rename this script class `EditModeUnitTests`.

22. Edit the `EditModeUnitTests.cs` script class, replacing its content with the following code:

```
using System;
using UnityEngine.TestTools;
using NUnit.Framework;
using UnityEngine;

public class EditModeUnitTests
{

    // inner unit test classes go here

}
```

23. Add the following class and basic tests to the `EditModeUnitTests` class in `EditModeUnitTests.cs`:

```
public class TestCorrectValues
```

```csharp
{
    [Test]
    public void DefaultHealthOne()
    {
        // Arrange
        Player player = new Player();
        float expectedResult = 1;

        // Act
        float result = player.GetHealth();

        // Assert
        Assert.AreEqual(expectedResult, result);
    }

    [Test]
    public void HealthCorrectAfterReducedByPointOne()
    {
        // Arrange
        Player player = new Player();
        float expectedResult = 0.9f;

        // Act
        player.ReduceHealth(0.1f);
        float result = player.GetHealth();

        // Assert
        Assert.AreEqual(expectedResult, result);
    }

    [Test]
    public void HealthCorrectAfterReducedByHalf()
    {
        // Arrange
        Player player = new Player();
        float expectedResult = 0.5f;

        // Act
        player.ReduceHealth(0.5f);
        float result = player.GetHealth();
```

```
        // Assert
        Assert.AreEqual(expectedResult, result);
    }
}
```

24. Add the following class and limit test to the `EditModeUnitTests` class in `EditModeUnitTests.cs`:

```
public class TestLimitNotExceeded
{
    [Test]
    public void HealthNotExceedMaximumOfOne()
    {
        // Arrange
        Player player = new Player();
        float expectedResult = 1;

        // Act
        player.AddHealth(1);
        player.AddHealth(1);
        player.AddHealth(0.5f);
        player.AddHealth(0.1f);
        float result = player.GetHealth();

        // Assert
        Assert.AreEqual(expectedResult, result);
    }
}
```

25. Add the following class and event tests to the `EditModeUnitTests` class in `EditModeUnitTests.cs`:

```
public class TestEvents
{
    [Test]
    public void CheckEventFiredWhenAddHealth()
    {
        // Arrange
        Player player = new Player();
        bool eventFired = false;

        Player.OnHealthChange += delegate
```

```
        {
            eventFired = true;
        };

        // Act
        player.AddHealth(0.1f);

        // Assert
        Assert.IsTrue(eventFired);
    }

    [Test]
    public void CheckEventFiredWhenReduceHealth()
    {
        // Arrange
        Player player = new Player();
        bool eventFired = false;

        Player.OnHealthChange += delegate
        {
            eventFired = true;
        };

        // Act
        player.ReduceHealth(0.1f);

        // Assert
        Assert.IsTrue(eventFired);
    }
}
```

26. Add the following class and exception tests to the `EditModeUnitTests` class in `EditModeUnitTests.cs`:

```
public class TestExceptions
{
    [Test]
    public void Throws_Exception_When_Add_Health_Passed_Less_Than_Zero()
    {
        // Arrange
        Player player = new Player();
```

```
        // Act

        // Assert
        Assert.Throws<ArgumentOutOfRangeException>(
            delegate
            {
                player.AddHealth(-1);
            }
        );
    }

    [Test]
    public void Throws_Exception_When_Reduce_Health_Passed_Less_Than_
Zero()
    {
        // Arrange
        Player player = new Player();

        // Act

        // Assert
        Assert.Throws<ArgumentOutOfRangeException>(
            () => player.ReduceHealth(-1)
        );
    }
}
```

27. Add the following class and logging tests to the `EditModeUnitTests` class in `EditModeUnitTests.cs`:

```
public class TestLogging
{
    [Test]
    public void Throws_Exception_When_Add_Health_Passed_Less_Than_Zero()
    {
        Debug.unityLogger.logEnabled = true;

        // Arrange
        Player player = new Player();
        HealthChangeLogger healthChangeLogger = new HealthChangeLogger();
        string expectedResult = "health = 0.9";
```

```
        // Act
        player.ReduceHealth(0.1f);

        // Assert
        LogAssert.Expect(LogType.Log, expectedResult);
    }
}
```

You can see that the inner classes allow us to group the **unit tests** visually in the **Test Runner** panel.

28. Click **Run All**.

29. You should see the results of your unit test being executed – if all the tests conclude successfully, there should be green ticks (checkmarks) next to each test:

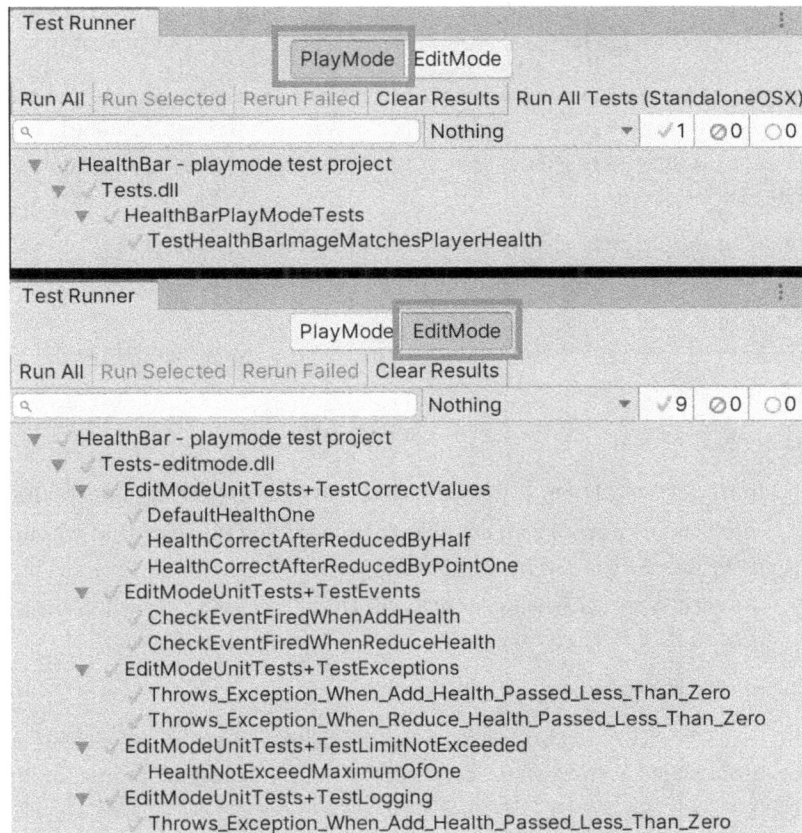

Figure 19.21: PlayMode and EditMode test results – all of which passed (green ticks)

How it works...

Let's take a look at how this recipe works in detail.

PlayMode testing

The PlayMode test called `TestHealthBarImageMatchesPlayerHealth()` loads the `HealthBar` scene, gets a reference to the instance object of **PlayerManager**, which is a component of the **PlayerManager** GameObject, and invokes the `ReduceHealth()` method. This method reduces the player's health by 0.1. So, from its starting value of `1.0`, it becomes `0.9`.

The **PlayerManager** GameObject also has as a component that's an instance object of the C# `HealthBarDisplay` script class. This object registers to listen to published events from the player class. It also has a public **UI image** variable that has been linked to the **UI image** of the health bar filler image in the scene.

When the player's health is reduced to 0.9, it publishes the `OnChangeHealth(0.9)` event. This event is received by the `HealthBarDisplay` object instance, which then sets the `fillAmount` property of the linked health bar filler image in the scene.

The `TestHealthBarImageMatchesPlayerHealth()` PlayMode test gets a reference to the object instance named `image-health-bar-filler`, storing this reference in the `healthBarFiller` variable. The test assertion that's made is that the `expectedResult` value of `0.9` matches the actual `fillAmount` property of the UI image in the scene:

```
Assert.AreEqual(expectedResult, healthBarFiller.fillAmount);
```

Unit tests

There are several unit tests that can be grouped by placing them inside their own classes, inside the `EditModeUnitTests` script class:

- `TestCorrectValues`:

  - `DefaultHealthOne()`: This tests that the default (initial value) of the player's health is 1.

  - `HealthCorrectAfterReducedByPointOne()`: This tests that when the player's health is reduced by 0.1, it becomes `0.9`.

  - `HealthCorrectAfterReducedByHalf()`: This tests that when the player's health is reduced by 0.5, it becomes `0.5`.

- `TestLimitNotExceeded`:

  - `HealthNotExceedMaximumOfOne()`: This tests that the value of the player's health does not exceed 1, even after attempts to add 1, 0.5, and 0.1 to its initial value of 1.

- TestEvents:

  - CheckEventFiredWhenAddHealth(): This tests that an OnChangeHealth() event is published when the player's health is increased.

  - CheckEventFiredWhenReduceHealth(): This tests that an OnChangeHealth() event is published when the player's health is decreased.

- TestLogging:

  - CorrectDebugLogMessageAfterHealthReduced(): This tests that a Debug.Log message is correctly logged after the player's health is reduced by 0.1 to 0.9.

- TestExceptions:

  - Throws_Exception_When_Add_Health_Passed_Less_Than_Zero(): This tests that an ArgumentOutOfRangeException is thrown when a negative value is passed to the AddHealth(...) player method.

  - Throws_Exception_When_Reduce_Health_Passed_Less_Than_Zero(): This tests that an ArgumentOutOfRangeException is thrown when a negative value is passed to the ReduceHealth(...) player method.

> These two tests illustrate one convention of naming tests that adds an underscore (_) character between each word in the method name in order to improve readability.

See also

You can learn more about the LogAssert Unity script reference in the Unity documentation: https://docs.unity3d.com/Packages/com.unity.test-framework@1.3/api/UnityEngine.TestTools.LogAssert.html.

The method for unit testing C# events has been adapted from a post on philosophicalgeek.com: http://www.philosophicalgeek.com/2007/12/27/easily-unit-testing-event-handlers/.

The delegate-event publishing of health change events in this health bar feature is an example of the **Publisher-Subscriber** design pattern.

Reporting Code Coverage testing

A useful tool in projects with code testing is to be able to analyze how much of a C# script class is being tested. For example, is every method being tested with at least one set of test data? Unity now offers a Code Coverage feature, which we'll explore in this code-testing recipe. As shown in the following screenshot, Unity allows us to create a set of HTML pages for documenting the code coverage of tests against C# code.

With this, we can see what percentage of our code is covered by tests, and even which lines of code are, and are not, covered by our tests:

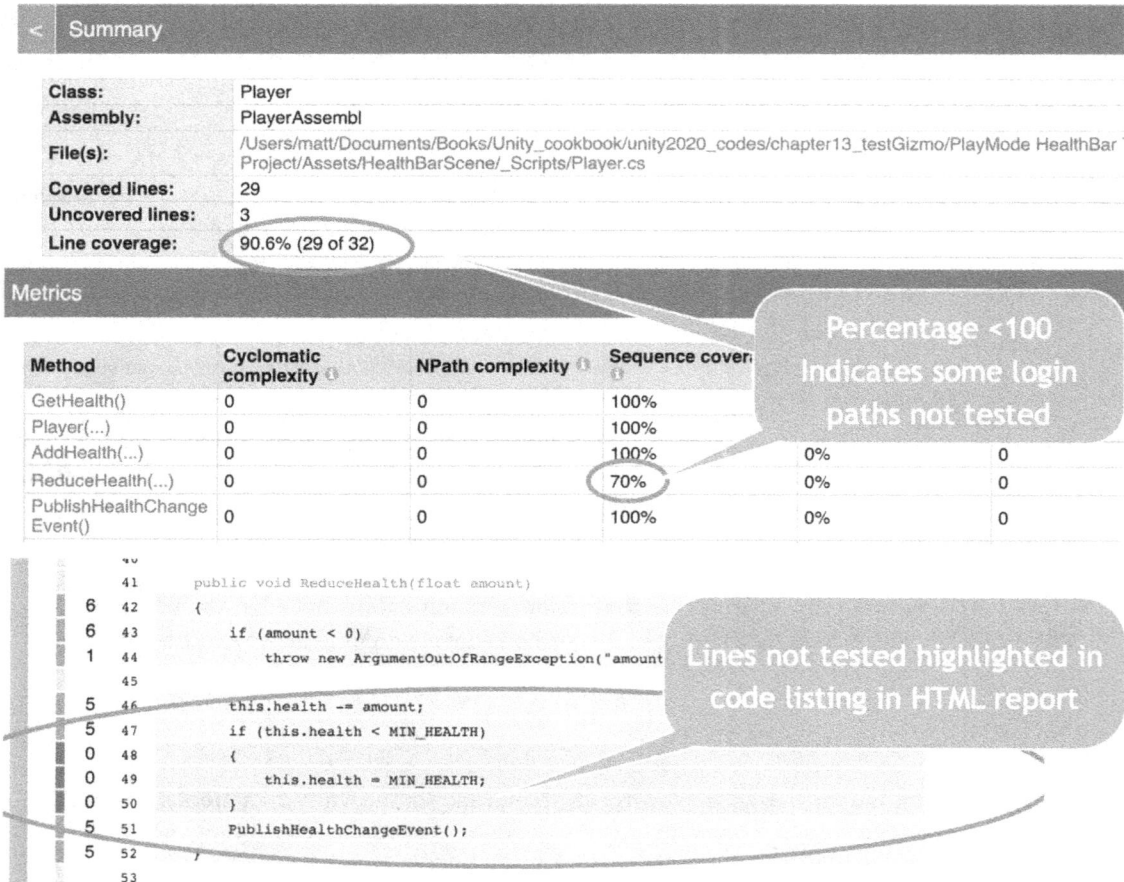

Summary	
Class:	Player
Assembly:	PlayerAssembl
File(s):	/Users/matt/Documents/Books/Unity_cookbook/unity2020_codes/chapter13_testGizmo/PlayMode HealthBar Project/Assets/HealthBarScene/_Scripts/Player.cs
Covered lines:	29
Uncovered lines:	3
Line coverage:	90.6% (29 of 32)

Metrics

Method	Cyclomatic complexity	NPath complexity	Sequence cover:		
GetHealth()	0	0	100%		
Player(...)	0	0	100%		
AddHealth(...)	0	0	100%	0%	0
ReduceHealth(...)	0	0	70%	0%	0
PublishHealthChange Event()	0	0	100%	0%	0

Percentage <100 Indicates some login paths not tested

```
       40
       41        public void ReduceHealth(float amount)
  6    42        {
  6    43            if (amount < 0)
  1    44                throw new ArgumentOutOfRangeException("amount
       45
  5    46            this.health -= amount;
  5    47            if (this.health < MIN_HEALTH)
  0    48            {
  0    49                this.health = MIN_HEALTH;
  0    50            }
  5    51            PublishHealthChangeEvent();
  5    52        }
       53
```

Lines not tested highlighted in code listing in HTML report

Figure 19.22: Code Coverage HTML report for the Player script class

Getting ready

This project builds on the previous one, so make a copy of what you made in that recipe and work on the copy.

How to do it...

To add Code Coverage reporting to a project with unit tests, follow these steps:

1. Open the **Code Coverage** window by going to **Window | Analysis | Code Coverage**.
2. In the **Code Coverage** window, click the **Switch to Debug Mode** button.
3. Next, check the **Enable Code Coverage** option. Then, ensure the **HTML Report** and **Auto Generate Report** options are checked under the **Report Options** section.

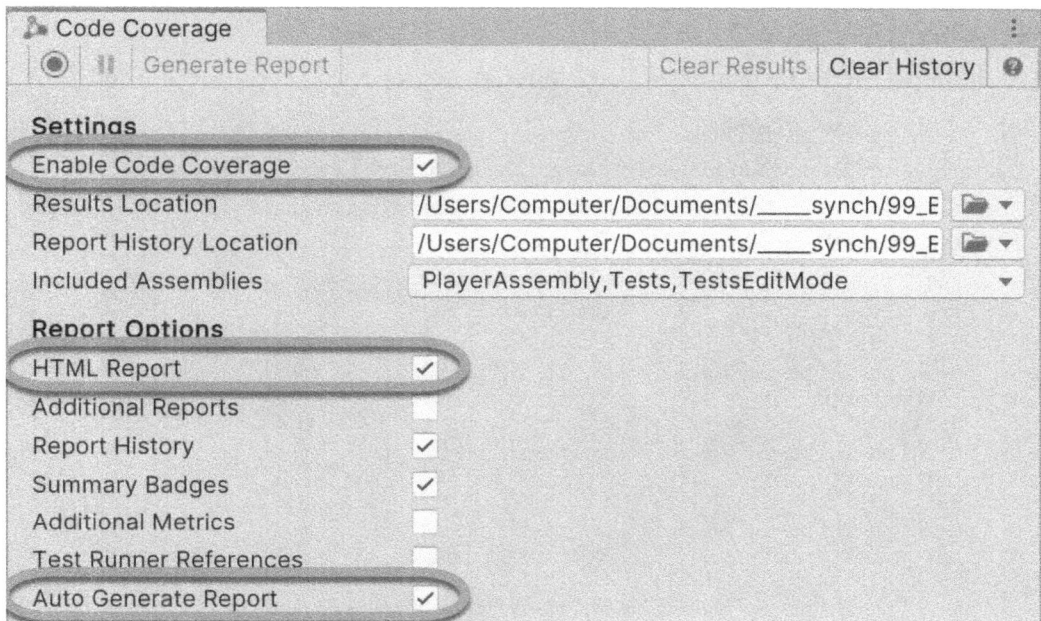

Figure 19.23: Code Coverage window options

4. In the **Test Runner**, choose **PlayMode** and click the **Run All Tests** button.

5. Once the testing is finished, an HTML Code Coverage report should automatically be created and opened in your computer's default web browser application. The report will be created in a new folder named Report inside a folder named Code Coverage in your Unity project folder.

How it works...

The Unity debugger monitors which lines of code are being executed as the tests are run. It uses this data to compile a report on how much of each method of each class has been executed for the assemblies involved in the testing. By adding the Code Coverage package and enabling it, debugging data will be collected and reported upon when you run Unity tests.

The Unity **Code Coverage** tool generates a set of web pages to inform us about how much and which lines of our code are being examined with our unit tests. While even 100% coverage does not guarantee the code is "correct," a high percentage of code coverage does indicate that the behavior of most of our code is being tested to some extent.

Running simple Python scripts inside Unity

Unity Python is a package that allows Python code to be executed as part of a Unity project. In this recipe, we'll install the package, test the **Python Script Editor** window with a traditional Hello World Python print statement, and create C# scripts to run Python based on the examples provided by Unity at https://docs.unity3d.com/Packages/com.unity.scripting.python@4.0/manual/inProcessAPI. html.

How to do it...

To run simple Python scripts inside Unity, follow these steps:

1. Create a new 2D project.
2. Open the **Package Manager** by navigating to **Window | Package Manager**.
3. Set the list of packages to those in the **Unity Registry** and search for **Python Scripting**. Then, click **Install**.

Figure 19.24: Adding the Python Scripting package to a project

4. Open the **Python Script Editor** panel by going to **Window | General | Python Script Editor**.
5. Enter print ('Hello World from Python') in the editor section (lower half) of the **Python Script Editor** panel and click the **Execute** button. You should see a message stating Hello World appear in the output (top) section of the window:

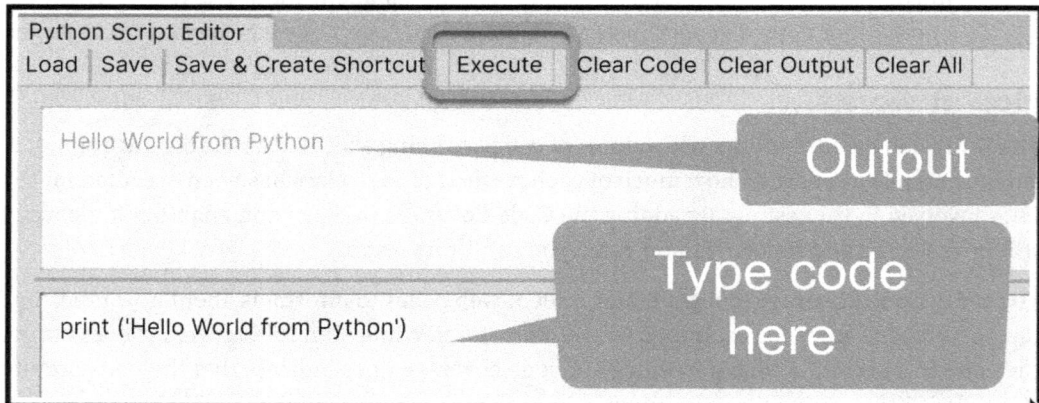

Figure 19.25: Executing a Hello World statement in the Python Script Editor window

6. Now, let's create a C# script that can execute Python in a string using the Python API. Create a _Scripts folder in the **Project** panel. Inside it, create a new C# script class file named HelloConsole.cs that contains the following code:

```
using UnityEditor.Scripting.Python;
using UnityEditor;
```

```
public class HelloConsole
{
[MenuItem("My Python/Hello Console")]
static void PrintHelloWorldFromPython()
{
    PythonRunner.RunString(@"
            import UnityEngine;
            UnityEngine.Debug.Log('hello console')
            ");
}
}
```

7. You should now see a new menu named **My Python** with an item called **Hello Console**. When you select this menu item, a message stating `hello console` should be output to the **Console** panel.

8. Now, let's create a text file in our _Scripts folder containing pure Python – you may need to create this text file outside the Unity Editor by navigating to the _Scripts folder and using a text editor application to create and save this new file. Create a new text file named `renamer.py` that contains the following code:

```
import UnityEngine

all_objects = UnityEngine.Object.FindObjectsOfType(UnityEngine.
GameObject)
for go in all_objects:
    if go.name[-1] != '_':
        go.name = go.name + '_'
```

Note: Indentation is very important in Python programming since indentation is used instead of braces (curly brackets) for grouping statements. Python uses four space characters for indentation (see code above) – so ensure any Python code you write uses four spaces wherever indentation is required.

9. Now, let's create a C# script that offers a menu item that will execute our Python script file. Create a new C# script class file called `InvokeRenamer.cs` that contains the following code:

```
using UnityEditor.Scripting.Python;
using UnityEditor;
using UnityEngine;

public class InvokeRenamer
```

```
    {
        [MenuItem("My Python/Underscore Renamer")]
        static void RunEnsureNaming()
        {
            PythonRunner.RunFile($"{Application.dataPath}/_Scripts/renamer.
py");
        }
    }
}
```

10. On the **My Python** menu item, there should now be a second menu item named **Underscore Renamer**. When you select this menu item, you should see that all the GameObjects in the **Hierarchy** panel now end with underscore characters.

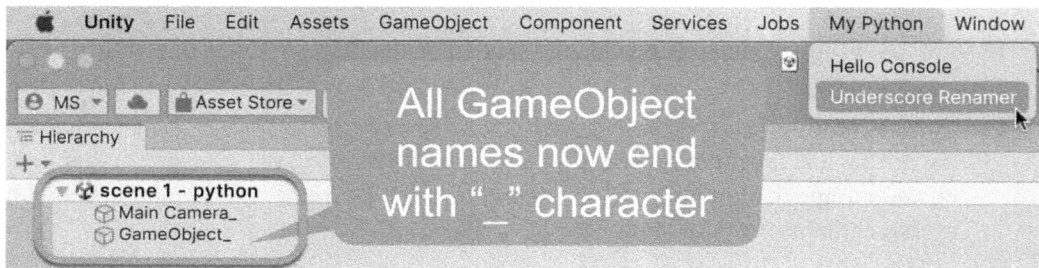

Figure 19.26: The My Python menu items and the renamed Hierarchy window GameObjects

How it works...

The Python Scripting package adds an API (code library) that we can access through our C# code, as well as the **Python Script Editor** window for testing and running simple Python statements. We tested **Python Script Editor** with a traditional Hello World from Python print statement.

Our C# script class HelloConsole demonstrated how we can use the PythonRunner.RunString($...) method in our C# scripts to execute a string containing Python code. Note that we must add a using statement in our Python code to import the UnityEditor.Scripting.Python library for this to work.

> **Note:** The @ sign in our **HelloConsole** C# script class allowed us to write several lines of Python code in a single-string variable declaration. This sign means that the string contents are treated as a verbatim string literal, meaning any special characters such as new lines or slashes are considered part of the string inside the two double quotation marks.

In both our scripts, we used the [MenuItem("<menu/item>")] Editor instruction to enable us to easily test out code from a menu item.

Our InvokeRenamer C# script class demonstrated how we can use the PythonRunner.RunFile($...) method in our C# scripts to execute a text file containing Python code. We created a file called renamer. py containing pure Python code in order to loop through all the GameObjects.

We added an underscore suffix if they didn't already have such a suffix. We followed the convention of using the `.py` file extension for files containing just Python code.

Notice how we indicated the location of the `renamer.py` file by writing `{Application.dataPath}/_Scripts/` before the file name. We could change this to indicate a different location for the Python file we wish to execute.

In both our Python string and file, we were able to make use of the **UnityEngine** library (assembly) by writing `import UnityEngine` before our Python statements – this works just like the C# `using` statement.

Further reading

You can learn more about gizmos at the following links:

- The Unity Gizmos manual entry at https://docs.unity3d.com/ScriptReference/Gizmos.html
- Unity Gizmos tutorial: https://learn.unity.com/tutorial/creating-custom-gizmos-for-development-2019-2

You can learn more about Unity testing at the following links:

- Unity Test Framework manual documentation: `https://docs.unity3d.com/Manual/testing-editortestsrunner.html`
- Unity Test Framework package documentation: `https://docs.unity3d.com/Packages/com.unity.test-framework@1.1/manual/index.html`
- Unity Test Framework how-to pages: `https://unity.com/how-to/unity-test-framework-video-game-development`
- A website for the book *The Art of Unit Testing* (and lots of other learning resources associated with testing): `http://artofunittesting.com/`
- A great dual article tutorial about Unity testing by Tomek Paszek from Unity (talking about the old Unity test tools, but most of the content is still very relevant): `https://blogs.unity3d.com/2014/06/03/unit-testing-part-2-unit-testing-monobehaviours/`
- The Infalliblecode YouTube channel, where you can learn lots about Unity testing (and other topics): `https://www.youtube.com/infalliblecode`
- CodeProject.com's introduction to TDD and NUnit: `https://www.codeproject.com/Articles/162041/Introduction-to-NUnit-and-TDD`
- A great tutorial about unit testing by Anthony Uccello on Kodeco (formally Ray Wenderlich): `https://www.kodeco.com/9454-introduction-to-unity-unit-testing`
- The Code Coverage features of the Unity Test tools: `https://docs.unity3d.com/Packages/com.unity.testtools.codecoverage@1.0/manual/`

- Testing in Unity usually involves **Unity Assemblies**. This is an approach to separating the components of a game into separate modules. A great introduction to Unity Assemblies by *Erdiizgi* can be found at `https://erdiizgi.com/why-modular-game-development-and-how-to-do-it-with-unity/`

- You can learn more about Unity Python by reading the official package documentation: `https://docs.unity3d.com/Packages/com.unity.scripting.python@4.0/manual/inProcessAPI.html`

Learn more on Discord

To join the Discord community for this book – where you can share feedback, ask questions to the author, and learn about new releases – follow the QR code below:

`https://packt.link/unitydev`

‹packt›

packt.com

Subscribe to our online digital library for full access to over 7,000 books and videos, as well as industry leading tools to help you plan your personal development and advance your career. For more information, please visit our website.

Why subscribe?

- Spend less time learning and more time coding with practical eBooks and Videos from over 4,000 industry professionals
- Improve your learning with Skill Plans built especially for you
- Get a free eBook or video every month
- Fully searchable for easy access to vital information
- Copy and paste, print, and bookmark content

At www.packt.com, you can also read a collection of free technical articles, sign up for a range of free newsletters, and receive exclusive discounts and offers on Packt books and eBooks.

Other Books You May Enjoy

If you enjoyed this book, you may be interested in these other books by Packt:

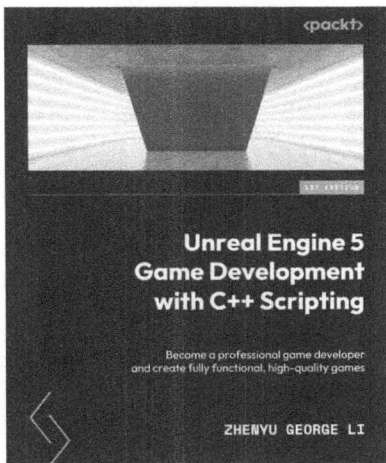

Unreal Engine 5 Game Development with C++ Scripting

ZHENYU GEORGE LI

ISBN: 9781804613931

- Develop coding skills in Microsoft Visual Studio and the Unreal Engine editor
- Discover C++ programming for Unreal Engine C++ scripting
- Understand object-oriented programming concepts and C++-specific syntax
- Explore NPC controls, collisions, interactions, navigation, UI, and the multiplayer mechanism
- Use the predefined Unreal Engine classes and the programming mechanism
- Write code to solve practical problems and accomplish tasks
- Implement solutions and methods used in game development

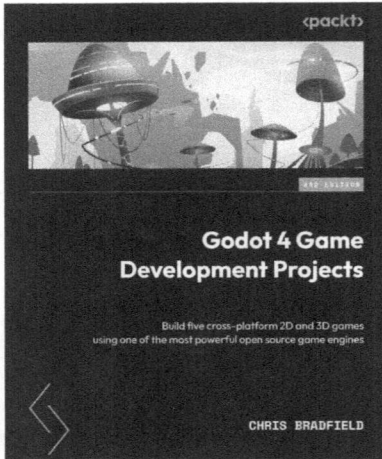

Godot 4 Game Development Projects - Second Edition

Chris Bradfield

ISBN: 9781804610404

- If you're new to Godot, get started with the game engine and editor
- Learn about the new features of Godot 4.0
- Build games in 2D and 3D using design and coding best practices
- Use Godot's node and scene system to design robust, reusable game objects
- Use GDScript, Godot's built-in scripting language, to create complex game systems
- Implement user interfaces to display information
- Create visual effects to spice up your game
- Publish your game to desktop and mobile platforms

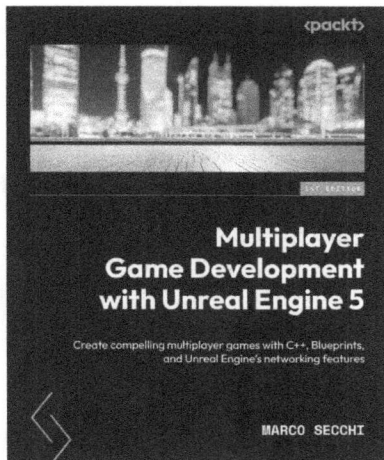

Multiplayer Game Development with Unreal Engine 5

Marco Secchi

ISBN: 9781803232874

- Get to grips with the basics of multiplayer game development
- Understand the main elements of a networked level
- Explore Unreal multiplayer features such as replication, RPCs, relevancy, and roles
- Debug and optimize code for improved game performance
- Deploy the game on LAN or online platforms
- Use Epic Online Services to elevate the player experience

Packt is searching for authors like you

If you're interested in becoming an author for Packt, please visit authors.packtpub.com and apply today. We have worked with thousands of developers and tech professionals, just like you, to help them share their insight with the global tech community. You can make a general application, apply for a specific hot topic that we are recruiting an author for, or submit your own idea.

Share your thoughts

Now you've finished *Unity Cookbook, Fifth Edition* we'd love to hear your thoughts! Scan the QR code below to go straight to the Amazon review page for this book and share your feedback or leave a review on the site that you purchased it from.

https://packt.link/r/1805123025

Your review is important to us and the tech community and will help us make sure we're delivering excellent quality content.

Index

Symbols

A

S

Y

Z

Download a free PDF copy of this book

Thanks for purchasing this book!

Do you like to read on the go but are unable to carry your print books everywhere? Is your eBook purchase not compatible with the device of your choice?

Don't worry, now with every Packt book you get a DRM-free PDF version of that book at no cost.

Read anywhere, any place, on any device. Search, copy, and paste code from your favorite technical books directly into your application.

The perks don't stop there, you can get exclusive access to discounts, newsletters, and great free content in your inbox daily

Follow these simple steps to get the benefits:

1. Scan the QR code or visit the link below

https://packt.link/free-ebook/9781805123026

2. Submit your proof of purchase
3. That's it! We'll send your free PDF and other benefits to your email directly

www.ingramcontent.com/pod-product-compliance
Lightning Source LLC
Chambersburg PA
CBHW081208220326
41598CB00037B/6710